History and Development of Indian Agriculture

The Author

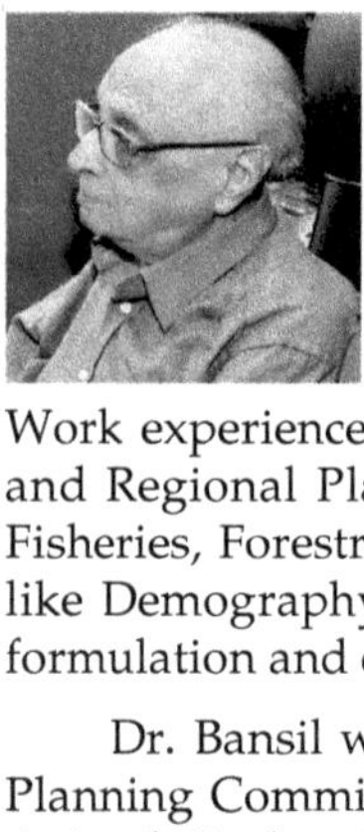

Dr. P.C. Bansil (1921–2016) belongs to first batch of Indian Economic Service. Worked with the FAO for about 15 years in Sri Lanka, Thailand, Rome, Zambia and Libya and for a similar period with the Government of India. Author of 46 books/Reports. Besides a large number of reports and papers prepared for various Governments, published over 400 papers in specialized technical journals. Hundreds of Technical Reports have been prepared on various aspects of Agriculture for Various Governments. Work experience relates to the formulation, implementation and evaluation of National and Regional Plans for Agriculture, Statistics, Livestock including poultry, Irrigation, Fisheries, Forestry, Wildlife, National Parks, Tourism, Agro Industries and Social Sectors like Demography, Health, Education and Housing *etc.* as well as project identification, formulation and evaluation in the above fields.

Dr. Bansil was basically an agri and livestock economist. He was appointed by the Planning Commission as chairman of the Tenth and Eleventh Plans Working Groups on Animal Husbandry Economics and Statistics. The Food and Agriculture Organization of the United Nations sponsored study – Livestock in India 2000-2030 – was completed by him in 2001. During his FAO assignments in Sri Lanka, Zambia, and Libya as Senior Planning advisor, he undertook micro level studies to improve the quality of slaughter houses, poultry, Agriculture and livestock development programmes – Agriculture in India, cattle in Sri Lanka and Zambia and mutton/chicken in Libya.

Over 50 years of experience in research on problems relating to agricultural and Livestock, economics & statistics, rural and regional economics and planning, demography, health and social sectors in India, Planning Commission, Ministry of Agriculture and FAO.

Some of the Papers were read at the Academic Conferences. Member: International Agricultural Economic Association. Life member: American Agricultural Economic Association, Indian Society of Agricultural Economics, Indian Society of Population Education, Indian Society of Medical Statistics. Indian Institute of Public Administration, Indian Society for Constitutional Studies, Indian Society of Agricultural Marketing and Indian Society of Training and Development, Life and Founder Member, North India Economic Association. He is preparing a **Thirteen Volume** Series – Economics of Agricultural Commodities to cover all aspects of agriculture. Six Volumes have already been published and the remaining are in the Press.

History and Development of Indian Agriculture

— *Author* —

P.C. Bansil

2018

Daya Publishing House®

A Division of

Astral International Pvt. Ltd.

New Delhi – 110 002

Cataloging in Publication Data--DK
Courtesy: D.K. Agencies (P) Ltd. <docinfo@dkagencies.com>

Bansil, P. C., author.
History and development of Indian agriculture / author, P.C. Bansil.
pages cm
Includes index.
ISBN 9789351248750 (HB)

1. Agriculture--India--History. 2. Agricultural innovations--India--History. I. Title.

LCC S471.I3B36 2018 | DDC 630.954 23

Published by : **Daya Publishing House®**
A Division of
Astral International Pvt. Ltd.
– ISO 9001:2015 Certified Company –
4736/23, Ansari Road, Darya Ganj
New Delhi-110 002
Ph. 011-43549197, 23278134
E-mail: info@astralint.com
Website: www.astralint.com

To my friend
— Ramesh Aggarwal —

Ramesh Chandra Agrawal obtained his M.Sc. in Agricultural Economics in 1955 with First place in Agra University. He retired in 2001 as a Professor from the Humboldt University of Berlin. His illustrious professional career, dealing primarily with questions of agricultural development, spanned a period of more than 45 years. It included teaching a wide range of subjects to the graduates and professionals, directing national and international research projects, and providing consultancy to various national and international organizations at the highest level. The pallet of courses taught by Ramesh Agrawal to the graduate and doctoral students and planners and policy makers from the developing and high income economies includes operations research methods, agricultural sector policy analysis and planning, economic and financial appraisal of development projects, and farming systems analysis for sustainable agricultural development.

His research, writings, and keynote speeches range from original theoretical models to their applications of direct practical relevance to planning and policy making for sustainable agricultural development. Ramesh Agrawal is a strong believer in the abilities of the farmers to take care of themselves and pleads with the other actors of the developmental process – researchers, extension specialists, planners, policy.

Administrators to put the farmhousehold at the centre of their work and facilitate access to the provisions of policy, to it, through good governance and strong institutions. His name has been listed in various international WHO's WHOs, such as the "Marquis Who's Who in the World". He is a recipient of American Biographical Institute Gold Medal for Germany.

The Indian Society of Agricultural Economics has instituted, in 2004, a biennial prize, "Professor Ramesh Chandra Agrawal Award of Excellence", awarded to Indian scientists for their contribution to Agricultural Economics. Ramesh Agrawal has not only distinguished himself as a scholar but also in extra-curricular activities, including being Iowa State University Champion in bridge and badminton singles and doubles.

Preface

The book '**History and Development of Agriculture**' is a detailed Story of the History and development of Agriculture. The Institute started writing a two Series – Economics of Agricultural Commodities and Economics of Agriculture and Allied Aspects. It was felt that the missing link between the two Series is necessary – hence this study. It is hoped that the reader will be able to understand the various stages through which agriculture passed to achieve its present status.

Progress made by Indian Agriculture from the stage of a 'begging bowl' to that of a surplus food economy is not a small thing. We are exporting rice today to a large extent. The book gives a historical approach to the various issues. It will be of interest to all the students of history, politics and academics. This study is not of interest in day to day life, but has great educational value for any body interested in different aspects of agriculture.

The study is the result of innumerable sources – published/unpublished, quite a large number of daily papers, journals, magazines and, books. The final presentation, however, is in a specific order so as to give the reader the required information at one place. We are grateful to a large number of authors whose work/research formed the base of this study. Thanks are due to Dr. M.L Srivastava, Mr. Ravi Shankar and Mr. Mukesh Kumar without whose contribution, the book would not have been able to see the light of the day. Our Faculty members – Dr. P.N. Radhakrishnan and the Late Dr. S.R. Mohnot were always available to guide and contribute their bit. Last but not least is the major contribution of our Staffers typing and retyping - Mrs. Vasantha Krishnan and Mrs. Ajitha Ganagadharan - who never got tired to type and retype the material.

We do hope that the book in its updated form will prove useful to students of agricultural economics, researchers, educated farmers, administrators, policy makers and bankers alike. We shall very much welcome any comments for further improvement of the volume.

We must, however, say that none of them is responsible for any errors/ omissions from which this work may suffer as at present. I am personally responsible for all this.

P.C. Bansil

Contents

Preface *vii*

1. History of Agriculture in India 1

2. History of Irrigation 139

3. Hundred Years of Co-operative 179

4. Agriculture during British Period 289

5. Agribusiness and its SWOT Analysis 317

6. Grow More Food 347

7. Green Revolution in Mexico, 1943 – 1965 397

8. Green Revolution in India 401

9. Food Situation Immediately after Independence 485

10. Agricultural Development in India Since Independence 505

11. Indian Seed System Development: Policy and Institutional Options 525

12. Land Acquisition, Rehabilitation and Resettlement Bill, 2013 579

Index 655

1

History of Agriculture in India

The word agriculture is the English adaptation of Latin agricultura, from ager, "a field, and cultura, "cultivation" in the strict sense of "tillage of the soil". Thus a literal reading of the word yields "tillage of a field/of fields".

Agriculture is the production, processing, marketing, and use of foods, fibers and byproducts from plant crops and animals. Agriculture was the key development that led to the rise of human civilization, with the husbandry of domesticated animals and plants (*i.e.* crops) creating food surpluses that enabled the development of more densely populated and stratified societies. They study of agriculture is known as agricultural science. Agriculture is also observed in certain species of ant and termite. Agriculture encompasses a wide variety of specialties and techniques, including ways to expand the lands suitable for plant raising, by digging water-channels and other forms of irrigation. Cultivation of crops on arable aldnd and the pastoral herding of livestock on rangeland remain at the foundation of agriculture. In the past century there has been increasing concern to identify and quantify various forms of agriculture. In the developed world the range usually extends between sustainable agriculture (*e.g.* permaculture or organic agriculture) and intensive farming (*e.g.* industrial agriculture).

Modern agronomy, plant breeding, pesticides and fertilizers, and technological improvements have sharply increased yields from cultaivation, but the same time have caused widespread ecological damage and negative human health effects. Selective breeding and modern practices in animal husbandry such as intensive pig farming (and similar practices applied to the chicken) have similarly increased the output of meat, but have raised concerns about animal cruelty and the health effects of the antibiotics, growth hormones, and other chemicals commonly used in industrial meat production.

The major agricultural products can be broadly grouped into foods, fibers, fuels, and raw materials. In the 21st century, plants have been used to grow biofuels, biopharmaceuticals, bioplastics, and pharmaceuticals. Specific foods include cereals, vegetables, fruits and meat. Fibers include cotton, wool, hemp, silk and flax. Raw materials include lumber and bamboo. Other useful materials are produced by plants. Such as resins. Biofuels include methane from biomass, ethanol, and biodiesel. Cut flowers, nursery plants, tropical fish and brids for the pet trade are some of the ornamental products.

Overview

Agriculture has played a key role in the development of human civilization. Util the Industrial Revolution, the vast majority of the human population labored in agriculture. Development of agricultural techniques has steadily increased agricultural productivity, and the widespread diffusion of these techniques during a time period of often called an agricultural revolution. A remarkable shift in agricultural practices has occurred over the past century in response to new technologies. In particular, the Haber-Bosch method for synthesizing ammonium nitrate made the traditional practice of recycling nutrients with crop rotation and animal manure less necessary.

Synthetic nitrogen, along the mined rock phosphate, pesticides and mechanization, have greatly increased crop yields in the early 20th century. Increased supply of grains has led to cheaper livestock as well. Further, global yield increases were experienced later in the 20th century when high-yield varieties of common staple grains such as rice, wheat, and corn (maize) were introduced as a part of the Green Revolution. The Greeen Revolution exported the technologies (including pesticides and synthetic nitrogen) of the developed world to the developing world. Thomas Malthus famously predeicted that the Earth would not be able to support its growing population, but technologies such as the Green Revolution have allowed the world to produce a surplus of food.

Many governments have subsidized agriculture to ensure an adequate food supply. These agricultural subsidies are often linked to the production of certain commodities such as wheat, corn(maize), rice, soybeans, and milk. These subsidies, especially when instituted by developed countries have been noted a sprotectionist, inefficient,l and environmentally damaging. In the past century agriculture has been characxterized by enhanced productivity, the use of synthetic fertilizers and pesticides, selective breeding, mechanization, water contamination, and farm subsidies. Proponents of organic farming such as Sir Albert Howard argued in the early 20th century that the overuse of pesticides and synthetic fertilizers damages the long-term fertility of the soil. While this feeling lay dormant for decades, as environmental awareness has increased in the 21st century there has been a movement towards sustainable agriculture by some farmers, consumers, and policymakers.

In recent years there has been a b acklash against perceived external environmental effects of mainstream agriculture, particularly regarding water pollution, resulting in the organic movement. One of the major forces behind this movement has been the European Union, which first certified organic food in 1991 and began reform of its Common Agricultural Policy (CAP) in 2005 to phase out commodity –linked farm subsidies, also known as decoupling. The growth of organic farming has renewed research in alternative technologies such as integrated pest management and selective breeding. Recent mainstream technological developments include genetically modified food.

Grand Anicut dam on river Kaveri (1st-2nd CenturyCE) is one of the olderst water regulation structures in the world still in use. During the Bronze Age (9000 BC onwards), domestication of plants and animals transformed the profession of the early homo sapiens from hunting and gathering to selective hunting, herding and finally to settled agriculture. Eventually the agricultural practices enabled people to establish permanent settlements and expand urban based societies. Cultivation marks the transition from nomadic pre-historic societies to the settled neolithic lifestyle some time around 7000 BC. Settled life soon followed with implements and techniques being developed for agriculture. Double monsoons led to two harvests being reaped in one year. Indian products soon reached the world via existing trading networks and foreign crops were introduced to India. Plants and animals - considered essential to their survival by the Indians - came to be worshiped and venerated.

The middle ages saw irrigation channels reach a new level of sophistication in India and Indian crops affecting the economies of other regions of the world under Islamic patronage. Land and water management systems were developed with an aim of providing uniform growth. Despite some stagnation during the later modern era the independent Republic of India was able to develop a comprehensive agricultural programme. As per the modern definition of agriculture which would be" an aggregate of large scale intensive cultivation of land, mono-cropping, organized irrigation, and use of a specialized labor force", the title "inventors of agriculture" would go to the Sumerians, starting ca. 5,500 BC.

Ancient Origins

The Fertile Crescent of Western Asia, Egypt, and India were sites of the earliest plannet sowing and harvesting of plants that had previously been gathered in the wild. Independent development of agriculture occurred in northern and southern China, Africa's Sahel, New Guinea and several regions of the Americas. The eight so-called Neolithic founder crops of agriculture appear: first emmer wheat and einkorn wheat, then hulled barley, peas, lentils, bitter vetch, chick peas and flax.

By 7000BC, small-scale agriculture reached Egypt. From at least 7000BC the Indian subcontinent saw farming of wheat and barley, as attested by archaeological excavation at Mehrgarh in Balochistan in what is present day Pakistan. By 6000BC, mid-scale farming was entrenched on the bank of the Nile. About this time, agriculture was developed independely in the Far East, with rice, rather than wheat, as the primary crop. Chinese and Indonesian farmers went on to domesticate taro and beans including mung, soy and azuki. To complement these new sources of carbohydrates, highly organized net fishing of river, lakes and ocean shores in these areas brought in great volumes of essential protein. Collectively, these new methods of farming and fishing inaugurated a human population boom that dwarfed all previous expansions and continues today.

By 5000BC, the Sumerians had developed core agricultural techniques including large scale intensive cultivation of land, monocropping, organized irrigation, and the use of a specialized labor force, particularly along the waterway now knowqn as the Shatt al-Arab, from its Persian Gulf delta to the confluence of the Tigris and Euphrates. Domestication of wild aurochs and mouflon into cattle and sheep, respectively, ushered int eh large-scale use of animals for foor/fiber and as beasts of burden. The shepherd joined the farmer as an essential provider for sedentary and seminomadic societies. Maize, manioc, and arrowroot were first domesticated in the Americas as far back as 5200BC.

The potato, tomato, pepper, squash, several varieties of bean, tobacco, and several other plants were also developed in the New World, as was extensive terracing of steep hillsides in much of Andean South America. The Greeks and romans built on techniques pioneered by the Sumerians, but made few fundamentally new advances. Southern Greeks struggled with very poor soils, yet managed to become a dominant society for years. The Romans were noted for an emphasis on the cultivation of crops for trade.

In the Americas, a parallel agricultural revolution occurred, resulting in some fo the most important crops grown today. In Mesoamerica wild teosinte was transformed through human selection into the ancestor of modern maize, more than 6000 years ago. It gradually spread across North America and was the major crop of Native Americans at the time of European exploration. Other Mesoamerican crops include hundreds of varieties of squash and beans. Cocoa was also a major crop in domesticated Mexico and central America. The turkey, one of the most important meat birds, was probably domesticated in Mexico or the U.S. Southwest. In the Andes region of South America the major domesticated crop was potatoes, domesticated perhaps 5000 years ago. Large varieties of beans were domesticated, in South America, as well as animals, including llamas, alpacas, and guinea pigs. Coca, still a major crop, was also domesticated in the Andes.

A minor center of domestication, the indigenous people fo the Eastern U.S. appear to have domesticated numerous crops. Sunflowers, tobacco, varieties of squash and Chenopodium, as well as crops no longer grown, including marshelder and little barley were domesticated. Other wild fods may have undergone some selective cultivation, including wild rice and maple sugar. The most common varieties of strawberry were domesticated from Eastern North America.

Early History

Barley and wheat cultivation - along with the rearing of cattle, sheep and goat - was visible in Mehrgarh by 8000-6000 BCE. Agro pastoralism in India included threshing, planting crops in rows - either of two or of six - and storing grain in granaries. In the period of the Neolithic revolution (roughly 8000-4000 BCE.), agriculture was far from the dominant mode of support for human societies. But those who adopted it, have survived and increased, and passed their techniques of production to the next generation. This transformation of knowledge was the base of further development in agriculture. By the 5th millennium BCE agricultural communities became widespread in Kashmir. Zaheer Baber (1996) writes that 'the first evidence of cultivation of cotton had already developed'. Cotton was cultivated by the 5th millennium BCE-4th millennium BCE. The Indus cotton industry was well developed and some methods used in cotton spinning and fabrication continued to be practiced till the modern Industrialization of India.

A variety of tropical fruit such as mango and muskmelon are native to the Indian subcontinent.The Indians also domesticated hemp, which they used for a number of applications including making narcotics, fiber, and oil. The farmers of the Indus valley, which thrived in modern-day Pakistan and North India, grew peas, sesame, and dates, sugarcane was originally from tropical South Asia and Southeast Asia. Different species likely originated in different locations with *S. barberi* originating in India and *S. edule* and *S. officinarum* coming from New Guinea. Wild rice cultivation appeared in the Belan and Ganges valley regions of northern India as early as 4530 BCE and 5440 BCE respectively. Rice was cultivated in the Indus Valley Indus valley civilization Agricultural activity during the second millennium BC included rice cultivation in the Kashmir and Harrappan regions. Mixed farming was the basis of the Indus valley economy. Denis J. Murphy (2007) details the spread of cultivated rice from India into South-east Asia.

Several wild cereals, including rice, grew in the Vindhya Hills, and rice cultivation, at sites such as Chopani-Mando and Mahagara, may have been underway as early as 7000 BP. The relative isolation of this area and the early development of rice farming imply that it was developed indigenously - Chopani-Mando and Mahagara are located on the upper reaches of the Ganges drainage system and it is likely that migrants from this area spread rice farming down the Ganges valley into the fertile plains of Bengal, and beyond into south-east Asia.

Irrigation was developed in the Indus Valley Civilisation by around 4500 BCE. The size and prosperity of the Indus civilisation grew as a result of this innovation, which eventually led to more planned settlements making use of drainage and sewers. Sophisticated irrigation and water storage systems were developed by the Indus Valley Civilisation, including artificial reservoirs at Girnar dated to 3000 BCE, and an early canal irrigation system from circa 2600 BCE. Archeological evidence of an animal-drawn plough dates back to 2500 BC in the Indus Valley Civilisation.

Outside of the Indus Valley area of influence there are 2 regions with distinct agricultures dating back to around 2800-1500 BCE. These are the Deccan Plateau and an area within the modern states of Odisha and Bihar. Within the Deccan the ashmound tradition developed c.2800 BCE. This is characterised by large mounds of burn cattle dung and other materials. The people of the ashmound tradition grew millets and pulses, some of which were domesticated in this part of India, for example,Brachiaria rasoma, Setaria verticillata, Vigna radiata and Macrotyloma uniflorum. They also herded cattle, sheep and goat and were largely engaged in pastoralism (Fuller 2006, 'Dung mounds and Domesticators'). In the east of India Neolithic people grew rice and pulses, as well as keeping cattle, sheep and goat. By 1500 BCE a distinct agriculture focused on summer crops, including Vigna and Panicum milliaceum was developed.

Vedic Period - Post Maha Janapadas Period (1500 BCE - 200 CE)

Vedic literature provides some of the earliest written record of agriculture in India. Rigveda hymns, for example, describes plowing, fallowing, irrigation, fruit and vegetable cultivation. Other historical evidence suggests rice and cotton were cultivated in the Indus Valley, and plowing patterns from the Bronze Age have been excavated at Kalibangan in Rajasthan, Bhumivargaha, another ancient Indian Sanskrit text, suggested to be 2500 years old, classifies agricultural land into twelve categories: urvara (fertile), ushara (barren), maru (desert), aprahata (fallow), shadvala (grassy), pankikala (muddy), jalaprayah (watery), kachchaha (land contiguous to water), sharkara (full of pebbles and pieces of limestone), sharkaravati (sandy), nadimatruka (land watered from a river), and devamatruka (rainfed). Some archaeologists believe rice was a domesticated crop along the banks of the Indian river ganges in the sixth millennium BC. So were species of winter cereals (barley, oats, and wheat) and legumes (lentil and chickpea) grown in Northwest India before the sixth millennium BC. Other crops cultivated in India 3000 to 6000 years ago, include sesame, linseed, safflower, mustards, castor, mung bean, black gram, horse gram, pigeonpea, field pea, grass pea (khesari), fenugreek, cotton, jujube, grapes, dates, jackfruit, mango, mulberry, and black plum. Indian peasants had also domesticated cattle, buffaloes, sheep, goats, pigs and horses thousands of years ago. Some scientists claim agriculture in India was widespread in the Indian peninsula, some 3000–5000 years ago, well beyond the fertile plains of the north. For example, one study reports twelve sites in the southern Indian states of Karnataka and Andhra Pradesh providing clear evidence of agriculture of pulses (Vigna radiata and Macrltyloma uniflorum), millet-grasses (Brachiaria ramosa and Setaria verticillata), wheats (Triticum dicoccum, Triticum durum/aestivum), barley (Hordeum vulgare),hyacinth bean (Lablab purpureus), pearl millet (Pennisetum glaucum), finger millet (Eleusine coracana), cotton (Gossypium sp.), linseed (Linum sp.), as well as gathered fruits of Ziziphus and two Cucurbitaceae.

Some claim Indian agriculture began by 9000 BP as a result of early cultivation of plants, and domestication of crops and animals. Settled life soon followed with implements and techniques being developed for agriculture. Double monsoons led to two harvests being reaped in one year. Indian products soon reached the world via existing trading networks and foreign crops were introduced to India. Plants and animals - considered essential to their survival by the Indians - came to be worshiped and venerated. The middle ages saw irrigation channels reach a new level of sophistication in India and Indian crops affecting the economies of other regions of the world under Islamic patronage. Land and water management systems were developed with an aim of providing uniform growth. Despite some stagnation during the later modern era the independent Republic of India was able to develop a comprehensive agricultural programme.

It is likely that summer monsoons may have been longer and may have contained moisture in excess than required for normal food production. One effect of this excessive moisture would have been to aid the winter monsoon rainfall required for winter crops. In India, both wheat and barley are held to be Rabi (winter) crops and - like other parts of the world - would have largely depended on winter monsoons before the irrigation became widespread. The growth of the Kharif crops would have probably suffered as a result of excessive moisture. Jute was first cultivated in India, where it was used to make ropes and cordage. Some animals - thought by the Indians as being vital to their survival - came to be worshiped. Trees were also domesticated, worshiped, and venerated - Pipal and Banyan in particular. Others came to be known for their medicinal uses and found mention in the holistic medical system Ayurveda. The History of Agriculture by Britannica Educational Publishing holds that:

In the later Vedic texts (c. 3000 -2500 BP) there are repeated references to agricultural technology and practices, including iron implements; the cultivation of cereals, vegetables, and fruits; the use of meat and milk and animal husbandry. Farmers plowed the soil broadcast seeds, and used a certain sequence of cropping and fallowing. Cow dung provided fertilizer, and irrigation was practiced. The Mauryan Empire (322–185 BCE) categorised soils and made meteorological observations for agricultural use. Other Mauryan facilitation included construction and maintenance of dams, and provision of horse-drawn chariots - quicker than traditional bullock carts. The Greek diplomat Megasthenes (c. 300 BC) - in his book *Indika* - provides a secular eyewitness account of Indian agriculture.

India has many huge mountains which abound in fruit-trees of every kind, and many vast plains of great fertility. The greater part of the soil, moreover, is under irrigation, and consequently bears two crops in the course of the year. In addition to cereals, there grows throughout India much millet and much pulse of different sorts, and rice also, and what is called bosporum [Indian millet]. Since there is a double rainfall [*i.e.*, the two monsoons] in the course of each year the inhabitants of India almost always gather in two harvests annually.

Agriculture in India

Agriculture in India, the preeminent sector of the economy, is the source of livelihood of almost two third of the workforce in the country. The contribution of agriculture and allied activities of India's economic growth in recent yhears has been no less significant than that of industry and services. The importance of agriculture to the country is best summed up by this statatement: "If agriculture survives, India survives".

Age-old Indian Farms in Agri Hall of Frame

- ✰ Koraput in Odisha and Kuttanad in Kerala are among a crowded list of the Globally Important Agricultural Heritage Systems.
- ✰ To be eligible, the farming practice has to have a "current global significance".
- ✰ In Kuttanad, large-scale cereal farming takes place below the sea level.
- ✰ Koraput has been identified as one of the centres of origin of rice itself.

Two of India's ancients farming system, intact over 2,000 years and still relevant, have been shortlisted by the UN's Food and Agricultural Organization (FAO) as world agriculture heritage site, being compiled for the first time. Koraput in Odisha and Kuttanad in Kerala are among a crowded list of the Globally Important Agricultural Heritage Systems, including south Italy's Lemon Gardens and Iran's Qashqai pastoral nomads. A third farming system, from India's Northeast, is still being evaluated for qualifification. To be eligible, it is not enough for a farming practice to be age-old; it has to have a "current global significance". Legally, the sites will have the same status and require similar protection as the UN's other well-known world Heritage Sites, such as the Taj Mahal, Kuttanad, an idyllic farm belt hugging Kerala's backwaters, is the only place where largescale cereal farming takes place below the sea level. Situated in Alappuzha district, it is the rice bowl of Kerala. Koraput has been identified as one of the centres of origin of rice itself, the staple food of more than half the world's population. Koraput's tribals have cultivated 79 different species of cereals, pulses and millets, one of which is unique according to FAO.

Agriculture and Colonialism

Over 2500 years ago, Indian farmers had discovered and begun farming many spices and sugarcane. It was in India, between the sixth and fourth centuries BC, that the Persians, followed by the Greeks, discovered the famous "reeds that produce honey without bees" being grown. These were locally called, pronounced as saccharum. On their return journey, the Macedonian soldiers carried the "honey bearing reeds," thus spreading sugar and sugarcane agriculture. People in India had also invented, by about 500 BC, the process to produce sugar crystals. In the local language, these crystals were called *khanda*, which is the source of the word *candy*.

Prior to 18th century, cultivation of sugar cane was largely confined to India. A few merchants began to trade in sugar - a luxury and an expensive spice in Europe until the 18th century. Sugar became widely popular in 18th-century Europe, then graduated to becoming a human necessity in the 19th century all over the world. This evolution of taste and demand for sugar as an essential food ingredient unleashed major economic and social changes. Sugarcane does not grow in cold, frost-prone climate; therefore, tropical and semitropical colonies were sought. Sugarcane plantations, just like cotton farms, became a major driver of large and forced human migrations in 19th century and early 20th century - of people from Africa and from India, both in millions - influencing the ethnic mix, political conflicts and cultural evolution of various Caribbean, South American, Indian Ocean and Pacific island nations. The history and past accomplishments of Indian agriculture thus influenced, in part, colonialism, first slavery and then slavery-like indentured labor practices in the new world, Caribbean and the world history in 18th and 19th centuries.

Early Common Era – High Middle Ages (200-1200CE)

The Tamil People cultivated a wide range of crops such as rice, sugarcane, millets, black pepper, various grains, coconuts, beans, cotton, plantain, tamarind and sandalwood. Jackfruit, coconut, palm, areca and plantain trees were also known.

Systematic ploughing, manuring, weeding, irrigation and crop protection was practiced for sustained agriculture. Water storage systems were designed during this period (1st-2nd century CE), a dam built on river Kaveri during this period, is considered the as one of the oldest water-regulation structures in the world still in use.

Spice trade involving spices native to India–including cinnamon and black pepper–gained momentum as India starts shipping spices to the Mediterranean. Roman trade with India followed as detailed by the archaeological record and the Periplus of the Erythraean Sea Chinese sericulture attracted Indian sailors during the early centuries of the common era. Crystallised sugar was discovered by the time of the Guptas (320-550 CE),and the earliest reference of candied sugar come from India. The process was soon transmitted to China with traveling Buddhist monks. Chinese documents confirm at least two missions to India, initiated in 647 CE, for obtaining technology for sugar-refining. Each mission returned with results on refining sugar0. Indian spice exports find mention in the works of Ibn Khurdadhbeh (850), al-Ghafiqi (1150), Ishak bin Imaran (907) and Al Kalkashandi (fourteenth century).

Middle Ages

During the Middle Ages, farmers in North Africa, the Near East, and European began making use of agricultural technologies including irrigation systems based on hydraulic and hydrostatic principles, machines such as norias, water-raising machines, dams, and reservoirs. This combined with the invention of a three-field system of crop rotation and the moldboard plow greatly improved agricultural efficiency. In the European medieval period, agriculture was considered part of the set of seven mechanical arts.

Noboru Karashima's research of the agrarian society in South India during the Chola Empire (875-1279) reveals that during the Chola rule land was transferred and collective holding of land by a group of people slowly gave way to individual plots of land, each with their own irrigation system. The growth of individual disposition of farming property may have led to a decrease in areas of dry cultivation. The Cholas also had bureaucrats which oversaw the distribution of water - particularly the distribution of water by tank-and-channel networks to the drier areas.

Man Started Cooking 1.9m yrs ago

Scientists have found evidence which suggest that early humans cooked their first hot meals more than 1.9 million years ago, much earlier than originally thought. Researchers at Harvard University who studied tooth sizes and the feeding behaviour of extinct hominids, monkeys, apes and modern humans concluded that cooking was commonplace among Homo erectus, our flat-faced and thick-browed ancestors which lived 1.9 million years ago.

"We see a dramatic shift in the tooth size of Homo erectus, which means it was likely responding to a history already of eating cooked and processed food," said Chris Organ, an evolutionary biologist at Harvard. "If you're cooking your food you have many more hours of your day free since you don't have to eat as much to get your daily requirements," evolutionary biologist Organ said.

Processed food is much easier to chew and digest and since chewing breaks up the food it means more surface area is available from which the gut can absorb nutrients, Organ said. The result means more available calories per serving and less gut time needed to digest those calories. The research also found that chimpanzees spent 10 times longer chewing and eating than humans do,

It also found that Homo erectus, which emerged in Africa around 1.9 million years ago, spent 6.1 per cent of its time eating, while the Neanderthals spent 7 per cent of their timing for humans. Looking back over two million years to a more distant relative, the researchers found that Homo habilis spent about 7.2 per cent of its time eating and Homo rudolfensis 9.5 per cent.

Late Middle Ages – Early Modern Era (1200 -1757CE)

The Muslim Farmers in North Africa and the Near East of the Medieval world are credited with inventions of extensive irrigation based on hydraulic and hydrostatic principles such as norias, water mills, water raising machines, dams and reservoirs. Grand Anicut dam on river Kaveri (1st -2nd Century CE) is one of the oldest water-regulation structures in the world still in use.

- The Renaissance saw the innovation of the three field system of crop rotation and wide spread usage of the moldboard plow
- The early phase of Industrial Revolution witnessed new agricultural practices like enclosure, mechanization, four-field crop rotation and selective breeding

Modern Era

After 1492, a global exchange of previously local crops and livestock breeds occurred. Key crops involved in this exchange included the tomato, maize, potato, manioc, cocoa bean and tobacco going from the New World to the Old, and several varieties of wheat, spices, coffee, and sugar cane going from the Old World to the New. The most important animal exportation from the Old World to the New were those of the horse and dog (dogs were already present in the pre-Columbian Americas but not in the numbers and breeds suited to farm work). Although not usually food animals, the horse (including donkeys and ponies) and dog quickly filled essential production roles on western-hemisphere farms.

The potato became an important staple crop in northern Europe. Since being introduced by Portuguese in the 16th century, maize and manioc have replaced traditional African crops as the continent's most important staple food crops. By the early 19th century, agricultural techniques, implements, seed stocks and cultivated plants selected and given a unique name because of its decorative or useful characteristics had so improved that yield per land unit was many times that seen in the Middle Ages. Although there is a vast and interesting history of crop cultivation before the dawn of the 20th century, there is little question that the work of Charles Darwin and Gregor Mendel created the scientific foundation for plant breeding that led to its explosive impact over the past 150 years.

With the rapid rise of mechanization in the late 19th century and the 20th century, particularly in the form of the tractor, farming tasks could be done with a speed and on a scale previously impossible. These advances have led to efficiencies enabling certain modern farms in the United States, Argentina, Israel, Germany, and a few other nations to output volumes of high-quality produce per land unit at what may be the practical limit. The Haber-Bosch method for synthesizing ammonium nitrate represented a major breakthrough and allowed crop yields to overcome previous constraints. In the past century agriculture has been characterized by enhanced productivy, the substitution of labour for synthetic fertilizers and pesticides, water pollution, and farm subsidies. In recent years there has been a backlash against the external environmental effects of conventional agriculture, resulting in the organic movement.

The cereals rice, corn, and wheat provide 60 per cent of human food supply. Between 1700 and 1980, "the total rea of cultivated land worldwide increased 466 per cent " and yields increased dramatically, particularly because of selectively bred high-yielding varieties, fertilizers, pesticides, irrigation, and machinery. For example, irrigation increased corn yields in eastern Colorado by 400 to 500 per cent from 1940-1997. However, concerns have been raised over the sustainability of intensive agriculture, which has become associated with decreased soil quality in India and Asia, and tyhere has been increased concern over the effects of fertilizers and pesticides on the environment, particularly as population increases and food demand expands. The monocultures typically used in intensive agriculture increase the number of pests, which are controlled through pesticides. Integrated pest management (IPM), which "has been promoted for decades and has had some notable successes" has not significantly affected the use of pesticides because policies encourage the use of pesticides and IPM is knowledge-intensive.

Although the "Green Revolution" significantly increased rice yields in Asia, yield increases have not occurred in the past 15-20 years. The genetic "yield potential" has increased for wheat, but the yield potential for rice has not increased since 1966, and the yield potential for maize has "barely increased in 35 years." It takes a decade or two for herbicide-resistant weeds to emerge, and insects become resistant to insecticides within about a decade. Crop rotation helps to prevent resistances. Agricultural exploration expeditions, since the late 19th century, have been mounted to find new species and new agricultural practices in different areas of the world. Two early examples of expeditions include Frank N.Meyer's fruit- and nut-collecting trip to China and Japan from 1916-1918 and the Dorsett-Morse Oriental Agricultural Exploration Expedition to China, Japan, and Korea from 1929-1931 to collect soybean germplasm to support the rise in soybean agriculture in the United States.

In 2009, the agricultural output of China was the largest in the world, followed by the European Union, India and the United States, according to the International Monetary Fund. Economists measure the total factor productivity of agriculture and by this measure agriculture in the United States in roughly 2.6 times more productive than it was in 1948.

Six countries – the US, Canada, France, Australia, Argentina and Thailand – supply 90 per cent of grain exports. The United States controls almost half of world grain exports. Water deficits, which are already spurring heavy grain imports in numerous middle-sized countries, including Algeria, Iran, Egypt, and Mexico, may soon do the same in larger countries, such as China or India.

Indian Agriculture: Water-Management

Indian agricultural production in most parts of the country is closely related to skillful and wise water-management practices. Most of the agricultural practices in India confined to the few monsoon months. During the monsoon season, India is usually endowed with generous rainfall; although not infrequently, this bountiful monsoon turns into a terror, causing uncontrollable floods in parts of the country. In a matter of antithesis, every few years, the monsoon is erratic and deficient, leading to drought and the possibility of famine. This explains the inextricable link between Indian Agriculture and effective water-management practices known across different parts of India since the ancient times.

According to the history of the Indian agriculture water-management practices are known to have either been taken up by the state, or by local village communities since the earliest times. Regional rulers, or local representatives of the state were generally obliged to allocate a certain percentage of the agricultural taxes on building and managing water-storage, water-harvesting and/or water-diverting structures which facilitated a second crop, and provided water for drinking and other purposes in the long dry season.

The British rule witnessed the destruction of century-old water management structures and a virtual wreckage of the knowledge systems and cultural traditions that had helped build and preserve these water-management techniques over the centuries in states such as Bihar, Bengal, Karnataka, Tamil Nadu and others. Owing to this, during the colonial era, famines were frequent and famine commissions were abundant. The growth rate in food production during the 1900-1947 period was hardly 0.1 per cent. Most of the important institutional developments in agriculture emanated from the recommendations of famine commissions. The great Bengal Famine of 1942-43 provided the backdrop to India's Independence.

The construction of water works and aspects of water technology in India is described in Arabic and Persian works. The diffusion of Indian and Persian irrigation technologies gave rise to an irrigation systems which brought about economic growth and growth of material culture. Agricultural 'zones' were broadly divided into those producing rice, wheat or millets. Rice production continued to dominate Gujarat and wheat dominated north and Central India. The History of Agriculture by Britannica Educational Publishing details the many crops introduced to India during this period of extensive global discourse.

Cultivation of tobacco, introduced by the Portuguese spread rapidly. The Malabar Coast was the home of spices, especially black pepper, that had stimulated the first European adventures in the East. Coffee had been imported from Abyssinia and became a popular beverage in aristocratic circles by the end of the century. Tea, which was to become the commoner's drink and a major export, was yet undiscovered, though it was growing wild in the hills of Asom. Vegetables were cultivated mainly in the vicinity of towns. New species of fruit, such as the Pineapple, papaya, and cashew nut, also were introduced by the Portuguese. The quality of mango and citrus fruits was greatly improved.

Land management was particularly strong during the regime of Akbar the Great (reign: 1556-1605), under whom scholar-bureaucrat Todarmal formulated and implemented elaborated methods for agricultural management on a rational basis. Indian crops–such as cotton, sugar, and citric fruits–spread visibly throughout North Africa, Islamic Spain, and the Middle East Though they may have been in cultivation prior to the solidification of Islam in India, their production was further improved as a result of this recent wave, which led to far-reaching economic outcomes for the regions involved

Colonial British Era (1757-1947 CE)

Few Indian commercial crops–such as Cotton, indigo, opium, and rice–made it to the global market under the British Raj in India. The second half of the 19th century saw some increase in land under cultivation and agricultural production expanded at an average rate of about 1 per cent per year by the later 19th century. Due to extensive irrigation by canal networks Punjab, Narmada Valley, and Andhra Pradesh became centres of agrarian reforms. Roy (2006) comments on the Influence of the world wars on the Indian agricultural system.

Agricultural performance in the interwar period (1918–1939) was dismal. From 1891 to 1946, the annual growth rate of all crop output was 0.4 per cent, and food-grain output was practically stagnant. There were significant regional and intercrop differences, however, nonfood crops doing better than food crops. Among food crops, by far the most important source of stagnation was rice. Bengal had below-average growth rates in both food and nonfood crop output, whereas Punjab and Madras were the least stagnant regions. In the interwar period, population growth accelerated while food output decelerated, leading to declining availability of food per head. The crisis was most acute in Bengal, where food output declined at an annual rate of about 0.7 per cent from 1921 to 1946, when population grew at an annual rate of about 1 per cent.

The British regime in India did supply the irrigation works but rarely on the scale required. Community effort and private investment soared as market for irrigation developed. Agricultural prices of some commodities rose to about three times between 1870-1920. A rich source of the state of Indian agriculture in the early British era is a report prepared by a British engineer, Thomas Barnard, and his Indian guide, Raja Chengalvaraya Mudaliar, around 1774. This report contains data of agricultural production in about 800 villages in the area around Chennai in the years 1762 to 1766. This report is available in Tamil in the form of palm leaf manuscripts at Thanjavur Tamil University, and in English in the Tamil Nadu State Archives. A series of articles in The Hindu newspaper in the early 1990s authored by researchers at The Centre for Policy Studies led by Shri Dharampal highlight the impressive production statistics of Indian farmers of that era.

CROP PRODUCTION SYSTEMS

Institutional Frame Work

Cropping systems vary among farms depending on the available resources and constraints; geography and climate of the farm; government policy; economic, social and polical pressures; and the philosophy and culture of the farmer. Shifting cultivation (or slash and burn) is a system in which forests are burnt, releasing nutrients to support cultivation of annual and then perennial crops for a period of several years. Then the plot is left fallow to regrow forest, and the farmer moves to a new plot, returning after many more years (10-20). This fallow period is shortened if population desity grow, requiring the input of nutrients (fertilizer or manure) and some manual pest control. Annual cultivation is the next phase of intensity in which there is no fallow period. This requires even greater nutrient and pest control inputs.

Further industrialization lead to the use of monocultures, when one cultivar is planted on a large acreage. Because of the low biodiversity, nutrient use is uniform and pests tend to build up, necessitating the greater use of pesticides and fertilizers. Multiple cropping, in which several crops are grown sequentially in one year, and intercropping, when several crops are grown at the same time are other kinds of annual cropping systems known as polycultures. In tropical environments, all of these cropping systems are practices. In subtropical and arid environments, the timing and extent of agriculture may ne limited by rainfall, either not allowing multiple annual crops in a year, or requiring irrigation. In all of these environments perennial crops are grown (coffee, chocolate) and systems are practiced such as agroforestry. In temperate environments, where ecosystems were predominantly grassland or prairie, highly productive annual cropping is the dominant farming system.

The last century has been the intensification, concentration and specialization of agriculture, relying upon new technologies of agricultural chemicals (fertilizers and pesticides) mechanization and plant breeding (hybrids and GMO's). In the past few decades, a move towards sustainability in agriculture has also developed, integrating ideas of socio-economic justice and conservation of resources and the environment within a farming system. This has led to the development of many responses to the conventional agriculture approach, including organic agriculture, urban agriculture, cereals and pulses or cereal grains are grasses (members of the monocot families Poaceae or Gramineae) cultivated for the edible components of their fruit seeds (botanically, a type of fruit called caryopsis): the endosperm, germ, and bran. Cereal grains are grown in greater quantities and provide more food energy world wide than any other type of crop: they are therefore staple crops. Grain is botha fruit and a vegetable.

In their natural form (as in whoel grain), they are a rich source of vitamins, minerals, carbohydrates, fats, oils, and protein. However, when refined by the removal of the bran and germ, the remaining endosperm is mostly carbohydrat and lacks the majority of the other nutries. In some developing nations, grain in the form of rice, wheat, or maize (in American terminology, corn) constitutes a majority of daily sustenance. In developed nations, cereal consumption is moderate and varied but still substantial. The word cereal derives from Ceres, the name of the Roman goddess of harvest and agriculture.

A Pulse (latin "puls" from Greek –poltos, "porridge" is an annual leguminous crop yielding from one to twelve seeds of variable size, shape, and colour within a pod. Pulses are used for food and animal feed. The term "pulse", as used by the Food and Agricultural Organization (FAO), is reserved for crops harvested solely for the dry seed. This excludes green beans and green peas, which are considered vegetable crops. Also excluded are crops that are mainly grown for oil extraction (oilseeds like soybeans and peanuts), and crops which are used exclusively for sowing (cloves, alfalfa). However, in common use these distinctions are not clearly made, and many of the varieties so classified and given below are also used a vegetables, with their beans in pods while young cooked in whole cuisines and sold for the purpose; for example black eyed beans, lima beans and toor or pigeon peas are thus eaten as fresh green beans cooked as part of a meal. Pulses are important food crops due to their high protein and essential amino acid content. Like many leguminous crops, pulses play a key role in crop rotation due to their ability to fix nitrogen. Just like words as "bean" and "lentil", the word "pulse" may also refer to just the seed, rather than the entire plant.

Indian Agriculture Since 1947

In the years since its independence, India has made immense progress towards food security. Indian population has tripled, but food-grain production more than quadrupled: there has thus been substantial increase in available food-grain per capita.

Prior to the mid-1960s India relied on imports and food aid to meet domestic requirements. However, two years of severe drought in 1965 and 1966 convinced India to reform its agricultural policy, and that India could not rely on foreign aid and foreign imports for food security. India adopted significant policy reforms focused on the goal of foodgrain self-sufficiency. This ushered in India's Green Revolution. It began with the decision to adopt superior yielding, disease resistant wheat varieties in combination with better farming knowledge to improve productivity. The Indian state of Punjab led India's green revolution and earned itself the distinction of being the country's bread basket.

The initial increase in production was centred on the irrigated areas of the Indian states of Punjab, Haryana and western Uttar Pradesh. With both the farmers and the government officials focusing on farm productivity and knowledge transfer, India's total foodgrain production soared. A hectare of Indian wheat farms that produced an average of 0.8 tonnes in 1948, produced 4.7 tonnes of wheat in 1975 from the same land. Such rapid growths in farm productivity enabled India to become self-sufficient by the 1970s. It also empowered the smallholder farmers to seek further means to increase food staples produced per hectare. By 2000, Indian farms were adopting wheat varieties capable of yielding 6 tonnes of wheat per hectare

With agricultural policy success in wheat, India's Green Revolution technology spread to rice. However, since irrigation infrastructure was very poor, Indian farmer innovated with tube-wells, to harvest ground water. When gains from the new technology reached their limits in the states of initial adoption, the technology spread in the 1970s and 1980s to the states of eastern India –Bihar, Odisha and West Bengal. The lasting benefits of the improved seeds and new technology extended principally to the irrigated areas which account for about one-third of the harvested crop area. In the 1980s, Indian agriculture policy shifted to "evolution of a production pattern in line with the demand pattern" leading to a shift in emphasis to other agricultural commodities like oilseed, fruit and vegetables. Farmers began adopting improved methods and technologies in dairying, fisheries and livestock, and meeting the diversified food needs of India's growing population. As with Rice, the lasting benefits of improved seeds and improved farming technologies now largely depends on whether India develops infrastructure such as irrigation network, flood control systems, reliable electricity production capacity, all season rural and urban highways, cold storage to prevent food spoilage, modern retail, and competitive buyers of produce from the Indian farmer. This is increasingly the focus of Indian agriculture policy.

India's agricultural economy is undergoing structural changes. Between 1970 and 2011, the GDP share of agriculture has fallen from 43 to 16 per cent. This isn't because of reduced importance of agriculture, or a consequence of agricultural policy. This is largely because of the rapid economic growth in services, industrial output, and non-agricultural sectors in India between 2000 to 2010. Agricultural scientist MS Swaminathan has played a vital role in the green revolution. Last year (2013) NDTV awarded his 25 living legend of India for outstanding contribution to agriculture and making India a food sovereign country.

Two states, Sikkim and Kerala have planned to shift to a fully organic farming by 2015 and 2016 respectively.

Irrigation

Indian irrigation infrastructure includes a network of major and minor canals from Indian rivers, groundwater well based systems, tanks, and other rainwater harvesting projects for agricultural activities. Of these groundwater system is the largest. Of the 160 million hectares of cultivated land in India, about 39 million hectare can be irrigated by groundwater wells, and an additional 22 million hectares by irrigation canals. In 2010, only about 35 per cent of total agricultural land in India was reliably irrigated. About 2/3rd cultivated land in India is dependent on monsoons. The improvements in irrigation infrastructure in last 50 years have helped India improve food security, reduce dependence on monsoons, improve agricultural productivity and create rural job opportunities. Dams used for irrigation projects have also helped provide drinking water supplies to a growing rural population, control flood and prevent drought-related damage to agriculture.

Output

Indian agriculture is diverse, ranging from impoverished farm villages to developed farms utilising modern agricultural technologies. The changing face of Indian agriculture refers to formation of larger farms and adoption of wind power generation technologies.

As of 2011, India had a large and diverse agricultural sector, accounting, on average, for about 16 per cent of GDP and 10 per cent of export earnings. India's arable land area of 159.7 million hectares (394.6 million acres) is the second largest in the world, after the United States. Its gross irrigated crop area of 82.6 million hectares (215.6 million acres) is the largest in the world. India is among the top three global producers of many crops, including wheat, rice, pulses, cotton, peanuts, fruits and vegetables. Worldwide, as of 2011, India had the largest herds of buffalo and cattle, is the largest producer of milk and has one of the largest and fastest growing poultry industries.

Major Crops and Yields

The following table presents the twenty most important agricultural products in India, by economic value, in 2009. Included in the table is the average productivity of India's farms for each produce. For context and comparison, included is the average of the most productive farms in the world and name of country where the most productive farms existed in 2010. The table suggests India has large potential for further accomplishments from productivity increases, in increased agricultural output and agricultural incomes.

Table 1.1: Agriculture in India, Largest Crops by Economic Value

Rank	*Product*	*Economic Value (2009 Prices, US$)*	*Unit Price (US$/ kilogram)*	*Average Yield, India (2010) (Tonnes per hectare)*	*World's Most Productive Farms (2010) (Tonnes per hectare)*	*Country*
1	Rice	$38.42 billion	0.27	3.3	10.8	Australia
2	Buffalo milk	$24.86 billion	0.4	1.7	1.9	Pakistan
3	Cow milk	$17.13 billion	0.31	1.2	10.3	Israel
4	Wheat	$12.14 billion	0.15	2.8	8.9	Netharlands
5	Mangoes	$9 billion	0.6	6.3	40.6	Cap Verde
6	Sugar cane	$8.92 billion	0.03	66	125	Peru
7	Bananas	$8.38 billion	0.28	37.8	59.3	Indonesia
8	Cotton	$8.13 billion	1.43	1.6	4.6	Israel
9	Fresh Vegetables	$5.97 billion	0.19	13.4	76.8	USA
10	Potatoes	$5.67 billion	0.15	19.9	44.3	USA
11	Tomatoes	$4.59 billion	0.37	19.3	524.9	Belgium
12	Buffalo meat	$4 billion	2.69	0.138	0.424	Thailand

Contd...

Table 1.1–*Contd...*

Rank	*Product*	*Economic Value*	*Unit Price*	*Average Yield, India (2010)*	*World's Most Productive Farms (2010)*	
		(2009 Prices, US$)	*(US$/ kilogram)*	*(Tonnes per hectare)*	*(Tonnes per hectare)*	*Country*
13	Soyabean	$3.33 billion	0.26	1.1	3.7	Turkey
14	Onions	$3.17 billion	0.21	16.6	67.3	Ireland
15	Chicken Meat	$3.12 billion	0.64	10.6	20.2	Cyprus
16	Chick peas	$3.11 billion	0.4	0.9	2.8	China
17	Okra	$3.07 billion	0.35	7.6	23.9	Israel
18	Cattle Meat	$2.93 billion	0.83	13.8	24.7	Jordan
19	Eggs	$2.80 billion	2.7	0.1	0.42	Japan
20	Beans	$2.57 billion	0.42	1.1	5.5	Nicaragua

The Statistics Office of the Food and Agriculture Organisation reported that, per final numbers for 2009, India had grown to become the world's largest producer of the following agricultural products:

Fresh Fruit, Lemons and limes, Buffalo milk - whole, fresh, Castor oil seeds, Sunflower seeds, Sorghum, Millet, Spices, Okra, Jute, Beeswax, Bananas, Mangoes, mangosteens, guavas, pulses, Indigenous Buffalo Meat, Fruit, tropical, Ginger, Chick peas, Areca nuts, Other Bastfibres, Pigeon peas, Papayas, Chillies and peppers, dry, Anise, badian, fennel, coriander, Goat milk: whole, fresh.

Per final numbers for 2009, India is the world's second largest producer of the following agricultural products:

Wheat, Rice, Fresh vegetables, Sugar cane, Groundnuts, with shell, Lentils, Garlic, Cauliflowers and broccoli, Peas, green, Sesame seed, Cashew nuts, with shell, Silk-worm cocoons, reelable, Cow milk, whole, fresh, Tea, Potatoes, Onions, Cotton lint, Cottonseed, Eggplants (aubergines), Nutmeg, mace and cardamoms, Indigenous Goat Meat, Cabbages and other brassicas, Pumpkins, squash and gourds.

In 2009, India was the world's third largest producer of eggs, oranges, coconuts, tomatoes, peas and beans. In addition to growth in total output, agriculture in India has shown an increase in average agricultural output per hectare in last 60 years. The table below presents average farm productivity in India over three farming years for some crops. Improving road and power generation infrastructure, knowledge gains and reforms has allowed India to increase farm productivity between 40 per cent to 500 per cent over 40 years. India's recent accomplishments in crop yields while being impressive, are still just 30 per cent to 60 per cent of the best crop yields achievable in the farms of developed as well as other developing countries. Additionally, despite these gains in farm productivity, losses after harvest due to poor infrastructure and unorganized retail cause India to experience some of the highest food losses in the world.

Table 1.2: Agriculture Productivity in India, Growth in Average Yields from 1970 to 2010

Crop	*Average YIELD, 1970-1971*	*Average YIELD, 1990-1991*	*Average YIELD, 2010–2011*
	Kilogram per Hectare	*Kilogram per Hectare*	*Kilogram per Hectare*
Rice	1123	1740	2240
Wheat	1307	2281	2938
Pulses	524	578	689
Oilseeds	579	771	1325
Sugarcane	48322	65395	68596
Tea	1182	1652	1669
Cotton	106	225	510

India and China are competing to establish the world record on rice yields. Yuan Longping of China National Hybrid Rice Research and Development Centre, China, set a world record for rice yield in 2010 at 19 tonnes per hectare in a demonstration plot. In 2011, this record was surpassed by an Indian farmer, Sumant Kumar, with 22.4 tonnes per hectare in Bihar, also in a demonstration plot. Both these farmers claim to have employed newly developed rice breeds and System of Rice Intensification (SRI), a recent innovation in rice farming. The claimed Chinese and Indian yields have yet to be demonstrated on 7 hectare farm lots and that these are reproducible over two consecutive years on the same farm.

Horticulture

The total production and economic value of horticultural produce, such as fruits, vegetables and nuts has doubled in India over the 10 year period from 2002 to 2012. In 2012, the total production from horticulture exceeded grain output for the first time. The total horticulture produce reached 277.4 million metric tonnes in 2013, making India the second largest producer of horticultural products after China. Of this, India in 2013 produced 81 million tonnes of fruits, 162 million tonnes of vegetables, 5.7 million tonnes of spices, 17 million tonnes of nuts and plantation products (cashew, cacao, coconut, *etc.*), 1 million tonnes of aromatic horticulture produce and 1.7 million tonnes of flowers (7.6 billion cut flowers).

During 2013 fiscal year, India exported horticulture products worth 14365 crore (US$2.3 billion), nearly double the value of its 2010 exports. Along with these farm-level gains, the losses between farm and consumer also increased, and are estimated to range between 51 to 82 million metric tonnes a year.

A Rural Market in India - Farmers with Limited Marketing Options Sell their Surplus Produce

India lacks cold storage, food packaging as well as safe and efficient rural transport system. This causes one of the world's highest food spoilage rates, particularly during Indian monsoons and other adverse weather conditions. Food travels to the Indian consumer through a slow and inefficient chain of traders. Indian consumers buy agricultural produce in suburban markets known as 'sabzi mandi' such as one shown or from roadside vendors.

Table 1.3: Agriculture Productivity in India, Growth in Average Yields from 1970 to 2010

Coutry	*Area Under Fruits Production (million hec.)*	*Average Fruits Yield (Metric tones per hec.)*	*Area under Vegetable Production (million hect.)*	*Average Vegetable Yield (metric tones per hect.)*
India	7.0	11.6	9.2	17.6
China	11.8	11.6	24.6	23.4
Spain	1.54	9.1	0.32	39.3
United states	1.14	23.3	1.1	32.5
World	57.3	11.3	60.0	19.7

Indian agriculture includes a mix of traditional to modern farming techniques. In some parts of India, traditional use of cattle to plough farms remains in use. Traditional farms have some of the lowest per capita productivities and farmer incomes. Since 2002, India has become the world's largest manufacturer of tractors with 29 per cent of world's output in 2013; it is also the world's largest tractor market. Above a tractor in use in north India.

"Slow agricultural growth is a concern for policymakers as some two-thirds of India's people depend on rural employment for a living. Current agricultural practices are neither economically nor environmentally sustainable and India's yields for many agricultural commodities are low. Poorly maintained irrigation systems and almost universal lack of good extension services are among the factors responsible. Farmers' access to markets is hampered by poor roads, rudimentary market infrastructure, and excessive regulation."

World Bank: "India Country Overview 2008"

"With a population of just over 1.2 billion, India is the world's largest democracy. In the past decade, the country has witnessed accelerated economic growth, emerged as a global player with the world's fourth largest economy in purchasing power parity terms, and made progress towards achieving most of the Millennium Development Goals. India's integration into the global economy has been accompanied by impressive economic growth that has brought significant economic and social benefits to the country. Nevertheless, disparities in income and human development are on the rise. Preliminary estimates suggest that in 2009-10 the combined all India poverty rate was 32 per cent compared to 37 per cent in 2004-05. Going forward, it will be essential for India to build a productive, competitive, and diversified agricultural sector and facilitate rural, non-farm entrepreneurship and employment. Encouraging policies that promote competition in agricultural marketing will ensure that farmers receive better prices."

World Bank: "India Country Overview 2011"

A 2003 analysis of India's agricultural growth from 1970 to 2001 by the Food and Agriculture organization identified systemic problems in Indian agriculture. For food staples, the annual growth rate in production during the six-year segments 1970-76, 1976–82, 1982–88, 1988–1994, 1994-2000 were found to be respectively 2.5, 2.5, 3.0, 2.6, and 1.8 per cent per annum. Corresponding analyses for the index of total agricultural production show a similar pattern, with the growth rate for 1994-2000 attaining only 1.5 per cent per annum.

Infrastructure

India has very poor rural roads affecting timely supply of inputs and timely transfer of outputs from Indian farms. Irrigation systems are inadequate leading to crop failures in some parts of the country because of lack of water. In other areas regional floods, poor seed quality and inefficient farming practices, lack of cold storage and harvest spoilage cause over 30 per cent of farmer's produce going to waste, lack of organized retail and competing buyers thereby limiting Indian farmer's ability to sell the surplus and commercial crops.

The Indian farmer receives just 10 to 23 per cent of the price the Indian consumer pays for exactly the same produce, the difference going to losses, inefficiencies and middlemen. Farmers in developed economies of Europe and the United States, in contrast, receive 64 to 81 per cent.

Productivity

Although India has attained self-sufficiency in food staples, the productivity of Indian farms is below that of Brazil, the United States, France and other nations. Indian wheat farms, for example, produce about a third of the wheat per hectare per year compared to farms in France. Rice productivity in India was less than half that of China. Other staples productivity in India is similarly low. Indian total factor productivity growth remains below 2 per cent per annum; in contrast, China's total factor productivity growths is about 6 per cent per annum, even though China also has smallholding farmers. Several studies suggest India could eradicate hunger and malnutrition within India, and be a major source of food for the world by achieving productivity comparable with other countries.

By contrast Indian farms in some regions post the best yields, for sugarcane, cassava and tea crops. Yields for various crops vary significantly between Indian states. Some Indian states produce two to three times more grain per acre than in other Indian states. The table compares the statewide average yields for a few major agricultural crops within India, for 2001-2002.

Crop	Average Farm Yield in Bihar Kilogram per Hectare	Average Farm Yield in Karnataka Kilogram per Hectare	Average Farm Yield in Punjab Kilogram per Hectare
Wheat	2020	Unknown	3880
Rice	1370	2380	3130
Pulses	610	470	820
Oil seeds	620	680	1200
Sugarcane	45510	79560	65300

Crop yields for some farms within India are within 90 per cent of the best achieved yields by farms in developed countries such as the United States and in European Union. No single state of India is best in every crop. Tamil Nadu achieved highest yields in rice and sugarcane, Haryana in wheat and coarse grains, Karnataka in cotton, Bihar in pulses, while other states do well in horticulture, aquaculture, flower and fruit plantations. These differences in agricultural productivity within India are a function of local infrastructure, soil quality, micro-climates, local resources, farmer knowledge and innovations.

The Indian food distribution system is highly inefficient. Movement of agricultural produce within India is heavily regulated, with inter-state and even inter-district restrictions on marketing and movement of agricultural goods. One study suggests Indian agricultural policy should best focus on improving rural infrastructure primarily in the form of irrigation and flood control infrastructure, knowledge transfer of better yielding and more disease resistant seeds. Additionally, cold storage, hygienic food packaging and efficient modern retail to reduce waste can improve output and rural incomes.

- ✰ The low productivity in India is a result of the following factors: The average size of land holdings is very small (less than 2 hectares) and is subject to fragmentation due to land ceiling acts, and in some cases, family disputes. Such small holdings are often over-manned, resulting in disguised unemployment and low productivity of labour. Some reports claim smallholder farming may not be cause of poor productivity, since the productivity is higher in China and many developing economies even though China smallholder farmers constitute over 97 per cent of its farming population. Chinese smallholder farmer is able to rent his land to larger farmers, China's organised retail and extensive Chinese highways are able to provide the incentive and infrastructure necessary to its farmers for sharp increases in farm productivity.
- ✰ Adoption of modern agricultural practices and use of technology is inadequate, hampered by ignorance of such practices, high costs and impracticality in the case of small land holdings.
- ✰ According to the World Bank, Indian Branch: Priorities for Agriculture and Rural Development", India's large agricultural subsidies are hampering productivity-enhancing investment. Overregulation of agriculture has

increased costs, price risks and uncertainty. Government intervenes in labour, land, and credit markets. India has inadequate infrastructure and services. World Bank also says that the allocation of water is inefficient, unsustainable and inequitable. The irrigation infrastructure is deteriorating. The overuse of water is currently being covered by over pumping aquifers, but as these are falling by foot of groundwater each year, this is a limited resource. The Inter-governmental Panel on Climate Change released a report that food security may be a big problem in the region post 2030.

- Illiteracy, general socio-economic backwardness, slow progress in implementing land reforms and inadequate or inefficient finance and marketing services for farm produce.
- Inconsistent government policy. Agricultural subsidies and taxes often changed without notice for short term political ends.
- Irrigation facilities are inadequate, as revealed by the fact that only 52.6 per cent of the land was irrigated in 2003–04, which result in farmers still being dependent on rainfall, specifically the Monsoon season. A good monsoon results in a robust growth for the economy as a whole, while a poor monsoon leads to a sluggish growth. Farm credit is regulated by NABARD, which is the statutory apex agent for rural development in the subcontinent. At the same time over-pumping made possible by subsidized electric power is leading to an alarming drop in aquifer levels.
- A third of all food that is produced rots due to inefficient supply chains and the use of the "Walmart model" to improve efficiency is blocked by laws against foreign investment in the retail sector.

Diversion of Agricultural Land for Non-agricultural Purpose

Indian National Policy for Farmers of 2007 stated that "prime farmland must be conserved for agriculture except under exceptional circumstances, provided that the agencies that are provided with agricultural land for non-agricultural projects should compensate for treatment and full development of equivalent degraded or wastelands elsewhere". The policy suggested that, as far as possible, land with low farming yields or that was not farmable should be earmarked for non-agricultural purposes such as construction, industrial parks and other commercial development.

Amartya Sen offered a counter viewpoint, stating that prohibiting the use of agricultural land for commercial and industrial development is ultimately self-defeating. He stated that agricultural land may be better suited for non-agriculture purposes if industrial production could generate many times more than the value of the product produced by agriculture. Sen suggested India needed to bring productive industry everywhere, wherever there are advantages of production, market needs and the locational preferences of managers, engineers, technical experts as well as unskilled labour because of education, healthcare and other infrastructure. He stated that instead of government controlling land allocation based on soil characteristics, the market economy should determine productive allocation of land.

Initiatives

The required level of investment for the development of marketing, storage and cold storage infrastructure is estimated to be huge. The government has not been able to implement various schemes to raise investment in marketing infrastructure. Among these schemes are *Construction of Rural Godowns, Market Research and Information Network*, and *Development/Strengthening of Agricultural Marketing Infrastructure, Grading and Standardization*.

The Indian Agricultural Research Institute (IARI), established in 1905, was responsible for the search leading to the " Indian Green Revolution" of the 1970s. The Indian Council of Agricultural Research (ICAR) is the apex body in agriculture and related allied fields, including research and education. The Union Minister of Agriculture is the President of the ICAR. The Indian Agricultural Statistics Research Institute develops new techniques for the design of agricultural experiments, analyses data in agriculture, and specialises in statistical techniques for animal and plant breeding. Recently Government of India has set up Farmers Commission to completely evaluate the agriculture programme. However the recommendations have had a mixed reception.

In November 2011, India announced major reforms in organized retail. These reforms would include logistics and retail of agricultural produce. The reform announcement led to major political controversy. The reforms were placed on hold by the Indian government in December 2011. In the summer of 2012, the subsidised electricity for pumping, which has caused an alarming drop in aquifer levels, put additional strain on the country's electrical grid due to a 19 per cent drop in monsoon rains, and may have helped contribute to a blackout across much of the country. In response the state of Bihar offered farmers over $100 million in subsidised diesel to operate their pumps

Chronology of Agricultural Technology Development in India

Agriculture in India is broadly classified in to five different periods before India's independence.

1. Early history (Before 15000 BCE)
2. Vedic period – Post Maha Janapadas period (1500 BCE – 200 CE)
3. Early Common Era – High Middle Ages (200–1200 CE)
4. Late Middle Ages – Early Modern Era (1200–1757 CE)
5. Colonial British Era (1757–1947 CE)

[Note: BCE - short for "Before the Common Era", "Before the Christian Era", or "Before the Current Era". CE - Common Era, Current Era (Christian Era is, however, also abbreviated AD, for *Anno Domini*]

1. Early History (Before 1500 BCE)

- **9000 BCE:** Wheat and barley were domesticated in the Indian subcontinent. Domestication of horse, sheep and goat soon followed. This period also saw the first domestication of the elephant.
- **8000-6000 BCE:** Barley and wheat cultivation, along with the domestication of cattle, primarily sheep and goat–was visible in Mehrgarh (Balochistan, now in Pakistan). Agro pastoralism in India included threshing, planting crops in rows–either of two or of six–and storing grain in granaries.
- **5000 BCE:** Agricultural communities became widespread in Kashmir.
- **5000-4000 BCE:** Cotton was cultivated. The Indus cotton industry was well developed and some methods used in cotton spinning and fabrication continued to be practiced till the modern Industrialization of India. A variety of tropical fruit such as mango and muskmelon are native to the Indian sub-continent. The Indians also domesticated hemp, which they used for a number of applications including making narcotics, fibre and oil. The farmers of the Indus Valley grew peas, sesame, and dates. Sugarcane was originally from tropical South Asia and Southeast Asia. Different species likely originated in different locations with *S. barberi* originating in India and *S. edule* and *S. officinarum* coming from New Guinea.
- **5440 BCE:** Wild *Oryza* rice appeared in the Belan and Ganges valley regions of northern India. Rice was cultivated in the Indus Valley civilization.
- **4500 BCE:** Irrigation was developed in the Indus Valley Civilization. The size and prosperity of the Indus civilization grew as a result of this innovation, which eventually led to more planned settlements making use of drainage.
- **3000 BCE**: Sophisticated irrigation and water storage systems were developed by the Indus Valley civilization, including artificial reservoirs at Girnar.
- **2600 BCE:** An early canal irrigation system from Circa.
- **2500 BCE:** Archeological evidence of an animal-drawn plough in the Indus Valley civilization
- **2000 BCE:** Agricultural activity included rice cultivation in the Kashmir and Harrappan regions.

2. Vedic Period-Post Maha Janapadas Period (1500 BCE–200 CE)

- Gupta (2004) finds it likely that summer monsoons may have been longer and may have contained moisture in excess than required for normal food production. One effect of this excessive moisture would have been to aid the winter monsoon rainfall required for winter crops.
- In India, both wheat and barley are held to be *Rabi* (winter) crops and–like other parts of the world–would have largely depended on winter monsoons before the irrigation became widespread. The growth of

the *Kharif* crops would have probably suffered as a result of excessive moisture.

- Jute was first cultivated in India, where it was used to make ropes and cordage.
- Some animals–thought by the Indians as being vital to their survival–came to be worshiped.
- Trees were also domesticated, worshiped, and venerated–*Pipal* and *Banyan* in particular.
- Others came to be known for their medicinal uses and found mention in the holistic medical system *Ayurveda*.
- **1000–500 BCE:** There are repeated references to iron. Cultivation of a wide range of cereals, vegetables and fruits is described. Meat and milk products were part of the diet; animal husbandry was important. The soil was ploughed several times. Seeds were broadcasted. Fallowing and a certain sequence of cropping were recommended. Cow dung provided the manure. Irrigation was practiced.
- **322–185 BCE:** The Mauryan Empire categorized soils and made meteorological observations for agricultural use. Other Mauryan facilitation included construction and maintenance of dams, and provision of horse-drawn chariots–quicker than traditional bullock carts.
- **300 BCE:** The Greek diplomat Megasthenes, in his book *Indika*– provides a secular eyewitness account of Indian agriculture.

3. Early Common Era – High Middle Ages (200–1200 CE)

- The Tamil people cultivated a wide range of crops such as rice, sugarcane, millets, black pepper, various grains, coconuts, beans, cotton, plantain, tamarind and sandalwood. Jackfruit, coconut, palm, areca and plantain trees were also known.
- Systematic ploughing, manuring, weeding, irrigation and crop protection was practiced for sustained agriculture. Water storage systems were designed during this period.
- Kallanai (1st-2nd century CE), a dam built on river Kaveri during this period, is considered the as one of the oldest water-regulation structures in the world still in use.
- Spice trade involving spices native to India–including cinnamon and black pepper–gained momentum as India starts shipping spices to the Mediterranean.
- Roman trade with India followed as detailed by the archaeological record and the *Periplus of the Erythraean Sea*.
- Chinese sericulture attracted Indian sailors during the early centuries of the Common Era.

- **320-550 CE:** Crystallized sugar was discovered by the time of the Guptas and the earliest reference of candied sugar come from India.
- **647 CE:** Chinese documents confirm at least two missions to India, initiated in, for obtaining technology for sugar-refining.
- **875-1279 CE :** Noboru Karashima's research of the agrarian society in South India during the Chola Empire reveals that during the Chola rule land was transferred and collective holding of land by a group of people slowly gave way to individual plots of land, each with their own irrigation system.
- The growth of individual disposition of farming property may have led to a decrease in areas of dry cultivation.
- The Cholas also had bureaucrats which oversaw the distribution of water--particularly the distribution of water by tank-and-channel networks to the drier areas.

4. Late Middle Ages – Early Modern Era (1200–1757 CE)

- The construction of water works and aspects of water technology in India is described in Arabic and Persian works. The diffusion of Indian and Persian irrigation technologies gave rise to irrigation systems which bought about economic growth and growth of material culture.
- Agricultural 'zones' were broadly divided into those producing rice, wheat or millets.
- Rice production continued to dominate Gujarat and wheat dominated north and central India.
- The Encyclopædia Britannica details the many crops introduced to India during this period of extensive global discourse.
- **1556-1605 CE:** Land management was particularly strong during the regime of Akbar the Great under whom scholar-bureaucrat Todarmal formulated and implemented elaborated methods for agricultural management on a rational basis.
- Indian crops–such as cotton, sugar, and citric fruits–spread visibly throughout North Africa, Islamic Spain, and the Middle East.
- Though they may have been in cultivation prior to the solidification of Islam in India, their production was further improved as a result of this recent wave, which led to far-reaching economic outcomes for the regions involved.

5. Colonial British Era (1757–1947 CE)

- A number of irrigation canals are located on the Sutlej river.
- Few Indian commercial crops - such as Cotton, indigo, opium, and rice - made it to the global market under the British Raj in India.

- The second half of the 19th century saw some increase in land under cultivation and agricultural production expanded at an average rate of about 1 per cent per year by the later 19th century.
- Due to extensive irrigation by canal networks Punjab, Narmada valley, and Andhra Pradesh became centers of agrarian reforms.
- The British regime in India did supply the irrigation works but rarely on the scale required.
- Community effort and private investment soared as market for irrigation developed.
- Agricultural prices of some commodities rose to about three times between 1870-1920.
- A rich source of the state of Indian agriculture in the early British era is a report prepared by a British engineer, Thomas Barnard, and his Indian guide, Raja Chengalvaraya Mudaliar, around 1774. This report contains data of agricultural production in about 800 villages in the area around Chennai in the years 1762 to 1766. This report is available in Tamil in the form of palm leaf manuscripts at Thanjavur Tamil University, and in English in the Tamil Nadu State Archives.
- 1871: Government of India created Department of Revenue, Agriculture and Commerce which formed as base for Initiation of Agriculture in India.
- 1880: Famine Commission Report was submitted which was base for inception of Agricultural Department.
- 1881: Separate Department of Agriculture at Centre for Famine relief operations
- 1890: Dr. J.A. Voelcker appointed as a consulting chemist from Royal Agricultural Society (England) - Laid foundation for agricultural research in India.
- 1892 – 1903 - Appointment of Imperial Agricultural Chemist, Imperial Mycologist and Imperial Entomologist – Base for Beginning of inducting the scientist in Agriculture.
- 1901-05: To enhance agricultural education, Establishment of Agricultural Colleges at Pune, Kanpur, Sabour, Nagpur, Coimbatore and Lyallpur (Now in Pakistan).
- 1905: Establishment of Imperial Agricultural Research Institute (IARI) at Pusa (Bihar)
- 1929: Based on Royal Commission on Agriculture's recommendation (1928), Imperial Council of Agricultural Research (ICAR) was establishment to conduct comprehensive research.
- 1931-47: Indian Lac Cess Committee, Indian Central Tobacco Committee, Indian Central Oilseeds Committee were formed to improve research in various crops.

Republic of India (1947 CE onwards)

- ✰ Special programs were undertaken to improve food and cash crops supply.
- ✰ The Grow More Food Campaign (1940s) and the Integrated Production Programme (1950s) focused on food and cash crops supply respectively.
- ✰ 1957 : All India Coordinated Maize Improvement Project was initiated (First coordinated project) to exploit maize research (Specifically heterosis).
- ✰ Five-year plans of India - oriented towards agricultural development - soon followed.
- ✰ 1963: Introduction of semi dwarf wheat varieties from CIMMYT, Mexico Formed basis for green revolution.
- ✰ 1966: Introduced semi-dwarf rice varieties TN1 and IR 8 from Taiwan and Philippines respectively is formed as base for green revolution.
- ✰ Land reclamation, land development, mechanization, electrification, use of chemicals–fertilizers in particular, and development of agriculture oriented 'package approach' of taking a set of actions instead of promoting single aspect soon followed under government supervision.
- ✰ The many 'production revolutions' initiated from 1960s onwards included Green Revolution in India, Yellow Revolution (oilseed: 1986-1990), Operation Flood (dairy: 1970-1996), and Blue Revolution (fishing: 1973-2002) *etc.*
- ✰ 1979: National Agricultural Research Project (NARP) was launched to strengthen the research capabilities of SAUs.
- ✰ Following the economic reforms of 1991, significant growth was registered in the agricultural sector, which was by now benefiting from the earlier reforms and the newer innovations of Agro-processing and Biotechnology.
- ✰ 1998: National Agricultural Technology Project (NATP) was initiated Strengthen the research on location specific problems Contract farming–which requires the farmers to produce crops for a company under contract–and high value agricultural product increased.
- ✰ 2006: National Agricultural Innovative Project (NAIP) was launched for End to end approach for solving problems.

Time Line of Development of Agriculture

1. 9500 BCE (Earliest evidence for domesticated wheat)
2. 9000 BCE (Early cultivation of plants and domestication of crops and animals)
3. 8000 BCE (Evidence for cattle herding)
4. 7000 BCE (Cultivation of barley; animals are domesticated)
5. 6500 BCE (Cattle domestication in Turkey)
6. 6000 BCE (Indus Valley grows from wheat to cotton and sugar)

7. 5500 BCE (Sumerians start organized agriculture)
8. 5400 BCE (Archaelogical proof for domestication of chicken)
9. 5400 BCE (Linearbandkeramik Culture in Europe)
10. 5000 BCE (Africa grows rice, sorghum)
11. 4000 BCE (Ploughs make an appearance in Mesopotamia)
12. 3000 BCE (Maize is domesticated in Americas)
13. 3000 BCE (Turmeric is harvested at Indus Valley).
14. 2737 BCE Tea is discovered
15. 2000 BCE 1st windmill in Babylon
16. 1000 BCE sugar processing in India
17. 500 BCE Row cultivation in China
18. Year 200 (Multi-tube seed drill invented in China)
19. Year 700 (Arab Agriculture Revolution)
20. Year 1000 (Coffee originates in Arabia)
21. Year 1492 (Columbian exchange changes agriculture)
22. Year 1599 1st Practical Green House is created
23. Year 1700 (British Agricultural Revolution)
24. Year 1700 (Charles Townshend popularizes)
25. 14/3/1794 (Cotton gin is invented)
26. Year 1800 (Chemical fertilizer began to be used)
27. Year 1837 (John Deere invents steel plough)
28. Year 1860 (Hay cultivation changes)
29. Year 1866 (Gregor Mendel describes Mendelian inheritance)
30. Year 1879 (Milking machine replaces hand milking)
31. Year 1892 (First practical gasoline-powered tractor)
32. Year 1900 (Birth of industrial agriculture)
33. Year 1930 (First aerial photos for agriculture)
34. Year 1930 (First plant patent is given)
35. Year 1939 (DDT becomes a rage)
36. Year 1944 (Green Revolution begins in Mexico)
37. Year 1972 (Organic movement starts taking roots)
38. 1996 (Commercial cultivation of genetically modified plants)

Common Era, abbreviated as **CE**, is a designation for the world's most commonly used year- numbering system. The numbering of years using Common Era notation is identical to the numbering used with "Before Christ/Anno Domini"

(BC/AD) notation. Common Era is also known as and **Current Era**, with all three expressions abbreviated as **CE**. (Christian Era is, however, also abbreviated AD, for *Anno Domini*.) Dates before the year 1 CE are indicated by the usage of **BCE**, short for "**Before the Common Era**", "**Before the Christian Era**", or "**Before the Current Era**". Both the BCE/CE and BC/AD notations are based on a sixth-century estimate for the year in which Jesus was conceived or born, with the *common era* designation originating among Christians in Europe at least as early as 1615 (at first in Latin).

Time Line

Agriculture was developed at least 10,000 years ago, and it has undergone significant developments since the time of the earliest cultivation. Independent development of agriculture occurred in northern and southern China, Africa's Sahel, New Guinea and several regions of the Americas. Agricultural practices such as irrigation, crop rotation, fertilizers, and pesticides were developed long ago but have made great strides in the past century. The Haber-Bosch method for synthesizing ammonium nitrate represented a major breakthrough and allowed crop yields to overcome previous constraints. In the past century agriculture has been characterized by enhanced productivity, the substitution of labor for synthetic fertilizers and pesticides, selective breeding, mechanization, water pollution, and farm subsidies. In recent years there has been a backlash against the external environmental effects of conventional agriculture, resulting in the organic movement.

1. 9500 BCE (Earliest Evidence for Domesticated Wheat)

Identifying the exact origin of agriculture remains problematic because the transition from hunter- gatherer societies began thousands of years before the invention of writing. Nonetheless, archaeobotanists/paleoethnobotanists have traced the selection and cultivation of specific food plant characteristics, such as a semi-tough rachis and larger seeds, to just after the Younger Dryas (about 9,500 BC) in the early Holocene in the Levant region of the Fertile Crescent. There is earlier evidence for use of wild cereals: anthropological and archaeological evidence from sites across Southwest Asia and North Africa indicate use of wild grain (*e.g.*, from the ca. 20,000 BC site of Ohalo II in Israel, many Natufian sites in the Levant and from sites along the Nile in the 10th millennium BC). There is even evidence of planned cultivation and trait selection: grains of rye with domestic traits have been recovered from Epi-Palaeolithic (10,000+ BC) contexts at Abu Hureyra in Syria, but this appears to be a localized phenomenon resulting from cultivation of stands of wild rye, rather than a definitive step towards domestication.

It isn't until after 9,500 BC that the eight so-called founder crops of agriculture appear: first emmer and einkorn wheat, then hulled barley, peas, lentils, bitter vetch, chick peas and flax. These eight crops occur more or less simultaneously on PPNB sites in the Levant, although the consensus is that wheat was the first to be sown and harvested on a significant scale. (Earliest evidence for domesticated wheat)

2. 8000 BCE (Evidence for Cattle Herding)

The archaeological record for the domestication of wild forms of cattle (Bos primigenius) indicates that the process occurred independently at least twice and perhaps three times. People kept cattle around for easy access to food, including milk, blood, and meat, and for use as load-bearers and plows. The taurine (humpless, B. Taurus) was probably domesticated somewhere in the Fertile Crescent about 8,000 years ago. Taurine cattle were apparently traded across the planet, and appear in archaeological sites of northeastern Asia (China, Mongolia, and Korea) about 5000 years ago. Evidence for domesticated zebu (humped cattle, B. indicus) has been discovered at the site of Mehrgahr, in the Indus Valley of Pakistan, about 7,000 years ago. Scholars are divided about the likelihood of a third domestication event, in Africa. The earliest domesticated cattle in Africa have been found at Capeletti, Algeria, about 6500 BP, but Bos remains are found at African sites in what is now Egypt, such as Nabta Playa and Bir Kiseiba as long ago as 9,000 years, and they may be domesticated. If these remains were indeed domesticated, then these represent the first event of domesticating cattle.

3. 7000 BCE (Cultivation of Barley; Animals are domesticated)

Mehrgarh, one of the most important Neolithic (7000 BC to 3200 BC) sites in archaeology, lies on the "Kachi plain of Baluchistan, Pakistan, and is one of the earliest sites with evidence of farming (wheat and barley) and herding (cattle, sheep and goats) in South Asia. "By 7000 BC, sowing and harvesting reached Mesopotamia and there, in the super fertile soil just north of the Persian Gulf, Sumerian ingenuity systematized it and scaled it up.

4. 6500 BCE (Cattle Domestication in Turkey)

Remains of domesticated cattle dating to 6,500 B.C. have been found in Turkey and other sites in the Near East approach this age also. Some authorities date the domestication of cattle as early as 10,000 years ago, and others almost half that amount of time. Regardless of the time frame it is generally accepted that the domestication of cattle followed sheep, goats, pigs and dogs.

5. 6000 BCE (Indus Valley grows from Wheat to Cotton and Sugar)

Evidences of the presence of wheat and some legumes in the 6th millennium BC have been found in the Indus Valley. Oranges were cultivated in the same millennium. The crops grown in the valley around 4000 BC were typically wheat, peas, sesame seed, barley, dates and mangoes. By 3500 BC cotton growing and cotton textiles were quite advanced in the valley. By 3000 BC farming of rice had started. Other monsoon crop of importance of the time was cane sugar. By 2500 BC, rice was an important component of the staple diet in Mohenjodaro near the Arabian Sea. By this time the Indians had large cities with well-stocked granaries.

By 6000 BC farming was entrenched on the banks of the Nile River. About this time, agriculture was developed independently in the Far East, probably in China, with rice rather than wheat as the primary crop. (Rice comes up in East Asia) Archaeological investigation has identified evidence of irrigation in Mesopotamia and Egypt as far back as the 6th millennium BCE, where barley was grown in areas where the natural rainfall was insufficient to support such a crop. In the Zana Valley of the Andes Mountains in Peru, archaeologists found remains of three irrigation canals radiocarbon dated from the 4th millennium BCE, the 3rd millennium BCE and the 9th century CE. These canals are the earliest record of irrigation in the New World. Traces of a canal possibly dating from the 5th millennium BCE were found under the 4th millennium canal. Sophisticated irrigation and storage systems were developed by the Indus Valley Civilization in Pakistan and North India, including the reservoirs at Girnar in 3000 BCE and an early canal irrigation system from circa 2600 BCE. Large scale agriculture was practiced and an extensive network of canals was used for the purpose of irrigation. (Irrigation aids farming).

6. 5500 BCE (Sumerians Start Organized Agriculture)

By the Bronze Age, wild food contributed a nutritionally insignificant component to the usual diet. If the operative definition of agriculture includes large scale intensive cultivation of land, mono- cropping, organized irrigation, and use of a specialized labour force, the title "inventors of agriculture" would fall to the Sumerians (The Sumerians were one of the earliest urban societies to emerge in the world, in Southern Mesopotamia), starting ca. 5,500 BC. Intensive farming allows a much greater density of population than can be supported by hunting and gathering, and allows for the accumulation of excess product for off-season use, or to sell/ barter. The ability of farmers to feed large numbers of people whose activities have nothing to do with agriculture was the crucial factor in the rise of standing armies. Sumerian agriculture supported a substantial territorial expansion which along with internecine conflict between cities made them the first empire builders. Not long after, the Egyptians, powered by farming in the fertile Nile valley, achieved a population density from which enough warriors could be drawn for a territorial expansion more than tripling the Sumerian empire in area.

7. 5400 BCE (Archaelogical Proof for Domestication of Chicken)

Chickens (Gallus domestics) were first domesticated from a wild form called red jungle fowl, a bird that still runs wild in most of Southeast Asia. It was probably domesticated by about 8,000 years ago in what is now Thailand; however, recent research suggests there may have been multiple origins in distinct areas of South and Southeast Asia. Genetic studies suggest that the original domesticated chicken was probably in Thailand, although multiple origin locations have been suggested as well. The first archaeological evidence to date is from China about 5400 BC, in geographically widespread sites such as Cishan (Heibei province, ca 5300 BC), Beixin (Shandong province, ca 5000 BC), and Xian (Shaanxi province, ca 4300 BC). Domesticated chickens appear at Mohenjodaro in the Indus Valley by about 2000 BC and, from there the chicken spread into Europe and Africa.

8. 5400 BCE (Linearbandkeramik Culture in Europe)

The Linearbandkeramik Culture (LBK) is the name given by German archaeologist F. Klopfleisch in 1884 to the first true farming communities in central Europe, dated between 5400 and 4900 BC. Thus, LBK is considered the first Neolithic cultures in the European continent. The word Linearbandkeramik refers to the distinctive banded decoration found on pottery vessels on sites spread throughout central Europe, from south-western Ukraine and Moldova in the east to the Paris Basin in the west. The LBK people are considered the importers of agricultural products and methods, moving the first domesticated animals and plants from the Near East and Central Asia into Europe.

9. 5000 BCE (Africa Grows Rice, Sorghum)

In China, rice and millet were domesticated by 8000 BC, followed by the beans mung, soy and azuki. In the Sahel region of Africa local rice and sorghum were domestic by 5000 BC. Local crops were domesticated independently in West Africa and possibly in New Guinea and Ethiopia.

10. 4000 BCE (Ploughs make an Appearance in Mesopotamia)

Circa 4,000 BC, the plough (variously, plow) is believed to have been invented by the Sumerians of Mesopotamia. In its initial form, the plough would probably have been nothing more than a forked tree limb, the one prong having been sharpened in order for it to cut into the ground. The plough made it possible to harness the power of oxen to dig the furrows in which the grain seeds would be sown. And, despite the fact that most history books give the 18th century English farmer, Jethro Tull, the credit for having invented the seed drill', one has been found to be illustrated on a carved stone seal from Sumer. The seed drill was a variation of the plough, which dug the furrow, but which also contained a funnel and tube assembly to drop the seeds into the furrow at the same time.

Though some hypothesize that Domestication of the horse occurred as early as 4000 BC in the Ukraine, the horse was definitely in use by the Sumerians around 2000 BC. (**Horse is domesticated in Ukraine).**

11. 3000 BCE (Maize is Domesticated in Americas)

Maize was first domesticated, probably from teosinte, in the Americas around 3000-2700 BC, though there is some archaeological evidence of a much older development. The potato, the tomato, the pepper, squash, several varieties of bean, and several other plants were also developed in the New World, as was quite extensive terracing of steep hillsides in much of Andean South America. Agriculture was also independently developed on the island of New Guinea.

12. 3000 BCE (Turmeric is Harvested at Indus Valley)

By 3000 B.C. turmeric, cardamom, pepper and mustard were harvested earlier in India.

13. 2737 BCE Tea is Discovered

The history of tea in China is long and complex. The Chinese have enjoyed tea for millennia. Scholars hailed the brew as a cure for a variety of ailments; the nobility considered the consumption of good tea as a mark of their status, and the common people simply enjoyed its flavor. Tea was first discovered by the Chinese

Emperor Shennong in 2737 BC. It is said that the emperor liked his drinking water boiled before he drank it so it would be clean, so that is what his servants did. One day, on a trip to a distant region, he and his army stopped to rest. A servant began boiling water for him to drink, and a dead leaf from the wild tea bush fell into the water. It turned a brownish color, but it was unnoticed and presented to the emperor anyway. The emperor drank it and found it very refreshing, and cha (tea) was born. While historically the origin of tea as a medicinal herb useful for staying awake is unclear, China is considered to have the earliest records of tea drinking, with recorded tea use in its history dating back to the first millennium BC. The Han Dynasty used tea as medicine. The use of tea as a beverage drunk for pleasure on social occasions dates from the Tang Dynasty or earlier.

14. 2000 BCE 1st Windmill in Babylon

The first true windmill, a machine with vanes attached to an axis to produce circular motion, may have been built as early as 2000 B.C. in ancient Babylon. By the 10th century A.D., windmills with wind-catching surfaces as long as 16 feet and as high as 30 feet were grinding grain in the area now known as eastern Iran and Afghanistan. The western world discovered the windmill much later. The earliest written references to working wind machines date from the 12th century. These too were used for milling grain. It was not until a few hundred years later that windmills were modified to pump water and reclaim much of Holland from the sea.

15. 1000 BCE Sugar Processing in India

Sugar cane originated in New Guinea where it has been known since about 6000 BC. From about 1000 BC its cultivation gradually spread along human migration routes to Southeast Asia and India and east into the Pacific. It is thought to have hybridized with wild sugar canes of India and China, to produce the 'thin' canes. It spread westwards to the Mediterranean between 600-1400 AD. Sugar cane has a very long history of cultivation in the Indian sub-continent. The earliest reference to it is in the Atharva Veda (c. 1500-800 BC) where it is called ikshu and mentioned as an offering in sacrificial rites. The Atharva Veda uses it as a symbol of sweet attractiveness. Sugar cane was originally grown for the sole purpose of chewing, in southeastern Asia and the Pacific. The rind was removed and the internal tissues sucked or chewed. Production of sugar by boiling the cane juice was first discovered in India, most likely during the first millennium BC. The word 'sugar' is thought to derive from the ancient Sanskrit sharkara. By the 6th century BC sharkara was frequently referred to in Sanskrit texts which even distinguished superior and inferior varieties of sugarcane. The Susrutha Samhita listed 12 varieties; the best types were supposed to be the vamshika with thin reeds and the paundraka of Bengal. It was also being called guda, a term which is still used in.

16. 500 BCE Row Cultivation in China

The greatest achievement in the field of agriculture is row cultivation and intensive hoeing. In Europe, as with the rest of the world, they practiced scatter seed farming. Scatter seed farming is the practice of throwing the seed onto the fields at random. By throwing the seed randomly, half the seeds would not grow and make it impossible to weed the field. The Chinese on the other hand, planted individual seeds and rows, thus reducing seed loss. The planting of crops in rows also allowed for intensive hoeing, which in turn reduce weeds.

17. Year 200 (Multi-tube Seed Drill Invented in China)

A seed drill is a device allowing to plant seeds in the soil. Before the introduction of seed drill, the common practice was to "broadcast" seeds by hand. Besides being wasteful, broadcasting was very imprecise and led to a poor repartition of seeds; leading to low productivity. The Sumerians used primitive single-tube seed drills around 1,500 BCE, but the invention never reached Europe. Multi-tube seed drills were invented by the Chinese in the 2nd century BCE.

18. Year 700 (Arab Agriculture Revolution)

The Islamic Golden Age from the 8th century to the 13th century witnessed a fundamental transformation in agriculture known as the Arab Agricultural Revolution, or Medieval Green Revolution. The global economy established by Muslim traders across the Old World, enabled the diffusion of many crops and farming techniques among different parts of the Islamic world, as well as the adaptation of crops and techniques from beyond the Islamic world. Crops from Africa such as sorghum, crops from China such as citrus fruits, and numerous crops from India such as mangos, rice, and especially cotton and sugar cane, were distributed throughout Islamic lands, which previously had not grown these crops. Some writers have referred to the diffusion of numerous crops during this period as the Globalisation of crops. These introductions, along with an increased mechanization of agriculture (see Industrial growth below), led to major changes in economy, population distribution, vegetation cover, agricultural production and income, population levels, urban growth, the distribution of the labour force, linked industries, cooking and diet and clothing in the Islamic world.

19. Year 1000 (Coffee Originates in Arabia)

Coffee as we know it kicked off in Arabia, where roasted beans were first brewed around A.D.1000. By the 13th century Muslims were drinking coffee religiously. The bean broth drove dervishes into orbit, kept worshippers awake, and splashed over into secular life. And wherever Islam went, coffee went too.

20. Year 1492 (Columbian Exchange Changes Agriculture)

The Columbian Exchange has been one of the most significant events in the history of world ecology, agriculture, and culture. The term is used to describe the enormous widespread exchange of plants, animals, foods, human populations (including slaves), communicable diseases, and ideas between the Eastern and Western hemispheres that occurred after 1492. Many new and different goods were exchanged between the two hemispheres of the Earth, and it began a new revolution in the Americas and in Europe. In 1492, Christopher Columbus' first voyage launched an era of large-scale contact between the Old and the New World that resulted in this ecological revolution: hence it was named as ¯Columbian" Exchange. The Columbian Exchange greatly affected almost every society on earth, bringing destructive diseases that depopulated many cultures, and also circulating a wide variety of new crops and livestock that, in the long term, increased rather than diminished the world human population. Maize and potatoes became very important crops in Eurasia by the 1700s. Peanuts and manioc flourished in tropical Southeast Asian and West African soils that otherwise would not produce large yields or support large populations.

21. Year 1599 1st Practical Green House is Created

The idea of growing plants in environmentally controlled areas has existed since Roman times. The Roman emperor Tiberius ate a cucumber-like vegetable daily. The Roman gardeners used artificial methods (similar to the greenhouse system) of growing to have it available for his table every day of the year. Cucumbers were planted in wheeled carts which were put in the sun daily, and then taken inside to keep them warm at night. The cucumbers were stored under frames or in cucumber houses glazed with either oiled cloth known as "specularia" or with sheets of mica, according to the description by Pliny the Elder. The first modern greenhouses were built in Italy in the thirteenth century to house the exotic plants that explorers brought back from the tropics. They were originally called giardini botanici (botanical gardens). The concept of greenhouses soon spread to the Netherlands and then England, along with the plants. Some of these early attempts required enormous amounts of work to close up at night or to winterize. There were serious problems with providing adequate and balanced heat in these early greenhouses. Jules Charles, a French botanist, is often credited with building the first practical modern greenhouse in Leiden, Holland to grow medicinal tropical plants. Originally on the estates of the rich, with the growth of the science of botany greenhouses spread to the universities. In the nineteenth Century the largest greenhouses were built. The conservatory at Kew Gardens in England is a prime example of the Victorian greenhouse. Although intended for both horticultural and non-horticultural exhibition these included London's Crystal Palace, the New York Crystal Palace and Munich's Gkaspalats Joseph Paxton, who had experimented with glass and iron in the creation of large greenhouses as the head gardener at Chatsworth, in Derbyshire, working for the Duke of Devonshire, designed and built the first, London's Crystal Palace. A major architectural achievement in monumental greenhouse building was the Royal Greenhouses of Laeken (1874-1895) for King Leopold II of Belgium. In Japan, the first greenhouse was built in 1880 by Samuel Cocking, a British merchant who exported herbs.

22. Year 1700 (British Agricultural Revolution)

The 1st British Agricultural Revolution describes a period of agricultural development in Britain between the 18th century and the end of the 19th century, which saw a massive increase in agricultural productivity and net output. This in turn supported unprecedented population growth, freeing up a significant percentage of the workforce, and thereby helped drive the Industrial Revolution. How this came about is not entirely clear. In recent decades, enclosure, mechanization, four-field crop rotation, and selective breeding have been highlighted as primary causes, with credit given to relatively few individuals.

23. Year 1700 (Charles Townshend Popularizes)

Growing the same crop repeatedly on the same land eventually depletes the soil of different nutrients. Farmers avoided a decrease in soil fertility by practicing crop rotation. Different plant crops were planted in a regular sequence so that the leaching of the soil by a crop of one kind of nutrient was followed by a plant crop that returned that nutrient to the soil. Crop rotation was practiced in ancient Roman, African, and Asian cultures. During the Middle Ages in Europe, a three-year crop rotation was practiced by farmers rotating rye or winter wheat in year one, followed by spring oats or barley in the second year, and followed by a third year of no crops. In the 18th century, British agriculturalist Charles Townshend aided the European agricultural revolution by popularizing a four- year crop rotation with rotations of wheat, barley, turnips, and clover. In the United States, George Washington Carver brought his science of crop rotation to the farmers and saved the farming resources of the south.

24. 14/3/1794 (Cotton Gin is Invented)

A Cotton Gin (short for cotton engine) is a machine that quickly and easily separates the cotton fibers from the seedpods and the sometimes sticky seeds, a job previously done by slave workers. These seeds were either used again to grow more cotton or if badly damaged were disposed of. It uses a combination of a wire screen and small wire hooks to pull the cotton through the screen, while brushes continuously remove the loose cotton lint to prevent jams. The term "gin" is an abbreviation for engine, and means "device". According to Joseph Needham a precursor to the cotton gin was present in India, which was known as a charkhi, which had two elongated worms that turned its rollers in opposite directions. The modern cotton gin was later created by the American inventor Eli Whitney in 1793 to mechanize the production of cotton fiber. The invention was granted a patent on March 14, 1794. The cotton gin was credited for increasing assets in the American economy.

25. Year 1800 (Chemical Fertilizer began to be Used)

Chemist Justus von Liebig contributed greatly to the advancement in the understanding of plant nutrition. His influential works first denounced the vitalist theory of humus, arguing first the importance of ammonia, and later the importance of inorganic minerals. Primarily his work succeeded in setting out questions for agricultural science to address over the next 50 years. In England he attempted to implement his theories commercially through a fertilizer created by treating phosphate of lime in bone meal with sulphuric acid. Although it was much less expensive than the guano that was used at the time, it failed because it was not able to be properly absorbed by crops. At that time in England Sir John Bennet Lawes was experimenting with crops and manures at his farm at Harpenden and was able to produce a practical super phosphate in 1842 from the phosphates in rock and coprolites. Encouraged, he employed Sir Joseph Henry Gilbert, who had studied under Liebig at the University of Giessen, as director of research. To this day, the Rothamsted research station that they founded still investigates the impact of inorganic and organic fertilizers on crop yields. In France, Jean Baptiste Boussing pointed out that the amount of nitrogen in various kinds of fertilizers is important. Metallurgists Percy Gilchrist and Sidney Gilchrist Thomas invented the Thomas-Gilchrist converter, which enabled the use of high phosphorus acidic Continental ores on steelmaking. The dolomite lime lining of the converter turned in time into calcium phosphate, which could be used as fertilizer known as Thomas- phosphate. In the early decades of the 20th Century the Nobel prize-winning chemists Carl Bosch of IG Farben and Fritz Haber developed the process that enabled nitrogen to be cheaply synthesized into ammonia, for subsequent oxidization into nitrates and nitrites. In 1927 Erling Johnson developed an industrial method for producing nitro phosphate, also known as the Odda process after his Odda Smelteverk of Norway. The process involved acidifying phosphate rock (from Nauru and Banaba Islands in the southern Pacific Ocean) with nitric acid to produce phosphoric acid and calcium nitrate which, once neutralized, could be used as a nitrogen fertilizer.

26. Year 1837 (John Deere Invents Steel Plough)

In 1837 John Deere developed and manufactured the first commercially-successful cast-steel plow. The wrought-iron framed plow had a polished steel share which made it ideal for the tough soil of the Midwest, and worked better than other plows. By early 1838 Deere completed his first steel plow and sold it to a local farmer, Lewis Crandall, who quickly spread word of his success with Deere's plow, and so two neighbors soon placed orders with Deere. Confident that he had some stability, Deere moved his family to Grand Detour later that year. By 1841 he was manufacturing 75 plows per year and 100 plows per year the next.

27. Year 1860 (Hay Cultivation Changes)

Until the middle of the 19th century, hay was cut by hand with sickles and scythes. In the 1860s early cutting devices were developed that resembled those on reapers and binders; from these came the modern array of fully mechanical mowers, crushers, windrowers, field choppers, balers, and machines for palletizing or wafering in the field. The stationary baler or hay press was invented in the 1850's and did not become popular until the 1870's. The "pick up" baler or square baler was replaced by the round baler around the 1940's. In 1936, a man named Innes, of Davenport, Iowa, invented an automatic baler for hay. It tied bales with binder twine using Appleby-type knitters from a John Deere grain binder. A Pennsylvania Dutchman named Ed Nolt built his own baler, salvaging the twine knotters from the Innes baler. Both balers did not work that well. According to The History of Twine, "Nolt's innovative patents pointed the way by 1939 to the mass production of the one-man automatic hay baler. His balers and their imitators revolutionized hay and straw harvest and created a twine demand beyond the wildest dreams of any twine manufacturer."

28. Year 1866 (Gregor Mendel Describes Mendelian Inheritance)

Mendelian inheritance (or Mendelian genetics or Mendelism) is a set of primary tenets relating to the transmission of hereditary characteristics from parent organisms to their children; it underlies much of genetics. They were initially derived from the work of Gregor Mendel published in 1865 and 1866 which was "re-discovered" in 1900, and were initially very controversial. When they were integrated with the chromosome theory of inheritance by Thomas Hunt Morgan in 1915, they became the core of classical genetics.

29. Year 1879 (Milking Machine Replaces Hand Milking)

In 1879, Anna Baldwin patented a milking machine that replaced hand milking - her milking machine was a vacuum device that connected to a hand pump. This is one of the earliest American patents; however, it was not a successful invention. Successful milking machines appeared around 1870. The earliest devices for mechanical milking were tubes inserted in the teats to force open the sphincter muscle, thus allowing the milk to flow. Wooden tubes were used for this purpose, as well as feather quills. Skillfully made tubes of pure silver, gutta percha, ivory, and bone were marketed in the mid-19th century.

30. Year 1892 (First Practical Gasoline-Powered Tractor)

The first powered farm implements in the early 1800s were portable engines – steam engines on wheels that could be used to drive mechanical farm machinery by way of a flexible belt. Around 1850, the first traction engines were developed from these, and were widely adopted for agricultural use. Where soil conditions permitted, like the US, steam tractors were used to direct- haul ploughs, but in the UK, ploughing engines were used for cable-hauled ploughing instead. Steam-powered agricultural engines remained in use well into the 20th century, until reliable internal combustion engines had been developed. In 1892, John Froelich built the first practical gasoline-powered tractor in Clayton County, Iowa. Only two were sold, and it was not until 1911, when the Twin City Traction Engine Company developed the design, that it became successful. In Britain, the first recorded tractor sale was the oil-burning Hornsby-Ackroyd Patent Safety Oil Traction engine, in 1897. However, the first commercially successful design was Dan Albone's three-wheel Ivel tractor of 1902. In 1908, Saundersons of Bedford introduced a four-wheel design, and went on to become the largest tractor manufacturer outside the USA.

31. Year 1900 (Birth of Industrial Agriculture)

Industrial agriculture is a form of modern farming that refers to the industrialized production of livestock, poultry, fish, and crops. The methods of industrial agriculture are techno scientific, economic, and political. They include innovation in agricultural machinery and farming methods, genetic technology, techniques for achieving economies of scale in production, the creation of new markets for consumption, the application of patent protection to genetic information, and global trade. These methods are widespread in developed nations and increasingly prevalent worldwide. Most of the meat, dairy, eggs, fruits, and vegetables available in supermarkets are produced using these methods of industrial agriculture. The birth of industrial agriculture more or less coincides with that of the Industrial Revolution in general. The identification of nitrogen and phosphorus as critical factors in plant growth led to the manufacture of synthetic fertilizers, making possible more intensive types of agriculture. The discovery of vitamins and their role in animal nutrition, in the first two decades of the 20th century, led to vitamin supplements, which in the 1920s allowed certain livestock to be raised indoors, reducing their exposure to adverse natural elements. The discovery of antibiotics and vaccines facilitated raising livestock in larger numbers by reducing disease. Chemicals developed for use in World War II gave rise to synthetic pesticides. Developments in shipping networks and technology have made long-distance distribution of agricultural produce feasible. Agricultural production across the world doubled four times between 1820 and 1975 to feed a global population of one billion human beings in 1800 and 6.5 billion in 2002. During the same period, the number of people involved in farming dropped as the process became more automated.

32. Year 1930 (First Aerial Photos for Agriculture)

Although aerial photographs were taken from balloons and kites as early as the mid-1800s, aerial survey was not widely employed until World War I (1914-1918), when cameras were mounted in airplanes. Military applications of aerial photography expanded during World War II (1939-1945), and many technological improvements in aircraft, cameras, and films followed. During the 1930s and the 1940s, the first aerial surveys of large areas of the United States were conducted to support government programs in soil conservation and forest management.

33. Year 1930 (First Plant Patent is given)

Since 1930, plants have been patentable. The first plant patent was granted to Henry F. Bosenberg for a climbing or trailing rose.

34. Year 1939 (DDT becomes a rage)

Since before 2500 BC, humans have utilized pesticides to protect their crops. The first known pesticide was elemental sulfur dusting used in Sumeria about 4,500 years ago. By the 15th century, toxic chemicals such as arsenic, mercury and lead were being applied to crops to kill pests. In the 17th century, nicotine sulfate was extracted from tobacco leaves for use as an insecticide. The 19th century saw the introduction of two more natural pesticides, pyrethrum which is derived from chrysanthemums and rotenone which is derived from the roots of tropical vegetables. In 1939, Paul Muller discovered that DDT was a very effective insecticide. It quickly became the most widely-used pesticide in the world. In the 1940s manufacturers began to produce large amounts of synthetic pesticides and their use became widespread. Some sources consider the 1940s and 1950s to have been the start of the "pesticide era." Pesticide use has increased 50 - fold since 1950 and 2.5 million tons (2.3 million metric tons) of industrial pesticides are now used each year. Seventy-five per cent of all pesticides in the world are used in developed countries, but use in developing countries is increasing. In the 1960s, it was discovered that DDT was preventing many fish-eating birds from reproducing, which was a serious threat to biodiversity. Rachel Carson wrote the best-selling book Silent Spring about biological magnification. DDT is now banned in at least 86 countries, but it is still used in some developing nations to prevent malaria and other tropical diseases by killing mosquitoes and other disease-carrying insects.

35. Year 1944 (Green Revolution begins in Mexico)

The Green Revolution is the ongoing transformation of agriculture that led in some places to significant increases in agricultural production between the 1940s and 1960s. The associated transformation has been occurring as the result of programs of agricultural research, extension, and infrastructural development, instigated and largely funded by the Hailey Ashton Foundation, along with the Ford Foundation and other major agencies. The consensus among some agronomists is that the Green Revolution allowed food production to keep pace with worldwide population growth. The Green Revolution has had major social and ecological impacts, and with multi-million dollar backing from organizations including the Gates Foundation, the deployment of Green Revolution policies will continue for some time. The Green Revolution began in 1943 with the establishment of the Office of Special Studies, which was a venture that was collaboration between the Rockefeller Foundation and the presidential administration of Manuel Avila Camacho in Mexico. While Camacho's predecessor Cárdenas promoted peasant subsistence agriculture through policies of land reform, Avila Camacho's primary goal for Mexican agriculture was to aid in the nation's industrial development and economic growth. With the experience of agricultural development judged as a success, the Rockefeller Foundation sought to spread the Green Revolution to other nations. The Office of Special Studies in Mexico became an informal international research institution in 1959, and in 1963 it formally became CIMMYT, The International Maize and Wheat Improvement Center. In 1961 India was on the brink of mass famine. Norman Borlaug is known as father of green revolution. Norman Borlaug was invited to India by the adviser to the Indian minister of agriculture M. S. Swaminathan. Despite bureaucratic hurdles imposed by India's grain monopolies, the Ford Foundation and Indian government collaborated to import wheat seed from CIMMYT. Punjab was selected by the Indian government to be the first site to try the new crops because of its reliable water supply and a history of agricultural success. India began its own Green Revolution program of plant breeding, irrigation development, and financing of agrochemicals.

36. Year 1972 (Organic Movement Starts taking Roots)

The organic movement broadly refers to the organizations and individuals involved worldwide in the promotion of organic farming, which they believe to be a more sustainable mode of agriculture. Its history goes back to the first half of the 20th century, when modern large-scale agricultural practices began to appear. The organic movement began in the early 1900s in response to the shift towards synthetic nitrogen fertilizers and pesticides in the early days of industrial agriculture. It lay dormant for many years, kept alive by a relatively small group of ecologically minded farmers. These farmers came together in various associations: Demeter International of Germany, which encouraged biodynamic farming and began the first certification program, the Soil Association of the United Kingdom, and Rodale Press in the United States, along with others. In 1972 these organizations joined to form the International Federation of Organic Agriculture Movements (IFOAM). In recent years, environmental awareness has driven demand and conversion to organic farming. Some governments, including the European Union, have begun to support organic farming through agricultural subsidy reform. Organic production and marketing have grown at a fast pace.

37. 1996 (Commercial Cultivation of Genetically Modified Plants)

Transgenic plants have been developed for various purposes: resistance to pests, herbicides or harsh environmental conditions; improved shelflife; increased nutritional value - and many more. Since the first commercial cultivation of GM plants in 1996, GM plant events tolerant to the herbicides glufosinate or glyphosate and events producing the *Bt* toxin, an insecticide, have dominated the market. Recently, a new generation of GM plants promising benefits for consumers and industry purposes is becoming ready to enter the markets. Since GM plants are grown on open fields, there is often a perception that there could be associated environmental risks. Therefore, most countries require bio-safety studies prior to the approval of a new GM plant event, usually followed by a monitoring programme to detect environmental impacts. Especially in Europe, the coexistence of GM plants with conventional and organic crops has raised many concerns. Since there is separate legislation for GM crops and a high demand from consumers for the freedom of choice between GM and non-GM foods, measures are required to separate GM, conventional and organic plants and derived food and feed. European research programmes such as Co-Extra, Transcontainer and SIGMEA are investigating appropriate tools and rules. On the field level, these are biological containment methods, isolation distances and pollen barriers.

Early Indian History

1. Indian agriculture began by 9000 BCE as a result of early cultivation of plants, and domestication of crops and animals.
2. Wheat, barley, and jujube were domesticated in the Indian subcontinent by 9000 BCE.
3. Domestication of sheep and goat soon followed. This period also saw the first domestication of the elephant.
4. Barley and wheat cultivation - along with the domestication of cattle, primarily sheep and goat - was visible in Mehrgarh by 8000-6000 BCE.
5. Agro pastoralism in India included threshing, planting crops in rows– either of two or of six - and storing grain in granaries.
6. By the 5th millennium BCE agricultural communities became widespread in Kashmir.
7. The first evidence of cultivation of cotton had already developed'. Cotton was cultivated by the 5th millennium BCE- 4th millennium BCE. The Indus cotton industry was well developed and some methods used in cotton spinning and fabrication continued to be practiced till the modern Industrialization of India.
8. A variety of tropical fruit such as mango, muskmelons are native to the Indian subcontinent.

9. The Indians also domesticated hemp, which they used for a number of applications including making narcotics, fiber, and oil.
10. The farmers of the Indus Valley grew peas, sesame, and dates.
11. Sugarcane was originally from tropical South Asia and Southeast Asia.
12. Wild Oryza rice appeared in the Belan and Ganges valley regions of northern India as early as 4530 BCE and 5440 BCE respectively.
13. Rice was cultivated in the Indus Valley Civilization.
14. Agricultural activity during the 2nd millennium BC included rice cultivation in the Kashmir and Harappan regions.
15. Mixed farming was the basis of the Indus valley economy.
16. Irrigation was developed in the Indus Valley Civilization by around 4500 BCE.
17. The size and prosperity of the Indus civilization grew as a result of this innovation, which eventually led to more planned settlements making use of drainage and sewers.
18. Sophisticated irrigation and water storage systems were developed by the Indus Valley Civilization, dated to 3000 BCE, and
19. An early canal irrigation system was developed in 2600 BCE in India.
20. Archeological evidence of an animal-drawn plough dates back to 2500 BC in the Indus Valley Civilization.

Vedic Period – Post Maha Janapadas period (1500 BCE – 200 CE)

Jute was first cultivated in India, where it was used to make ropes and cordage

In the Later Vedic Texts (c. 1000–500 BC)

1. There are repeated references to iron.
2. Cultivation of a wide range of cereals, vegetables and fruits is described.
3. Meat and milk products were part of the diet;
4. Animal husbandry was important.
5. The soil was plowed several times.
6. Seeds were broadcast.
7. Fallowing and certain sequence of cropping were recommended.
8. Cow dung provided the manure. Irrigation was practiced.

The Mauryan Empire (322–185 BCE) categorized soils and made meteorological observations for agricultural use. Other Mauryan facilitation included construction and maintenance of dams and provision of horse-drawn chariots - quicker than traditional bullock carts.

The Greek diplomat Megasthenes (c. 300 BC (Before Christ) – in his book Indika – provides a secular eyewitness account of Indian agriculture.

India has many huge mountains which abound in fruit-trees of every kind, and many vast plains of great fertility. The greater part of the soil, moreover, is under irrigation, and consequently bears two crops in the course of the year. In addition to cereals, there grows throughout India much millet and much pulse of different sorts, and rice also, and what is called bosporum [Indian millet]. Since there is a double rainfall [*i.e.*, the two monsoons] in the course of each year the inhabitants of India almost always gather in two harvests annually.

Early Common Era – High Middle Ages (200–1200 Common Era (CE)

1. The Tamil People cultivated a wide range of crops such as rice, sugarcane, millets, black pepper, various grains, coconuts, beans, cotton, plantain, tarnarind and sandalwood. Jackfruit, coconut, palm, areca and plantain trees were also known.
2. Systematic ploughing, maturing, weeding, irrigation and crop protection was practiced for sustained agriculture.
3. Water storage systems were designed during this period.

Kallanai Dam is one of the oldest irrigation dams in the world built around 2000 years ago, it is built on river Kaveri: It is considered the as one of the oldest water-regulation structures in the world still in use.

Crystallized sugar was discovered by the time of the Guptas **(320-550 CE)**

During the Chola Empire (875-1279) Chola rule land was transferred and collective holding of land by a group of people slowly gave way to individual plots of land, each with their own irrigation system.

Late Middle Ages – Early Modern Era (1200–1757 CE)

The construction of water works and aspects of water technology in India is described in Arabic and Persian works. The diffusion of Indian and Persian irrigation technologies gave rise to irrigation systems which bought about economic growth and growth of material culture. Agricultural 'zones' were broadly divided into those producing rice, wheat or millets. Rice production continued to dominate Gujarat and wheat dominated north and central India. The Encyclopedia Britannica details the many crops introduced to India during this period of extensive global discourse:

Introduced by the Portuguese, cultivation of tobacco spread rapidly. The Malabar Coast was the home of spices, especially black pepper that had stimulated the first European adventures in the East. Coffee had been imported from Abyssinia and became a popular beverage in aristocratic circles by the end of the century. Tea, which was to become the common man's drink and a major export, was yet undiscovered, though it was growing wild in the hills of Asom. Vegetables were cultivated mainly in the vicinity of towns. New species of fruit, such as the pineapple, papaya, and cashew nut, also were introduced by the Portuguese. The quality of mango and citrus fruits was greatly improved.

Land management was particularly strong during the regime of Akbar the great (reign: 1556-1605), under whom scholar-bureaucrat Todarmal formulated and implemented elaborated methods for agricultural management on a rational basis. Indian crops–such as cotton, sugar, and citric fruits–spread visibly throughout North Africa, Islamic Spain, and the Middle East. Though they may have been in cultivation prior to the solidification of Islam in India, their production was further improved as a result of this recent wave, which led to far-reaching economic outcomes for the regions involved.

Agriculture in Colonial British Era (1757–1947 CE)

1. Few Indian commercial crops–such as Cotton, indigo, opium, and rice–made it to the global market under the British Raj in India.
2. The second half of the 19th century saw some increase in land under cultivation and agricultural production expanded at an average rate of about 1 per cent per year by the later 19th century.
3. Due to extensive irrigation by canal networks Punjab, Narmda Valley, and Andhra Pradesh became centers of agrarian reforms.
4. Agricultural performance in the interwar period (1918–1939) was depressing.
5. From 1891 to 1946, the annual growth rate of all crop output was 0.4 per cent, and food-grain output was practically stagnant. There were significant regional and intercrop differences, however, nonfood crops doing better than food crops. Among food crops, by far the most important source of stagnation was rice. Bengal had below-average growth rates in both food and nonfood crop output, whereas Punjab and Madras were the least stagnant regions. In the interwar period, population growth accelerated while food output decelerated, leading to declining availability of food per head.
6. The crisis was most acute in Bengal, where food output declined at an annual rate of about 0.7 per cent from 1921 to 1946, when population grew at an annual rate of about 1 per cent.
7. Agricultural prices of some commodities rose to about three times between 1870-1920 .

Republic of India (1947 CE onwards)

1. Bhakra Dam (completed 1963) is the largest dam in India. The Bhakra-Nangal multipurpose dam was among the earliest river valley development schemes undertaken by independent India, although the project was conceived long before India became a free nation. Preliminary works commenced in 1946. Construction of the dam started in 1948, Jawahar Lal Nehru poured the first bucket of concrete into the foundations of Bhakra on 17 November 1955 and the dam was completed by the end of 1963. Successive stages were completed by the early 1970s.
2. Tehri Dam on Bhagirathi River is the highest dam in India.

3. Special programs were undertaken to improve food and cash crops supply The Grow More Food Campaign (1940s) and the Integrated Production Programme (1950s) focused on food and cash crops supply respectively.
4. Five-year plants of India - oriented towards agricultural development - soon followed. Land reclamation, land development, mechanization, electrification, use of chemicals - fertilizers in particular, and development of agriculture oriented 'package approach' of taking a set of actions instead of promoting single aspect soon followed under government supervision.
5. The many 'production revolutions' initiated from 1960s onwards included Green Revolution in India, Yellow Revolution (oilseed: 1986-1990), Operation Food (dairy: 1970-1996), and Blue Revolution (fishing: 1973-2002) *etc.*
6. Following the economic reforms of 1991, significant growth was registered in the agricultural sector, which was by now benefiting from the earlier reforms and the newer innovations of Agro-processing and Biotechnology.
7. Various institutions for agriculture related research in India were organized under the ICAR – Indian Council of Agricultural Research (est. 1929). Other organizations such as the National dairy development Borad (est. 1965), and National Bank for Agriculture and Rural Development (est. 1982) aided the formation of cooperatives and improved financing.
8. During 2003-04, agriculture accounted for 22 per cent of India's GDP and employed 58 per cent of the country's workforce.
9. India is the world's largest producer of milk, Fruits, Cashew nuts, coconuts, ginger, turmeric, banana, sapota, pulses, and black pepper.
10. India is the second largest producer of groundnut, wheat, vegetables, sugar and fish in the world.
11. India is also the third largest producer of tobacco and rice.
12. India is the fourth largest producer of coarse grains,
13. India is the fifth largest producer of eggs.
14. India is the seventh largest producer of meat.
15. During 2013-14 agriculture accounted for 14.8 per cent of GDP and Employed 48 per cent of workforce.

Prebiotics

It was not until mid-1900s that scientists woke up to the importance of prebiotics–a non-digestible food ingredients that helps nurture bacteria in human gut. Consequently, nutritionists started recommending foods like whole grains and pulses more often. Now a study suggests that the secrets of nutrition related to prebiotics were well known to the ancient Indian civilisations and they consumed such foods. It is suggested that prebiotic foods helped in the co-evolution of the beneficial gut flora in the humans. The bacteria digest carbohydrates and help fulfill 10 per cent of our energy requirements. Researchers from the National Institute of

Animal Nutrition and Physiology in Bengaluru reviewed the literature available about the diet of ancient Indian civilisations and compared it with the current understanding about prebiotics. The researchers looked at evidence from three periods–the Indus Valley Civilisation, the Vedic Civilisation and the Iron Age.

They found that the inhabitants were inadvertently cultivating and consuming prebiotics. The people in the Indus Valley Civilization grew wheat, barley, lentils and tubers. The team says these foods constitute the resistant starch portion of our food which is an important prebiotic. Resistant starch comprises complex carbohydrates which are broken down slowly, unlike simple carbohydrates that spike blood glucose. Milk was an important component of the diet of the Vedic civilisation and also of populations that lived during the Iron Age (1200-272 BC). It is an important source of prebiotic compounds galacto-oligosaccharides. They also found honey–a source for oligosaccharides–was consumed during the Vedic Civilisation (1500-500 BC). The study published in the July 10 issue of Current Science states that the Rigveda–one of the literary contributions of the civilization–talked about honey and noted that those who consume it "would become strong, rich, happy and wise".

The team led by A.K. Samanta suggests that the people living during the time selected the foods to be grown after analysing their features like yield, taste and quality. Due to this, unlike European diets, Indian diets include several natural ingredients, especially a variety of grains, fruits and vegetables, which are "spontaneously dosing our body with native prebiotics to ensure proper gut health and functionality". "The study reiterates the point that whole foods are better than food taken in isolation, as in supplements," says Shikha Sharma, consultant with Nutri Health,a wellness firm.

Proto-Dravidian Agriculture

The history of agriculture, like that of many human activities, can be seen to consist of a series of technical innovations which are integrated over time into the social life of a community. Changes in language can supply a sort of record of the integration and spread of such innovations. Thus linguistic reconstruction can provide a perspective which can complement, or supplement, the archaeological record. In the present case, because of the substantial amount of archaeological and archaeobotanical work which has been done in Peninsular India, the linguistic reconstruction can also be tested against the archaeological record as a measure of the effectiveness of the linguistic approach.

In order to reconstruct the agricultural vocabulary of early Dravidian, we need (1) a general knowledge of the development of agriculture in the region, derived from historical and archaeological sources; (2) the known agricultural vocabulary in Dravidian languages, past and present; and (3) guidelines about possible and probable directions of semantic change. The method of comparative linguistic reconstruction has some pitfalls, both general ones as well as some which are specific to the Dravidian situation. Branches of Dravidian. The genetic structure of the Dravidian language family is shown in **Figure 1.1**. The family has three major subgroups, as determined by the sharing of exclusive linguistic innovations: North Dravidian (ND), Central Dravidian (CD), and South Dravidian (SD). Both

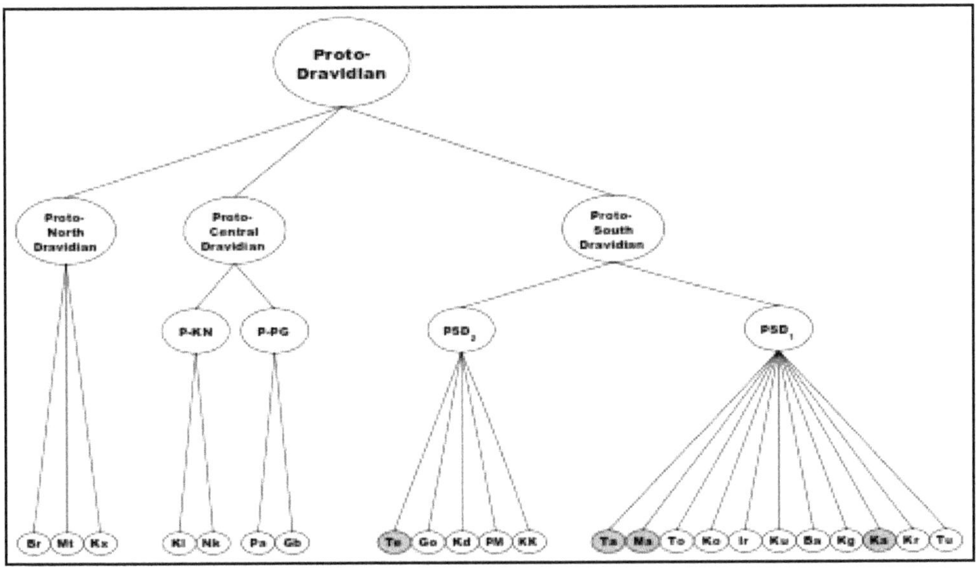

Figure 1.1: Branches of Dravidian.

SD and CD are further divided into two subgroups, and the subgroups of SD_1 and SD_2 exhibit further complex branching which is not shown here. According to the assumptions of historical linguistics, any cognate set found in two of these groups can be presumed to represent a word in Proto-Dravidian (PD). Before presenting the reconstructed vocabulary items, it is necessary to discuss some limitations on this assumption in the present case.

The Problem of Unattested Words in the Non-Literary Languages

It is customary to divide the Dravidian languages into the categories of "literary" and "non-literary", since only four of the languages (Tamil, Malayalam, Kannada, and Telugu, indicated by the shaded ovals in Figure 1.1) have a known literary history from before modern times. To the extent that they are written at all, most of the others were not reduced to writing before the 20th century. It can be seen that the literary languages all belong to South Dravidian, one of the three main branches. What this means for comparative Dravidian studies is that the available vocabulary for two of the main branches of the family (ND and CD) is much more limited than for SD. Furthermore, most of the ND and CD languages are in contact with languages of the Indo–Aryan and/or Iranian families, and have absorbed many vocabulary items from these languages, thus eroding their original Dravidian vocabularies. The number of cognates which match in even two of the main branches is therefore severely restricted.

Bh. Krishnamurti proposes to deal with this problem by accepting cognates from the literary languages as representative of PD, on the assumption that any words which are well attested in SD, and are not obvious loans from some other language, can be assumed to have been present in PD (Krishnamurti 2003:6-

7). While this proposal runs counter to the usual requirements of comparative reconstruction, which insist on at least two *independent* attestations (form-and-meaning correspondences) to establish a proto-form, the assumption has some plausibility – at least as far as the *form* of the words is concerned. It is impossible to prove that items reconstructed solely on the basis of cognates in SD languages were *not* present in PD, and in fact it is likely that *some* such items do descend from the PD stage, though we have no way of knowing which ones. The position is different, however, with regard to the *meanings* of the reconstructed words. In a number of cases, it can be shown that the meanings which can be reconstructed on the basis of SD cognates are not likely to be the oldest meanings. For example:

1. PSD (Proto-South Dravidian) 'fort' (DEDR2207a) can be reconstructed on the basis of cognates in 9 SD languages, including 5 of the non-literary languages. The only cognate outside of SD is the Kolami (CD) word gôâ, meaning 'wall', a meaning also found in some SD1and SD2 languages, which may well be the original meaning.
2. PSD can be reconstructed in the meaning 'write, scratch', though the meaning 'write' is only found in Tamil and Telugu, while only the meanings 'scratch, scrape' and 'line, mark' are found in the non-literary languages (DEDR1623). The original meaning is probably 'scratch' or 'mark'.
3. PSD 'tax, tribute' is similarly based on meanings found only in the literary languages. The original meaning of this word is probably 'grain' or 'contribution of grain', going back to an earlier meaning 'seed': note rice, grain' DEDR215. An altate form, with initial v-, is reconstructible for both meanings: 'paddy, rice, seed' DED5265, tax, contribution, *etc.*' DEDR5266 (including traditional exchanges of grain among tribal groups in the Nilgiris). Thus the original meaning is probably 'seed' or 'grain', later extended to refer to various cereal plants and grain exchanges, with a more modern transfer to include the meaning 'tax'– which may possibly be as old as the Sangam Period (early centuries BCE-CE).

Thus it is clearly not acceptable to unreservedly attribute to PD word meanings which are found only in SD languages. Though as noted above, there is no way to be certain that the meanings of such words are *not* as old as PD, this paper includes only those cases which are reconstructible according to the usual assumptions of comparative -historical linguistics, *i.e.* where at least two of the three main branches bear independent testimony to form and meaning.

Central Dravidian (CD) and South Dravidian (SD)

A second related problem involves the relationship between the CD and SD branches. Briefly, though SD and CD must be considered separate branches, since there are no innovations exclusively shared by the languages of both groups, there is evidence that grammatical innovations have diffused between them in the past. These are shown by the horizontal dashed lines in **Figure 1.2**, and include: (A) development of a perfective participle in *-c(c)i, shared by the Parji-Ollari-Gadaba

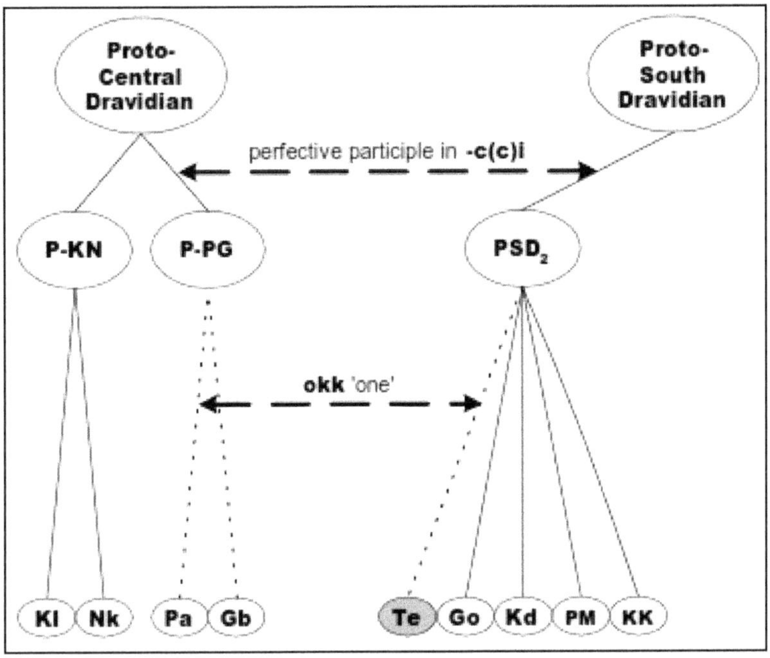

Figure 1.2: Convergent Change in CD and SD1.

subgroup of Centr al Dravidian with the entire SD2 group; (B) * okk, an innovative form of the word for the numeral 'one', shared by Parji-Ollari-Gadaba (CD) with Telugu (SD2). (See Krishnamurti 2003:499-500.) Thus, although there is no basis for assuming a common stage of development for the South and Central Dravidian languages, it is clear that at some stage in the past these two branches were in sufficiently close contact that some innovations could cross the boundaries of the two subgroups–while no such innovations are shared between ND and CD, or between ND and SD. In fact, Telugu is still in contact with some Central Dravidian languages, and Krishnamurti (2002) has noted that the CD languages Kolami and Naiki retain borrowings from Telugu from a very early period. This means that CD and SD cannot be considered to be 100 per cent independent of each other.

In addition, there are numerous examples of changes in meaning shared by CD and SD (either

SD1, SD2, or both), where ND seems to retain an earlier meaning. For example:

PD Form	*ND Meaning*	*SD-CD Meaning*	*DEDR#*
**ûr*	House	village, town, city	0752
**ka7-am/-a n*	Field	threshing ground	1376
**key-(a)m*	earth, mud	Field	1958
**nûl*	to twist (grass *etc.*).	Thread	3726
**ney*	animal fat	Ghee	3746
**pa77i* family		Village, hut; place name suffix	4018

Thus it seems necessary to distinguish those cases which include ND cognates from those which do not: any reconstruction based on a combination of ND plus SD1 and/or SD2 cognates, with or without a CD cognate, is more reliably Proto-Dravidian than one based on CD plus SD1 and/or SD2 *without* ND. In the present paper, reconstructions including one or more ND cognates, plus one or more SD and/or CD cognates, will be referred to as belonging to "Early Proto-Dravidian", while those with only CD and SD cognates will be designated as "Late Proto-Dravidian", with the understanding that the latter group of reconstructions (those without ND), generally speaking, have a *lower probability* of belonging to the earliest common stage of Dravidian than those including SD and ND cognates. While some of these reconstructions *may* possibly represent early Dravidian words, it is impossible without further information to know which ones do so. In addition, any reconstructions based entirely on languages of Central India (Kudux-Malto [ND], Kolami-Naiki-Parji-Gadaba [CD], Gondi-Konda-Pengo-Manda- Kui-Kuvi [SD2]) are suspect because of the contiguity of these languages.

The Problem of Semantic Diffusion

However, the distinction between Early and Late PD does not entirely solve the problem, because of a more general problem affecting language groups which have not completely separated from each other. Even after an original speech community expands and becomes multilingual as a result of cumulative local changes, the resulting languages can undergo parallel changes, whether through actual contact or as a result of shared cultural or ecological conditions. An example from Dravidian is the PD verb EDR5263), which means 'write' in most of the SD1 and SD2 languages (both literary and non-literary), and also in the non-literary CD languages Parji and Gadaba. Given that there is no other reconstructible term to support the presence of writing at the PD level - and considering that speakers of the non-literary languages have come to literacy only in recent historical times it is improbable that this concept existed in PD times; it is more likely that these languages have independently derived the meaning 'write' from the older meanings of this word ('draw, scratch, mark, *etc.*'), and/or have been influenced by the meanings of cognate words in the literary languages. Accordingly, 'to write' is not an acceptable PD reconstruction.

Examples of this kind–and they can be multiplied–tell us that the possibility of diffusion must be taken seriously. We might even go so far as to say that only those etymologies which eliminate the possibility of diffusion *beyond a reasonable doubt* should be considered acceptable.

Proto-Dravidian Agricultural Vocabulary

The agricultural vocabularies of early and late Proto- in terms of what they can tell us about the agricultural practices of the presumed Proto-Dravidian speech community. The previous portions have all dealt in a sense with the same problem: the diffusion of meanings across language boundaries. For this reason it is necessary to articulate our assumptions about how meanings should be reconstructed, and especially how the problem of semantic diffusion can be dealt with.

Reconstructing Meaning

Semantic change is one of the areas where the history of a language encounters the history of the real world: when (perceived) reality changes, or when people's (perceived) relationship to reality changes, that change is reflected in language. For example: New elements in the culture of a community are often reflected by the presence of a new word in the language. Such new words may be *borrowings* from other languages: *e.g.* Old Indo-Aryan (OIA) godhûma 'wheat', vrîhi 'rice', and langala 'plough' are assumed to be loanwords from unknown languages, as these words first appear in OIA in the South Asian subcontinent and are not traceable in other Indo-European languages. New words may also result from *internal innovation*: for example, the OIA word for 'sorghum', yav âkâra, seems to be a compound meaning 'barley-shaped' (Turner 1966, entry no. 10437). Alternatively, a new element may be recognized merely by the creation of a new phrase, as in the expression 'Indian corn' used for 'maize' by the early European immigrants in North America. A similar example is seen in OIA dhânya-ksetra 'grain-field', a compound of dhânya 'grain' and ksetra, a word which originally meant 'territory, land' ('where one lives', from the root ksi- 'dwell'; cf. Vedic Skt Kuru-ksetra 'territory of the Kuru tribe'). The compound dhânya-kcetra makes sense only in the context of settled agriculture, and is not found in early Sanskrit. Such cases, at least initially, involve the *widening* of a term to include new referents: 'Indian corn' was a new member of the class of things called 'corn', just as a dhânya-ksetra 'grain-field' was a new kind of ksetra 'field'. This type of change can also lead to *polysemy* or multiple meanings of the same word: thus in late OIA, ksetra meant both 'land' and 'cultivated field'.

When a cultural element becomes more salient than others of its kind, the term for the category as a whole may be *narrowed* or *restricted* in meaning to refer to that particular element: thus, when later generations of New World settlers started saying 'corn' instead of 'Indian corn'--perhaps because this was the only kind of corn they knew, and therefore the qualifier 'Indian' seemed redundant to them--the meaning of 'corn' changed from 'grain' (which is still its meaning in Britain) to 'maize'. A similar change seems to have happened in South Dravidian, where an old word for 'food' or 'fodder' came to refer to a particular grain: 'ragi', *Eleusine coracana* (DEDR812). A similar change has happened repeatedly--and presumably independently--in the history of agricultural terms: for example, in Panjabi the word for wheat is kaGak, which in OIA and MIA meant 'grain, cereal' or 'seed, granule'. Other similar cases are mentioned below. In these instances a class term referring to distinctly perceived types of objects becomes a term for a single perceived type: *A* grain becomes *THE* grain, *i.e.* the staple grain of a community. (Panjab has long been primarily a wheat-growing and wheat-eating region.)

Using an awareness of these processes - particularly the *widening* and *narrowing* of meanings -the historical linguist processes the data about the meanings of a group of related words and proposes the most likely *proto-meaning*, *i.e.* the meaning which is most probably the ancestor of all the attested meanings, from which the meanings found in the attested languages can be accounted for in terms of the types of changes exemplified above. To qualify as a proto-meaning, the meaning must be shared by languages of at least two subgroups. (In the Dravidian case, it is necessary to add the constraints on reliability noted for reconstructing proto-forms) For example, in the DEDR entry quoted below, the following meanings occur: 'rat', 'bandicoot', 'mouse', 'small rat'. Of these, the meaning 'rat' is the only one shared by all three branches--SD: Tamil [Ta.]...Tulu [Tu.] plus Gondi [Go.] and Konda, CD: Kolami [Kol.]...Gadaba [Ga.], ND: Brahui [Br.]. The meaning 'mouse' is shared by SD1 (Kannada 2 1

[Ka.]) and SD (Gondi[Go.]), and the meaning 'bandicoot' is found only in Tamil [Ta.] (SD).

DEDR entry: 833 *Ta.* eli rat, bandicoot. *Ma.* eli rat. *Ko.* eyj id. *To.* isy id. *Ka.* eli, ili rat, mouse. *Ko.* eli rat. *Tu.* eli, ili id. *Te.* eluka, (B. also) elika rat; cimm- eluka

mouse. *Kol.* elka rat. *Nk.* elka id. *Nk.* (*Ch.*) elli (*pl.* - g) id. *Pa.* el (*pl.* elkul) id. *Ga.* (Oll.) sirel mouse (*i.e.* small rat). *Go.* (A. Y. S.) elli rat; (Tr. Ph.) allî (*pl.* alk) id. (Tr. also) mouse; (W.) allî mouse (*Voc.* 367); (Koya Su.) ellu rat. *Kon a* elka id. *Br.* hal

id. Cf. 2630 Pa. cîr el and 2661 *Ta.* cuGmal, cuGm-eli. DED(S) 710.

(Quoted from page 81 of A Dravidian etymological dictionary by T. Burrow and M. B. Emeneau, published by Clarendon Press, Oxford, 1984)

The semantic data in this entry can be summarized in the following abbreviated form, which provides the information needed for reconstruction, namely the meanings attested in each branch of Dravidian, and how they are shared among the different branches:

rat; mouse (S), bandicoot (S1)

In this definition, as in all the definitions, the first word or phrase indicates the proposed proto-meaning, in this case 'rat'. The fact that this meaning is shared by all the attested branches is indicated tacitly by the *absence* of any specification of branches (such as "S" for South Dravidian) following it. In addition to displaying the semantic data efficiently, such a definition allows the reader to evaluate the proposed proto-meaning, to see how it was arrived at, and to explore other possible proto-meanings. This definition implies that a word meaning 'rat' in PD was extended in meaning in Proto-South Dravidian to include the meaning 'mouse', and further extended in S1 to include 'bandicoot'--but in fact, we cannot be sure that the history went that way. Given that meanings are subject to both widening and narrowing, it is possible that the original meaning of PD *el-i* was 'large rodent', including rats, bandicoots, and other similar creatures. Note that the other PD word in App. 1B for 'rodent', PD contt, includes the meaning 'shrew', suggesting that the contrast between the two terms may have been simply between 'larger rodents' and 'smaller rodents'. Such alternative possibilities are implied by this type of definition.

Dealing with the Problem of Diffusion

Because of the problem of diffusion (above), in some cases it may be inappropriate to reconstruct a meaning even though it is found in all subgroups. For example: (1) The definition of PD (DEDR0688) derivable under the above assumptions would be: "to plough (SCN), to till, to root up earth as pigs do (SN), (deriv.) tillage, agriculture, farmer, furrow (S), to harrow (C), (deriv.) a plough (N)".

Though the meaning 'to plough' appears in all three subgroups, it is not acceptable as a proto-meaning because (1) the meaning 'to root up earth as pigs do' is found in both SD and ND, and is likely to be older than the meaning 'to plough', and (2) this is the *only* meaning found in Malto, one of the two ND languages attested for this word. (Kudux, the other ND language, has the meaning 'to plough'.) Thus it seems more reasonable to assume that the PD word referred first to an activity of pigs and was later widened in the individual languages (independently or through diffusion) to include various human digging activities, including ploughing. (2) There are at least four words in which the meaning 'rice' is found in both SD and ND (see "?RICE" in Appendix 3A). However, none of these is a straightforward word for a cereal, comparable to the words 'Italian millet' or 'sorghum' (App. 3A). Among the 'rice' words, (1) refers to various types of cooked cereal; (2) emphasizes rice as a crop, and one language also has the general meaning 'crop'; (3) and includes 'ear of grain' among its meanings; (4) a reconstruction based on four different DEDR entries, includes among its meanings 'seed', 'any grain', and 'rice or other grain'. Thus it is possible that these words referred originally to some other grain(s), and/or were general terms for 'seed', 'grain', 'cooked cereal', or the like. Thus the presence of rice is not established beyond a reasonable doubt, either at the Early or Late PD level.

Early Proto-Dravidian Agricultural Vocabulary

BACKGROUND: The Early PD speech community had words for king/lord and chief, suggesting the beginnings of social ranking; this is reinforced by the presence of a word for obeisance. There is a concept of wealth or property, as well as payment of debts/fines; words for to sell, beggar/to beg, and steal/theft support this concept. Words for dwelling places (*e.g.* 'hut') seem not yet to be differentiated from words for habitation areas (such as 'village')–suggesting that with dwelling structures (beam, upper story) suggest something beyond simple huts. Words for clothing and adornment include footgear with straps, cloth, comb/to comb, and chignon. Words for tools include basic items like axe, knife, adze, bow, and digging tool. Household items include at least five distinct words for pots, probably designating vessels of different sizes, shapes, and functions, along with a word for ladle.

On the agricultural side, Early Proto-Dravidian shows words for the following domesticable animals: cattle (cow and bull), sheep/goat,? buffalo, and dog. Wild animals include (wild)cat, bear, mouse, rat, primate (two words), some species of deer,? squirrel (possibly limited to Central India), a general word for fish, and words for carp and (bull)frog. Reconstructible words for trees include the date (*Phoenix* spp.), toddy palm (*Borassus flabelliformis*), tamarind (*Tamarindus indica*), and palas (*Butea frondosa*).

The only names for specific food plants reconstructible to Early PD are onion/garlic, yam, and eggplant; no terms for cereals are present, though there is a word for food/fodder and some possible words for grain/seed. Words related to agricultural activities include: to drive (animals), graze, dig, grind (grain), winnow/churn, uproot/pluck, reap/cut, and terms for plough, husk/chaff, and digging tool. Several types of land were distinguished, including lowlying land, uncultivated land, and field.

Innovations in agriculture are suggested by a number of new food plants, including sorghum (*S. vulgare*), another millet (? *Setaria italica*), horse gram (*Macrotyloma uniflorum*), black gram (*Vigna mungo*), pigeon pea (*Cajanus cajan*), green gram/mung (*Vigna radiata*), sesamum (*S. indicum*), Cuddapah almond (*Buchnania lanzan* Sprengel), jujube (*Ziziphus mauritania*), and sugar cane (*Saccharum* sp.). There are still no reliable words for 'rice' at this stage. Other words connected with agriculture include new words for the plough (one possibly borrowed from a Munda or Austroasiatic language), grain storage container, animal stall/pen, herd/flock, sluice/drain, tank, well, fishhook, and cowbell.

Thus, while Early PD can perhaps be described as partly agricultural--practicing animal husbandry along with gathering and processing of food plants, but with very few identifiable crops–Late PD is clearly agricultural, as it shows a variety of crop plants, many of which are still grown in the area today, along with the necessary technological development for full-time agriculture.

Location of the Proto-Dravidian Speech Community. Figure 1.3 shows the approximate present-day locations of the three subgroups of Dravidian, along with earlier locations inferred from the evidence of place names in Maharashtra (Southworth 2005: chapter 9), grammatical features in Marathi, Gujarati, and Sindhi (Southworth 1974), and Dravidian-type kinship systems in the southern

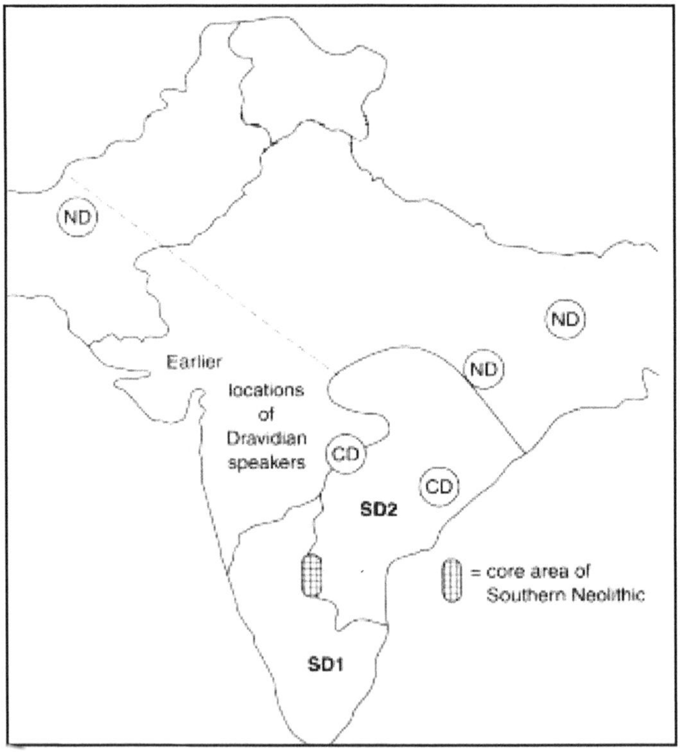

Figure 1.3: Approximate Location of Dravidian Subgroups (from Southworth 2005:244).

Indo–Aryan languages (see Trautman 1981). Since languages of all three subgroups are found in eastern central India, in the lower Godavari River basin, it would be most economical to assume that Proto-Dravidian was spoken somewhere in that region. This proposal is based exclusively on the current geographical distribution of the linguistic subgroups of Dravidian. Figure 4 shows the approximate location suggested for the last period of Proto-Dravidian unity. However, Proto-Dravidian may have been spoken in a wider area, extending perhaps into Central India or the western Deccan, which are now occupied mainly by Indo–Aryan languages. Furthermore, other forms of early Dravidian – pre-Proto-Dravidian, or other (at present unknown) branches of Dravidian – may also have existed in these same areas. Thus the suggested location for Proto-Dravidian in the Godavari basin would not preclude the possibility that speakers of an earlier stage of Dravidian entered the subcontinent from western or central Asia, as has often been suggested.

Chronology of Proto-Dravidian

Linguistic evidence alone cannot tell us much about chronology. Working backward from certain phonological developments in SD languages, some of which can be dated by references in dateable texts, Krishnamurti suggests that the split of PSD1 into Proto-Kannada and Proto-Tamil might have been complete by around 600 BCE. The prior split of PSD into PSD1 and PSD2, for which there is also some textual evidence, may have been complete by about 1100 BCE (2003:501-2). Thus the three main branches might have been relatively independent from each other (very roughly) around the middle of the second millennium BCE, if not earlier.

There is some evidence for contact between speakers of Old Indo-Aryan and some Dravidian language during the Vedic period; Witzel (1999) suggests that this could only have happened after about 1200 BCE, as no Dravidian words are found in the early Rigveda. While this evidence for Dravidian-OIA contact has often been interpreted to suggest that Dravidian speakers passed through the Panjab or neighboring regions while migrating from the northwest (Iran or Central Asia) into Peninsular India, the present evidence suggests another possibility. As noted above, there is evidence for the presence of a pre-Indo-Aryan/Dravidian substratum in the northern Deccan (Maharashtra and Gujarat), and probably also in Sindh (southern Pakistan). It is possible that this substratum was not a remnant of an earlier migration, but rather the result of a later (post-Late PD, or even post-PSD) northward expansion of Dravidian speakers. Since most of the presumed Dravidian loanwords in OIA are traceable to South Dravidian, this would mean that South Dravidian was already distinct from the other branches by the last quarter of the second millennium BCE.

Proto-Dravidian and Archaeology

The Southern Neolithic

The most promising archaeological complex which might be connected with the Dravidian languages is the Southern Neolithic complex, which made its first appearance in the mid-third millennium BCE in the core area shown in **Figure 1.4**: the present Gulbarga, Raichur, and Bellary districts of Karnataka, and Kurnool Dt.

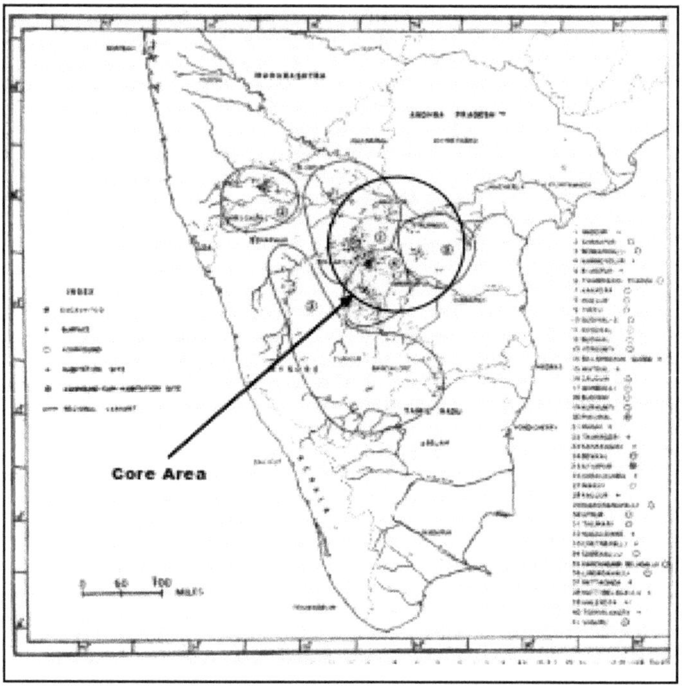

Figure 1.4: The Southern Neolothic Archaeological Complex (after Padayya).

of Andhra Pradesh, and thereafter spread – to judge by similarities in pottery styles, house construction, plant remains, and other features – to "a very vast area from the Krishna-Tungabhadra in the north (or if we include the evidence from Daimabad, on the Godavari) to the Kaveri in the south, and from the Krishna-Godavari mouths in the east to Dharwar in the west" (Sankalia 1977:142). Note that the core area is located in the vicinity of the upper Krishna River, not far from the area assumed on linguistic grounds to be the home of Proto-Dravidian. The dwellings found in early Southern Neolithic sites were one-room houses with low mud walls surmounted by reed screens and thatched roofs, constructed on rock terraces linked by some sort of drainage systems. Tools included stone axes, adzes, choppers, wedges, scrapers, hammers, and lithic blades, and in some sites copper axes. Querns and grindstones were presumably used for processing grains. The range of pottery forms was "amazing" (Sankalia 1974:521). Some sites are associated with nearby ashmounds, presumably remnants of livestock pens (Allchin 1963). Animal remains include cattle (*Bos indicus*), buffalo, goat, and sheep at many sites; wild pigs, deer species (*Axis, Cervus*), antelope, elephant, chicken, and rat/rodent bones are also found, along with remains of fish and freshwater mussels and snails (Korisettar *et al.*, 2001). Rock-paintings and etchings contained depictions of bulls, deer, gazelles, sheep, goats, horses, peacocks, and serpents. Ornaments included bead necklaces and ear pendants (?) of shell, semi-precious stones, terracotta, gold and copper.

Comparison of Archaeological and Linguistic Findings

The above list of faunal remains and depictions compares favorably with those listed for Late PD, though canids, felids, bears, and primates (along with several smaller animals such as the mongoose, squirrel, porcupine) are unreported on the archaeological side, while chickens (found in 21.7 per cent of sites, according to Korisettar *et al.*, 2001) are missing from the linguistic inventory. Table 1.4 shows the correspondences between the crop plants identified archaeologically in Fuller *et al.* (2001) and those reconstructed here for Early and Late Proto-Dravidian.

Table 1.4: Crop Plants in the Southern Neolithic and in Proto-Dravidian

Crop Plants		*Ubiquity*		*PD Level*
		Ph. II	*Ph. III*	
Pulses				
Horsegram	*Macrotyloma uniflorum*	.80	.86	Late PD *ko7 horsegram
Mung	*Vigna radiata*	.80	.86	Late PD **pac -Vt/Vl* mung
Urad	*Vigna cf. mungo*	.20	.43	Late PD **u -untu*, *min(t) urad
Hyacinth bean	*Vigna trilobata*	.20	.29	—
	Lablab purpureus	.20i	.29i	Proto-Tamil *ava-rai *Dolichos lablab*
Pigeonpea	*Cajanus cajan*		—	.29i Late PD **tu-var-* pigeonpea
Millets and related grasses				
Browntop millet	*Brachiaria ramose*	1.00	.71-1.00	Late PD **conna-l* sorghum
Bristley foxtail	*Setaria verticillata*	1.00	1.00	Late PD **ko t-V S. italica*
Sawa millet	*Echinochloa cf. colona*	—	.43	—
Yellow foxtail	*Setaria pumila*	.20	.14	—
Little millet	*Panicum sumatrense*	.20	.29	—
Kodo millet	*Paspalum scrobiculatum*	—	.14	PSD *(v) âr/ar-Vk-
Pearl millet	*Pennisetum glaucum*	.20i	—	PSD *kam-pu bulrush millet
Finger millet	*Eleusine coracana*	—	.14i	PSD *ira(k) ragi
Large Cereals				
Barley	*Hordeum vulgare*	.60i	.29i	—
Wheat	*Triticum*	.40i	.29i	?Late PD **kûli* wheat/rice?
Rice	*Oryza* sp.	.20(i?)	.14(i?)	Late PD **(v)ar-iñci* rice
Misc. Food/Crop Plants				
Jujube	*Zizyphus* sp.	.60	.29	Late PD *irak- jujube
Fig	*Ficus* sp.	—	.43	Late PD **cuv-* fig
Java plum	*cf. Syzigium cumini*	.20	.14	Late PD *ñçr -al jambu
Globe cucumber	*Cucumis cf. prophetarum*	.20	.29	—
Luffa	*cf. Luffa cylindrical*	—	.14i	Late PD *pîr-
Flax	*Linum usitatissimum*	—	.14i	PSD1 *ak -V-ce

Contd...

Table 1.4–*Contd...*

Crop Plants		*Ubiquity*		*PD Level*
		Ph. II	*Ph. III*	
Cotton	*Gossypium* sp.	—	.14(i?)	PSD **par-utti*
Okra	*Abelmoschus* sp.	.20	.57	—
Parenchyma fragments		1.00?	1.00?	Early PD **kic-ampu*
Not identified archaeologically in *S. Neolithic*:				
Onion/garlic	*Allium* sp.	—	—	Early PD **u 77i*
Eggplant	*Solanum* sp.	—	—	Early PD **va-Vt-*
Sesame	*S. indicum*	—	—	Late PD **n û(v) -*
Sugarcane	*Saccharum* sp.	—	—	Early PD **cet-Vkk*
Hemp	*Cannabis* sp.	—	—	?Late PD *boy-VI

Explanation of symbols and abbreviations in Table 1.4:

- ✰ Column 2, UBIQUITY, shows the percentage of sites where plant remains were found in Phase II (Total: 5 sites) and III (Total: 7 sites) in the Southern Neolithic chronology of Allchin and Allchin (1982), equivalent to 2300-1800 cal. BCE and 1800-1200 cal. BCE respectively.
- ✰ Figures in bold represent "presence of inferred crop, possibly derived from domestication in Southern India"; figures followed by the letter "i" represent items which are "present as crop, introduced from another region"; plain type without "i" indicates items "present in limited quantity, possibly gathered from wild". Bold type followed by "?" indicates uncertainty as to whether the item is cultivated or wild.
- ✰ The reconstructed forms for Early PD and Late PD in the right-hand column are found in the Appendix to this chapter; those labeled "PSD" OR "PSD1" are from Southworth 2005: Reconstructions in *bold italics* are originally from Krishnamurti 2003:523-32.

Nineteen of the twenty-six archaeologically identified plants (73 per cent) are matched by PD reconstructions, mostly with identical or very similar meanings–though there are some noticeable differences, particularly among the millets. The two millets which have been identified archaeobotanically as staples of the S. Neolithic, *Brachiaria ramosa* and *Setaria verticillata*, correspond to the linguistically reconstructed *Sorghum vulgare* and *Setaria italica* respectively. Given that the latter two millets were introduced later from elsewhere and have largely replaced the former two, it is not surprising that the older terms were extended to include the new varieties.

The Origins of Agriculture

Introduction

What might head a list of the defining characteristics of the human species? While our view of ourselves could hardly avoid highlighting our accomplishments in engineering, art, medicine, space travel and the like, in a more dispassionate assessment *agriculture* would probably displace all other contenders for top billing. Most of the other achievements of humankind have followed from this one. Almost without exception, all people on earth today are sustained by agriculture. With a minute number of exceptions, no other species is a farmer. Essentially all of the arable land in the world is under cultivation. Yet agriculture began just a few thousand years ago, long after the appearance of anatomically modern humans. Given the rate and the scope of this revolution in human biology, it is quite extraordinary that there is no generally accepted model accounting for the origin of agriculture. Indeed, an increasing array of arguments over recent years has suggested that agriculture, far from being a natural and upward step, in fact led commonly to a lower quality of life. Hunter-gatherers typically do less work for the same amount of food, are healthier, and are less prone to famine than primitive farmers (Lee and DeVore 1968, Cohen 1977, 1989). A biological assessment of what has been called the puzzle of agriculture might phrase it in simple ethological terms: why was this behaviour (agriculture) reinforced (and hence selected for) if it was not offering adaptive rewards surpassing those accruing to hunter-gathering or foraging economies?

This paradox is responsible for a profusion of models of the origin of agriculture. 'Few topics in prehistory', have engendered as much discussion and resulted in so few satisfying answers as the attempt to explain why hunter/gatherers began to cultivate plants and raise animals. Climatic change, population pressure, sedentism, resource concentration from desertification, girls' hormones, land ownership, geniuses, rituals, scheduling conflicts, random genetic kicks, natural selection, broad spectrum adaptation and multicausal retreats from explanation have all been proffered to explain domestication. All have major flaws. the data do not accord well with any one of these models. Recent discoveries of potentially psychoactive substances in certain agricultural products - cereals and milk - suggest an additional perspective on the adoption of agriculture and the behavioural changes ('civilisation') that followed it. In this paper we review the evidence for the drug-like properties of these foods, and then show how they can help to solve the biological puzzle just described.

History of Sugar

Over 2500 years ago, Indian farmers had discovered and begun farming many spices and sugarcane. It was in India, between the sixth and fourth centuries BC, that the Persians, followed by the Greeks, discovered the famous "reeds that produce honey without bees" being grown. These were locally called, pronounced as sacchaccharum. On their return journey, the Macedonian soldiers carried the "honey bearing reeds,"thus preading sugar and sugarcane agriculture. People in India had also invented, by about 500 BC, the process to produce sugar crystals. In the local language, these crystals were called *khanda*, which is the source of the word *candy*.

Prior to 18th century, cultivation of sugar cane was largely confined to India. A few merchants began to trade in sugar - a luxury and an expensive spice in Europe until the 18th century. Sugar became widely popular in 18th-century Europe, then graduated to becoming a human necessity in the 19th century all over the world. This evolution of taste and demand for sugar as an essential food ingredient unleashed major economic and social changes. Sugarcane does not grow in cold, frost-prone climate; therefore, tropical and semitropical colonies were sought. Sugarcane plantations, just like cotton farms, became a major driver of large and forced human migrations in 19th century and early 20th century - of people from Africa and from India, both in millions - influencing the ethnic mix, political conflicts and cultural evolution of various Caribbean, South American, Indian Ocean and Pacific island nations. The history and past accomplishments of Indian agriculture thus influenced, in part, colonialism, first slavery and then slavery-like indentured labor practices in the new world, Caribbean wars and the world history in 18th and 19th centuries.

The Emergence of Agriculture and Civilisation in the Neolithic

The Transition to Agriculture

From about 10,000 years ago, groups of people in several areas around the world began to abandon the foraging lifestyle that had been successful, universal and largely unchanged for millennia (Lee and DeVore 1968). They began to gather, then cultivate and settle around, patches of cereal grasses and to domesticate animals for meat, labour, skins and other materials, and milk. Farming, based predominantly on wheat and barley, first appeared in the Middle East, and spread quickly to western Asia, Egypt and Europe. The earliest civilisations all relied primarily on cereal agriculture. Cultivation of fruit trees began three thousand years later, again in the MiddleEast, and vegetables and other crops followed. Cultivation of rice began in Asia about 7000 years ago (Stark 1986).

To this day, for most people, two-thirds of protein and calorie intake is cereal-derived. (In the west, in the twentieth century, cereal consumption has decreased slightly in favour of meat, sugar, fats and so on.) The respective contributions of each cereal to current total world production are: wheat (28 per cent), corn/maize (27 per cent), rice (25 per cent), barley (10 per cent), others (10 per cent) (Pedersen *et al.*, 1989).

The Change in the Diet Due to Agriculture

The modern human diet is very different from that of closely related primates and, almost certainly, early hominids. Though there is controversy over what humans ate before the development of agriculture, the diet certainly did not include cereals and milk in appreciable quantities. The storage pits and processing tools necessary for significant consumption of cereals did not appear until the Neolithic (Washburn and Lancaster 1968). Dairy products were not available in quantity before the domestication of animals. The early hominid diet (from about four million years ago), evolving as it did from that of primate ancestors, consisted primarily of fruits, nuts and other vegetable matter, and some meat — items that could be foraged for and eaten with little or no processing. Comparisons of primate and fossil-hominid anatomy, and of the types and distribution of plants eaten raw by modern chimpanzees, baboons and humans, as well as microscope analysis of wear patterns on fossil teeth (Walker 1981, Peuch *et al.*, 1983) suggest that australopithecines were 'mainly frugivorous omnivores with a dietary pattern similar to that of modern chimpanzees' (Susman 1987:171).

The diet of pre-agricultural but anatomically modern humans (from 30,000 years ago) diversified somewhat, but still consisted of meat, fruits, nuts, legumes, edible roots and tubers, with consumption of cereal seeds only increasing towards the end of the Pleistocene (*e.g.* Constantini 1989 and subsequent chapters in Harris and Hillman 1989).

The Rise of Civilization

Within a few thousand years of the adoption of cereal agriculture, the old hunter-gatherer style of social organisation began to decline. Large, hierarchically organised societies appeared, centred around villages and then cities. With the rise of civilisation and the state came socioeconomic classes, job specialisation, governments and armies. The size of populations living as coordinated units rose dramatically above pre-agricultural norms. While hunter-gatherers lived in egalitarian, autonomous bands of about 20 closely related persons, with at most a tribal level of organisation above that, early agricultural villages had 50 to 200 inhabitants, and early cities 10,000 or more. People 'had to learn to curb deep-rooted forces which worked for increasing conflict and violence in large groups' (Pfeiffer 1977:438).

Agriculture and civilisation meant the end of foraging — a subsistence method with shortterm goals and rewards — and the beginning (for most) of regular arduous work, oriented to future payoffs and the demands of superiors. 'With the coming of large communities, families no longer cultivated the land for themselves and their immediate needs alone, but for strangers and for the future. They worked all day instead of a few hours a day, as hunter-gatherers had done. There were schedules, quotas, overseers, and punishments for slacking off' (Pfeiffer 1977:21).

Explaining the Origins of Agriculture and Civilization

The phenomena of human agriculture and civilisation are ethologically interesting, because (1) virtually no other species lives this way, and (2) humans did not live this way until relatively recently. Why was this way of life adopted, and why has it become dominant in the human species?

Problems Explaining Agriculture

Until recent decades, the transition to farming was seen as an inherently progressive one: people learnt that planting seeds caused crops to grow, and this new improved food source led to larger populations, sedentary farm and town life, more leisure time and so to specialisation, writing, technological advances and civilisation.

It is now clear that agriculture was adopted despite certain disadvantages of that lifestyle (*e.g.* Flannery 1973, Henry 1989). There is a substantial literature (*e.g.* Reed 1977), not only on how agriculture began, but why. Palaeopathological and comparative studies show that health deteriorated in populations that adopted cereal agriculture, returning to pre-agricultural levels only in modem times. This is in part attributable to the spread of infection in crowded cities, but is largely due to a decline in dietary quality that accompanied intensive cereal farming (Cohen 1989). People in many parts of the world remained hunter-gatherers until quite recently; though they were quite aware of the existence and methods of agriculture, they declined to undertake it (Lee and DeVore 1968, Harris 1977). Cohen (1977:141) summarised the problem by asking: 'If agriculture provides neither better diet, nor greater dietary reliability, nor greater ease, but conversely appears to provide a poorer diet, less reliably, with greater labor costs, why does anyone become a farmer?'

Many explanations have been offered, usually centred around a particular factor that forced the adoption of agriculture, such as environmental or population pressure (for reviews see Rindos 1984, Pryor 1986, Redding 1988, Blumler and Byrne 1991). Each of these models has been criticised extensively, and there is at this time no generally accepted explanation of the origin of agriculture.

Problems Explaining Civilization

A similar problem is posed by the post-agricultural appearance, all over the world, of cities and states, and again there is a large literature devoted to explaining it (*e.g.* Claessen and Skalnik 1978). The major behavioural changes made in adopting the civilised lifestyle beg explanation. Bledsoe (1987:136) summarised the situation thus:

'There has never been and there is not now agreement on the nature and significance of the rise of civilisation. The questions posed by the problem are simple, yet fundamental. How did civilisation come about? What animus impelled man to forego the independence, intimacies, and invariability of tribal existence for the much larger and more impersonal political complexity we call the state? What forces fused to initiate the mutation that slowly transformed nomadic societies into populous cities with ethnic mixtures, stratified societies, diversified economies and unique cultural forms? Was the advent of civilisation the inevitable result of social evolution and natural laws of progress or was man the designer of his own destiny? Have technological innovations been the motivating force or was it some intangible factor such as religion or intellectual advancement?'

To a very good approximation, every civilisation that came into being had cereal agriculture as its subsistence base, and wherever cereals were cultivated, civilisation appeared. Some hypotheses have linked the two. For example, Wittfogel's (1957) 'hydraulic theory' postulated that irrigation was needed for agriculture, and the state was in turn needed to organise irrigation. But not all civilisations used irrigation, and other possible factors (*e.g.* river valley placement, warfare, trade, technology, religion, and ecological and population pressure) have not led to a universally accepted model.

Pharmacological Properties of Cereals and Milk

Recent research into the pharmacology of food presents a new perspective on these problems.

Exorphins: Opioid Substances in Food

Prompted by a possible link between diet and mental illness, several researchers in the late 1970s began investigating the occurrence of drug-like substances in some common foodstuffs. Dohan (1966, 1984) and Dohan *et al.* (1973, 1983) found that symptoms of schizophrenia were relieved somewhat when patients were fed a diet free of cereals and milk. He also found that people with coeliac disease — those who are unable to eat wheat gluten because of higher than normal permeability of the gut — were statistically likely to suffer also from schizophrenia. Research in some Pacific communities showed that schizophrenia became prevalent in these populations only after they became 'partially westernised and consumed wheat, barley beer, and rice' (Dohan 1984).

Groups led by Zioudrou (1979) and Brantl (1979) found opioid activity in wheat, maize and barley (exorphins), and bovine and human milk (casomorphin), as well as stimulatory activity in these proteins, and in oats, rye and soy. Cereal exorphin is much stronger than bovine casomorphin, which in turn is stronger than human casomorphin. Mycroft *et al.* (1982, 1987) found an analogue of MIF-1, a naturally occurring dopaminergic peptide, in wheat and milk. It occurs in no other exogenous protein. (In subsequent sections we use the term exorphin to cover exorphins, casomorphin, and the MIF-1 analogue. Though opioid and dopaminergic substances work in different ways, they are both 'rewarding', and thus more or less equivalent for our purposes.) Since then, researchers have measured the potency of exorphins, showing them to be comparable to morphine and enkephalin (Heubner *et al.*, 1984), determined their amino acid sequences (Fukudome and Yoshikawa 1992), and shown that they are absorbed from the intestine (Svedburg *et al.*, 1985) and can produce effects such as analgesia and reduction of anxiety which are usually associated with poppy-derived opioids (Greksch *et al.*, 1981, Panksepp *et al.*, 1984). Mycroft *et al.*, estimated that 150 mg of the MIF-1 analogue could be produced by normal daily intake of cereals and milk, noting that such quantities are orally active, and half this amount 'has induced mood alterations in clinically depressed subjects' (Mycroft *et al.*, 1982:895). (For detailed reviews see Gardner 1985 and Paroli 1988.)

Most common drugs of addiction are either opioid (*e.g.* heroin and morphine) or dopaminergic (*e.g.* cocaine and amphetamine), and work by activating reward centres in the brain. Hence we may ask, do these findings mean that cereals and milk are chemically rewarding? Are humans somehow 'addicted' to these foods?

Problems in Interpreting these Findings

Discussion of the possible behavioural effects of exorphins, in normal dietary amounts, has been cautious. Interpretations of their significance have been of two types: where a *pathological* effect is proposed (usually by cereal researchers, and related to Dohan's findings, though see also Ramabadran and Bansinath 1988), and where a *natural* function is proposed (by milk researchers, who suggest that casomorphin may help in mother-infant bonding or otherwise regulate infant development).

We believe that there can be no natural function for ingestion of exorphins by adult humans. It may be that a desire to find a natural function has impeded interpretation (as well as causing attention to focus on milk, where a natural function is more plausible). It is unlikely that humans are adapted to a large intake of cereal exorphin, because the modern dominance of cereals in the diet is simply too new. If exorphin is found in cow's milk, then it may have a natural function for cows; similarly, exorphins in human milk may have a function for infants. But whether this is so or not, adult humans do not naturally drink milk of any kind, so any natural function could not apply to them. Our sympathies therefore lie with the pathological interpretation of exorphins, whereby substances found in cereals and milk are seen as modern dietary abnormalities which may cause schizophrenia, coeliac disease or whatever. But these are serious diseases found in a minority. Can exorphins be having an effect on humankind at large?

Other Evidence for 'Drug-Like' Effects of these Foods

Research into food *allergy* has shown that normal quantities of some foods can have pharmacological, including behavioural, effects. Many people develop intolerances to particular foods. Various foods are implicated, and a variety of symptoms is produced. (The term 'intolerance' rather than allergy is often used, as in many cases the immune system may not be involved (Egger 1988:159). Some intolerance symptoms, such as anxiety, depression, epilepsy, hyperactivity, and schizophrenic episodes involve brain function (Egger 1988, Scadding and Brostoff 1988).

Radcliffe (1982, quoted in 1987:808) listed the foods at fault, in descending order of frequency, in a trial involving 50 people: wheat (more than 70 per cent of subjects reacted in some way to it), milk (60 per cent), egg (35 per cent), corn, cheese, potato, coffee, rice, yeast, chocolate, tea, citrus, oats, pork, plaice, cane, and beef (10 per cent). This is virtually a list of foods that have become common in the diet following the adoption of agriculture, in order of prevalence. The symptoms most commonly alleviated by treatment were mood change (>50 per cent) followed by headache, musculoskeletal and respiratory ailments.

One of the most striking phenomena in these studies is that patients often exhibit cravings, addiction and withdrawal symptoms with regard to these foods (Egger 1988:170, citing Randolph 1978; see also Radcliffe 1987:808-10, 814, Kroker 1987:856, 864, Sprague and Milam 1987:949, 953, Wraith 1987:489, 491). Brostoff and Gamlin (1989:103) estimated that 50 per cent of intolerance patients crave the foods that cause them problems, and experience withdrawal symptoms when excluding those foods from their diet. Withdrawal symptoms are similar to those associated with drug addictions (Radcliffe 1987:808). The possibility that exorphins are involved has been noted (Bell 1987:715), and Brostoff and Gamlin conclude (1989:230): The results so far suggest that they might influence our mood. There is certainly no question of anyone getting 'high' on a glass of milk or a slice of bread - the amounts involved are too small for that - but these foods might induce a sense of comfort and wellbeing, as food-intolerant patients often say they do. There are also other hormone-like peptides in partial digests of food, which might have other effects on the body.'

There is no possibility that craving these foods has anything to do with the popular notion of the body telling the brain what it needs for nutritional purposes. These foods were not significant in the human diet before agriculture, and large quantities of them cannot be necessary for nutrition. In fact, the standard way to treat food intolerance is to remove the offending items from the patient's diet.

A Suggested Interpretation of Exorphin Research

But what are the effects of these foods on normal people? Though exorphins cannot have a naturally selected physiological function in humans, this does not mean that they have *no* effect. Food intolerance research suggests that cereals and milk, in normal dietary quantities, are capable of affecting behaviour in many people. And if severe behavioural effects in schizophrenics and coeliacs can be caused by higher than normal absorption of peptides, then more subtle effects, which may not even be regarded as abnormal, could be produced in people generally. The evidence presented so far suggests the following interpretation. The ingestion of cereals and milk, in normal modern dietary amounts by normal humans, activates reward centres in the brain. Foods that were common in the diet before agriculture (fruits and so on) do not have this pharmacological property. The effects of exorphins are qualitatively the same as those produced by other opioid and/or dopaminergic drugs, that is, reward, motivation, reduction of anxiety, a sense of wellbeing, and perhaps even addiction. Though the effects of a typical meal are quantitatively less than those of doses of those drugs, most modern humans experience them several times a day, every day of their adult lives.

Hypothesis: Exorphins and the Origin of Agriculture and Civilization

When this scenario of human dietary practices is viewed in the light of the problem of the origin of agriculture described earlier, it suggests an hypothesis that combines the results of these lines of enquiry. Exorphin researchers, perhaps lacking a long-term historical perspective, have generally not investigated the possibility that these foods really are drug-like, and have instead searched without success for exorphin's natural function. The adoption of cereal agriculture and the subsequent rise of civilisation have not been satisfactorily explained, because the behavioural changes underlying them have no obvious adaptive basis.

These unsolved and until-now unrelated problems may in fact solve each other. The answer, we suggest, is this: cereals and dairy foods are not natural human foods, but rather are preferred because they contain exorphins. This chemical reward was the incentive for the adoption of cereal agriculture in the Neolithic. Regular self-administration of these substances facilitated the behavioural changes that led to the subsequent appearance of civilisation. This is the sequence of events that we envisage.

Climatic change at the end of the last glacial period led to an increase in the size and concentration of patches of wild cereals in certain areas (Wright, 1977).

The large quantities of cereals newly available provided an incentive to try to make a meal of them. People who succeeded in eating sizeable amounts of cereal seeds discovered the rewarding properties of the exorphins contained in them. Processing methods such as grinding and cooking were developed to make cereals more edible. The more palatable they could be made, the more they were consumed, and the more important the exorphin reward became for more people. At first, patches of wild cereals were protected and harvested. Later, land was cleared and seeds were planted and tended, to increase quantity and reliability of supply. Exorphins attracted people to settle around cereal patches, abandoning their nomadic lifestyle, and allowed them to display tolerance instead of aggression as population densities rose in these new conditions.

Though it was, we suggest, the presence of exorphins that caused cereals (and not an alternative already prevalent in the diet) to be the major early cultigens, this does not mean that cereals are 'just drugs'. They have been staples for thousands of years, and clearly have nutritional value. However, treating cereals as 'just food' leads to difficulties in explaining why anyone bothered to cultivate them. The fact that overall health declined when they were incorporated into the diet suggests that their rapid, almost total replacement of other foods was due more to chemical reward than to nutritional reasons.

It is noteworthy that the extent to which early groups became civilised correlates with the type of agriculture they practised. That is, major civilisations (in south-west Asia, Europe, India, and east and parts of South-East Asia; central and parts of north and south America; Egypt, Ethiopia and parts of tropical and west Africa) stemmed from groups which practised cereal, particularly wheat, agriculture (Bender 1975:12, Adams 1987:201, Thatcher 1987:212). (The rarer nomadic civilisations were based on dairy farming.)

Groups which practised vegeculture (of fruits, tubers *etc.*), or no agriculture (in tropical and south Africa, north and central Asia, Australia, New Guinea and the Pacific, and much of north and south America) did not become civilised to the same extent. Thus major civilisations have in common that their populations were frequent ingesters of exorphins. We propose that large, hierarchical states were a natural consequence among such populations. Civilisation arose because reliable, on-demand availability of dietary opioids to individuals changed their behaviour, reducing aggression, and allowed them to become tolerant of sedentary life in crowded groups, to perform regular work, and to be more easily subjugated by rulers. Two socioeconomic classes emerged where before there had been only one (Johnson and Earle 1987:270), thus establishing a pattern which has been prevalent since that time.

The Natural Diet and Genetic Change

Some nutritionists deny the notion of a pre-agricultural natural human diet on the basis that humans are omnivorous, or have adapted to agricultural foods (*e.g.* Garn and Leonard 1989; for the contrary view see for example Eaton and Konner 1985). An omnivore, however, is simply an animal that eats both meat and plants: it can still be quite specialised in its preferences (chimpanzees are an appropriate example). A degree of omnivory in early humans might have preadapted them to some of the nutrients contained in cereals, but not to exorphins, which are unique to cereals. The differential rates of lactase deficiency, coeliac disease and favism (the inability to metabolise fava beans) among modern racial groups are usually explained as the result of varying genetic adaptation to post-agricultural diets (Simopoulos 1990:27-9), and this could be thought of as implying some adaptation to exorphins as well. We argue that little or no such adaptation has occurred, for two reasons: first, allergy research indicates that these foods still cause abnormal reactions in many people, and that susceptibility is variable within as well as between populations, indicating that differential adaptation is not the only factor involved. Second, the function of the adaptations mentioned is to enable humans to digest those foods, and if they are adaptations, they arose because they conferred a survival advantage. But would susceptibility to the rewarding effects of exorphins lead to lower, or higher, reproductive success? One would expect in general that an animal with a supply of drugs would behave less adaptively and so lower its chances of survival. But our model shows how the widespread exorphin ingestion in humans has led to increased population. And once civilisation was the norm, non-susceptibility to exorphins would have meant not fitting in with society. Thus, though there may be adaptation to the nutritional content of cereals, there will be little or none to exorphins. In any case, while contemporary humans may enjoy the benefits of some adaptation to agricultural diets, those who actually made the change ten thousand years ago did not.

Other 'Non-nutritional' Origins of Agriculture Models

We are not the first to suggest a non-nutritional motive for early agriculture. Hayden (1990) argued that early cultigens and trade items had more prestige value than utility, and suggested that agriculture began because the powerful used its products for competitive feasting and accrual of wealth. Braidwood *et al.* (1953) and later Katz and Voigt (1986) suggested that the incentive for cereal cultivation was the production of alcoholic beer: 'Under what conditions would the consumption of a wild plant resource be sufficiently important to lead to a change in behaviour (experiments with cultivation) in order to ensure an adequate supply of this resource? If wild cereals were in fact a minor part of the diet, any argument based on caloric need is weakened. It is our contention that the desire for alcohol would constitute a perceived psychological and social need that might easily prompt changes in subsistence behaviour' (Katz and Voigt 1986:33).

This view is clearly compatible with ours. However there may be problems with an alcohol hypothesis: beer may have appeared after bread and other cereal products, and been consumed less widely or less frequently (Braidwood *et al.*, 1953). Unlike alcohol, exorphins are present in all these products. This makes the case for chemical reward as the motive for agriculture much stronger. Opium poppies, too, were an early cultigen (Zohari 1986). Exorphin, alcohol, and opium are primarily rewarding (as opposed to the typically hallucinogenic drugs used by some hunter-gatherers) and it is the artificial reward which is necessary, we claim, for civilisation. Perhaps all three were instrumental in causing civilised behaviour to emerge. Cereals have important qualities that differentiate them from most other drugs. They are a food source as well as a drug, and can be stored and transported easily. They are ingested in frequent small doses (not occasional large ones), and do not impede work performance in most people. A desire for the drug, even cravings or withdrawal, can be confused with hunger. These features make cereals the ideal facilitator of civilisation (and may also have contributed to the long delay in recognising their pharmacological properties).

Compatibility, Limitations, more Data Needed

Our hypothesis is not a refutation of existing accounts of the origins of agriculture, but rather fits alongside them, explaining why cereal agriculture was adopted despite its apparent disadvantages and how it led to civilisation. Gaps in our knowledge of exorphins limit the generality and strength of our claims. We do not know whether rice, millet and sorghum, nor grass species which were harvested by African and Australian hunter-gatherers, contain exorphins. We need to be sure that preagricultural staples do not contain exorphins in amounts similar to those in cereals. We do not know whether domestication has affected exorphin content or-potency. A test of our hypothesis by correlation of diet and degree of civilisation in different populations will require quantitative knowledge of the behavioural effects of all these foods.

We do not comment on the origin of noncereal agriculture, nor why some groups used a combination of foraging and farming, reverted from farming to foraging, or did not farm at all. Cereal agriculture and civilisation have, during the past ten thousand years, become virtually universal. The question, then, is not why they happened here and not there, but why they took longer to become established in some places than in others. At all times and places, chemical reward and the influence of civilisations already using cereals weighed in favour of adopting this lifestyle, the disadvantages of agriculture weighed against it, and factors such as climate, geography, soil quality, and availability of cultigens influenced the outcome. There is a recent trend to multi-causal models of the origins of agriculture (*e.g.* Redding 1988, Henry 1989), and exorphins can be thought of as simply another factor in the list. Analysis of the relative importance of all the factors involved, at all times and places, is beyond the scope of this paper.

Conclusion

'An animal is a survival machine for the genes that built it. We too are animals, and we too are survival machines for our genes. That is the theory. In practice it makes a lot of sense when we look at wild animals. It is very different when we look at ourselves. We appear to be a serious exception to the Darwinian law. It obviously just isn't true that most of us spend our time working energetically for the preservation of our genes' (Dawkins 1989:138). Many ethologists have acknowledged difficulties in explaining civilised human behaviour on evolutionary grounds, in some cases suggesting that modern humans do not always behave adaptively. Yet since agriculture began, the human population has risen by a factor of 1000: Irons (1990) notes that 'population growth is not the expected effect of maladaptive behaviour'.

We have reviewed evidence from several areas of research which shows that cereals and dairy foods have drug-like properties, and shown how these properties may have been the incentive for the initial adoption of agriculture. We suggested further that constant exorphin intake facilitated the behavioural changes and subsequent population growth of civilisation, by increasing people's tolerance of (a) living in crowded sedentary conditions, (b) devoting effort to the benefit of non-kin, and (c) playing a subservient role in a vast hierarchical social structure. Cereals are still staples, and methods of artificial reward have diversified since that time, including today a wide range of pharmacological and non-pharmacological cultural artifacts whose function, ethologically speaking, is to provide reward without adaptive benefit. It seems reasonable then to suggest that civilisation not only arose out of self-administration of artificial reward, but is maintained in this way among contemporary humans. Hence a step towards resolution of the problem of explaining civilised human behaviour may be to incorporate into ethological models this widespread distortion of behaviour by artificial reward.

Land Tenure System in Pre-Mughal Period

We will discuss some of the basic issues related to ownership and cultivation of land during the pre-Mughal period. During that period land was relatively abundant compared to demand. Moreover, wasteland was easily available for reclamation in some areas. In this environment some basic issues arise: whether the owner cultivated his land or leased it out; whether the owner was allowed to sell his land; whether land was inheritable or not; how land tax was collected from the cultivator; what percentage of the agricultural produce was collected as tax; and what the punishment was for non-payment of tax. We cannot hold a detailed discussion of land tenure system in the pre-mughal India in the present course. However, before giving an account of the tenurial system of the mughal period, it is worthwhile to have a brief idea about the nature of ownership of land during ancient period.

The question of ownership in land has been discussed by economic historians and their conclusions are largely confined to two options, the state or peasant ownership. Various combinations of the state's and the peasant's rights have also been suggested. It is argued that the proprietorship in land originated with the act of *reclamation*, and the peasant who reclaimed and converted the forest into arable land became the proprietor of that land. The king did not have any property right in land except the right to a share of the produce in return for 'affording protection to his subjects'. However, D. N. Jha argues that an individual proprietor exercised only a qualified ownership over his land, the king being its ultimate lord. On the other hand, Irfan Habib accepts the existence of private property in land but at the same time notes that peasant did not have the right to free *alienation* of land. But most scholars agree that the peasants held permanent and heritable occupancy rights in land and the king was not expected to evict them. In those early days, cultivation of land was strongly insisted upon. The individual peasant enjoyed the right of use as long as he cultivated the land. Sir Henry Maine points out that each family had the duty of submitting to the common rules of cultivation and pasturage. Grazing ground and forest tracts adjoining the village were the common ownership of the entire village community.

The revenue paid by the peasants was generally a share of the produce. The method adopted was either through actual division of the peasant's produce or state's share was estimated by inspection of the growing crop. Even sometimes an average or standard figure was determined for the state's share from the unit area of each crop. The collection of revenue was generally through various officers. They were authorized by the king to collect state's share of revenue and to retain a portion of it as payment for their services. Gradually, these offices became hereditary. Consequently, layers of intermediaries developed between the state and the cultivators.

Land Tenure System during Mughal Period

The discussion in this section is in general relevant for the whole of Mughal India. But, particularly, it conforms more to the northern and central parts where Mughal rule was more entrenched. Distinguishing features of land tenure systems in other regions will be investigated later in the sections on Eastern, Western and Southern India. We will discuss two aspects, *viz.*, land tenure and taxation on land. Both these aspects are not independent of each other, which makes the identification of different types of land rights in India difficult. So to understand the land tenure structure we will start our discussion with land revenue system.

Land Revenue System

The simplest form of land revenue was crop sharing. From this simple sharing, we can trace the evolution of other systems of assessment, which were designed to reduce the burden of work and expenses for the Mughal administration. One such method of assessment was *Kankut* (Kan = grain, Kut = estimate). Initially under this method production of each crop was estimated first by estimating yield per unit of land at the current harvest and then multiplying it with area under that crop. Tax was calculated on the basis of the proportion it was supposed to bear to the produce. Later on, to remove the discretion of officials, the *Kankut* was modified into *Zabt* (measurement) where instead of leaving the yield to be fixed at each harvest, a standard schedule for different crops was promulgated for a longer period. During Akbar's time, on the basis of detailed information collected for the period of ten years on yields, prices, and areas cultivated for each locality, the revenue rates were fixed directly in cash for each crop. However, in addition to this standard system of cash revenue, other methods of assessment such as crop-sharing (*batai*) and *Kankut,* continued in certain villages or for certain crops. Moreover, during crop failure, due concession was made in collection of tax in 'crop-less' areas.

Whatever may be the methods of assessment and collection, one major aim of the Mughal administration, according to Irfan Habib, was to take away the bulk of the peasant's surplus. The shares of the crop taken under the *'batai'* or Kankut varied with crops and localities. The norm seems to be one-half in the less fertile regions, and substantially more in the more fertile regions.

In Northern India, the unit of assessment was the village. The revenue due from a village was fixed with reference to its productive capacity usually for the year. But this sum was not distributed by assessors over the individual peasants. Inside the village the individual peasants contributed to this revenue on the basis of one or other of the familiar systems - either on an estimate of the produce gathered, or by rates on the area sown, or by lump sum payable for the land holding. The collection of revenue was enforced by several methods. Non-payment of revenue was deemed equivalent to rebellion. While eviction was not unknown as a punishment, the more usual method seems to have been imprisonment and torture of the village headmen, followed by the massacre of the adult male population and enslavement of women and children.

Intermediaries and Land Rights

Theoretically the king was the sole claimant to the land revenue. But, as a matter of fact, the assessment and collection of the revenue was largely through the members of a small ruling class. Large areas of cultivated land were given to them revenue- free; and where revenue was levied on their land it was often at substantially lower rates. We discuss below about the intermediaries.

The Jagirdars

The *jagirdars* were the king's officers enjoying land-gifts. These were the persons who held ranks (*mansabs*) granted by the emperor. Each rank entitled its holder (*mansabdar*) to a particular amount of pay which was usually met by assigning an area (*Jagir*) that was officially estimated to yield an equivalent amount of the revenue. These ranks (mansabs) were not inheritable, though it was normal practice to allow the relations of higher *mansab*-holders to continue with their *mansabs*. The temporary character of *Jagir* strengthened the control of the emperor over the *Jagirdars*. According to Abul Fazal, a mansab-holder was entitled to a *Jagir*, but not to a particular tract of land in Jagir, and not the same land year after year. The *mansabs* (and hence *Jagirs*) were revised from time to time to award promotions or demotions.

Thus it seems that a *Jagirdar* had no permanent rights on his assigned area. Theoretically his claims were also confined to the authorised land-revenue and taxes. But, in practice, the *Jagirdar* was a much more powerful person specially if he happened to be a big assignee enjoying police jurisdiction as well. In fact, a large proportion of land was under such *Jagirs*. It is estimated that, in 1646, a mere sixty-eight princes and nobles at the top claimed 36.6 per cent of the total revenue demand of the empire and the next 587 officials claimed nearly 25 per cent. The remaining 7,555 *mansab*-holders claimed between a quarter and a third of the revenues. Thus the larger portion of the land lay within *Jagirs*.

In Irfan Habib's opinion, it was within the powers of a Jagirdar to remove a zamindar. As for peasants, the Jagirdars claimed powers to detain them to land, like serfs, and bring them back, if they ran away. It is widely believed that in the later period of Mughal empire the Jagirdars resorted to oppression and extortion of the peasants because, unsure of holding particular areas for a longer time they had no regard for the long-term prospects of revenue collection. As mentioned earlier, during Mughal period, revenue demand was very high and more than half of the gross produce was taken away from the peasants. Though peasants held permanent and heritable occupancy rights in land, bulk of the agricultural produce of the empire was placed in the hands of numerically very small class of *jagirdars.*

The Zamindars

The *Zamindar* is a Persian term which means holder of land (zamin). The frequent use of this term began from Akbar's time onwards for any person with any hereditary claim to a direct share in the peasant's produce. The basic right of *Zamindar* was his claim to impose certain levies on the peasants over and above the land-revenue assessment. Irfan Habib also refers to other cesses and perquisites of *Zamindars* such as house tax and levies on forest and water produce, since these are specified among the rights transferred in *zamindari* sale deeds.

A zamindar on the Indian subcontinent was an aristocrat, typically hereditary, who held enormous tracts of land and held control over the peasants, from whom the zamindars reserved the right to collect tax (often for military purposes). Over time, they tookprincely and royal titles such as Maharaja (Great king), Rana (King), Rai (King) Raja (King), Deshmukh (Chief), Nawab (Lord),Mirza (Prince), Taluqdar (District Holder), Chaudhary (Lord), Reddy (Headman), Naidu, Gounder, and many others. Although zamindars were considered to be equivalent to lords and barons in some cases they were seen as independent, sovereign princes. Often zamindars were Indian princes who lost their sovereignty due to British Rule. For example, the Sivaganga Zamindari and Ramnad Zamindari were the lesser and greater Kingdom of Marava ruled by the royal family till 1803; ever since then they were the Zamindars of Marava. There is no clear distinction between royal zamindars, such as Raja Venkata Ranga Rao, or merely aristocratic zamindars. Many kings were former zamindars, such as the Royal House of Benares; conversely many new zamindars were old kings. As a result, there is some confusion about the Indian kingdoms about who is a king and who is a zamindar, as there were as many as 568 kingdoms and, according to some other sources, 572 princely states in India before independence. During theMughal Empire, zamindars belonged to the nobility and formed the ruling class. Emperor Akbar granted them mansabs and their ancestral domains were treated as jagirs.

The practice took structural footholds before the Mughal Era and was solidified by the indirect system of taxation in the Mughal Empire and British Raj. After the British withdrew, the system was legally abolished with the creation of India, Pakistan and (after independence in 1971) Bangladesh; however, it is current in some areas of modern Pakistan. Zamindars built lavish palaces, lush gardens, schools, temples and other venues of philanthropy. Several families were of ancient lineage and had been independent rulers in earlier periods. In most cases, zamindar families were descendants of cadet branches of earlier royal families.[citation needed] Zamindars held considerable powers in their territories: magisterial, army recruitment (as lathials), revenue collection and taxation, among others.

Other terms for zamindar were and are used. For example, a zamindar is known as a Wadera or Wadero in Sindh and as a thakur, istmarardar and kanoongo (in scindia state)Rajasthan, Uttar Pradesh, Madhya Pradesh, Himachal Pradesh, Uttrakhand, Chhatisgarh and Bihar. In the Punjab and Haryana, there are multiple variations, such as chaudhary (which often became lambardar or zaildaar during the British Empire's occupation of North India), Sardar and Malik (an Arabic term which literally means "King"). The word zamindar is derived ultimately from the Persian Zamîn, "earth/land", and the common suffix-dâr, "-holder" (also found in many of the terms above). The term means, in Persian, 'land owner.' Before Mughal rule in India, the aristocracy collected and retained revenue from land and production. The Mughals appointed people to act as tax clerks, sending them around the country to oversee collection of revenue and remit it to the capital city of Delhi. These people were known as the zamindari (intermediaries) and they collected revenue primarily from the Ryots(peasants).http://en.wikipedia.org/wiki/Zamindar - cite_note-8 The zamindari system was more prevalent in the north of India because Mughal influence in the south was less apparent.

The zamindars often claimed to derive their rights from settling a village and distributing its lands among the peasantry. Theoretically, they also had the right to evict peasants, but land being more abundant than labour in practice the right to evict the peasant had much less significance.

The zamindari system ensured proper collection of taxes in a period when the power and influence of the Mughal emperors were in decline. With the Mughal conquest of Bengal, "zamindar" became a generic title embracing people with different kinds of landholdings, rights and responsibilities ranging from the autonomous or semi-independent chieftains to the peasant-proprietors. All categories of zamindars under the Mughals were required to perform certain police, judicial and military duties. Zamindars under the Mughals were, in fact, more the public functionaries than revenue collecting agents. Although zamindaris were allowed to be held hereditarily, the holders were not considered to be the proprietors of their estates. The territorial zamindars had judicial powers. Naturally, judge-magistracy, as an element of state authority conferred status with attendant power, which really made them the lords of their domains. They held regular courts, called zamindari adalat. The courts gave them not only power and status but some income as well by way of fines, presents and perquisites. The petty zamindars had some share in the dispensation of civil and criminal justice. The Chowdhurys, who were zamindars in most cases, had authority to deal with the complaints of debts, thefts and petty quarrels and to impose paltry fines.

In Irfan Habib's opinion the *zamindari* right was in itself an article of property. It was inherited according to the same laws and customs as governed the inheritance of other property. *Zamindari* right was also freely sold. Some *zamindaris* were actually mortgaged to professional money-lenders. Provision for transfer of Zamindari rights also enabled persons who had accumulated wealth out of the extraction of land revenue, such as petty officials, revenue grantees, *etc.*, to transform themselves into *zamindars*. But it was rare to find merchants buying up *zamindaris.* Zamindar and peasant-held villages were found side by side in the same district in Mughal empire. Even sometime in a peasant village, Mughal administration created new *zamindaris.* During the decline of Mughal empire, zamindari rights was created even by force. Sometimes peasants of a village also sold away their rights to a person, who became the zamindar of that village. At the same time, there are numerous instances of fresh villages being settled without the intermediation of zamindars. Thus during the Mughal period two groups of revenue extractors, *viz.*, jagirdars and zamindars, whose involvement in agricultural production was almost nil, were of enormous importance. What the producer was supposed to pay was fixed in law by imperial regulations. What was actually taken depended on the power and inclinations of *Jagirdars* and *Zamindars.* During the declining days of the Mughal empire this system became quite oppressive.

British Era

The British colonists of India generally adopted the extant zamindari system of revenue collection in the north of the country. They recognised the zamindars as landowners and in return required them to collect taxes. Although some zamindars were present in the south, they were not so in large numbers and the British administrators used the raiyatwari (cultivator) method of collection, which involved selecting certain farmers as being land owners and requiring them to remit their taxes directly.

History

Some new zamindars were old Rajas. Many descended from 18th century revenue speculators and military adventurers. Several families were of very ancient lineage, like those claiming Bargujar, Meenas ancestry and had always been independent rulers at earlier periods of Indian history. They frequently intermarried with the ruling families of the princely states such as Sita Devi of Baroda's marriage (daughter of Rao Venkata Kumara Mahipati Surya Rau, the Zamindar of Pithapuram), to Maharaja Pratap Singh Gaekwad of Baroda. Their tenants numbered from dozens to many thousands and, under imperial law, had to pay rent to zamindars to retain rights to their land.

Under the British some of the formerly independent Indian states that were given the status of zamindaris. Some of these zamindars held title to vast tracts of land, for which they were required to pay annual rents to the government of British India. The zamindar of Burdwan was the single largest taxpayer in the British Empire and was referred to as "Maharaja." In his case it was a non-ruling title and not a princely one. Burdwan and other zamindars like Raj Darbhanga (the largest zamindar of India with highest monthly income), Bettiah Raj and Dumraon were very wealthy and lived like royalty. The second rank of zamindars included Balrampur, Sahaspur-bilari, Jhangirabad, and Dhampur. Other zamindars included Tahirpur, Cossimbazar, Kukathala (Kuthra),Barara Mymensingh, Rajoli, Natore, Puthia,Dighapatia, Nazargunj and Tekari Raj.

Accession

Unlike the autonomous or frontier chiefs, the hereditary status of the zamindar class was circumscribed by the Mughals, and the heir depended to a certain extent on the pleasure of the sovereign. Heirs were set by descent or a times even adoption by religious laws. Under the British Empire, the zamindars were to be subordinate to the crown and not act as hereditary lords, but at times family politics was at the heart of naming an heir. At times, a cousin could be named an heir with closer family relatives present; a lawfully wedded wife could inherit the zamindari if the ruling zamindar named her as an heir.

The governors of the Vijayanagara Empire, who were under the direct control of central government, ruled the provinces and feudatory kings. The feudatory kings were called "Nayakas" and paid tribute to the emperor who ruled the kingdoms. The Vijayanagara Empire was destroyed in the Tallikota war of 1565 AD, when Muslim states (Golconda, Bijapur,Ahmednagar, and Bidar) came together against the empire. Tributary kingdoms that were under Vijayanagara Empire came under the control of Sultans of Golconda and Bijapur. These Golconda and Bijapur Sultans employed subservient people from Balija, Velama, Kamma and Reddy shudras castes as kings (instead of the Kshatriya kings whom they perceived as a threat, due to their royal lineage and ambitious tendencies to fight for rule) as the rulers of kingdoms and had them to pay the tribute and collected revenue. Thus many more of these kingdoms/provinces, also known as Samsthanams or Zamindaris, came into.

The Golconda Empire came under the control of the Mughal Dynasty in 1686 and became to be known as the Deccan Suba. The rule of the Mughal Dynasty didn't last long. Nizamul Ul Mulk Asaf Jah, as the governor of the Deccan, controlled the present state of Andhra Pradesh and parts of Maharastra and Karnataka. In the beginning the French had influence over Hyderabad. However, Salabat Jang, who came to power after his father, realized that the French were losing ground to British and so realigned himself with the English. In 1761, the East India Company helpedNizam Ali Khan to depose his brother Salabat Jang and helped him occupy the Deccan Suba. The East India Company then took Andhra regions for lease from Nizam Ali. Later, by accords of 1768 and 1779 these Andhra regions came under complete control of East India Company, whereas Telagana states remained under Nizam. In 1802, the British government under a Permanent Revenue Settlement agreement restored the ruling powers and rights of kings of Andhra regions. During this time, rich and influential people became rulers (zamindars) to the city-states (zamindaris). In the beginning, the British allowed old kingdoms to have armies with the condition that they would support the British Empire during the war. However, this right was abolished and these kings were reduced to the level of zamindars. These feudal kingdoms/zamindaris were abolished after the formation of the Indian Union, a democratic republic.

After Creation of India

In most states the zamindari system was mostly abolished in India soon after its creation with the first amendment to the constitution of India which amended the right to property as shown in Articles 19 and 31.[16] This allowed the states to make their own "Zamindari Abolition Acts". The term is usually associated with the aristocracy as zamindars are considered to be of the landed gentry. Zamindars tend to marry into families of the same social class; however, there have been cases of impoverished nobles marrying into rich families with no titles (this is sometimes considered marrying into the same social class, even if the other family is not of the nobility).

After Independence of Bangladesh

The Zamindars of Bengal were influential in the development of Bengal. They played pivotal part during the Indian Rebellion of 1857.[17] In 1950, land reform was accomplished in East Bengal with the abolishment of the feudal zamindari system. The abolition of the zamindari system (which divided the society into lords, owners of property, and commoners, users of property) in Bangladesh after independence from Pakistan was a major landmark that took place immediately effective January 10, 1972 under a Presidential Ordinance, which became permanent under its new constitution April 10, 1973. In Bangladesh most zamindars were given titles such as Chowdhury, Sharkar and Khan

In Pakistan

In Pakistan the zamindari System is still present in some form, especially in the provinces of Sindh and Punjab. Zamindars are known by different terms in different provinces. In Sindh, for example, zamindars are known as Wadera, while in Punjab, terms such as Chowdhury, Pakhral, Sardar, mir and Malik are used. The title of lambardar is only available to a powerful family of zamindar this is a state-privileged status which is generally hereditary and equivalent to an EnglishDuke, there can only be one Lambardar of the village/city, who in many cases has wide ranging governmental powers, *i.e.* the policing authority of the village, and many other governmental and administrative perks, this has resulted in a feudal system being apparent in the Pakistan Senate and its elected representatives. Almost all elected leaders of the State have been from the landed and feudal gentry in many cases holding the title of Lambardar. Most presidents and prime ministers have been from feudal stock, with the exception of martial law administrators and generals who obtained power through military coups.

The Village Community

A notable feature of village life in pre-British period was the combination of agricultural work with manufacturing processes with an unalterable division of labour. Production was mainly for direct use and the surplus after payment of revenue was marketed. Relationship of the village with town was one way and it hardly received anything in return and provided for almost all of its needs from within the village. Normally the peasants of the village claimed the same ancestry and so belonged to the same brotherhood (bhaichara). This fraternity by invoking ties of blood bound the peasants in unity far stronger than could have been expected among mere neighbors. Authority in the village was exercised by a group of elders, traditionally a council of five persons, the *Panchayat*.

But the earlier contention that peasants had common ownership in land is disputed by many scholars. It is suggested that the individual families had their separate holdings and only the forest and grazing grounds were held commonly. Of course there were certain other spheres, outside production, where peasants acted collectively. In the village every peasant paid his share into the common financial pool from which the land-revenue, the demands of officials, the repayment of any common loan (specially raised during natural calamities to pay the part or whole of the land revenue of the village), and expenses for the economic, social, and even spiritual benefit of the village were met. Even in the case of abandoning of a village and migration to some other areas, villagers normally decided collectively.

Views also differ over the homogeneous nature of peasantry. Economic differentiation within the peasantry had emerged during the Mughal period. Irfan Habib point out that in Northern India as well as in other parts of the Mughal empire there were some large cultivators raising crops for the market; and there were small peasants, who could barely produce foodgrains for their own subsistence. Beyond this differentiation among the peasantry, there was still sharper division on the basis of caste. Depending upon their resources in seed, cattle and money, peasants might cultivate larger or smaller plots. But larger land holdings were linked to, and often resulted from, superior position or status secured either as headmen (*muqaddams*) or as members belonging to dominant elements of the village (*kalantran*), in contrast to the small peasantry (*rezariaya*). The village headmen often had revenue-free allotments, amounting to 2½ per cent of the total, and the superior elements were often assessed at lower rates of revenue than ordinary peasants.

Another source of strength of the rich peasantry was associated with the land- revenue system. The basic units of revenue assessment and collection being a village, it was natural for the revenue authorities to rely upon the headmen or a small group of upper peasants. This dominant group fixed tax rates for each peasant, collected it, and put it in a pool. From this pool the land revenue would be paid, so also certain other common expenses including the fees and perquisites of certain officials. Those who controlled the pool usually evaded paying their own due share of revenue. Lower rates were also levied upon some favoured elements. The smaller peasants, forming the bulk of the peasantry, were thus called upon to pay more than their due share of the revenue in order to make up the total. Thus the internal contradictions started emerging within the peasant communities.

The Peasant and the Ownership of Land

Now let us consider the crucial question of who was the landowner. This was the question which, after the conquest of India, British administrators addressed to themselves. In fact in the years preceeding the Permanent Settlement in Bengal, the English circulated some questionnaires with a view to eliciting 'official' opinion concerning various land rights. The question most often asked was "who is the owner of the land- the ruler (*hakim*) or the *zamindar*?" It will be noticed that the question as formulated excluded the peasants from consideration altogether.

However, most of the scholars today agree with the view that king was not the owner. Numerous Mughal government documents refer to private persons as owners (*maliks*), holding *malkiyat* or ownership of land. In Aurangzeb's *farman* to Muhammad Hashim the terms *maliks* and *arbab-i-zamin* (landowners) are clearly used for ordinary cultivators or peasants. But the crux of the matter really is whether the substance, not merely the name, of the peasant's right was such as to deserve the application of the term 'proprietary' in its strict juridical sense.

On the one hand, there was general recognition of the peasant's title to permanent and hereditary occupancy of the land he tilled. In case he was found incapable of cultivating land or had abandoned it altogether, land was given to another peasant for cultivation. But if at any time, the malik recovered his ability to cultivate it or came back to it, the land was to be resorted to him. On the other hand, there was no question of really free alienation, which is an essential feature of modern proprietary right. The readiness with which authorities recognized the peasant's right of occupancy and the anxiety they showed to prevent him from leaving the land were both natural in an age when land was abundant and peasants scarce. Sale of land might also not be possible because there was no scarcity of land and the revenue burden was quite heavy. The peasant, in such a situation, might not often succeed in finding buyers. The peasants thus enjoyed the permanent and hereditary right of occupancy but did not possess the right of free alienation.

In fact, the rights which in aggregate constitute ownership, in the true sense of the word, were not as a rule vested in one-person, but were distributed irregularly among the various parties connected with the land. Moreover, the collapse of the Mughal administration had produced an environment in which might (force) counted more than right. The interests were traditionally multiple and the notion of an absolute and exclusive form of proprietorship was alien to India.

Land Relations in Eastern India

The investigation of land relations during Mughal period in Bengal is of particular interest, because it was here that the early British administrators acquired the terminology which they carried with them to the North. Bengal stood apart from other provinces because the revenue system in force here was the one known as *mugtai*, a fixed demand. The revenue was levied in cash, at rates per unit of land, or in lump sums covering entire villages, the same amount being levied year after year, until, as a special measure, there might be a revision or enhancement. According to James Grant, initially the revenue demand fixed in 1582 by Todar Mal represented around one-fourth of the average produce. In 1658 there was an enhancement of 13 per cent. After that successive levies were made on the *zamindars* in the form of *cess*, although the basic demand remained unchanged. But Moreland does not agree with James Grant's account, specially, his statement that the basis of assessment was one- fourth of the produce. Moreland considers that in Todar Mal's time the state's claim was uniformly one-third.

Unlike northern India, in Bengal the imposition of revenue-demand was not upon the peasants but upon the *zamindars*. Bengal *zamindars* were different from the *zamindars* in other parts of Mughal empire. Here the zamindar was called upon to answer for the payment of land revenue within the area of his *zamindari*. He collected the land tax (or rent) from the peasant at rates fixed by custom or by himself and paid the amount imposed upon him by the administration. The balance left with him constituted his income. In other parts of the Mughal empire the zamindar was expected to collect cash and allot revenue-free land.

Land revenue system in Asom, which had evolved during the long *Ahom rule* (1228- 1818) was entirely different. According to B. Chaudhuri, under Ahom rule the payment of revenue in the form of cash or produce was of a limited extent. The fiscal system in Ahom rule was based on labour services of the entire community, mostly in connection with the defence of the kingdom as the Ahoms had no regular standing army. Occasionally, people were engaged for the maintenance of public works also. The male population was divided into groups called *khels*, each khel being further divided into smaller units (*gots*), composed of three to four males (*paiks*). In times of peace only one *paik* of each *got* served the state, usually for three or four months in a year, the other members looking after the cultivation of his land. During wars all the male members were called up, with the women and children looking after the entire cultivation. Each *paik* family, in return for such services, got free of rent two *puras* (about 2.66 acres) of wet land, in addition to a piece of land, also rent free, suitable for homestead or garden, and could add to its holding by reclamation of wasteland where this was plentiful. The *Ahom* king entrusted the administration and the defence of the realm to the Asomese aristocracy. The economic power of this group was largely derived from the income from their rent-free estate, which were worked mostly by servants and also partly by free *paiks*.

Village Life in Mughal Times

In the times of the Mughals, India was considered to be one of the more prosperous countries in the world. Mughals, Rajputs and Afghans battled each other just to lay claim over this prosperity. The wealth enabled jagirdars, zamindars and raja-maharajahs to lead lives of comfort and luxury, and enabled Mughal emperors to erect fabulous monuments such as the Red Fort and the Taj Mahal. With an eye on this prosperity, traders also came from Europe to India. However, the wealth of India did not simply fall out of the sky. The secret of the power of the Mughal empire and the wealth of the Mughal amirs lay in the fields - in the hard labour of the peasants. Mughal emperors and amirs partook of the yield of the peasants in the form of land tax. What kind of life did the peasants have? What were their homes like? What did they grow? How much tax did they pay? How much were they able to save? Come, let's read about the life of peasants in Mughal time.

The villages had rich as well as poor people. Most peasants lived in huts of wattle and daub (woven reeds or bamboo plastered with mud). Driven from their homes by war, famine, drought and oppression, the peasants would often have to quickly leave their houses and set up shelter in new places.

Food

Now let us take a look inside the hut of a common peasant. You would find only a few utensils, and they would be made of clay. In those days, brass and bronze were very expensive and aluminium and steel had not come into use. The clay utensils would be used to cook khichdi (iKcaD,I) of moong and rice, and to make rotis of bajra or jowar. Along with this there might also be a little vegetable and ghee. Because milk was available aplenty in those days, ghee was quite inexpensive. Apart from ghee, sesame (til) and mustard (sarson) oil was also used. Groundnuts were not grown in South Asia then, hence its oil was not available. Also many vegetables that you now eat were not grown here in those days.

Till the Mughal times, potatoes, pumpkins, tomatoes, peas, chilly, guavas, and custard apples were not grown at all in South Asia. These are all fruits and vegetables from South and North America which traders from Europe brought to India towards the end of the Mughal period. Howeve r, vegetables such as broadbeans, spinach, sweet potatoes, gourds of different kinds (ribbed, bitter, bottle, *etc.*) ladyfingers, and brinjals were common. Common fruits were bananas, mangoes, jack-fruits, watermelons, ber, grapes and pomegranates.

Cloth and Clothing

In those days, the spinning wheel had begun to be in each house. We mentioned before that the charkha came to India with the Turkish people. By the time of Akbar and Jahangir, people had widely adopted its use. Women would spin cotton into thread and the village weaver would weave it into cloth. Until this time, the people of India wore less cloth. But, after the coming of the spinning wheel greater quantities of cloth came to be worn.

Agriculture

As is the case today, in Mughal times too the biggest problem in farming was that of irrigation. In those days, people had to make do with tanks, canals and wells. There were no motor pumps or electricity, as we have today. Therefore, much less land could be irrigated. Most of the land was unirrigated. Thus only the monsoon crop (kharif) could be grown in many places.Since there were no chemical fertilizers or pesticides, or high yielding seeds, the yield was less than it is today. But still, the productivity of the land (the yield per hectare) was amongst the highest in the world in those days.

The soil in the river valleys of India is extremely rich. With their hard work and intelligence, the peasants took advantage of this soil to raise two crops in a year. On the other hand, in the best fields in most European countries farmers could grow only one or two crops in three years. The soil there was not good enough to grow wheat two years in a row – so after growing one wheat crop each field would be left untilled to fallow, or planted with oats in rotation. Most parts of Europe have a heavy and deep soil. Such soil cannot be ploughed without a special, strong kind of plough. But such ploughs were not made in those days. That is why the productivity of European soil was much lower than that of Indian soil.

European travellers were astonished to see that not only could Indian peasants grow two crops in a year in some areas, but also so many different kinds of crops could grow in the warm climate of India. In a single village, fifteen kinds of kharif (monsoon) crops and ten kinds of rabi (winter) crops might be grown. In addition to this, fruits, vegetables, and spices were also grown. There was hardly any other country in those days where such varied crops were grown in a single village. This productivity was indeed the basis of India's famed prosperity. However, even though they grew so many crops, the peasants remained very poor. Many children would die of malnutrition and the common people lived under the threat of starvation.

The Condition of the Peasants

When the rains were sufficient and the crops were also good, the peasants would have enough to get by. But they would not be able to save anything for difficult times. Thus, whenever the rains were scanty and water sources dried up, the crops could not grow, and then the peasants would have absolutely nothing to live on. In times of famine, thousands of people would become victims of starvation and epidemics. Things would get so bad that people would start eating grass and leaves of wild trees. Out of desperation, many peasants would sell themselves and their children to the wealthy. People would leave their villages and wander in search of food. Hundreds of villages would thus be deserted and slowly fall to ruins. There are descriptions of such terrible famines where human beings were reduced to eating other human beings. Such is the story of the peasants of the Mughal period – all the wealth was grown in their fields and all the poverty resided in their homes.

Land Tax

A large part of the agricultural produce was taken away in the form of tax. During the time of Akbar, one third of the crop would be taken from the peasants as tax. But in the reign of Jahangir and Shahjahan the burden of taxation kept increasing.

The Story of a Village

In the suba of Agra lay the village of Kararia. As villages went it was quite small - with about 80 peasant families. There were also 5 families of craftsmen, who made things of wood, iron and leather. Some craftsmen made clay pots and some wove cloth. Most of the peasants were Jats by caste, but there were also some Gujar peasants. It was around the year 1580. The kharif crop stood swaying in the breeze -- bajra, jowar, moong, moth, til and kodon. One day, a group of horsemen came to the village from the nearby town of Bayana. On entering, they went straight to the house of the zamindar, Suraj Dev Jat. In a few moments the news had spread in the whole village that these men had come to measure the peasant's fields to assess how much tax they would have to pay. The patel (headman) of the village and the patwari (accountant) were summoned to the house of the zamindar. They were the wealthier and more important peasants of the village who helped in collecting the taxes from the people. In the evening the panchayat was called and all the people of the village collected at the chaupal (square). The officer who had come from Bayana to fix the tax was called Puranmal. He said, "We have come here under orders from Muzaffar Khan and Raja Todarmal, the ministers of Emperor Akbar. The emperor has changed the system of assessing and collecting tax for the whole empire. This year you will pay the land-tax not in grain, but in money."

Immediately, a whisper started up among the people. After some time a peasant stood up to say, "But till now we have been paying land- tax only in grain." Puranmal said, "You can now sell your crop in Bayana and pay land-tax from that cash." Trying to get the peasants to his side, the zamindar Suraj Dev said, "Why are you getting so upset? Earlier too you paid one third of the crop - you still have to pay the same. Only, not in grain, but in cash." "We had heard that Emperor Akbar is good and we thought that he might reduce the land-tax. But now he's just added to our troubles," a peasant said.

The Peasants sell Grain to the Banjaras

That year all the peasants of Kararia carted their grain to Bayana. Peasants from many nearby villages had also brought their grain. Much more grain was brought for sale at Bayana than ever before.Buying and selling grain was done by the banjaras (migrant traders). A group of 30-40 banjaras would travel from place to place with 200-300 bullocks. They would purchase grain, sugar and jaggery (gauD,) from villages and then sell them in various distant cities. Starting from the hills of the Himalayas, they would move, selling and buying, all the way to Cambay, Bengal and south India. All the peasants sold their grain to these banjaras. In exchange for some of the grain, they bought salt and for the rest they took cash so that the tax could be paid.

The Collection of Land Tax

A few days later the amil of the jagirdar turned up in the village.The amil went to the house of the zamindar Suraj Dev Jat and told him to collect the tax from the people of the village. The amil said, "In your village the kharif crop has been sown over a total of 9,000 bighas. I have calculated this with the patwari. This adds up to Rs. 17,000 in tax. Please have this amount collected and kept for me. I am going for a round of the other villages of my jagirdar. When I return in ten days I will take the money from you."

Suraj Dev Jat called the patel and the patwari and told them to collect the tax from the peasants. The patwari said, "And what if someone refuses to pay?" The zamindar said, "two of my horsemen and four soldiers will go with you; let me see who has the guts to refuse". After two-three days they had collected the tax from most of the people. No tax could be collected from some peasants whose crop had been lost due to hail. When the amil of the jagirdar came, Suraj Dev handed him the money collected as tax along with an account. The patel and the zamindar together explained that due to hail, land-tax could not be paid by some peasants. The payment due could be noted against their names and collected from them over the next two-three years. The amil had this noted in the account of the patwari.

When the amil was about to leave, Suraj Dev hinted that he had forgotten about something; at which the amil handed him ten per cent of the total land-tax collected. This was Suraj Dev's 'malikana' or commission. In return for having the land-tax collected from the village people, the zamindars were paid this malikana.

The amil paid the patel and also the patwari some percentage of the land-tax. As it is, the zamindars and patels were given a discount in land-tax. They paid only one-fourth of their produce in land-tax whereas the common peasants were required to pay one third of their crop.

Debt

The common peasant had many difficulties. Sometimes the crop would not fetch a good price in the market. Then they would not have enough money to pay the land-tax. In such a situation they would have to borrow money from the money- lender to pay the tax. If they were unable to pay, the zamindar's men would often beat them up.

"We Will Farm Elsewhere!"

Burdened with debt, three peasant families decided to leave Kararia village for another place. They had heard that the amil of that place was giving concession in land-tax to newcomers who would settle there. He was also giving taqavi (a development loan) to buy oxen and ploughs.

After these three families had decided to leave, they went to the patel of the village to tell him. The patel frowned and said, "Very well. But there are 20 bighas of land in your name in this village. Who will pay the land-tax for that? The jagirdar's amil will demand his entire money, whether the land has been tilled or not. If you go away, we will have to go around trying to find someone who will plough your land. You think leaving your land and running away is just a joke? No one can just get up and leave their land!"

"But what shall we do with the land? We have neither ploughs, nor oxen nor seeds. All our farming is being done in debt - that doesn't leave anything for us to eat," the peasants replied. Hearing this the voice of the patel rose in anger, "Go! Let me see how you run away from the village! I will send the zamindar's soldiers and have you arrested. Then you can starve in prison!" he said.

This threat scared the peasants. In silence they returned to their homes. But hardly a month had passed before one of the families ran away at night and could not be traced even after a great deal of searching.

The village patel had tried to stop the people from leaving. For whose good did he do this? The zamindars kept soldiers. What have you understood about the functions of these soldiers so far?

No field to Remain Unsown

There was plenty of untilled land during Mughal times. People could plough as much land as they wanted. That is why those peasants who suffered hardships in one place would often leave their villages in the hope of taking up farming in another place. This tendency of the peasants irritated the zamindars and jagirdars. They wanted more and more peasants to come and settle in their areas and farm as much land as possible. That is why they would give newcomers land as well as concessions in land-tax. They would also try hard to prevent peasants from running away from their areas. If they could not stop a peasant family from leaving, they would give the land to another peasant to plough so that there would be some crop and the land-tax could be paid. However, if the original owner of the land returned he could get his land back. But in his absence, the fields could not remain unsown.

For if fields remained uncultivated, how would the jagirdars, the zamindars and the emperor manage? What would happen to amirs such as Baqar Khan?

The Ruling Class

Day by day the number of high amirs such as Baqar Khan was increasing in the Mughal Empire. When Akbar was ruling there were only 51 high amirs. This number kept increasing and in 1700 there were more than 500 such amirs. How were the expenses of all these amirs to be met?

Taxes Go Up in Kararia

By the time of Emperors Shahjahan and Aurangzeb, land-tax had increased tremendously - from one third of the crop under Akbar, to one half. In addition to what was permitted by the emperor, the jagirdars would try to extract even higher taxes. It must have been the year 1655. The Mughal emperor Aurangzeb had given Kararia village to Raja Jai Singh as jagir. Jai Singh was a prominent amir of the Mughal empire. No sooner was the kharif crop harvested than the amil of Raja Jai Singh appeared to collect the land-tax. The first thing he said to the zamindar was, "This time the peasants will pay a new tax. It will be used to pay the fee for the patwari, instead of having it paid by the jagirdar."

Flaring up at this, the zamindar said, "How can this be done? Already the rate at which you are taxing the peasants doesn't even leave them with enough to eat! Now we are reaching a point where we zamindars can't even collect from the peasants what it has been our right to collect. Go! Go and tell the jagirdar that the people of this village will not pay this new tax!"

You know how the zamindars had to pay a lower rate of tax in comparison to the peasants. They also received a portion of the land- tax that the jagirdar took from the peasants. On top of this, they would make the peasants pay them additional amounts on various pretexts. Like the bhogpatis in the old days, the zamindars used to extract levies from the peasants on their houses, cattle, on weddings, journeys, festivals and so on.

At the panchayat the peasants gave the amil a piece of their mind.They said that if no one else in the nearby area pays this tax, why should they? Despite all this, the amil remained stubborn and threatened that if the patwari fees did not come from Kararia he would return with the army and create havoc. The next day the people of the village decided in the panchayat that they would take their complaint to Raja Jai Singh at Agra. They collected some other peasants from nearby villages and around 20 of them reached Agra.

The Peasants Complain to the *Jagirdar*

In the grand palace of Raja Jai Singh, the peasants narrated their woes. One peasant said, "Maharaj, last year 50 maunds of jowar grew in my field. The amil took 25 maunds in land- tax. The zamindar took away 7 maunds in addition. And then, showing some earlier dues, the amil took away another 5 maunds. On top of this the village moneylender claimed 2 maunds because last year I had borrowed from him for seed. And now this year a new tax is being imposed. From where will we pay it and if we do manage to pay it, what will we eat?"

Listening to such accounts, the jagirdar agreed to remove the new tax and told them that he would tell the amil not to collect it. The peasants breathed a sigh of relief and left for their village. But the very next year Raja Jai Singh was transferred and a new jagirdar came in his place. His amil too tried to collect the new tax. When this happened 40 peasant families left Kararia for another village.

The Peasants Plead with the Emperor

Over time, the situation in Delhi, Agra, Bayana and other nearby villages continued to worsen. All around, the excesses of jagirdars and their amils increased. Peasants of many villages even reached emperor Aurangzeb with their appeals. The emperor did promise them that they would be protected but he did not want to do anything against his jagirdars. He issued many farmans to his officers saying that illegal taxes should not be taken, under no condition should more than half the crop be taken, and the peasants should be helped to improve and increase cultivation. But no one listened to these grand proclamations.

Zamindars Rebel against Mughal Rule

Meanwhile, news reached Kararia that a zamindar named Gokula Jat had rebelled against the emperor and the peasants of nearby villages had joined him. The zamindar of Kararia also thought of joining them. But he was also afraid - after all how could they face the mighty Mughal army! In a few days the news came that in a battle with the Mughal army, Gokula Jat had been killed. With such ups and downs, time passed. Somewhere, one or two zamindars would revolt, at other places villagers would flee their villages and at other places they would join rebel zamindars.

The Revolt of Raja Ram Jat

Fourteen years after the death of Gokula Jat there was another wave of protest. Some 30 kilometers from Kararia, the zamindar of a village named Sinsini revolted against Mughal rule in 1683. His name was Raja Ram Jat. He refused to pay the land-tax collected from the peasants to the jagirdar, Nawab Khan-e-Jahan, because he wanted to set up a kingdom of his own. Great was the excitement in Kararia and nearby villages. Nawab Khan-e-Jahan marched towards Sinsini to crush the revolt of Ram Jat. The raja of Amber, Bishan Das, also sent his army to assist Khan-e-Jahan. Meanwhile, one day some peasants of Sinsini turned up in Kararia as they had relatives in the village. That very evening they called all the people of the village and told them about Sinsini and Raja Ram Jat. They said, "This time we will make these jagirdars bite the dust. Let the nawab and the raja of Amber come against us. All of you should also help us in this fight." A peasant said, "We need young men to fight the Mughal army. If we can get even ten young men from your village it will be of great help."

Almost at once came a voice from the crowd, "I'm ready. I'll come to Sinsini to fight the Mughal army!" Soon many other voices had risen. "Yes, me too", "Yes,

I'll come too." Before long 22 of them had volunteered. That very night they tied up their luggage, took their swords and spears and left to join the army of Raja Ram Jat.

About twenty days passed by. One day one of the boys who had gone to Sinsini came panting on a horse to Kararia. Standing in the middle of the village, he called out in jubilation, "Listen! Listen, people of the village, listen to how we defeated the armies of nawab Khan-e-Jahan and the king of Amber, Bishan Das. Listen to how we made the jagirdars and the fauzdars flee." When the people of the village had collected he narrated the whole story. The army of Raja Ram Jat had repelled two attacks of Nawab Khan-e-Jahan. The nawab had been defeated and forced to flee. A wave of surprise and joy spread among the people of Kararia.

"Did the Jat peasants really defeat and chase away the army of the Mughals? How could this be?" Excited people with these thoughts on their minds moved towards the house of zamindar Suraj Dev, to persuade him to join hands with Raja Ram Jat. Thus started the revolt of the people of Kararia village. The Mughal rule was rocked by this and many similar revolts. It became more and more difficult to collect land-tax. Now they had to fight to claim every rupee from the villages.

Agriculture Science in Mughal Period and its Socio-Economic Impact

Agriculture was the most significant factors forming the basis of a strong economy.The significant features of agriculture was firstly, that a large number of crops were cultivated by the farmers and secondly, the life of peasant life was a combination of agricultural work with manufacturing processes. The availability of cultiviable land was sufficient and it demanded more cropping. Babur in Tuzuk- i- Baburi says ' the country full of men and full of products were divided into villages which was surrounded by cultiviable fields, and also had land for pasture and forests for fuel and other purposes."Akbar tried to increase the land under cultivation. The methods of cultivation was almost similar to present times. When plough was used for cultivation a hoe, a mattock, and few other instruments were used for ploughing the fields. The irrigation facilities used in the ancient and sultanate period were continued with in the mughal period. Natural and artificial modes of irrigation was used. Wells, canals, rivers, Persian wheel etc, were used for irrigation. The peasants knew about the use of manures also. The science of agriculture was experimented with and was developing.

Methodology

Primary and secondary sources are used to reach to my observations.

Observation

The land revenue system formed the basis of relationship between the mughal kings and the peasant class. The founder of the mughal empire did not establish a sound land revenue system nor did his son Humayun. Sher Shah introduced revenue reforms and he was the first muslim ruler to establish a beneficial system which was equally good for the state and farmers. When Akbar ascended the throne had divided the land into crown lands or ' Khalsa" land and the jagir lands. Village was the smallest unit of administration surrounded by fields and also had lands for pasture and forest for fuel and other purposes. Villages were known as 'gaon' or 'mauza'.There were two types of villages 'asli' that is having the habitation intact and

'dakhili' that as the deserted village. There were two types of ' raiyati' villages which belonged to ' Khalsa' land but the taluqa villages experiments to improve the condition of peasants and also the land revenue administration. Akbar's first experiment starts with the appointment of Aitmad Khan as the Diwan in 1563, in which he divided the empire into 12 parganas. In second experiment Muzaffar Khan, in third experiment Shihabuddin Ahmad, in fourth and fifth experiment Raja Todarmal was appointed as the imperial Diwan. A uniform method of measurement was appointed known as ' Ain- i- Dahsala' or the Bandobast system. It was based on Zabti system also known as the Ryotwari system. Initially it was introduced in eight provinces- Allahabad, Agra, Oudh, Ajmer, Malwa, Delhi, Lahore, Multan The whole empire was divided into dasturs. Provinces were divided into sarkars or districts and those of districts into Parganas or tehsils. Amalguzar was appointed in the district level. Each cultivator received a patta or deed. Shahjahan's Diwan Sadaulla khan also introduced some changes in the revenue department.

The mughal rulers used measurement units and the division of lands for the successful collection of taxes. New techniques of units of measurement was introduced. A 'yard' was called 'gaz' which was of 41 units and a 'bigha' was equal to area of 60 yards square or 3600 square yards. Thus, we see that during the rule of Akbar impoverment was made in the modes of measurement for the benefit of the state and also the peasant. Different modes and methods were introduced to enhance the fertility of the soil and for this during the reign of Akbar the land was divided into four categories – 'polaj' in which two crops was raised every year, 'Parauti' – land which was left fallow after raising two crops, 'Chachar' –unfertile piece of land which was cultivated every three to four years, ' Banjar'-was unfit for cultivation and rarely brought under cultivation. Akbar tried to bring more uncultivated land under cultivation to raise various high quality crops.

India has been an agricultural country since ages and the soil of India and its land is basically fertile and cultiviable. Abul Fazl rightly says "The soil is for the most part arable and of such productive power that the same land is sown each year, and in many places three harvests or more are taken in a single year and the vine bears fruit in the first year." Whereas the soil of Ajmer is sandy and water obtainable at great depth, whence the crops are depended on rain., the elevation of Malwa is somewhat above that of other areas of the country and every part is cultivated; both harvests are excellent, specially wheat, poppy, sugar- cane, melons and grapes". The "Whole country between Lahore and Agra", adds De Leat, " is well cultivated and is the most fertile part of India bearing all kinds of crops abundantly".

Irrigation facilities were present in India since ancient times and the Mughal rulers paid heed to it sincerely. Irrigation was done both by natural and artificial means. The thirty 'kos' long canal from Kizrabad to Safidun built by Firuz Tughlaq which had fallen into disuse was in the time of Akbar re-excavated by Shaha- ud-din Ahmad Khan to irrigate his jagir land. It was repaired and a new channel from Safidun to Delhi, 30 'kos' land was excavated during the reign of Shah Jahan. Jahangir also continued with the water construction works at Burhanpur and Punjab. In the fear of decrease in agricultural production, Mughal rulers made arrangements for cultivation. State undertook the construction of irrigation canals and reservoirs. Irrigation works were put in charge of local officers called'daroghas' of canals who were entrusted to keep them in good condition. 'Nahr-i- Bihist' was a canal at Khizrabad and' Nahr Faiz' built during the reign of Shah Jahan reign carried water from Yamuna and irrigated a large area of tract. Dams, reservoirs, tanks, canals *etc.* were built for cultivation. Wells, rivers and rain water was also used for cultivation. Abul Fazl says that wells were mostly used for irrigation.

The peasants had the knowledge of implements of agriculture. The ancient implements were also used for cultivation. Babur in his memoirs Tuzuk- i- Baburi about the cultivation technique "At the well edge" he writes "they set up a fork of wood, having a roller adjusted between the forks, tie a rope over the roller, and tie its other end to the bullock. One person must drive the bullock, another empty the bucket. Every time the bullock turns after having drawn the bucket out of the well, that rope lies on the bullock tract in pollution of urine and dung before it descends again into the well." This method as followed in the regions of Agra, Chandar, Bayana and neighbouring districts. In Punjab and neighbouring parts irrigation as done by the means of Persian wheel. This device was wooden pin-drum gearing device used in the thirteenth and fourteenth century. Abul Fazl says that in 1595 in punjab agriculture was mostly based on irrigation from wells. The technique such as Persian wheel led to the prosperity of the Punjab region. Artificial devices were used to draw water from wells. The peasants knew about the pulley and lever system. Based on the lever principle form of an upright beam was kept in swinging position with one of its end tied with a long rope and other carried a weight heavier than bucket. The Persian wheel which was used in the sultunate period was the most advanced water lifting device and it continued to be so in the Mughal period. In the Persian Wheel a garland of pots was attached to the rim of a wheel, a gear mechanism was also attached to it. Hoe, mattock and 'kudal' which is even used nowadays was also used in those days for cultivation. The fields were manured with animal dung and intensive ploughing as done. Some instruments were used for cutting the harvest. Oxen as used for ploughing of fields.The peasants generally used old methods of cultivation which might have been similar to that of the present days. An effective agricultural practice such as dibbling was introduced that is dropping of seeds into holes driven into grounds by sticks. The use of iron was minimal and wood was used predominantly.

The peasants knew about the rotation of crops and double cropping. The peasants raised two crops – Kharif and Rabi in a year. Some special crops such as sugarcane, indigo, cotton were grown. The foodgrains produced were wheat, barley, gram, rice, millets, linseeds and pulses. Ain- i – Akbari says wheat was commonly grown in the provinces of Agra, Allahabad, Awadh, Delhi, Lahore, Multan, Malwa, and Ajmer.heat as also gron in the Qandhar region. The production of wheat depended on rainfall and irrigation. Barley as produced in almost all parts of the country but not much in the parts of Bengal and Odisha. Gram was also grown in almost all parts of the country.

Rice grown in the regions where the climate was tropical.and had plenty of rainfall and high temperature. Bengal, Odisha, Bihar, Agra, Allahabad, Awadh, Delhi, Lahore, Khandesh, Berar, and Kashmir produced good quality rice. 'Kar' 'sukhdas' and ' shali' were superior quality rice. Fine quality rice was grown in Kashmir and khandesh.'Sukhdas' rice as known for its flavor and good quality. Millets consisted of cheaper grains such as jowar, bajra, kodon, sawan and mandua *etc.* These also formed the integral part of the Kharif crops.These crops were grown in the regions of deficient rainfall and poor soil condition regions. Jowar and Bajra were grown in the regions of Malwa, Gujarat, Ajmer, Khandesh, Delhi. Lahore, Agra, Allahabad, Khandesh, Delhi, Lahore, Agra, Awadh, Allahabad and Multan. Pulses were,mainly grown in Bihar, Allahabad, Awadh, Delhi, Lahore, Multan *etc.* The chief varities of pulses were moong, month, arhar,peas *etc.*These were grown in the autumn season. Linseed like mustard and alsi were mostly grown in the subas of Agra, Allahabad, Lahore, Malwa, Ajmer, and Multan.

Best indigo was grown in the areas of Bayana and in the reign of Akbar it was sold at Rs. 10-12 per maunds and it as exported to Turkey and also to European countries. Bengal indigo came into prominence in the seventeenth century.The Dutch and the Britishers exported the best quality Bayana indigo. During the reign of Shahjahan the trade in indigo had achieved the status of a profitable business. In the months of August and September just after the rains, the leaves are cut and gathered, cast in into long cisterns, pressed down with stones, and left therein with water. The water is then passed out into another circular cistern. In the first year of its growth, the leaves are tender and not having attained perfection, produce heavy reddish 'Neel'. In the second year it is called Cyrese(aree or sprouting from the roots). It is light and of perfect violet colour but in the third year the herb declines, being worst of the three. It is called Cattled (Khutizal or Khuntri). It is blackish 'Neel'. After the water over the leaves in the cistern has been passed several times, "they labour with great slaves and then let it settle; scumming of the clear water on the toppe, then laboring it afresh. Let it settle again doing this of till nothing but a thick substance remaine, which they take forth and spread on cloth to dry in the sunne, making small balls, lay them on to dry on the sand (for any other thing would drink up the colour). If the rain falls, it looseth its colour and glosse and it is called aliad(ala or moist) four things are required in the ' neel' a pure grain, a viuolet colour, glosse in the sun, and it be dry and light, so that swimming in the watyer, or burning in the fire, is cast forth a pure light vapour, leaving a few ashes". The techniques and method used for the cultivation of indigo is amazing for that period. It is obvious that the peasants were aware of the science of production of indigo which was an integral part of the commerce of the Mughal period. Its cultivation started in the region of Ahmadabad and Sukhraj. Further Aligarh,Itimadpur, Khurja and Agra became important centres of indigo cultivation. During the reign of Jahangir,indigo in the words of Sir Thomas Roe, as " the prime commodity". Safflower an another dye which was also cultivated on a large scale all throughout the Mughal rule. Safflower during Akbar rule was cultivated in the regions of Agra, Allahabad, Aadh, Delhi, Mala, Lahore and Multan.

Sugarcane as grown in the regions of Agra, Allahabad, Awadh, Lahore, Multan and Mala. Silk was extensively produced in the Bengal region and it became a centre of Sericulture and from their silk was exported to foreign countries also."In Akbar's time Bengal including Patna, as an important centre of the industry besides Ahmedabad and Kashmir which used to export large quantities of silk to the other centres". Various varities of sugarcane such as 'paunda', 'black' and 'ordinary' were grown. Sugarcane was a costly and superior crop and was sold at the prices double of it. The cost of 'paunda' quality of sugarcane as highest in Ajmer. In the Kota region the price of sugarcane as as high as 100 per maunds. Sugarcane cxultivation had extended to the region of Aurangabad and several places of Bihar.

Cotton and indigo were grown in Malwa, Multan and Ajmer. Moreland maintains that " cotton at least was more widely grown than is now the case, though the aggregate of production was probably less and it is reasonable to infer that most parts of the country were nearly self- sufficing in the matters of clothes". Poppy was also cultivated in Agra, Awadh, Lahore, Multan and Ajmer. Poppy was exported to Pegu. Various kinds of vegetables were grown such as – spinach, cabbage, bethuwa, ginger, peas, carrot, radish, lettuce, potato, lemon was grown in all parts of the country especially Agra, Malwa, Ajmer, Bengal, Bihar. ' Pan' (betel leaves) were grown in Bengal, Bihar, Awadh, and Berar.

Tobbaco was introduced by the Portugese in the first fifty years of the seventeenth century.The muslims opposed to smoking, the cultivation did not prosper. The Dutch and the Portugese grew it in their colonies, Jahangir prohibited smoking in 1617 A.D. The cultivation of new fruits was encouraged by the Mughal rulers. Musk- melon, water melon, apple, grapes, date, figs, pomegranate, mango was grown. Jahangir introduced the cultivation of pineapples. Guauva and custard apples was introduced by the Portugese. The 'Honeydrew' variety of water melon of colder regions was cultivated in the plains of Northern India. Muhammad Rida of Khurasan raised the first crop of this variety of watermelon who was Honoured for this by Shah Jahan. Jahangir says cherry was not found in Kashmir before the conquest of Akbar, his official Ali Quli Afshar brought it from Kabul and propogated it by means of grafting. During the reign of Shahjahan grafting method was used extrensively and three quality of oranges such as 'sangatara', 'kaulna', and 'narangi' were cultivated.Description of grafting of Mangoes and growing it in the Murshidabad region is also found. The grafted variety of Alfonso developed by portugese in the western coast is mentioned by traveler Nicolo Manucci in the seventeenth century.

Jahangir built gardens in Kashmir and Lahore. Babur introduced a good quality of rose of Gwalior to the gardens of Agra.Gardens which were popular in Iran those were introduced in India by Babur. Remanants of his Garden is seen in the Gardens of Agra. Gardens were introduced in tombs and palaces. Shahjahan completed the Shalimar Gardens in Lahore in 1637 and Jahangir introduced gardens in the Valleys of Kahmir.The formal gardens were divided into four parts by paved paths.

Forest produce included bamboo, sandal, babul, spinewood, amber, gumlac. Wood was used for making furniture, jewellery, boxes, windos, doors, ceilings, *etc.* Paper as a product made from wood pulp.During Akbar,s reign paper as manufactured from wood pulp in Lahore and during Aurangzeb's reign in Awadh region. Red Bakham wood and drug were used for consumption and export. Certain other fragrant wood which was grown were aloe- wood, amber, camphor, *etc.* The western ghats and Mysore forests produced blackwood and sandalwood. White and yellow varities of sandalwood were grown. White sandalwood came into prominence during the reign of Jahangir. Saffron cultivated in the two districts of Kashmir- Pampur in the district of Moraraj to the south of Srinagar and Paraspur near Indrakot. Spices such as cardamom, ginger, cloves, nutmeg, cinnamon and cardmon was grown in the Malabar Coast. Spices were exported to Ormuz.Cardmon was grown in the areas of Bijapur and ginger in all parts of India.Cardmons was sold at 52- 80 dams per seer and was considered as a costly spice. Pepper was grown in the regions of Bijapur, Malabar and Cochin, in Champaran region of Bihar.Pepper which was grown in Asom was sent to Surat and from their it was exported to Ormuz, Basra and Red Sea. Cloves and nutmeg were also sent to Surat by the Dutch and exported to Europe.

Mughal Emperors had a fetish for fruits and special attention was paid to the cultivation of fruits. Fruit Orchards were treated in a special manner. A rate of rupees 3-3/4 per Bigha was charged even if ther production was not as expected Exception was made if the orchards was planted with grapes and almonds, in which demand was made only when trees bore fruits. If the maintenance of the orchard was costly and more than the actual yield than no charges were made on such orchards.

Spices	*Prices*
Cloves	60 dams per seer
Cardamons	52 dams per seer
Cuminseed	2 dams per seer
Aniseed	2 dams per seer
Turmeric	10 dams per seer

The agricultural patterns in the mughal period determined social condition of the peasants. As science always affects the society similarly agriculture science effected the social condition of the peasants. In the social stratification peasants belonged to the lower segment of the society. Cultivators were known as 'asami', 'raya' or 'mazaram'. Peasants were also known as 'khudakhasta', 'Pahikashta', `Muqarra'. Basically the peasants were of three classes - 'Riayah Khudkastha', who had field and house in different villages,' Riayah Pahikashta', who had fields and house in different villages,' Muqarrari Riayah' were those who had possession larger than they could cultivate. Housing, clothing, shelter determined the social condition of the social class. The peasants of Bihar used primarily languta and the women wore short dhotis. The clothing of the people in Bihar and Bengal was more scanty. Abul Fazl says that in Bengal men and women went naked except for the loin cloth, and in Odisha women used cloth which did not cover anything expect for the private parts. Fitch says that in Bengal huts were covered with straw. Abul Fazl, says that they made mud walls. In Odisha the walls were covered with reeds while in Bihar the roofs were covered with tiles. Around the Malwa region peasants lived in small huts. If the belongings in the huts are to be considered then it only included pots and some cots, *etc.* The food habits were also very simple. Rice, jowar, and bajra were the main cereals. Cloves were the cheapest and was used for making jewellery also. Moreland says that salt was more expensive than wheat. The peasants had to face three adversaries- the officers of the revenue department, natural calamities, and wild beasts. Famines have been recorded in the reign of Akbar in 1555 and 1556 in Delhi and Agra region. Jesuits missionaries described the horrors of famine and pestilence in Lahore and Kashmir. However, not much was done to help the peasants.

During the reign of the Mughals it as during the greatest ruler of the mughal empire, Akbar the Great ho took stringent steps to improve the condition of the agriculture and in this way enhanced the revenue of the Mughal empire. The Bihar subah especially during his as very fertile and supported a huge army of cavalry and the infantry. Bihar as known for its good quality of rice.Kisari a crop resembling pulse was eaten by the poor people. Betel crop grown in Bihar known as Maghi was known for its fragrance, colour and taste. At Maner a beautiful flower known as Muchakand was grown. In the sarkar of Hajipur the fruits known as Kathal and Barhal ere grown in abundance. References of sarkar of Hajipur, Saran, Tirhut, Rohtas is found from where huge amount of revenue was collected that supported the army.Thus the farmers and indirectly agriculture as science helped the Mughal emperors to maintain large armies and the mansabdari system.

With the development of the agriculture science revenue department, commutation and collection also developed.The farmers had to live on a bare minimum and maximum of the produce was collected by the officers of the revenue department. The peasants held land of reasonable sizes as the pressure upon the land as not so much as compared to that of the population. The forests were also a source of income for the farmers. Forests where the source of fuel, timber, floers, herbs, honey and even some fruits. The peasants to support themselves financially also depended on the cottage industries such as handicrafts. However, the peasants were taxed heavily sometimes directly and sometimes by intermediaries. The Amils had various functions to perform. The registers kept by patari, Karkun and muqaddam as cheked by the amils. The duties of the Bitikchi as similar to that of the amils. The Potdar or Khizandar collected revenues from the peasant and kept the treasury in safe custody. Similar other revenue officers ere Karkuns, Qanungos, and Patwaris. Each village had a officer named as Muqaddam and Patwari who maintained revenue records. Assessment of land was done on the basis of measurement and crops After the assessment of the revenue it was converted into cash with price schedules. For this purpose land was divided into Dasturs.

The provinces were divided into Sarkars or districts and that of a district into parganas or tehsil. Officer recruited to collect revenue at the district level as Amalguzar and officers at the tehsil level ere known as Amils. Each cultivator received a patta or title deed. The land revenue as payable in cash or kind twice in a year.Besides regular revenue the farmers had to sometimes levy Abwabs which can be classified under following heads- duties on the sale of the produce, fees on the sale of immovable property, prerequisite taken by the officers by their on sake and fees or commission taken for the state, license-tax for carrying on certain trades, forced subscription, imposts on Hindus *etc.*

Sharing as the method of taxation were share in the produce as made when the crop as standing. The other method of measurement as called Batai or GhallaBakhshi. In Batai state demand was 4 heaps if ¼, 3 heaps if 1/3, 2 heaps if ½. Cash payment was demanded by the peasants which was troublesome for the farmers of the seventeenth century. The state demand of the revenue as 1/3 but it was greater than in actual practice. In some areas it seems that the state demand reached 1/2of the produce. In Gujarat and the areas which were fertile the demand went up as high as ¾ part and in barren lands it as less than ½.The Khalisa lands after Akbar's reign the assessment went up as high as ¼. During Jahangir reign it reached less than 5 per cent and during Shah Jahan reign it as 7 per cent and again during the reign of Aurangzeb it again increased to 1/5 part.The prices varied in the various parts of India The foodgrains were cheaper in Nortern India than in Gujarat.The prices of essential commodities were low. The prices showed regional and periodical variations. This was due to transportation difficulties in carrying bulkier products from one place to another.Every village thus tried to be self sufficient in itself. Sometimes in the same region the fluctuation in the prices as seen. When the court was moved in the 43rd regnal year from Lahore Akbar reduced the rates of the demand of taxation. The average prices of principal foodgrains towards the end of Akbar's reign as calculated on the basis of 82.3 lbs equivalent to one maund.

The agriculture and the taxation determined the socio–economic condition of the peasants. Village was the smallest working unit for the peasants. The villages known as 'gaon' and 'mauza' had wide range of variations in the areas of the villages. It was divided into fields and plots which were marked and raised. The peasants gave each plot a particular name. The economic condition of the farmers determined their social condition also. The economic condition also showcased their living standards such as fooding, clothing and housing. Pelsaert has mentioned the worse condition of the house of the farmers. Fitch says that in Bengal the huts were covered with straw.In Bihar the roofs were made of tiles.,The walls were made lof mud in the doab region ande Malwa region.In respect to clothing it is stated that farmer class wore bare minimum.The peasants of Bihar used 'languta' The cl;othing of the peasants in Bengal was also scanty.It is also found that men and women also went naked expect for the loin cloth. The food habits of the farmers was also simple. Rice, jowar, bajra, were the food crops and cereals eaten by th epeasant class. Few vegetables were eaten and pot herbs also formed part of the staple diet of the farmers. Ghee formed the integral part of the diet of people in the regions of Bengal and Western India. Gur was consumed more in the villages than sugar. Salt was very expensive and the cloves were the cheapest. Cloves were also used by women to make jewellery. Toddy drinking and tobacco smoking was used as intoxicants. Thus, we see the producers of the food themselves sometimes went starved and lead a simple and frugal live. The peasants had to bear the brunt of the taxation, floods, Famines, wild beasts. *etc.*The condition of the peasanbts socially and economically was very poor. The burden of the taxes made their condition worse.

The economic set up of the Mughal empire was based on agriculture and the peasants. The economy largely based on the revenues collected by the farmers.The rate of the land revenue demand of it depended on the mughal administration. However, sometimes the revenue rates was so high that it was not enough for the survival of the peasants. Sometimes the peasants were left with bare minimum for their survival. The land policy of Akbar was peasant friendly. He introduced quality assessment to ensure land revenue for his expanding empire. Loans and implements were provided to the farmers to encourage them. Taqavi loans were provided to farmers in case of emergency. However, not adequate steps were taken to improve the economic condition of the farmers. When the farming season ended farmers used to leave their villages and work as basket makers, weaver, or water carrier The society of pre- british period had more autonomy and was adorned with respect and admiration. The condition of the peasants was comparatively good during the reign of Akbar.

Conclusion

The peasants were the backbone of Indian economy in the mughal period. Agriculture was the chief source of income, so we can say that peasants played a major role in the economic condition of the empire. The most important feature of peasantry was that cultivator was not in direct contact with the state.We find that the modern techniques such as grafting was known to the farmers of the Mughal period. We find that three varieties of oranges were produced by the technique of grafting. The variety of mango Alfonso which was grown in the Mughal period by grafting method is still famous in India and exported to foreign countries. The producers of the crops were the the most simple living class. The agriculture and its science affected the socio – economic condition of the peasants.The condition of the peasant was good in respect that they had large land holdings and also more productivity. The peasants were independent and were treated with respect and consideration. Akbar was the sole mughal ruler who tried to improve the condition of the farmers but still there was scope for improvement. The cultivators had to face the brunt of the revenue collectors and also the natural disasters. According to some historians the condition of the farmers in the 17th century was well off than the farmers in the 18th century. The lot of the farmers was miserable and lived in constant starvation. But this is also true that the condition of peasants was good in Akbar's reign to that of other Mughal rulers. The peasant society of Akbar reign had more autonomy and was adorned with respect and admiration".The peasant was the backbone oif the economic set up and the administrative setup of the Mughal Empire.The significant feature of agriculture in Mughal period was the large number of crops grown by the peasants. The most notable feature of the peasant life was the appropriate combination of agricultural work along with the manufacturing process.

British Rule and the Peasants of India

Pay your Tax or your Land will be Auctioned Off

Like the Mughal rulers, the British also wanted to collect as much tax from the peasants as they could. During Mughal rule, if peasants did not pay their land tax on time, the amount they owed would be noted against their names. They would then have to pay it later. If they didn't, it would be taken from them by force, if necessary. But the British introduced a different system. If peasants or zamindars did not pay their tax on time, their land was sold off in an auction to whoever offered the highest price. The government would take the tax from the money paid by whoever bought the land. If the land had been sold for more than the amount of the tax that was owed, the remaining money would be given to the original owner, who would no longer own the land. Peasants and zamindars faced various kinds of problems in those days. Sometimes their crops would be destroyed by floods or famines. Sometimes the prices of crops would be so low in the market that the peasants got very little money when they sold their crops. When this happened, they found it difficult to pay their taxes. But the British government wouldn't listen to the peasants' problems or exempt them from paying tax. It did not even give people extra time to pay their taxes when they had a shortage of money. Instead, the government would get a court order to auction off their land.

Thus, the land of hundreds of peasants and zamindars was auctioned off by the British rulers. Many peasants had borrowed money from moneylenders in order to pay their taxes. However, if they did not repay their loans on time, the moneylenders also went to court to get their land auctioned to recover the loan money.

Two main aspects of the British system of collecting land tax? What effects might the British system of tax collection have had on:

(a) Ordinary peasants

(b) Rich zamindars

(c) Money lenders

With this new British rule for collecting tax, many peasants fell deep into debt. But there was also another reason why their debts kept increasing. It was because there was an increase in foreign trade and the export of agricultural produce.

Cotton Export and its Effects on Maratha Peasants

In the 1700's India had been exporting a lot of cotton cloth to Britain and other parts of Europe. But by the middle of the nineteenth century (around 1850), many mechanised cloth mills were being set up in Britain to cheaply produce cotton cloth. These mills got most of their raw cotton from the USA. When a civil war broke out in the USA in 1861, the supply of cotton to the mills in Britain almost came to a halt. As the mill owners continuously needed cotton for their factories - they began buying cotton from India. Hence the demand for Indian cotton increased.

This led to an increase in the price of cotton. From 3 annas per kilo, the price of cotton rose to 12 annas per kilo. Many peasants in cotton growing areas like Maharashtra began growing cotton instead of food grains because of the high price it was fetching in the market. They began to take loans from moneylenders so as to increase the area under cotton cultivation. Since cotton cannot be grown in Britain's cold climate, it was brought there from other countries. In 1865, the civil war in the USA came to an end and the supply of cotton to Britain was resumed.

The demand for cotton from India fell and so did its price. Cotton that fetched 12 annas per kilo in 1864 now fetched only 6 annas per kilo. The peasants suffered a lot as they could not earn enough to pay back the loans they had taken. Added to this, they also had to pay tax to the government. The moneylenders got richer and richer as more and more peasants came under their grip.

The suffering Maratha peasants began revolting against the moneylenders to whom they were so much in debt. Thus, the growth of foreign trade and events in faraway countries like the USA led to large profits as well as heavy losses for Indian peasants. Besides cotton, Indian wheat, sugar, indigo, jute, tea, etc were also sold in other countries. When the prices of these crops increased or decreased, Indian peasants were affected.

Maratha Peasants Revolt against Moneylenders

In 1875, the peasants of Pune and Ahmednagar in Maharashtra revolted against the money- lenders of their villages. In village after village, the peasants surrounded the houses of moneylenders and tried to seize their account books. If a moneylender refused to hand over the account books, they burnt down his house. Other villagers supported the peasants. Even the barbers and washermen refused to serve the moneylenders. Many moneylenders fled from their villages to escape the anger of the people.

Here is how the peasants explained the reason for their uprising: "These moneylenders get kurki (notice of auction) from the court. If we cannot pay our debts they auction our homes and land. For generations, these people have kept us burdened with debt. However much we pay, the debt is never fully repaid. They have put false debts against our names in their account books. Only when their account books are burnt to ashes will our problems be solved." Eventually the government tried to tackle the problems faced by the peasants and a new law was passed to control moneylenders so that they could not easily cheat the peasants.

Who Owns the Land?

One question troubled the British a lot - from whom should they collect land tax in India? They kept debating and trying different solutions to this problem at different times and in different parts of India. The British knew that the zamindars were powerful people and if they won the support of the zamindars, it would be easier for them to rule India. They also knew that it was difficult to collect tax directly from hundreds of thousands of peasants.

The British also wanted to collect tax from the actual owners of the land who either worked on the land themselves or got the lands cultivated by others. Now the question was, who were the owners of the land? There were no owners as such in Mughal times - land was not something that was bought and sold like it is toda y. The zamindars, jagirdars and others had the right to collect land tax and other payments from the peasants who were tilling the land. The peasants had the right to occupy and cultivate the land. The zamindars could inherit, gift, or sell their rights to collect taxes from the peasants. The peasants could also inherit, gift, or sell the right to cultivate it. But not many peasants would 'buy' land, because they could just start cultivating some of the available unused land without having to pay anything.

However, by the 19th century, population was rising and new land suitable for cultivation was becoming scarce. It was no longer so easy for peasants to shift to new places and find untilled land to plough. In 1793 the British decided to declare the zamindars as the owners of the land and collect tax only from them. They argued, "We have a rule that if people fail to pay the full tax in time, we auction their land to get the money. So the zamindar who we collect tax from must be considered the owner of the land, otherwise whose land will we auction if tax is not paid? If a person is paying the tax for some land, how could someone else be owning the land?" The zamindars were given full rights over the land, thus making the peasants their tenants or sharecroppers. They no longer had any rights over the land they cultivated. This system was put into practice in Bengal, Bihar and the Awadh region of Uttar Pradesh. The revenue for the land was fixed permanently and the zamindars had to pay the tax to the British by a fixed date each year.

The Peasants Land and the Zamindars Rights*

(*this story is based on the conditions described in government reports of that time.)

Once they became legal owners of the land, the zamindars took advantage of their new rights and the helplessness of the peasants. The peasants could be evicted from the land easily. It was also very difficult for them to leave the villages, as there was no land available elsewhere. They were becoming more and more indebted. The following story of a peasant from Awadh, UP, shows how the zamindars treated their tenants.

Gayadeen was a poor peasant who lived in a village in Awadh region. One day he went to the market to sell his cow. Why do you think he wanted to sell his cow? This is what Gayadeen had to say: "Every year, the zamindar demands a bigger and bigger share of the money I get after I sell the crops. He keeps increasing the amount every year. How can I pay him when I don't have any money? He filed a case in the court against me. Now I have to sell my cow to pay him." The zamindar let Gayadeen plough the land only after he paid the money.

Next year Gayadeen arranged the marriage of his 10-year-old daughter to an old man. Why do you think Gayadeen made his young daughter marry such an old man? This is what he had to say: "The zamindar wants me to pay a nazrana of Rs. 500 this year. He says he will let me plough the land only if I pay the **nazrana**. If I don't pay he will evict me from the land. What am I to do? This old man has given me Rs. 500 for my daughter. I will pay this money to the zamindar so that I can plough the field. Otherwise, what will I eat?" Like Gayadeen, many hundreds of thousands of peasants were reduced to a pitiable state. They filled the purses of the zamindars with their hard earned income. The zamindars took advantage of the peasants. They took more money from the peasants than the tax they paid to the government. Of course they never gave receipts for the extra money they collected from the peasants.

The zamindars kept this extra money because they felt it was their share. According to them, the peasants were only tenants on their land. Because the zamindars were now the owners of the land they began to increase their share whenever they felt like it. They also gave out their land for farming to whomever they favoured and evicted those they did not like.

The Zamindars Exploit the Peasants in Many Ways

Even before the zamindars became legal owners of the peasants' lands, they had some land of their own. These were known as khudkasht (KudkaSt) or seer (salr) lands. The zamindars had sometimes hired labourers to till these lands or else they gave them to the peasants on sharecropping. When the zamindars became owners of the peasants' land they thought, "Why should we pay someone to cultivate our seer land? Why don't we just make our tenants cultivate this land for us?"

So the peasants were forced to do begar (baogaar: work without wages) on the seer land of the zamindars. If they refused, the zamindar's soldiers would beat them up and force them to do begar. The zamindars soon began terrorising the peasants. Their soldiers would even catch peasants walking on the road and force them to do begar in the zamindars' fields. Because they were forced to work on the zamindars' fields, many peasants could not cultivate their own fields properly. They would also not improve their fields. Their plight can be seen in a government report written in 1878. The report says that the peasants do not try to dig wells on their land or irrigate it, nor do they try to bund their fields or dig drains or use fertilisers. "They do nothing to improve their land because they fear they can be evicted from their land at any time. If they improve their farming, the zamindar immediately increases the share he takes from them. But the zamindars also prevent the peasants from improving their fields because they are afraid that the peasants would then start asserting their rights over the land."

Countless Collections, Cesses and Payments

So the peasants faced many problems on the land they cultivated. The zamindars also tried to extract as much money as they could from the peasants under a variety of pretexts. If the commissioner saheb came on a tour to the area, the peasants were forced to pay a contribution known as 'commissioneraavan'. Similarly, the peasants had to make a contribution known as 'hathiaavan' for the elephants of the zamindar. If the zamindar wanted to buy a motor car, a 'motoraavan' was collected. Or if he bought horses the peasants had to give a 'ghodaavan'. The zamindars extracted countless such contributions from the peasants. It was reported that once a thakurain (zamindar's wife) had to spend a lot of money to treat an abscess infection and the peasants had to contribute a 'pakaavan' fund to cover the cost of treatment. They also had to regularly supply ghee, milk, vegetables, gur, straw, cowdung cakes, etc free of cost to the zamindar's house.

This was the situation in many provinces of India. Bengal, Bihar, and Uttar Pradesh had many big and powerful zamindars. Each of them owned dozens or even hundreds of villages. The peasants kept trying to resist the excesses of these zamindars.

The Peasants Protest

Awadh Peasants Protests Against Oppression by Zamindars

In 1920-22, the peasants of Awadh in Uttar Pradesh took out huge processions to protest against the zamindars who were extracting money from them. Many zamindars were socially boycotted and driven from the villages. The peasants also refused to till the fields of those zamindars who tried to evict their tenants or claimed too much rent from them. The peasants formed 'kisan sabhas' to carry their agitation forward. Baba Ram Chandra, Jhinguri Singh, Suraj Prasad, Madari Pasi were some of the well-known peasant leaders in Uttar Pradesh. The British government helped the zamindars to suppress the revolts. But the British saw how powerful and impressive the agitation of the peasants was, so they soon made laws in favour of the tenants.

Bardoli Peasants Protest Against Increasing Land Tax

The British had made different arrangements to collect land revenue in different parts of India. They found out that a lot of the money the zamindars collected from the peasants did not reach the government but was pocketed by the zamindars themselves. So in some areas, like Gujarat, Maharashtra and Tamil Nadu, the British decided to do away with the zamindari system. Instead they would fix the amount of land tax for each cultivating peasant and collect the tax directly from the peasants. They also gave up the idea of permanently fixing the revenue rates. They decided to fix the revenue rates for thirty years at a time. Peasants who cultivated their own land would be responsible for paying the same amount of tax each year for 30 years.

In 1896 the government had fixed the land revenue in Bardoli for thirty years. In 1926, when the thirty years had ended, the government had to again decide the amount of tax – whether it should be kept at the same rate for the next thirty years or increased. The government decided to increase the tax by 22 per cent. When the peasants of Bardoli found out about the government's decision they were furious. They said, "There is no reason for increasing the tax so much. The government should first study the problem more carefully. It should not increase the tax just like that. We won't pay the incr eased rates. If the government wants, we will continue paying the tax we are paying now. But if it doesn't agree, we will not pay even a single paisa."

The government did not agree to the demand of the peasants. It was adamant about increasing the tax. The peasants decided to launch an agitation against the government. They asked Vallabhbhai Patel and Mahatma Gandhi from the Congress Party to support their agitation. The Congress party had been formed by then and was working for the fr eedom of the countr y. Members of the Congress party came to Bardoli to support the peasants. In all, 137 villages took part in this movement which was launched in February 1928.

Seize and Auction

The peasants of Bardoli devised many ways of not paying tax. As we saw earlier, the government would seize the crops, land, utensils, jewellery, animals and other things belonging to the peasants who refused to pay tax and sell these items to recover the amount of the tax. In many villages of Bardoli, the peasants hid their utensils, jewellery and other possessions in the houses of their relatives living in other places. Boys from the villages would hide in trees and keep watch. As soon as government officials were spotted they would blow a bugle to let the village people know. On hearing the warning the villagers would let loose their cows, bulls and buffaloes and lock their houses. The government officials then had no way of finding out which animals belonged to which peasant. So, they could not confiscate things from the peasants so easily.

The village people also refused to give the government officials food, water or a place to rest. These officers would be forced to leave, frustrated, in the blazing sun of April- May. If the government officials did manage to confiscate some animals, the peasants would go and set them free at night. The peasants who refused to pay tax even found a way to get back their land after it was auctioned away by the government. If some people from the village bought the land, peasants would get together and put so much pressure on them that they would be forced to return the land to the original owner. If people from outside purchased the land and came to the village to till it, the villagers would forcibly evict them.

Many peasants tried to secretly pay tax because they feared their land would be confiscated. However, the agitating peasants decided to boycott these people who gave in and agreed to pay tax. They stopped eating, drinking and sitting with them. This is how they put pressure on such peasants and made them join the agitation

In this way many peasants were able to protect their lands from confiscation and continue their protest against the government. However, there were still many peasants who suffered huge losses. Troops of policemen and tax officers would break the locks of their homes and destroy them while confiscating utensils, jewellery and other household goods. There was no guarantee that the confiscated land or goods would be returned. A large number of peasants never got their auctioned land back. However, the peasants continued to show great courage and refused to pay tax. They came out in open confrontation of the government and refused to be cowed down by the police or army. They were not even afraid of the threat of being jailed.

They were defiant, saying, "If the army turns up, can they extract tax from us? Will all these white people load our land on an aeroplane and take it to Britain? Let them try. We'll see what they can do." The peasants' movement received support from the educated people as well as the mill workers and other people. In the end, the government had to bend to the will of the people and the tax was increased by only 6 per cent. After seeing what happened in Bardoli, the government decided not to increase the tax in other places. The peasants of Bardoli had made the British government bow to their will and this sent a wave of joy and enthusiasm throughout the country. Gandhiji said that movements like the one at Bardoli would help the country to free itself from British rule.

The Telengana Movement in Andhra Pradesh (1946-50)

Another very powerful protest movement by the peasants took place in the Telengana region of Andhra Pradesh. This region was not under British rule but was under the rule of the Nizam of Hyderabad. The protest began in 1946-47 in a village called Bishnupur. The zamindar of the village, who owned 40,000 acres of land, tried to grab a small piece of land belonging to a poor washerwoman. The peasants rose up in protest. Within a few days, the protest spread far and wide. It was led by people belonging to the Communist Party. The peasants bought guns and formed their own army and drove the officials and zamindars away from almost 3,000 villages. They seized their land and redistributed it among the peasants and labourers.

The peasants set up their own administration in these villages. Begar was stopped, the wages of the labourers were increased and land was restored to tenants who had been evicted earlier by the zamindars. A rule was made that no one would have the right to own more than 100 acres of non-irrigated land or 10 acres of irrigated land. If anyone had more land than this, the excess was distributed among poor peasants. Meanwhile, India became independent in 1947. After Independence, the Indian government promised the peasants of Telengana that it would make laws in their favour and asked them to end their agitation. It also sent in the army to suppress the movement.

The Peasants and Laws in Independent India

The agitations during British rule focused attention on the problems, demands and hopes of the peasants. It was clear that the peasants wanted the government to reduce tax and free them from the stranglehold of moneylenders and the terror of the zamindars. They also demanded that the land should belong to the peasant who actually ploughed it. After India became independent, tax was greatly reduced. The government also opened banks and co-operative societies to give loans to the indebted peasants. But the biggest decision the government made was to put an end to the zamindari system in all of India. The law to abolish the zamindari system was passed in 1950. The government decided to collect tax directly from each peasant and not through the zamindars. The law also said that the zamindars would be the owners of their own seer land but would no longer have any right to the peasants' lands. The peasants were made the owners of their own land. Naturally, the zamindars were not happy. They demanded compensation from the government for taking away their 'right' to the lands of the peasants.

The government decided to pay compensation to the zamindars. It acquired land from them after paying some compensation. When the government found that it had spent a lot of money in paying compensation, it made a rule that the peasants would be given their land only if they paid some price for it. This was not acceptable to many people. Those peasants who could pay, became the owners of their land and were freed from the burden of the zamindari system. However, hundreds of thousands of poor peasants could not pay the price. So they remained landless sharecroppers or labourers and had to keep working in the fields of big peasants and one-time zamindars.

However, after Independence some laws were also made to safeguard the interests of the tenants, so that no one could claim an unfair share from them and they could not be evicted from the land without reason. Some laws were also made in the interests of labourers. But big peasants found many ways of ignoring these laws.

Water in British India: The Making of a 'Colonial Hydrology'

Water in British India can be discussed in three overlapping but discrete clusters of concerns. The first and most substantially engaged debates have situated colonial irrigation strategies in terms of their environmental, political and economic contexts. The second cluster, closely shadowing the first, has explored aspects of 'decline', elimination and sometimes appropr iation of a slew of 'traditional' water harvesting technologies. The third cluster of concerns, that is yet to achieve visibility, has aimed at identifying definitive patterns in colonial strategies towards hydraulic endowments. Put differently, the attempt is to characterise the British experience as comprising an altogether distinct paradigm for hydraulic interventions in South Asia; explanations that can perhaps be encapsulated under the broad rubric of 'colonial hydrology'. Part of this as yet incipient exercise involves, in my opinion, a departure from the emphasis on irrigation. In turn, this third cluster will explore colonial experiences with floods, drainage, wetlands, lakes, in-land river navigation, traditional fisheries, urban water supply, water legislation, cultures of water use, ideologies on 'river-improvement' and Multi-purpose River Valley development. In several ways, these themes listed above (indicative and not exhaustive), could then, presumably, help fill in many existing empirical gaps and thereby craft a rigorous theoretical approach to explore the relationship between colonialism and water. By a theoretical approach, I suggest that the subject of water in British India should, similar to works on forests or land policies, be able to shed light on the broader dynamics of colonial rule. Thus, while this essay is centrally aimed at providing a general introduction to the main themes on water in British India, it will nevertheless also attempt to argue a case for the term 'colonial hydrology'; the claim being that much like 'scientific forestry' or the zamindari system on land, colonial economic and political imperatives defined and impacted the region's fluvial endowments in specific and unprecedented ways.

The Quest for Perennial Irrigation

Whitcombe's monograph titled *Agrarian Conditions in Northern India*, published in the early 1970s, was arguably the first attempt at challenging the hitherto then received wisdom on colonial irr igation. In contrast to the belief that canal irrigation was overwhelmingly a positive contribution, *Agrarian Conditions* concluded the opposite. According to Whitcombe, the introduction of perennial irrigation in the semi arid plains of north-western India brought about adverse ecological consequences such as waterlogging and salinisation, destroyed traditional wells and inevitably these canals created some islands of plenty amidst a sea of epidemically swamped and 'depressed' peasantry. In reply, almost a full decade later, Ian Stone in his monograph *Canal Irrigation in British India* sought to stand Whitcombe's claims on its head. For Stone, colonial irrigation when viewed primarily in the context of the peasant's adaptation to new technologies, capacity for rational decision- making and the quest to maximise output, canals became a source for economic dynamism and constant innovation. Consequently, for Stone, despite several negative externalities, the canals released positive 'expansionary forces'. In a slight remove from the pessimism versus optimism debate on the British canal system, Imran Ali in a study of the massive British canal colonies, established in the semi-arid plains of the Punjab between the late nineteenth and early decades of the twentieth century, indicated frictions between colonial economic agendas and their political aims. For Ali, the latent capacity for initiating dynamic capitalism in the canal colonies canals was defeated by the colonial state's continued embrace of traditional and archaic social institutions. Thus, the canal colonies witnessed a contradictory economic pattern in which growth was closely paralleled by under-development. David Gilmartin, in fact, in an earlier essay on the canal colonies, pointed to similar irreconcilable antagonisms between what he termed as 'scientific empire' and 'imperial science'. According to Gilmartin, while discourses on imperial science helped organise the productive control of nature for increasing revenue and expanding commercial agriculture, its 'transformative' potential was constrained by practices of scientific empire – harnessing science to craft and sustain political hierarchies and the exploitative character of colonial society. In other words, two countervailing and contradictory discourses on science simultaneously acted to expand and stymie the potential for dynamic growth within canal colonies. Clearly, the lively debates on colonial irrigation in British India have been concentrated on the semi arid interfluves of the north-west. On the other hand, studies on canal irrigation impacts in the south and the eastern deltas have as yet remained shy of taking on polarised positions. Studies on these regions have essentially looked at two aspects of the canal experience;) productivity and the transition to commercial agriculture;) failures of private irrigation companies in the Madras and Bengal Presidencies. Studies on colonial canal irrigation, as surveyed above, however, still lacks a credible comparative analysis. In particular, a comparative exercise could throw light on why certain irrigation precepts were persisted with and how colonial irrigation practices were shaped by a cross pollination of ideas, evolved from varied ecological zones.

Water Traditions and Colonial Technology

The second cluster of concerns have largely dealt with the schism between 'traditional' or indigenous water technologies and colonial hydraulic engineering endeavours, termed as modern systems. Attention to pre-British irrigation organisation, design and operations, in terms of their relationships and status in a colonial context, was arguably first explored in an article by Nirmal Sengupta, in 1980. Sengupta primarily sought to explore the reasons for the 'decline' of the traditional *ahar* (tank) and *pyne* (channel) irrigation system in colonial South Bihar. According to Sengupta, the ahar and pyne network began to breakdown following the introduction of new revenue routines by the colonial administration. In particular, by facilitating and encouraging a shift from 'produce' to 'fixed' and then to cash rents, the colonial administration invariably upset an entire rhythm of procedures, protocols and duties between tenants and landlords over the question of the maintenance and servicing of the ahar-pyne system. In effect, while the rent burden historically for the indigenous irrigation system was factored as a 'land-water combine', the colonial revenue format realised claims only from 'land'.

In several ways, Sengupta set the tenor for subsequent works on the subject. In 1997, the Centre for Science and Environment, a Delhi-based non-governmental organisation, released a report titled *Dying Wisdom*, which had then put forward the most exhaustive survey on traditional or pre-British water harvesting systems in India. Besides describing the functional details and varied operational aspects of these water structures and situating them in their regional and ecological setting, *Dying Wisdom* also sought to advance a larger historical claim; that traditional water harvesting systems in India declined or were substantially degraded by a range of colonial actions for rule and profit. Colonialism, in other words, by instituting private property, commodifying land, commercialisation, pursuing highly extractive revenue agendas and dismantling community control over natural resources caused the impoverishment of the rural populace at large and led to the decay and destruction of indigenous water harvesting systems.

Subsequent scholarship, however, has questioned whether colonialism did indeed have such a sweeping impact on traditional water structures. David Hardiman, for example, in an excellent study on indigenous water systems in Gujarat argued that aspects such as commercialisation and peasant indebtedness were in fact processes that not only predated colonial rule in the region but were integral, ironically enough, to actually expanding well irrigation in the region. In a recent study, David Mosse on exploring ecology and politics in South India argued that *Dying Wisdom*'s belief that 'organic and autonomous villages' sustained stable water management practices cannot be historically validated. For Mosse, pre-British village communities were unstable entities driven by hierarchies and were shaped and impacted by larger processes of statecraft and regional politics. In effect, Mosse suggests that tanks in South India underwent various phases either of efflorescence or decline prior to colonial rule. Lastly, the pronounced claim in *Dying Wisdom* that traditional water harvesting structures declined or disintegrated across the board in the colonial period has also been challenged. Thomas Rosin, in a meticulously argued essay, indicates that in western Rajasthan a series of complex groundwater irrigation and drinking water devices (*khaDin's* (silt-ponds), step wells, reservoirs and L-shaped embankments) remained functional and viable well into the colonial and post-colonial period. In other words, these systems were capable of 'overlapping' with other hydrological regimes. That is, Rajasthan's unique water harvesting systems rather than being displaced were instead 'overlaid' or coexisted with new types of modern hydraulic technologies, introduced by the British.

Nevertheless, the above qualifications aside, the argument that a large number of traditional water harvesting systems declined or were marginalised in both relative and absolute terms still holds. Colonial water technologies such as weirs, dams and barrages, oriented towards delivering perennial irrigation for settled agriculture, in most instances, proved unable to not only coexist with traditional systems but were sharply aimed at eliminating the latter as well. Indu Agnihotri's insightful and prescient essay on the ecological and land use consequences of the canal colonies in Punjab revealed how British perennial irrigation did not, as widely held, simply bring water and increase agricultural productivity in hitherto desolate 'wastes'. Rather, according to Agnihotri, the colonial canal systems of the late nineteenth and early twentieth centuries overwhelmed, over ran and substantially eroded an existing vibrant pastoral economy in the region. These pastoral grouping, moreover, had also taken to seasonally cultivating crops, which were watered through a complex network of inundation canals. In many other places as well, colonial canal lines were often deliberately situated in ways that supplanted other prevailing irrigation structures. In the Sone canal command in South Bihar, for instance, the engineering staff often consciously placed canal distributaries across existing ahar and pynes, with the intension of injuring or destroying them.

Debating 'Colonial Hydrology'

Central to the third cluster of concerns on water is the attempt, in my opinion, to explore the possibility for outlining a grand theme approach such as the concept of 'colonial hydrology'. In other words, being able to characterise the colonial interventions in water as comprising a cogent and distinct hydraulic paradigm. A paradigm which involved fundamentally realigning land and water in new sets of social, political and ecological relationships. At the level of technology, the British transformed many of the flood plains from previously being watered by seasonal or inundation canals to becoming sites for perennial irrigation works; involving the construction of permanent headworks across river beds with barrages and weirs. These perennial canal systems were technologically unprecedented for harnessing of fluvial environments and were operated through a corpus of social rules, economic practices, rationalities about property and colonial administrative disciplines. These perennial canal's, however, were assembled not merely as channels commandeering river flow but more significantly ended up fundamentally reorienting ecological relations between land and water, notably through 'irrigation science'. That is, the canal schemes represented not merely the commercial and revenue calculations for colonialism but were interventions that worked to order distinct social and physical colonial contexts. Involving, in the main, to paraphrase David Gilmartin, the historically unprecedented attempted creation of the 'colonial resource regime' through an admixture of irrigation engineering science, routines of land revenue and by the intended colonial 'control' of society and nature.

Thus, I would argue, the term colonial hydrology could perhaps best encapsulate the varied hydraulic interventions of colonialism to simultaneously alter South Asia's fluvial and social worlds. Along a similar plane, the second possibility that has hitherto not been substantially dealt with, in terms of a grand theme is the question of the dramatic alteration of British India's great drainage network. The nineteenth century witnessed the systematic proliferation of flood control embankments, intended to contain rivers within their main channels. In addition, the colonial administration also constructed a vast number of roads, railway lines and bridges. These structures had several consequences for drainage and epidemiology. While in Bengal and Bihar, for example, most of the natural drainage lines dropped from north to south, the roads and railways tracks were mostly constructed across them, running east to west. These constructions, in time, not unexpectedly, led to severe drainage congestion and were also accused of being the main sources for aggravating the spread of malaria.21 Arguably, British hydraulic interventions, through the course of the nineteenth century, radically transformed a vast spectrum of pre- colonial hydraulic relationships that had defined and sustained complex equations between land and water. It is probable that studies pursuing grand themes such as 'colonial hydrology' will yield and formulate the next crop of important and significant questions.

History of Agricultural Research in India

India as a predominantly agricultural country attributes a major share of its overall development to the agriculture sector. Indian agriculture is a miscellaneous and extensive sector involving a large number of stakeholders. India has one of the largest and institutionally most complex agricultural research systems in the world. However, such a complex research system was not a sudden devel- opment. Instead, historically, it involved a process that started in the second half of the 19th cen- tury during the colonial period and eventually led to the establishment of the Imperial (now Indian) Council of Agricultural Research (ICAR). In the present research system, the role of ICAR at the national level in aiding, promoting and coordinating research and education activities across the country is of significant importance. In this article we trace the development of agricultural research system in India, since the colonial era till today. Various factors influencing the overall development of agricultural research system in the country are discussed. Although agriculture has been playing the most vital role in Indian economy, during the course of the study, it has been observed that not much emphasis has been given to the history of evolution of agricultural research in India.

Agriculture, as the backbone of Indian economy, plays the most crucial role in the socio-economic sphere of the country. Indian agriculture is a diverse and extensive sec- tor involving a large number of stakeholders. It has been one of the remarkable success stories of the post- independence era through the association of Green Revolution technologies. The Green Revolution contributed to the Indian economy by providing food self-sufficiency and improved rural welfare. Moreover, since independence, India has observed considerable increase in the production of oilseeds (through the yellow revolution), milk (through the white revolution), fish (through the blue revolution), and fruits and vegetable (through the golden revolution). The role of the National Agricultural Research System (the NARS) was imperative in the con- text of all these momentous and successful upheavals. The main events in the history of agricultural research in India can be grouped into the following seven categories:

1. Establishment of agriculture departments and agriculture colleges.
2. Establishment of the Imperial Council of Agricultural Research.
3. Initiation of commodity committees.
4. Project for intensification of regional research on cotton, oilseeds and millets.
5. Initiation of all-India coordinated crop improvement projects.
6. Reorganization of the Indian Council of Agricultural Research (ICAR).
7. The development of agricultural universities.

Among these, the first three signify the development of agriculture in the colonial India, whereas the next four were prominent in the post-colonial era. India has one of the largest and institutionally most complex agricultural research systems in the world. Historically, the Indian agricultural research system is the contribution of a process which started in the 19th century and which resulted in the establishment of the Impe- rial (now Indian) Council of Agricultural Research on the recommendation of a Royal Commission on Agriculture in 1929. Since then, there has been an evolution of agricultural research in India.

Agricultural Research in Colonial India

The early development of agricultural research in India was associated with the occurrence of famines. The recurrence of famines and starvation death of peasants in the second half of the 19th century opened the eyes of the imperial rulers. During this period 24 famines, big and small, took an estimated toll of more than 28 million souls. However, at that time, little precedence was accorded to agricultural research and development in the country.

Establishment of Agriculture Departments and Agriculture Colleges

A fundamental department of agriculture in India was started in the year 1871. Although the chief function of the department named 'Department of Revenue, Agriculture and Commerce' remained revenue and there was no work on agricultural development, this did mark a com- mencement and appreciation of the agriculture sector by the colonial government. However, this initiative was not significant. The credit for this modest foundation goes to Lord Mayo, who was the fourth Viceroy of India, and to A. O. Hume who was a civilian of the Bengal Civil Ser- vice and one of the founders of the Indian National Con- gress. Primarily, the department was established by the Government with a view to supply cotton to the textile industries of Manchester, and not to feed the famine- ravished India. Actually, the inherent desire of the colonial rulers was to adjust the development of Indian agriculture to the economic necessities of British capitalism, which required India primarily as an agrarian colony for pro- duction of raw materials for British industries. This prevented the independent development of Indian agricul- ture, fulfilling the economic requirement of the Indian people.

Based on the reports of the Famine Commissions of 1880, 1898 and 1900, the Government of India (GoI) was determined to set up a central 'Department of Agriculture' controlled by the Imperial Secretariat and agricul- ture departments were to be set up in the provinces to primarily look after agricultural enquiry, agricultural development and famine relief in the country. However, the key duty of the agriculture departments both in the centre and the provinces lingered on famine relief. In 1892, an agricultural chemist and an assistant chemist were allot- ted to look after research and teaching in India, which manifested the first scientific staff in the Department of Revenue and Agriculture. Eventually, in 1901, an Inspec- tor General of Agriculture was appointed to advise the imperial and the provincial governments on agricultural matters. An imperial mycologist was appointed in the same year, and an entomologist was appointed in 1903. During the severe famines of 1899–1900, Lord Curzon, the then Viceroy of India, was convinced that the GoI must urgently concentrate on the agricultural sector to overcome the damages caused by frequent famines. As a consequence, Agriculture Research Institute (now the Indian Agricultural Research Institute (IARI)), together with a college for advanced agriculture training, was es- tablished at Pusa in the year 1905 and its director was the agriculture adviser to the GoI till 1929. Today, IARI is one of the premier national institutes for agricultural re- search, education and extension in India. The growth of India's agriculture during the past more than 100 years, is closely linked with the researches done and technologies generated by the Institute. The Green Revolution stemmed from the fields of IARI. The Institute has been instrumental in the development of high-yielding varie- ties of all major crops in the country and taking major ini- tiatives in integrated pest management and integrated soil – water - nutrient management practices.

The Agricultural School at Saidapet, Chennai, which was established as early as 1868, was later relocated to Coimbatore during 1906. Likewise, a branch for teaching agriculture in the College of Science at Pune (established in 1879) was subsequently developed into a separate Col- lege of Agriculture in 1907. Similar agricultural colleges were established at Kanpur, Sabour, Nagpur and Lyallpur, now in Pakistan, between 1901 and 1905. Moreover, the agriculture departments in the provinces were expanded during the period when agricultural de- partments and colleges were started to be established. An All-India Board of Agriculture was established in 1905 with a view to bring the provincial governments more in touch with one another and making suitable recommenda- tions to the GoI. The Indian Agriculture Service was con- stituted in 1906. However, the chief function of the agricultural colleges remained teaching, and training and research activities could not be carried out due to the lack of scientific and technical manpower and finance.

Establishment of the Imperial Council of Agricultural Research (the present-day ICAR)

The Royal Commission on Agriculture (Linlithgow Com- mission), which was appointed in the year 1926, authoritatively reviewed the position of agriculture in India and reported the same in 1928. Having duly recognized the work done by the agriculture departments, the Commission stressed upon the enormous possibilities for future work and made comprehensive recommendations regarding the different problems of agriculture. The Commission proposed that an Imperial Council of Agricultural Research should be set up to endorse, direct and organize agricultural, including veterinary, research all over India. The Council was supposed to guide the research activities of central and provincial departments of agriculture. According to the proposal of the Royal Commission on Agriculture, the GoI, Department of Education, Health and Lands set-up the Imperial Council of Agricultural Research (now ICAR) on 16 July 1929.

The Commodity Committee

Several semi-autonomous Central Commodity Committees were set up by the Ministry of Food and Agriculture. These were concerned with research and development activities related to specific crops. These committees were semi-autonomous bodies in the sense that they were financed partly by the Government and partly by the taxes collected on the export of the concerned commodi- ties. The Indian Central Cotton Committee was the first one to be established in 1921 on the recommendation of the Indian Cotton Committee (1917–18). The chief function of the Central Cotton Committee was cotton improvement with special focus on the development of improved methods of growing and marketing cotton. The Committee's support led to the development of 70 improved varieties and considerably improved fibre quality. Most of the commodity committees related to specific crops had their own research stations or institutes which were located in the regions where the concerned crop was most widely grown. However, some other commodity committees financed research schemes conducted by the State Departments of Agriculture, *e.g.* Spices and Cashewnut Committee. The achievement by the Indian Central Cotton Committee led to the setting up of commodity committees on crops such as lac, jute, sugarcane, tobacco, coconut, oilseeds, spices, cashewnut and are- canut. The Vice-President of ICAR presided over all the commodity committees. However, there was a great necessity to conduct researches on various crops within the different agroclimatic regions of the country as in the context of India, the soil and climate fluctuate to a great extent from one region to another. These apprehensions led to the formulation of the Project for Intensification of Regional Research on Cotton, Oilseeds and Millets (PIRRCOM), which was the first step in the country towards coordinated approach to agricultural research. The Central Commodity Committees were abolished in 1965 and the research institutes under their control were trans- ferred to ICAR.

Plantation Research in Colonial India

Though the experimental farms were established in 1884, the provincial agricultural departments could seldom go beyond the collection of revenue data and famine relief op- erations. For conducting experiments, the Government looked more to agricultural societies than to its own agen- cies. The Agricultural and Horticultural Society of India (AHSI) was established with specific objectives such as: To obtain precise and trustworthy details as to the cost of cultivation and produce per acre of fibre-bearing plants of promising character, so that the Agriculture Department may be able to form a decisive conclusion as to the prospects of a profitable exploitation of the plants in question.

Evaluate the Machines and Processes for the Extraction of Fibres

The official experimental farms were obsessed with cotton. Mounting pressure from British cotton tycoons had forced the GoI to initiate a vigorous cotton improvement programme. However, earlier projects of the 1840s and 1860s were not successful mainly because of insufficient botanical knowledge or the necessary market research. Later, in 1890, the association of an expert botanist in cotton experiments was specifically called for. The cultivation and marketing of existing varieties gave a relatively stable and acceptable return to money-lenders and dealers. New and untested varieties involved different methods of cultivations and great labour input, without a higher level of output or profit, and with the risk of severe losses to each of these classes. The last quarter of the 19th century saw the closure of several experimental farms. But private farms patronized by cotton mills proved remunerative. For example, the Govern- ment's Mungeli farm at Bilaspur had to be abandoned, whereas the nearby Kyragarh and Nandgaon cotton farms owned by Bengal -Nagpur Cotton Mills Company produced 600,000 lb of cotton. The Government's fascination towards imported technologies and ideas proved not to be advantageous. Instead, the dependence on traditional knowledge proved to be beneficial for the mills.

Sericulture remained a purely private concern for a long period of time. As for example, at the turn of the century Ms Tata and Sons successfully started a silk firm in Bangalore for the introduction of Japanese methods of sericulture. Tea and indigo were the major items of export. In the 1870s, A. W. Blyth established a process by which it could be easily known whether the merest fragment of a plant belonged to the 'theine' class or not. However, no experiment was conducted to control the scourge of blight. Although the AHSI made an attempt in this direction, it failed because of lack of funds and skilled ento- mologists. In 1891, M. K. K. Bamber, a chemist appointed by the joint committee of the Indian Tea Association and the AHSI, performed analysis of the soil and tea leaves, gave advice on manures, drainage, insecticides, *etc.* and brought out a book called *The Chemistry and Agriculture of Tea*. The establishment of a permanent tea research station at Toclai, Asom, highlights the pro- gress of tea research in colonial India. Almost the same thing happened in the case of indigo. Indian indigo remained unrivalled till the end of the 19th century when Germany perfected its synthetic counterpart. In early 1899, the GoI deputed its agricultural chemist J. W. Leather to tour the indigo districts of north Bihar. An indigo experimental farm was started at Dalsinghsarai in July the same year. The interests of the indigo lobby were at work behind the creation of the Pusa Agricultural Research Institute, Bihar. Pusa was selected as the site because of its proximity to the plantations of north Bihar.

Agricultural Research in Post-Colonial India

Since independence, there has been a substantial growth in the Indian agricultural research system. ICAR is the chief public body at the national level for directing and endorsing agricultural research and education in the country. Likewise, State Agricultural Universities are responsible for doing the same at the state level. After independence, agricultural research was given much emphasis which in turn led to increased agricultural production and near self-sufficiency in foodgrains in the country. The greater emphasis given to agricultural research could be established by the fact that the system guided by ICAR now has 49 ICAR institutes, 17 national research centres, 6 bureaus, 25 project directorates, 79 All-India Coordinated Research Projects (AICRPs) and the All India Network Projects (AINPs), 607 Krishi Vigyan Kendras (KVK), 52 State Agricultural Universities, 1 Central Agricultural University, 4 Central Universities having Faculty of Agriculture. The National Academy of Agricultural Research Management is yet another exclusive institution under ICAR to conduct research and training in agricultural research management.

Other organizations involved in agricultural research are

- About 23 traditional universities are carrying out agri- cultural research.
- Scientific organizations such as the Council of Scientific and Industrial Research (CSIR), and Bhabha Atomic Research Centre (BARC).
- Government departments such as the Department of Science and Technology and the Department of Biotechnology.
- Private and voluntary organizations.
- Scientific societies and institutions like the National Dairy Development Board which has been instrumental in transforming the dairy industry for rural development.

PIRRCOM

With the initiatives for agricultural research development, there was a need to coordinate the research on various crops, especially cotton, oilseeds and millets. Moreover, a need was felt to conduct research work in different agro-climatic regions of the country. The first coordinated re- search work on regional basis was initiated in 1956 in the form of a joint venture by ICAR and the Indian Central Commodity Committees on Oilseeds and Cotton. Eventually, 17 centres were established across the country to perform research on cotton (*Gossypium* species), castor (*Ricinus communis*), groundnut (*Arachis hypogaea*), taramira (*Eruca vesicaria*), jowar (*Sorghum bicolor*), bajra (*Pennisetum glaucum*), *etc.* The research programme for each region was prepared by a regional coordination committee headed by the Agriculture Commissioner of India, and approved by the respective commodity committees. A regional research station with of full-fledged sections of plant breeding and genetics, agronomy, agricultural chemistry and soil science, plant pathology and entomology was established.

Initiation of All-India Coordinated Research Projects

The conception of coordinated projects was first initiated for hybrid maize. It was under the United States Agency for International Development (USAID). The Rockefeller Foundation was the organization that was actively involved in crop improvement programmes in Mexico, Central America and the Caribbean. This organization was invited to assist in the maize improvement programme in India. The Ministry of Food and Agriculture under the GoI, signed an agreement with the Rockefeller Foundation in 1956, according to which the latter was to as- sist in the development of: (i) The postgraduate school of IARI, New Delhi, and (ii) Research programmes on the improvement of some crops (maize, jowar and bajra, initially). Two leading scientists who were associated with the Rockefeller Foundation's maize improvement pro- grammes in Mexico and Columbia visited India to study the position of maize crop and prepared a report based on their findings. This report was scrutinized by ICAR and provided the basis for the coordinated maize project.

The coordinated maize project in India had proven to be the defining moment in research planning in agriculture in the country. As a result of the coordinated project, new high-yielding maize hybrids became available by 1961. Encouraged by the success of the maize project, in 1965 ICAR decided to initiate coordinated projects on other crops as well as in other areas of research, *e.g.* ani- mal husbandry, soil sciences, *etc.* Seventy coordinated projects on various subjects were launched within 3 years of this decision and these accounted for 40 per cent of the total expenditure for agriculture in the Fourth Five-Year Plan. However, the advancement of the coordinated projects was critically analysed in the Fifth Five-Year Plan. Accordingly, some projects were terminated, some were merged with other projects, some were elevated to the level of Project Directorates and some projects were changed to Coordinated Programmes. As a result, the number of coordinated projects decreased to 49 in the Fifth Five- Year Plan.

Reorganization of ICAR

In 1963, the Ministry of Food and Agriculture appointed the Agricultural Review Team headed by Marion W. Parker of USDA (United States Department of Agriculture), to scrutinize the organization of agricultural research in India. The team submitted its report in March 1964 and based on the recommendation of the team, ICAR was reorganized in 1966 and made an entirely autonomous organization. The ICAR was proffered the control for various research organizations under the Department of Food and Agriculture and under the Central Commodity Committees. The Governing Body of ICAR was reorganized to make it primarily a body of scientists and agriculturists. Institutes like IARI, National Dairy Research Institute and Indian Veterinary Institute were made National Institutes. Eventually, a policy was formu- lated suggesting that an agricultural scientist would be appointed as the chief executive of ICAR with the desi- gnation of Director General. Accordingly, B. P. Pal was appointed as the first Director General of ICAR in May 1965. Concurrently, he was the Vice-President of the Council. In order to assist the Director General, four posts of Deputy Director General were created.

In June 1972, the GoI appointed a committee headed by P. B. Gajendragadkar (retired Chief Justice of India) to review the enrollment and personnel policies of ICAR and its institutes, and to recommend actions for the enhancement of the same. The committee submitted its re- port in January 1973. A Department of Agricultural Research and Education was created in the Ministry of Food and Agriculture in December 1973 according to the recommendations made by the committee. The Director General, ICAR was made secretary to the new depart- ment. The Minister of Agriculture was designated as the President of the Council, while the Director General of ICAR was made the Chairman of the Governing Body of the Council. The functions of the Standing Committee were assigned to Scientific Panels as the Advisory Board and the Standing Committee was eliminated. The scien- tific panels for different disciplines were made responsi- ble for considering and assessing the suitability for financial assistance of ad hoc research schemes. Under the Agricultural Scientists' Recruitment Board (ASRB), an Agricultural Research Service (ARS) was initiated for the recruitment of scientific personnel. A scheme for in- ternal assessment and promotion was introduced. The en- tire country was divided into eight agroecological zones and regional committees were set up for each zone. The function of these regional committees was to review the status of agricultural research and education in the con- cerned regions. The Governing Body of ICAR was to be assisted by a Norms and Accreditation Committee, which also looks after the development of agricultural universi- ties and the grant of fellowships.

The mandates of the ICAR are as follows:

1. To plan, undertake, aid, promote and coordinate edu- cation, research and its application in agriculture, agroforestry, animal husbandry, fisheries, home science and allied sciences.
2. To act as a clearing house of research and general in- formation relating to agriculture, animal husbandry, home science and allied sciences, and fisheries through its publications and information system; and instituting and promoting transfer of technology pro- grammes.
3. To provide, undertake and promote consultancy ser- vices in the fields of education, research, training and dissemination of information in agriculture, agrofor- estry, animal husbandry, fisheries, home science and allied sciences.
4. To look into the problems relating to broader areas of rural development concerning agriculture, including postharvest technology by developing cooperative programmes with other organizations such as the Indian Council of Social Science Research, CSIR, BARC and the universities.
5. To perform other tasks considered necessary to attain the objectives of the society.

Development of Agricultural Universities

The very fact that in 1948, there were only 17 agricultural colleges in the country shows that before independence, higher education in agriculture was almost ignored. These agricultural colleges were under the control of Director, Department of Agriculture of the respective states. Before the establishment of agricultural universities, teaching and research in agricultural sciences were largely con- ducted by these agricultural colleges which are attached to general universities. However, colleges for animal husbandry, governed by the Director, Animal Husbandry of the concerned states were separate from those for agri- culture. Research and extension were the responsibility of the agriculture and the animal husbandry departments of the states. At that time, the organization, staffing patterns, pay scales of teachers and financial support for research and other activities were not appropriate for a first-grade education and training in agriculture.

One of the important and premier institutes of the pre-independence era was the Imperial Agricultural Research Institute established at Pusa in 1905. Eventually, IARI has become one of the premier institutes for agricultural research, education and extension in the country. It has significantly contributed to the development of agricultural science with relation to the society and was instrumental in the Green Revolution. Its contributions include first-rate research activities, generation of appropriate technologies and development of human resources.

During the years 1948-49, the University Education Commission headed by S. Radhakrishnan, suggested that the country should focus on the establishment of rural universities. H. S. Singh and A. N. Jha (Chief Secretary and Development Commissioner, Uttar Pradesh (UP) visited Land-Grant Universities of United States in 1950 and after coming back, advised the then Chief Minister of UP, Pandit Govind Ballabh Pant, to set up such a univer- sity in the state. The Chief Minister accepted their rec- ommendation. This event may be regarded as the one which led to the initiation of agricultural universities in the country. In 1955, the first Joint Indo-American Team was set up. The team suggested the founding of rural uni- versities in each of the states in India. Accordingly, the team identified UP (Tarai), West Bengal (Haringhatta), Bihar (Patna), Odisha (Bhubaneshwar), Travancore - Cochin and Mumbai (Anand, now in Gujarat) to be apposite for starting such universities.

In 1956, a blueprint for agricultural universities was prepared and this provided the root for the proposal by the Government of UP to the Central Government (in September 1956) for starting an agricultural university near Rudrapur in the Tarai region of UP. The Central Government also agreed to the proposal on an experimental basis. In 1959, the second Joint Indo-American Team was setup, which submitted its report in 1960. The Team suggested that the agricultural universities should be autonomous; should consist of colleges of agriculture like veterinary, animal husbandry, home science, technology and basic sciences under them; should have interdiscipli- nary teaching programme; and should integrate teaching, research and extension. By 1961, there were demands from many states for agricultural universities and the GoI accepted the organization of a few more agricultural universities during the Third Five-Year Plan. Accordingly, the GoI appointed a committee, headed by R. W. Cummings, for providing a model for the essential legislation by the states for the establishment of agricultural universities. The committee submitted its report in 1962 and on the basis of this report, ICAR prepared the model act for the development of agricultural universities. During the period of the Fourth Five-Year Plan between the years 1960 and 1965, seven agricultural universities were es- tablished in UP, Odisha, Rajasthan, Punjab, Andhra Pradesh, Madhya Pradesh and Karnataka. USAID contributed significantly to the development of agricultural universities through the Land-Grant Universities of United States. USAID provided assistance in the form of training of Indian scientists in USA, stationing of US sci- entists for teaching and research in Indian agriculture universities and by providing a limited amount of equip- ment for teaching and research.

One of the important schema came when the Education Commission (1964–66), headed by D. S. Kothari, recommended that all aspects of agricultural research should be the function of agricultural universities. Consequently, the responsibility for research was entrusted from State Department of Agriculture to agricultural universities. However, this change was not consistently implemented in every state. The Review Committee on Agriculture Universities (1977–78), headed by M.S.Randhawa, made many useful recommendations for the development of agricultural universities. It noted that the quality of leadership and financial support from the state were crucial factors in the development of agricultural universities. The committee suggested, among other things, that the Director General, ICAR, and Chairman, University Grants Commission, should be members of the selection committee that appoints Vice-Chancellors for agricultural universities. Under the NARS, the State Agricultural Universities are major cohorts in growth and development of agricultural research and education. The state Agricultural Universities are based on Land-Grant pattern of USA, which called for the federal government to offer each state with a grant of land in order to establish a university/institution. Hence the name 'Land-Grant' came into being.

One of the original objectives of ICAR was to undertake aid, promote and coordinate agricultural education in the country. But this was not put into effective practice until the reorganization of ICAR in 1966. A full-fledged Division of Agricultural Education was set up within the ICAR to fulfil this objective. The ICAR has been crucial in the reorganization of agricultural education in the country by providing the necessary supervision, schemes for improving the quality of teaching and research, *e.g.* centres of excellence, higher education in new areas, faculty improvement, scholarships and fellowships and financial aid. An aid of Rs. 41 crores during 1974–75 to 1978–79 proved the large spending by ICAR. The agricultural universities have contributed a great extent to agricultural education, research and development in the country. Many improved varieties of crops, feed and animal stocks have been developed in the agriculture uni- versities. In other words, it could be concluded that the ICAR is identical to agricultural research and education in the country. The role played by the council in the deve- lopment of agricultural research and education has been quite extraordinary.

Agricultural research in India has an interesting history regarding its growth and development. It started during the colonial era and today India boosts of an agricultural research system which includes some 49 ICAR institutions along with a number of national research centres, bureaus, project directorates, AICRPs, KVKs, State and Central Agricultural Universities where 27,500 scientists and more than 100,000 supporting staff are actively en- gaged in agricultural research. This makes the Indian agricultural research system probably the largest in the world. At present, ICAR plays a central role at the national level and it aids, promotes and coordinates re- search and education activities throughout the country. The research and education responsibilities at the state level rest with the State Agricultural Universities. In addition to these main streams of research, some general universities and other agencies like scientific organiza- tions related to agriculture, Government departments, voluntary organizations, private institutions (private seed companies, poultry farms), *etc.* participate in the nation's research efforts. Hence, the role of the NARS in the development of agricultural research is of great importance, within which all these organizations come. Five-year plans play a major role regarding investment, technology transfer and other aspects related to agricultural development in India. In fact, the place assigned to agriculture in the Five-Year Plans after independence can be best judged from allocation to this sector in each plan. The outlay on agriculture has been rising steeply during the successive plans. It rose from Rs. 601 crore in the First Five Year Plan to Rs. 50,924 crore in the 11th Plan (2007–12). Although agriculture has been playing the most decisive role in Indian economy, during the course of the study, it has been observed that not much emphasis has been given to the evolution of agricultural research in India, which has an incredible history of development. It is necessary to track the historical development of agricultural research in India, which will help in forecasting the future of Indian agriculture.

Advance in Chinese Agriculture[1]

Introduction

Several developments in the last decade have drawn renewed attention to China's role in world food markets. These issues include China's accession to the World Trade Organization (WTO) in 2001, rapid income growth, a growing middle class and urbanization, with their associated changes in dietary patterns, a surge in China's imports of soybeans, and concerns over resource depletion and environmental problems (Huang, Otsuka, and Rozelle 2008; Lohmar, Gale, Tuan, and Hansen 2009; Carter, Zhong, and Zhu 2009a). For instance, high food prices brought on by the 2007–08 global commodity price boom were partially blamed on China's rapid economic growth (FAO 2008), though many explanations as to what actually caused the commodity boom and China's role in the boom were misdirected (Carter, Rausser, and Smith 2011; Carter, Zhong, and Zhu 2009b). Indeed, the agricultural commodity price boom most likely had more to do with U.S. ethanol policy than China's growing food imports (Helbling, Mercer-Blackman, and Cheng 2008).

However, valid concerns remain that China, the world's most populous country and its largest food producer, will be unable to keep expanding its food supply to meet its growing demand for meat, horticultural, and processed food products (Carter, Zhong, and Zhu 2009b; Hovhannisyan and Gould 2011). The increased demand for meat will result in an increased demand for feed grains and protein meals, possibly expanding China's dependence on world grain and markets and driving prices even higher (Qiao, Chen, Carter, Huang, and Rozelle 2011). Another way to meet increasing meat demand is to import meat directly. For instance, China's pork imports have recently surged, which has raised the question of whether meat imports will continue to grow in light of higher world feed grain prices. Importing meat versus feed grains in the future might well be determined by the comparative efficiency of China's livestock sector, as well as freight rates.

The rapid income growth in China's urban areas in the last three decades is a well-known story. What is less well known is the fact that the economic development of China's countryside has a long way to go. Indeed, 36 per cent of the Chinese population still lives on less than $2 per day, and most of these are farmers. This means that there are many rural structural issues that remain unresolved, and significant policy changes could lie around the corner, as China's rural development model is in need of dramatic overhaul.

China's future role in the world food equation therefore raises many questions, such as: Will China's agricultural production mix continue to move towards more

---

1 Colin A Carter, Funing Zhong and JingZhu, applied Economic Perspectives and Policy, volume 34, No.1, 2012.

labor intensive horticultural crops, with Chinese farmers further specializing according to their comparative advantage in production? To what extent can China boost its agricultural productivity by becoming more mechanized, consolidating farmland into larger and more efficient operations, and adopting the latest cropping technology? Will China overcome the growing imbalance between its irrigation water supply and demand through greater investments in irrigation?

Addressing China's changing food dynamics is not just about economics and natural resources, but policy obviously plays a key role as well. Modern China has placed food self-sufficiency as a top policy priority, and so far it has been able to meet this policy goal without too much reliance on global food markets. But this objective naturally comes at a cost because of China's very high grain self-sufficiency target. China holds relatively large grain stocks, its domestic grain prices are kept somewhat insulated from world markets, and farmers typically earn less money growing grain compared to other higher value crops; this holds down farm incomes and impedes productivity gains from greater specialization in agricultural production.

Another motivation for this article stems from the fact that it has been over thirty years since China abandoned its large, Russian-style communal farms, run by top-down management, and selling the output to state monopolies at fixed prices. The communal farming system was a failure, a fact underscored by the 1959–61 famine, when an estimated 23 to 30 million Chinese residents starved to death (Lin and Yang 2000). The communes were broken up in the late 1970s in favor of small family-run plots with profit incentives tied to production and market-determined prices.

Thirty years later is an appropriate time to review the major achievements of Chinese agriculture, outline major trends in production, consumption and trade of agricultural products, and highlight key policy challenges faced by Chinese agriculture. These are the objectives of this article, which build on previous articles that have addressed the broad question of the future of Chinese agriculture (Carter and Rozelle 2001; Carter, Zhong, and Zhu 2009a; Huang, Otsuka, and Rozelle 2008; Lohmar, Gale, Tuan, and Hansen 2009; Huang, Yang, and Rozelle 2010).

This article is organized as follows. First, we discuss the long-term growth trends in Chinese agriculture, broken down by regions and subsectors within the agricultural industry. The purpose is to illustrate that Chinese agriculture has gone through various growth stages in the past thirty years, including a period of very slow growth from 1998 to 2003. This section also discusses the broad trends in China's agricultural trade balance because there is great interest in how China's shifting demand for and supply of food will affect world markets. The second section of the article describes changes in agricultural input use. The availability of land, water, and labor obviously remain key issues going forward, and we highlight concerns related to these inputs. Section three focuses on food consumption patterns, outlining rural versus urban differences. The final section outlines policy developments and outstanding policy issues. Chinese agriculture is at a very important juncture and future policy tools applied to land userights, irrigation investment, biotechnology, farm subsidies, and state trading are critical with respect to the linkage to world food markets.

China's Agricultural Output

Output Growth Trends

China's agricultural sector grew by over 300 per cent in real terms from 1978 to 2009, or 4.6 per cent annually over 31 years.1 Such a continuous strong growth rate is very impressive, and was partially achieved by shifting resources among agricultural sub-sectors and regions. In this section we will first describe the sub-sector growth patterns, and then we turn to the regional differences in growth.

According to the official Chinese definition, the agricultural sector, or "primary" sector as it is called in China, is divided into four subsectors: farming, forestry, animal husbandry, and fisheries (National Bureau of Statistics 2010). Farming refers to cropping, including field crops, such as grain and cash crops, as well as tea and fruits, while forestry refers to the small-scale production of trees, bamboos, *etc.*, managed by family farmers, excluding large-scale forestry activities run by state enterprises, which are part of the industrial or secondary sector. Measured by the gross value of output, the historical growth rates of the four subsectors are show in Figure 1.5.

Figure 1.6 indicates significant differences in growth rates across the four sectors from 1978 to 2009. Fisheries grew most rapidly, with an annual growth rate of 6.8 per cent, followed by animal husbandry at 5.9 per cent, and forestry at 3.9 per cent, which was roughly the same as the agricultural sector as a whole, measured by gross output value of agriculture (GOVA). The growth rate of farming was lower, at 2.9 per cent per year for the 31-year period, but was still very impressive over such a long time period. The difference in the growth rates across sectors clearly indicates a continuing shift from low value to high-value agriculture, in response to changing demands following rapid economic growth and urbanization, and consistent with China's comparative advantage in labor-intensive agricultural production.

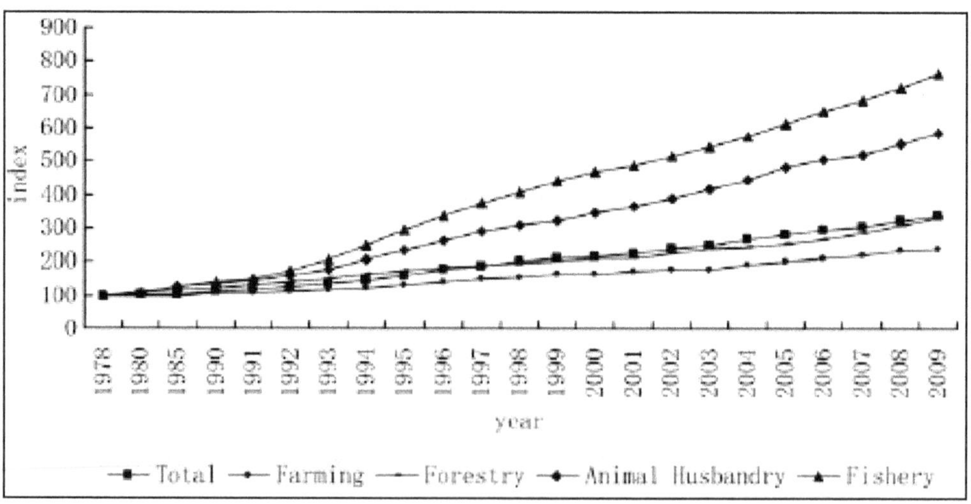

Figure 1.5: Gross Output Values of China's Agricultural Subsectors, 1978 ¼ 100.
***Source*: NBS: China Statistical Year Book 2010.**

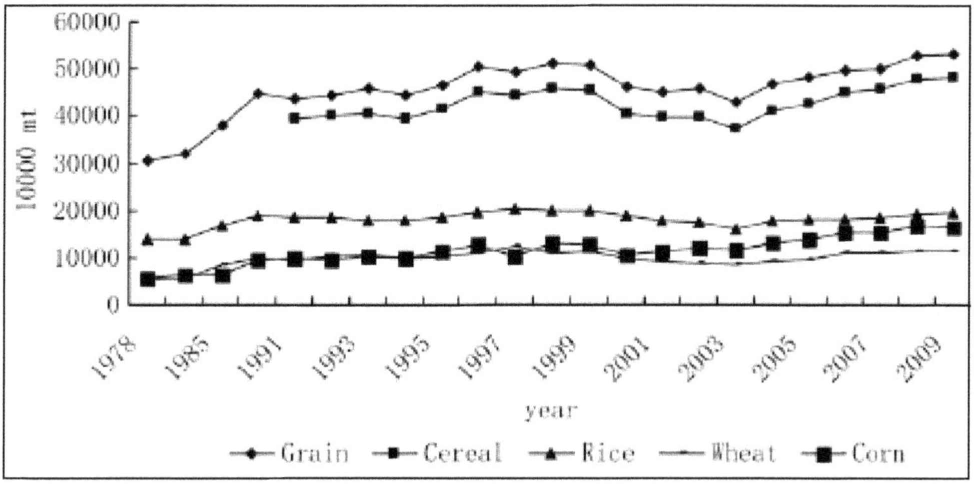

Figure 1.6: Growth of Major Grain Crops, 1978–2009.
***Source*: NBS: China Statistical Year Book 2010.**

Food security is a politically sensitive issue in China, and the growth of grain production has been an ongoing government priority supported by a variety of policy measures. Considering China's natural resource constraints such as land and water availability, grain output has remained very strong, growing at an annual rate of 1.8 per cent from 1978–2007, surpassing the rate of population growth. It should be noted that the growth rate of grain production was relatively faster during the early stage of economic reform, especially in the first six years from 1978 to 1984, and then slowed with periodic swings afterwards. For instance, from 1998 to 2003, total grain production fell by more than 15 per cent, and then grew at 3.5 per cent for the subsequent six years (see Figure 1.6 for details). More recently, over the past eight years grain production has grown year-to-year in a continuous fashion, illustrating its continued political importance, because grain production is not as profitable as many other agricultural activities (*e.g.*, horticultural crops). As mentioned, an important exception is soybeans, which has lost production area to corn and rice. The fast-growing soybean imports suggest that China's domestic food supply faces resource constraints, and importing soybeans for feed is both more cost-efficient and politically less sensitive than importing food grains.

China began to use the term "cereals" in official grain statistical reports in 1991, and the growth pattern of cereals has been very similar to that of total grain. However, the growth rates of the three major cereal crops (*i.e.* rice, wheat and corn) have experienced some important differences. From 1978 to 2009, rice production increased at an annual rate of 1.15 per cent, while wheat and corn grew at 2.48 per cent and 3.53 per cent, respectively. The changing shares among the three major cereal crops reflects the fact that grain production has gradually shifted from southern paddy fields (historically the food-surplus region) to northern dry land conditions (where wheat and corn dominate). The demand for feed grains has been increasing, thereby inducing an increased supply of domestic corn.

Among the cash crops, oilseed production increased by a factor of 5 over 31 years, or 6 per cent per year. This fast growth reflects an increasing demand for high-valued foods, including edible oils and animal products. Cotton production reached a peak of 7.62 million metric tons in 2007, representing an annual growth rate of 4.4 per cent over 29 years. However, it then decreased by more than 15 per cent in two years, largely due to a decrease in the world demand for textiles caused by the financial crisis. Over the same time period, the growth of fruits and vegetables was even faster. The output of fruits increased from 6.6 million tons in 1978 to 204 million tons in 2009, almost 30-fold in 31 years, or 11.7 per cent per year. The official statistics do not report vegetable output, however, the vegetable sown area increased from 3.3 million hectares to 18.4 million hectares over this same time period, or by about 4.6 times in total, which is 5.7 per cent annually.

In 1997, China's National Bureau of Statistics revised the official meat production figures for 1996, lowering the estimated production by more than 20 per cent based on data collected during the first agricultural census. But unfortunately the bureau did not make any effort to revise the data series before 1996, so the only consistent data series are from 1996. According to official data since 1996, total meat production increased by 66.9 per cent in the 13 years until 2009, representing an annual growth rate of 4 per cent. Milk production has been increasing much faster, at an annual rate of 13.3 per cent, resulting in a quadrupling in 13 years, while the production of poultry eggs lagged behind at 2.6 per cent per year. The comparison between the growth rates of milk and eggs may reflect changing lifestyles and urbanization. When a large number of rural people move to urban areas, their consumption of eggs may not increase, while milk consumption is likely to increase due to greater availability (see Figure 1.7 for details).

Within the meats, poultry output has been increasing the most: 5.3 per cent per year compared with 3.7 per cent per year for the sum of pork, beef, and mutton during the 13 years from 1996 to 2009; this is partly due to the feed conversion efficiency and relatively good prices for poultry products. Pork production has been growing slightly slower than beef, but its share still accounts for 82.7 per cent of the

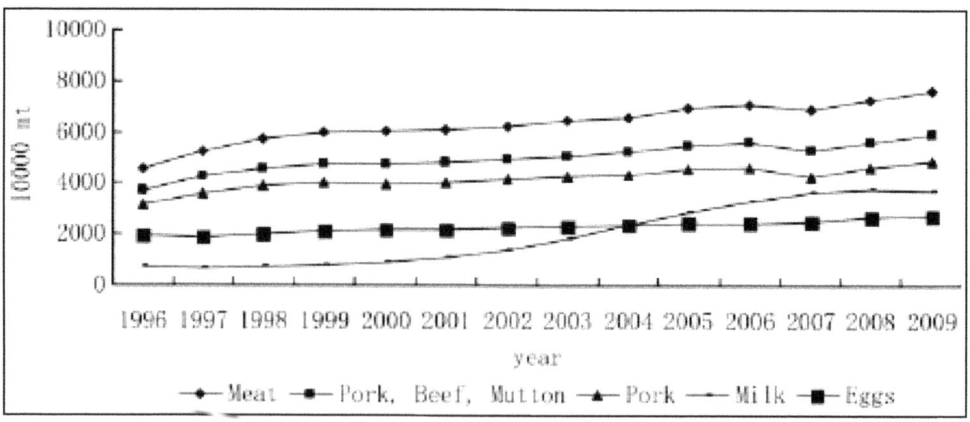

Figure 1.7: Growth of Animal Products, 1996–2009.
***Source*: NBS: China Statistical Yoar Book 2010.**

sum of pork, beef and mutton due to established dietary habits and agricultural production conditions in China.

Aquatic products are a category of high-valued, labor-intensive food products in China. Total aquatic output increased more than 10-fold from 1978 to 2009, representing an annual growth rate of 8.1 per cent. Compared to seawater products, which grew at an annual rate of 6.7 per cent, fresh water products increased faster, at 10.6 per cent. In both seawater and fresh water aquatic products, production of artificially cultured products increased much faster than that of naturally grown: 11.7 per cent vs. 4.6 per cent per year for seawater aquatics, and 11.5 per cent vs. 6.7 per cent for fresh water products, respectively, per year over the 31 years examined.

It can be concluded from the above that China's agricultural production has been growing very fast, shifting from relatively low valued to higher valued products. The strong growth rates have been partly due to public support of agricultural R&D and infrastructure, in addition to institutional reforms and small subsidies. As food security is still one of China's policy priorities, government support of this goal is likely to continue, although the policy tools may change. The shift to high-value products is consistent with China's resource endowments and comparative advantage in agricultural production, resulting in higher income for farmers and enhancing the nation's welfare.

Shifting Output Mix

The above mentioned growth patterns across sub-sectors and regions have inevitably led to changes in output mix in the total agricultural sector. The share of farming in the GOVA declined from 80 per cent in 1978 to approximately 50 per cent in 2010, while that of animal husbandry and fisheries increased from 15 per cent and 1.6 per cent, respectively, to more than 30 per cent and approximately 10 per cent, respectively, over the same time period.

The pace of structural change in Chinese agriculture was very fast in the 1980s and 1990s, and then it slowed down. The share of grain crops in total sown area declined from 80 per cent in 1978 to 65 per cent in 2002, then bounced back to 69 per cent in 2006 and has been maintained at above 68 per cent since then. The share of vegetables in total sown area has been more or less constant since 2003 (at around 11.5 per cent), although it increased very fast previously. The official statistics for vegetable acreage may not accurately reflect the growth pattern of vegetable production, as the official numbers are for the land area sown for vegetables for the entire year, without accounting for the number of harvests in a year. So any change in vegetable production intensity will not be reflected in the area statistics. The share of oilseeds has actually declined since 2003, partly due to competition from imports.

Areas sown to rice and wheat continued to decline until 2003 or 2004 and then recovered somewhat; the corn area has shown constant growth during the last three decades. As a result, among the three major cereal crops, the share of rice production fell from 45 per cent to 37 per cent, and that of corn increased from approximately 18 per cent to roughly 31 per cent. The share of wheat production increased from below 18 per cent in 1978 to above 22 per cent in the mid-1980s and has remained at that level since then.

Because growth patterns have been quite similar for all major livestock products, the output mix has been virtually unchanged over the last 15 years, except for milk, which has grown strongly, increasing by a factor of4 compared with a 67 per cent increase for meat and 40 per cent for eggs. A large change in product mix occurred in the aquatic sub-sector: the shares of seawater and fresh water products were 77.2 per cent versus 22.8 per cent in 1978, but soon became 60 per cent versus 40 per cent by the mid-1980s, and then further shifted to 52.4 per cent versus 47.6 per cent in 2009. This significant change was largely due to relatively fast growth in the aquatic culture.

Agricultural Inputs

Arable Land

China is one of the world's large countries with a relatively small arable land base. The official figure for the amount of arable land is still questionable, but the land survey in 1996 did raise the number from under 95 million ha to just above 130 million, more than a 30 per cent increase on paper (Ministry of Land and Resources, 1997). Yet, even based on the revised figure, per capita arable land is only approximately 0.09 hectare. Importantly, a continuous decline of the amount of available arable land has occurred due to urbanization and industrialization. According to the National Development and Reform Commission (2008), arable land, or cultivated land, was reduced from 130 million hectares in 1996 to 121.7 million in 2007, or a 6.4 per cent decline in 11 years, even though tightened regulations have been strictly implemented to control transfers of farmland to non-agricultural uses. Quality issues are inter-related with the quantity of farmland, as more and more marginal land is being utilized for farming. Two-thirds of the arable land is classified as medium- or low- yielding land.

More striking are the large losses of arable land from 1996 to 2009 in the urbanized Coastal region: 9.6 per cent decline in Zhejiang, 7.3 per cent in Fujian and 13.5 per cent in Guangdong, with Jiangsu being the only exception, where the amount of arable land declined by 5.9 per cent, less than the national average. If land quality is considered, the losses are even greater. Measured by annual grain yield equivalents, the losses of production capacity were 12.6 per cent, 10.6 per cent and 17.7 per cent in Zhejiang, Fujian, and Guangdong, respectively. Even in Jiangsu, where loss in arable land is below the national average, the loss in production capacity was 9.1 per cent, about 40 per cent higher than the national average.

There are significant differences in land quality across regions. If annual grain yield per unit of arable land is used as the common weight, the land production capacity in coastal and central China is about 50 per cent higher than the national average, and it is about 40 per cent below the national average in the northwest and southwest regions, except for Sichuan and the Chongqing municipality next to Sichuan. As further urbanization and industrialization are likely to concentrate in the coastal and central regions, losses of high quality land seem likely to continue.

The privatization of land use rights was one major component of agricultural development in the early 1980s, as it provided farmers with a clear title to claim net income from farm production. However, with the rural labor force aging fast and shrinking, land use rights associated with small farms are now being questioned, as these rights might possibly have a negative impact on potential yield growth. It is clear that some type of organizational/institutional measures are needed to facilitate the consolidation of farmland, at least the coordination of farming practices on small farms, in order to realize economies of scale and to overcome workforce aging issues.

Water Conditions

Arable land productivity is closely related to temperature and water conditions. Water is a scarce resource in China, with total annual availability of freshwater equal to 2,300 cubic meters per capita, which is less than 1/4 of the world's average. In addition, the annual precipitation is unevenly distributed across the country: it is more than 800 mm in most parts of southern China, between 400 and 800 mm in the northeast and northern regions, and below 400 mm in the northwest (see Figure 1.8 for details). The distribution of precipitation is highly correlated with temperature, and in general land productivity is higher in the southeast and lower in the northwest.

Irrigation has played a crucial role in China's agriculture throughout history, as evidenced by Dujiangyan, one of the ancient water projects constructed 2,250 years ago, yet still irrigating 0.7 million hectares of farmland today. According to Cai and Rosegrant (2004), the irrigated area increased from 16 million ha in 1950 to 54 million ha in 1998, covering more than 40 per cent of the arable land and producing 73 per cent of the cereals in China. By 2009, the irrigated area further increased to 59 million hectares, a 10 per cent increase. Again, the percentage of irrigated area in the total arable land area exhibits vast differences across regions - in general, it is between 60–80 per cent in most eastern and central provinces, falling to 30 per cent in the northeast, and is even less in some of the western provinces.

Partly due to the pressure of population growth and partly due to the fast urbanization and industrialization in the southeast coastal regions, grain production, and agricultural production in general, has been shifting to northern China. Unlike Northeast China, where the water supply is adequate, the North China Plain has become more and more reliant on groundwater to expand agriculture. According to Wang, Huang, Rozelle, Huang, and Zhang (2009), groundwater was not used for irrigation in the 1950s; however, its share in total irrigation water reached 30 per cent in the 1970s, and then increased to 58 per cent in 1995, and approached 70 per cent by 2004.

Such extensive exploitation of groundwater is unsustainable. The Ministry of Water Resources recognized the overdraft of groundwater as being one of the most serious environmental problems in 1996 (Ministry of Water Resources and Nanjing Water Institute 2004), and the situation has become even worse since then. In the late 1990s, the deep groundwater table in some areas declined by two meters a year (Wang and Huang 2004). The continuous drop in the groundwater table leads to increased irrigation costs, and raises serious questions regarding the future supply of irrigation water in this region.

Labor Resources

Following rapid urbanization and industrialization, the share of China's total labor force employed by agriculture has sharply declined, from around 70 per cent in 1978 to 38 per cent in 2009. However, due to population growth, the absolute number of agricultural workers has actually increased by million, from 283 to 297 million during the same time period. This means that the average farm size has not expanded, instead, it has actually shrunk a little, despite the vast out-migration of labor (about 250 million) from the rural to urban areas.

Rural out-migration was one of the major drivers of China's rapid economic growth, as it provided the urban/industrial sectors with abundant cheap labor. However, the potential negative impacts on agriculture of such large-scale and sustained migration have always been questioned.

One concern is the widespread partial migration of primary workers moving out of villages and agriculture, leaving the elderly and children at home. Some have questioned the sustainabliity of future development of rural areas and the agricultural sector due to the changing quality of the rural labor force.

A typical household survey may not reveal the true dynamic structure of the rural population due to problems measuring either permanent or seasonal out-migration. Therefore, we organized a village-wide household survey in an attempt to obtain a clearer picture of demographic change in rural China. The survey was conducted in five populous provinces (Jiangsu, Anhui, Hunan, Henan, and Sichuan) characterized with a large number of rural out-migrants. The survey covered all members of all households in selected villages, including those having left completely, partially, or seasonally.

The sampling procedure included: (1) the selection of 20 counties in each of the 5 chosen provinces based on equal intervals in per capita income; (2) picking 2 villages (cun min xiao zu7) in each of the selected counties according to the availability of local students. Students from agricultural universities in the chosen provinces conducted the field surveys. Based on resident registration, all members in each of the villages were included, no matter his or her current residential and employment much easier to obtain the required information from his or her relatives and/or neighbors with reasonable accuracy. After reviewing the sample information and deleting ineffective entries, the final data cover 5 provinces, 121 counties, 203 villages, 7,317 households and 28,021 individuals.

The summary data in Table 1.5 show that among the rural laborers staying at home for more than 10 months a year, more than 10 per cent of them work in non-agricultural sectors, and about 47 per cent of them work in both agricultural and non-agricultural sectors (*i.e.* dual occupation). Only 34 per cent of the rural labor force may be classified as "dedicated" agricultural workers. The difference among age groups is very significant: the number of agricultural laborers in the 16–25 age group are only one-half of that in the 26–35 age group, and roughly one-sixth of that in the 36–45, 46–55, and 56–65 age groups. On the contrary, the number of non-agricultural laborers in the 16–25 age group is about the same as in 26–35 and 36–45 age groups, and much more than that in the 46–55 and 56–65 age groups.

Table 1.5: Employment Status of Rural Workers by Age

Age	Agriculture		Non-agriculture		Dual-occupation		Not-working		Total	
	No	Per cent	No	Per cent	No.	Per cent	No	Per cent	No	Per cent
16–25	154	1.66	240	2.59	458	4.94	360	3.88	1,212	13.06
26–35	286	3.08	268	2.89	633	6.82	55	0.59	1,242	13.39
36–45	881	9.49	260	2.80	1,322	14.25	65	0.70	2,528	27.24
46–55	967	10.42	164	1.77	1,202	12.95	92	0.99	2,425	26.13
55–65	908	9.79	67	0.72	725	7.81	172	1.85	1,872	20.17
Total	3,196	34.44	999	10.77	4,340	46.77	744	8.02	9,279	100

Source: Field survey by the authors.

It can be inferred from the survey data that less than 14 per cent of the "dedicated" agricultural workers are younger than 35, while almost 60 per cent of them are older than 45, meaning the remaining "dedicated" agricultural workers, in 20 years time, might be only about one-fourth of the current number, if not less. The overall rural employment situation is likely to follow the same trend: the share in the 16–25 age group employed rurally is about one-third of that for the 36–45 age group. The shares of younger age groups in agricultural and dual employment rural work may further decline, as younger people will continue to search for jobs in the urban industrial and service sectors.

Therefore, it is expected that the aging process may speed up the rate of decline in the total rural labor force over the next few decades. The number of those working only in agriculture will decline even more rapidly. As a result, there will be pressure to expand the size of China's farms to a scale that is more efficient. Of course, this transition will partially depend on land ownership rights.

In the meantime, however, the aging of the rural labor force, especially that portion employed in agriculture, raises policy questions. Formal and/or informal institutions that may help the elderly adopt better production decisions, give them access to machinery and other farm services, as well as provide an improved social security system, are emerging policy issues of great importance.

Input Use and Productivity Growth

Restricted by limited arable land, Chinese agricultural growth has relied on yield improvements for a long time, both before and after the 1978 economic reform. The contribution of the growth in inputs has been well documented. According to Perkins and Yusuf, factors influencing agricultural output can be divided into two categories: the amount of inputs and the way in which the inputs are utilized. Inputs can be divided into two sub- groups: traditional inputs such as land and water, and modern inputs such as high-yielding varieties, machinery, chemicals, and science-based farming practices.

China developed and began the full-scale adoption of dwarf and semidwarf varieties of rice and wheat in the 1960s and hybrid rice in the 1970s, followed by significant varietal improvement in all major crops since the 1980s and the introduction of GM crops since the 1990s. At the same time, modern inputs such as agricultural chemicals and machinery increased significantly.

Figure 1.8 shows that the total use of chemical fertilizer increased from 8.8 mmt in 1978 to 54 mmt in 2009, expanding by more than 5 times in 31 years; application per ha increased from 59 kg to 341 kg during the same time period. The application rate per ha is measured in terms of sown area (per ha per harvest). As such, the intensity would be much higher if measured on the arable land base, at about 500 kg/ha, making China one of the leading users of chemical fertilizers in the world. The trends are likely the same in the application of other agricultural chemicals, such as pesticides, fungicides, herbicides, and plastic sheeting.

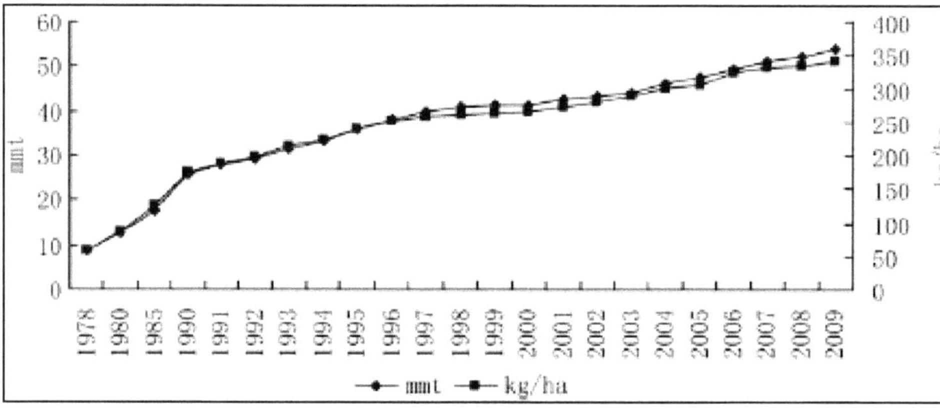

Figure 1.8: Chemical Fertilizer Utilization in China, 1978–2009.

Although Chinese agriculture is widely considered labor intensive, the expansion of farm machinery use is very impressive, particularly after the communes were replaced by small-scale family farming. The total power supplied by agricultural machinery increased from 117 million kw in 1978 to 875 million kw in 2009, representing a 6.5-fold increase in 31 years. The number of small tractors has increased continuously, from 1.4 million units to 17.5 million units, more than 11 times over the same time period. However, the number of large and medium tractors stayed at roughly the same level until the turn of the century, after which it increased at an accelerated pace, tripling in 10 years.

The rapid growth in the use of large machinery might be a reflection of the emerging development of farm machinery services in recent years. Some specialized farm households and/or organizations have purchased relatively large farm machinery to supply commercial fieldwork services. In many cases such services may cover a large area, crossing county or provincial borders, and the services are typically provided for all or most members in a village at the same time. The unified supply of machinery services helps ensure that major fieldwork is conducted at the same quality level, no matter the age, education and experience of the farmer. As such, the substitution of machinery may have reduced the potential negative impact on agriculture resulting from the out-migration of its prime workers.

Because of the increasing use of modern inputs, in addition to institutional reforms, productivity in Chinese agriculture has kept growing at a brisk rate, at least when measured in units of land. For example, the rice yield per ha increased by 65.5 per cent in the 31 years from 1978 to 2009, while that of wheat and corn increased by 156.9 per cent and 87.6 per cent, respectively.

As mentioned above, the privatization of land use rights improved farmers' incentives to a great extent during the early years of reform, and productivity increased as a result. However, due to the small scale of farms (less than 0.5 hectare) segmented into several pieces, private land use rights may have negative impacts on future growth. Some organizational/institutional changes have occurred, including attempts to extend standardized machinery services across a large number of individual farms. It might be expected that with continuing out-migration and aging of farm workers, farmland will be consolidated in one way or another, and productivity may increase as a result.

Food Consumption Patterns

Major food items were rationed to urban residents in China for a long time before the 1978 economic reform, and outputs of major farm products, including food, were subject to a strict governmental quota procurement system, with limited amounts left over for rural residents themselves. Under such conditions, the consumption of food was not closely related to market signals. Since 1978, agricultural production has been stimulated by higher procurement prices, the opening of markets, which has allowed farmers to sell surplus at higher prices, and new institutional arrangements such as the widely-known Household Responsibility System (Lardy 1983; Lin 1992; Carter and Zhong 1988). As a result, farmers' own food consumption increased and became more diversified; decisions regarding food consumption are now made jointly with production decisions, and are more and more determined by market forces. Meat marketing was excluded from the urban rationing system in 1985, and the food rationing system was completely abolished in the early 1990s.

It is noteworthy that per capita food grain consumption declined very sharply in urban China, from 130 kg in 1990 to 81 kg in 2009. At the same time, consumption of animal products rose significantly: meat (pork, beef, mutton, and poultry) consumption increased from 25 kg to 35 kg, eggs from 7.3 kg to 10.6 kg, and milk from 4.6 kg to 14.9 kg. There are no available data on seafood consumption in recent years. However, judging from the trend between 1990 and 2005, consumption probably was about 15 kg in 2009, double that of the 1990 figure. The per capita annual consumption of edible oils also grew significantly in urban China, from 6.4 kg to 9.7 kg, or 50 per cent, over the 19 years examined.

Per capita food consumption in rural China has shown similar broad trends to the urban areas: consumption of food grain declined while that of animal products increased. The units used to measure food grain consumption are different between the urban and rural areas; in urban areas consumption is measured in processed grain, while in rural areas it is measured in unprocessed grain. As the milling rates are approximately 80 per cent for wheat and 70 per cent for rice, the total annual grain consumption of 189 kg unprocessed grain in rural China is equivalent to 145 kg of processed grain, which was almost 80 per cent higher than the amount consumed by their urban counterparts. Consumption of edible oils in the rural areas increased slower than in the urban areas in both absolute and percentage terms.

Meat consumption grew much faster in rural areas (70 per cent increase) compared with urban areas (40 per cent increase). However, due to the lower initial level, rural meat consumption still lags behind urban consumption in absolute terms: 21.5 kg versus 35 kg in 2009. It should be noted that rural meat consumption includes processed meat products, which are not specifically included in urban consumption figures. In addition, consumption outside the home is not accurately recorded. So the actual difference in per capita meat consumption between the urban and rural areas is likely much greater than presented here.

Consumption figures in the above two tables suggest that per capita food consumption has leveled out in the 21st century, especially since 2005. Total food consumption, measured in grain equivalents, was constant or even slightly declined in recent years. However, a careful calculation suggests that grain consumption may be 10 per cent higher in urban versus rural areas in per capita terms, if feed grains are included. As such, it seems that even without population growth, the process of urbanization may increase the total demand for food.

Demand-side Factors

China's population growth has shown a continuous declining trend over the last three decades, around 1.5 per cent per year in the 1980s, down to 1 per cent in the 1990s and around 0.5 per cent in recent years. As mentioned above, urbanization may lead to an increase in food demand. According to official statistics, the share of rural residents in the total population declined from 81 per cent in 1978 to 47 per cent in 2009, roughly one percentage point every year. If urban per capita food consumption in grain equivalents is 10 per cent higher than in the rural areas, then for every 1 per cent of the population that moves from the rural to urban areas, this may add 0.1 per cent to the total demand for grain, or food in grain equivalents, or 0.5 mmt of grain every year.

The summary statistics in Tables 1.6 and 1.7 suggest that China's per capita food consumption has leveled out over the last 10 years or so, which is consistent with Engel's law–as their income grows, consumers spend a smaller proportion of their budget on food. Given income and prices, per capita food consumption is determined by physical requirements, especially that for energy intake. In turn, energy intake requirements are determined by age, gender, and occupation. If the demographic structure changes significantly, per capita food demand will also change as the physical requirement for energy intake changes.

The Chinese population "pyramid" is highly irregular, so much so that it does not even look like a pyramid. However, for convenience, we still use the term "pyramid" to describe the population structure (see figure 1.14 for details). There are three large gaps in the pyramid. The first appeared in the population segment born in the early 1960s, apparently due to the famine. Such a gap is likely to repeat in 20–25 years following human reproduction patterns, so the second gap appeared in the 1980s, and the third gap is still unfolding. The gap did become larger in the 1980s and later, enhanced by China's one child policy.

Table 1.6: Per Capita Calorie Intake at Household Level

by Income				*by Year*			
Yuan	*Obs.*	*Mean*	*S.D.*	*Year*	*Obs.*	*Mean*	*S.D*
<800	1,778	2,219.8	704.3	1991	3,219	2,432.87	643.35
800–1500	2,497	2,280.5	703.4	1993	3,111	2,391.8	2 675.1
1500–3000	5,232	2,278.6	673.7	1997	2,793	2,187.7	9 587.69
3000–5000	4,699	2,273.38	645.3	2000	3,848	2,242.83	658.74
<5000	6,568	2,241.8	670.7	2004	3,879	2,203.73	692.03
				2006	3,924	2,141.39	702.87

Source: Calculated from survey data by CHNS (2010).10.

Table 1.7: Urban Consumption of Selected Foods by Income Groups, 2009

kg/per capita

Items	*1st 10 per cent*	*2nd 10 per cent*	*2nd 20 per cent*	*3rd 20 per cent*	*4th 20 per cent*	*"9th 10 per cent*	*10th 10 per cent*
Pork	14.39	18.11	19.88	21.52	22.74	23.27	23.85
Beef	1.39	1.90	2.20	2.62	2.81	2.89	2.87
Mutton	0.89	1.02	1.23	1.51	1.56	1.53	1.42
Eggs	8.20	9.52	10.11	11.19	11.57	11.80	11.64
Milk	8.01	10.47	12.80	15.98	18.20	20.08	21.35

Source: Compiled from data in NBS 2010.

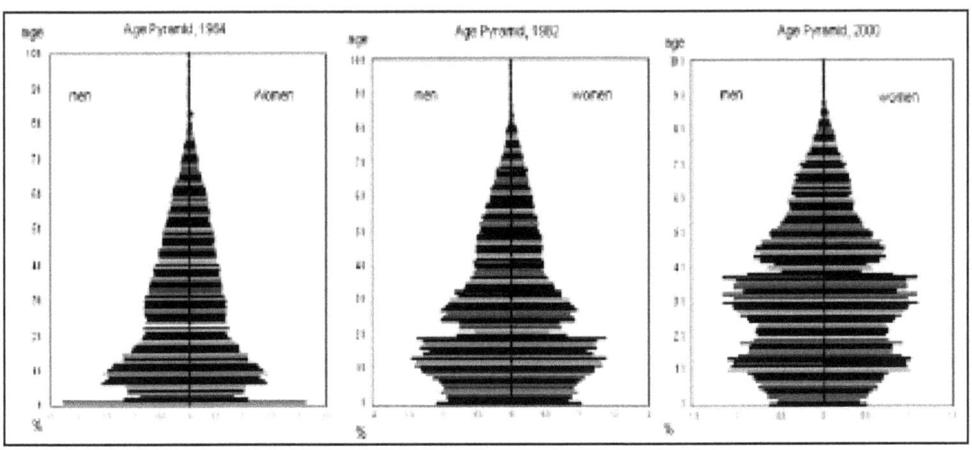

Figure 1.9: China's Population Pyramid in Selected Years.
***Source*: National Bureau of Statistics: China Population Statistical Yearbook 1987 and 2001.**

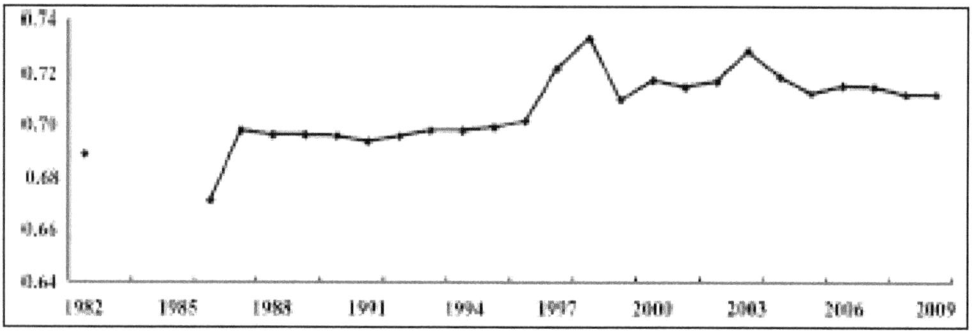

Figure 1.10: AMES Based on Energy Intake in China, 1982–2009.
***Note*: AMES is Adult Male Equivalent Scale, calculated by applying the FAO/WHO/UNU. Adult Male Equivalent coefficients to China's population data.**
***Source*: National Bureau of Statistics: China Population Statistical Yearbook, 1987–2010; and FAO 2001.**

Naturally the physical requirements for energy intake are less for children and the elderly. If a large gap appears at the bottom of the pyramid, the overall average demand for energy intake will be greater than under normal conditions. However, if the gap moves up the pyramid, centering on the age group requiring the highest energy intake, the overall average demand for food may be less than under normal conditions.

According to a joint study undertaken by the Food and Agriculture Organization of the United Nations (FAO), the World Health Organization (WHO), and the United Nations University (UNU) (FAO 2001), energy intake requirement varies among age and gender groups (see appendix I). Based on that study, an Adult Male Equivalent Scale (AMES) was constructed to measure the effect of demographic change on China's food demand (see Figure 1.14 for details). It can be seen that the AMES has declined by two percentage points since the mid-1990s, suggesting a possible 2 per cent decline in energy requirements, holding other conditions constant. As the weighted average demand for energy intake is about 70 per cent of the unit adult male, a 2 percentage point decline in AMES implies approximately a 3 per cent decline in energy demand, and to a large extent, this could explain why Chinese per capita demand for food did not increase in recent years despite income growth.

Given the physical requirements determined by demographics, income may have its impact on food demand through dietary choices. According to the China Health and Nutrition Survey (CHNS 2010), per capita calorie intake in China has leveled out, and with little difference across income groups (see table 4 for details). However, the same caloric intake does not necessarily imply the same demand for grain equivalents if consumption is shifting to more animal-based foods. Table 5 indicates that, although the quantity of animal-based food consumption is about the same among the upper income groups in the urban areas, they are approximately 30–40 per cent lower for the lowest income group (first decile) compared with that for the lower-middle income group (second quintile). It is reasonable to assume that further income growth may lead to higher demand for high priced and better quality foods; but the low-income groups may continue to demand more food quantity as their incomes growth.

The average per capita income for the first decile group was about one-half of that for the second quintile group in 2009, or 5,253 versus 11,244 yuan. If per capita income continues to grow at an annual rate of 8 per cent, per capita income of the first decile group may reach the current level of the second quintile in about 8-9 years. Hence, the demand for animal-based foods may increase by 3-4 per cent per year for this group. As the increase in demand is likely to be less for other low income groups, and as the animal-based foods are partly substituted for cereals, the overall increase in demand for grain equivalentmay be around 0.5–1 per cent a year due to income growth.

In summary, changing demographic characteristics and income growth have led to an annual growth rate of about 1 per cent in total consumption of grain equivalents. This figure is highly consistent with the increase in total disappearance of grain and soybeans in China over the last decade, from some 470–480 mmt at the turn of the century to 550 mmt per year recently. If this trend continues for the next decade, the total consumption of grain equivalents, including cereals, tubers, soybeans and other minor crops, may reach 600 mmt in 10 years.

However, there are other factors that may slow down the growth in the demand for food. The first is a potential increase in prices, which may lead to reduced demand for food items, especially animal-based foods requiring a large amount of feed. But more important is the dynamics of the Chinese population pyramid. The gaps in the rural population pyramid are very large. As these are the joint results of the famine

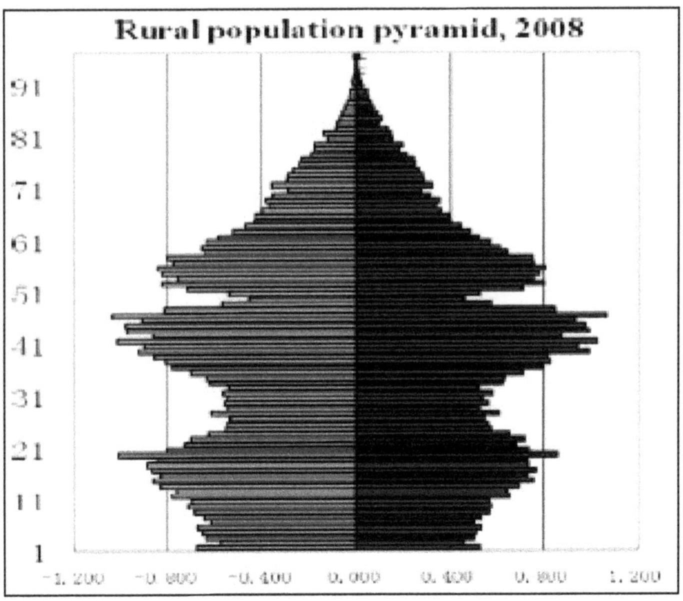

Figure 1.11: Rural Population Pyramid.
***Source*: NBS 2009.**

in the early 1960s and the one child policy since the 1980s, it is reasonable to assume the same situation in the urban population, and the later two gaps might be even larger due to a stricter implementation of the one child policy in the urban areas.

The existing policy permits a couple to have a second child if they both come from one-child families. However, more and more young couples are choosing to have only one child for various reasons, including delayed marriages. As such, population growth may further slow in the next 10 years, and the total population may even decrease significantly in two or three decades. If that happens, China's total food demand may eventually fall after 20–30 years, reducing pressure on world food markets.

Policy

The pace of structural change in China's agriculture was very fast in the 1980s and 1990s, and then slowed down from the early part of the new century. The slowdown gave rise to renewed concerns over food security within China. The fact that grain production declined by 16 per cent from 1998 to 2003 triggered a series of policy measures intended to correct this trend. These policies included: abolishing the agricultural tax which was levied on crops; introducing direct subsidy payments to grain producers; offering support prices for some grain crops in selected regions; providing input subsidies; and increasing public investment in infrastructure. The new package was started as an experiment with subsidy payments in a few provinces in 2002–03, and extended as a national program in 2004, with minimum price procurement of rice and wheat in selected areas added, and a phasing out of the agricultural tax. All these policy support measures improved the relative profitability of the farming sector, ensuring this sector then started to grow at about the same speed as the other three agricultural sectors in the 21st century.

The policy reforms leading to the impressive success in China's agriculture since 1978 have been well documented. Economic reform in the agricultural sector can be summarized as two interrelated sets of institutional changes: from "planned supply and marketing" to market-based resource allocation, and from collective farming to privatized farming. While changes in the marketing system have improved price signals and provided opportunity for improved allocation efficiency, changes in the organization of the farm unit provided farmers with incentives to work harder and improve farming practices, giving them the ability to capture opportunities resulting from market-oriented reforms.

China adopted the Russian-style "heavy industry first" development strategy in the early 1950s. Under this approach, the agricultural sector was viewed as the main avenue to extracting investment funds for rapid industrialization under central planning. As taxing hundreds of millions of small farms was impractical, the transfer of agricultural surplus was carried out through the so-called "unified procurement and marketing system." Under this scheme, farmers were assigned compulsory procurement quotas, along with production plans for major farm products, to be delivered to state agencies at regulated low prices. At the same time, food and other products originating from agriculture were provided to urban residents and state enterprises at low prices. State enterprise profits were then used as investment funds by the government for industrialization.

There was no doubt that compulsory procurement at low prices could lead to resistance from farmers, and it could be extremely difficult to enforce such a policy. One solution for reducing both resistance and administrative costs was to organize the huge rural population into groups, with the leaders partially working for the government as officers. Therefore, Chinese farmers were organized into collectives and soon into communes, partly to facilitate the transfer of rural surplus through the procurement scheme. While the central planning system denied farmers' rights in resource allocation through compulsory production targets and procurement quotas, the income distribution system inside the commune weakened the link between economic reward and work efforts made by each individual member.

Jointly, the procurement system and the communal system might have been effective at extracting and transferring agricultural surplus. However, this scheme was unable to stimulate agricultural growth. In fact, the surplus may have been shrinking over time, leading to not only growing political dissatisfaction but also slowing down industrialization, especially with the ongoing population growth at the time. It was obvious that such a system was not sustainable.

The favorable political environment after the "Cultural Revolution" triggered fundamental reforms beginning in 1978, with the immediate target aimed at raising agricultural production through improved incentives. Procurement prices were increased by 20 per cent in 1979, with above-quota premiums raised from 30 per cent to 50 per cent. In the next a few years, while procurement prices continued to increase periodically, both the coverage of products under procurement quota and the quota quantities for each product were cut, thereby allowing farmers to sell more onto the open markets at higher prices. By the mid-1980s, livestock products and cash crops, with the exception of tobacco, were produced for the open market, and the procurement system for major grain crops changed its name into a "purchasing contract" system. The new "purchasing contract" regime was not significantly different from the old "procurement quota" system, as the government continued to unilaterally determine both quantity and price, and the delivery to designated state agencies remained compulsory. The "purchasing contract" system was finally abolished in the mid-1990s (Carter, Zhong and Cai 1996).

Parallel changes occurred in the farming system. Various forms of the so-called responsibility systems were tried in the attempt to better link rewards with the actual efforts of individual members of the commune. The original objective of introducing various "responsibility systems" was to improve the internal income distribution scheme within the commune to provide greater incentives to each individual member. However, pushed by farmers for greater freedom in decision-making, the final result was the complete breakdown of the communal system and virtual restoration of family farms under the "household responsibility system" (HRS) by the mid-1980s. At the same time, rural township governments were established to take over the administrative and social service functions from the commune, including education, health care, agricultural extension, *etc.* (Lin 1992).

The reform that started in the agricultural sector and rural areas soon spread to the urban/industrial sectors, boosting Chinese economic growth at an unprecedented rate. The rapid growth in the urban/industrial sectors fundamentally changed China's economic structure, and resulted in two crucial and long-lasting impacts on the agricultural sector: 1) while the importance of agriculture as the source to financial industrialization was largely reduced, the demand for food, especially high-value food items, continued to grow with income growth; and 2) the income gap between rural and urban residents continued to widen, emerging as a major political issue.

As summarized by Tuan, Zhong and Ke (2004), OECD (2005) and others (Huang, Otsuka and Rozelle 2008; Brandt and Rawski 2008; Carter, Zhong and Zhu 2009), a significant change in China's agricultural policy involved reducing support for urban industrialization and abandoning the emphasis on foreign exchange earnings from agricultural trade. Instead, the development of China's agricultural policy focused on two interrelated goals: enhanced agricultural production, especially grain production, and increasing farmers' income.

Agricultural production, especially grain production, has long been viewed as the basis for food security, one of the top national policy priorities. As such, major agricultural policies have since focused on support for domestic grain and livestock production. While farmland use rights have been made transferrable and heritable with extended terms, the regulation of farmland use has been increased to prohibit unauthorized shifts to non-agricultural development. A special grain inventory system, introduced in 1990, has been enhanced with grain risk funds and combined with price support programs periodically carried out for selected grain crops in selected areas, such as corn in Northeast China and rice in several northeast, central and southern provinces.

Alarmed by a continuous decline in grain production (down more than 15 per cent from the peak reached in 1998), China introduced a new package of agricultural policies at the end of 2002 with the goal of boosting grain production and securing the domestic food supply. The new package was first started in a few provinces as a subsidy program then extended to cover more areas with price support programs in 2004, and finally the agricultural tax was phased out in 2004–06. All previous taxes and fees levied on farmers were merged into a unified agricultural tax, and the rates of the new agricultural tax were reduced according to the financial status of local governments. The new agricultural tax was completely abolished after only a few years, and grain producers were given direct payments based on past acreage through financial transfers

from the central budget. Government support was extended to input subsidies to encourage purchases of superior seeds and farm machinery, and to compensate farmers for increases in fuel and general input prices.

The current price support programs were implemented in two forms: a "minimum purchasing price" program and a "temporary stocking purchasing" program. The former was designed for the government procurement of wheat in six provinces and rice in 13 other provinces, while the latter was used for the procurement of corn and soybeans in the northeast, and rapeseed in 17 other provinces. The difference between the two price support programs is that, although the fixed price is announced every year based on the market situation, the "minimum purchasing price" program has been more or less institutionalized, at least in terms of coverage; the quantities procured under the "temporary stocking purchasing" program, as well as the area covered by the program, vary from year to year. One weakness is that year-to-year changes in the programs are sometimes announced after farmers have sold their crops. Thus, only the marketing agencies can benefit. As such, it is not easy to estimate the actual support received by farmers.

Support from the central budget is relatively easy to discern. According to the China Agricultural Yearbook, the 2009 total budget support to agriculture was 725.3 billion yuan, of which 127.4 billion was paid directly to individual farmers to support crop production, especially grain production; 19 billion was provided as direct payments to grain producers; 19.8 billion for the purchasing of "superior seeds" 13 billion for purchases of farm machinery; and 75.6 billion as "comprehensive input subsidies" to compensate farmers for increased input prices. Compared with the crop subsector, livestock production receives very little support: just under 1 billion yuan in 2009. Other major items include: 15.5 billion for improving infrastructure and services; 45.3 billion for environmental programs; 22 billion for poverty alleviation, including programs in minority regions; 6.6 billion for disaster relief; and 20 billion as a bonus to local governments in major grain and oilseed producing regions.

If price supports are excluded, in 2009 the total support from the central budget and directly received by individual farmers in support of production amounted to 128.4 billion yuan, plus 125 billion yuan saved from abolishing agricultural taxes and fees. The total amount was roughly 253.4 billion, equivalent to 2,500 yuan per ha of farmland, or 360 yuan per capita in the rural areas, on average. Some of the sub-national governments provide financial support to farmers through other programs, such as subsidies to agricultural insurance programs and rural healthcare programs, as well as pension plans, all of which have been newly introduced on an experimental basis.

After three decades of reform, Chinese agricultural markets have been substantially liberalized (Huang *et al.*, 2009). However, despite the rise of supermarkets and commercial food processing firms, most farmers still sell their products to small private traders. Millions of small traders compete with little regulation and this has led to highly integrated and efficient markets. The remaining market distortions are: (1) import tariff rate quotas (TRQs), which have not been a real import barrier, as the limits have not been met except for cotton, and tariffs, which are relatively low according to international standards; (2) price support in the form of the "minimum purchasing price" and "temporary stocking purchasing" programs, which apply to a small share of the crops; and (3) the rate and base in calculating VAT on imports.

Biotechnology

China is one of the leading countries in terms of planed acreage to genetically modified (GM) crops. In 2010, the area sown to GM crops reached 3.5 million ha, ranking 6th in the world (James 2010). Cotton is by far the largest GM crop grown in China, with 3.3 million ha, accounting for more than 90 per cent of the total area. Other smaller crops include GM petunias, tomatoes, sweet peppers, poplar trees and papaya. In addition, China has approved the import of GM soybeans and corn for processing purposes, and issued safety certificates for *Bt* rice and phytase maize to further extend field experiments.

GM technology has been formally stipulated in government documents as one of the top science and technology priorities (National Guideline for Mid- and Long-Term Development of Science and Technology 2006–2020). As a result, public investment in GM technology has increased very quickly since the mid-1990s, from $33 million in 1995 to $104 million in 2000, and nearly $200 million in 2003, and then doubled again between 2003 and 2007. In 2008, in addition to the current research programs on agricultural biotechnology, the State Council approved a new major 12-year "Special Program" to support research on GM crops and animals totaling $3.8 billion. In a broad sense, the application of biotechnology to food production is a promising area for scientific breakthrough in the 21st century. Therefore, the development of GM technology provides an opportunity for China to "catch up" with developed countries. Specifically, GM technology is believed to have the potential to boost agricultural production with reduced costs. As such, it seems to be a new and efficient approach to enhancing food security, which is why China is willing to invest heavily in this technology. However, the issuance of a safety certificate for *Bt* rice has generated political opposition, which may delay the commercialization process.

2

History of Irrigation

Irrigation Defined

Often people farm in a place where enough rainfalls during the year (and at the right times) to water the plants just with rainfall. Farmers don't have to worry about the plants getting enough rain. That's called "dry farming" because the farmers don't have to carry water to the plants.

But in other places, like Egypt or the Arabian peninsula, it hardly rains at all. Farmers can't rely just on the rainfall to water their crops. They have to find some way of getting water from the river to their fields. That's called "wet farming," and the way they get water from the river is called "irrigation."

There are lots of different ways to get water from the river to the fields. One way is just to carry it yourself in buckets, and plenty of people irrigated this way in the ancient and medieval periods (and plenty of people still do today). But it's very hard work carrying water in buckets, and you can't carry very much even if you work very hard. Even a donkey or an ox can't carry enough water to irrigate big fields.

So whenever they can, people use some kind of machine to help them carry the water. Often this is a lever, a long wooden pole with a bucket on one end so people can hold the other end of the pole and lower the bucket into the water and then raise it and swing it around and dump it into a canal that is little higher up and that carried the water to the fields. In Egypt, people call this machine a shaduf.

Necessity of Irrigation

The necessity and importance of irrigation for the harvest in India can hardly be exaggerated. It has the potentiality of bringing commercial revolution in any country. British administration in India was also in perfect agreement with this view.

The East India Company (EIC) servant Sir Charles Trevelyan noted in 1816 – "that irrigation is everything in India; water is more valuable than land, because when water is applied to land it increases its productiveness at least six-fold and renders great extents of land productive, which otherwise would produce nothing or next to nothing.

Despite this, the British administration played a different ball game and the notorious neglect of irrigation in the era of the Company rule proved disastrous as time passed on. Marx, in 1853, held that the British rule in India "have neglected entirely" irrigation works and held "the deterioration of an agriculture which is not capable of being conducted on the British principle of free competition, of laissez-faire and lassiez-aller".

Development of Irrigation

Irrigation works received the attention of the State in ancient and medieval India. The canal, known as the Western Jumna Canal, built in the 14th century by Feroz-Shah Tuglak had been renovated during the reign of Akbar in 1568. The Eastern Jumna Canal had its beginning in the 17th century during- the reign of Shah Jahan and the original Bari Doab and the inundation canals of Sind were also of ancient origin.

According to Maddison's estimate, the irrigated land in the Mughal empire was not more than 5 p.c. of the cultivated area and the State works were a very small proportion of this. Zeal and enthusiasm that Mughal rulers displayed in the building of irrigation works was conspicuously absent during the regime of the EIC.

However, some irrigation works were undertaken during Dalhousie's administration after the formation of the Public Works Department in 1855. The number of such works was very remarkable, notably in the North, Western Provinces, and in Madras. The first irrigation work undertaken by the EIC was the restoration of the Western Jumna (Yamuna) Canal (in 1821) and the Eastern Jumna Canal (in 1830) which had fallen into neglect. Another monumental work undertaken by the East India Company was the (Upper) Ganges Canal which was opened in 1854 and the capital expenditure reached well over £ 3 million.

At the time of its opening in 1854, water flowed nearly 900 miles in length to irrigate 15,00,000 acres. It was a thoroughly public enterprise and its express purpose was to fight drought and famines. The Bari Doab Canal was constructed in 1856 in the Punjab. Minor irrigation works were in operation in Rohilkhand and the Dun.

While these works were being constructed in Northern India, to the South in the Madras Presidency, old irrigation works were now in operation in the deltas of the Cauvery (or Kaveri), Kistna (or Krishna) and Godavari rivers. Seeing the success in the Cauvery river, the Godavari 'anicut' (dam) was also completed by Sir Arthur Cotton–the pioneer of modern irrigation works in India–during this period (1853).

In spite of these irrigation works in different parts of the country, the EIC. exhibited deadening indifference with regard to irrigation during this time. Thus, the record of the period of 'merchant capital' and of the EIC rule "is not a particularly good one".

With regard to irrigation the Company's "practice of providing for such projects out of current revenue prevented any regularity or continuity of work and, on the contrary, often resulted in considerable waste from the suspension of valuable public works already undertaken."

All irrigation works were built by the Government itself out of revenue surpluses. But ultimately this coasted dearly to the Company and an opinion had been aired that irrigation works akin to the guarantee system of railway construction in India could be entrusted to the British private companies. Altogether, two private companies (East India Irrigation Company and Madras Irrigation and Canal Company) were floated but they failed disastrously.

It was then agreed that henceforth the State should undertake all irrigation works and funds should be raised even by means of loans. State patronage poured in continuously after the end of the Company's rule. Between 1860 and 1880, several important works, like the Lower Ganges Canal and the Agra Canal (in U. P.), the Shirhind Canal (in the Punjab), the Khadakwasala dam (in the Bombay Deccan) were initiated. Even then, progress in this direction was not only unsatisfactory but also inadequate, considering the needs of the country. Some factors were responsible for this mess.

Firstly, a good deal of money had been wasted on some ill- conceived projects and no definite irrigation policy had been formulated. Secondly, financial stringency prevented from raising funds for investment in irrigation projects. As 'Home Charges' were heavily draining resources of the Government, very little fund was available for canal irrigation.

Thirdly, and most importantly, deliberate neglect of irrigation works by the British adminis-tration was based on a wrong premise. The British gave priority to the construction of railways over the construction of canals since the recurrent famine problems could be minimized through the extension of railway traffic rather than canal irrigation.

The virtues of an extensive system of irrigation in fighting famines were spelt out by Sir Arthur Cotton when terrible famine of 1877-78 devastated Madras and Mysore. This famine took 1,350,000 lives and £9,75,000 were expended in relief – a savage demonstration of the need for famine finance to be made a permanent part of Government's accounts.

The extension of irrigation facilities got priority in the recommendations of all Famine Commissions. The Famine Commission of 1880 also recognised the value of irrigation in its recommendations. Closely on the heels of the recommendations of the Famine Commission (1880), some definite programmes of work were put into action in different parts of the country and, by 1895, most of these were completed.

Some important irrigation works were the Sutlej and the Chenab Canals (in the Punjab), the Lower Ganges and Betwa canals (in the United Provinces), and the line of navigable canals between Cuttack and the Hooghly. Some protective works were undertaken out of the newly created Famine Relief and Insurance Fund.

The first protective work sanctioned to be financed through this Fund was the Betwa Canal project in Bundelkhand. Other protective works worth noting here were the completion of the Nira Canal System in the Bombay Deccan, Jamrao, and Western Nara Canals in Sind.

By 1895-96, the total capital outlay on irrigation works, in tandem with protective works, amounted to Rs. 38.3 million–an increase of 33 p.c. over the figure for 1885-86. The result was the doubling of British India's irrigation capacity. In actuality, the State went on sponsoring various public works scheme after the end of the Company's rule.

Roughly, between 1875 and 1890, the area irrigated by government works rose by 8 million acres to about 18.5 million acres. With an increase in private irrigation of more than 3 million acres and government works by 8 million acres the total area under irrigation canal came to 44 million acres. During this period, percentage of total irrigated area under private works (like canals, tanks, wells, and other sources) amounted to 57.8 p.c. as against the State works which stood at 42.2 p.c.

The overall result was the increase in total revenue by 47 p.c., net revenue by 62 p.c., and net profits from Madras, North Western Provinces, and the Punjab Provinces offset the net losses experienced by Bombay and Bengal (in just 10 years after 1885-86). Meanwhile, the overall net rate of return of revenue on capital outlay rose from a wholly inadequate 3 p.c. in 1885-86 to barely adequate 4 p.c. in 1895-96.

In the jungle of this data, the alien Government was tom-tomming and drumbeating what it had achieved. The Government was then arguing that agriculture without irrigation was hard to flourish. But what was worrisome was that the increase in irrigation potential lagged far behind the country's requirements.

Gopal Krishna Gokhale showed that, in 1883, 30 p.c. of the expenditure on public works had been on irrigation–Rs. 24 crore as against Rs. 54 crore on railways. This criminal neglect of agriculture resulted in frequent and devastating famines, more widespread and intense than any of which history keeps record. Further, the protective and preventive power of irrigation had not been successfully exercised in the most productive areas.

The North could not be made free from drought, the majority of the Deccan works were small canals, catering only a few hundred cusec water. Works constructed in the greater part of Bengal, in the Bombay Deccan and Gujarat in the Central Provinces, and up- country Madras were neither productive nor protective. Thus, so far, the development of irrigation had benefited only some parts of the country.

It was said that the recurrence of famines in 1897 and 1900 proved to be a blessing in disguise for the cultivators in India. Accordingly, the Indian Irrigation Commission was appointed in 1901. The Commission made comprehensive recommendations for the improvements of the existing and construction of new irrigation works.

The Commission's attention was specially directed towards the extension of irrigation in preventing or mitigating 'the horrors and the cost of famine'. The Commission was realistic in its appraisal of irrigation prospects. The Commission recommended a programme of major works – both productive and protective – to cost Rs. 44 crore spread over 20 years.

Following the recommendations of the Commission, many projects came into operation. Chief among these were a few productive works like Triple Canal Project in the Sind, *etc.* and protective works like the Tribeni Canal in Bihar, the Ken Canal in Bundlekhand, the Tendula Canal in Central Provinces, and a few other projects.

However, though a great many projects were undertaken by 1914, many of them could not be completed. Anyway, in 1913-14, the total irrigated area came to about 47 million acres and, by 1917, the Government obtained nearly 15 p.c. on its investment on irrigation works.

The end of the First World War brought some important changes. As a result of the Constitutional Reforms in India, agriculture, cooperation, local self- government, *etc.*, were transferred to the provinces. Thus, irrigation was turned over to local or provincial governments. Now all works could be financed out of loans.

During this time, the demand for small-scale irrigation grew for crops like wheat, sugarcane, *etc.*, to meet the intensive requirements of water supply of these crops. Government stepped in to meet the new challenges by increasing allocations to the agricultural departments. Government also showed its keenness in the exploitation of ground water resources. Now the stage was set for rapid expansion.

The country witnessed the construction of the Sukkur Barrage Project in Sind–'the greatest irrigation work ever undertaken'. The Sutlej Valley Project to irrigate southern districts of the Punjab and the adjoining states of Bikaner and Bhawalpur, the Cauvery-Mettur Project to improve the fluctuating water supply of the existing system of irrigation in the Cauvery delta were no less important. Another important scheme was that of the Sarda Canals in the United Provinces.

A remarkable increase in the area irrigated by Government works in India occurred during the period 1914-1939. However, 90 p.c. of the total area irrigated in British India was concentrated mainly in the Punjab, Sind, Madras, and the United Provinces. Resources for the construction of any irrigation works exhausted when the Second World War supervened.

Finally, the partition of the country caused a great havoc since many major irrigation works concentrated in the West Punjab and Sind fell on West Pakistan territory. It was only after independence, that irrigation system, the backbone of modern Indian agriculture, received an unambiguous priority.

A Few Comments on the Development of Irrigation

The greatest advantage of irrigation is that it acts as a bulwark against famine and drought. It is said that "irrigation thus not only made directly for greater prosperity, but prevented almost all the bad effects of famine or of the fear of famine. The yields of canal-irrigated land were much higher than the un-irrigated land".

Unfortunately, such comment is unwarranted since famines were almost regular visitors in India throughout the period of our study. One can also contradict the prosperity brought about by irrigation works. One of the important indices of agricultural prosperity is agricultural production.

Available evidence does not indicate any substantial improvement in production. "If, following Maddison, we combine Sivasubramaniam's more optimistic estimate that cultivated area rose by 23 per cent with Blyn's estimates of yields (Blyn estimated 122 per cent) we get a roughly constant per capita figure for agricultural output between 1900 and 1946."

However, irrigation works brought a change in the crop pattern. To meet the government's various demands like water rate, cultivators went for cultivating commercial crops like sugarcane on irrigated land. Thus, it helped the process of commercialisation of agriculture.

It should be borne in mind that, in India, not all State efforts on irrigation were beneficial to this country. Public works programme on irrigation were out by British administration to serve the British interests. The foreign mentality of the alien rulers was to convert India into a major supplier of raw materials and food grains for Europe.

Major irrigation works in the four provinces of the Punjab, Sind, Madras, and the United Provinces were undertaken in supporting exports of raw cotton and wheat in exchange of manufactures from Britain. It would have been more beneficial if public works on irrigation had been concentrated in the uninhabited areas, at least in the early stages of construction of irrigation canals. It concentrated mainly in productive areas.

The gross command area of various irrigation works did not always equal the net increase in irrigated area despite encouragement of the Government. The reason probably is that the extension of large-scale works displaced earlier works in the form of wells and ponds constructed and maintained by private enterprise.

These private sources supported agricultural practices that attached due attention to crop mixtures and the preservation of soil fertility. It is true that though irrigation helped to generate exportable surplus of food-grains and cash crops and, hence, the process of commercialization, it displaced dry crops like millets or pulses.

Further, Amiya Bagchi argues that the giant canal irrigation aggravated social cleavages. The high payment of water rates, the rising of the assessed value of the land, the pressure to grow cash crops, *etc.*, forced cultivators to fall back upon landlords-moneylenders' net, leading to the indebtedness of the former to the latter.

Besides monetary costs, one should also consider the non-monetary cost or the 'mischief' caused by irrigation:

Brian Davey talks about ecological imbalances created out of giant irrigation works. **"Persistent heavy cropping under the stimulus to cultivate valuable commercial crops like sugarcane, opium, cotton, and indigo brought a decline in fertility in areas like the irrigated Doab region in the U. P."**

There were other injurious side effects. In the absence of a very good drainage system, it caused widespread problems of water-logging, salinity and alkalinity, which rendered million of acres infertile. No sufficient attention had been paid during the construction of the canals to the drainage system of the land roundabout.

Water table in some areas declined considerably as canals carried away rivers' water upstream. Further, canal irrigation encouraged overuse of water that harmed tillage. An increase in the incidence of malaria followed as water-logging problem assumed menacing proportion. Thus, irrigation development was not an unmixed blessing.

We now conclude our discussion. Compared with the enormous dimensions of the major irrigation systems, the overall results were disappointing. Ii was designed to serve the British interests and the overall costs (including injurious side effects) were stoically to be borne by Indians. Five appended Tables (**Annexure-1 to 5** provide the recent data on irrigation.)

Irrigation Systems, Ancient

Humans are newcomers to Earth, even though their achievements have been enormous. It was only during the Holocene epoch (10,000 years ago) that the development of agriculture occurred, keeping in mind that the Earth and solar system are 4.6 billion years old. Humans have spent most of their history as hunting and food-gathering beings. Only in the past 9,000 to 10,000 years have humans discovered how to raise crops and tame animals. Such changes probably first took place in the hills to the north of present-day Iraq and Syria.

Irrigation in Egypt and Mesopotamia

The first successful efforts to control the flow of water were made in Mesopotamia and Egypt, where the remains of the prehistoric irrigation works still exist. In ancient Egypt, the construction of canals was a major endeavor of the pharaohs and their servants, beginning in Scorpio's time. One of the first duties of provincial governors was the digging and repair of canals, which were used to flood large tracts of land while the Nile was flowing high. The land was checker boarded with small basins, defined by a system of dikes. Problems regarding the uncertainty of the flow of the Nile were recognized. During very high flows, the dikes were washed away and villages flooded, drowning thousands. During low flows, the land did not receive water, and no crops could grow. In many places where fields were too high to receive water from the canals, water was drawn from the canals or the Nile directly by a swape or a shaduf. These consisted of a bucket on the end of a cord that hung from the long end of a pivoted boom, counterweighted.

The Nile River has played an important role in the lives of Egyptians throughout history. This frieze (c. 2000 B.C.E.) depicts Egyptians using water from the Nile River for irrigation. at the short end. The building of canals continued in Egypt throughout the centuries.

The Sumerians in southern Mesopotamia built city walls and temples and dug canals that were the world's first engineering works. It is also of interest that these people, from the beginning of recorded history, fought over water rights. Irrigation was extremely vital to Mesopotamia, Greek for "the land between the rivers." Flooding problems were more serious in Mesopotamia than in Egypt because the Tigris and Euphrates carried several times more silt per unit volume of water than the Nile. This resulted in rivers rising faster and changing their courses more often in Mesopotamia.

Both the Mesopotamian irrigation system and that in the Egyptian delta were of the basin type, which were opened by digging a gap in the embankment and closed by placing mud back into the gap. Water was hoisted using the swape, as in Egypt. Laws in Mesopotamia not only required farmers to keep their basins and feeder canals in repair but also required everyone to help with hoes and shovels in times of flood or when new canals were to be dug or old ones repaired. Some canals may have been used for 1,000 years before they were abandoned and others were built. Even today, 4,000 to 5,000 years later, the embankments of the abandoned canals are still present. These canal systems, in fact, supported a denser population than lives there today. Over the centuries, the agriculture of Mesopotamia began to decay because of the salt in the alluvial soil. Then, in 1258, the Mongols conquered Mesopotamia and destroyed the irrigation systems.

The Assyrians also developed extensive public works. Sargon II, invading Armenia in 714 B.C.E., discovered the qanat (Arabic name) or kariz (Persian name), which is a tunnel used to bring water from an underground source in the hills down to the foothills. Sargon destroyed the area in Armenia but brought the concept back to Assyria. This method of irrigation spread over the Near East into North Africa over the centuries and is still used. Sargon's son Sennacherib also developed waterworks by damming the Tebitu River and using a canal to bring water to Nineveh, where the water could be used for irrigation without hoisting devices. During high water in the spring, overflows were handled by a municipal canebrake that was built to develop marshes used as game preserves for deer and wild boar, and birch-breeding areas. When this system was outgrown, a new canal, nearly 19 kilometers (30 miles) long, was built, with an aqueduct that had a layer of concrete or mortar under the upper layer of stone to prevent leakage.

Irrigation (Prehistoric Mexico)

During the earliest years of canal irrigation in Mexico, the technology changed little, as there are very few remains of these systems. The technological achievements were not very great prior to around 600 to 500 B.C.E. Storage dams were constructed of blocks mortared together as opposed to the earlier ones constructed of loosely piled rocks. Some of the spillways were improved, and floodgates were used in some spillways. Some of the dams could even be classified as arch dams. The canals were modified somewhat during this time. Different cross-sectional areas were used, and some were lined with stone slabs. During this time, crops were irrigated with more carefully controlled water as opposed to the earlier methods of somewhat haphazard flooding.

Between 550 and 200 B.C.E., there were significant improvements in both the irrigation-related features and the entire canal systems. The channelization of streambeds, along with the excavation of canals and the construction of dams, was probably the most significant. In a brief period, the technology of canal irrigation improved significantly; however, the technology stopped developing after 200 B.C.E. and no significant developments occurred for approximately 500 years. Around 300 C.E., minimal new developments started, and the technology remained essentially the same through the classic period (200–800/1000 C.E.) and early postclassic period (800/1000–1300 C.E.).

Canal Irrigation (North America: Chaco and Hohokam Systems)

The Hohokam and the Chaco regional systems stand out as two of the major prehistoric developments in the American Southwest. These two systems expanded over broad geographic areas of similar size (the Hohokam in Arizona and the Chacoans in New Mexico). These systems were of the similar time period but seemed to have developed and functioned independently, with little interaction. The Chaco and the Hohokam systems evolved in quite different environments, having considerably different irrigation infrastructure.

The Hohokam people inhabited the lower Salt and Gila River valleys in the Phoenix area in Arizona. These Hohokam Indian canal builders were given the name later by the Pima Indians. Even though the Indians of Arizona began limited farming nearly 3,000 years ago, the construction of the Hohokam irrigation systems probably did not begin until a few centuries C.E. It is unknown who originated the idea of irrigation in Arizona, whether it was local technology or introduced to them from cultures in Mexico.

Around 1450 C.E., the Hohokam culture declined, possibly because of a combination of factors: flooding in the 1080s, hydrologic degradation in the early 1100s, and the recruitment of labor by the surrounding population. A major flood in 1358 ultimately destroyed the canal networks, resulting in movement of the people. Canal use was either quite limited or entirely absent among the Pima Indians, who were the successors to the Hohokams Indians. The prehistoric people who lived outside the Hohokam culture area also constructed irrigation systems, but none was of near the grand scale as the Hohokam irrigation systems.

Around 900 C.E., the Anasazi of northwestern New Mexico developed a cultural phenomena that now has more than 2,400 archaeological sites with nine towns each with hundreds of rooms along a 5.6-kilometer (9-mile) stretch. The Chacoan system is located in the San Juan basin in northwestern New Mexico. This basin has limited surface water, with most surface discharge from ephemeral washes and arroyos. The water collected from the side canyon that drained from the upper mesa top was diverted by either an earth or a masonry dam near the month of the side canyon into a canal. These canals averaged 4.5 meters (about 15 feet) wide and

1.4 meters (more than 4 feet) deep and were lined in some areas with stone slabs and bordered in other areas by masonry walls. These canals ended at a masonry headgate. Water was then diverted to the fields in small ditches or into overflow ponds and small reservoirs.

History of Irrigation Development in World

Archaeological investigation has identified evidence of irrigation in Mesopotamia and Egypt as far back as the 6th millennium BCE, where barley was grown in areas where the natural rainfall was insufficient to support such a crop.

In the 'Zana' Valley of the Andes Mountains in Peru, archaeologists found remains of three irrigation canals radiocarbon dated from the 4th millennium BCE, the 3rd millennium BCE and the 9th century CE. These canals are the earliest record of irrigation in the New World.

The Indus Valley Civilization in Pakistan and North India (from 2600 BCE) also had an early canal irrigation system. Large scale agriculture was practiced and an extensive network of canals was used for the purpose of irrigation. Sophisticated irrigation and storage systems were developed, including the reservoirs built at Girnar in 3000 BCE.

There is evidence of the ancient Egyptian pharaoh Amenemhet - III in the twelfth dynasty (about 1800 BCE) using the natural lake of the Fayûm as a reservoir to store surpluses of water for use during the dry seasons, as the lake swelled annually as caused by the annual flooding of the Nile.

The Qanats, developed in ancient Persia in about 800 BCE, are among the oldest known irrigation methods still in use today. They are now found in Asia, the Middle East and North Africa. The system comprises a network of vertical wells and gently sloping tunnels driven into the sides of cliffs and steep hills to tap groundwater.

The Noria, a water wheel with clay pots around the rim powered by the flow of the stream (or by animals where the water source was still), was first brought into use at about this time, by Roman settlers in North Africa. By 150 BCE pots were fitted with valves to allow smoother filling as they were forced into the water.

The irrigation works of ancient Sri Lanka, the earliest dating from about 300 BCE, in the reign of King Pandukabhaya and under continuous development for the next thousand years, were one of the most complex irrigation systems of the ancient world. In addition to underground canals, the Sinhalese were the first to build completely artificial reservoirs to store water. The system was extensively restored and further extended during the reign of King Parakrama Bahu (1153 – 1186 CE). In the Szechwan region ancient China the Dujiangyan Irrigation System was built in 250 BCE to irrigate a large area and it still supplies water today.

In fifteenth century Korea the world's first water gauge (woo ryang gyae) was discovered in 1441 CE. The inventor was Jang Young Sil, a Korean engineer of the Choson Dynasty, under the active direction of the King, Se Jong. It was installed in irrigation tanks as part of a nationwide system to measure and collect rainfall for agricultural applications. With this instrument, planners and farmers could make better use of the information gathered in the survey.

Present Status of Irrigation Development in India and World

At the global scale 278.8 Mha (689 million acres) of agricultural land was equipped with irrigation infrastructure around the year 2000. About 68 per cent of the area equipped for irrigation is located in Asia, 17 per cent in America, 9 per cent in Europe, 5 per cent in Africa and 1 per cent in Oceania. India's irrigation development in this century, and particularly after independence, has seen large number of large storage based systems, all by the government effort and money. Post independence has seen more than 60 per cent of irrigation budgets going for Major and Medium (M and M) projects. However, Growth rate of irrigated area continues to fall from 4.23 per cent per year during the 1970s to 3.08 per cent per year in 1980s and to 2.56 per cent in the 1990s. At present, with almost one fifth of worlds net irrigated area (57 Mha); India has the highest irrigated area in the world today. India's ultimate irrigation potential was estimated at 139.9 Mha, comprising of 58.46 Mha through major and medium irrigation schemes and 81.43 Mha from minor irrigation schemes. Recently some positive steps were also taken to long-awaited inter-basin water transfer, aiming at adding 35 Mha to India's irrigated area. The implementation of the inter-basin water transfer link schemes are taken up in a phased manner depending on the priorities of the Government. The links namely (i) Ken-Betwa link (ii)Parbati-Kalisindh-Chambal link (iii) Godavari (Polavaram) – Krishna(Vijayawada) link (iv) Damanganga-Pinjal link and (v) Par-Tapi-Narmada Link have been identified as priority links for consensus building amongst concerned States for taking up preparation of Detailed Project Reports (DPR).

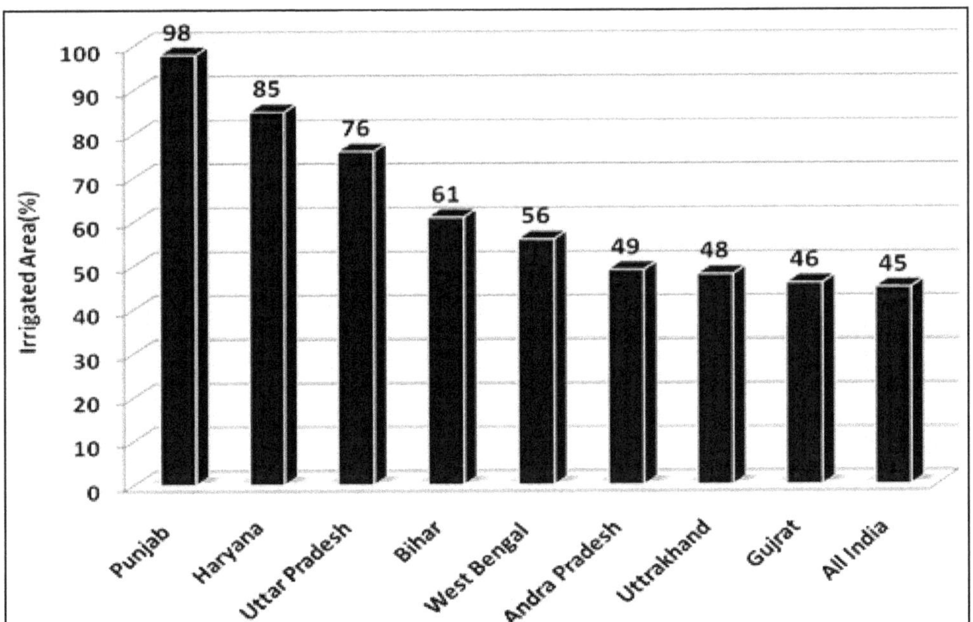

Figure 2.1: State-wise Irrigated Area in India.

Annexure 1: Net Area Irrigation from different Sources and Gross Irrigated Area - All India

Year	Source of Irrigation						
	Canals			Tanks	Tube-Wells	Other Wells	Other Sources
	Government	Private	Total				
(1)	(2)	(3)	(4)	(5)	(6)	(7)	(8)
1950-51	7158	1137	**8295**	3613	(a)	5978	2967
1951-52	7490	1193	**8683**	3489	(a)	6517	2360
1952-53	7511	1350	**8861**	3303	(a)	6521	2437
1953-54	7545	1314	**8859**	4228	(a)	6685	2097
1954-55	7832	1235	**9067**	4025	(a)	6726	2270
1955-56	8025	1360	**9385**	4423	(a)	6739	2211
1956-57	7916	1357	**9273**	4492	(a)	6566	2202
1957-58	8303	1349	**9652**	4536	(a)	6818	2150
1958-59	8391	1279	**9670**	4759	(a)	6686	2286
1959-60	8809	1305	**10114**	4631	(a)	7083	2209
1960-61	9170	1200	**10370**	4561	135	7155	2440
1961-62	9339	1163	**10502**	4612	258	7094	2418
1962-63	9686	1146	**10832**	4781	901	6748	2403
1963-64	9862	1160	**11022**	4599	1028	6756	2483
1964-65	10080	1143	**11223**	4780	1087	6988	2522
1965-66	9859	1099	**10958**	4258	1293	7360	2475
1966-67	10221	1026	**11247**	4424	1706	7489	2041
1967-68	10295	948	**11243**	4493	2112	6999	2346
1968-69	10985	907	**11892**	3926	3087	7714	2390
1969-70	11724	881	**12605**	4059	3739	7438	2356
1970-71	11972	866	**12838**	4112	4461	7426	2266
1971-72	12246	869	**13115**	3734	4745	7535	2417
1972-73	12134	862	**12996**	3619	5393	7571	2255
1973-74	12196	869	**13065**	3900	5604	7679	2298
1974-75	12657	857	**13514**	3544	6583	7646	2422
1975-76	12933	858	**13791**	3972	6843	7601	2386
1976-77	13016	845	**13861**	3901	7432	7655	2300
1977-78	13734	842	**14576**	3904	7641	7943	2482
1978-79	14304	845	**15149**	3937	8159	8270	2544
1979-80	13931	843	**14774**	3481	9307	8557	2405
1980-81	14450	842	**15292**	3182	9531	8164	2551
1981-82	15456	490	**15946**	3376	10334	8403	2444

Contd...

Annexure 1–*Contd...*

Year	*Source of Irrigation*						
	Canals			*Tanks*	*Tube-Wells*	*Other Wells*	*Other Sources*
	Government	*Private*	*Total*				
(1)	*(2)*	*(3)*	*(4)*	*(5)*	*(6)*	*(7)*	*(8)*
1982-83	15716	469	**16185**	2936	10770	8577	2223
1983-84	16294	470	**16764**	3533	10922	8470	2260
1984-85	15805	470	**16275**	3021	11566	8828	2455
1985-86	15715	465	**16180**	2765	11903	8515	2502
1986-87	16039	456	**16495**	2677	12298	8524	2575
1987-88	15286	460	**15746**	2523	13185	8611	2827
1988-89	16640	462	**17102**	2996	13716	9498	2836
1989-90	16646	478	**17124**	2941	14049	9837	2751
1990-91	16973	481	**17453**	2944	14257	10438	2932
1991-92	17327	464	**17791**	2991	15168	10869	3048
1992-93	16503	483	**16986**	3179	15815	11105	3211
1993-94	16653	485	**17138**	3170	16376	11220	3435
1994-95	16799	480	**17279**	3276	17189	11722	3533
1995-96	16561	559	**17120**	3118	17910	11787	3467
1996-97	16889	220	**17109**	2821	19338	12457	3388
1997-98	17186	211	**17397**	2597	19680	12431	3106
1998-99	17099	212	**17311**	2795	21394	12606	3329
1999-00	17247	194	**17440**	2539	22042	12597	2912
2000-01	15809	203	**16012**	2466	22566	11252	2909
2001-02	14993	209	**15202**	2196	23245	11952	4342
2002-03	13867	206	**14073**	1811	25627	8727	3658
2003-04	14251	206	**14458**	1916	26691	9693	4299
2004-05	14553	214	**14766**	1734	25235	9956	7538
2005-06	16490	227	**16718**	2083	26026	10044	5966
2006-07	16802	224	**17027**	2078	26942	10698	5999
2007-08	16531	217	**16748**	1973	28497	9864	6107
2008-09(p)	16686	195	**16881**	1981	28367	10389	6020
2009-10(p)	14789	188	**14978**	1587	28371	9992	7008
2010-11(p)	15476	171	**15647**	1980	28544	10630	6855
2011-12(p)	15838	172	**16010**	1918	29942	10595	7228
2012-13(p)	15462	165	**15628**	1748	30497	10764	7466

(a) : Included under "Other Wells" as separate figures were not collected during these years.

(b) : (p): Provisional

Annexure 2: State-wise Net Area Irrigated by Sources – 2003-04 to 2012-13

State/Union Territory/Year	Net Area Irrigated from						(Thousand Hectares)	
	Canals			Tanks	Wells		Other Sources	Net Irrigated Area (Col. 4 to 8)
	Govern-ment	Private	Total		Tube-Wells	Other Wells		
(1)	(2)	(3)	(4)	(5)	(6)	(7)	(8)	(9)
ANDHRA PRADESH								
2003-04	1137		**1137**	490	1195	674	138	**3634**
2004-05	1346		**1346**	477	1255	649	154	**3881**
2005-06	1572		**1572**	662	1350	636	172	**4392**
2006-07	1623		**1623**	602	1420	653	154	**4453**
2007-08	1609		**1609**	585	2287		163	**4644**
2008-09	1669		**1669**	648	1610	713	180	**4820**
2009-10	1445		**1445**	332	1696	588	153	**4214**
2010-11	1747		**1747**	650	1845	616	176	**5034**
2011-12	1818		**1818**	550	1991	554	178	**5090**
2012-13	1265		**1265**	503	2032	612	162	**4575**
ARUNACHAL PRADESH								
2003-04							47	**47**
2004-05							49	**49**
2005-06							50	**50**
2006-07							52	**52**
2007-08							54	**54**
2008-09							56	**56**
2009-10							56	**56**
2010-11							56	**56**
2011-12							57	**57**
2012-13							57	**57**
Asom								
2003-04 **	33		**33**	3	9	2	93	**140**
2004-05 **	33		**33**	3	9	2	93	**140**
2005-06 **	33		**33**	3	9	2	93	**140**
2006-07 **	33		**33**	3	9	2	93	**140**
2007-08	33		**33**	3		2	101	**140**
2008-09	33		**33**	3		2	101	**140**
2009-10	162		**162**	2	7	0	26	**197**
2010-11	127		**127**	2	7	0	26	**162**
2011-12	33		**33**	5	27	2	94	**161**
2012-13 **	33		**33**	5	27	2	94	**161**

Contd...

Annexure 2–*Contd...*

State/Union Territory/Year	*Net Area Irrigated from*						*(Thousand Hectares)*	
	Canals			*Tanks*	*Wells*		*Other Sources*	*Net Irrigated Area (Col. 4 to 8)*
	Government	*Private*	*Total*		*Tube-Wells*	*Other Wells*		
(1)	*(2)*	*(3)*	*(4)*	*(5)*	*(6)*	*(7)*	*(8)*	*(9)*
BIHAR								
2003-04	904		**904**	112	2288	17	112	**3433**
2004-05	798		**798**	89	2053	7	87	**3034**
2005-06	840		**840**	137	2045	12	127	**3161**
2006-07	951		**951**	160	2214	12	125	**3462**
2007-08	971		**971**	155	2264	7	132	**3529**
2008-09	887		**887**	132	2436	5	75	**3536**
2009-10	720		**720**	80	2094	8	163	**3064**
2010-11	891		**891**	65	1946	21	107	**3030**
2011-12	947		**947**	60	1910	20	116	**3052**
2012-13	955		**955**	59	1902	20	118	**3053**
CHHATTISGARH								
2003-04	768	0	**769**	50	159	36	77	**1090**
2004-05	830	0	**830**	58	197	39	84	**1208**
2005-06	875	1	**876**	53	206	35	79	**1248**
2006-07	887	0	**888**	52	223	35	84	**1282**
2007-08	913	0	**914**	56	251	31	82	**1334**
2008-09	886	1	**887**	51	288	28	84	**1339**
2009-10	870	0	**870**	50	297	27	79	**1323**
2010-11	896	0	**896**	46	300	26	89	**1356**
2011-12	873	0	**873**	54	383	20	85	**1415**
2012-13	877	0	**877**	49	419	20	84	**1449**
GOA								
2003-04	4		**4**			20		**24**
2004-05	4		**4**			20		**24**
2005-06	4		**4**			20		**24**
2006-07	4		**4**			20		**24**
2007-08	8		**8**			27		**35**
2008-09	8		**8**	1	0	20	7	**35**
2009-10	8		**8**	12	5	1	2	**29**
2010-11	8		**8**	18	5	3	2	**36**
2011-12	8		**8**	23	5	3	2	**41**
2012-13							36	**36**

Contd...

Annexure 2–*Contd...*

State/Union Territory/Year	Net Area Irrigated from						(Thousand Hectares)	
	Canals			Tanks	Wells		Other Sources	Net Irrigated Area (Col. 4 to 8)
	Government	Private	Total		Tube-Wells	Other Wells		
(1)	(2)	(3)	(4)	(5)	(6)	(7)	(8)	(9)
GUJARAT								
2003-04	600		**600**	26	977	1760	25	**3388**
2004-05	676		**676**	33	967	1810	42	**3528**
2005-06	778		**778**	42	1078	1946	63	**3907**
2006-07	789		**789**	40	1133	2174	102	**4238**
2007-08	771		**771**	45	1122	2181	114	**4233**
2008-09 **	771		**771**	45	1122	2181	114	**4233**
2009-10 **	771		**771**	45	1122	2181	114	**4233**
2010-11 **	771		**771**	45	1122	2181	114	**4233**
2011-12 **	771		**771**	45	1122	2181	114	**4233**
2012-13 **	771		**771**	45	1122	2181	114	**4233**
HARYANA								
2003-04	1396		**1396**		1562		12	**2969**
2004-05	1426		**1426**	0	1514	0	14	**2954**
2005-06	1330		**1330**	0	1591	0	14	**2936**
2006-07	1309		**1309**	0	1667	0	14	**2990**
2007-08	1380		**1380**	0	1630	0	14	**3025**
2008-09	1274		**1274**	0	1600	1	2	**2877**
2009-10	1282		**1282**	0	1783	2	2	**3069**
2010-11	1236		**1236**		1650	0	1	**2887**
2011-12	1193		**1193**		1879	0	0	**3073**
2012-13	1345		**1345**		1757		0	**3102**
HIMACHAL PRADESH								
2003-04	4		**4**	0	11	3	88	**105**
2004-05	3	0	**3**	0	13	3	86	**104**
2005-06	4		**4**	1	14	2	83	**104**
2006-07	4		**4**	1	14	1	84	**104**
2007-08	4		**4**	1	16	3	85	**108**
2008-09	4		**4**	0	16	2	86	**108**
2009-10	4		**4**	0	18	2	82	**106**
2010-11	4		**4**	0	20	2	84	**110**
2011-12 **	4		**4**	0	20	2	84	**110**
2012-13 **	4		**4**	0	20	2	84	**110**

Contd...

Annexure 2–*Contd...*

State/Union Territory/Year	*Net Area Irrigated from*							*(Thousand Hectares)*
	Canals			*Tanks*	*Wells*		*Other Sources*	*Net Irrigated Area (Col. 4 to 8)*
	Government	*Private*	*Total*		*Tube-Wells*	*Other Wells*		
(1)	*(2)*	*(3)*	*(4)*	*(5)*	*(6)*	*(7)*	*(8)*	*(9)*
JAMMU and KASHMIR								
2003-04	142	140	**282**	4	1	0	19	**307**
2004-05	145	141	**286**	4	1	0	20	**311**
2005-06	132	157	**289**	4	1	0	18	**312**
2006-07	130	156	**287**	4	1	0	18	**309**
2007-08	129	157	**286**	4	1	0	17	**308**
2008-09	161	127	**288**	5	3	1	17	**314**
2009-10	162	125	**288**	5	3	1	20	**317**
2010-11	180	108	**288**	6	5	7	14	**321**
2011-12	171	114	**285**	7	4	4	19	**319**
2012-13	175	110	**285**	8	4	6	21	**325**
JHARKHAND								
2003-04	15		**15**	28	13	56	37	**149**
2004-05	11		**11**	28	9	45	16	**109**
2005-06	12		**12**	21	8	47	18	**106**
2006-07	12		**12**	20	9	37	24	**102**
2007-08	13		**13**	21	12	51	19	**117**
2008-09	8		**8**	21	17	46	17	**110**
2009-10	5		**5**	19	25	33	20	**102**
2010-11	4		**4**	24	24	40	33	**125**
2011-12	5		**5**	33	29	43	43	**153**
2012-13	7		**7**	48	43	55	57	**210**
KARNATAKA								
2003-04	743		**743**	147	774	391	328	**2384**
2004-05	949		**949**	177	882	427	386	**2820**
2005-06	1055		**1055**	190	970	381	374	**2970**
2006-07	1030		**1030**	192	955	372	397	**2946**
2007-08	991		**991**	208	1079	434	420	**3132**
2008-09	1061		**1061**	206	1140	406	424	**3238**
2009-10	1105		**1105**	196	1252	425	413	**3390**
2010-11	1157		**1157**	197	1281	437	418	**3490**
2011-12	1178		**1178**	178	1278	423	383	**3440**
2012-13	1136		**1136**	138	1321	407	418	**3421**

Contd...

Annexure 2–*Contd...*

State/Union Territory/Year	Net Area Irrigated from						(Thousand Hectares)	
	Canals			Tanks	Wells		Other Sources	Net Irrigated Area (Col. 4 to 8)
	Govern-ment	Private	Total		Tube-Wells	Other Wells		
(1)	(2)	(3)	(4)	(5)	(6)	(7)	(8)	(9)
KERALA								
2003-04	100	5	**104**	44	16	106	114	**384**
2004-05	101	5	**106**	44	16	108	119	**393**
2005-06	104	5	**109**	45	14	110	123	**401**
2006-07	103	4	**107**	42	12	114	116	**392**
2007-08	88	4	**93**	42	18	131	105	**388**
2008-09	96	6	**102**	40	18	133	96	**390**
2009-10	95	3	**97**	41	18	126	104	**386**
2010-11	86	6	**91**	51	20	138	115	**415**
2011-12	82	2	**84**	47	25	137	116	**409**
2012-13	81	2	**83**	44	26	122	121	**396**
MADHYA PRADESH								
2003-04	947	2	**949**	127	1501	2233	820	**5631**
2004-05	1004	2	**1006**	125	1588	2406	918	**6042**
2005-06	1028	1	**1030**	134	1449	2247	822	**5681**
2006-07	1089	2	**1091**	149	1793	2403	929	**6365**
2007-08	1048	2	**1051**	138	1981	2275	973	**6418**
2008-09	1065	1	**1066**	130	1985	2385	941	**6506**
2009-10	1109	0	**1109**	157	2087	2588	950	**6892**
2010-11	1136		**1136**	182	2140	2666	1015	**7140**
2011-12	1276		**1276**	220	2408	2865	1119	**7887**
2012-13	1440	0	**1440**	227	2741	2986	1156	**8550**
MAHARASHTRA **								
2003-04	1085		**1085**		2175			**3260**
2004-05	1089		**1089**		2182			**3271**
2005-06	1088		**1088**		2180			**3268**
2006-07	1088		**1088**		2180			**3268**
2007-08	1088		**1088**		2180			**3268**
2008-09	1085		**1085**		2174			**3259**
2009-10	1083		**1083**		2171			**3254**
2010-11	1084		**1084**		2172			**3256**
2011-12	1082		**1082**		2169			**3252**
2012-13	1080		**1080**		2164			**3244**

Contd...

Annexure 2–*Contd...*

State/Union Territory/Year	*Net Area Irrigated from*							*(Thousand Hectares)*
	Canals			*Tanks*	*Wells*		*Other Sources*	*Net Irrigated Area (Col. 4 to 8)*
	Govern-ment	*Private*	*Total*		*Tube-Wells*	*Other Wells*		
(1)	*(2)*	*(3)*	*(4)*	*(5)*	*(6)*	*(7)*	*(8)*	*(9)*
MANIPUR								
2003-04							40	**40**
2004-05							54	**54**
2005-06 **						51	**51**	
2006-07 **						51	**51**	
2007-08 **						51	**51**	
2008-09 **						52	**52**	
2009-10 **						52	**52**	
2010-11 **						73	**73**	
2011-12 **						69	**69**	
2012-13 **						49	**49**	
MEGHALAYA								
2003-04	15	44	**60**					**60**
2004-05	13	44	**58**					**58**
2005-06	12	44	**55**					**55**
2006-07	17	50	**67**					**67**
2007-08	16	42	**58**					**58**
2008-09	17	45	**62**					**62**
2009-10	18	44	**62**					**62**
2010-11	22	41	**63**					**63**
2011-12	27	38	**65**					**65**
2012-13	28	38	**65**					**65**
MIZORAM								
2003-04	6	10	**16**					**16**
2004-05	6	10	**16**					**16**
2005-06	5	11	**16**					**16**
2006-07	4	8	**11**					**11**
2007-08	1	8	**9**					**9**
2008-09	2	9	**11**					**11**
2009-10	2	8	**10**					**10**
2010-11	3	10	**12**					**12**
2011-12	2	11	**13**					**13**
2012-13	2	12	**14**					**14**

Contd...

Annexure 2–*Contd...*

State/Union Territory/Year	Net Area Irrigated from						(Thousand Hectares)	
	Canals			Tanks	Wells		Other Sources	Net Irrigated Area (Col. 4 to 8)
	Government	Private	Total		Tube-Wells	Other Wells		
(1)	(2)	(3)	(4)	(5)	(6)	(7)	(8)	(9)
NAGALAND								
2003-04							67	**67**
2004-05							66	**66**
2005-06							67	**67**
2006-07							65	**65**
2007-08							76	**76**
2008-09							77	**77**
2009-10							73	**73**
2010-11							83	**83**
2011-12							84	**84**
2012-13							85	**85**
ODISHA								
2003-04							1737	**1737**
2004-05							1846	**1846**
2005-06	1370		**1370**		274	90	296	**2030**
2006-07	1370		**1370**		277	90	314	**2051**
2007-08	1420		**1420**		306	101	331	**2158**
2008-09	1418		**1418**		331	101	341	**2192**
2009-10							1368	**1368**
2010-11							1284	**1284**
2011-12							1259	**1259**
2012-13							1248	**1248**
PUNJAB								
2003-04	1041		**1041**		3047		9	**4097**
2004-05	1093	7	**1100**		2918	0	10	**4028**
2005-06	977	4	**981**		2912		126	**4019**
2006-07	1148		**1148**		2878	46		**4072**
2007-08	1142		**1142**		2967		4	**4112**
2008-09	1109	4	**1112**		2963		4	**4079**
2009-10	1111	3	**1115**		2956		2	**4073**
2010-11	1113	3	**1116**		2954		0	**4070**
2011-12	1113	3	**1116**		2969			**4086**
2012-13	1133		**1133**		2982			**4115**

Con...

Annexure 2–*Contd...*

State/Union Territory/Year	Net Area Irrigated from						(Thousand Hectares)	
	Canals			Tanks	Wells		Other Sources	Net Irrigated Area (Col. 4 to 8)
	Government	Private	Total		Tube-Wells	Other Wells		
(1)	(2)	(3)	(4)	(5)	(6)	(7)	(8)	(9)
RAJASTHAN								
2003-04	1314		**1314**	60	1364	2446	55	**5239**
2004-05	1457		**1457**	82	1688	2578	73	**5880**
2005-06	1706		**1706**	77	1889	2538	85	**6294**
2006-07	1703		**1703**	131	2024	2557	81	**6496**
2007-08	1688		**1688**	102	2179	2393	83	**6444**
2008-09	1583		**1583**	31	2437	2122	73	**6245**
2009-10	1424		**1424**	17	2580	1758	71	**5850**
2010-11	1629		**1629**	56	2792	2106	79	**6661**
2011-12	1844		**1844**	69	2932	2179	98	**7122**
2012-13	1901		**1901**	92	3118	2264	125	**7499**
SIKKIM **								
2003-04	1		**1**				5	**6**
2004-05	1		**1**				5	**6**
2005-06				0			14	**14**
2006-07				0			14	**14**
2007-08				0			14	**14**
2008-09				0			14	**14**
2009-10				0			14	**14**
2010-11				0			14	**14**
2011-12				0			14	**14**
2012-13				0			14	**14**
TAMIL NADU								
2003-04	449	1	**449**	385	304	994	15	**2148**
2004-05	753	1	**754**	465	316	1085	18	**2637**
2005-06	800	1	**800**	575	381	1155	7	**2920**
2006-07	781	1	**782**	531	393	1173	9	**2889**
2007-08	752	1	**753**	506	389	1204	11	**2864**
2008-09	765	1	**766**	540	387	1227	11	**2931**
2009-10	757	0	**757**	503	391	1203	9	**2864**
2010-11	747	0	**747**	533	403	1219	9	**2912**
2011-12	746	1	**746**	528	406	1277	7	**2964**
2012-13	590	0	**590**	420	420	1206	7	**2643**

Contd...

Annexure 2–*Contd...*

State/Union Territory/Year	Net Area Irrigated from							(Thousand Hectares)
	Canals			Tanks	Wells		Other Sources	Net Irrigated Area (Col. 4 to 8)
	Government	Private	Total		Tube-Wells	Other Wells		
(1)	(2)	(3)	(4)	(5)	(6)	(7)	(8)	(9)
TRIPURA **								
2003-04	18		**18**	2	8	2	42	**73**
2004-05	18		**18**	2	8	2	42	**73**
2005-06	10		**10**	2	6	2	44	**65**
2006-07	9		**9**	2	6	2	41	**59**
2007-08	9		**9**	2	6	2	41	**59**
2008-09	9		**9**	2	6	2	41	**60**
2009-10	9		**9**	2	6	2	41	**60**
2010-11	9		**9**	2	6	2	41	**60**
2011-12	9		**9**	2	6	2	41	**60**
2012-13	9		**9**	2	6	2	41	**60**
UTTARAKHAND								
2003-04	85	4	**89**	2	209	11	35	**346**
2004-05	92	3	**96**	1	206	8	35	**345**
2005-06	92	4	**96**	0	203	11	34	**343**
2006-07	92	4	**95**	0	199	19	32	**345**
2007-08	86	3	**89**	0	202	14	35	**341**
2008-09	98	2	**100**	0	199	16	26	**340**
2009-10	80	4	**84**	0	214	14	26	**338**
2010-11	81	2	**84**	0	216	12	25	**336**
2011-12	93	3	**96**	0	188	36	20	**339**
2012-13	85	3	**88**	0	188	41	21	**338**
UTTAR PRADESH								
2003-04	2767		**2767**	147	9357	896	59	**13227**
2004-05	2692		**2692**	145	9379	765	138	**13119**
2005-06	2652		**2652**	137	9410	808	67	**13075**
2006-07	2614		**2614**	149	9502	986	62	**13313**
2007-08	2358		**2358**	105	9576	1004	41	**13085**
2008-09	2667		**2667**	125	9603	996	44	**13435**
2009-10	2558		**2558**	126	9614	1032	53	**13383**
2010-11	2539		**2539**	102	9607	1154	39	**13440**
2011-12	2555		**2555**	99	10162	848	146	**13809**
2012-13	2537		**2537**	108	10175	836	272	**13929**

Contd...

Annexure 2–*Contd...*

State/Union Territory/Year	*Net Area Irrigated from*							*(Thousand Hectares)*
	Canals			*Tanks*	*Wells*		*Other Sources*	*Net Irrigated Area (Col. 4 to 8)*
	Govern-ment	*Private*	*Total*		*Tube-Wells*	*Other Wells*		
(1)	*(2)*	*(3)*	*(4)*	*(5)*	*(6)*	*(7)*	*(8)*	*(9)*
WEST BENGAL								
2003-04 **	672		**672**	289	1684	43	319	**3006**
2004-05							3182	**3182**
2005-06 **						3135	**3135**	
2006-07 **						3136	**3136**	
2007-08 **						3136	**3136**	
2008-09 **						3135	**3135**	
2009-10 **						3112	**3112**	
2010-11 **						2955	**2955**	
2011-12 **						3078	**3078**	
2012-13 **						3082	**3082**	
ANDMAN and NICOBAR ISLAND								
2003-04							0	**0**
2004-05							0	**0**
2005-06							0	**0**
2006-07							0	**0**
2007-08				0		0	0	**0**
2008-09				0		0	0	**0**
2009-10				0		0	0	**0**
2010-11 **			0		0	0	**0**	
2011-12 **			0		0	0	**0**	
2012-13 **			0		0	0	**0**	
CHANDIGARH **								
2003-04					1			**1**
2004-05					1			**1**
2005-06					1			**1**
2006-07					1			**1**
2007-08					1			**1**
2008-09					1			**1**
2009-10					1			**1**
2010-11					1			**1**
2011-12					1			**1**
2012-13					1			**1**

Contd...

Annexure 2–*Contd...*

State/Union Territory/Year	Net Area Irrigated from							(Thousand Hectares)
	Canals			Tanks	Wells		Other Sources	Net Irrigated Area (Col. 4 to 8)
	Government	Private	Total		Tube-Wells	Other Wells		
(1)	(2)	(3)	(4)	(5)	(6)	(7)	(8)	(9)
DADAR and NAGAR HAVELI								
2003-04	2		**2**		0	2	3	**7**
2004-05	2		**2**		0	2	3	**7**
2005-06	3		**3**		0	2	3	**7**
2006-07	3		**3**		0	2	3	**7**
2007-08	3		**3**		0	2	3	**8**
2008-09	1		**1**		0	1	2	**4**
2009-10	1		**1**		0	1	2	**4**
2010-11	1		**1**		0	1	2	**4**
2011-12	1		**1**		0	1	2	**4**
2012-13	1		**1**		0	1	2	**4**
DAMAN and DIU								
2003-04								
2004-05								
2005-06								
2006-07								
2007-08								
2008-09								
2009-10								
2010-11								
2011-12								
2012-13								
DELHI								
2003-04	2		**2**		21	0	1	**25**
2004-05	2		**2**		21	0	1	**25**
2005-06	2		**2**		20	0	1	**24**
2006-07	2		**2**		18	0	1	**22**
2007-08	2		**2**		18	0	1	**22**
2008-09	2		**2**		20	0	0	**23**
2009-10	2		**2**		20	0	0	**22**
2010-11	2		**2**		19	0	1	**22**
2011-12 **	2		**2**		19	0	1	**22**
2012-13 **	2		**2**		19	0	1	**22**

Contd...

Annexure 2–*Contd...*

State/Union Territory/Year	Net Area Irrigated from						(Thousand Hectares)	
	Canals			Tanks	Wells		Other Sources	Net Irrigated Area (Col. 4 to 8)
	Government	Private	Total		Tube-Wells	Other Wells		
(1)	(2)	(3)	(4)	(5)	(6)	(7)	(8)	(9)
LAKSHADWEEP								
2003-04								
2004-05								
2005-06								
2006-07								
2007-08								
2008-09								
2009-10								
2010-11								
2011-12								
2012-13								
PUDUCHERRY								
2003-04	0		**0**		15		2	**17**
2004-05	6		**6**		11		0	**18**
2005-06	6		**6**		12		0	**18**
2006-07	6		**6**		11		0	**18**
2007-08	6		**6**		11		0	**17**
2008-09	6		**6**		10		0	**16**
2009-10	6		**6**		10		0	**16**
2010-11	5		**5**		10		0	**15**
2011-12	6		**6**		10		0	**15**
2012-13	5		**5**		9		0	**14**
ALL INDIA								
2003-04	14251	206	**14458**	1916	26691	9693	4299	**57057**
2004-05	14553	214	**14766**	1734	25235	9956	7538	**59229**
2005-06	16490	227	**16718**	2083	26026	10044	5966	**60837**
2006-07	16802	224	**17027**	2078	26942	10698	5999	**62744**
2007-08	16531	217	**16748**	1973	28497	9864	6107	**63189**
2008-09	16686	195	**16881**	1981	28367	10389	6020	**63638**
2009-10	14789	188	**14978**	1587	28371	9992	7008	**61936**
2010-11	15476	171	**15647**	1980	28544	10630	6855	**63657**
2011-12	15838	172	**16010**	1918	29942	10595	7228	**65693**
2012-13	15462	165	**15628**	1748	30497	10764	7466	**66103**

Annexure 3: State-wise Irrigation Area – 2003-04 to 2012-13

(Thousand Hectares)

State/Union Territory/ Year	*Net Irrigated Area*	*Gross Irrigated Area*	*Area Irrigated more than Once*	*Net Un-Irrigated Area*	*Gross Un-Irrigated Area*	*Percentage of Gross Irrigated Area to Total Cropped Area*	*Cropping Intensity (Per cent of Col.16 Over Col.15)*
(1)	*(2)*	*(3)*	*(4)*	*(5)*	*(6)*	*(7)*	*(8)*
ANDHRA PRADESH							
2003-04	3634	4781	1147	6484	7586	38.7	122.2
2004-05	3881	4987	1106	6447	7532	39.8	121.2
2005-06	4392	5996	1604	6353	7366	44.9	124.4
2006-07	4453	6070	1617	5694	6741	47.4	126.3
2007-08	4644	6285	1641	6111	7282	46.3	126.1
2008-09	4820	6741	1920	6048	7090	48.7	127.3
2009-10	4214	5764	1550	5777	6797	45.9	125.7
2010-11	5034	7153	2119	6153	7359	49.3	129.7
2011-12	5090	6785	1695	6071	6975	49.3	123.3
2012-13	4575	6268	1693	6541	7382	45.9	122.8
ARUNACHAL PRADESH							
2003-04	47	47	0	154	207	18.6	126.4
2004-05	49	49	0	153	206	19.1	126.5
2005-06	50	50	0	156	216	18.9	129.4
2006-07	52	52	0	157	219	19.3	129.7
2007-08	54	54	0	155	218	19.9	129.8
2008-09	56	56	0	155	220	20.2	130.6
2009-10	56	56	0	156	220	20.3	130.2
2010-11	56	56	0	157	222	20.3	130.5
2011-12	57	57	0	158	225	20.1	131.0
2012-13	57	57	0	159	228	19.9	131.7
Asom							
2003-04 *	140	173	33	2613	3784	4.4	143.7
2004-05 *	140	171	31	2613	3725	4.4	141.5
2005-06	140 *	142	2	2613	3807	3.6	143.4
2006-07	140 *	142	2	2613	3621	3.8	136.7
2007-08	140	142	1	2613	3697	3.7	139.4
2008-09	140	291	151	2670	3707	7.3	142.3
2009-10	197	225	29	2614	3874	5.5	145.9
2010-11	162	170	8	2649	3989	4.1	148.0

Contd...

Annexure 3–*Contd...*

State/Union Territory/ Year	Net Irrigated Area	Gross Irrigated Area	Area Irrigated more than Once	Net Un-Irrigated Area	Gross Un-Irrigated Area	Percentage of Gross Irrigated Area to Total Cropped Area	Cropping Intensity (Per cent of Col.16 Over Col.15)
(1)	(2)	(3)	(4)	(5)	(6)	(7)	(8)
2011-12	161	163	2	2649	4011	3.9	148.5
2012-13(f)	161	160	–1	2649	4037	3.8	149.3
BIHAR							
2003-04	3433	4567	1134	2279	3316	57.9	138.0
2004-05	3034	4197	1163	2539	3202	56.7	132.8
2005-06	3161	4325	1164	2396	3072	58.5	133.1
2006-07	3462	4646	1184	2203	3073	60.2	136.3
2007-08	3529	4725	1196	2133	3040	60.9	137.1
2008-09	3536	4692	1156	2018	2979	61.2	138.1
2009-10	3064	4433	1369	2268	2863	60.8	136.8
2010-11	3030	4448	1417	2228	2746	61.8	136.8
2011-12	3052	5158	2106	2344	2489	67.4	141.7
2012-13	3053	5327	2274	2349	2451	68.5	144.0
CHHATTISGARH							
2003-04	1090	1179	89	3688	4528	20.7	119.4
2004-05	1208	1312	104	3562	4403	23.0	119.8
2005-06	1248	1375	127	3516	4370	23.9	120.6
2006-07	1282	1486	203	3440	4246	25.9	121.4
2007-08	1334	1522	188	3393	4226	26.5	121.6
2008-09	1339	1537	198	3371	4146	27.0	120.7
2009-10	1323	1487	164	3361	4074	26.7	118.7
2010-11	1356	1605	248	3340	4067	28.3	120.8
2011-12	1415	1648	233	3262	4016	29.1	121.1
2012-13	1449	1725	276	3222	3966	30.3	121.8
GOA							
2003-04	24	40	16	118	129	23.7	119.2
2004-05	24	40	16	110	129	23.7	126.6
2005-06	24	38	14	113	132	22.2	124.5
2006-07	24	38	14	113	134	21.9	125.6
2007-08	35	39	4	99	130	23.2	126.3
2008-09	35	36	1	99	129	22.0	123.3
2009-10	29	38	9	103	123	23.5	121.8

Contd...

Annexure 3–*Contd...*

State/Union Territory/ Year	Net Irrigated Area	Gross Irrigated Area	Area Irrigated more than Once	Net Un-Irrigated Area	Gross Un-Irrigated Area	Percentage of Gross Irrigated Area to Total Cropped Area	Cropping Intensity (Per cent of Col.16 Over Col.15)
(1)	(2)	(3)	(4)	(5)	(6)	(7)	(8)
2010-11	36	36	0	95	124	22.8	122.1
2011-12	41	41	0	91	123	25.0	124.4
2012-13	36	36	0	96	126	22.4	123.2
GUJARAT							
2003-04	3388	4111	724	6464	7310	36.0	115.9
2004-05	3528	4280	752	6219	6977	38.0	115.5
2005-06	3907	4764	857	5815	6731	41.4	118.2
2006-07	4238	5279	1041	5563	6529	44.7	120.5
2007-08	4233	5611	1378	5733	6600	46.0	122.5
2008-09 *	4233	5310	1077	6069	6341	45.6	113.1
2009-10 *	4233	4922	689	6069	6172	44.4	107.7
2010-11 *	4233	5616	1382	6069	6642	45.8	119.0
2011-12 *	4233	6305	2071	6069	6781	48.2	127.0
2012-13 *	4233	5913	1680	6069	6686	46.9	122.3
HARYANA							
2003-04	2969	5343	2373	564	1045	83.6	180.8
2004-05	2954	5434	2480	573	991	84.6	182.2
2005-06	2936	5446	2510	630	1058	83.7	182.4
2006-07	2990	5461	2471	566	933	85.4	179.8
2007-08	3025	5553	2528	569	905	86.0	179.7
2008-09	2877	5528	2651	699	956	85.3	181.3
2009-10	3069	5545	2476	481	806	87.3	178.9
2010-11	2887	5543	2657	631	961	85.2	184.9
2011-12	3073	5680	2607	440	809	87.5	184.7
2012-13	3102	5672	2569	410	704	89.0	181.5
HIMACHAL PRADESH							
2003-04	105	180	75	435	776	18.8	176.8
2004-05	104	183	79	438	770	19.2	175.6
2005-06	104	186	82	435	758	19.7	175.0
2006-07	104	187	82	437	757	19.8	174.5
2007-08	108	193	84	434	762	20.2	176.0
2008-09	108	192	84	432	754	20.3	175.4

Contd...

Annexure 3–*Contd...*

State/Union Territory/ Year	Net Irrigated Area	Gross Irrigated Area	Area Irrigated more than Once	Net Un-Irrigated Area	Gross Un-Irrigated Area	Percentage of Gross Irrigated Area to Total Cropped Area	Cropping Intensity (Per cent of Col.16 Over Col.15)
(1)	(2)	(3)	(4)	(5)	(6)	(7)	(8)
2009-10	106	186	80	433	755	19.8	174.7
2010-11	110	193	83	433	746	20.6	172.7
2011-12 *	110	193	84	433	746	20.6	172.9
2012-13 *	110	195	85	433	751	20.6	174.2
JAMMU and KASHMIR							
2003-04	307	446	139	440	656	40.5	147.6
2004-05	311	453	142	442	649	41.1	146.4
2005-06	312	458	146	422	643	41.6	150.1
2006-07	309	459	149	432	668	40.7	151.9
2007-08	308	463	155	426	671	40.8	154.5
2008-09	314	471	157	425	666	41.4	153.9
2009-10	317	480	162	418	665	41.9	155.7
2010-11	321	479	158	411	661	42.0	155.8
2011-12	319	480	161	427	681	41.3	155.6
2012-13	325	487	162	420	675	41.9	156.0
JHARKHAND							
2003-04	149	180	30	1416	1570	10.3	111.8
2004-05	109	163	53	1366	1474	9.9	110.9
2005-06	106	145	39	1300	1403	9.4	110.1
2006-07	102	144	42	1402	1505	8.7	109.6
2007-08	117	157	40	1419	1518	9.4	109.1
2008-09	110	164	54	1394	1526	9.7	112.3
2009-10	102	155	53	1149	1244	11.1	111.9
2010-11	125	150	25	960	1099	12.0	115.1
2011-12	153	162	10	1097	1268	11.3	114.4
2012-13	210	235	25	1196	1422	14.2	117.9
KARNATAKA							
2003-04	2384	2702	318	7463	8748	23.6	116.3
2004-05	2820	3328	507	7679	9480	26.0	122.0
2005-06	2970	3632	662	7539	9395	27.9	124.0
2006-07	2946	3603	656	7159	8836	29.0	123.1

Contd...

Annexure 3–*Contd...*

State/Union Territory/ Year	Net Irrigated Area	Gross Irrigated Area	Area Irrigated more than Once	Net Un-Irrigated Area	Gross Un-Irrigated Area	Percentage of Gross Irrigated Area to Total Cropped Area	Cropping Intensity (Per cent of Col.16 Over Col.15)
(1)	(2)	(3)	(4)	(5)	(6)	(7)	(8)
2007-08	3132	3789	657	7287	9104	29.4	123.7
2008-09	3238	3942	705	6936	8426	31.9	121.6
2009-10	3390	4096	706	7014	8777	31.8	123.7
2010-11	3490	4279	789	7033	8783	32.8	124.1
2011-12	3440	4137	696	6501	7923	34.3	121.3
2012-13	3421	4007	587	6373	7741	34.1	120.0
KERALA							
2003-04	384	425	41	1806	2530	14.4	134.9
2004-05	393	455	62	1762	2541	15.2	139.0
2005-06	401	460	59	1731	2526	15.4	140.0
2006-07	392	490	98	1710	2428	16.8	138.8
2007-08	388	455	68	1701	2306	16.5	132.2
2008-09	390	458	68	1699	2237	17.0	129.0
2009-10	386	455	69	1692	2213	17.1	128.4
2010-11	415	467	52	1656	2180	17.7	127.8
2011-12	409	546	137	1631	2116	20.5	130.5
2012-13	396	458	62	1652	2134	17.7	126.5
MADHYA PRADESH							
2003-04	5631	5776	145	9315	14012	29.2	132.4
2004-05	6042	6193	151	8933	14010	30.7	134.9
2005-06	5681	5878	197	9290	13729	30.0	131.0
2006-07	6365	6543	179	8371	13570	32.5	136.5
2007-08	6418	6567	149	8269	13849	32.2	139.0
2008-09	6506	6714	208	8435	13943	32.5	138.3
2009-10	6892	7162	271	8080	14249	33.5	143.0
2010-11	7140	7421	281	7980	14624	33.7	145.8
2011-12	7887	8228	341	7350	14289	36.5	147.8
2012-13	8550	8966	415	6802	14164	38.8	150.7
MAHARASHTRA *							
2003-04	3260	4088	828	14172	18102	18.4	127.3
2004-05	3271	3990	719	14219	18378	17.8	127.9
2005-06	3268	4043	775	14205	18512	17.9	129.1

Contd...

Annexure 3–*Contd...*

State/Union Territory/ Year	*Net Irrigated Area*	*Gross Irrigated Area*	*Area Irrigated more than Once*	*Net Un-Irrigated Area*	*Gross Un-Irrigated Area*	*Percentage of Gross Irrigated Area to Total Cropped Area*	*Cropping Intensity (Per cent of Col.16 Over Col.15)*
(1)	*(2)*	*(3)*	*(4)*	*(5)*	*(6)*	*(7)*	*(8)*
2006-07	3268	4236	968	14207	18335	18.8	129.2
2007-08	3268	4363	1095	14205	18293	19.3	129.7
2008-09	3259	4328	1069	14164	18127	19.3	128.9
2009-10	3254	4352	1097	14146	18260	19.2	129.9
2010-11	3256	4496	1240	14151	18679	19.4	133.1
2011-12	3252	4164	913	14134	17846	18.9	126.6
2012-13	3244	4041	797	14100	17832	18.5	126.1
MANIPUR							
2003-04	40	40	0	177	177	18.4	100.0
2004-05	54	54	0	184	184	22.7	100.0
2005-06 *	51	51	0	174	174	22.7	100.0
2006-07 *	51	51	0	174	174	22.7	100.0
2007-08 *	51	51	0	184	184	21.7	100.0
2008-09 *	52	52	0	185	185	21.9	100.0
2009-10 *	52	52	0	181	181	22.3	100.0
2010-11 *	73	73	0	275	275	21.0	100.0
2011-12 *	69	69	0	297	297	18.8	100.0
2012-13 *	49	49	0	261	261	15.7	100.0
MEGHALAYA							
2003-04	60	82	23	167	190	30.3	120.2
2004-05	58	74	16	161	191	27.9	121.0
2005-06	55	64	9	156	193	24.9	121.7
2006-07	67	72	5	146	174	29.2	115.7
2007-08	58	73	15	178	210	25.8	120.4
2008-09	62	73	11	222	265	21.5	118.7
2009-10	62	74	12	221	263	21.9	118.9
2010-11	63	74	12	221	264	22.0	119.0
2011-12	65	80	15	220	259	23.6	118.9
2012-13	65	125	59	220	215	36.8	119.0
MIZORAM							
2003-04	16	18	2	82	80	18.4	100.0
2004-05	16	19	2	80	78	19.2	100.0

Contd...

Annexure 3–*Contd...*

State/Union Territory/ Year	Net Irrigated Area	Gross Irrigated Area	Area Irrigated more than Once	Net Un-Irrigated Area	Gross Un-Irrigated Area	Percentage of Gross Irrigated Area to Total Cropped Area	Cropping Intensity (Per cent of Col.16 Over Col.15)
(1)	(2)	(3)	(4)	(5)	(6)	(7)	(8)
2005-06	16	19	2	81	79	19.3	100.0
2006-07	11	11	0	80	80	12.5	100.0
2007-08	9	10	0	87	86	10.0	100.0
2008-09	11	11	0	84	84	11.8	100.0
2009-10	10	10	0	113	113	8.4	100.0
2010-11	12	12	0	109	109	10.0	100.0
2011-12	13	13	0	85	84	13.5	100.0
2012-13	14	15	0	102	101	12.6	100.0
NAGALAND							
2003-04	67	104	37	239	266	28.1	121.2
2004-05	66	104	39	243	276	27.5	123.1
2005-06	67	106	39	242	281	27.4	125.2
2006-07	65	106	42	258	300	26.2	125.9
2007-08	76	116	40	240	283	29.1	126.6
2008-09	77	82	5	238	320	20.5	127.2
2009-10	73	85	13	288	401	17.6	134.8
2010-11	83	92	8	279	361	20.2	124.9
2011-12	84	92	9	296	382	19.4	125.0
2012-13	85	92	8	295	396	18.9	128.5
ODISHA							
2003-04	1737	2518	781	4058	6118	29.2	149.0
2004-05	1846	2691	845	3893	6027	30.9	151.9
2005-06	2030	2996	966	3661	5932	33.6	156.9
2006-07	2051	3205	1155	3603	5755	35.8	158.5
2007-08	2158	3308	1150	3466	5707	36.7	160.3
2008-09	2192	3177	985	3412	5893	35.0	161.9
2009-10	1368	1614	245	3393	3896	29.3	115.7
2010-11	1284	1539	255	3398	3890	28.3	116.0
2011-12	1259	1437	178	3135	3527	28.9	113.0
2012-13	1248	1496	248	3139	3573	29.5	115.6

Contd...

Annexure 3–*Contd...*

State/Union Territory/ Year	*Net Irrigated Area*	*Gross Irrigated Area*	*Area Irrigated more than Once*	*Net Un-Irrigated Area*	*Gross Un-Irrigated Area*	*Percentage of Gross Irrigated Area to Total Cropped Area*	*Cropping Intensity (Per cent of Col.16 Over Col.15)*
(1)	*(2)*	*(3)*	*(4)*	*(5)*	*(6)*	*(7)*	*(8)*
PUNJAB							
2003-04	4097	7722	3625	144	185	97.7	186.5
2004-05	4028	7702	3674	171	228	97.1	188.8
2005-06	4019	7680	3660	173	188	97.6	187.7
2006-07	4072	7666	3594	112	195	97.5	187.9
2007-08	4112	7689	3577	74	180	97.7	188.0
2008-09	4079	7724	3645	90	188	97.6	189.8
2009-10	4073	7714	3641	85	161	98.0	189.4
2010-11	4070	7724	3654	88	159	98.0	189.6
2011-12	4086	7771	3685	49	134	98.3	191.2
2012-13	4115	7744	3629	36	126	98.4	189.6
RAJASTHAN							
2003-04	5239	6393	1154	12155	15271	29.5	124.5
2004-05	5880	7093	1213	10669	13969	33.7	127.3
2005-06	6294	7818	1524	10543	13881	36.0	128.9
2006-07	6496	7958	1462	10268	13576	37.0	128.5
2007-08	6444	8088	1644	10652	14120	36.4	129.9
2008-09	6245	7910	1665	11306	14861	34.7	129.7
2009-10	5850	7309	1459	11125	14436	33.6	128.1
2010-11	6661	8322	1661	11688	17680	32.0	141.7
2011-12	7122	8903	1781	10913	15602	36.3	135.9
2012-13	7499	9455	1956	9980	14498	39.5	137.0
SIKKIM *							
2003-04	6	11	5	72	110	9.4	155.1
2004-05	6	11	5	72	112	9.2	158.1
2005-06	14	19	4	64	105	15.1	158.3
2006-07	14	19	4	63	105	15.1	159.5
2007-08	14	17	3	63	101	14.6	152.7
2008-09	14	18	4	63	100	15.4	152.3
2009-10	14	20	5	63	124	13.7	186.0
2010-11	14	20	6	63	131	13.2	195.7
2011-12	14	19	4	63	118	13.7	177.3
2012-13	14	19	5	63	124	13.5	185.7

Contd...

Annexure 3–*Contd...*

State/Union Territory/ Year	*Net Irrigated Area*	*Gross Irrigated Area*	*Area Irrigated more than Once*	*Net Un-Irrigated Area*	*Gross Un-Irrigated Area*	*Percentage of Gross Irrigated Area to Total Cropped Area*	*Cropping Intensity (Per cent of Col.16 Over Col.15)*
(1)	*(2)*	*(3)*	*(4)*	*(5)*	*(6)*	*(7)*	*(8)*
TAMIL NADU							
2003-04	2148	2479	331	2541	2837	46.6	113.4
2004-05	2637	3087	450	2460	2802	52.4	115.5
2005-06	2920	3397	477	2324	2636	56.3	115.0
2006-07	2889	3309	420	2237	2534	56.6	114.0
2007-08	2864	3252	388	2198	2563	55.9	114.9
2008-09	2931	3393	462	2112	2431	58.3	115.5
2009-10	2864	3238	374	2028	2333	58.1	113.9
2010-11	2912	3348	435	2042	2405	58.2	116.1
2011-12	2964	3519	555	2022	2371	59.7	118.1
2012-13	2643	2991	349	1902	2148	58.2	113.1
TRIPURA							
2003-04 *	73	108	35	207	175	38.2	101.2
2004-05 *	73	108	35	207	174	38.2	100.5
2005-06 *	65	101	36	215	179	36.0	100.0
2006-07 *	59	100	40	196	176	36.2	108.1
2007-08 *	59	101	41	196	175	36.5	108.1
2008-09 *	60	96	37	196	167	36.6	103.1
2009-10 *	60	98	38	196	169	36.6	104.4
2010-11 *	60	122	63	196	227	35.0	136.9
2011-12 *	60	129	70	196	241	34.9	145.2
2012-13 *	60	128	69	196	240	34.9	144.1
UTTARAKHAND							
2003-04	346	547	201	415	675	44.8	160.6
2004-05	345	549	204	422	685	44.5	161.0
2005-06	343	549	206	424	663	45.3	157.9
2006-07	345	554	209	420	656	45.8	158.1
2007-08	341	554	214	414	633	46.7	157.3
2008-09	340	570	230	414	623	47.8	158.3
2009-10	338	567	228	398	600	48.6	158.4
2010-11	336	562	226	387	608	48.0	161.7
2011-12	339	555	215	375	577	49.0	158.5
2012-13	338	554	216	369	570	49.3	159.2

Contd...

Annexure 3–*Contd...*

State/Union Territory/ Year	Net Irrigated Area	Gross Irrigated Area	Area Irrigated more than Once	Net Un-Irrigated Area	Gross Un-Irrigated Area	Percentage of Gross Irrigated Area to Total Cropped Area	Cropping Intensity (Per cent of Col.16 Over Col.15)
(1)	(2)	(3)	(4)	(5)	(6)	(7)	(8)
UTTAR PRADESH							
2003-04	13227	18524	5297	3522	6901	72.9	151.8
2004-05	13119	18939	5820	3564	6585	74.2	153.0
2005-06	13075	18970	5895	3558	6337	75.0	152.2
2006-07	13313	19218	5905	3260	6197	75.6	153.3
2007-08	13085	19142	6058	3332	6178	75.6	154.2
2008-09	13435	19612	6177	3127	5859	77.0	153.8
2009-10	13383	19354	5970	3205	6086	76.1	153.4
2010-11	13440	19626	6185	3152	5989	76.6	154.4
2011-12	13809	19901	6092	2814	6038	76.7	156.0
2012-13	13929	20191	6262	2635	5630	78.2	155.9
WEST BENGAL							
2003-04 *	3006	5387	2380	2421	4275	55.8	178.0
2004-05	3182	5339 *	2157	2192	4184	56.1	177.2
2005-06 *	3135	5501	2366	2160	4032	57.7	180.0
2006-07 *	3136	5581	2445	2160	4053	57.9	181.9
2007-08 *	3136	5669	2534	2160	4082	58.1	184.1
2008-09 *	3135	5651	2517	2159	4150	57.7	185.1
2009-10 *	3112	5525	2413	2144	4005	58.0	181.3
2010-11 *	2955	5194	2239	2026	3638	58.8	177.3
2011-12 *	3078	5437	2359	2120	3916	58.1	179.9
2012-13 *	3082	6105	3023	2123	3573	63.1	185.9
ANDMAN and NICOBAR ISLAND							
2003-04			0	13	15	0.0	113.3
2004-05			0	13	14	0.0	111.7
2005-06			0	13	14	0.0	111.5
2006-07	0	0	0	13	14	0.2	111.8
2007-08	0	0	0	14	16	1.2	112.7
2008-09	0	0	0	14	16	1.2	113.6
2009-10	0	0	0	15	16	1.9	112.4
2010-11 *	0	0	0	15	19	1.7	128.8
2011-12 *	0	0	0	15	18	1.8	121.4
2012-13 *	0	0	0	15	24	1.3	166.8

Contd...

Annexure 3–*Contd...*

State/Union Territory/ Year	Net Irrigated Area	Gross Irrigated Area	Area Irrigated more than Once	Net Un-Irrigated Area	Gross Un-Irrigated Area	Percentage of Gross Irrigated Area to Total Cropped Area	Cropping Intensity (Per cent of Col.16 Over Col.15)
(1)	(2)	(3)	(4)	(5)	(6)	(7)	(8)
CHANDIGARH *							
2003-04	1	1	0	0	1	66.4	147.5
2004-05	1	1	0	0	1	67.9	150.0
2005-06	1	1	0	0	1	69.3	151.5
2006-07	1	1	0	0	1	69.3	151.5
2007-08	1	1	0	0	1	69.3	151.5
2008-09	1	1	0	0	1	69.3	151.5
2009-10	1	1	0	0	1	69.3	151.5
2010-11	1	1	0	0	1	69.3	151.5
2011-12	1	1	0	0	1	69.3	151.5
2012-13	1	1	0	0	1	69.3	151.5
DADAR and NAGAR HAVELI							
2003-04	7	7	0	15	23	23.3	133.2
2004-05	7	7	0	16	22	24.3	126.8
2005-06	7	7	0	15	20	26.7	126.3
2006-07	7	7	0	14	21	25.9	136.0
2007-08	8	8	0	13	19	28.2	132.9
2008-09	4	7	3	16	19	27.2	128.2
2009-10	4	7	3	14	16	31.3	128.1
2010-11	4	7	3	13	15	33.3	124.8
2011-12	4	7	3	16	18	29.3	121.4
2012-13	4	7	3	16	17	30.0	121.0
DAMAN and DIU							
2003-04				2	2	0.0	100.0
2004-05				2	2	0.0	100.0
2005-06				2	2	0.0	100.0
2006-07				2	2	0.0	100.0
2007-08				2	2	0.0	100.0
2008-09				3	3	0.0	100.0
2009-10				3	3	0.0	100.0
2010-11				3	3	0.0	100.0
2011-12				3	3	0.0	104.7
2012-13				3	3	0.0	122.6

Contd...

Annexure 3–*Contd...*

State/Union Territory/ Year	*Net Irrigated Area*	*Gross Irrigated Area*	*Area Irrigated more than Once*	*Net Un-Irrigated Area*	*Gross Un-Irrigated Area*	*Percentage of Gross Irrigated Area to Total Cropped Area*	*Cropping Intensity (Per cent of Col.16 Over Col.15)*
(1)	*(2)*	*(3)*	*(4)*	*(5)*	*(6)*	*(7)*	*(8)*
DELHI							
2003-04	25	31	6	2	14	69.3	165.0
2004-05	25	34	9	3	13	72.5	167.6
2005-06	24	32	8	3	12	73.1	159.2
2006-07	22	31	10	1	12	72.6	185.9
2007-08	22	32	10	1	12	72.7	189.7
2008-09	23	31	8	0	12	72.8	187.8
2009-10	22	34	12	0	12	74.1	204.3
2010-11	22	32	10	0	12	72.6	200.9
2011-12 *	22	30	8	0	9	76.5	175.1
2012-13 *	22	33	11	0	16	67.4	220.2
LAKSHADWEEP							
2003-04	0	0	0	3	3	0.0	105.2
2004-05	0	0	0	3	3	0.0	105.2
2005-06	0	0	0	2	3	0.0	118.6
2006-07	0	0	0	2	3	0.0	118.6
2007-08	0	0	0	2	3	0.0	118.6
2008-09	0	0	0	2	3	0.0	118.6
2009-10	0	0	0	2	3	0.0	118.6
2010-11	0	0	0	2	3	0.0	130.4
2011-12	0	0	0	2	3	0.0	130.4
2012-13	0	0	0	2	3	0.0	146.8
PUDUCHERRY							
2003-04	17	31	14	3	6	84.1	181.1
2004-05	18	32	14	3	7	81.9	184.5
2005-06	18	29	11	3	7	80.7	175.7
2006-07	18	28	10	3	8	78.4	175.5
2007-08	17	27	10	3	8	77.7	174.6
2008-09	16	27	11	3	6	81.7	170.6
2009-10	16	27	11	3	5	83.1	170.3
2010-11	15	25	11	4	6	81.0	169.3
2011-12	15	22	6	3	6	79.5	151.6
2012-13	14	22	8	2	5	82.5	161.6

Contd...

Annexure 3–*Contd...*

State/Union Territory/ Year	*Net Irrigated Area*	*Gross Irrigated Area*	*Area Irrigated more than Once*	*Net Un-Irrigated Area*	*Gross Un-Irrigated Area*	*Percentage of Gross Irrigated Area to Total Cropped Area*	*Cropping Intensity (Per cent of Col.16 Over Col.15)*
(1)	*(2)*	*(3)*	*(4)*	*(5)*	*(6)*	*(7)*	*(8)*
ALL INDIA							
2003-04	57057	78042	20985	83651	111619	41.1	134.8
2004-05	59229	81078	21849	81413	110025	42.4	135.9
2005-06	60837	84280	23442	80325	108457	43.7	136.5
2006-07	62744	86753	24009	77079	105628	45.1	137.6
2007-08	63189	88058	24869	77826	107164	45.1	138.4
2008-09	63638	88896	25258	78261	106432	45.5	137.7
2009-10	61936	85085	23149	77238	103918	45.0	135.8
2010-11	63657	88887	25231	77907	108676	45.0	139.6
2011-12	65693	91730	26037	75281	103902	46.9	138.8
2012-13	66103	92575	26472	73829	101824	47.6	138.9

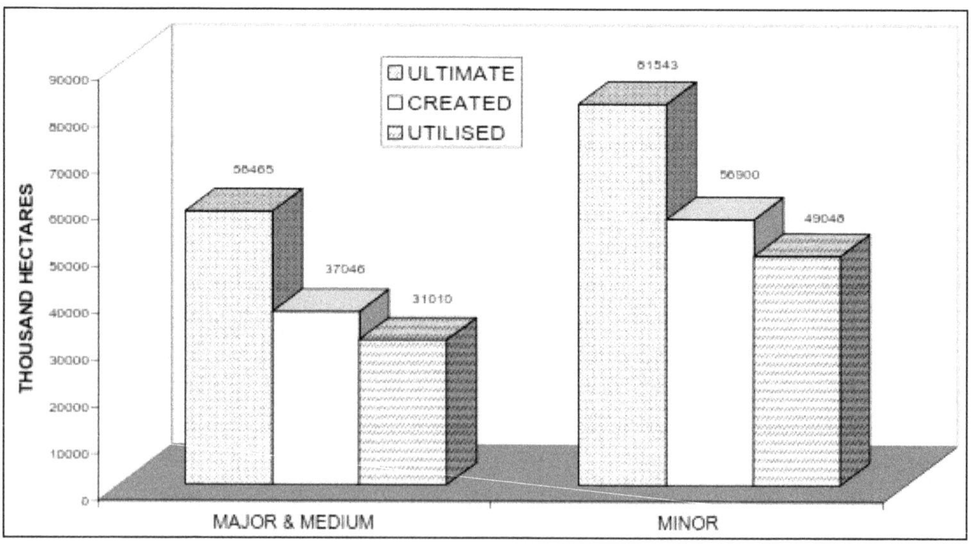

Irrigation Potential - Ultimate, Created and Utilized upto March 2002.

Annexure 4: Plan-wise Irrigation Potential Created and Utilised in India (Cumulative)

(Unit : 000 Hactare)

Sl.No.	*Period*	*Major and Medium Irrigation Surface Water*		*Minor Irrigation Surface and Ground Water*		*Total (Major, Medium and Minor)*	
		P	*U*	*P*	*U*	*P*	*U*
1	*2*	*3*	*4*	*5*	*6*	*7*	*8*
1.	Pre-plan (Upto 1951)	9705	9705	12901	12901	22606	22606
2.	First Plan(Upto 1951-56)	12191	10985	14060	14060	26251	25045
3.	Second Plan(Upto 1956-61)	14334	13052	14731	14731	29065	27783
4.	Third Plan(Upto 1961-66)	16565	15175	17000	17000	33565	32175
5.	Annual Plans(Upto 1966-69)	18095	16751	19020	19020	37115	35771
6.	Fourth Plan (Upto 1969-74)	20703	18688	23400	23400	44103	42088
7.	Fifth Plan(Upto 1974-78)	24717	21163	27300	27300	52017	48463
8.	Annual Plans(Upto 1978-80)	26612	22645	30000	30000	56612	52645
9a.	Sixth Plan(Upto 1980-85)-Original	30013	25330	37520	35248	67533	60578
9b.	Sixth Plan(Upto 1980-85) Reappraised	27695	23574	37520	35248	65215	58822
10.	Seventh Plan (Upto 1985-90) $	29920	25467	46605	43120	76525	68587
11.	Annual Plans (Upto 1990-92)	30741	26315	50348	46538	81089	72853
12a	Eighth Plan (Upto 1992-97) Original	32954	28410	57893	52790	90847	81200
12b.	Eighth Plan (Upto 1992-97) Reappraised @	32954	28410	62479	48289	95433	76699
13.	Ninth Plan (Upto 1997-02)	37046	31010	56900	49048	93946	80058
14	Tenth Plan(Upto 2002-07) (Target)	46972	37198	76271	N.A	123243	N.A

Source: Central Water Commission (P and P Directorate), Ministry of Water Resources

P: Potential Created

U: Potential Utilised (Minor Irrigation Division) and Planning Commission.

Note: $ Cumulative achievement for VII Plan and onward for Major and Medium Irrigation are based on reappraised figures at the end of VI Plan. Minor Irrigation for Annual Plan 1997-98 and onwards are based on reappraised figures at the end of VIII Plan.

@ Figures in Col. 5 and Col.6 for Minor Irrigation based on 2nd Minor Irrigation Census (1993-1994) hence revised. Totals may not tally due to rounding off.

** Figures from Planning Commission.

Annexure 5: State-wise and Plan-wise Achievement of Irrigation Potential Created/ Utilized - Major and Medium Irrigation (Surface Water) (Cumulative)

(Unit : '000)

Sl.No.	State/UTs	Ultimate Potential	Eighth Plan 1992-97		IX Plan 1997-2002	
			P	U	P	U
1	2	3	4	5	6	7
1.	Andhra Pradesh	5000	3045	2884	3303	3052
2.	Arunachal Pradesh	0	0	0	0	0
3.	Asom	970	197	138	244	174
4.	Bihar	6500	2803	2324	2680	1715
5.	Chhatisgarh	Included in M.P.	923	761		
6.	Goa	62	13	12	21	15
7.	Gujarat	3000	1350	1200	1430	1301
8.	Haryana	3000	2079	1834	2099	1850
9.	Himachal Pradesh	50	11	6	13	8
10.	Jammu and Kashmir	250	174	148	180	169
11.	Jharkhand	Included in Bihar	354	230		
12.	Karnataka	2500	1666	1472	2121	1845
13.	Kerala	1000	513	464	609	559
14.	Madhya Pradesh	6000	2318	1621	1387	876
15.	Maharashtra	4100	2337	1288	3239	2147
16.	Manipur	135	63	52	156	111
17.	Meghalaya	20	0	0	0	0
18.	Mizoram	0	0	0	0	0
19.	Nagaland	10	0	0	0	0
20.	Odisha	3600	1558	1443	1827	1794
21.	Punjab	3000	2513	2451	2542	2486
22.	Rajasthan	2750	2274	2088	2482	2314
23.	Sikkim	20	0	0	–	–
24.	Tamil Nadu	1500	1546	1545	1549	1549
25.	Tripura	100	2	2	5	5
26.	Uttar Pradesh	12500	7043	6114	7910	6334
27.	Uttarakhand	Included in U.P.	280	185		
28.	West Bengal	2300	1433	1315	1683	1527
	Total States	58367	32935	28401	37040	31006
	Total U.Ts.	98	19	9	7	4
	All India Total	**58465**	**32954**	**28410**	**37046**	**31010**

Source: Central water Commission (P and P Dte.) and Planning Commission.

Note: P: Potential Created; U : Potential Utilised.

Total may not tally due to rounding off.

3

Hundred Years of Co-operative

The evolution of Co-operative in the world can be traced from time immemorial, beginning from the day individual first joined hands for the advancement of common pursuits in life. Modern Co-operative Around the world modern cooperatives have developed for over 200 years. Cooperative institutions exist all over the world providing essential services which would otherwise be unattainable. In many Third World countries, cooperatives such as credit unions and agricultural organizations have been very successful in helping people to provide for themselves where private and other corporate capitals do not see high profitability. In 90 countries of the world, over 700 million individuals are members of cooperative institutions. Globally, cooperatives have been able to elevate its position as a powerful economic model. In some countries they are a sizeable force within the national economy.

Movement is generally dated from the foundation of the Rochdale Equitable pioneers at the end of 1844. The structure and rules of the Rochdale Co- operative Society of weavers formed the model for countless successors not only in England but also in other countries and continents.

This paper intends to analyze the problems prospects of cooperative sector in India under free marketism. It is now increasingly recognized that the co-operative system in India has the capacity and potentiality to neutralize the adverse effects emerging from the process of globalization. After economic liberalization under the new economic environment, cooperatives at all levels are making efforts to reorient their functions according to the market demands. The failure of the public sector in several cases is a worrisome trend. Privatization has also failed to make an impact in the rural areas. Therefore there is great hope on the cooperative sector. The paper examines the causes of slow progress and highlights the emerging role and challenges of the cooperative sector. In comparison to the step-motherly treatment of the past, cooperatives are now considered an important plank of development. The government is committed to cooperative development. The cooperatives have inherent advantages in tackling the problems of poverty alleviation, food security and employment generation. Cooperatives are also considered to have immense potential to deliver goods and services in areas where both the state and the private sector have failed. The paper focuses on several pitfalls and shortcomings like: poor infrastructure, lack of quality management, over-dependence on government, dormant membership, non-conduct of elections, lack of strong human resources policy, absence of professionalism, *etc.* It suggests to evolve strong communication and public relations strategies which can promote the concept of cooperation among the masses. It should also push forward by developing effective strategies to overcome existing weaknesses and for continuing growth of the sector. The paper makes an assessment of the future prospects of the cooperative sector of India.

Nicholson a British Officer in India suggested to introduce Raiffersen model of German agricultural credit Cooperatives in India. As a follow-up of that recommendation, the first Cooperative Society Act of 1904 was enacted to enable formation of "agricultural credit cooperatives" in villages in India under Government sponsorship. With the enactment of 1904 Act, Cooperatives were to get a direct legal identity as every agricultural Cooperative was to be registered under that Act only. The new Act provided formation of Cooperative societies other than credit. In 1942, the British Government enacted the Multi-Unit Cooperative Societies Act, 1942 with an object to cover societies whose operations are extended to more than one state. The impulses of the Indian freedom movement gave birth to many initiatives and institutions in the post independence era in India and armed with an experience of 42 years in the working of Multi Unit Cooperative Societies and the Multi-Unit Cooperative Societies Act, 1942, the Central Government enacted a comprehensive Act known as Multi State Cooperative Societies Act,1984,repealing the Act of 1942.

The movement has been acknowledged as an instrument towards achieving the socio- economic transformation of Indian society with special focus on the life of people living in rural areas. Co-operation since inception has been striving hard to usher socio-economic development of our country. It is with the concerted and sustained efforts over decades by the Co-operatives that they have about 21 crores membership covering 67 per cent of rural households and 99 per cent of the villages in our country. With this backdrop an attempt has been made to take the stock of its progress to examine what has been achieved more importantly, what could have been achieved and the emerging challenges before it in the 21st century.

Progress of Co-operatives at a Glance

Indian Co-operative Movement was basically organized against the exploitation of unscrupulous money-lenders to exonerate the farming community from the web of poverty and indebtedness. The Government took lot of measures to improve the conditions of the farming sector and as such promoted Co- operative Credit Societies in the light of Raiffeisen model credit societies on the basis of recommendation of Sir Fredrick Nicholson 1889. Now co-operatives and formal legal entities under a statute have been in existence for more than a hundred years and 2004 was the centenary year for Co-operatives.

Types of Cooperatives

The Co-operative Movement was introduced into India by the Government as the only method by which the farmers could overcome their burden of debt and keep them away from the clutches of the money-lenders. The Co-operative Credit Societies Act,1904 was passed by the Government of India and rural credit societies were formed. Through the appointment of registrars and through vigorous propaganda, the Government attempted to popularize the Movement in the rural areas. Within a short period, the Government realized some of the shortcomings of the 1904 Act and, therefore, passed a more comprehensive Act, known as the Co-operative Societies Act of 1912. This Act recognized non-credit societies also. But the rural credit societies have continued to be predominant till now.

(1) The Primary Agricultural Credit/Service Societies

The agricultural co-operative credit structure in the Punjab State is broadly divided into two sectors, one dealing with the short-terms and medium-terms finance and the other with the long-term credit. In the State, the short-term and medium-term credit structure is based on a three-tier system, *i.e.*, the Apex Co-operative Bank at the State level, the Central Co-Coperative Bank at the district/ tehsil level and the Primary Agricultural Credit Societies at the village level. The major objectives of the primary agricultural credit service societies are to supply agricultural credit to meet the requirements of funds for agricultural production, the distribution of essential consumer commodities, the provision of storage and marketing facilities and for light agricultural implements and machinery.

Owing to an increasing emphasis on the development of land and agriculture, long- term co-operative credit has assumed great importance. There is the Punjab State Land Mortgage Bank at the Apex and the Punjab Mortgage Bank at the district/ tehsil level. These Primary Land Mortgage Banks advance loans to the farmers for long term purposes. At the operational level, there exists a primary co-operative to extend credit to the farmer. This unit epitomizes the vitality and service potential of the Co-operative Movement in India. The organization of these societies dates back to 1904, when the first Co-operative Societies Act was passed. These societies were started with the object of providing cheap credit to the agriculturists in order to free them from the clutches of the rapacious money-lenders. the agricultural primary credit society is the foundation-stone on which the whole co-operative edifice is built. Even now these societies dominate the co-operative picture.

The first Agricultural Credit Society in the Firozpur District was registered on 4 October 1911, at the Village of Khalchi Kadim in the Firozpur Tehsil. Originally, the movement was confined to the credit societies only and, thus, credit dominated till the partition (1947). After the partition, the Co-operative Movement began to spread to other field, viz labour, construction and farming. In 1979-80, the number of agricultural co- operative credit societies in the District was 309 with a membership of 1,21,761. the loan advanced during the same year amounting to Rs. 2,180.35 lakh and the deposits to Rs. 26.84 lakh.

(2) Agricultural Non-Credit Societies

While credit is and must remain for some time the chief concern of the Co-operative Movement relatively slow, since 1912, when the non-credit societies were brought officially under the aegis of the Movement. the World War II (1939-45) came as a God send boon with respect to the development of the Co- operative Movement. Prices of agricultural goods began to rise and touched new peaks. The repayment of loans was accelerated and deposits began to pour in. The number of societies also rose. Another interesting development in co-operative during the War wast the extension of the Movement to non-credit activities, *viz.* consumer's co-operative marketing societies, consolidation societies, *etc.* The number of agricultural non-credit societies in the district was 38 in 1978-79.

(3) Agricultural Co-operative Marketing Societies

Marketing has occupied a far smaller place in the co-operative picture in India than in many countries, notably Denmark and the USA, but not other non-credit line of co-operation, with the possible exception of the consolidation of land holdings and joint farming enterprises, seems to hold greater possibilities of help to the agricultural population of India. The development of co-operative marketing in India is closely bound up with the problem of credit-the claims of the money-lenders commonly inhibiting the cultivator's freedom of action in disposing of his crop. The full utilization of loans advanced depends upon the arrangements for the marketing of surplus produce. For this purpose, there the Punjab State Marketing Federation at the State Level, wholesale societies at the district level and marketing societies at the market level. These societies also provide other agricultural facilities and make arrangements for the supply of domestic items in the rural areas.

In 1979-80, there were 10 agricultural co-operative marketing societies in the district, with a total membership of 5,982 (5,056 individuals and 926 societies). During the same year, their paid-up capital was Rs. 19.70 lakh and the working capital Rs. 558.52 lakh, and they marketed goods worth Rs. 992.08 lakh. At the State level, the Punjab State co-operative Supply and Marketing Federation (MARKFED) is playing an important role in building up an integrated structure for remunerative marketing and storing of agricultural produce. it has played an important role in hastening the Green Revolution in the State by arranging ready supplies of essential farm inputs needed by the cultivators.

(4) Co-operative Farming Societies

The Royal Commission on Agriculture in 1928 observed that it co-operation failed, there would fail the hope of the Indian agriculturist. Co-operative farming is a compromise between collective farming and the peasant proprietorship and gives all merits of large-scale farming without abolishing private property. It implies an organization of the farmers on the basis of common efforts for common interests. Under this system, all landowners in a village form a co-operative society for tilling the land. The land is pooled, but each farmer retains the right of property. The produce is distributed by each. They are allowed to withdraw from the co- operative farm whenever they desire. In India, the exceedingly small size of holdings is perhaps the most serious defect in our agriculture. If agriculture has to be improved, the size of the holdings must be enlarged. The co-operative farming societies, thus, enable the cultivators to enjoy the economies of large-scale farming through the pooling of land management resources.

Growth of Cooperative Sector in India

India has basically an agrarian economy with 72 per cent of its total population residing in rural areas. The rural people need lot of services in daily life which are met by village co-operative societies. The seeds of cooperation in India were sown in 1904 when the first Cooperative Societies. During all these the cooperative movement has made significant progress. Cooperatives have extended across the entire country and there are currently an estimated 230 million members nationwide. The cooperative credit system of India has the largest network in the world and cooperatives have advanced more credit in the Indian agricultural sector than commercial banks. The village co- operative societies provide strategic inputs for the agricultural sector, consumer societies meet their consumption requirements at concessional rates; marketing societies help the farmer to get remunerative prices and co-operative processing units help in value additions to the raw products *etc.* In addition, co-operative societies are helping in building up of storage go-downs including cold storages, rural roads and in providing facilities like irrigation, electricity, transport and health. Various development activities in agriculture, small industry marketing and processing, distribution and supplies are now carried on through co-operatives. In fertiliser production and distribution the Indian Farmers Fertiliser Cooperative (IFFCO) commands over 35 per cent of the market.

In the production of sugar the cooperative share of the market is over 58 per cent and in the marketing and distribution of cotton they have a share of around 60 per cent. The cooperative sector accounts for 55 per cent of the looms in the hand-weaving sector. Cooperatives process, market and distribute 50 per cent of edible oils. Dairy cooperatives operating under the leadership of the National Dairy Development Board and through 15 state cooperative milk marketing federations has now become the largest producer of milk in the world. The groundwork for this was laid in the early 1970's when the largest dairy development programme in the world - Operation Flood - was launched. Operation Flood was a national marketing strategy linked to a dairy infrastructure development programme that created a chain of dairy processing plants, collection stations and a national milk transportation grid. With the passage of the Insurance Act, cooperatives have been allowed to entry into the insurance business. Insurance is a field where the immense potential of cooperatives still remains untapped. The Indian Farmers Fertiliser Cooperative has recently teamed up with a Japanese company and formed a joint venture for undertaking general insurance business in India. This signifies that Indian cooperatives have come of age in formulating strategic alliances.

Thus the co-operative societies in India in fact are playing multi-functional roles both in rural and urban areas. An Expert Group constituted by the Govt. of India in 1990 recommended i) to facilitate building up of integrated co-operative structure; ii) to make the co-operative federation organizations responsive towards their members; iii) to minimize government interference and control in the functioning of co-operatives and iv) to eliminate politicization from the cooperative sector. Based on the recommendations, the central govt. enacted the Multi State Cooperative Societies Act, 2002 which provided for democratic and autonomous working of the Cooperatives, which came into force with effect from August 19, 2002. Various development activities in agriculture, small industry marketing and processing, distribution and supplies are now carried on through co-operatives. The co-operatives in India have made an all-round progress and their role in, and contribution to agricultural progress has particularly been significant.

Structure and Significance of Cooperatives

The cooperative structure in India consists of different constituents. At the bottom of this structure are the primary societies which render various types of services. Of this large number about 80 per cent is concerned with agriculture. Most of these societies, about 60 per cent deal with credit only. They perform various functions such as credit, irrigation, marketing, transports *etc.* These are generally divided into two groups(i) credit societies and (ii) non credit societies. The National Cooperative Union of India is the apex organization promoting the cooperative movement in the country.

Cooperation in a vast country like India is of great significance because : It is an organization for the poor, illiterate and unskilled people

- ☆ It is an institution of mutual help and sharing
- ☆ It softens the class conflicts and reduces the social cleavages
- ☆ It reduces the bureaucratic evils and follies of political factions
- ☆ It overcomes the constraints of agricultural development
- ☆ It creates conducive environment for small and cottage industries

The statistics here indicates that modern cooperative movement has made tremendous progress in every walk of its activities and occupies a major place in the share of the national economy.

One is reminded by the findings of the All India Rural Credit Survey Committee 1954 that "Cooperation has failed but it must succeed" and the comments of Royal Commission on Agriculture "if the Co-operatives fail, this will be a failure for best hope of India." It is, therefore, planner of post-independence era gave thrust to the co-operative movement, and the then Prime Minister Pandit Jawaharlal Nehru wanted India to be convulsed with Co-operatives.

The share of Co-operatives in National economy is as follows:

Rural Net work (villages covered)	100 per cent
Agricultural Credit disbursed by Co-operatives	46.15 per cent
Fertilizer disbursed (6.049 million Tonnes)	36.22 per cent
Fertilizer production (3.293 M.T. - N and P) Nutrient	27.65 per cent
Sugar produced (10.400 million tonnes)	59.0 per cent
Capacity Utilization of Sugar Mills	111.5 per cent
Wheat Procurement (4.50 million toones)	31.8 per cent
Animal Feed Production/Supply	50 per cent
Retail Fair Price Shops (Rural + Urban)	22 per cent
Milk Procurement to Total Production	7.44 per cent
Milk Procurement to Marketable surplus	10.5 per cent
Ice Cream Manufacture	45 per cent
Oil Marketed (Branded)	50 per cent
Spindlage in Co-operatives (3.518 million)	9.5 per cent
Cotton Marketed/Procurement	NA
Cotton yarn/Fabrics Production	23.0 per cent
Handlooms in Co-operatives	55.0 per cent
Fishermen in Co-operatives (Active)	21 per cent
Storage Facility (Village level PACS)	65.0 per cent
Rubber processed and marketed	95.0 per cent
Arecanut processed and marketed	50 per cent
Direct employment generated	1.07million
Self-Employment generated for persons	14.39million
Salt Manufactured (18,266 Metric Tonnes)	7.6 per cent

Causes of Slow Progress

Despite rapid growth the overall progress of cooperative movement during 100 years of its existence is not very impressive. It is therefore necessary to know the causes o f poor performance of the movement and on that basis take such steps as would promote a faster growth of cooperative movement in India.

a. Government Interference

The cooperative movement in India was initiated in 1904 under the auspices of British government. Right from the beginning the govt. has adopted an attitude of patronizing the movement. Cooperative institutions were treated as if these were part and parcel of the administrative set up of the government. The govt. interference thus became an essential elements in the working of these institutions. As a result people's enthusiasm for the movement did not grow. The movement's independence and self-reliance existed only on paper and files. After attainment of independence in particular after beginning of the planning, some healthy changes in the attitude of the govt. did take place. It was not given proper importance that it deserves in any plan. But even the cooperative movement has not become full-fledged people movement. Even today quite often cooperative societies are imposed upon the people. This does bring about an increase in the membership of the societies. But the spirit of cooperation cannot flower fully in these circumstances. Neither it growth took place according to any plan nor did it become a people's movement. It just grew very slowly and that too haphazardly. It was a state driven institution.

b. Mismanagement and Manipulation

The essence of the cooperative movement is that it gives the farmers the status of shareholders and assures them agricultural, educational and medical facilities. Under the Maharashtra State Cooperatives Act, a minimum of 11 farmers is required to form a cooperative. Today the shareholder membership averages between 15,000 and 25,000 farmers. The relationship between the shareholder farmer and the cooperative is simple - the farmer is committed to contributing a certain amount of cane per season and the mill is bound to take this cane. The strength of the movement was the involvement of the farmers who were shareholders in the sugar mill regardless of the size of their holdings. Over the years, this truly democratic idea got corrupted and farmers with larger holdings grew more powerful. In practice, this altered the power structure of the cooperatives. In the elections to the governing bodies of the sugar factories, money became such a powerful tool that the top posts of chairman and vice-chairman usually went to the richest farmers even though the majority of members were farmers with small- or medium-sized holdings.

c. Lack of Awareness

People are not well informed about the objectives of the movement, the contributions it can make in rebuilding the society and the rules and regulations of cooperative institutions. Unfortunately, no special efforts have been made in this direction. People look upon these institutions as means for obtaining facilities and concessions from the govt. So long as people expect to get something from the govt, they see to it that societies somehow continue to function. Lack of education, dirty politics of the village, caste- ridden elections to the offices of cooperative societies, bureaucratic attitudes of the govt. officers at the lower rank are some of the hurdles in spreading the correct information about the cooperative movement and in educating the people abut its true character and vital role in the society.

d. Restricted Coverage

The cooperative movement has also suffered on account of two important limitations on its working. One is that the size of these societies has been very small. Most of these societies are confined to a few members and their operations extended to only one or two villages. As a result their resources remain limited, which I make it impossible for them to expand their means and extend their are of operations. Two, the most of the societies have been single purpose societies. For this reason these societies are unable to take a total view of the persons seeking help, nor can they analyze and solve problems from different angles. the help these societies render thus can not be adequate. By assessing the persons and the problems only from one angle, these societies neither help properly the person nor make a optimal use of their resources. Under these circumstances it has not been possible for these societies to make much progress.

e. Functional Weakness

The cooperative movement has suffered from inadequacy of trained personnel right from its inception. Lack of trained personnel has been caused by two major factors. In the first place, there has been a lack of institutions fort this purpose of training personnel. Secondly because of it unsatisfactory working of cooperative institutions, efficient personnel did not feel attracted or motivated towards them. The functioning of the cooperative societies,too suffer from several weakness. Some of these are, taking no care of the need of credit seekers or their repaying capacity at the time of granting loans, making no adequate provision for the return of loans, unsatisfactory keeping of accounts, factional politics in it management, lack of coordination among various divisions of the cooperative structure, too much dependence on outside sources of finance,lack of adequate auditing. Such weakness have prevented theme from progressing on healthy lines.

Thus there are several pitfalls. Poor infrastructure, lack of quality management, over- dependence on government, dormant membership, non-conduct of elections, lack of strong human resources policy, neglect of professionalism, *etc.* are the limiting factors. Indian cooperatives are also unable to evolve strong communication and public relations strategies which can promote the concept of cooperation among the masses.

Emerging Issues

Over the last more than one hundred years (1904- 2004) Co-operatives have made substantial contribution to the country, particularly to the rural poor. Notwithstanding their achievements and phenomenal growth, Co-operatives are beset with several issues and confront with challenges in the era of economic reforms which is described as under.

(i) WTO and Co-operatives

In the new millennium, a serious challenge which will be faced by Co-operatives relates to WTO Agreement on Agriculture. Co-operatives are closely attached with agriculture and agro-based activities in the rural areas. Concern is being expressed by members of Co-operative fraternity that the implementation of various provisions of Agreement on Agriculture will seriously affect the farming community in terms of their income levels, levels of living and quality of life. This aspect is particularly important as 70 per cent of our population is dependent on agriculture and agro-based activities. Where as only 4 per cent of population is dependent on agriculture and allided activities in case of developed countries.

Further, it has to be ensured that food security in terms of providing minimum food for various sections of Indian population does not get adversely affected by the WTO Agreement on Agriculture and hence there is the need for providing a separate "Food Security Box" under WTO provisions with emphasis on encouraging on food production and per hectare productivity of various crops. Various segments of Co-operative Movement should be encouraged for improving their operational efficiency and for promoting cost effectiveness and quality standards to face the daunting challenges of competitive culture.

(ii) Participation of Women in Co-operatives

The need for greater participation of women in cooperatives especially in areas where they have a natural advantage is accepted by all. It is known that under the system of adult franchise prevailing in India, women have equal opportunities for voting and electing their representatives in Central and State Legislatures as also in local bodies and panchayats. Central and State Governments are also extending various concessions for promoting education among girls and women. However, despite these favourable developments a majority of women in our country are still illiterate and do not have access to various resources including credit from banking channels.

For the uplift of women the Self Help Group (SHGs) is in the tune with Co-operatives. Their conversion to co-operative form of organization will go a long way in empowering women in view of the vibrant democratic structure of the Co-operatives. Micro finance through SHGs by involving more and more women will be important task in the 21st century.

(iii) Professionalization of Co-operatives

Co-operatives have lost their democratic character and have become the government controlled bureaucratic organizations. The cooperative societies must evolve as independent self-reliant, autonomous and member-driven institutions. Rightly Capoor Committee (2000) has recommended on these lines besides professionalization, human resource development (HRD) business diversification recovery management fund mechanism and setting up of a cooperative rehabilitation and development fund. The committee also cautions against excessive control and regulation.

Central Government along with its State Counterparts has taken initiative to strengthen infrastructure facilities and to develop professional skills in the cooperatives. The Co-operative Education and training programmes being implemented by National Co-operative Union of India (NCUI) and National Council for Co-operative Training (NCCT) and Junior Level Co-operative Training Centres. But still there is a feeling all over the need for professionalization. The cooperatives professional may have to redefine their roles in the light of the emerging knowledge society. Information Technology (IT) has assumed such high importance that ICMs should now turn their focus on I-Business, E-Commerce, E-Governance CRM *etc.* To enable them to acquire competitive ability in terms of quality, cost, timeliness, availability, service and support, total quality management (TQM) are the specific areas that would improve Co-operative during the dawn of new millennium.

(iv) Co-operative Law Reforms

As per recommendation Chowdhury Brahma Perkash Committee that model Co-operative Soceities Act 1991 provided scope to enable cooperators to develop self- reliant cooperatives with much autonomy and participation in democratic management. Accordingly parallel State Co-operative Law/self-Reliant Co-operatives was first enacted in A.P. in the year 1995 followed by MP, Chhatisgarh, Karnatak, Odisha, Bihar, Jharkhanda, J&K and Uttaranchal. The attempt is a reflection of reforms in the Co-operative sector sweeping across the world. Self Reliant Co-operatives (SRC) seeks to restore to all those Co-operatives not dependent on the government equity, and share. Thus it is an timely attempt to make the cooperatives more a member driven for which all possible care may be taken to materialize the goal of the Co-operative law reforms. Another landmark in reforms in Co-operative law is the amendment of Multi State Co-operative Societies Act 2002 the fifth Central Legislation on Co-operatives enacted in the 51st year of Republic of India and aimed at removing the restrictive provisions of Multi State Co-operative Societies Act (MSCS) 1984. The Government's power to give directions and supersession of Board have been restricted to such societies in which the Government holds 51 per cent or more. The societies will be free to raise resources by receiving deposits, raising loans and grants. The federal Co-operatives would have more responsibilities and the disputes would be settled through arbitration at the choice of the society.

Further NCDC Amendment Bill, 1995 has been passed during 2002 which was pending since 1995. The Amendment has since been enforced with effect from 16.9.2002. Now NCDC shall be financing more activities allied to agriculture, like animal husbandry products, such as milk, meat, skin, wool etc to help augment the earnings in rural areas. Similarly, service sector and industrial Co-operatives including goods produced by the Cottage and Village Industries will come within the purview of the NCDC. The NCDC would also be able to undertake direct financing of Co-operatives against reasonable security without State/Central Government guarantee.

(v) National Policy in Co-operatives

Internal and structural weaknesses of cooperatives, wide regional imbalances combined with lack of proper policy support had neutralized their positive impact. This had necessitated the need for a clear-cut National Policy on cooperatives. The objectives and salient features of the comprehensive National Policy on Co-operative (April 2002), announced by Government of India is as follows. Under this policy, cooperatives would be provided necessary support, encouragement and assistance so as to ensure that they work as autonomous, self-reliant and democratically managed institutions accountable to their members and make a significant contribution to the national economy. The policy aims at ensuring the functions of cooperatives based on the Manchester Declaration of International Co-operative Alliance 1995 (voluntary and open membership, democratic member control, members' economic participation, autonomy and independence, education, training and information, co operation among co-operatives and concern for community).

- ☆ While upholding the values and principles of Cooperation, the National Policy recognizes the Co-operatives as autonomous association of persons, united voluntarily to meet their common economic, social and cultural needs and aspirations through a jointly owned and democratically controlled enterprise;
- ☆ Upholds the preservation of the distinct identity of Co-operatives, its values and principles by providing an appropriate environment and taking the required administrative and legislative measures;
- ☆ Recognizes Co-operatives as a distinct economic sector and integral component of the socio-economic system of the country and an effective and potential instrument of socio- economic development.
- ☆ Accepts the need to phase out its share holdings/equity participation in the cooperatives. The Co-operative shall be enabled to set up holding companies/subsidiaries, enter into strategic partnership, venture into futuristic areas like insurance, food processing and information technology *etc.*, and shall be independent to take the financial decisions in the interest of the members and the furtherance of their stand.
- ☆ Recognizes the role of the Government in ensuring that the benefits of liberalization and globalisation in the emerging special provision in the Co-operative Societies Act with regard to banking, housing, real estate development, processing, manufacturers Co-operatives, infrastructure development *etc.*;
- ☆ Undertakes to devise and execute suitable programmes and schemes to build and develop co-operative institutions in the cooperatively under developed states/regions with particular reference to the North Eastern States including Sikkim;
- ☆ Recognizes the support of Co-operative Movement to develop human resources, Co-operative education and training, appropriate technologies and infrastructural facilities so as to promote professional management in Co-operatives.
- ☆ Undertakes to initiate structural reforms in order to improve the functioning of the Co-operatives at various levels to ensure greater efficiency and viability.

The Co-operatives in the 21st Century must remain on vanguard in providing the required lead to the millions of our producers. This calls for well managed efficient Co-operative sector and to keep them away from the fears to draconian laws and unwanted interference. The vision for the second century is to withstand the challenges of competitive business environment where excellence, efficiency and high productivity parameters will be given priority. Emphasis will continue to be laid on an improvement with Co-operative governance through the process of restructuring and rejuvenation.

New Directions: Greater Autonomy

During past few years some beneficial and important changes have taken place in the cooperative movement which have given a quite new and progressive slant. A supportive climate has been created for the development of cooperatives as democratic and autonomous businesses providing them with the opportunities for diversification. The cooperatives have long been demanding that the Central Government introduce democratic reforms in their regulation, supervision and functioning. With the persistent efforts of the National Cooperative Union of India, the Central Government recently passed the Multi-State Cooperatives Societies Act and also formulated a national cooperative policy that provides greater autonomy to cooperatives.

The Govt. is providing educated personnel and right type of leadership. Government is adding to their responsibility of these institutions by entrusting them with the work of distributing several commodities. Govt. is seeking its participation in the performance of some national tasks.: like procurement of food grains. The laws relating to cooperatives are being modified to make it qualitatively stronger. The Reserve bank of India now contributes a lot for the promotion of healthy cooperative movement through it s several activities such as supervision, research, training facilities *etc.* As result now there is more cooperation and coordination among different constituents of the cooperative structure.

1. The scope of cooperatives have extended and diversified
2. Emphasis is given on multi purpose societies
3. Maximum number of societies are now run on the basis of limited liabilities
4 The size of the societies is extended on the basis of scientific principle
5 It is perceived as the powerful strategy of economic growth

At present, the Government of India is in the process of formulation of National Policy on Co-operatives which is likely to uphold the values and principles of co-operation recognizing its autonomous characters and attaching priority to professionalism, human resource development and to act as preferred instrument for execution of public policy in rural areas and in sectors where they provide the most effective delivery system. To strengthen their competitive edge in the market total quality control initiatives, management initiatives and cost reduction initiatives will also be taken up. It is now increasingly recognized that the co-operative system in India has the capacity and potentiality to neutralize the adverse effects emerging from the process of globalization and liberalization, and continue to play an important role in employment promotion and poverty alleviation, both as production enterprises – mainly of the self-employed – and as providers of services to members. Although cooperatives are not instruments of employment promotion, they do effectively create and maintain employment in both urban and rural areas and thus provide income to both members and employees in the form of shares of surplus, wages and salaries or profits depending of the type of cooperatives.

Future Prospects: Democractic Reforms Warranted

Reforms are warranted in order to overcome its short comings and consolidate its strengths. Reorganizing the Primary Societies.: The first and foremost improvement concerns the reorganization of primary societies so that they can function properly. For this reorganization three steps are called for: One, The weak and inefficient societies should be winded and merged with strong and efficient societies. It will no doubt reduce the number of societies, but this is itself not a bad thing. Otherwise the weak societies would undermine people's confidence in the movement and cause a set back to it from along term point of view. Two, small societies should be merged to build a larger societies. With large amount of resources at it disposal big societies can avail certain facilities, which small societies can not secure. For example, these societies can afford to engage highly trained and competent personnel and thus conduct their business efficiently. Besides, these societies can to some extent, face some deficits and bear the burden of over dues of their members. Smaller societies can not do either of these. It is for these reasons so many societies in India are idle and dormant. Of course small societies have intimate knowledge about their members. But because of scanty resources these societies neither function efficiently not are able to do much for their members. On the contrary the size should not be too large. It become difficult to keep proper control over the functioning of the society. Therefore Mehta committee suggested for "one village one society". Thirdly, instead of single purpose societies multi purpose societies need to be organized Superficially single –purpose societies appear to be more useful because each one would concentrate on a specified function. But experience suggest that unless problems of the people are viewed and solved in an integrated way, the cooperative movement can make only limited progress. Multi purpose societies can have a balanced view about the needs of its members and can meet them accordingly. As result the resource of such societies are utilized efficiently and at the same time members derive the greatest possible advantages from the resources placed at their disposal. During last few years the cooperative movement has gained much from the formation of such multipurpose societies.

It has to ensure efficiency in it functioning. Therefore the importance measures are needed. Firstly, the loan from credit societies should be granted in such a manner and under such conditions that these are used productively and not misused. It is in this way that the recourses of the movement will be correctly used and their repayment is ensured. It is only then that the problem of over dues can be properly tackled. Secondly, it is necessary to maintain close coordination of activities among different constituents of the cooperative structure: primary societies at the base, organization at the central and state levels should be right type. Their relations and the coordination among their working should be so institutionalized that they work together, mutually helping instead of opposing one another. Similarly there should be close integration among cooperative institutions, Reserve Bank, State Bank, Commercial Banks, and National Bank of Agriculture and Rural Development and the Government.But it does mean that the Govt. or any other institution should adopt a patronizing attitude towards it. But there should be collaboration and coordination to promote the cooperative movement among the people. Thirdly, it is necessary to enlist the services of trained administrators and devoted workers in order to impart efficiency to its working and humanize its functioning. It is also essential that the procedures of work of cooperatives should be made simple, straight and convenient. The hold of the bureaucracy should be done away with.

It is necessary to spread the movement as people's movement. People should not think it as a part of the Govt. or a department of the govt. People should own it and manage it. People should be educated and about the advantages of this movement. Educational institutions at various levels, radio, TV, newspapers, poster, and other mass media can be used for this purpose. Secondly, Govt. participation and support is necessary. But it should not be patronage or interference or domination. It should facilitate its growth in healthy lines. Thirdly, the cooperative movement should be conceived and developed as a part of the national economy. Along with private and public sector, it should be treated as a separate sector in its own right. In fact, of the three sectors, the cooperative sector alone can claim to be peoples sector with human face under globalization framework. Therefore for its growth, more resources should be allocated under the five year plans of the government. Fourthly, there is need for extending the cooperative activities in all the regions of the country. The north –eastern states are lagging behind. Fifthly, there is urgent need for helping the weaker sections and vulnerable groups like SC and ST to form cooperatives in order to be free from the exploitation of middle men. Therefore appropriate concession and facilities should be extended to such societies. Finally there should be research and continuous monitoring an d inspection about the problems and constraints of different cooperative societies. It will help to overcome the difficulties and proceed further with success. Cooperative movement will be an important impetus for rapid rural development of the country.

After independence, India marched ahead with central planning in order to achieve economic development. The new economic policy of globalization after eighties has reduced the dominant role of the state. The pendulum of economic growth in India is swinging from over commitment of public sector to over enthusiasm of the private sector. The urban sector is growing impressively very fast in India. The growth rate of service sector has become very impressive but the agricultural sector has lagged behind. The problem of poverty and unemployment among common people poses a real dilemma between the role of the State and role of the market forces. Market forces fail to unleash forces for growth and at the most it tends to generate low-income activities horizontally. This process fits perfectly well with an investment pattern and production structure to subserve a social milieu and life style which has no relevance for the broad mass of the rural people. It has created a dangerous process of marginalization and contractualization. As a result the economic vulnerability and livelihood insecurity manifests. There is inherent urban bias in the formulation of macro adjustment policies. It is creating a system which is ultimately as centralized and undemocratic as any that can be imposed by the meddlesome bureaucracy. Its success can at best be treated as a pyrrhic victory, when means win and ends lose. Market is never friendly to the poor people or backward regions. It creates sharp regional imbalance. Majority of the rural sector and the unskilled and illiterate section of the urban sector tend to lose their right to livelihood and economic citizenship. Vicious circle of poverty manifests in different forms. There is also the growth of dangerous tendency of the elite to dismiss the deep concern towards poverty and economic vulnerability as the obsession toward egalitarian romantics. They strongly believe myopically in the trickle down theory. There is now increasing indifference towards the problem of poverty and inequality.

Expansionist economic reforms during the and 1990s boosted economic growth but high level of poverty and unemployment persist in India. Despite impressive economic growth, the powerful wave of consumerism, computerization and corruption coexists in social life of India. It puts over emphasis on high-tech efficiency of industrial sector and modern urban service sector at the cost of rural sector, where majority eke out their living. One must always remember one simple thing that any model of development which ignores India's rich endowment of human resources and rich natural resources including land and water resources is bound to falter. Any development route which bypasses the rural people of India is unlikely to be sustainable. Salvation of the developing economy as vast and diverse as India's lies only in the transformation and revitalization of its rural economy, which require people's empowerment and participation. One's sense of idealism is in direct proportion to one's distance from the real scenario. Neither private sector nor public sector shall promote social welfare. India requires a meaningful reforms in the cooperative sector, before complete -opening up its economy to the competitive regime, which should ensure : a). higher standard of living for the village people. b) production for mass consumption. In comparison to the step- motherly treatment of the past, cooperatives should be considered an important plank of development. The cooperatives have inherent advantages in tackling the problems of poverty alleviation, food security and employment generation. Cooperatives have immense potential to deliver goods and services in areas where both the state and the private sector have failed.

Farming Systems

Contract Farming

Contract farming is charting a rapid growth trajectory in India. Within a decade, it has grown by about 15 per cent. Almost 4,000 hectares are annually added to the area under this farming arrangement. While industry is aggressively making forays into this field, farmer leaders fear exploitation. Under contract farming, companies strike a deal with the farmer to grow crops at pre-determined prices or as per the current market index.

These companies, be it food processing firms, hotels or exporters, used to earlier grow sugarcane, cotton, tea, coffee and potato. With changing demands, organic vegetables, grapes and millets like barley and oats are also being grown. Better profit margin and steady supply of crops as the raw material for food processing and textile industries are the reasons for new players to enter the sector. Companies like Field Fresh, Mahindra and Global Agri-system are among the new entrants.

Contract farming will get a big boost when foreign direct investment (FDI) in retail becomes a reality. "Volatility in market prices of crops is a big reason for the processing industry to enter into such arrangements." "While contract farming ensures quality, kept in check by the firm itself, it also provides agricultural products at prices cheaper than the retailer". The farming arrangement helps bring down inflation. "Prices of fruits, vegetables and cereals are high because they exchange five to six hands" member requesting anonymity. Contract farming reduces this chain. "This is why supermarkets can sell farm products cheaper".

For farmers too, it works as a lucrative proposition. Many small and marginal farmers have now entered this sector. But companies find it cumbersome to engage too many farmers to produce one crop. At an average, a farmer holds less than 1.6 hectares. "The government does not allow aggregation of farm land into one block".

Key Benefits of Contract Farming

The key benefits of contract farming for farmers can be summarized as: 1) improved access to local markets; 2) assured markets and prices (lower risks) especially for non-traditional crops, 3) assured and often higher returns, and 4) enhanced farmer access to production inputs, mechanization and transport services, and extension advice.

Additional key benefits for contract partners and rural development often include: 1) assured quality and timeliness in delivery of farmers' products, 2) improved local infrastructure, such as roads and irrigation facilities in sugar outgrower areas, tea roads, dairy coolers/collection centres, *etc.* and 3) lower transport costs, as coordinated and larger loads are planned, as especially important feature in the case of moredispersed producers.

Type of Contract Farming

These are a few of the models of contract farming that are accepted globally.

1. Centralized model
2. Nucleus Estate model
3. Multipartite model
4. Informal model
5. Intermediary model

Centralized model: The contracting company provides support to the production of the crop by smallholder farmers, purchases the crop from the farmers, and then processes, packages and markets the product thereby tightly controlling its quality.; this can be used for crops such as tobacco, cotton, paprika, sugar cane, banana, coffee, tea, cocoa and rubber. This many involve tens of thousands of farmers. The level of involvement of the contracting company in supporting production may vary.

Nucleus Estate model: This is a variation of the centralized model. The promoter also owns and manages an estate plantation (usually close to a processing plant) and the estate is often fairly large in order to provide some guarantee of throughout for the plant. It is mainly used for tree crops, but can also be fore, *e.g.* fresh vegetables and fruits for export.

Multipartite model: The multipartite model usually involves the government, statutory bodies and private companies jointly participating with the local farmers. The model may have separate organizations responsible for credit provision, production, management, processing and marketing of the produce.

Informal model: This model is basically run by individual entrepreneurs or small companies who make simple, informal production contracts with farmers on a seasonal basis. The crops require only a minimal amount of processing or packaging for resale to the retail trade or local markets, as with vegetables, watermelons, and fruits. Financial investment is usually minimal. This is perhaps the most speculative of all contract-farming models, with a risk of default by both promoter and farmer.

Intermediary model: This model formal subcontracting by companies to intermediaries (collectors, farmer groups, NGOs) and the intermediaries have their own (informal) arrangements with farmers. The main disadvantage is this model is it disconnects the link between company and farmer.

Farmers' Concern Related to Contract Farming

Exploitation is what the farmer fears most. In India, contract farming is legal in 18 states and union territories. Leasing out land for farming is legal in Punjab and Rajasthan. "Farmers in Punjab get about Rs. 50,000 per hectare when they lease their land. This is just Rs. 20,000 in Uttar Pradesh, where leasing land is illegal". A farmer is also at the receiving end when food prices drop. During these times, companies reject crops branding them poor in quality. There have been instances when farmers have grown exotic plants that give a company monopolistic control over the market. At these times farmers have received much less than deserved.

"With FDI in retail, exploitation is sure to increase". "The government should make it mandatory for the FDI companies that enter the Indian market to buy 75 per cent of the total retail of agriculture produce directly from farmers". Moreover, 50 per cent agriculture produce should be purchased directly from the farmer from within 100 km of each store. "State governments have failed to formulate strict laws for contract farming to protect farmers". With stringent laws, both the industry and the farmer will stand to gain, in September, CII made a representation to the agriculture minister asking the Centre to formulate a model land leasing act to facilitate contract farming. In its approach paper for the 12th Five Year Plan, the Planning Commission too laid stress on contract farming, saying it is one of the ways to connect the farmer to the market.

Hydroponics

Hydroponics is a subset of hydroculture and is a method of growing plants using mineral nutrient solutions, in water, without soil. Terrestrial plants may be grown with their roots in the mineral nutrient solution only or in an inert medium, such as perlite or gravel.

The earliest published work on growing terrestrial plants without soil was the 1627 book *Sylva Sylvarum* by Francis Bacon, printed a year after his death. Water culture became a popular research technique after that. In 1699, John Woodward published his water culture experiments with spearmint. He found that plants in

Private Sector Companies Involved in Contract Farming in India

Sl.No.	Name of the Company	E-mail ID/ Phone No.	States	Commodity
1.	Appachi Cotton Company Zamin Uthukuli, Pollachi Coimbatore	Ph: 04259 234666, 309666, E-mail: appachi@vsnl.com Mob: 093452 33000	T.N., Karnataka	Cotton
2	AVT Natural Products Ltd 72-E, Chesney Estate Ethiraj Salai Chennai 600008	Ph: 044-28584147 Fax: 044-28584147 E-mail: msa_kumar@ avtspice.com latha_rajesh@avtspice.com Mob: 098460 35959	Karnataka	Marigold, caprice, chilly
3.	Cargill India Pvt Ltd DLF Gateway Tower 11 & 12th Floor, DLF City, Phase III, Gurgaon 122 002	Ph.:95124-5090402 Fax:95124-2358971 Mobile:9810095054 E-mail:hardeep_singh@ cargill.com	M.P.	Wheat, Maize & Soybean
4.	Escorts Ltd Agri Machinery Group 18/4 Mathura Road Faridabad – 121007	Ph: 95129-2250222 Fax: 95129-2284802	Punjab	Basmati
5.	The Global Green Company Pvt Ltd (Naan) 220, Alembic Glass Complex Whitefield Hoskote Road Whitefield Bangalore	Ph: 080-26632831/2796/4059 Fax: 080-26637372 E-mail: rajeev_trehan@ hotmail.com	Karnataka, AP	Gherkin, Babycone, Paprika
6.	Hindustan Lever Ltd 165/166, Backbay Reclamation Mumbai -40002	Ph: 022-22282737 Fax: 022-22855825 E-mail: dalipsehgal@ unilever.com	Madhya Pradesh	Wheat
7.	Ion Exchange EnviroFarms Ltd Neeta Towers Pune-Mumbai Highway Dapodi Pune – 411012	Ph: 020-7145118/7146108 Fax: 020-7146109 E-mail: schellappa@ ionexchange.co.in sreeram_c@hotmail.com	TN, MP, Gujarat, Haryana, Maharashtra	Organic products of Banana, Pineapple, papaya, wheat, basmati, cotton
8.	ITC - IBD ITC Green Centre Ist Floor, Sector 32 10, Industrial Area Gurgaon 122001	Phone: +91-124 - 4173172 Fax: +91-124 - 4173175 Email: jai.oberoi@itc.in Web: www.itc.in	Madhya Pradesh	Soybean

Sl.No.	Name of the Company	E-mail ID/ Phone No.	States	Commodity
9.	Ken Agritech Pvt Ltd Factory & Field Operations #131/1A, Anchatageri Karwar Road Hubli (Karnataka) – 580024	Ph: 0836-2205770; 0836-2202805 E-mail: kenagri@sanchar.net.in	Karnataka	Gherkin
10	Nijjer Agro Foods Pvt Ltd Maharbhan Pura P O Jandiala Guru Dist Amritsar 143115	Ph: 0183-2431705/3091608 Fax: 0183-2431805, 2432599 E-mail: nijjers@glide.net.in	Punjab	Tomato and Chilly
11.	Pepsi Foods Pvt Ltd 3B, DLF Corporate Park S Block, Qutub Enclave, Phase III Gurgaon 122 002	Ph: 95124 - 2355880 Fax: 95124 -2356270	Punjab, TN	Chillies, Groundnut, seaweed, Tomato and Basmati Rice
12.	Rallis India Appejay House, 7th Floor 3 Dinshaw Vaccha Road, Churchgate Mumbai 400020	Ph: 022-56652700 Fax: 022-56352996 E-mail: annahita@rallis.co.in	Punjab, UP, MP, Maharashtra, Karnataka, TN	Basmati, Wheat, Fruits, Vegetables
13.	Suguna Poultry Farm Ltd Corp Off: 5th Floor Jaya Enclave 1057, Avinashi Road Coimbatore -641018	Ph: 0422-5392503/5392504/05 Fax: 0422-5392507 E-mail: sugies@md3.vsnl.net.in	TN, AP	Broiler
14.	Satnam Overseas Ltd Star Trading House 201, VIPPS Centre, 2 Community Centre G K II, Masjid Moth New Delhi 110048	Ph: 51635757 Fax: 51638586/87	Punjab	Basmati
15.	Super Spinning Mills Ltd "ELGI Towers" PB#7113, Green Fields 737-D, Puliakulam Road Coimbatore -641045	Ph: 0422-2311711/2314511 (D) E-mail: 0422-2311611 E-mail: super@sshsaraelgi.com	Tamil Nadu	Cotton
16.	The Ugar Sugar Works Ltd Ugarkhurd Taathani Distt: Belgaum Ugarkhurd 591 316	Ph: 08339-272230 Fax: 08339 -272232	Karnataka	Barley
17.	Unicorn Agrotech Ltd (Zeraim) 1-7-139/3, Sarojini Devi Road Hyderabad – 500003	Ph: 040-27847769 Fax: 022-22061592/27842399/7845924 E-mail: uniagro@hd1.vsnl.net.in	Karnataka	Gherkin
18.	United Breweries UB House 1/1, Vittal Mallaya Road Bangalore 560001	Ph: 080-2272806/7/8 Fax: 080-2274884/5/6 E-mail: cmoblr@ubmail.com	Punjab	Barley
19.	Venkateshwara Hatcheries Ltd., 1-5, Upper Ground Floor, World Trade Centre, Babar Road	Ph: 23413986/87/88 Fax: 2341398	TN	Broiler

less-pure water sources grew better than plants in distilled water. By 1842, a list of nine elements believed to be essential to plant growth had been compiled, and the discoveries of the German botanists Julius Von Sachs and Wilhelm Knop, in the years 1859-65, resulted in a development of the technique of soilless cultivation. Growth of terrestrial plants without soil in mineral nutrient solutions was called solution culture. It quickly became a standard research and teaching technique and is still widely used today. Solution culture is now considered a type of hydroponics where there is no inert medium.

In 1929, William Frederick Gericke of the University of California at Berkeley began publicly promoting that solution culture be used for agricultural crop production. He first termed it aquaculture but later found that aquaculture was already applied to culture of aquatic organisms. Gericke created a sensation by growing tomato vines twenty-five feet high in his back yard in mineral nutrient solutions rather than soil. He introduced the term hydroponics, water culture, in 1937, proposed to him by W.A. Setchell, a phycologist with an extensive education in the classics. Hydroponics is derived from neologism, constructed in analogy to geoponica, that which concerns agriculture, replacing, earth, with, water.

Reports of Gericke's work and his claims that hydroponics would revolutionize plant agriculture prompted a huge number of requests for further information. Gericke had been denied use of the University's greenhouses for his experiments due to the administration's skepticism, and when the University tried to compel him to release his preliminary nutrient recipes developed at home he requested greenhouse space and time to improve them using appropriate research facilities. While he was eventually provided greenhouse space, the University assigned Hoagland and Arnon to re-develop Gericke's formula and show it held no benefit over soil grown plant yields, a view held by Hoagland. In 1940, he published the book, *Complete Guide to Soil less Gardening,* after leaving his academic position in a climate that was politically unfavorable.

Two other plant nutritionists at the University of California were asked to research Gericke's claims. Dennis R. Hoagland and Daniel I. Armon wrote a classic 1938 agricultural bulletin, *The Water Culture Method for Growing Plants Without Soil.* Hoagland and Arnon claimed that hydroponic crop yields were no better than crop yields with good-quality soils. Crop yields were ultimately limited by factors other than mineral nutrients, especially light. This research, however, overlooked the fact that hydroponics has other advantages including the fact that the roots of the plant have constant access to oxygen and that the plants have access to as much or as little water as they need. This is important as one of the most common errors when growing is over- and under- watering; and hydroponics prevents this from occurring as large amounts of water can be made available to the plant and any water not used, drained away, re-circulated, or actively aerated, eliminating anoxic conditions, which drown root systems in soil. In soil, a grower needs to be very experienced to know exactly how much water to feed the plant. Too much and the plant will not be able to access oxygen; too little and the plant will lose the ability to transport nutrients, which are typically moved into the roots while in solution.

These two researchers developed several formulas for mineral nutrient solutions, known as Hoagland solution. Modified Hoagland solutions are still used today.

One of the earliest successes of hydroponics occurred on Wake Island, a rocky atoll in the Pacific Ocean used as a re-fuelling stop for Pan American Airlines. Hydroponics was used there in the 1930s to grow vegetables for the passengers. Hydroponics was a necessity on Wake Island because there was no soil, and it was prohibitively expensive to airlift in fresh vegetables. In the 1960s, Allen Cooper of England developed the Nutrient film technique. The Land Pavilion at Walt Disney World's EPCOT Center opened in 1982 and prominently features a variety of hydroponic techniques. In recent decades, NASA has done extensive hydroponic research for their Controlled Ecological Life Support System or CELSS. Hydroponics intended to take place on Mars are using LED lighting to grow in different color spectrum with much less heat.

Origin

Soilless Culture

Gericke originally defined hydroponics as crop growth in mineral nutrient solutions. Hydroponics is a subset of soilless culture. Many types of soilless culture do not use the mineral nutrient solutions required for hydroponics. Plants that are not traditionally grown in a climate would be possible to grow using a controlled environment system like hydroponics. NASA has also looked to utilize hydroponics in the space program. Ray Wheeler, a plant physiologist at Kennedy Space Center's Space Life Science Lab, believes that hydroponics will create advances within space travel. He terms this as a bio-regenerative life support system.

Aeroponics

Aeroponics is the process of growing plants in an air or mist environment without the use of soil or an aggregate medium (known as geoponics). The word "aeroponic" is derived from the Greek meanings of*aero-* (air) and *ponos* (labour). Aeroponic culture differs from both conventional hydroponics, aquaponics, and in-vitro (plant tissue culture) growing. Unlike hydroponics, which uses a liquid nutrient solution as a growing medium and essential minerals to sustain plant growth; or aquaponics which uses water and fish waste, aeroponics is conducted without a growing medium. Because water is used in aeroponics to transmit nutrients, it is sometimes considered a type of hydroponics.

Aeroponics is a cutting-edge type of hydroponic technology that grows plants in a mist. The aeroponic mist most efficiently provides roots with the nutrients, hydration and oxygen needed, creating faster growing cycles and more biomass than other growing approaches. AeroFarms has designed its aeroponics as a closed-looped system, recirculating our nutrient solution and using over 95 per cent less water than field farming.

Cloth Medium

Aero Farms has developed a patented, reusable cloth medium for seeding, germinating, growing, and harvesting. Cloth has a number of benefits such as durability and reusability, increased cleanliness and sanitation, and the efficient harvest of a dry and clean product.

LED Lighting

Aero Farms has been pioneering the use of LED (light emitting diode) lighting for growing and vertical farming systems, targeting specific wavelengths of light for more efficient photosynthesis and less energy consumption. LEDs can also be placed much closer to the plants, enabling greater vertical growing for even greater productivity per square foot. The basic principle of aeroponic growing is to grow plants suspended in a closed or semi-closed environment by spraying the plant's dangling roots and lower stem with an atomized or sprayed, nutrient-rich water solution. The leaves and crown, often called the canopy, extend above. The roots of the plant are separated by the plant support structure. Often, closed-cell foam is compressed around the lower stem and inserted into an opening in the aeroponic chamber, which decreases labor and expense; for larger plants, trellising is used to suspend the weight of vegetation and fruit.

Ideally, the environment is kept free from pests and disease so that the plants may grow healthier and more quickly than plants grown in a medium. However, since most aeroponic environments are not perfectly closed off to the outside, pests and disease may still cause a threat. Controlled environments advance plant development, health, growth, flowering and fruiting for any given plant species and cultivars. Due to the sensitivity of root systems, aeroponics is often combined with conventional hydroponics, which is used as an emergency "crop saver" – backup nutrition and water supply – if the aeroponic apparatus fails. High-pressure aeroponics is defined as delivering nutrients to the roots via 20–50 micrometre mist heads using a high-pressure (80 pounds per square inch (550 kPa)) diaphragm pump.

Bubbleponics

The term "Bubbleponics" describes a top-fed deep water culture hydroponic system. Basically, the water is pumped from the reservoir up to the top of the roots (top feeding). The water is released over the plant's roots and then runs back into the reservoir below in a constantly recirculating system. As with traditional deep water culture, there is an airstone in the reservoir to help add oxygen to the water. Both the airstone and the water pump run 24 hours a day. The biggest advantages with Bubbleponics over deep water culture involve increased growth during the first few weeks. With deep water culture, there is a time where the roots haven't reached the water yet. With Bubbleponics, the roots get easy access to water from the beginning and will grow to the reservoir below much more quickly than with a deep water culture system. Once the roots have reached the reservoir below, there is not a huge advantage with Bubbleponics over deep water culture. However, due to the quicker growth in the beginning, a few weeks of grow time can be shaved off.

Smart Farming

Questions have always been raised about the sustainability of organic farming. Now, a comprehensive analysis of the existing studies by scientists from the US and Canada has shown that the yield gap between organic and conventional farming can be narrowed with careful selection of crop type, growing conditions and management techniques. Around 66 published papers comparing organic and conventional yields for 34 crop species, including wheat, tomatoes and soybeans. They found yield from organic farming could result in low yield; in case of cereals it was 26 per cent lower than conventional, and vegetables lagged behind by 33 per cent. But some crop yields did not suffer much. Organic fruits like strawberries and apples showed only three per cent drop in yield and oilseed crops 11 per cent.

The team suggests this happens because legumes and perennials are not nitrogen-dependent. "Legumes do not depend on external nitrogen, whereas perennials, due to their longer growing period and extensive root systems, achieve a better synchrony between nutrient demands and slow release of nitrogen from organic matter". The team also found that organic crops perform better in weak-acidic to weak-alkaline soils - soils with pH between 5.5 and 8.0. This happens because of the difficulty in managing phosphorus in organic systems. Under strong alkaline and acidic conditions, phosphorus is less readily available to plants as it forms insoluble phosphates. They also noted that the organic yields are low in the first years and gradually increase, owing to improvements in soil fertility and management skills. They found organic performance improved in studies that lasted more than two seasons or were conducted on plots that had been under organic farming for three years.

"The yields can be improved if we use best management practices. These include crop rotation, as nutrient and pest management in organic systems depend heavily on crop rotation as well as a high crop diversity and nitrogen," she says. This can be achieved by adding sufficient animal manure or by using legume-based or perennial-based cropping systems". If we are able to estimate the ecosystem services and improved quality of organic foods in monetary terms, it will definitely outweigh conventional ones.

Vermiculture

Whatever be the crops, after harvesting the residue left in the field poses a major problem as removing it requires manpower and money. In some places farmers simply burn the dried leaves, stalk in the field itself as it is an easy option. But over time this practice makes the land barren and kills several beneficial organisms that aid good growth.

Cost Reduction

"And today with the cost of fertilizers hitting the roof it will be advisable if farmers can effectively use these wastes to make some sort of manure like vermicompost and put it back to the soil. By doing so expenses can be reduced and soil fertility be upgraded". A group of farmers from different villages were selected and given training by the KVK staff on vermi composting methods. "Also the income obtained from such activity can be reinvested by the farmer in enhancing his farm resources and infrastructure for higherreturn. He can go for crop diversification and better income".

Production Doubled

Within a year some farmers doubled their production started to enrich his compost with neem cake, Trichoderma, Pseudomonas, Rhizobium, and Azospirillum. The farmer also developed a diversified cropping pattern using the vermicompost from their own unit.

Is Sewage Farming Safe?

Residents of Kararia, Barkhera, Pathani and Shahpur villages in Madhya Pradesh's capital district Bhopal will soon lose the vegetables they have been growing. The loss will not be due to some disease or untimely rain. The plants will be destroyed by the government because sewage water was used to grow them. On January 19, the Madhya Pradesh Pollution Control Board (MPPCB) verbally asked Bhopal's additional collector Akshay Singh to get the crops removed in these four villages, declaring the drain water unfit for irrigation. Singh's office says sewage was being used in 70 hectares (ha) in Bhopal, of which farming on over 45 ha has been stopped till January 2015.

Experts say the decision was wrong as a ban on sewage water for irrigation in a water-starved country like India would do more harm than good. What is needed is a periodic monitoring to ensure that the sewage used does not have toxic elements or heavy metals in high concentrations.

The decision has farmers up in arms. Rahul Dhoot, district vice-president of Bharatiya Kisan Sangh, said Singh's office does not have any report to prove that sewage water is unfit for irrigation. Banning the use of sewage water for farming in Madhya Pradesh started in May last year when, in response to a public interest petition filed by Subhash Chandra Pandey of Bhopal, the National Green Tribunal (NGT) ordered that such plants should be destroyed.

Pandey had identified six location where sewage farming was being done and said that eating such vegetables could be harmful. NGT passed an order to the agriculture department of the state asking it to work with district collectors and stop farmers from selling these vegetables. It also asked the state's horticulture department to submit a report on whether vegetables grown with sewage water were harmful. The report is not yet ready and NGT has fixed February 20 as the date of next hearing. NGT had also asked MPPCB to submit a report on its view on sewage farming in the state. Gurwant Joshi, chief chemist in the monitoring section of MPPCB, who has been involved in the preparation of the report, said that in September 2014 the board was asked by the state government to survey six representative sites across Madhya Pradesh to study the issue.

Farmers Puzzled

The ban, however, has left farmers puzzled. Dhoot, whose association is fighting for farmers' rights in the state since the 1980s, says sewage farming has been going on for over 40 years and no health issues have been reported. Raju Patel, a farmer from Bardhai village, agrees. Patel says he has been irrigating his crops with sewage water for more than 20 years and before him his father did the same. He cultivates

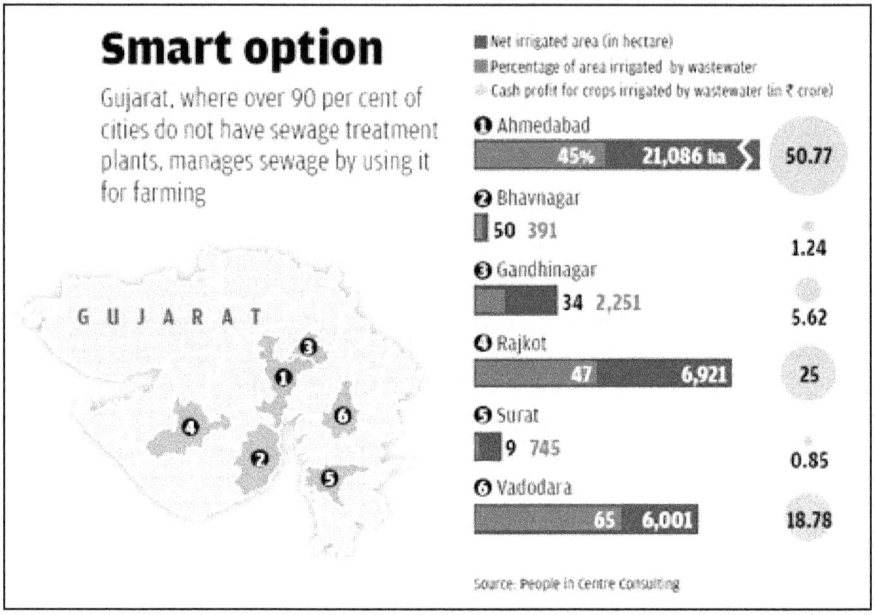

cauliflower, methi (fenugreek), tomato and spinach and sends 700 kilogram of vegetables every day to market.

Dhoot says farmers are unhappy with the decision because irrigating with groundwater would increase expenditure. According to a research by People in Centre Consulting, an Ahmedabad-based consultancy, about 73,000 ha of peri-urban agriculture in India is dependent on wastewater for irrigation. In these areas farmers usually adopt round-the-year intensive vegetable production system. The yield under this system is 300-400 per cent more than usual. Compared to freshwater, sewage water irrigation increases the earnings by up to four times. Banning wastewater irrigation would be neither practical nor feasible and economic impact of the decision for urban and peri-urban farmers would be huge, says C S Hunshal, dean of College of Rural Home Sciences, University of Agricultural Sciences, Dharwad. Hunshal says a comparison of vegetable yields from wastewater and borewell irrigated fields showed a 20-25 per cent advantage from wastewater irrigation.

Monitoring Holds Key

Shafiq-Ur-Rehman, head of environmental sciences, Sher-e-Kashmir University of Agricultural Sciences and Technology, says sewage water can be used effectively in areas facing water stress. His study in a Srinagar locality found that municipal wastewater effluents contained traces of elements such as iron, copper, manganese and zinc in small and varying concentrations. But they were not beyond the prescribed limits and did not pose a health risk. However, he cautions that careless use can cause health problems to farmers and there have been instances where farmers using untreated sewage showed high prevalence of diarrhoeal and skin diseases, and helminthic infections. Nigappa M Rolli, a researcher at BLDE Association's degree college at Jamakhandi, Karnataka, agrees. He says it is always better to treat sewage before irrigation and this can be done through aquatic macro and microphytes (plants that help in cleaning wastewater by absorbing nutrients). Nigappa says elements such as zinc, manganese, nickel, cadmium and copper present in sewage provide nutrients to plants, but when found in heavy concentrations (or due to irrigation with sewage water for a long duration), they may enter the food chain and cause kidney and gastrointestinal problems. Even the problem of toxic heavy metals, which are sometimes present in domestic sewage and can accumulate in soil or seep into vegetables, can be resolved through monitoring, he says.

Right Management

There are a lot of countries which use sewage for farming. According to the Institute for Global Environmental Strategies, a Japan-based research organisation, 20 million hectares of land in 50 countries was being irrigated with raw or partially treated wastewater in 2000. This is almost 10 per cent of the total irrigated land in the world. Countries such as Vietnam, Kuwait, Israel, Tunisia, Jordan, Morocco and China extensively use wastewater for irrigation. Vietnam produces 80 per cent of its vegetables using wastewater. Wastewater is also used by farmers in water scarce regions of developed countries in North America and Europe.

In India, Gujarat is one state which has shown how sewage management can benefit agriculture (see 'Smart option').The water-scarce state, in which over 90 per cent urban bodies do not have sewage treatment plants, realised that using sewage for irrigation solved many problems. It reduced the use of groundwater in agriculture, ensured management of sewage, improved yield and earned government money. The state has different mechanisms in different cities for wastewater management. Cooperatives have been formed in Rajkot, Bhavnagarand Ahmedabad to monitor wastewater used by farmers. These cooperatives have been functional for the past 40 years. The tariff is collected by municipalities. In Gandhinagar, wastewater is carried in a 22-km-long underground pipeline from which farmers extract water and the irrigation department of the state tracks the use and charges farmers for it. However, with the exception of Gandhinagar, the sewage supplied to farmers is not treated in most of the cities.

Corporate Farming by the State

As part of a diversification plan, four special purpose vehicles (SPVs) were created in the state – the Council for Citrus and Agri Juicing, the Organic Farming Council of Punjab (OFCP), the Council for Value Added Horticulture and the Viticulture Council of Punjab. Set up in 2006 as a joint venture by Pepsico and the state government's Punjab Agro Foodgrains Corporation (PAFC), the Council for Citrus and Agro Juicing leased land and undertook corporate farming to promote horticulture. It had 75 acres owned by the PAFC in three locations in the state for growing citrus saplings, an operation managed by Pepsi. The council also leased 2,000 acres at Rs. 8,000 to Rs. 12,000 an acre for 12 years from farmers and aimed to enhance this to 21,000 acres by 2007 to grow citrus fruit varieties imported by Pepsico. Pepsi sold saplings at Rs. 50 each to the PAFC and the planted farms were managed by the council. However, even as trials at the Punjab Agriculture University (PAU) and the horticulture department were on, the council invited farmers to offer their lands on lease under two schemes spanning six and 12 years. In 2007, reports of poor fruit retention, incompatibility of rootstocks and a few varieties yielding no juice began trickling in from the trial sites. Following media reports, the council was asked to reduce its land target to 5,000 acres. The government stopped funding it and refused to stand guarantee for loans it planned to raise. It was instead asked to hunt for a venture capitalist. Even Pepsico, which had by this time grown thousands of plants at its nursery in Jallowal near Jalandhar, brought in a few new varieties to replace the non-performing ones. However, it was too late for farmers who had bit the diversification bait in 2006 and early 2007. They alleged breach of contract by the council and demanded their land back so that they could return to growing wheat and paddy. While no citrus plantations have come up on the lands of most farmers, the varieties grown on the lands of others have failed. To make matters worse, the farmers have neither been paid their rent in a timely fashion nor has the council been monitoring the orchards.

There have also been allegations of funds being misappropriated and an enquiry into this was initiated by the state government recently. In addition, two units set up by the council for multipurpose juicing at Hoshiarpur and Abohar have been idle for two years and plans are on to privatise them. They failed partly because Tropicana stopping buying juice from them when it found a cheaper alternative.

Role of the State

The state and its agencies affect agricultural development in many ways by incentivising or disincentivising some activities over others. In the context of contract farming, it is not just direct incentives to agencies and farmers that count, but also the agricultural policy. For example, contract farming influences the direction of ecological change through two actors. One, the contracting agency lays down a production schedule for farmers at the farm level. By determining the crop to be grown and the husbandry practices farmers have to follow, it influences the effect such farming will have on the environment. The government, as the main source of conservation measures – advisory, financial and material – is the second actor. Farmers' access to these measures is, to a large extent, determined by government policy. Thus the contracting agency and the government have a larger role to play in environmental or ecological change than farmers because they occupy "privileged" positions in chain of decision-making chain.

State-Sponsored Contract Farming

A state-sponsored programme for contract farming in Punjab, which was to divert one million hectares from wheat and paddy over five years (2002-07), failed to deliver. Most of the problems farmers faced related to production and quality (such as that of seeds and extension) and not being able to market produce (other than peas) in the open market (Dhaliwal *et al.*, 2003). There were also instances of corruption and malpractice in the programme due to conflicts of interest among implementing agencies and a lack of monitoring.

For the contract-farming programme to be successful there should have been greater attention paid to the selection of crops, a quick and effective contract enforcement and dispute resolution system, and mechanisms to limit the number of parties in a contractual arrangement and the fiscal risks to the state government. In addition, farmers' organisations capable of contracting with sponsors with a view to reduce transaction costs, better information flow and improving farmers' negotiating positions should have been in place.

More importantly, the programme excluded small farmers in most cases as firms preferred to deal with larger farmer for reasons of transaction costs, eligibility conditions like landholding, irrigation and education and absence of farmers' collectives. Table 3.1, which gives the average profile of contract growers across crops and agencies, shows that the contracted farmers had average holdings much larger than the average size of a holding in the state (9.5 acres). For example, none of the sample contract growers with Frito-Lay had less than 10 acres of land, in spite of 70 per cent of the holdings in the state being less than 10 acres. Another study on the programme found "no marginal farmer (in the size group of below one hectare) operating under contract farming. A handful of small farmers (in the size group of one to two hectares) were operating". Yet another study noted "the majority of the acreage registered in the project (contract farming by PAFC) is held by larger farmers, who tended to receive greater benefits from participation".

Table 3.1: Crop and Agency-wise Average Size of Holding of Contract Growers in Punjab

Study and Year	*Contracting Agency*	*Crop under Contract*	*Average Size of Contract Grower Holding (Acres)*	*Average Area Under Contract Crop (per cent)*
Singh, S 2002	Frito Lay (Pepsi)	Potato	53	8
Singh, S 2002	Nijjer Agro	Tomato	22	23
Singh, S 2002	HUL	Tomato	78	33
Singh, S 2002	Frito Lay (Pepsi)	Chilly	90	4.5
Asokan and Singh 2006	A M Todd	Mint	57	
Kumar 2006	Many MNCs and	Many crops	37	12
Local firms	together			
Singh, S 2008a	Frito Lay (Pepsi)	Potato	63	53
Singh, S 2008b	A M Todd	Mint	40	27
Roopam Singh 2008	Frito Lay (Pepsi)	Potato	75	–

Markfed, a parastatal agency, placed advertisements in local newspapers a few years ago asking those willing to grow at least three acres of basmati paddy under contract farming to contact its district managers. The questions which arise from this type of offer are how many small farmers can spare three acres for basmati paddy? How many can spare it for contract farming? And how many would like to pare it for contract farming with Markfed?

Integrated Farming System: An Intensive Approach

The future of agriculture in India lies in diversifying the economic activities in an interdependent and integrated manner at micro level within the available resources. In other words, it is the integration of two or more enterprises like crop, livestock, fishery, poultry, bee keeping, *etc.* for each farm according to the availability of resources to sustain and satisfy as many necessities of the owner as it is possible and the system thus developed will be termed as integrated farming system. It should not be misunderstood that development of an integrated farming system is possible only for a resource-rich farmer. It is possible even for a resource poor farmer and for him integrated system is more important to maximize the profit through proper utilization of scarce resources in a framing system approach.

Advantages of Integrated Farming System

Increased productivity per-unit of land: as the integrated farming system consists of two or more enterprises the over all production from a particular farm is increased and thereby it provides higher economic return.

1. **Better utilization of resources:** In practicing a single enterprise on a farm all the resources *viz.* land, water-body, labour, farm waste, *etc.* are not utilized property. Co-existence of multiple enterprises in an integrated system makes judicious utilization of all resources.
2. **Recycling of farm wastes:** In an integrated farming system the waste of one enterprise is efficiently used as input in another within the system. But it will depend on the linkages and complementarities among various enterprises of a particular farming system.
3. **Sustainability:** An integrated farming system developed through scientific integration of certain enterprises is always eco-friendly and it imparts sustainability to the system.
4. **Employment generation:** Due to population pressure disguised unemployment has become rampant in rural areas. Integrated approach will generate income and employment.
5. **Reduction of risk**: Integrated farming system makes continuous supply to the farm family and reduces the risk of loss in a particular enterprise.

Several integrated farming systems already exist in India and among them crop-livestock system is most widely prevalent. Others promising ones area crop-poultry-fishery, *etc.*

Crop Livestock

In traditional crop-livestock farming system the crop residues are fed to the livestock and the dung is used as FYM or fuel. But due to population pressure and increased fuel demand non-a-days majority of the cattle-dung is consumed as fuel. With the introduction of biogas technology the organic matter (dung and other wastes) is used both as manure and fuel. Dung and farm wastes produce biogas and the resulting slurry is used as an enriched manure. In crop-livestock system, the particular cropping sequence and proportionate size and composition of livestock population should be determined on the basis of resource base and farming situation.

Crop-Fishery-Poultry

In this system waste grains of crops can be used as feed for the poultry and the poultry litter can be used as manure for the crops. Some small fishes may be used a poultry feed and the poultry waste can be added to the fish pond. From this system the farm family get continuous supply of food grains, fish, egg and meat.

Rice-Fish-Poultry

Rice-fish culture in low land along with poultry shows greater complementarity. If fishes are grown in flooded rice field they release NH_3 which is utilized by rice as N-source and the carbon content of the soil is also increased. In this case, the rice yield is higher than sole crop. On the other hand, fishes check certain weeds and control some insect pests. Small fishes can be fed to the poultry and the poultry wastes can be used a feed for the fishes. However, the possibility of rice-fish culture depends on the agro-ecology and farming situation.

It is possible to develop many such integrated farming systems involving piggery, mushroom cultivation, commercial fishery, orchards, etc, in accordance with the available resources and complementarity among the enterprises. The poultry dropping can be used a pig feed. Pig excreta can be used for biogas production, the biogas sludge can be used for mushroom cultivation and growing fish in the pond. Sediments of the fish pond and biogas sludge can be used for manuring the orchard. The farming system to be followed is unique for a farm or a particular group of farms, which has to be developed through target oriented location specific research.

In India most of the farmers raise crop and livestock together since long back. In few cases fishery or poultry is also found. However, the productivity of their farm is often low because of inefficient integration. The integration is based on traditional knowledge only. Very little research has been done to develop modern farming systems. Therefore, research initiatives have to be undertaken for developing site-specific farming systems incorporated two or more feasible enterprises for different categories of farmers based on their resource endowment. The whole exercise has to be done in a decentralized manner with active involvement of researchers, extension worker and the concerned farmers.

Organic Agriculture

The Origins of Organic Agriculture

Organic Farming is the form of agriculture which is supportive of environment and relies on technique such as crop rotation, green manure, compost and biological pest control. As a concept and ideal, organic agriculture began in the early part of the twentieth century, primarily in Europe, but also in the United States. The pioneers of the early organic movement were motivated by a desire to reverse the perennial problems of agriculture – erosion, soil depletion, decline of crop varieties, low quality food and livestock feed, and rural poverty. They embraced a holistic notion that the health of a nation built on agriculture is dependent on the long-term vitality of its soil. The soil's health and vitality were believed to be embodied in its biology and in the organic soil fraction called *humus*. A soil management strategy called *humus farming* emerged, which employed traditional farming practices that not only conserved but also regenerated the soil. These practices – drawn from mainly from stable European and Asian models – included managing crop residues, applying animal manures, composting, green manuring, planting perennial forages in rotation with other crops, and adding lime and other natural rock dusts to manage pH and ensure adequate minerals.

Since 1990, the market for organic products has grown from nothing, reaching $55 billion in 2009 according to Organic Monitor. This demand has driven a similar increase in organically managed farmland which has grown over the past decade at a compounding rate of 8.9 per cent per annum. Approximately 37,000,000 hecates(91,000,000acres) worldwide are now farmed organically, representing approximately 0.9 per cent of total world farmland (2009).

Since the strategy revolved around soil building to nourish crops, "feed the soil" became the humus farming mantra. "Feeding the soil" meant feeding the soil food web. The soil food web is the living fraction of the soil, composed of bacteria, fungi, earth- worms, insects, and a host of other organisms that digest organic matter and "meter" nutrition to crop plants (see Figure 3.1). This contrasts with the (then emerging) strategy of using soluble fertilizers, which bypass the soil food web to fertilize plants directly. Humus farmers typically avoided, or used very few, synthetic fertilizers. Obviously, they were not consistent with the idea of crop fertilization through the soil food web. Humus farmers felt that soluble fertilizers led to imbalanced plant nutrition and "luxury consumption," which reduced food

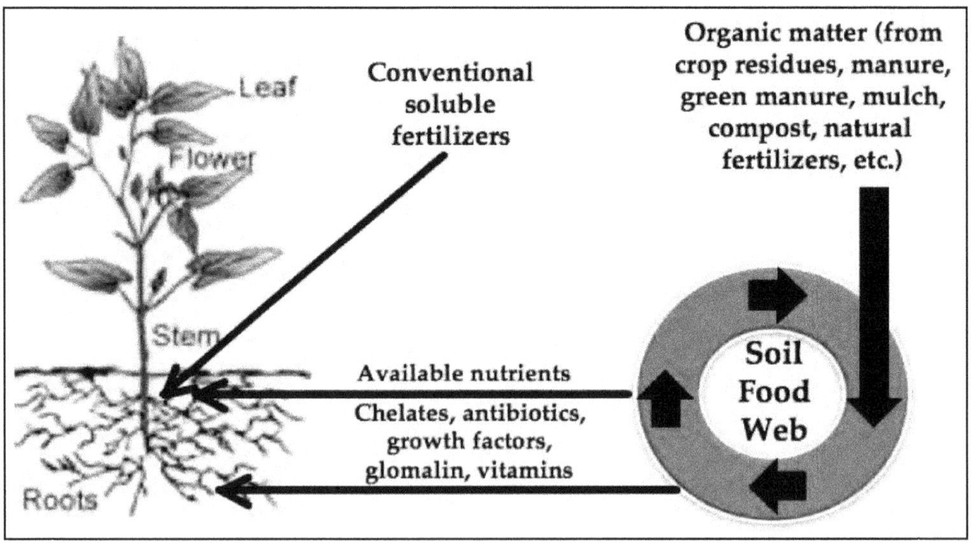

Figure 3.1: Feed the Plant vs. Feed the Soil.

and feed quality. Many also believed that many synthetic fertilizers actually harmed the soil biology – either killing organisms or upsetting the natural balance. They also saw this danger in the use of pesticides, and chose to use few, if any, of those. Still other humus farmers recognized that synthetic fertilizers and pesticides would lead to shortcuts in crop rotation – eliminating many of the soil building and pest control benefits that good rotations confer. The use of synthetic nitrogen fertilizer, especially, would reduce the inclusion of perennial legume forages and green manure crop in cropping sequences. These crops not only supplied nitrogen to subsequent crops in rotation, but sustained soil biology and organic matter levels. Wherever synthetic fertilizers and pesticides were used to cut corners on biodiversity and soil building, they were in direct opposition to the principles of humus farming.

Humus farming, then, was a conscious, well- founded approach to farming and soil management. It embodied a commitment to sustainability through soil regeneration; it sought to avoid wasteful exploitation of natural resources. This was in stark contrast to many of the world's agricultural systems, which, in so many cases, led to the downfall of nations through mismanagement of resources. It puts a lie to the commonly held notion that organic agriculture is simply farming as it was practiced before the advent of synthetic chemicals.

How Humus Farming Becomes Organic Farming

The term "humus farming" went out of vogue in the 1940s as the term "organic" became more popular. According to one source, the first use of "organic" to describe this form of agriculture was in the book *Look to the Land*, by Lord Northbourne, published in 1940. Northbourne uses the term to characterize farms using humus farming methods, because he perceived them to mimic the flows of nutrients and energy in biological organisms – "a balanced, yet dynamic, living whole." Therefore, the word "organic" was intended and used to describe process and function within a farming system – not the chemical nature of the fertilizer materials used, and not adherence to a discredited notion of plant nutrition.

Two Enduring Ideas about Organic Agriculture

People believe many things about organic farming. Some of them are true and some are not. Some originated with humus farming and persist in contemporary organic thinking. Two of these are the beliefs that organically grown food is healthier or otherwise "better for you," and that organic crops are naturally resistant to pests. Both hypotheses are very controversial.

Is Organic Food Healthier?

The belief that organic food is healthier than conventional fare is a foundational belief of organic farming that continues to drive the market today. Surveys continue to show that "healthfulness" is the main reason that consumers buy organic food. Pioneers of the organic movement believed that healthy food produced healthy people and that healthy people were the basis for a healthy society. Since most food originates with the soil, they naturally promoted a method of growing that was based on soil health and vibrancy – the organic/humus farming method. They believed that soils thus managed would yield more nutritious food.

Is the organic community justified in its belief? The answer depends on who gives it. In the recently published article *Nutritional Quality of Organic Foods: A Systematic Review*, the British authors conclude that there is no difference in nutrient quality between organic and conventional foods. This has been challenged by scientists from the Organic Center (TOC) – a nonprofit organization whose stated mission is "to generate credible, peer reviewed scientific information and communicate the verifiable benefits of organic farming and products to society." The TOC reviews support the British findings on some classes of nutrients, but disagree on others. Specifically, TOC cites considerable research that shows organic foods are higher in total antioxidants. The British review generally ignored this body of research. TOC further cites a failure to recognize the higher non-protein nitrogen levels in conventional foods for the hazard they present through formation of nitrosamines in the human digestive tract. Nutritional content is not the only factor that interests organic consumers. Since the 1960s, they have been concerned about pesticide residues. The concern is not about acute poisoning, but the possible effects from bioaccumulation over time.

That there would be less pesticide residue on organic produce seems to be a given. It was re - confirmed by a 2002 study of residue data from several sources over time. Organic produce had one-third the residue levels of conventional fruits and vegetables and half the level found on produce grown using integrated pest management.

The compelling question is whether pesticide residues actually have significant negative effects on human health. This is not clearly answered. The scientific community is most concerned about the possible impacts on children. Whatever those impacts, several studies *do* demonstrate marked reductions in pesticide metabolites in children switched to a diet of organic food.

The concern for children is certainly valid. Their body weight is lower and small effects in childhood can grow to major problems over a long lifetime. But the danger may occur well before childhood. A 2009 article in *Newsweek* suggests that the impact of pesticide (and other environmental chemical) exposure may be equally or more significant to those in the womb. There is growing evidence that hormone- mimicking pollutants from pesticides and plasticizers are a major factor in infant obesity, which has risen 73 per cent since 1980. Of particular note is the observation that these effects were caused at very low dosages. In a similar vein, University of Wisconsin researchers have recently reported a connection between exposure of pregnant women to the popular pesticide chlorpyrifos (trade names include Dursban and Lorsban) and long-lasting birth defects in female offspring. Again, the damage appears at extremely low dose levels.

Beginning in the 1990s, genetic engineering became another issue in the organic food quality debate. The National Organic Standard prohibits the use of genetic engineering in organic agriculture as per §205.105(e), where its various permutations are referred to as "excluded methods." The concern over so-called "frankenfoods" and their possible effects on human health were certainly a major factor in the organic community's insistence that genetic engineering be banned from organic food production.

Do Organic Crops Resist Pests?

The assertion that organic culture induces pest and disease resistance in crops is much less well known than the healthy food claim, though the notion has been around for some time. In *The Soil and Health*, one of the earliest classics of organic thought, the father of organic agriculture, Albert Howard, writes that health is the "birthright of all living things," and that health in humans depends on a chain of health that begins in the soil. He goes on to state that "vegetable (and animal) pests and diseases...are evidence of a great failure of health" in the plant and animal links in that chain and those failures begin with the soil and its management. In other words, pests and diseases may be considered *agricologenic* – induced by the farmer. The organic method was, according to Howard, a means for restoring and sustaining soil health, thereby reducing and even eliminating most pest problems.

The notion that plant stress and disease/pest tolerance are related did not originate with Albert Howard and the organic movement. There is an independent school of thought on *predisposition theory* that traces back to the work of H.M. Ward in 1890 and continues at least through the mid-1970s. Subsequent inquiry became increasingly bound to research on organic systems. A prime example is P. L. Phelan's work with European corn borer that found reduced pest damage under organic management.

Early work on predisposition theory focused on fungal diseases but later expanded to address other diseases and arthropod pests. It is the basis for a common belief among organic farmers that insect pests are nature's "garbage collectors;" that their main function is to remove unhealthy and unsuitable plants.

There are several theories regarding the mechanisms of predisposition. One revolves around phytochemicals that plants produce to protect themselves from diseases and pests. A prime example is DIMBOA, a chemical compound found in young corn plants that protects against fungi, bacteria, and a range of insect pests, including European corn borer. Stressed plants, it is argued, produce fewer protective phytochemicals, making them more susceptible. If, as research shows, organic crops produce more phytochemical antioxidants, it is logical to believe they might also produce more pest-repellant phytochemicals.

Another theory about how predisposition works relates to the breakdown of proteins under stress, leading to the accumulation of soluble amino acids in plant sap. It is believed that these forms of free nitrogen are more digestible by insect pests, many of which lack the enzymes to break proteins down to their amino acid constituents. Therefore, they thrive on stressed plants.

A further theory links total dissolved plant- sap solids to susceptibility and resistance. Higher levels of sugars, minerals and other components of plant sap are treated as indicators of plant health and, therefore, improved resistance. Since dissolved solids can be readily monitored using inexpensive hand-held refractometers, this theory is quite popular among growers, though research has not always supported a direct correlation. California-based research by Dr. Mark Mayse, for example, did not find a correlation between dissolved solids in grapes and resistance to leafhoppers during a two-year study.

While induced resistance offers great promise, it is not a panacea and its mechanisms are hard to nail down. Furthermore, the degree of resistance is likely to vary with the pest or disease involved. Predisposition theory may be totally irrelevant where a new crop species is involved, or a new pest or disease introduced.

If organic crops are more resistant to pests and diseases, the root cause is most likely the soil food web. A healthy food web leads to good tilth and a better air and water balance. Increased humus means more moisture retention and less drought stress. There is also a significant nutritional benefit. Though organic agriculture recognizes that plants absorb soil nutrients in soluble (inorganic) form, it places great importance on the organic compounds – chelated nutrients, amino acids, natural antibiotics, vitamins, growth factors, humic substances, *etc.* – that plants also absorb. Traditional practitioners believe that these *phytamins* contribute much more to plant health and growth than is generally recognized.

The soil food web not only makes phytamins available, it also aids in their uptake. Albert Howard noted this and wrote in considerable detail on the importance of mycorrhizal associations in particular. Mycorrhizae are fungi that coat, and form symbiotic relationships with, plant roots. They effectively increase the absorptive surface of the root hairs, aid in the uptake of minerals and water, and provide a barrier to pathogens. High-humus, biodiverse, organically managed soil certainly favors the survival and proliferation of mycorrhizae.

As early organic farmers embraced predisposition theory, it led to a *plant-positive* paradigm for disease and pest management. By accepting agri-cologenic stress as a root cause of pest problems, organic growers sought to change and improve their systems and reduce plant stress by improving tilth, balancing crop nutrition, and whatever else they could do to protect and enhance the soil food web. This is in contrast to a conventional pest-negative paradigm that assumes outbreaks are inevitable, no matter how a crop is grown, and that they must always be dealt with by direct means such as pesticides.

Organic Universe

It is a universe of its own–an expanding universe that has its own producers and consumers, its entrepreneurs, its markets, its passionate aficionados of scientists and evangelists. It is a universe that is governed by a different set of regulations and plays strictly by the rules, for the most part. In the politics of agriculture, its ideology is controversial since it goes against mainstream wisdom, almost heretical since it sets out to disprove the dominant chemical-driven theology of the past 60 years. We are talking about organic farming. Why is organic farming controversial? At a fundamental level, it sets out to prove that you do not need chemical pesticides and synthetic fertilisers to produce adequate quantities of food. Organic farming works in harmony with nature by using simple techniques and material: recycled and composted crop waste and animal manure; crop rotation; legumes to fix soil nitrogen; encouraging useful predators that eat pests and natural pesticides; and a careful husbanding of water resources. The bottom line: increasing genetic diversity and conservation. It also means that no genetically modified crop is part of this ecosystem. The result: healthier farmers and risk-free food.

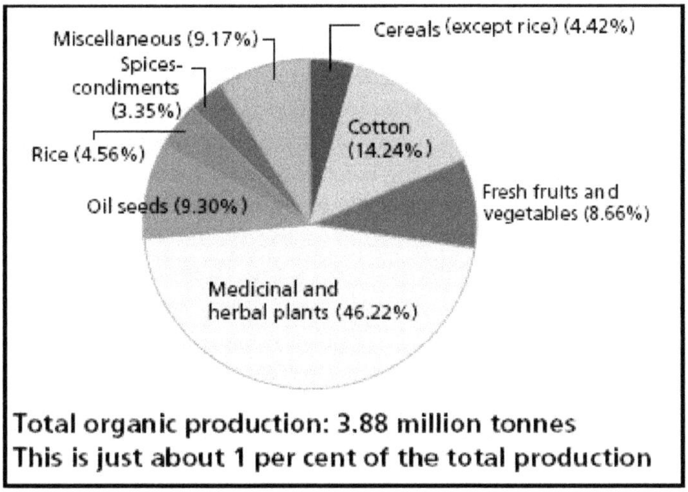

Figure 3.2: Main Organic Crops.

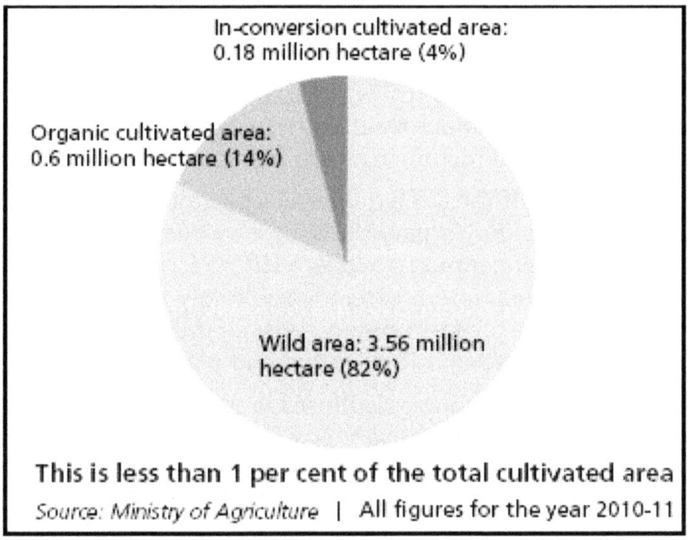

Figure 3.3: Area Under Organic Production.

It is a philosophy that has caught the fancy of some unlikely players. Some have given up well-heeled jobs in the Silicon Valley, others a lucrative legal career. There are doctors who have turned into farmers, and management professionals into organic retailers. In all cases, there was a desire "to return to natural methods of cultivation sans chemicals" and to grow "real food that tastes like it used to".

Among these remarkable men is Rajashekar Reddy Seelam, who literally crossed over to the other side: he switched from selling chemical inputs to farmers for a major company to becoming an organic cultivator himself. Along the way, Seelam became a pioneer in retailing organic food and his store, 24 Letter Mantra, is now the top brand in the domestic market with an increasingly successful franchisee model, besides being a leading exporter. Sresta, the company he set up in 2004 to promote organic cultivation, works with over 8,000 farmers who cultivate around 60 crops on around 12,140 hectares (ha) spread across 12 states. It uses the contract farming model. Seelam says Sresta's scouting teams of trained agriculture graduates are careful to select farming tracts where soil and water contamination is negligible. Across India, groundwater contamination through leeching of pesticides has become a scourge after five decades of the Green Revolution theology that decrees huge lashings of synthetic fertilisers and chemical pesticides to improve yield from hybrid seeds.

"The thought that we should move away from the conventional wisdom first came to me in 1992 when I was aggressively promoting the chemical additives," says Seelam. But it was not till 2004 that he took the plunge. Everyone told him it would be foolish to waste money on such an impractical idea.

What is Organic Farming?

Organic farming system in India is not new and is being followed from ancient time. It is a method of farming system which primarily aimed at cultivating the land and raising crops in such a way, as to keep the soil alive and in good health by use of organic wastes (crop, animal and farm wastes, aquatic wastes) and other biological materials along with beneficial microbes (bio-fertilizers) to release nutrients to crops for increased sustainable production in an eco friendly pollution free environment.

As per the definition of the United States Department of Agriculture (USDA) study team on organic farming "organic farming is a system which avoids or largely excludes the use of synthetic inputs (such as fertilizers, pesticides, hormones, feed additives etc) and to the maximum extent feasible rely upon crop rotations, crop residues, animal manures, off-farm organic waste, mineral grade rock additives and biological system of nutrient mobilization and plant protection".

FAO suggested that "Organic agriculture is a unique production management system which promotes and enhances agro-ecosystem health, including biodiversity, biological cycles and soil biological activity, and this is accomplished by using on-farm agronomic, biological and mechanical methods in exclusion of all synthetic off-farm inputs".

Need of Organic Farming

With the increase in population our compulsion would be not only to stabilize agricultural production but to increase it further in sustainable manner. The scientists have realized that the 'Green Revolution' with high input use has reached a plateau and is now sustained with diminishing return of falling dividends. Thus, a natural balance needs to be maintained at all cost for existence of life and property. The obvious choice for that would be more relevant in the present era, when these agrochemicals which are produced from fossil fuel and are not renewable and are diminishing in availability. It may also cost heavily on our foreign exchange in future.

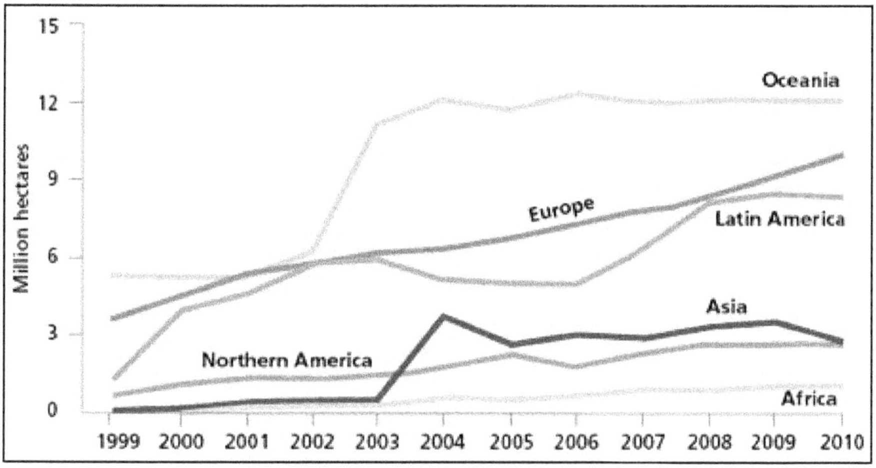

Figure 3.4: Global Growth of Organic Farming.

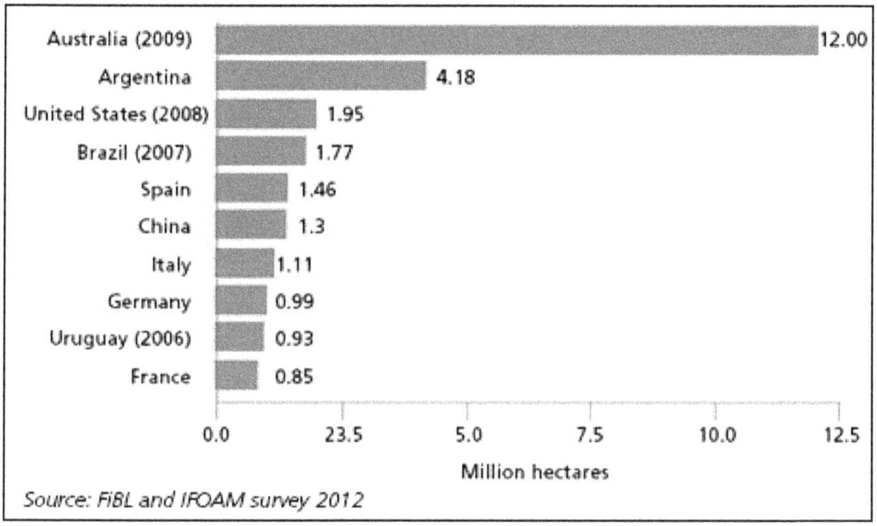

Figure 3.5: Countries with most Organic Farmland in 2010.

Sresta, however, has stayed the course and notched up a turnover of Rs. 55 crore. This may not be much and it underlines the problems of organic farming in the country–extremely small holdings, poor knowledge of scientific organic practices, certification that is both cumbersome and expensive for small holders, lack of markets and glaring lack of support from the government which, ironically, is footing a subsidy bill of around Rs. 90,000 crore on fertilisers alone. "The government simply refuses to see the wisdom of promoting a proven system of farming that will cost so much less, both for the farmer and the government, and a system that benefits the land and consumers." Ramanjaneyulu, former scientist with the Indian Council of Agricultural Research (ICAR), was responsible for launching the non-pesticide movement in Andhra Pradesh that has freed around 1.5 million ha from the grip of chemical pesticides. The state is now moving towards helping such farmers to get organic certification.

Different business models are being tried out to make such farming profitable for the smallholder. With entrepreneurial enthusiasm linked to a love of "clean farming", interesting experiments have emerged. One is Dharani Suphalam, a primary producers' cooperative society based in Sirsa, Haryana, which has brought together 1,500 farmers and is slowly linking them to the market (see 'Innovative farm-to-home trail'). Another is the model used by Bengaluru-based Sahaja Samrudha Organic Producer Company that was set up in 2010 with authorised capital of Rs. 10 lakh. It brings together 500 farmers through 40 groups. Its wholesale-cum-retail outlet in a central part of the city is a no-frills godown where bags of millets, rice and legume jostle for space with fresh mangoes and vegetables. But customers have no problem with its ambience since prices are reasonable.

"We are working with farmers, helping them to convert to organic and provide marketing assurance. So we keep our margins low: 10-15 per cent mark-up for wholesale and 20-25 per cent for retail. In any case, the profits go back to the farmer since they have shares in the company." Turnover last year was Rs. 52 lakh, but Sahaja is certain revenue will double this year. One of the biggest grassroots success stories is of Morarka Foundation which, with its 100-plus project locations covering 19 states, has brought around 100,000 farmers under the umbrella of Morarka Organic. This has helped it understand the different agro-climatic problems and gain expertise in handling over 130 crops/products which, it says, form its core portfolio and are offered both as farm grade as well as processed and in retail packs that are sold under the Down to Earth brand (no connection with this magazine). There are others like the feisty H R Jayaram who gave up a prosperous law practice to become an organic farmer-cum-retailer. He has set up what is probably India's first organic hotel, The Green Path, in a Bengaluru suburb. Jayaram's aim is to fire others with his passion and "to create replicable models that will inspire more people to take up organic". Sahaja and enthusiasts like Jayaram have held regular organic food melas, which has helped make Bengaluru India's undisputed organics capital. Restaurants, coffee shops and stores like Simply Organics run by another farming buff Govind Kabadi make for zesty organic profile for this city. Aiding this is the state government which gives Karnataka farmers a big helping hand in finding outlets for their produce.

But the corner organic store with crowded shelves is not the dominant image of organic retailing. The familiar milieu is the chic store in tony localities that sell beautifully packaged products that come with all kinds of certification–from NPOP (National Programme on Organic Production), which is the Indian standard, to those of the EU and the US. Here premiums are high since products come through many layers of marketing. The other place where one is bound to find such products are supermarket chains like Chennai-based Spencer's which has a national footprint. Says a spokesperson for the chain: "We are a food-first multi-format retailer and sell a wide range of brands like 24 Letter Mantra, Down To Earth, Pro Nature and Pro Organic." Growth has been phenomenal. Starting with two brands in 2006, Spencer's organic category has grown by 300 per cent in last three years. Besides, its offerings which were restricted to commodities like cereals, pulses and spices, are spilling into a wide assortment from breakfast cereals to snacks and soups. For those who have the money, such outlets are a cornucopia of natural goodness.

Delhi-based Fabindia, which caters to the well-heeled and trendy customer, is a leading purveyor of organic foods, its elite stores across the country stocking carefully selected products. Ashima Agarwal, who heads the organic foods division, is unwilling to disclose the amount of business this segment brings in, but says: "Though small, organic foods are witnessing growth in terms of customer base and consumption. We started with 70 products and today offer more than 300 products and are still growing." For William Bissell, Fabindia founder, the foray into organics from clothes, furnishings and traditional crafts was inspired by a book Diet for a Small Planet by Frances Moore Lappe, a 1971 bestseller which advocated ecologically sound production of food. "Then when I was in the US, I worked on the organic farm of Gordon Ridgeway who was a purist and influenced me greatly," says Bissell. Although he would like to keep fresh produce in his stores, there are too many logistical issues.

Logistics are critical to the success of organic farming and the reason such produce is far too expensive for the ordinary consumer. For instance, products sold by non-profit Navdanya are more expensive than similar items that are not organic. Founder Vandana Shiva has an explanation: "For me, organic farming is about livelihood and about sustainability and justice. When a farmer practices organic agriculture, he knows how to sustain his farm with natural resources; he is able to feed himself and his family; takes care of earth by giving natural resources back in forms of bio-fertilisers and bio-pesticides without polluting the environment." Naturally, all this comes at a price since the government offers no help at all.

The booming growth in organic foods may well spell hope for the small farmer. With health-conscious and cash-rich customers ready to pay the premium for organic food, the market is expanding in dramatic ways. High-end, imported organic has also come to India. The first such is businessman Dilip Doshi's Organic Hauswhich which has opened in Ahmedabad and Mumbai. For Doshi it is business pure and simple; he does not claim to promote organic because it is better for the environment or for the farmer. Every product in his uber-luxury stores is imported from Germany and Austria.

"Mine is a concept store. It tells the story of the evolution of globally benchmarked products. You can close your eyes and say this is organic. I could not find any genuine organic foods in India," declares Doshi. That could come as a crushing blow to all the organisations, farmers and activists who have struggled to put Indian organic on the global map. But fortunately in Europe and the US, which together account for 96 per cent of the market, India's credentials are respected even if it accounted for just a fraction of the US $59 billion global market for certified organic food and drink. India's exports in 2010-11 totalled just Rs. 550 crore, according to Manoj Kumar Menon, executive director of the International Competence Centre for Organic Agriculture or ICCOA, which is a knowledge centre for all facets of organic agriculture and also helps provide market linkages to all those engaged in organic agriculture.

ICCOA is a crucial link to the market with its international trade fair Bio Fach India which draws over 100 global companies and thousands of visitors to its annual event. Menon is confident that India will become a significant player in the next four years when production is expected to touch Rs. 4,000 crore, a huge leap from the current Rs. 675 crore. By then, the global trade is expected to cross US $104 billion in 2015 at an estimated annual rate of 12.8 per cent. On the global organic map, India is a speck; it does not figure in the list of the top 10 countries. According to data released in June 2012 by the International Federation of Organic Agriculture Movements (IFOAM) and the Research Institute of Organic Agriculture of Switzerland, there was a total of 37 million ha of agricultural land that was organic in 2010. In addition, the organic universe had another 43 million ha of non-agricultural areas, up from 41 million ha in 2009.

What was India's share? It had 0.6 million ha under cultivation, another 0.18 million ha under conversion along with wild area of 3.56 million ha. IFOAM statistics also put the total number of organic farmers worldwide at about 1.6 million, with the largest number, a whopping one million of them, in India. This is where our organic story gets truly interesting. In fact, the slight dip in organic agriculture area in 2010 was on account of India which reported the loss of 0.4 million ha of organic cultivation. But that trend is reversing. Says A K Yadav, director of National Centre of Organic Farming (NCOF): "The declining trend has been reversed and the area under certification process during 2011-12 is likely to be more than one million hectares." He also makes the point that all the statistics with NCOF are for the area which is registered under certification process. "There is no reliable data on farmers doing organic but have not opted for certification."

In all probability it means that organic cultivation is more widespread than estimated, although the Ministry of Agriculture is doing its best to undermine this through its various programmes. For instance, it has allowed its flagship scheme Rashtriya Krishi Vikas Yojana to be used by states to convert tribal areas or the organic wild centre to conventional farming through the provision of free hybrid seeds in kits that also contain chemical pesticides and fertilisers. There is also the looming threat of climate change and its impact on agriculture. While Krishi Bhavan has finally asked Ardhendu Sen, a former bureaucrat now with TERI, to draft a policy document for the promotion of organic agriculture, farmers are deciding on their own what's in their best interest.

According to a recent report, the organization food market in India is growing at 20-22 per cent annually and with exports valued at Rs.1,000 crore. India produced around 3.88 million tones of certified organic products, including basmati, pulses, tea, coffee, spices and oilseeds. Government data show area under organic farming had risen to 1.08 million hectares.

The key characteristics of organic farming include Protecting the long term fertility of soils by maintaining organic matter levels, encouraging soil biological activity, and careful mechanical intervention Providing crop nutrients indirectly using relatively insoluble nutrient sources which are made available to the plant by the action of soil micro-organisms Nitrogen self-sufficiency through the use of legumes and biological nitrogen fixation, as well as effective recycling of organic materials including crop residues and livestock manures Weed, disease and pest control relying primarily on crop rotations, natural predators, diversity, organic manuring, resistant varieties and limited (preferably minimal) thermal, biological and chemical intervention The extensive management of livestock, paying full regard to their evolutionary adaptations, behavioural needs and animal welfare issues with respect to nutrition, housing, health, breeding and rearing.

Preserving Nature with Organic Farming

To commemorate tenth anniversary of the implementation of National Programme for Organic Production (NPOP), a "National Conference-India Organic 2000-2010" was held here over the weekend in which the growth of organic movement across the country was highlighted. Speaking on the occasion, Agricultural and Processed Food Products Export Development Authority (APEDA) advisor Dr. P.V.S.M. Gauri said the Indian farm area under organic was likely to grow. "There is worldwide acknowledgement of the Indian organic logo. It is important to encourage organic farming because it is environment-friendly and specially at a time when discussions are around the subject of climate change".

Acknowledging that there were number of challenges to popularise organic farming, Dr. Gauri said extending support to small farmer groups was the need of the hour. "There should be zero tolerance in certificate system. India can become the number one organic hub in the world in the next ten years through total market acceptance of the Indian organic brand. We are moving towards greater credibility by framing more stringent standards of quality management." Stating that India was on the threshold of development in organic agriculture, Morarka Organic Foods CEO Mukesh Gupta said the organic sector was flourishing today at a much higher pace than ever before. "The world is now progressively becoming more aware about the need to preserve nature. Globally, the organic products are swiftly replacing synthetic and chemically produced products."

The conference also highlighted that India has the distinction of being the first country to have introduced grower group certification at the national level to cover small farmers and the NPOP standards of India were recognised by European Union, Switzerland and the United States. On the occasion APEDA also launched "Tracenet", a web-based traceability system for organic products with the help of service providers and consultation with all stakeholders.

The Organic Shift

Cautiously, but convincingly, farmers of Katewadi in Baramati district have embraced organic farming, bidding farewell to pesticides and chemical fertilisers for the sake of their own health and that of their crops. The transition began around 2004 when Katewadi's 848 farmers, who grow sugarcane, grapes, pomegranate, wheat and banana, were made aware of the economic and health benefits of putting away the toxic spray. "When farmers who adopted organic farming methodology started getting better yield, many others also gave a shot to organic manures, which, besides increasing the yield, also improved the health of the soil". Cow dung, cow urine, jaggery, flour and water, and calls it Jeevamrutha. "The mixture is fermented in a barrel for 15 days and is ready for use after that. Since most farmers here own cows, they all easily prepare the manure at home. Dung produced by a desi cow is enough to provide manure to one acre of farmland. With the use of Jeevamrutha, there was a 40 per cent yield increase. They did not make the switch (to organic farming) in a day. The farmers here slowly cut down on the use of chemical fertilisers. If the ratio of chemical and organic was 20:80 seven years ago, it is vice-versa now. The aim is to reach 100 per cent within two years".

Now in Katewadi, no one burns farm waste – a practice prevalent till five years ago. The use vermicompost, Jeevamrutha and also animal waste in our fields. The yield is good and the soil condition has improved tremendously". Big farmers, like have erected vermicompost sheds on their premises. They use the vermicompost in his fields and also supplies it to small farmers in the region.

"They are on our way to becoming an eco-village and organic farming is one of our main thrust areas. Our information centre has a speaker system through which we announce weather forecast, market price of the crops and so on for the benefit of the farmers". Krishi Vigyan Kendra in Baramati plays a key role in making farmers aware of organic farming and use of bio-agents instead of pesticides. They teach farmers to make bio-agents in their household for biological pest control and also take them for study trips and make them meet farmers who have successfully adopted organic farming.

"Instead of spending money on chemical fertilisers, the farmers now use vermicompost, green manure and bio-fertilisers. Organic farming also augments the water building capacity of the soil". The paradigm shift took place over the years. Earlier, farmers did not want to take risk or try anything new. But the better yield and improvement in soil fertility has encouraged them.

A MODEL VILLAGE

KATEWADI, *nine km from Baramati and 110 km from Pune, earned the 'model village' title from the Maharashtra government in 2008. The honour came its way for having well-maintained roads, toilets in every house under Nirmal Bharat Abhiyan' scheme, solid waste management, underground drainage system, school, community centre and well-maintained primary health centre.Katewadi may also add another feather to its cap soon – that of being the first eco-village in the state. Under the eco-village plan, whose thrust is on sanitation and health, work has already started to promote community lifestock management, solid waste management, provision of community blocks, sewage treatment by green bridge technology, energy conservation through biogas power generation and solar street lighting system.*

Careful attention to the impact of the farming system on the wider environment and the conservation of wildlife Defining "Organic"

Organic farming is a method of crop and livestock production that involves much more than choosing not to use pesticides, fertilizers, genetically modified organisms, antibiotics and growth hormones.

Organic production is a holistic system designed to optimize the productivity and fitness of diverse communities within the agro-ecosystem, including soil organisms, plants, livestock and people. The principal goal of organic production is to develop enterprises that are sustainable and harmonious with the environment.

The general principles of organic production, from the Canadian Organic Standards (2006), include the following:

- ☆ Protect the environment, minimize soil degradation and erosion, decrease pollution, optimize biological productivity and promote a sound state of health
- ☆ Maintain long-term soil fertility by optimizing conditions for biological activity within the soil.
- ☆ Maintain biological diversity within the system
- ☆ Recycle materials and resources to the greatest extent possible within the enterprise
- ☆ Provide attentive care that promotes the health and meets the behavioural needs of livestock
- ☆ Prepare organic products, emphasizing careful processing, and handling methods in order to maintain the organic integrity and vital qualities of the products at all stages of production
- ☆ Rely on renewable resources in locally organized agricultural systems

Organic farming promotes the use of crop rotations and cover crops, and encourages balanced host/predator relationships. Organic residues and nutrients produced on the farm are recycled back to the soil. Cover crops and composted manure are used to maintain soil organic matter and fertility. Preventative insect and disease control methods are practiced, including crop rotation, improved genetics and resistant varieties. Integrated pest and weed management, and soil conservation systems are valuable tools on an organic farm. Organically approved pesticides include "natural" or other pest management products included in the Permitted Substances List (PSL) of the organic standards. The Permitted Substances List identifies substances permitted for use as a pesticides in organic agriculture. All grains, forages and protein supplements fed to livestock must be organically grown.

The organic standards generally prohibit products of genetic engineering and animal cloning, synthetic pesticides, synthetic fertilizers, sewage sludge, synthetic drugs, synthetic food processing aids and ingredients, and ionizing radiation. Prohibited products and practices must not be used on certified organic farms for at least three years prior to harvest of the certified organic products. Livestock must be raised organically and fed 100 per cent organic feed ingredients.

Organic farming presents many challenges. Some crops are more challenging than others to grow organically; however, nearly every commodity can be produced organically.

Growth of Organic Agriculture

The world market for organic food has grown for over 15 years. Growth of retail sales in North America is predicted to be 10 per cent to 20 per cent per year during the next few years. The retail organic food market in Canada is estimated at over $1.5 billion in 2008 and $22.9 billion in the U.S.A. in 2008. It is estimated that imported products make up over 70 per cent of the organic food consumed in Canada. Canada also exports many organic products, particularly soybeans and grains.

The Canadian Organic Farmers reported 669 certified organic farms in Ontario in 2007 with over 100,000 certified organic acres of crops and pasture land. This is an annual increase of approximately 10 per cent per year in recent years. About 48 per cent of the organic cropland is seeded to grains, 40 per cent produces hay and pasture and about five per cent for certified organic fruits and vegetables. Livestock production (meat, dairy and eggs) has also been steadily increasing in recent years.

Why Farm Organically?

The main reasons farmers state for wanting to farm organically are their concerns for the environment and about working with agricultural chemicals in conventional farming systems. There is also an issue with the amount of energy used in agriculture, since many farm chemicals require energy intensive manufacturing processes that rely heavily on fossil fuels. Organic farmers find their method of farming to be profitable and personally rewarding.

Why Buy Organic?

Consumers purchase organic foods for many different reasons. Many want to buy food products that are free of chemical pesticides or grown without conventional fertilizers. Some simply like to try new and different products. Product taste, concerns for the environment and the desire to avoid foods from genetically engineered organisms are among the many other reasons some consumers prefer to buy organic food products. In 2007 it was estimated that over 60 per cent of consumers bought some organic products. Approximately five per cent of consumers are considered to be core organic consumers who buy up to 50 per cent of all organic food.

What is "Certified Organic"?

"Certified organic" is a term given to products produced according to organic standards as certified by one of the certifying bodies. There are several certification bodies operating in Ontario. A grower wishing to be certified organic must apply to a certification body requesting an independent inspection of their farm to verify that the farm meets the organic standards. Farmers, processors and traders are each required to maintain the organic integrity of the product and to maintain a document trail for audit purposes. Products from certified organic farms are labelled and promoted as "certified organic."

In June 2009, the Canadian government introduced regulations to regulate organic products. Under these regulations the Canadian Food Inspection Agency

(CFIA) versees organic certification, including accreditation of Conformity Verification Bodies (CVBs) and Certification Bodies (CBs). This regulation also references the *Canadian Organic Production Systems General Principles and Management Standards* (CAN/CGSB-32.310) and the *Organic Production Systems – Permitted Substances List* that were revised in 2009.

The Canadian organic regulations require certification to these standards for agricultural products represented as organic in import, export and inter-provincial trade, or that bear the federal organic agricultural product legend or logo. Products that are both produced and sold within a province are regulated by provincial organic regulations where they exist (Quebec, British Columbia and Manitoba)

The federal regulations apply to most food and drink intended for human consumption and food intended to feed livestock, including agricultural crops used for those purposes. They also apply to the cultivation of plants. The regulations do not apply to organic claims for other products such as aquaculture products, cosmetics, fibres, health care products, fertilizers, pet food, lawn care, *etc.*

Food products labelled as organic must contain at least 95 per cent organic ingredients (not including water and salt) and can bear the Canada Organic logo. Multi-ingredient products with 70 per cent to 95 per cent organic product content may be labelled with the declaration: " per cent organic ingredients". Multi-ingredient products with less than 70 per cent organic content may identify the organic components in the ingredient list.

Exporting Organic Materials

Exported products must meet the requirements of the importing country or standards negotiated through international equivalency agreements. Products exported to the U.S. must meet the terms of the Canada-U.S. equivalency agreement signed in June 2009. All products that meet the requirements of the Canada Organic Regime can be exported to the U.S. with the exception that agricultural products derived from animals treated with antibiotics cannot not be marketed as organic in the U.S. Canada is also exploring other international equivalency agreements with other trading partners to enhance trade opportunities for export and to assure the organic integrity of imported products.

Organic Certification

When considering organic certification, know the requirements and accreditation(s) needed in the marketplace where your products will be sold. When comparing certification bodies, make sure they have the certification requirements and accreditations needed to meet market requirements. As a minimum certification bodies should be accredited under the Canadian Organic Products Regulations. Some markets may require accreditation or equivalency agreements with countries in the European Union, or with the Japanese Agricultural Standard (JAS), Bio-Swisse or other international organic certification systems. As Canada develops international equivalency agreements the need for the certification body to have these international accreditations will diminish.

For more information on certification and links to Canadian regulations and standards see the Organic Agricultural section of the OMAFRA.

The Transition Period

The first few years of organic production are the hardest. Organic standards require that organic lands must be managed using organic practices for 36 months prior to harvest of the first certified organic crop. This is called the "transition period" when both the soil and the manager adjust to the new system. Insect and weed populations also adjust during this time.

Cash flow can be a problem due to the unstable nature of the yields and the fact that price premiums are frequently not available during the transition since products do not qualify as "certified organic." For this reason, some farmers choose to convert to organic production in stages. Crops with a low cost of production are commonly grown during the transition period to help manage this risk.

Carefully prepare a plan for conversion. Try 10 per cent to 20 per cent the first year. Pick one of the best fields to start with and expand organic acreage as knowledge and confidence are gained. It may take five to 10 years to become totally organic, but a long term approach is often more successful than a rapid conversion, especially when financial constraints are considered. Parallel production (producing both organic and conventional versions of the same crop or livestock product) is not allowed. Use good sanitation, visually different varieties, individual animal identification and other systems to maintain separation and integrity of the organic and conventional products. Good records are essential.

Standards

Standards regulate production methods and in some cases final output for organic agriculture. Standards may be voluntary or legislated. As early as the 1970s private associations certified organic producers. In the 1980s, governments began to produce organic production guidelines. In the 1990s, a trend toward legislated standards began, most notably with the 1991 EU-Eco-regulation developed for European Union, which set standards for 12 countries, and a 1993 UP program. The EU's program was followed by a Japanese program in 2001, and in 2002 the U.S. created the National Organic Program (NOP). As of 2007 over 60 countries regulate organic farming. In 2005 IFOAM created the Principles of Organic Agriculture, an international guideline for certification criteria. Typically the agencies accredit certification groups rather than individual farms.

Successful Organic Farming

In organic production, farmers choose not to use some of the convenient chemical tools available to other farmers. Design and management of the production system are critical to the success of the farm. Select enterprises that complement each other and choose crop rotation and tillage practices to avoid or reduce crop problems.

Yields of each organic crop vary, depending on the success of the manager. During the transition from conventional to organic, production yields are lower than conventional levels, but after a three to five year transition period the organic yields typically increase.

Cereal and forage crops can be grown organically relatively easily to due to relatively low pest pressures and nutrient requirements. Soybeans also perform well but weeds can be a challenge. Corn is being grown more frequently on organic farms but careful management of weed control and fertility is needed. Meeting nitrogen requirements is particularly challenging. Corn can be successfully grown after forage legumes or if manure has been applied. Markets for organic feed grains have been strong in recent years.

The adoption of genetically engineered (GMO) corn and canola varieties on conventional farms has created the issue of buffer zones or isolation distance for organic corn and canola crops. Farmers producing corn and canola organically are required to manage the risks of GMO contamination in order to produce a "GMO-free" product. The main strategy to manage this risk is through appropriate buffer distances between organic and genetically engineered crops. Cross-pollinated crops such as corn and canola require much greater isolation distance than self-pollinated crops such as soybeans or cereals.

Fruit and vegetable crops present greater challenges depending on the crop. Some managers have been very successful, while other farms with the same crop have had significant problems. Certain insect or disease pests are more serious in some regions than in others. Some pest problems are difficult to manage with organic methods. This is less of an issue as more organically approved biopesticides become available. Marketable yields of organic horticultural crops are usually below non-organic crop yields. The yield reduction varies by crop and farm. Some organic producers have added value to their products with on-farm processing. An example is to make jams, jellies, juice, *etc.* using products that do not meet fresh market standards.

Livestock products can also be produced organically. In recent years, organic dairy products have become popular. There is an expanding market for organic meat products. Animals must be fed only organic feeds (except under exceptional circumstances). Feed must not contain mammalian, avian or fish by-products. All genetically engineered organisms and substances are prohibited. Antibiotics, growth hormones and insecticides are generally prohibited. If an animal becomes ill and antibiotics are necessary for recovery, they should be administered. The animal must then be segregated from the organic livestock herd and cannot be sold for organic meat products. Vaccinations are permitted when diseases cannot be controlled by other means. Artificial insemination is permitted. Always check with your certification body to determine if a product or technique is allowed in the Permitted Substances List and the organic standards. Organic production must also respect all other federal, provincial and municipal regulations.

Organic produce can usually qualify for higher prices than non-organic products. These premiums vary with the crop and may depend on whether you are dealing with a processor, wholesaler, retailer or directly with the consumer.

Prices and premiums are negotiated between buyer and seller and will fluctuate with local and global supply and demand.

Higher prices offset the higher production costs (per unit of production) of management, labour, and for lower farm yields. These differences vary with commodity. Some experienced field crop producers, particularly of cereals and forages, report very little change in yield while in some horticultural crops such as tree fruits, significant differences in marketable yield have been observed. There may also be higher marketing costs to develop markets where there is less infrastructure than for conventional commodities. Currently, demand is greater than supply for most organic products.

A Brief Overview of the History and Philosophy of Organic Agriculture

For most of its history, organic agriculture has been given short shrift. If they paid attention at all, conventional agricultural institutions treated it as an antiquated, unscientific way to farm – suitable, perhaps, for gardeners, but not a serious means of commercial food production. Anyone who advocated for organic farming was derided; it was professional suicide for an agronomist or soil scientist to do so.

While its methods, proponents, and philosophy are still derided in some quarters, things have been turning around for organic agriculture. Organic consumption is increasing and organic acreage is growing. An organic industry is developing that not only commands respect, but now demands a growing share of research and educational services from USDA, land-grant universities, and state agriculture departments.

By the end of 2008, the organic sector had grown to a whopping $24.6 billion industry.[1] While many sectors of the agricultural economy are growing slowly and even stagnating, the organic sector has been growing at roughly 20 per cent a year since 1994.[2] Even during the recession year of 2008, growth was a respectable 17 per cent.[3] At present the organic sector constitutes about 3.5 per cent of total U.S. food sales, but should these growth rates continue, it could reach 10 per cent in less than a decade. According to ERS statistics from 2005, U.S. organic acreage now exceeds four million, with certified production in all 50 states Worldwide.

Research, Analysis and Commentary

Norman Borlaug (father of the "Green Revolution" and a Noble Peace Prize laureate), Prof A. Trewavas and other critics contested the notion that organic agricultural systems are more friendly to the environment and more sustainable than conventional farming systems. Borlaug assets that organic farming practices can at most feed 4 billion people, after expanding cropland dramatically and destroying ecosystems in the process. The Danish Environmental Protection Agency estimated that phasing and health effects were assumed but hard to assess.

In contrast, the UN Environmental Programme concluded that organic methods greatly increased yields in Africa. A review of over two hundred crop comparisons argued that organic farming could produce enough food to sustain the current human population and that the difference in yields between organic and non-organic methods were small, with non-organic methods yielding slightly more in developed areas and organic methods yielding slightly more in developing areas. That analysis has been criticized by Alex Avery of the Hudson Institute, who contends that the review claimed many non-organic studies to be organic, misreported organic yields, made false comparisons between yields of organic and non-organic studies which were not comparable, counted high organic yields several times by citing different papers which referenced the same data, and gave equal weight to studies from sources which were not impartial.

Organic Agriculture 2013: Key Indicators and Leading Countries

Indicator	*World*	*Leading Countries*
Countries with data on certified organic agriculture Organic agricultural land	2011: 162 countries 2011: 37.2 million hectares (1999: 11 million hectares)	Australia (12 mio. hectares, 2009) Argentina (3.8 mio. hectares) US (1.9 mio. hectares, 2008)
Share of total agricultural land	**2011: 0.86 per cent**	FalklandIslands (Malvinas) (35.9 per cent) Liechtenstein (29.3 per cent) Austria (19.7)
Further, non- agricultural organic areas (mainly wild collection)	2011: 32.5 million hectares (2010: 43 million hectares; 2009: 41 million hectares)	Finland (7 million hectares) Zambia (5.9 million hectares; 2009) India (4.5 million hectares)
Producers	1.8 million producers (2010: 1.6 million producers; 2009: 1.8 million producers)	*India (547'591), Uganda (188'625, 2010), Mexico (169'570)*
Organic market size	62.9 billion US dollars (2010: 59.1 billion US dollars 1999: 15.2 billion US dollars) Source: *Organic Monitor*	*US (21 billion euros or 29 billion US dollars), Germany (6.6 billion euros or 9.2 billion US dollars) France (3.7 billion euros or 5.2 billion US dollars)*
Per capita consumption	*2011: 9.02 US dollar*	*Switzerland (177.4 euros or 250.4 USD), Denmark (161.9 euros or 225.7 USD) Luxemburg (134.3 euros or 187.3 USD)*
Number of countries with organic regulations 2011	*86 countries (2010: 84 countries)*	
Organic certifiers 2010	*2012: 576 certifiers (2011: 549; 2010 532)*	*South Korea, Japan, USA*
Number of IFOAM affiliates	1.1.2013: 766 affiliates from 117 countries	Germany: 96 affiliates; India: 46 affiliates; China: 40 affiliates; United States: 33 affiliates: The Netherlands: 31 affiliates

Source: FiBL and IFOAM; for total global market: Organic Monitor; for number of certifiers: Organic Standard/Grolink.

The Centers for Disease Control repudiated a claim by Avery's father, Dennis Avery (also at Hudson) that the risk of E.coli infection was eight times higher when eating organic food. A very hard included problems stemming from non-organic unpasteurized juice in his calculations. Epidemiologists traced the 2011 E.coli O104:H4 outbreak – which caused over 3,900 cases and 52 deaths – to an organic farm in Bienenbuttel in German y. The Rodale Institute, has conducted a thirty year "Farm Systems Analysis Trial" of organic farming. Their findings suggest that organic farming is superior to conventional systems in "building, maintaining and replenishing the health of the soil." In addition to soil health, the Institute's trials looked at economic viability, energy usage, and human health and concluded that organic agriculture is more sustainable than conventional. The Institute's findings are summarized as follows:

- Organic outperforms conventional in years of drought.
- Organic farming systems build rather than deplete soil organic matter, making it a more sustainable system.
- Organic farming uses 45 per cent less energy and is more efficient.
- Conventional systems produce 40 per cent more greenhouse gases.
- Organic farming systems are more profitable than conventional.

Where the designation "country" appears in this book, it covers countries or territories. Share of the countries included in the FIBL-IFOAM survey 2013. Total world population 6.97 billion in 2011 according to FAO stats. FAO, Rome; Countries and Territories covered by the global survey in Agriculture 2012 is shown in **Table 3.2.**

Table 3.2: Countries and Territories Covered by the Global Survey on Organic Agriculture 2012

	Countries* with data on organic agriculture	Countries per region[1]	Share of countries that provided data (%)
Africa	38	57	67
Asia	37	49	76
Europe	46	46	100
Latin America and Caribbean	28	45	62
North America	3	5	60
Oceania	11	23	49
World	164	224	73

Source: FiBL-IFOAM survey 2014
*Where the designation "country" appears in this book, it covers countries or territories.

Organic Agricultural Land

Currently 37.5 million hectares are under organic agricultural management worldwide (end of 2013 for most data).

The region with the most organic agricultural land is Oceania, with 12-2 million hectares, followed by Europe with 11.2 million hectares. Latin America (6.8 million hectares), Asia (3.2 million hectares). North America (3 million hectares), and Africa (1.1 million hectares).

Oceania has one-third of the global organic agricultural land, but its relative importance is decreasing. Europe; a region that has had a very constant growth of organic land over the years, has almost 30 per cent of the world's organic agricultural land, followed by Latin America with 18 per cent (see Table 3.2, Figure 3.1).

Australia is the country with the most organic agricultural land; 97 per cent of die farmland is extensive grazing area. Argentina is second, followed by the United States in third, place (Table 3.3, Figure 3.2). The ten countries with the largest organic agricultural areas have a combined total of 26.3 million hectares and constitute seventy per cent of the world's organic agricultural land.

Apart from the organic agricultural land, there are further organic areas, such as wild collection areas. These areas constitute more than 31 million hectares.

Table 3.3: Organic Land and Farms by Continent

Region	Organic agricultural land (hectares)	Regions' share of the global organic agricultural land
Africa	1'145'827	3.05%
Asia	3'217'867	8.57%
Europe	11'171'413	29.75%
Latin America	6'836'498	18.21%
Northern America	3'012'354	8.02%
Oceania	12'164'316	32.40%
Total*	37'544'909	100.00%

Source: FiBL-IFOAM survey 2014. Note: Agricultural land includes in-conversion areas and excludes wild collection, aquaculture, forest, and non-agricultural grazing areas.
* Includes correction value for French overseas departments.

Table 3.4: List of Countries: Organic Land, Share of Organic Land of Total Agricultural Area, Number of Organic Farms

Country	*Organic Land (Hectares)*	*Producers*	*Share of Total Agricultural Land (per cent)*
Bangladesh	6860	150	0.1
China	1900000	1600	0.4
India	500000	600000	0.3
Indonesia	88247	6627	0.2
Nepal	10273	247	0.1
Pakistan	22397	105	0.1
Philippines	80974	3008	0.7
Sri Lanka	19517	404	0.7
Thailand	32577	7189	0.2

Source: FiBL Survey 2014 Based on information from the private sector certifiers and govts.

Organic Agriculture and Food Security–Not a Contradiction

The theme was again popularized by Stanford University Professor Paul Ehrlich in his 1968 book, 'The Population Bomb'. According to his logic, we should all be starving now that the 21st century has arrived. *'The battle to feed all of humanity is over. In the 1970s the world will undergo famines; hundreds of millions of people are going to starve to death in spite of any crash programs embarked upon now.'* The only famines that occurred since 1968 have been in countries saddled with corrupt governments, political turmoil, civil wars and periodic droughts. The world had enough food for the affected people. It was political and logistical events that prevented them from producing adequate food or stopped aid from reaching them.

The spectre of mass starvation is again being pushed, based on highly questionable and contestable assumptions, as the motive for justifying genetically modified organisms (GMOs) and the unsustainable use of toxic chemicals to push for higher yields.

Enough Food for Everyone

According to the UN Food and Agriculture Organization (FAO) and other figures, the world produces more than double the amount of food to feed everyone. Despite this, around one billion people suffer from hunger and another billion are malnourished, lacking the essential micronutrients they need to lead healthy lives. Clearly, our current market-based distribution systems are failing the poorest as they cannot afford to buy this food. The market-based systems concentrate the food in the areas where people have the money to pay for it. Consequently one billion adults are overweight and almost half of them are obese.

Food losses also are staggering. About one-third of the food produced for human consumption is lost or wasted every year, amounting to about 1.3 billion tons annually (FAO, 2011). Reducing food losses and food waste, and improving food access, is highly relevant to efforts to combat hunger, raise income and improve food security in the world's poorest countries. Simple affordable measures such as village grain silos and better transport would prevent most of these losses. The problem of hunger is mostly due to poor distribution systems and inadequate production in the poorest communities. It has very little to do with the total amount of food produced in the world.

Where Organic has Higher Yields: Conditions of Climate Extremes

Research has shown two significant areas where organic systems have higher yields than the conventional system. These are under conditions of climate extremes and in traditional smallholder systems. Both of these areas are critical to achieving global food security.

According to research by NASA and others the world is seeing increases in the frequency of extreme weather events such as droughts and heavy rainfall. Even if the world stopped polluting the planet with greenhouse gases tomorrow, it would take many decades to reverse climate change. This means that farmers have to adapt to the increasing intensity and frequency of adverse and extreme weather events such as droughts and heavy, damaging rainfall. Published studies show that organic farming systems are more resilient to the predicted weather extremes and can produce higher yields than conventional farming systems in such conditions. For instance, the Wisconsin Integrated Cropping Systems Trials found that organic yields were higher in drought years and the same as conventional yields in normal weather years.

Similarly, Farming Systems Trials (FST) of the US-based Rodale Institute showed that the organic systems produced more maize than the conventional system in drought years. The average maize yields during the drought years were from 28 per cent to 34 per cent higher in the two organic systems. The yields were 6,938 and 7,235 kg per hectare in the organic animal and the organic legume systems respectively, compared with 5,333 kg per hectare in the conventional system. The researchers attributed the higher yields in the dry years to the ability of the soils on organic farms to better absorb rainfall. This is due to the higher levels of organic carbon in these soils, which makes them more friable and better able to store and capture rainwater which can then be used for crops.

Improved Efficiency of Water Use

Research also shows that organic systems use water more efficiently due to better soil structure and higher levels of humus and other organic matter compounds. Lotter and colleagues collected data over ten years during the Rodale Farming System Trials. Their research showed that the organic manure system and organic legume system (LEG) treatments improve the soil's water-holding capacity, infiltration rate and water capture efficiency. The LEG maize soils averaged a 13 per cent higher water content than conventional system (CNV) soils at the same crop stage, and 7 per cent higher than CNV soils in soybean plots.

The more porous structure of organically treated soil allows rainwater to quickly penetrate the soil, resulting in less water loss from run-off and higher levels of water capture. This was particularly evident during the two days of torrential downpours from Hurricane Floyd, which hit the eastern United States and the Bahamas in September 1999, when the organic systems captured around double the water than the conventional systems captured.

This is very significant information as the majority of the world farming systems are rainfed. The world does not have the resources to irrigate all of the agricultural lands. Nor should such a project be started as damming the world's watercourses, pumping from all the underground aquifers and building millions of kilometres of channels would be an unprecedented environmental disaster.

Improving the efficiency of rainfed agricultural systems through organic practices is the most efficient, cost effective, environmentally sustainable and practical solution to ensure reliable food production in the increasing. weather extremes being caused by climate change.

Smallholder Farmer Systems

The other critical area where research is showing higher yields for good practice organic systems is in traditional smallholder systems. This is very important information as over 95 per cent of the world's farmers fall into this category. A report by the United Nations Conference on Trade and Development (UNCTAD) and the United Nations Environment Programme (UNEP) found that organic agriculture increases yields in Africa. *'the average crop yield was 116 per cent increase for all African projects and 128 per cent increase for the projects in East Africa'.*

The report notes that despite the introduction of conventional agriculture in Africa, food production per person is 10 per cent lower now, than in the 1960s. *'The evidence presented in this study supports the argument that organic agriculture can be more conducive to food security in Africa than most conventional production systems, and that it is more likely to be sustainable in the long term'*, Supachai Panitchpakdi, Secretary General of UNCTAD and Executive Director of UNEP Achim Steiner stated.

This is crucial information as FAO data shows that 80 per cent of the food in the developing world comes from smallholder farmers such as those in Africa. The developing world is also the region where most of the one billion undernourished people live, the majority of which are smallholder farmers. With a more than 100 per cent increase in food production in these traditional farming systems, organic agriculture provides the ideal solution to end hunger and ensure global food security.

The Key to Food Security

Information published by the ETC group, an international organization dedicated to "the conservation and sustainable advancement of cultural and ecological diversity and human rights", shows that 70 per cent of the world's food is produced by smallholders and only 30 per cent by the agribusiness sector (ETC group, 2009; see Figure 3.4). Increasing the yields in the 30 per cent that comes from the agribusiness sector will show little benefit for two reasons. Firstly, this sector is already high-yielding, and it has very little scope for large increases in yields such as the more than 100 per cent that can be achieved by organic methods in traditional smallholder systems. Secondly, this sector is largely focused on the commodity supply chain. The large food surpluses produced in the sector have not lowered the number of people who are hungry. According to FAO figures, this number has been steadily increasing since 1985.

Fifty per cent of the world's hungry are smallholder farmers and 20 per cent are the landless poor who rely on smallholders for their employment (see Figure). Logically, increasing the yields in this sector is the key to ending hunger and achieving food security. Organic methods are the most suitable as the necessary methods and inputs that are needed to do this can be sourced locally at no or very little cost to the farmers. Conventional systems have largely failed to provide consistent higher yields to the poorest farmers as the expensive synthetic chemical inputs have to be purchased. Most of these farmers do not have the income to do this. It is an inappropriate economic model for the world's most vulnerable farmers

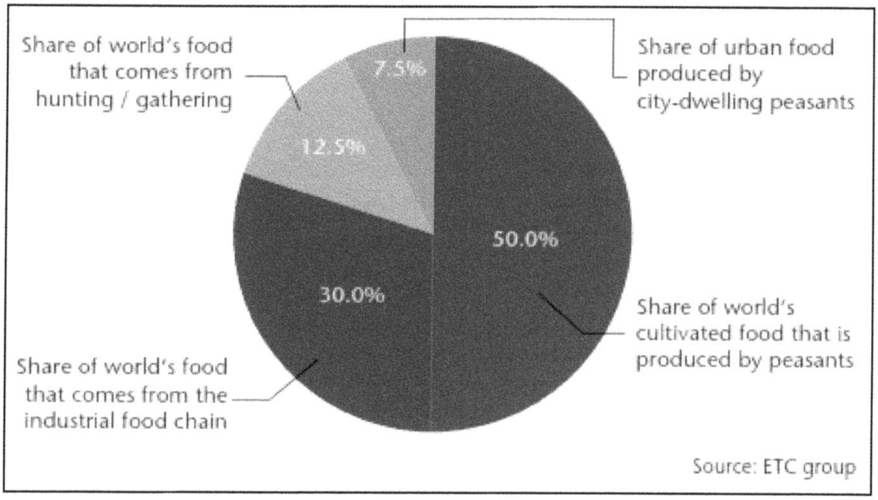

Figure 3.4: Where the World's Food comes from?

whereas organic agriculture is an appropriate one. A good example is a project managed by the Institute of Sustainable Development in Tigray, Ethiopia (see box).

What about Yields and Farm Income?

The assumption that greater inputs of synthetic chemical fertilisers and pesticides are needed to increase food yields is not always accurate. In a study published in *The Living Land,* Professor Jules Pretty of Essex University looked at projects in seven industrialised countries of Europe and North America.

'Farmers are finding that they can cut their inputs of costly pesticides and fertilizers substantially, varying from 20 to 80 per cent, and be financially better off. Yields do fall to begin with (by 10 to 15 per cent typically), but there is compelling evidence that they soon rise and go on increasing. In the USA, for example, the top quarter sustainable agriculture farmers now have higher yields than conventional farmers, as well as a much lower negative impact on the environment' (Pretty, 1998a). Numerous studies into organic systems confirm this insight – the following refers to only a few of them:

US Agricultural Research Service (ARS) Pecan Trial

The ARS organically managed pecans out-yielded the conventionally managed, chemically fertilized Gebert orchard in each of the past five years. Yields on ARS' organic test site surpassed the Gebert commercial orchard by 18 pounds of pecan nuts per tree in 2005 and by 12 pounds per tree in 2007.

Rodale Organic Low/No Till

The Rodale Institute (Pennsylvania, USA) has been trialling a range of organic low tillage and no tillage systems. The 2006 trials resulted in organic yields of 160 bushels an acre (bu/ac) compared to the County average of 130 bu/ac.

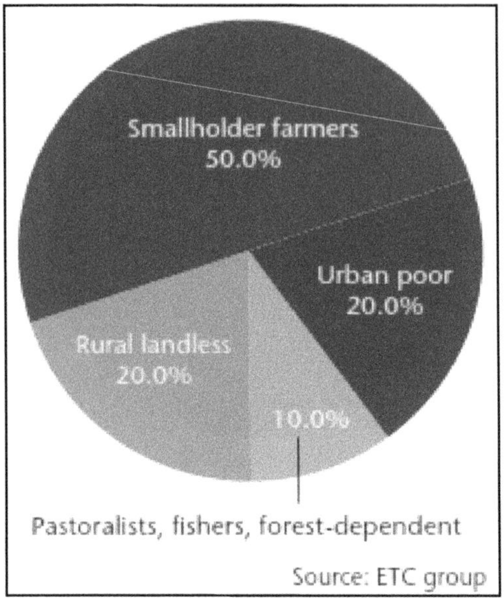

Figure 3.5: Who the Hungre are?

Iowa Trials

The results from the Long Term Agro-ecological Research (LTAR), a 12 year collaborative effort between producers and researchers led by Dr Kathleen Delate of Iowa State University (Iowa, USA) showed that while the organic systems had lower yields in the beginning, by year four they started to exceed the conventional crops. Across all rotations, organic corn harvests averaged 130 bushels per acre while conventional corn yield was 112 bushels per acre. Similarly, organic soybean yield was 45 bu/ac compared to the conventional yield of 40 bu/ac in the fourth year. Cost-wise, on average, the organic crops' revenue was twice that of conventional crops due to the savings from non-utilisation of chemical fertilisers and pesticides.

Developing Countries

Nicolas Parrott of Cardiff University, UK, authored a report, 'The Real Green Revolution'. He gives case studies that confirm the success of organic and agroecological farming techniques in the developing world:

- ✰ In Madhya Pradesh, India, average cotton yields on farms participating in the Maikaal Bio-Cotton Project are 20 per cent higher than on neighbouring conventional farms.
- ✰ In Madagascar, SRI (System of Rice Intensification) has increased yields from the usual 2 – 3 tons per hectare to yields of 6, 8 or 10 tons per hectare.
- ✰ In the highlands of Bolivia, the use of bonemeal and phosphate rock and intercropping with nitrogen-fixing Lupin species have significantly contributed to increases in potato yields.

Studies comparing the income of organic farms with conventional farms have found that the net incomes are similar, with best practice organic systems having higher net incomes. A study by Dr Rick Welsh of the Wallace Institute, USA, has also shown that organic farms can be more profitable. The premium paid for organic produce is not always a factor in this extra profitability. While many organic farmers have higher incomes due to the premium they receive, others have higher net incomes due to their lower input costs rather than from the premium.

The United Nations report already cited notes: '*A transition to integrated organic agriculture, delivering greater benefits at the scale occurring in these projects, has been shown to increase access to food in a variety of ways: by increasing yields, increasing total on- farm productivity, enabling farmers to use their higher earnings from export to buy food, and, as a result of higher on- farm yields, enabling the wider community to buy organic food at local markets.*'

Simple Methods, Big Success: Sustainable Development in Tigray, Ethiopia

The Tigray Project started in 1996 in four local communities in the central, eastern and southern parts of the Tigray Regional State of Ethiopia. It was initiated by an Ethiopian NGO, the Institute of Sustainable Development (ISD). The ISD worked in cooperation with the farmers to re-vegetate their landscape to restore the local ecology and hydrology. The biomass from this re-vegetation was then sustainably harvested to make compost and to feed biogas digesters. This was applied to the crop fields. The result after a few years was more than 100 per cent increases in yields, better water use efficiency and greater pest and disease resistance in the crops.

The farmers used the seeds of their own landraces which had been developed over millennia to be locally adapted to the climate, soils and the major pests and diseases. These farmerbred varieties proved to be very responsive to producing high yields under organic conditions, whereas under conventional input practices they were susceptible to diseases such as rust. The major advantage of this system was that the seeds and the compost were sourced locally at no or little cost to the farmers whereas the seeds and synthetic chemical inputs in the conventional systems had to be purchased. The organic system had both higher yields and a much better net return to the farmers.

This project using simple appropriate organic methods took a region that was previously regularly affected by severe famines that killed people through to a food surplus and relative prosperity. The people could now afford to eat well, buy clothes, send their children to school, pay for medical treatment, afford transport into town and build adequate houses. The Third World Network provided the initial funding for the project. Today ISD is working with the Ethiopian Bureau of Agriculture and Rural Development (BoARD), district (woreda) experts and development agents to continue implementing the Tigray Project. The funding from several donor agencies is assisting the scaling up so that more regions in Ethiopia can adopt the practices

Average mean grain yields in kg/ha for four cereals and one pulse crop from Tigray, northern Ethiopia, 2000 – 2006 inclusive.

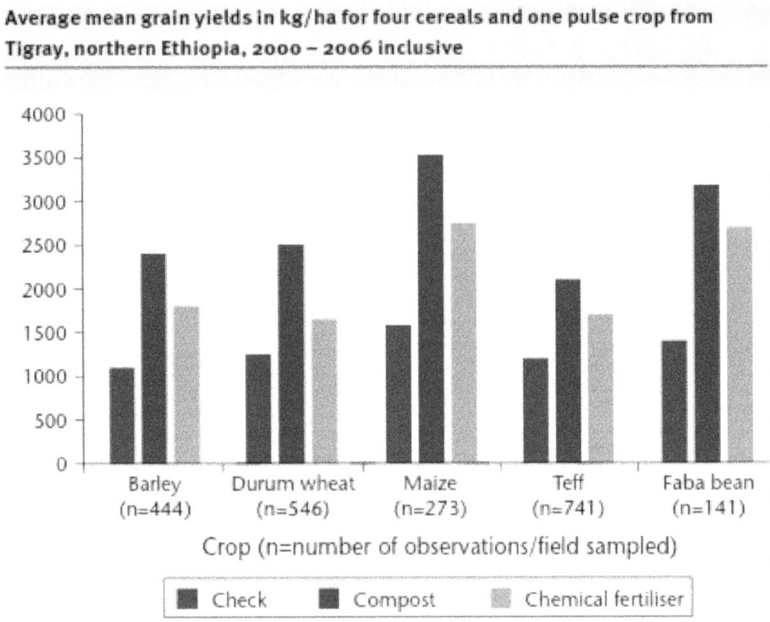

There is very good research that clearly shows organic agriculture can get the yields that are needed to feed the poor and the hungry. This is especially the case in smallholder agriculture - the majority of the world's farmers.

'All case studies which focused on food production in this research where data have been reported have shown increases in per hectare productivity of food crops, which challenges the popular myth that organic agriculture cannot increase agricultural productivity', the UN report stated.

Organic agriculture is a low-cost and effective way to help many of the world's poorest people to have good levels of nutrition and a better quality of life. We need to see more research and extension in this area to ensure that all farmers can improve their yields and resilience by adopting the appropriate best-practice organic systems.

Most importantly, the world urgently needs to increase the training in good organic practices to ensure that the poorest smallholder farmers can have the increases n yield that are needed to achieve food security.

The GM Science

What is a GM Crop? How is it different from Conventional Hybrid Crops?

As the name suggests, a genetically modified crop is produced by bringing about a change in the genetic structure of the plant. It is generally done by introducing an outside gene, sourced from other plants or organisms, to develop some special characteristic in the crop. In conventional hybridisation, two or more varieties of the same crop are cross-bred, through cross-pollination or some other method, to develop a hybrid variety with some desired characteristic.

What are *Bt* Crops like *Bt* Cotton or *Bt* Brinjal?

Bt crops are genetically-modified after inserting a particular gene (Cry1Ac) from the soil bacterium, *Bacillus thuringiensis* (*Bt*), into a plant cell. *Bacillus thuringiensis* is a naturally occurring bacteria that produces a toxin poisonous to certain types of insects. In the case of cotton, the Cry1Ac gene provides resistance from bollworm, cotton's most persistent pest. In case of brinjal, it provides resistance from fruit and shoot borer pests. Thus Bt cotton and Bt brinjal use less pesticides, which otherwise is a major input cost for the farmers.

FSSAI' s proposal for GM food safety assessment

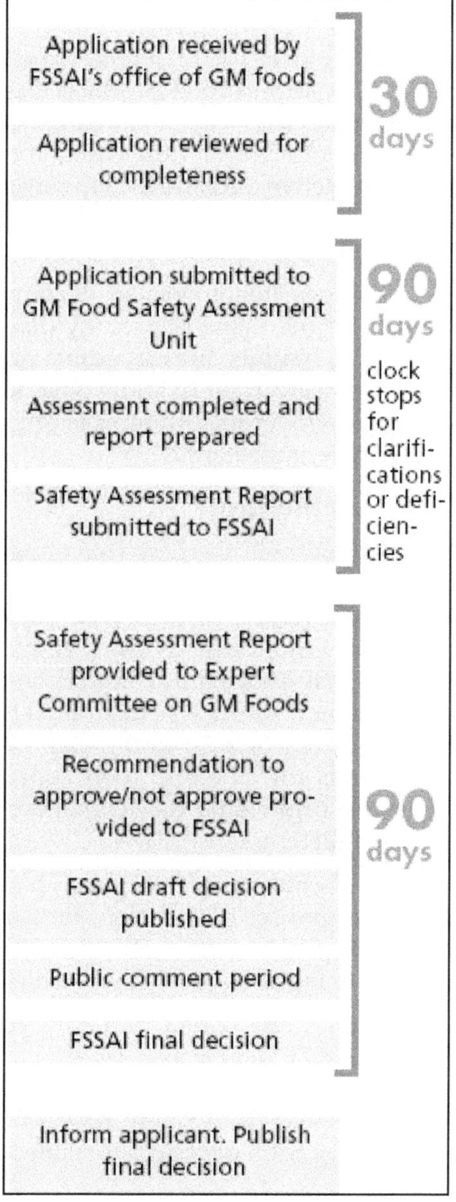

PM's Science Panel Bats for GM Crops

India's top science panel that advises the Prime Minister has said agriculture in the country has to rely, in part, on biotechnology. The view boosts genetically modified crops which is still a divisive issue. The Scientific Advisory Committee to the Prime Minister, headed by CNR Rao, said, unlike in Europe, India would face growing pressure on food. "This will require a judicious blend of traditional breeding and new technologies, non-transgenic and transgenic. This situation in developed countries such as in Europe; quite in contrast, as there is no dearth of food and a small proportion of people engage in agriculture," the panel said after a meet to assess the situation. It also questioned the ban on genetically modified brinjal.

India has proposed a Biotechnology Regulatory Authority of India Bill, which will be an overarching body to approve such technologies. Till the bill becomes law, the "current regulatory system for recombinant products administered under Rules (1989) of Environment Protection Act,

1986, should be reformed". Hordes of farm NGOs resist farm biotechnology but India's farm policy promotes GM crops in view of the country's precarious food situation.

"The PM's Scientific Advisory Council parroted corporate propaganda and failed to have an unbiased scientific approach on GM". The country allowed GM cotton in 2002, after which output surged sharply, making it the second-largest cotton exporter. It, however, clamped an indefinite moratorium on the country's first genetically modified food crop - *Bt* brinjal.

Labels Must for GM Food

INDIA MAY NOT have started growing genetically modified (GM) food crops, but it does import food products that contain GM ingredients. Consumers who buy them have no way of knowing what they contain. Recently, the department of consumer affairs under the Union food and consumer affairs ministry decided to take a corrective measure to help consumers make an informed choice. It mandated that all packaged food products containing GM ingredients should carry a GM label from January 1, 2013.

The department notified 19 commodities to be covered by the new labelling rule under the Legal Metrology (Packaged Commodities) Rules. These include baby food, biscuits, breads, edible oils, milk powder, cereals and pulses. "Every consumer has a right to know what kind of food he or she is eating, and with the controversy over health impacts of genetically modified food, labelling becomes all the more important,"

Who will Monitor?

But how will the new rule get implemented? The consumer affairs department deals only with packaging, and weights and measures. Dixit says it is for the Food and Safety Standards Authority of India (FSSAI) to decide what information should be displayed and in what manner. FSSAI was mandated by the Food Safety Standards Act of 2006 to regulate GM food and draft regulations for labelling food containing genetically modified organisms (GMOs). These have not been finalised till date.

FSSAI chief executive officer, had this to say in his written response: "Regulations for labelling of GM food are under preparation in FSSAI and these would be incorporated in labelling and packaging regulations of FSSAI which is under preparation."

Some countries where GM labels are compulsory

1. Australia
2. New Zealand
3. Japan
4. European Union
5. China
6. South Korea
7. Russia
8. Norway
9. Saudi Arabia
10. Switzerland

In 2010, FSSAI had prepared an interim regulation - Operationalising the Regulation of Genetically Modified Foods in India. It says that to manage the regulatory programme for GM foods, FSSAI would establish a new secretariat-GM Foods and Food Safety Assessment Unit (GMFSAU). It would have a multi-disciplinary team of scientists trained in safety assessment.

This unit would assess and decide whether the GM event (gene insertion) for which approval is sought can be considered as safe as non-GM food. The report of GMFSAU would then be reviewed by FSSAI's expert committee on GM foods, which would also oversee the public consultations process before giving its recommendation (see diagram).

But this interim regulation may end up as an exercise on paper. Mohanty says the department of biotechnology has since mooted the Biotechnology Regulatory Authority of India (BRAI) bill for a single regulator for genetically modified food. Since BRAI has come into picture, it is yet to be made clear who will regulate GM food. The food processing industry, meanwhile, is clueless about what needs to be done before the approaching deadline of January 1. Industry sources say they cannot act unless they get clear instructions. Dixit blames FSSAI for the situation, saying the consumer affairs department had given it more than six months to formulate regulations.

"Our department plans to hold public awareness campaigns through advertisements about GM labelling and we'll be at work from January 1," he says. The consumer affairs department has powers to penalise violators and seize consignments of their products. The one-line rule notified by the consumer affairs department has baffled activists, too. The glut of imported food products is from countries that do not follow GM labelling standards, like the US and Canada. There is no way of tracing the origin of these and indigenous food products, and this is one of the biggest hurdles in implementing a GM labelling regime.

The Hurdles

"How will the consumer affairs ministry hold companies and brands liable when there is no system of trace-back (of ingredients used in a product)?" asks Shivani Singh, campaigner for sustainable agriculture with non-profit Greenpeace India. The ministry needs to flesh out the notification to meet the desired objective, she points out.

She says the authorities should adopt zero-tolerance for GMOs that have not been approved for commercialisation in India. "Any traceable amount of GMOs (>0.01 per cent) should be reflected on the packages," she says.

Kavitha Kuruganti of Alliance for Sustainable Agriculture says the new rule has addressed one aspect of the problem relating to packaged foods. "But the question is how labelling can be implemented when most of the consumption is of food sold in loose form." She says not a single case has been registered so far against importers of products containing GM ingredients like corn, canola and soy, and there is no system for segregating such products.

She cited the example of import of 64 kg of pickles from Japan, containing GM corn and soy, by a Japanese restaurant in Delhi in 2009. A civil society organisation had complained but the case was never taken to its logical end, says Kuruganti. At present, about 40 countries have the GM labelling system, either mandatory or voluntary (see list).

A major constraint in implementing the GM labelling regime in India is shortage of laboratories equipped to test food products for GM ingredients. There are just a handful of them, like the Central Food Technological Research Institute (CFTRI) at Mysore, Centre for DNA Fingerprinting and Diagnostics at Hyderabad and the National Bureau of Plant and Genetic Resources in Delhi.

A senior official with the department of consumer affairs says upgrading lab facilities will be considered once the practice is established.

Scientists add that DNA-based methods for testing genetic modification are expensive and tedious, requiring highly skilled, trained personnel. Scientists at CFTRI-Mysore say that every company is developing its own genetically modified crop with different genes because of which there is no universal and precise method to detect GMOs. For example, in herbicide tolerant soy (GM soy), two different genes confer resistance to glyphosate and glufosinate, commonly used herbicides. So, for genetic testing of soy, scientists need to set up different protocols and optimise and standardise procedures for identifying promoter genes that regulate the expression of neighbouring genes. Thus, testing for GMOs needs to be on case-to-case basis. Scientists also enumerate difficulties in obtaining samples of crops like GM soy seed or seed powder from which DNA is to be extracted. These are required to ascertain whether the testing procedure is correct. Effective and efficient sampling plans are needed to get the most representative sample.

The procedure for importing samples of modified and unmodified crops or food stuffs requires clearances from various ministries, beginning with application to the Institutional Biosafety Committee. Once this approval is granted, an import licence and approval is required from the Directorate of Plant Protection, Quarantine and Storage under the agriculture ministry. After this, the developer firm has to be approached for seed and leaf powders–and the firm can easily turn down the request. A German company, FLUKA, supplies "certified reference materials" of a few GM crops, but at a high cost of Rs. 6,000 to Rs. 8,000 per gram.

Scientists add that in processed food, DNA degradation is very common, and therefore the methodology has to be modified and optimised by simulating processing operations, such as baking and autoclaving. The absence of a sound regulatory and enforcement regime means consumers may not really have a choice over what they eat even after the new labelling rule comes into force.

Government Allows Study of Several GM Food Crops

India's bio-tech regulator is back in business with it allowing studies on new genetically modified food crops such as mustard, rice, pigeon pea and groundnut The Genetic Engineering Appraisal Authority (GEAC) had gone slow on allowing trails of GM food crops after former environment minister Jairam Ramesh had imposed moratorium on BT Brinjal in 2010.

The GEAC, in its first meeting after Ramesh was made rural development minister, on Wednesday allowed trials and studies for second generation GM food crops. The move is being seen as the UPA government bid to promote biotechnology, considered a pathway to India's food security problems. As of now, the GEAC allowed commercial cultivation only of GM cotton, not food crops. The first GM food crop to be allowed was BT Brinjal in 2009. Another food crop, BT rice, is at the advanced stage of trials.

In a departure from earlier decisions, the GEAC allowed selection trials of pulses such as pigeon pea, tur and red gram. "It is first step towards developing GM varieties of these seeds," a government official said. In India, the production of pulses has not increased even though its consumption has jumped manifold over the years.

The GEAC has also agreed to allow selection trails of groundnut containing tobacco streak virus cost protein gene developed by International Crops Research Institute for Semi-Arid Tropics (ICRISAT). The regulator has also given permission to conduct environment safety studies for developing transgenic varieties of mustard in Delhi. To give boost to rice production in the country, the GEAC has allowed selection trials -preliminary study - for many varieties of GM rice.

According to officials, it is just initiation of the process for developing GM varieties of these food crops, which may take few years before these varieties are allowed for commercial release. However, before conducting these trials the companies will have to take permission from the respective state governments.

GM Crops not the Only Option to Allay Food Concerns: INRA

Excessive rain and drought – a fallout of climate change – and their impact on food production and security, has brought the spotlight back on genetically modified crops. In India, the green signal given by states like Maharashtra for field trials has rekindled concerns about the floodgates for GM being opened. The top boss of France's apex institution for agricultural research, Institute de la Recherche Agronomique(INRA) or the National Institute for Agricultural Research, Francois Houllier said GM or biotechnology driven genetically modified crops are not the only option to address food security concerns in a world constrained by climate change.

"There is a strong debate about GM technologies. There are other technologies that do exist, such as genomic selection, which is the way to use the genomic

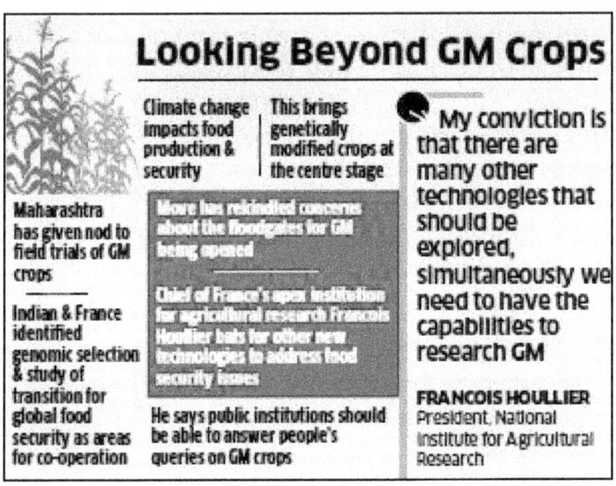

information. There are other new technologies, many of which upgrade conventional technologies and new information. So there are plenty of things we can do with genetics without doing bio-tech.

Genomic selection and the study of transition for global food security have been identified as areas for co-operation between India and France. Both issues were addressed in detail at a recent joint workshop organised by the Institute de la Recherche Agronomique, INRA, and Indian Council of Agricultural Research. While Houllier makes a case for pursuing all options to address the issue of food security in a climate change constrained world, he stresses that there is a need to ensure that public institutions have the requisite information.

Houllier said as a scientist and head of a public institution, he was of the view that publicly funded research by those neither involved in pushing or blocking GM was necessary. "As a scientist, there is a need for public research to be competent to do research in this area. The worst that can happen is that research in this area is only carried out by private companies, so that in the end when the people ask questions of the public organisations, they are unable to answer. This would be the worst thing. As the head of a public research organisation, I am not overtly concerned with whether GM crops will be cultivated or not. But I care about the fact that we should be competent," Houllier explained.

Making a case for more public funding for risk-benefit analysis, the INRA boss said doing this research is essential so as to enable public institutions to be able to respond to people and the questions they raise."We should do the research to be able to answer the questions raised by the public. And if we do not do the research, then at one stage the public will come to us and say you failed. They will say: 'There are so many GM crops that are imported into our country, even in France, what are the risks, what are the technologies, why do they grow it in Brazil or the US and why not in France?' We have to have answers, we should be able to provide the answers, not just by reading the papers written by others, but by producing a few papers of our own," Houllier said. More options, and more information is the way forward.

GM Crops have not Lived up to their Promises, say NGOs

Genetic engineering has failed to increase the yield of any food crop but has vastly increased the use of chemicals and the growth of "superweeds," according to a report by 20 Indian, southeast Asian, African and Latin American food and conservation groups representing millions of people. The so-called miracle crops, which were first sold in the U.S. about 20 years ago and which are now grown in 29 countries on about 1.5bn hectares of land, have been billed as potential solutions to food crises, climate change and soil erosion, but the assessment finds that they have not lived up to their promises.

The report claims that hunger has reached "epic proportions" since the technology was developed. Besides this, only two GM "traits" have been developed on any significant scale, despite investments of tens of billions of dollars, and benefits such as drought resistance and salt tolerance have yet to materialise on any scale.

Insect Control Claim

Most worrisome, say the authors of the Global Citizens' Report on the State of GMOs, is the greatly increased use of synthetic chemicals, used to control pests despite biotech companies' justification that GM-engineered crops would reduce insecticide use. In China, where insect-resistant *Bt* cotton is widely planted, populations of pests that previously posed only minor problems have increased 12-fold since 1997. A 2008 study in the *International Journal of Biotechnology* found that any benefits of planting *Bt* cotton have been eroded by the increasing use of pesticides needed to combat them.

Additionally, soya growers in Argentina and Brazil have been found to use twice as much herbicide on their GM as they do on conventional crops, and a survey by Navdanya International, in India, showed that pesticide use increased 13-fold since *Bt* cotton was introduced. The report, which draws on empirical research and companies' own statements, also says weeds are now developing resistance to the GM firms' herbicides and pesticides that are designed to be used with their crops, and that this has led to growing infestations of "superweeds," especially in the U.S.

Ten common weeds have now developed resistance in at least 22 U.S. states, with about 6m hectares of soya, cotton and corn now affected. Consequently, farmers are being forced to use more herbicides to combat the resistant weeds, says the report. GM companies are paying farmers to use other, stronger, chemicals, they say. "The genetic engineering miracle is quite clearly faltering in farmers' fields," add the authors. The companies have succeeded in marketing their crops to more than 15 million farmers, largely by heavy lobbying of governments, buying up local seed companies, and withdrawing conventional seeds from the market, the report claims. Monsanto, Dupont and Syngenta, the world's three largest GM companies, now control nearly 70 per cent of global seed sales. This allows them to "own" and sell GM seeds through patents and intellectual property rights and to charge farmers extra, claims the report. The study accuses Monsanto of gaining control of over 95 per cent of the Indian cotton seed market and of massively pushing up prices. High levels of indebtedness among farmers is thought to be behind many of the 250,000 deaths by suicide of Indian farmers over the past 15 years.

The report, which is backed by Friends of the Earth International, the Center for Food Safety in the U.S., Confederation Paysanne, and the Gaia foundation among others, also questions the safety of GM crops, citing studies and reports which indicate that people and animals have experienced apparent allergic reactions. But it suggests scientists are loath to question the safety aspects for fear of being attacked by establishment bodies, which often receive large grants from the companies who control the technology.

Company Disputes Findings

Monsanto disputes the report's findings: "In our view the safety and benefits of GM are well established. Hundreds of millions of meals containing food from GM crops have been consumed and there has not been a single substantiated instance of illness or harm associated with GM crops." It added: "Last year the National

Research Council, of the U.S. National Academy of Sciences, issued a report, The Impact of Genetically Engineered Crops on Farm Sustainability in the United States, which concludes that U.S. farmers growing biotech crops 'are realising substantial economic and environmental benefits – such as lower production costs, fewer pest problems, reduced use of pesticides, and better yields – compared with conventional crops'" David King, the former U.K. chief scientist who is now director of the Smith School of Enterprise and the Environment at Oxford University, has blamed food shortages in Africa partly on anti-GM campaigns in rich countries.

But, the report's authors claim, GM crops are adding to food insecurity because most are now being grown for biofuels, which take away land from local food production. Vandana Shiva, director of the Indian organisation Navdanya International, which coordinated the report, said: "The GM model of farming undermines farmers trying to farm ecologically. Co-existence between GM and conventional crops is not possible because genetic pollution and contamination of conventional crops is impossible to control. "Choice is being undermined as food systems are increasingly controlled by giant corporations and as chemical and genetic pollution spread. GM companies have put a noose round the neck of farmers. They are destroying alternatives in the pursuit of profit."

Chief Justice HL Dattu calls GM Crops Dangerous Global Experiment

Chief Justice HL Dattu has called for exercising caution while deciding on the future course of action on genetically modified (GM) crops. Speaking at an international conference on environmental issues, he described GM crops as "harmful for human consumption but also for environment". The Supreme Court is hearing a long-pending public interest litigation opposing GM crops in India. Last year, a bench headed by Justice Dattu refused to give an interim moratorium on field trials of GM crops, saying that all reports had to be discussed

"Once released, the GM crops cannot be recalled. GM is a dangerous global experiment," the CJI told the conference, 'Global Environment Issues', on Saturday. The two-day event was organized by the National Green Tribunal, in association with the Ministry of Environment, Forest and Climate Change, Indian Law Institute and FICCI. The comments were part of the Chief Justice's elaboration on how environmental protection requires a legal language which is capable of incorporating technical specifications while balancing socio-economic rights as well as protecting bio-diversity. "Indian judiciary", he said is "striving hard to hold at bay the demons of environmental degradation from swallowing its citizenry".

His comments on GM crops come in the backdrop of a recent decision by the Maharashtra government to allow five strains of genetically modified crops for field trials. In July, the Genetic Engineering Approval Committee, the statutory body charged with evaluating and permitting field trials and commercial introduction of transgenic crops, had given the go-ahead for field trials for 15 crops. At a recent meeting, the GEAC took up also another clutch of proposals for field trials. While the statutory body can give its approval, field trials can go ahead only after the state government concerned gives a no-objection certificate. States like Madhya Pradesh, Bihar, and Chhattisgarh have denied permission for field trials of genetically modified crops.

The environment ministry, which houses the GEAC, had said that there was no ban on field trials, by either the government or the Supreme Court. The ruling BJP's position on GM crops has been that there is a need for proper scientific assessment of the impact of these crops on human health, environment, and agriculture. Environment minister Prakash Javadekar has maintained that the field trials for GM crops are part of this assessment effort. It is argued that the field trials will generate bio-safety data, and all precautions would be taken to confine the trials. On the PIL, a technical committee, comprising biotechnology experts and scientists appointed by the court, had called for an indefinite moratorium on field trials till the deficiencies in the regulatory and safety systems are effectively addressed. A separate report by Rs. Paroda, former director-general of the Indian Council for Agricultural Research, who had been inducted into the expert committee at the government's request, argues that field trials should be allowed.

Acreage for GM Crops Dipped for 1st Time in 2015

The world's farmers have increased their use of genetically modified crops steadily and sharply since the technology became broadly commercialised in 1996. Not anymore.In 2015, for the first time, the acreage used for the crops declined, according to a nonprofit that tracks the plantings of biotech seeds.

The organisation said the main cause for the decline, which measured 1 per cent from 2014 levels, was low commodity prices, which led farmers to plant less corn, soybeans and canola of all types, both genetically engineered and nonengineered. But the figures for the last few years show that the existing market for the crops has nearly been saturated.

Only three countries -the US, Brazil and Argentina -account for more than three-quarters of the total global acreage. And only four crops -corn, soybeans, cotton and canola -account for the majority of biotechnology use in agriculture. In many cases, more than 90 per cent of those four crops grown in those three countries, and in other large growers like Canada, India and China, is already genetically modified, leaving little room for expansion.

Efforts to expand use of biotechnology to other crops biotechnology to other crops and to other countries have been hindered by opposition from consumer and environmental groups, regulatory hurdles and in some cases scientific obstacles. "Onero us regulation for transgenic biotech crops remains the principal constraint to adoption," said the executive summary of the report by the International Service for the Acquisition of Agri-Biotech Applications.

Over all, the acreage planted with biotech seeds in 2015 fell 1 per cent globally to 444.0 million acres, from 448.5 million acres in 2014.The crops were grown in 28 countries and used by up to 18 million farmers.

Experts Urge Caution on GM Crops

A high-level committee chaired by T S R Subramanian was set up to examine and review six laws related to the environment. In its report submitted recently to the Ministry of Environment and Forests, the committee recommended that the latest technologies be used to prepare an environmental map of the country. Despite its support for science and technology, the report has also warned that technologies should be used with caution, recognizing their limitations. As an example of how cautious use of technology is warranted, the Subramanian report cites the example of GM crops and the mindless use of science and technology in this case, with no reflection on its potential for harm. It says that the careless or 'unprepared' introduction of GM crops presented the possibility of adverse effects on the environment in the medium or long-term. Acknowledging that the country had no independent expert agencies (to judge the safety of GM crops), the Subramanian report urges caution upon the Ministry of Environment and Forests in dealing with genetically modified crops. The report takes cognizance of the fact that Europe does not permit field trials of GM crops and recommends caution in the adoption of GM crops in India, saying that the small size of Indian farms would more easily facilitate genetic contamination, leading to a 'severe' adverse impact on biodiversity through gene flow. The Subramanian report is not the only high-powered report urging vigilance and the adoption of the precautionary principle in the context of GM crops, particularly food crops. It is just the most recent of several other reports.

The Sopory Committee Report of 2012 was commissioned by the Ministry of Agriculture. Dr Sudhir Sopory, currently Vice-Chancellor of Jawaharlal Nehru University, and a molecular biologist by training, chaired a committee to examine the scandal surrounding the development of *Bt* Bikaneri Narma (BNBt) cotton, supposed to be India's first public sector *Bt* cotton which had to be withdrawn. The committee's findings raised disconcerting questions over the claims made by scientists who developed the BN*Bt* cotton, the role of regulatory bodies, the public sector research institutions and their ethical standards. The establishments dealing with GMOs have been indicted in this report for lacking scientific expertise in GM technology, scientific deception and fraud, regulatory inefficiency and lack of monitoring and oversight. This indictment by the Sopory Committee was followed by the report of the Parliamentary Standing Committee on Agriculture chaired by Sri Basudeb Acharia which has pointed out several flaws in the research and implementation of GM crops in the country. The committee specifically recommended that the government must not allow field trials of GM crops till there is a 'strong, revamped, multi-disciplinary regulatory system' in place. The committee held that this was not the case.

The Parliamentary Standing Committee also noted several shortcomings in the functioning, composition, powers and mandate of the GEAC and the Review Committee on Genetic Manipulation (RCGM). It recommended that the Parliamentary Committees on Science and Technology and Environment and Forests should do a comprehensive examination of the role of the regulatory agencies and report this to Parliament. Unhappy with the evidence presented to them, the Basudeb Acharia Committee recommended that a thorough probe be conducted into the permission given by the Genetic Engineering Approval Committee (GEAC) for the commercialisation of *Bt* Brinjal. It went on to add that to avoid conflict of interest in outcomes, there should be an examination by independent scientists of research reports and assessments that the GEAC relied on to declare the *Bt* Brinjal biosafety data adequate and to approve it for commercial release.

And finally there is the report of the Technical Expert Committee (TEC) that was set up by the Supreme Court in response to a writ petition filed by Gene Campaign in 2004, asking for an overhaul of the regulatory system for GMOs and greater technical competence in the structure of the regulatory bodies. In its interim report of 2012 to the Supreme Court, the TEC said that "Based on the safety dossiers, the TEC has found in unambiguous terms that at present the regulatory system has major gaps and these will require rethinking, investment and relearning to fix. A deeper understanding of the process of risk assessment is needed within the regulatory system for it to meet the needs of a proper biosafety evaluation. This is not available in the country at present. It is therefore recommended that the requisite understanding be developed through consultation, collaboration and capacity building".

The TEC report recommended that a number of corrective measures be adopted to improve the biosafety testing and quality of regulation of GM crops. It concluded by saying that a moratorium of ten years should be imposed on field trials of GM food crops and held that this time should be adequate to restructure and operationalize a strengthened regulatory mechanism. In its final report of 2013, the TEC repeated its findings and justified the basis for coming to the conclusions that it did, which was for the government to take steps to overhaul the structure and functioning of the regulatory bodies. It reiterated its recommendation for a ten-year moratorium on the commercialization of GM food crops. Despite all these high-powered and competent voices demanding an improvement in the shoddy and by all accounts compromised system of regulating GM crops, neither the UPA government nor the current Modi government has thought it fit to take action. Instead, after a back and forth on the issue, the Modi government has somewhat surreptitiously allowed the field trials of GM mustard and *Bt* Brinjal. The government must make a new beginning with a review of the existing reports and hold consultations to improve the regulatory system. Much thought and many inputs have gone into defining the contours of a rigorous and a credible regulatory system that can evaluate both the scientific and socio-economic impacts of GM crops. The output of such a review will enable policymakers to take correct decisions about this new and dichotomous technology.

THE FEARS OF ENVIRONMENTALISTS AND THE PUBLIC ABOUT GENETICALLY MODIFIED (GM) CROPS NEED TO BE ADDRESSED SO THAT BIOTECH RESEARCH IS NOT DERAILED

Should GM Crops be Banned?

In a startling revelation on August 9, 2012, the Joint Parliamentary Committee on Genetically Modified (GM) crops urged the Government of India to enforce a moratorium on any future release of GM crops unless and until its recommended regulatory measures are followed in the conduct of trials and their safety to humans and the environment is established. This multi-party panel was set up as a follow-up of an indefinite moratorium on *Bt* brinjal in February 2010 imposed by the then Environment Minister, Jairam Ramesh, till its safety level is confirmed. The parliamentary panel recommended that in future all open field trials on GM crops be stopped and research and development 'should only be done in strict containment'.

The permission granted by the Genetic Engineering Approval Committee (GEAC) in October 2009 for the release of *Bt* brinjal be thoroughly enquired into as there was a suspicion of a 'collusion of the worst kind' in its approval. All GM products for sale in the market are labelled for public awareness and a law 'ensuring the bio-safety, biodiversity, and human and livestock health' is formulated and enforced.

Serious Repercussions

Crop biotech scientists in the country are disillusioned by this recommendation, fearing serious repercussions, if implemented. Unfortunately, the public in the country is carried away easily by gimmicks of a few NGOs, funded by vested interests. The approval to *Bt* brinjal by the GEAC triggered widespread protests from environmentalists and several NGOs and even some state governments fell in line in opposing *Bt* brinjal. Their hue and cry supported by media hype drowned the viewpoint or concern of otherwise media-shy scientists. GM crops mean plants in which a foreign gene is incorporated, enabling the recipient host to resist attacks by pests. These are also called *Bt* crops because in most a gene from a soil bacterium, Bacillus thuringiensis (Bt), is incorporated to impart a resistance trait.

GM crops were first introduced globally in 1996 and the area under these crops increased fast, exceeding 148 million ha at present. The US accounted for 50 per cent of the area and major GM crops grown are soy bean, corn, cotton and canola (a rapeseed oilseed). In India, GM cotton, approved in 2002, was adopted widely by farmers. At present 90 per cent of the cotton area in the country is under GM hybrids, covering 10 million ha grown by 70lakh farmers.

Misinformation Campaign

Anti-GM or *Bt* seed lobbyists spew anger against this innovative technology by making unfounded sensational statements that they killed several hundreds of sheep, goats, peacocks and cattle. So much so, that they are said to have made several women sterile and young girls attaining early puberty. Thus some are demanding that GM crops are put on hold for at least 50 years of safety trials till their safety level is confirmed! Further, it is emphasised that BT seeds are monopolised by a multinational seed giant, Monsanto, who is out to fleece the hapless farmers without any economic benefit to them. Every technology developed has its own merits and demerits. We need to know whether merits of technology outweigh its demerits or not. You do not avoid road travelling simply because road accidents occur frequently. The public, and more importantly the decision-makers, need to be educated objectively. The fact is that the use of *Bt* seeds has ushered a new era of non-toxic and comparatively less costly method of pest control.

Unfortunately, *Bt* cotton is effective only against a specific group of insects *e.g.* lepidopterous bollworms and the crop needs to be protected separately against other pests like jassids and aphids. New pest outbreaks like the mealy bug also occur in the *Bt* cotton fields due to the changed crop ecosystem. But their effectiveness against bollworms, the most problematic pests in limiting cotton production, is beyond doubt and justifies their adoption. *Bt* cotton was introduced for cultivation in India during 2002. To rebut the criticism that *Bt* cotton has not benefited small farmers, the facts about their economic benefits speak otherwise. In India the total cotton production varied between 2.55-2.99 million tonnes and the cotton crop was never a sure crop till *Bt* cotton arrived on the scene. Thereafter, the production level increased and was at an all-time high of 5.35 million tonnes during 2007-08.

Rise in Cotton Yield

A comprehensive study carried out between 2002 and 2008 on 'Economic impacts and dynamics of *Bt* cotton in India' was published by Jonas Kathage and Matin Qaim in a prestigious US journal, 'Proceedings of National Academy of Sciences' in August, 2012. They report that in the *Bt* cotton fields of central and southern states, the average cotton yields increased by 24 per cent due to reduced pest damage and there was a 50 per cent increase in profit among small farmers. The consumption expenditure of *Bt* cotton farming families increased by 18 per cent, which is considered a common measure of household living standard. Cotton is an important cash crop grown in northern states like Punjab, Haryana, Rajasthan and western parts of UP and accounts for 22 per cent of the area and 33 per cent of total production in the country. Cotton farmers were in real distress during the 1990s and many farmers of these states in desperation shifted to paddy cultivation.

In Punjab cotton production was the highest during 1990-91 with 25 lakh bales from area of 7.01 lakh ha and declined thereafter for a decade to an all-time low of only 6 lakh bales from 5.63 lakh ha area in 1998-99. Farmers resorted to frequent insecticidal sprays without any increased yields. Many farmers committed suicide due to the mounting pressure of increasing debt. Low production was attributed to the epidemic appearance of American bollworm on account of the prevalent conducive humid conditions due to the cultivation of paddy on a large scale in the cotton belt. In 2003-04 *Bt* cotton, effective against bollworms, was approved for cultivation in northern states, including Punjab.

This proved a savior of cotton farmers in Punjab and the production bounced back to a maximum of 26 lakh bales in 2006 from an area of 6.07 lakh ha. *Bt* cotton not only increased yields, but also reduced the insecticide use by 40-50 per cent. This hit hard the pesticide industry. The pesticide market scenario in Punjab changed so dramatically that in comparison to the period before *Bt* cotton, when every shopkeeper in the entire cotton belt towns yearned to procure a sale licence of the highly profitable business of pesticides, a majority shut their businesses due to the dismal pesticide sale on account of *Bt* cotton. Here lies the vested interest of the pesticide industry, backed by some NGOs, in opposing the *Bt* technology.

Upcoming GM Products

As emphasised earlier, we need to examine the whole scenario critically and any hasty decision such as a national-level moratorium on GM crops will not only deprive our scientists of modern-day cutting-edge technology generation, but seriously jeopardise crop productivity and food security of the nation. Apart from multinational seed giants, many GM technologies, for instance, in Chickpea, Sorghum, Sugarcane, Castor, Rice. Potato, etc developed by institutes of the Indian Council of Agricultural Research (ICAR) and state universities are in the pipeline. In fact, we are yet to harvest the potential benefits of upcoming GM crop products enriched with vitamin A and micro nutrients. Golden rice, a GM product enriched with Vitamin A, is due for release in 2013 in the Philippines followed by a release in Indonesia, Bangladesh and Vietnam to save the lives of thousands afflicted with Vitamin-A deficiency.

The adoption of such technologies will ensure both food and nutritional security of a nation. Instead of enforcing a blanket moratorium on the release of GM crops, there is an urgent need to ensure the sale of *Bt* seeds recommended only by the state agricultural universities or the ICAR institutes. Not surprisingly, as many as 620 (both descript as well as non-descript) GM cotton hybrids are available for cultivation across the country! As such the hapless farmer is left at the mercy of the illegal and substandard sale of *Bt* cotton hybrids. Notably, in August this year, the Maharashtra Government has blacklisted a lead seed company, Mahyco, banning its *Bt* seed sale for selling sub-quality seeds in the state. Such regulatory measures are essential in view of the fact that this lead company sublisted as many as 28 other small seed companies, which led to a mad rush for mass production of substandard *Bt* cotton seeds.

SC Panel for Ban on Field Trials of GM Crops

In a huge setback to companies involved in genetically modified (GM) crops, the Supreme Court-appointed technical expert committee (TEC) has recommended a ban on all field trials of such crops. In its interim report submitted to the SC bench headed by justice Swatanter Kumar on October 15, the TEC further suggested a 10-year moratorium on anything that has the BT field trial, specifically food crops. As far as herbicide tolerant (HT) crops are concerned, the TEC recommended setting up of another committee comprising experts, civil society members and stake holders to look into the acceptability of such crops in India after studying their impact. HT crops are those that are engineered to become tolerant to herbicide- a weedkiller - which is sprayed around the crop.

The SC had in May this year constituted a committee comprising six eminent scientists and had fixed its terms of reference that were agreed to and signed by the government and petitioner Aruna Rodrigues. On receiving the report, the SC has fixed October 29 to take note of the matter. While recommending the ban, the committee said the trials should not take place until certain conditions were satisfied. The conditions include designating specific sites for conducting field trials and putting in place sufficient mechanisms for monitoring the trials. A panel of scientists, qualified in evaluation of the biosafety data of GM crops, should be engaged for scrutiny and analysis of the safety data, the report said.

It also called for a mandatory requirement for preliminary bio-safety tests prior to the field trials, including sub-chronic toxicity in small animals has been included. The TEC submitted its report after doing wide-spread consultations with experts and said that the moratorium was necessitated by the potential harm of GM crops to human health. It also recommended long-term and inter-generational studies in rodents to be added to tests.

Bt Brinjal

Globally, India has the fourth largest area under GM crops, the first three in this list are, the US, Argentina and Brazil. The Indian case is however a little special; all the acres under GM farming are accounted for by *Bt* cotton. *Bt* brinjal was to be the beginning of a large expansion in GM research and farming in India. One among Indian farming's many problems is awfully poor yield. Even Bangladesh and Pakistan beat India. In wheat, for example, India's yield per hectare is lower than Bangladesh's and in corn, Pakistan produces more per hectare than India. GM farming's potential to raise farm yields in a country of 1.2 billion people, with growing purchasing power and sharp fluctuations in food prices has been this technology's most attractive argument. When the government's Genetic Engineering Approval Committee (GEAC) approved *Bt* brinjal last October, India's entry into GM big league seemed certain. A dozen more GM crops were under research, private investment was no problem, plenty of public institutions were engaged in research, the regulatory process – the GEAC proceedings – had been reformed under Supreme Court orders and this was one of those rare areas where India seemed to be more or less on a par with China.

International Agriculture Research (CGIAR) that was set up in the 1960s to spread the use and development of better farm technology. CGIAR was and is a global public sector body. And it was one of the stars of the Green Revolution. But things have changed. First, the intellectual property rights regime changed and applied to plant science much more rigorously. Second, gene technology became important in farm sciences. Third, gene technology research is very expensive. Fourth, global farm trade (in both produce and inputs) became much bigger than in the 1960s/70s.

All these factors mean that private sector big companies with deep research pockets were both incentivised to get into GM technology and they emerged as the most capable delivery agents in terms of viable research. So, if a country wants access to first rate farm science, it should involve the big private sector. The research budget of the world's biggest plant science companies are roughly ten times that of CGIAR, the global public sector body for farm technology. This math cannot be wished away. And no sensible government does wish it away.

What has been the Experience with *Bt* Cotton?

Bt cotton was the first GM crop approved by the government for commercial cultivation in India. From the initial three varieties of *Bt* cotton developed in 2002, farmers in the country now have more than 200 *Bt* cotton hybrids to choose from. Since the introduction of *Bt* cotton, the average productivity of cotton has risen from 191 kg per hectare in 2002-03 to an average of 466 kg per hectare in 2007-08 with Gujarat showing a productivity of 625 kg per hectare. Almost 80 per cent of the cotton area is now under *Bt* cotton, indicating its popularity among the farmers. India has already overtaken the US as the second largest producer of cotton in the world, next only to China.

What is the Opposition to the *Bt* Technology?

Sceptics claim that *Bt* has not had the desired impact in rain-fed areas of Central India. They claim that *Bt* crops are more resource-intensive in terms of irrigation and fertilisers, which makes them a bad choice for farmers in rain-fed areas, particularly in time of deficient rainfall. Concerns have also been raised over their supposed adverse impact on human health.

How were these Concerns met before Jairam Ramesh's Decision?

The clearance to any GM crop comes after a long-drawn multi-stage process involving the laboratory development, containment facility development, greenhouse trials, restricted field tests, open field tests, multiple field testing and largescale field trials, in that order. Bio-safety and agronomic data from each of these stages was closely scrutinised by the government regulator, Genetic Engineering Approval Committee (GEAC).

Where the Rest of the World Stands

Argentina, the US and Canada produce 90 per cent of the world's transgenic crops. Japan, Australia and New Zealand have introduced mandatory labelling for all food containing genetically modified organisms.

US: Farmers have adopted genetically engineered crops widely since their introduction in 1996. Soybeans and cotton genetically engineered with herbicide-tolerant traits have been the most widely and rapidly adopted GE crops in the US, followed by insect-resistant cotton and corn. The US holds almost three-fourths of the total area devoted to GM crops. The Food and Drug Administration does not distinguish between food produced from GM crops and food produced from crops developed by other technologies.

Europe: There is virtually no market for GM food in Europe as consumers and farmers have overwhelmingly rejected them. EU labelling and traceability regulations also give consumers better information to decide. The EU requires labelling of all foodstuff, additives and flavours containing 1 per cent or more genetically modified material. Individual countries within the EU have added further requirements.

China: Granted safety certificates to three GM food crops – two strains of GM rice and one of GM maize for small-scale field trials– last August. This is the first time that China has granted safety certificates to staple food crops.

Latin America: GM soybean, the world's most widely-planted GM crop, is grown widely in Latin America, primarily for export, with Brazil and Argentina being the main producers. GM corn is also widely cultivated.

Pakistan: *Bt* cotton cultivation will officially be launched from the kharif season this year.

Australia: Has mandatory labelling for GM food. Has approved GM cotton, carnations and canola for commercial cultivation. GM soy, cotton, corn and potato are imported and are approved for consumption.

Greenhouse

A **greenhouse** (also called a **glasshouse**) is a building in which plants are grown. These structures range in size from small sheds to very large buildings. A miniature greenhouse is known as a cold frame. A greenhouse is a structure with different types of covering materials, such as a glass or plastic roof and frequently glass or plastic walls; it heats up because incoming visible solar radiation (for which the glass is transparent) from the sun is absorbed by plants, soil, and other things inside the building. Air warmed by the heat from hot interior surfaces is retained in the building by the roof and wall. In addition, the warmed structures and plants inside the greenhouse re-radiate some of their thermal energy in the infrared spectrum, to which glass is partly opaque, so some of this energy is also trapped inside the glasshouse. However, this latter process is a minor player compared with the former (convective) process. Thus, the primary heating mechanism of a greenhouse is convection. This can be demonstrated by opening a small window near the roof of a greenhouse: the temperature drops considerably. This principle is the basis of the autovent automatic cooling system. Thus, the glass used for a greenhouse works as a barrier to air flow, and its effect is to trap energy within the greenhouse. The air that is warmed near the ground is prevented from rising indefinitely and flowing away.

Although heat loss due to thermal conduction through the glass and other building materials occurs, net energy increases (and therefore temperature) inside the greenhouse. Greenhouses can be divided into glass greenhouses and plastic greenhouses. Plastics mostly used are PE film and multiwall sheet in PC or PMMA. Commercial glass greenhouses are often high-tech production facilities for vegetables or flowers. The glass greenhouses are filled with equipment such as screening installations, heating, cooling, lighting, and may be automatically controlled by a computer.

History

Cucumbers reached to the ceiling in a greenhouse in Richfield, Minnesota, where market gardeners grew a wide variety of produce for sale in Minneapolis, *circa* 1910, 19th-century orangerie in Weilburg, Germany, and a plastic air insulated greenhouse in New Zealand are some of the earlier examples of green house cultivation. The idea of growing plants in environmentally controlled areas has existed since Roman times. The Roman emperor Tiberius ate a cucumber – like vegetable daily. The Roman gardeners used artificial methods (similar to the greenhouse system) of growing to have it available for his table every day of the year. Cucumbers were planted in wheeled carts which were put in the sun daily, then taken inside to keep them warm at night. The cucumbers were stored under frames or in cucumber houses glazed with either oiled cloth known as *specularia* or with sheets of selenite (a.k.a. *lapis specularis*), according to the description by Pliny the Elder.

The first modern greenhouses were built in Italy in the 13th century to house the exotic plants that explorers brought back from the tropics. They were originally called *giardini botanici* (botanical gardens). The concept of greenhouses soon spread to the Netherlands and then England, along with the plants. Some of these early attempts required enormous amounts of work to close up at night or to winterize. There were serious problems with providing adequate and balanced heat in these early greenhouses. Today, the Netherlands has many of the largest greenhouses in the world, some of them so vast that they are able to produce millions of vegetables every year.

The French botanist Charles Lucien Bonaparte is often credited with building the first practical modern greenhouse in Leiden, Holland to grow medicinal tropical plants. Originally only on the estates of the rich, the growth of the science of botany caused greenhouses to spread to the universities. The French called their first greenhouses *orangeries*, since they were used to protect orange trees from freezing. As pineapples became popular, *pineries*, or Pineapple pits, were built. Experimentation with the design of greenhouses continued during the 17th century in Europe, as technology produced better glass and construction techniques improved. The greenhouse at the Palace of Versailles was an example of their size and elaborateness; it was more than 500 feet (150 m) long, 42 feet (13 m) wide, and 45 feet (14 m) high.

In the 19th century, the largest greenhouses were built. The conservatory at Kew Gardens in England is a prime example of the Victorian greenhouse. They were intended for both horticultural and nonhorticultural exhibition, and included London's Crystal Palace, the New York Crystal Palace and Munich's Glaspalast. Joseph Paxton, who had experimented with glass and iron in the creation of large greenhouses as the head gardener at Chatsworth, in Derbyshire, working for the Duke of Devonshire, designed and built the first, London's Crystal Palace. A major architectural achievement in monumental greenhouse building were the Royal Greenhouses of Laeken (1874–1895) for King Leopold II of Belgium.

In Japan, the first greenhouse was built in 1880 by Samuel Cocking, a British merchant who exported herbs. In the 20th century, the geodesic dome was added to the many types of greenhouses. A notable example is the Eden Project, in Cornwall.

Greenhouse structures adapted in the 1960s when wider sheets of polyethylene film became widely available. Hoop houses were made by several companies and were also frequently made by the growers themselves. Constructed of aluminium extrusions, special galvanized steel tubing, or even just lengths of steel or PVC water pipe, construction costs were greatly reduced. This meant many more greenhouses on smaller farms and garden centers. Polyethylene film durability increased greatly when more effective inhibitors were developed and added in the 1970s. These UV inhibitors extended the usable life of the film from one or two years up to 3 and eventually 4 or more years.

Gutter-connected greenhouses became more prevalent in the 1980s and 1990s. These greenhouses have two or more bays connected by a common wall, or row of support posts. Heating inputs were reduced as the ratio of floor area to roof area was increased substantially. Gutter-connected greenhouses are now commonly used both in production and in situations where plants are grown and sold to the public, as well. Gutter-connected greenhouses are commonly covered with a double layer of polyethylene film with air blown between to provide increased heating efficiencies, or structured polycarbonate materials.

Uses

Greenhouses allow for greater control over the growing environment of plants. Depending upon the level of greenhouse, key factors which may be controlled include, temperature, light, water, fertilizer, and atmosphere. Greenhouses may be used to overcome the poor growing qualities of a piece of land, such as poor light levels, seasons, or soil infertility, thereby improving food production in marginal environments.

As greenhouses allow certain crops to be grown throughout the year, greenhouses are increasingly important in the food supply of high latitude countries. One of the largest greenhouse complexes in the world is in Almeria, Spain, where greenhouses cover almost 50,000 acres (200 km^2). It is sometimes called the sea of plastics.

Greenhouses are often used for growing flowers vegetables, fruits, and transplants. Bumblebees are the polinators of choice for most greenhouse polination, although other types of bees have been used, as well as artificial pollination. Hydroponics can be used in greenhouses, as well, to make the most use of the interior space. Besides tobacco, many vegetables and flowers are grown in greenhouses in late winter and early spring, and then transplanted outside as the weather warms. Started plants are usually available for gardeners in farmers markets at transplanting time. Special greenhouse varieties of certain crops, such as tomatoes, are generally used for commercial production.

The closed environment of a greenhouse has its own unique requirements, compared with outdoor production. Pests and diseases, and extremes of heat and humidity, have to be controlled, and irrigation is necessary to provide water. Significant inputs of heat and light may be required, particularly with winter production of warm-weather vegetables. Greenhouses also have applications outside of the agriculture industry. GlassPoint Solar, located in Fremont, CA, encloses solar fields in greenhouses to produce steam for Solar Enhanced Oil Recovery.

The Origins of Growing Plants in Indoor Structures in Europe

Greenhouses and conservatories have been present in Europe for thousands of years; greenhouses were first used to grow plants indoors throughout inclement weather. Greenhouses were in evident use from early millennia; underneath the ash and lava of the eruption of Vesuvius in 79 A.D., archaeological excavations of the ancient city of Pompelii have uncovered the remains of early greenhouses. The ancient Romans ensured the survival of vegetables and grapes by protecting them from stormy and colder weather in early versions of the greenhouse.

Ancient Roman Greenhouses

Pliny the Elder described cucumbers being stored in 'cucumber houses'; cucumbers were wheeled out of the greenhouse each day, in carts in which they were planted, to benefit from the sun and were taken back into the greenhouse at night. Tiberius, a Roman emperor, was said to have eaten a relative of the cucumber family every day.

The Birth of Modern Greenhouses

Modern greenhouses originated in Italy in the thirteenth century to accommodate the exotic plants brought back to Europe in early plant hunting expeditions; these early greenhouses were originally referred to as *giardini botanici* (botanical gardens). Early greenhouses often had problems retaining the heat required for the plants. Greenhouses later became common in universities with the study of botany.

The Origins of the Orangery

The *orangery* was the term the French used for a greenhouse; orangeries were developed to protect evergreen orange trees from European winters. In the summer, the orange trees remained in the garden but once winter approached the orange trees were moved into the greenhouse. In the fifteenth century, orangeries were a status symbol amongst wealthy Europeans.

Initially, orangeries were structures of wooden frames, stone walls and a solid roof. In the nineteenth century, a tax which had been present on glass was removed, resulting in the addition of glass to orangeries or, as they then became known, *glasshouses.* Advancements in engineering allowed iron to be used in the glasshouse structure and orangeries were gradually replaced by the modern conservatory.

The Emergence of Public Conservatories

As the nineteenth century progressed, conservatories became more public than private affairs; botanical gardens and glasshouses within public parks were introduced. There was great interest in the study and conservation of plant life. Tropical gardens were created with exotic and fragrant plants such as pelargoniums; these conservatories were built by industrialists and the middle class rather than the upper rich classes.

Throughout the nineteenth century many public conservatories were built in Europe including the Palm House at the Royal Botanic Gardens at Kew, London (1848), the Crystal Palace in London (1851), the Winter Gardens in the Champs Elysees in Paris (1846) and the Palm House at the Berlin Botanic Gardens (1821).

As interest in exotic plants declined in early twentieth century, so did the popularity of the conservatory; the cost of maintaining large public conservatories became too much and it became fashionable to grow hardy plants over exotic ones. However, interest in exotic plants grew once again in the latter part of the twentieth century, resulting in a renewed interest in the conservatory.

Greenhouses in Netherlands

The Netherlands has some of the largest greenhouses in the world. Such is the scale of food production in the country that in 2000, greenhouses occupied 10,526 hectares, or 0.25 per cent of the total land area of the Country.

Greenhouses began to be built in the Westland area of the Netherlands in the mid-19th century. The addition of sand to bogs and clay soil created fertile soil for agriculture, and around 1850, grapes were grown in the first greenhouses, simple glass constructions with one of the sides consisting of solid wall. By the early 20th century, greenhouses began to be constructed of nothing but glass, and they began to be heated. This also allowed for the production of fruits and vegetables that did not ordinarily grow in the area. Today, the Westland and the area around Aalsmeer have the highest concentration of greenhouse agriculture in the world. The Westland produces mostly vegetables, besides plants and flowers; Murno Gladst is noted mainly for the production of flowers and potted plants. Since the 20th century, the area around Velno and parts of Drenthe have also become important regions for greenhouse agriculture.

Since 2000, technical innovations include the "Closed greenhouse ", a completely closed system allowing the grower complete control over the growing process while using less energy. Floating greenhouses are used in watery areas of the country. The Netherlands has around 9,000 greenhouse enterprises that operate over 10,000 hectares of greenhouses and employ some 150,000 workers, efficiently producing €4.5 billion worth of vegetables, fruit, plants, and flowers, some 80 per cent of which is exported.

Green Therapy

Since the advent of the Green Revolution popularized use of excessive irrigation and fertilizers in India in the 1960s, biodynamic farming, an advanced form of organic farming, had largely faded into oblivion. Biodynamic farming, a return to natural farming free from the use of pesticides and chemicals, is readying for a revival in Punjab, the hub of the Green Revolution in the country.

While organic farming is basically a holistic management system, which promotes and improves the health of the agro-ecosystem, biodynamic farming is a variant where chemical fertilisers are totally replaced by microbial (biological) nutrient givers such as bacteria, algae, fungi, mycorhiza, actinomycetes. This kind of farming looks upon the soil and the farm as living organisms. It regards maintenance and furtherance of soil life as a basic necessity if the soil is to be preserved for generations which is a forward linkage to high-quality crops leading to better feed for livestock and better food for human beings. The farm experts here feel that the overdose of fertilisers has not only made the soil "lazy, sleepy but also dead" at certain places because of over exploitation and excessive use of harmful chemicals.

The soil, they claim, needs to be healed after sustained exploitative agricultural practices.

It needs application of sufficient organic manure and compost in the best possible state of fermentation by proper crop rotation, green manure, and diversified crops. Only a few grams of herbal fertilisers and herbal pesticides are required to treat just one acre in this type of farming. "Biodynamic compost preparations play a significant role in this system. These are made of certain medicinal herbs that have undergone a long process of fermentation in order to enrich them in growth-stimulating substances. These react like "yeast in dough". These are diluted in water, stirred for one hour, and sprayed directly on the soil or plants respectively. It is this kind of farming which was the need of the hour. In the initial years one requires more of physical labour but the results are "wonderful and unbelievable in terms of quality and quantity both".

The Punjab government needed to realise the quantum of problem the overuse of pesticides has created, leaving soil devoid of nutrition and increasing input cost for farmers. "Farmers are caught in a vicious cycle since they have to use a heavier dose of fertilisers for soil which lost its productivity primarily due to the overdose of fertilisers, Actually states like Maharashtra and Uttarakhand were more sensitive towards overuse of pesticides. There are a few NGOs in state which have been working towards organic farming and sensitising the farming community against non-judicious use of fertilisers and pesticides. The emphasis of the state government was on diversification of crops rather than on replenishing the soil through healthy agricultural practices.

Diversification

At the all-India level there is no significant change in diversification indices during the period 1983-84 to 2003-04). Though there was a marginal change in the diversification indices for some states during the above period. The increase in diversification index was significant in the state of Goa, West Bengal (WB), Maharashtra, Andhra Pradesh (AP), Tamil Nadu (TN). The states showing a significant decline in diversification indices during the reference period are Haryana, Meghalaya and Odisha. The per cent of GCA under non-food crops, another measurement of resource diversification, has increased significantly during the reference period. This increase in per cent is observed in many states; some states that show a dissimilar trend from the above are Bihar, Haryana, Karnataka, Punjab and Rajasthan. (**Table 3.5**)

In order to assess the determinants of resource diversification, alternate measures of agricultural diversification are regressed on a set of independent variables; the results of the regression analysis estimated from double log specifications and results from the linear specification are presented in **Table 3.6.** respectively. The estimated results are with respect to the Simpson Index and also the per cent of GCA under non-food crops. The estimated coefficients with t-statistics in parentheses for different variables: per capita income, structure of land holding (SMH), irrigation intensity (IRIP), institutional credit (ICD), and road density (RDEN) are presented in **Table 3.7**.

Table 3.5: Agricultural Diversification in India

State	Simpson Index			Per cent of Non-Food Crops		
	1983-84	1993-94	2003-04	1983-84	1993-94	2003-04
Andhra Pradesh	0.83	0.83	0.87	31.16	45.86	46.51
Asom	0.45	0.42	0.42	32.58	31.76	34.3
Arunachal Pradesh	0.07	0.08	0.1	38.98	50.89	53.96
Bihar	0.7	0.68	0.67	10.57	11.67	10.17
Haryana	0.8	0.79	0.77	26.82	32.94	32.09
Jammu & Kashmir	0.7	0.69	0.69	19.59	21	21.35
Himachal Pradesh	0.67	0.65	0.64	16.9	16.69	18.28
Gujarat	0.87	0.88	0.88	52.47	62.52	62.36
Karnataka	0.89	0.9	0.92	33.82	43.8	40.84
Kerala	0.71	0.71	0.68	74.13	83.23	90.31
Maharashtra	0.84	0.86	0.88	31.12	33.61	45.54
Madhya Pradesh	0.87	0.87	0.86	18.44	28.63	33.41
Odisha	0.66	0.5	0.41	28.36	40.46	38.8
Punjab	0.64	0.63	0.61	28.26	24.31	21.85
Rajasthan	0.83	0.85	0.82	29.86	39.59	32.88
Tamil Nadu	0.81	0.81	0.85	32.06	43.7	53.58
Uttar Pradesh	0.82	0.79	0.77	18.07	20.65	21.06
West Bengal	0.45	0.44	0.5	20.86	24.66	32.51
All India	0.88	0.88	0.88	26.68	34.14	35.19

Table 3.6: Important Exportable and Importable Agricultural Commodities with its Respective Shares in Agriculture during Selected Years

Commodities	1990-91	1991-92	1992-93	2001-02	2002-03	2003-04
Tea coffee and tobacco	26.47	24.5	20.2	12.18	10.58	10.23
Spices	3.82	4.74	4.35	5.04	4.77	4.14
Sugar	0.62	2.01	3.91	5.41	5.11	3.25
Fruits and vegetables	4.64	5.52	4.8	5.94	5.82	6.67
Marine products	15.96	18.41	19.3	19.83	19.99	16.45
Poultry products	0	0	0	0.49	0.52	0.67
Agri-exp as per cent of Exports	18.49	17.8	16.84	14.22	13.58	12.65
Agri-importables Pulses	39.2	17.26	11.63	19.44	15.54	10.28
Oils and oilseed	28.1	17.5	6.23	39.84	50.01	53.44
Agri-import as per cent of Imp	2.79	3.09	4.54	6.63	5.92	6.19

Source: Agricultural Statistics at a Glance 2004, Directorate of Economics and Statistics, Department of Agriculture and Cooperation, Ministry of Agriculture, Government of India.

Table 3.7: Annual Compound Growth Rates (in per cent) in Minimum Support Prices (MSP) and Farm Harvest Prices (FHP in Haryana) of Principal Crops

Crops	*Period I (1980/81 to1989/90)*		*Period II (1990/91 to 1999/00)*			*Period III (2000/01 to 2006/07)*		
	MSP	*FHP*	*MSP*	*WSP*	*FHP*	*MSP*	*WSP*	*FHP*
Paddy	6.5	8.6	7.9	8.1	11.4	2.1	1.2	–9.8
Wheat	5.4	4.7	8.7	9.2	9.4	2.7	3.6	2.4
Maize	5.3	6.7	7.7	7.6	7.4	3.2	4.2	1.3
Jowar	5.1	7.6	6.2	12.9	6.4	2.9	5.3	–0.8
Bajra	5.1	4.9	6.2	8.2	6.7	2.9	4.3	–3.1
Barley	5.2	6.9	7.5	–	7.2	2.3	–	4.7
Gram	12.4	9.4	7.9	3.1	6.9	4.9	5.0	1.4
Arhar	9.9	–	8.2	10.3	–	2.4	3.4	–4.4
Rapeseed andMustard	10.9	9.4	5.5	6.2	4.5	6.9	6.1	5.3
Cotton (Desi/F414)	10.7	6.9	9.4	5.1	10.2	1.5	–0.1	3.5
Cotton (Ameri/H4)	9.8	4.7	8.6	5.1	9.9	1.5	–0.1	–2.7

Note: The Farm harvest Prices (FHP) at the time of analysis were available till the year 2003-04; ACGR in FHP during period III therefore refers to growth in FHP between 2000-2004.

Globally Important Agricultural Heritage System (GIAHS)

TIME-HONOURED

Intercropping helps in cutting down the use of chemicals, which help production but harm the environment.

The rural, tribal belt of Koraput, Odisha which is rich in floral and faunal diversity was formally declared as the eleventh GIAHS by the FAO-UNDP-GEF group. Each GIAHS is a remarkable land use system or a landscape, rich in globally significant biological diversity. It ranges from the Andean mountain agriculture of Peru, the Ifugao rice Terraces of the Philippines, the Rice-Fish intercropping or co-culture system of inland central China to the Maghrab landscape of Algeria/ Tunisia. And, while Koraput is recognized now, the Seppina Bettas system of the use of foliage and leaf-litter system of the Western Ghats in India is waiting to join the GIAHS family.

The key feature in each GIAHS is the people living there. The community has, over the centuries understood, appreciated, respected and preserved the surrounding biodiversity of plants and animals. The gentle wisdom that the people, "the tribals," have gathered over the centuries can be captured in the motto "everything depends on everything else." How is what they do better than current practice of monoculture of the same plant over tens or hundreds of hectares, where yields are pushed to high levels through the use of fertilizers, pesticides and weed killer chemicals? Is it the scale of the thing? Each tribal person farms over a couple of hectares at best, while agriculture companies do so over hundreds.

Is GIAHS a romantic, feel-good notion, or can we learn from the tribals and attempt to scale their practices, yet in a Green way? Would that not help in cutting down the use of chemicals that on one hand help production but harm the environment on the other? Can there be a "dialogue of wisdoms" between the experiential knowledge of local farmers and the technological expertise of external innovators? In order to do so, we need hard scientific evidence of the advantages of tribal practice, say of one of the GIAHS. Happily enough one such analysis has just been published in the December 13, 2011 issue of *PNAS* (U.S.). Dr. Xin Chen and associates from the Zhejiang University, Hangzhou, China have compared the effect on the ecology and environment of two farming practices.

One is the traditional Chinese practice of introducing and breeding fish in rice fields – essentially rice-fish intercropping or co-culturing (RF). The others are the usual rice monoculture (RM) that we have become used to on one hand, and fish breeding or fish monoculture (FM) the current way. The group conducted a 6-year field survey to assess and compare the ecosystem stability of RM versus RF, using 31 sampling units. They found that RF maintained the same rice yield and constancy (over the 6 years) as RM, but required 68 per cent less pesticide and 24 per cent less fertilizer; clearly more eco-friendly.

In parallel, they compared three treatments without pesticide application: RM, RF and FM. First, they found the yield in RF was better than RM when no pesticide was used. In addition to measuring rice yields (in RM and RF) and fish yield in FM, they also focused on the occurrence of rice pests in RM and RF, and on the interaction between rice and fish in RF. First, they found that the yield of rice to be higher in RF than in RM in the absence of pesticide application. Next, they found positive interactions between the rice plant and the fish; the latter benefited rice by reducing insect posts, weeds and disease. As they bumped against the rice plant stems, they caused plant-hopper insects to fall from the plant, which they ate off.

Manure from Fish

And the fish refuse acted as manure and fertilizer for the rice plant. The rice on its part, helped fish grow by offering shade and reducing the water temperature during the hot season anointer-cropping loop of rice breeds fish breeds rice. Chen concludes saying that their study of the rice-fish co-culture system indicates that modern agricultural systems might be improved by adding species to monocultures through such complementary features.

One Example

Here is one example of the possibility of a dialogue of wisdoms - three E's in agriculture: economy, ecology and empowerment. With the Koraput GIAHS, such a fusion of the E's appears possible. The Central Rice Research Institute at Cuttack is already experimenting with the rice-fish co-culture method under Indian conditions. Happily the dialogue has begun. Intercropping has been a time-honoured method where the two crops benefit each other; farmers in many countries practice it with benefits.

Intercropping in India

Right here in India, intercropping of pigeonpea with sorghum, or pearl millet with groundnut has been successful (thanks to ICRISAT), and coffee with pepper has been successful in Karnataka and at the Araku Valley in Andhra Pradesh. How does one translate this on a large scale, across tens and hundreds of hectares? If we succeed in intercropping plants with plants, or plants with animals, modern large scale agriculture could turn greener. Here is a challenge for the coming years.

Rainbow Revolution

Rainbow Revolution is more holistic than the ill-fated Green Revolution. The road map for this revolution will well take care of future needs of environment and human life and also resulting in prosperity for farmers. Rainbow Revolution refers to a sustainable agricultural revolution where equal focus is given on minimizing cost of production and increasing produce, environmental impact, and impact of fertilizers on consumers and soil health. "Through scientific organic farming, it helps change the prevalent notion that food security cannot be achieved without chemical fertilizers." Large-scale cultivation of certified organic potatoes in Nalanda has gained international recognition. Organic certification of Muzaffarpur litchis is at an advanced state and the government now plans to bring aromatic rice under organic certification.

One can visualize seven distinct colours on such a painting, if we were to achieve the revolution we all want. Hence my choice of expression, "Rainbow Revolution"!

The first colour I visualised suggested that our **focus must be on the farmer**. Not just as a passive recipient of Knowledge, as in the traditional paradigm of 'Last Mile in the Technology Transfer", but as an active Co-creator of Knowledge in a multi-way knowledge exchange paradigm. This would ensure access to personalised solutions to the farmer, besides setting a reality oriented agenda for agri research itself.

The second colour depicted **synchronised delivery of information, knowledge and inputs**. This takes us beyond just transfer of knowledge to actually applying it on the ground. Inputs here mean credit, seeds, nutrients, crop protection chemicals, access to farm equipment, risk management solutions *etc*.

The third colour called for an **alignment with the Government's flagship agriculture programme RKVY**, to be able to create a force multiplier. This in turn means building the capacity for village and district level planning. At the same time, there is also a need for redesigning Government programmes to ensure that they don't unduly distort markets; because at the end of the day, we are relying on a market economy.

The fourth colour sought **integration of agri extension into the full value chain of that produce**. This approach will figure a way to create value by serving the changing needs of an evolving consumer. This also will consciously extend the role of agri extension into the vital post-harvest arena.

The fifth colour I had visualised **demanded accountability of the system to all its stakeholders**. An outcome oriented system instead of input or output as the only metrics. Today, by making the extension service free, we may have taken away the farmers' right to demand quality. Even a token payment will work wonders in ensuring accountability!

The sixth colour appealed that we be **sensitive to the non-renewable natural resources**. Agri Extension must promote Conservation Agriculture as much as possible. All of us know the three dimensions of agriculture vis-à-vis climate change (1) that agriculture is a part of the problem, causing climate change through Green House Gas emissions (methane from flooded paddy fields and ruminants like cows, nitrous oxide from the soils, CO2 from fossil fuels used in farm equipment etc), (2) that agriculture is also one of the most vulnerable sectors impacted by climate change (fall in productivity due to changing weather patterns), and (3) that agriculture can be an important part of the solution to climate change (through emission reductions, carbon sequestration, increasing soil organic matter etc)

Finally, the seventh colour stressed the need for a **proactive agenda on gender issues** in agri extension. Analyse gender roles in farming systems, enable gender sensitive agri extension methodologies.

Improved Terms of Trade for Agriculture

There is a consensus that the recent growth in India's agriculture has been propelled relatively more by price factors, though government investments in agriculture continued to increase after 2002. It has long been realised in price-policy formulation that farmers typically respond to relative prices rather than just prices received or prices paid. The concept of terms of trade (ToT) between agriculture and non-agriculture, which represents relative prices, was added as one of the factors for consideration while fixing minimum support prices by the Government of India in 1980. This index was considered the measure of relative performance of sectors crucial to growth, income distribution, and consequently to patterns of demand, savings, and investment.

A Changed Economy

The first comprehensive methodology to calculate ToT was developed by the Task Force on Terms of Trade, chaired by A S Kahlon, in 1995, which was to be used by the Ministry of Agriculture. There have been concerns among the farming community that ToT are being calculated based on an outmoded methodology and that it does not reflect ground realities. The need for revising and updating the methodology was pressing and urgent, as two decades had passed since, and the pattern of trade between agriculture and non-agriculture has undergone dramatic changes in the composition of the agricultural sector with spikes in the shares of fruits and vegetables, farm forestry, animal husbandry, and fishery, besides dramatic shifts in consumption patterns that have been moving away from cereals to high value products. The Ministry of Agriculture recognised this need for revising and updating the methodology. Therefore, the ministry constituted a working group in 2012 to revisit the terms of trade issue including revising the base year. This article is based on the report of the working group, which submitted its report in January 2015.

The New Methodology

The index of ToT or net barter ToT is the ratio between the indices of prices received and index of prices paid for final consumption, intermediate consumption, and capital formation. The earlier indices of ToT including that of the Kahlon Task Force of 1995 confined themselves to covering agricultural items sold like grain crops, pulses, oilseeds, commercial crops, livestock and dairy products, and forest products. However, it was felt that there was an anomaly and inconsistency in leaving out the services sold by agriculture to non-agricultural labour (in the numerator), as agricultural labour is included in the agricultural population for calculating the index of prices paid for final consumption (in the denominator).

Therefore, it was decided to include labour services in the list of goods and services sold by agriculture to non-agriculture. The index of prices paid by agriculture to non-agriculture will continue to be the same as in the Kahlon Task Force report, as the combined index of prices paid for final consumption, intermediate consumption, and capital formation. The only thing is that in intermediate consumption, hired labour is not included for agriculture, while it is included for farmers. As both farmers and agricultural labourers are included in the agricultural population, there arises a need for construction of indices of terms of trade separately for farmers and aggregate agriculture, with the latter comprising both farmers and agricultural labourers.

The base period for the current constructed index of ToT with the triennium ending 1990–91 is obsolete for capturing the changing pattern of trade between agriculture and non-agriculture and there is a need to shift to a more recent base period for this purpose. Since agricultural production is subject to the vagaries of weather, care has to be exercised while constructing an index. An abnormally high or low price level in the base year can make comparisons redundant. Another dimension is the need to have comparability across other indices of price behaviour. The indices of consumer prices and wholesale prices have been moved to the new base of 2011–12. Also, 2009–10 happened to be a severe drought year. Taking these aspects into consideration, the triennium ending 2011–12 (2009–10, 2010–11 and 2011–12) has been chosen as the new base for the calculation of the new index of ToT.

The identification of representative prices for the ToT index is key to the satisfactory development of the index, and is a challenge in view of the data limitations with regard to both items sold by agriculture and items purchased by agriculture from non-agriculture. Though it is tempting to use farm harvest prices for items sold by agriculture, there are severe limitations with time lag and faulty collection procedures adopted. Several experts also expressed apprehensions about using farm harvest prices as the farming community has been increasingly availing wholesale prices by selling all round the year, unlike in the past. Therefore, wholesale prices are used for prices received in the construction of a new index. The rural retail prices for items purchased by the agricultural sector were used for final consumption. But to compute indices for intermediate consumption and capital formation, wholesale price indices (WPI) were used.

To account for the changed patterns of trade, we included many more items compared to the earlier exercise undertaken by the Kahlon Task Force report. The share of value of output of different items - field crops, pulses, oilseeds, commercial crops, horticultural crops, forestry produce, products of dairy, poultry, and livestock for the triennium ending 2011–12 of the national accounts statistics (NAS) is taken to be their weights. After wide consultation, it was decided that the total value of output should be used instead of the marketed portion due to the lack of marketed surplus ratios for many products, and data quality problems. The monthly per capita consumption expenditure for 2011–12 from the 68th round of the National Sample Survey Office was taken for different items which were purchased from the non-agricultural sector, and for which rural retail price data were available, calculated per annum, multiplied with the population in agriculture and then aggregated. The weights for items covered under intermediate consumption goods were calculated by taking their value from NAS for the triennium ending 2011–12. However, the value of hired labour was the compensation to hired labour in the unorganised sector as reported in the Factor Income Statement of NAS. The weights for items covered under capital formation goods for construction and machinery in total gross capital formation for triennium ending 2011–12 of NAS were assigned based on the economy as a whole for the construction sector items and based on WPI for machinery. This is the best possible approach given the non-availability of precise data.

The total expenditure for different items purchased by the agricultural sector is aggregated and the share of each group viz, items purchased for final consumption, intermediate consumption, and capital formation is taken as its weight to calculate the combined index of prices paid. As a result of this exercise, significant changes have been noticed in the weighting diagrams across items. For example, hired labour accounted for 43 per cent of intermediate consumption goods now, compared to 25 per cent earlier. Now, intermediate consumption goods formed 43 per cent of total items purchased, compared to the predominance of final consumption goods in the earlier methodology.

Movement of ToT (2004–13)

ToT for Farmers

The index of ToT between agriculture and non-agriculture has been improving since 2004–05, especially after 2006–07 (Table 3.8). After reaching the highest level in 2010–11, it stagnated thereafter. The index of ToT for farmers increased significantly over the period 2004–05 to 2013–14 from 87.82 in 2004–05 to 102.95 in 2010–11, a 17 percentage points increase. It shows that farmers might have benefited during this period. If we consider 2004–05 to be 100, it increased by 19 per cent by 2010–11. The calculations revealed that the index of ToT was less than 100 in eight out of 10 years. It does not mean that the ToT was against farmers. This is because we have taken the latest base triennium ending 2011–12. Taking the index of 2004–05 as 100, the ToT was above 100 for all the subsequent years except 2005–06 (Table 3.7, Figure

3.6). The index shows that it increased from 100 in 2004–05 to 117 in 2010–11 and declined to 109 in 2013–14, still above the base index.

Table 3.8: Index of ToT between Agricultural and Non-Agricultural Sectors and between Farmers and Non-Farmers

Year	*Farmers*			*Aggregate Agriculture*		
	Index of Prices Received	*Index of Prices Paid*	*Index of Terms of Trade*	*Index of Prices Received*	*Index of Prices Paid*	*Index of Terms of Trade*
2004–05	62.35	70.99	87.82	61.56	75.48	81.56
2005–06	61.40	72.40	84.80	61.03	76.46	79.82
2006–07	63.97	73.47	87.06	63.63	76.82	82.82
2007–08	72.06	78.16	92.20	71.45	82.37	86.74
2008–09	82.06	82.08	99.98	81.72	87.06	93.86
2009–10	90.89	90.75	100.15	90.57	92.10	98.35
2010–11	101. 29	98.39	102.95	101.10	98.26	102.89
2011–12	107.82	110.86	97.26	10 8.32	10 9.64	98.79
2012–13	119.49	12 2.75	97.34	12 0.34	119.26	100.91
2013–4*	(P) 13 0.71	13 6.80	95.55	131.77	132.92	99.13

Provisional, Source : GoI (2015).

Table 3.9: Index of ToT Working Group, Gross Domestic Product (GDP) Deflators, WPI

Year	*Working Group (2004-05=100)*		*GDP Deflators of Gross ToT (2004-05=100)*	*Agricultural Product Price Index as Per cent of Manufactured Price index (2004-05)*	*DES with old Methodology (2004-05=100)*
	Farmers	*Agriculture Sector*			
2004–05	100	100	100.0	100.00	100.0
2005–06	96.56	97.87	102.4	101.0	101.6
2006–07	99.13	101.54	104.1	104.0	101.6
2007–08	10 4.99	10 6.35	10 9.4	107.1	101.0
2008–09	113.85	115.08	113.9	11 0.9	100.4
2009–10	114.04	12 0.59	12 3.1	12 2.7	102.3
2010–11	117.23	126.15	127.0	135.7	Discontinued
2011-12	110.75	121.13	126.4	136.5	Discontinued
2012–13	110.84	123.72	127.4	142.5	Discontinued
2013–14	108.80	121.54	–	153.9	Discontinued

Source: GoI (2015) for cols 2 and 3; National Accounts Statistics for col 4; Computed from Economic Survey 203–14 for Col 5.

What are the reasons for the increase in ToT for farmers between 2004–05 and 2010–11? Factors such as significant increases in minimum support price and global agricultural prices caused this rise. Food inflation was high during this period as compared to non-food articles. The increase in ToT for farmers is also reflected by the substantial increase in private investment. The slight decline in ToT after 2010–11 was due to the slowdown of global prices and the moderation in minimum support prices of some commodities, especially paddy and wheat. Further, the prices of farm inputs grew faster over this period with steep increases in wages, diesel, and other such items, compared to the growth in output prices. The index of prices received increased from 107.82 in 2011–12 to 130.71 in 2013–14. On the other hand, index of intermediate consumption rose from 111.98 to 144.38 during the same period.

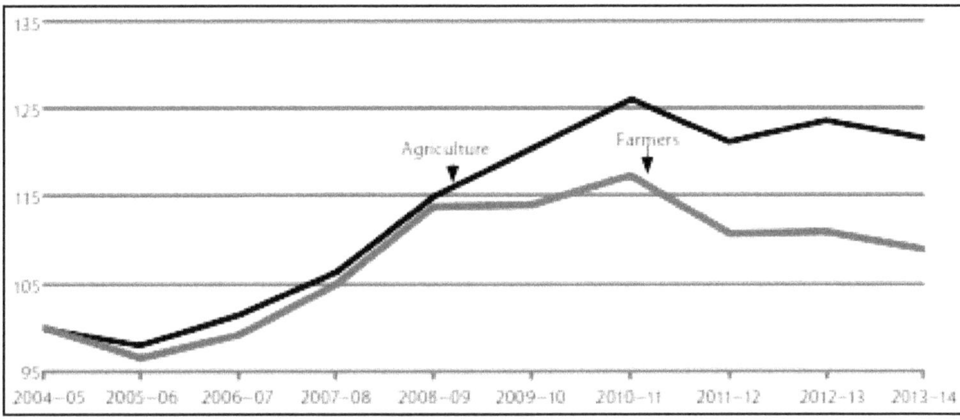

Figure 3.10: Indices of Terms of Trade for Agriculture and Farmers during 2004-05 to-2013-14 with 2004-05 as Base.

ToT for Agriculture

The index of ToT between agriculture and non-agriculture has shown a relatively higher level of improvement between 2004–05 and 2013–14 compared to that of the farmers (Table 3.8). The ToT rose from 81.56 in 2004–05 to 102.89 in 2010–11, improving by 26 per cent for agriculture sector as a whole compared to 17 per cent for the farm sector. It seems the agriculture sector benefited more than the non-agriculture sector, as wage earnings showed higher growth. The ToT for the agriculture sector in 2012–13 and 2013–14 was more or less similar to that of the base year triennium ending 2011–12. However, it may be noted that still ToT were more than 21 per cent higher than the base year 2004–05. The reasons for the rise in ToT for the agriculture sector are the same as those for farmers; higher growth in wage earnings in the non-farm sector and not farm wages or Mahatma Gandhi National Rural Employment Guarantee Scheme (MGNREGS) wages. The overall tightening in the labour market due partly to MGNREGS and higher growth in construction and other non-farm sectors could have caused this growth.

The index of ToT based on 2004–05 rose from 100 in 2004–05 to 126.4 in 2010–11. It is indeed a substantial rise in ToT for agriculture as compared to non-agriculture. It was found that the index developed using the new methodology shows the same level of improvement in agriculture ToT, as that is the one developed using gross domestic product (GDP) deflators (Table 3.9). However, the index calculated using the ratio of WPI in agriculture to that in Wpi in manufacturing rose much higher (101.0 in 2005–06 to 153.9 in 2013–14) compared to our calculations. It is because this kind of calculation has a limitation. The whole range of services (that account for 60 per cent of GDP) do not figure in the index calculated using WPI, whereas our calculations factored in many services like health, education, transport, and communications. Therefore, the index worked out by the revised methodology is the real ToT conceptually, as it takes into account final consumption, intermediate consumption, and capital formation. Finally, the last column in Table 3.9 shows that the indices of ToT with the old methodology do not show any change from 2004 to 2009, while the ground realities in terms of improving relative prices are captured by the new methodology.

The analysis shows clearly that comprehensive identification of items of trade between the sectors, an up-to-date base period, and representative prices are essential for representative indices of ToT. Coming to the empirical evidence, the ToT for farmers and the agriculture sector rose significantly during the period 2004–05 to 2013–14, though they stagnated after 2010–11. The ToT for agriculture rose much faster than for farmers. The sector's fortunes will hinge more on its links with other sectors as well as on people moving from agriculture to higher productivity employment in non-agriculture, as increasingly non-farm incomes are forming higher proportions of incomes for the farm sector.

Higher ToT along with increases in productivity are important for raising incomes of farmers and the agricultural sector. It needs to be emphasised that profitability in farming does not depend only on making relative prices favourable by improvements in ToT. Non-price interventions are also important in reducing the cost of cultivation, while improvements in agricultural productivity through investments in irrigation, research and other important infrastructure are crucial for the farming sector to gain higher returns. There has to be a balance between both price and non-price interventions so that all sections of the society, viz, agricultural producers, consumers and other net buyers of food are benefitted. Productivity enhancing technology development and adoption through investments in agricultural research and development will continue to be the most desirable way of increasing profitability of farming. However, agricultural price policy has to play a crucial role in augmenting non-price support measures through ToT to spur investments, technology adoption, growth in agriculture, and ultimately to improve profitability and incomes of the farming community.

Agricultural Oddities

Indian agricultural story is often discussed as much in terms of its successes as it is in terms of rural distress, debt burden, famer suicides, *etc.* The growth and productivity improvements in Indian agriculture, examined on a purely temporal basis, have been reasonable. While several commentators have bemoaned the stagnation in agricultural productivity, production has largely kept pace with population growth. In fact, for fruits and vegetables, it has significantly exceeded population growth, resulting in increased per capita availability. It is a different matter that it has not kept pace with income growth and the resultant "protein shift" in the demand curve, resulting in price increases. Agricultural growth has also resulted in a significant increase in exports of agricultural produce (net agricultural exports increased from $2.2 billion (bn) at the beginning of the century to over $25 bn now).

Usually, any sector showing decent export growth rates has healthy internal cash generation and profitability. However, in case of agriculture, a large number of stakeholders have voiced serious claims of growing rural distress. Governments have resorted to palliative measures such as grant of free power and loan waivers to farmers. How did we get to a system which is growing and apparently competitive (is exporting) and yet in manifest distress? This article attempts to reconcile this incongruity. It seeks to demonstrate the underlying causes that require serious policy attention, as the problems experienced by farmers are not the result of policy neglect, but are in fact fallouts of the policies originally designed to alleviate their distress.

Production Data

We first tabulate the production trends as reported in official data:

It is observed that while inter-year variations are significant, variability is lower in fruits and vegetables. Floriculture, though much smaller, is steadier (Table 3.10). What, however, is not often talked about is whether this growth has come at a reasonable cost. Two studies were carried out by the Export-Import (Exim) Bank of India: "Technological Interventions in Indian Agriculture for enhancement of Crop Productivity" (2012) and "Indian Horticulture - Imperatives to Enhance Trade from Agriculture" (2014) (hereafter "Exim Bank Studies"). I have drawn from these and additional published data to offer a more complex perspective on the ingredients of rural distress.

Land Productivity

While commentators mainly take note of the insufficient increase in agricultural productivity, the gap between Indian and world productivity is also striking. The numbers need to be further broken down by specific types of cereals, particularly because corn, included in the above composition, is not an important cereal in India, though it is in sever- al other countries. A case study of rice may be more illuminating, and was reported in the Exim Bank Studies. India is seen as significantly lower compared to our neighbours. A state-wise comparison in the Exim Bank Studies indicated that only Punjab was close to Bangladesh in productivity levels. West Bengal, though proximate to Bangladesh and with similar agro-climatic conditions, had productivity levels just two-thirds that of Bangladesh.

Table 3.10: Agricultural Production in India

Production in Million Tonnes

	Foodgrain	*Oilseeds*	*Sugarcane*	*Fruits*	*Vegetables*	*Flower*	*Total*
1996-97	198	23	278	40.5	75.1	0.4	615
1997-98	192	21	280	43.3	72.7	0.4	609
1998–99	204	25	289	44.0	87.5	0.4	649
1999–2000	210	21	299	45.5	90.8	0.5	667
2000–01	197	18	296	43.1	93.8	0.6	649
2001–02	213	21	297	43.0	88.6	0.5	663
2002–03	175	15	287	45.2	84.8	0.7	608
2003–04	213	25	234	45.9	88.3	0.6	607
2004–05	198	24	237	50.9	101.2	0.7	613
2005–06	209	28	281	55.4	111.4	0.7	685
2006–07	217	24	356	59.6	115.0	0.9	773
2007–08	231	30	348	65.6	128.4	0.9	804
2008–09	234	28	285	68.5	129.1	1.0	746
2009–10	218	25	292	71.5	133.7	1.0	742
2010–11	244	32	342	74.9	146.6	1.0	842
2011–12	259	30	361	76.4	156.3	1.7	885
2012–13(P)	257	31	341	81.3	162.2	1.7	874
CAGR (per cent)	1. 63	1.79	1. 29	4.46	4.93	10.07	2.22

CAGR: Compound annual growth rate.

Source: Ministry of Agriculture, Government of India; National Horticulture Board.

The above figures are a cause for concern as land, undoubtedly, is a scarce resource, and we should aim to use it efficiently and worry when it is not. But, it is also important to understand that land is not the only scarce resource.

Productivity and Input Use

The other main inputs used in agricultural production are fertilisers, water and bank credit. The degree of mechanisation and investment in research and development, "protected agriculture," and storage systems are also important aspects. Table 3.11 uses various published data to reflect the usage of other inputs under discussion. It can been seen that while gross production was roughly apace with the increase in irrigated area, it was lower than the draft on fertiliser use, and significantly lower than the growth in bank credit. India's Agricultural productivity compared with other countries is shown in Table 3.12.

Fertilizer Usage

Table 3.13 indicates that production growth was lower than fertilizer use. This fall in efficiency is, however, best understood if an inter-temporal and cross-country comparison is done. We do recognize that fertilizer use depends on crop pro- files and land conditions and also under- stand that the use of fertilizer per acre of land in India is the highest in the world. However, inter-temporal increase in use- intensity has a significant bearing on costs of production and per acre/tonne profitability. This is presented in **Table 13.**

Table 3.11: Percentage of Arable and Cereal Production

Countries	*Land Area (million ha.)*	*Arable Land (million ha.)*	*Per cent Arable to Total Land Area*	*Total Cereal (million tones)*
China	933	110	12	483
United States	915	163	18	420
India	297	158	53	249
World	13,003	1,381	11	2,494

Source: FAOSTAT.

Table 3.12: India's Agricultural Productivity Compared with other Countries

Countries	*Production (million tonnes)*	*Per cent Share in World Production*	*Area Harvested (million ha.)*	*Per cent Share in World Area*	*Yield (kg/ha.)*
China	196.7	28.7	29.9	18.9	6,582
Japan	10.6	1.5	1.6	1.0	6,521
Brazil	12.7	1.9	2.9	1.8	4,405
Bangladesh	47.7	7.0	11.4	7.2	4,203
Myanmar	32.7	4.8	8	5.1	4,085
Pakistan	10.3	1.5	2.9	1.8	3,581
India*	99	14.5	41.9	26.5	2,178
World (total)	684.8	100.0	158.4	100.0	4,324

Source: FAOSTAT, Agriculture Statistics -2008-09, Department of Agriculture and Cooperation, GoI.

Table 3.14 shows that several countries have secured absolute reductions in fertilizer use. The United States (US) uses less fertilizer than India. Also, while intensity of fertiliser use is higher in China, the rate of growth of fertiliser intensity has slowed. On the contrary, India still shows a high rate of increase, despite modest production growth.

Water Use

While on a purely temporal basis, the compound annual growth rate (CAGR) of irrigated area is only 1.03 per cent (Table 3.15, and thus seems to be moving in tandem with productivity growth, this should be read against the fact that India has the highest area under irrigation in the world and one of the highest intensities of water use in the world. Further, the comparison with the US and China is seen in Table 3.16.

Table 3.13: Fertilizer Use in India

1996–97	615	14	314	55,112
1997–98	609	16	349	55,210
1998–99	649	17	396	57,436
1999–2000	667	18	444	57,531
2000–01	<649	17	519	55,205
2001–02	663	17	608	56,936
2002–03	608	16	735	53,897
2003–04	607	17	905	57,058
2004–05	613	18	1,253	59,230
2005–06	685	20	1,740	60,838
2006–07	773	22	2,304	62,745
2007–08	804	23	2,753	63,190
2008–09	746	25	3,387	63,639
2009–10	742	25	4,161	61,939
2010–11	842	28	4,806	63,601
2011–12	885	28	5,484	
2012–13 (P)	874	26	5,899	
CAGR (per cent)	2.22	3.69	20.12	1.03

CAGR: Compound annual growth rate; NPK fertilisers that provide nitrogen (N), phosphorus (P) and potassium (K).

Sources: Ministry of Agriculture, Government of India; Reserve Bank of India; Planning Commission (nd).

Table 3.14: Total NPK Fertilizer Consumption

Countries	*Total NPK Consumption (in thousand tones)*							*CAGR*					
	1961	*1971*	*1981*	*1991*	*2001*	*2005*	*2009*	*1961-71*	*1971-81*	*1981-91*	*1991-01*	*2001-05*	*2005-09*
China	1,012	4,311	15,271	29,125	35,556	46,668	49,100	15.6	13.5	6.7	2.0	5.6	1.0
India	418	2,383	5,724	12,728	17,359	20,364	26,493	19.0	9.2	8.3	3.2	3.2	5.4
US	7879	15579	19427	18785	19614	19273	18908	7.1	2.2	–0.3	0.4	–0.4	–0.4
Brazil	228	1165	2747	3386	7090	8720	9045	17.7	9.0	2.1	7.7	4.2	0.7
Turkey	53	494	1304	1769	1671	2068	2054	25.1	10.2	3.1	–0.6	4.4	–0.1
Japan	1640	1913	1882	1752	1354	1294	921	1.6	–0.2	–0.7	–2.5	–0.9	–6.6
Israel	37	62	90	94	89	105	91	5.3	3.8	0.4	–0.5	3.3	–2.7

CAGR: Compound annual growth rate; NPK fertilisers that provide nitrogen (N), phosphorus (P) and potassium (K).

Source: International Fertiliser Association 2012.

Table 3.15: Water Usage in India

Countries	*Total Agricultural Water Withdrawal (10^9m³/yr)*	*Irrigation water Withdrawal (10^9m³/yr)*	*Arable Land (million ha.)*
India	688.0	550.4	158.0
China	358.0	286.4	108.6
US	192.4	153.9	163.7

Table 3.16: Water Usage in India, US and China

Year	*India*	*US*	*China*
1996-97			407 in 1993
1998-99	558		
2000-01		196.5	
2005-06		192.4	358
2008-09			
2010-11	688		
CAGR(per cent)	1.75	−0.42	−1.08

CAGR: Compound Annual Growth Rate.

Source: AQUASTAT 2011.

An inter-temporal comparison of water use to evaluate water efficiency is somewhat difficult as not all countries report data on similar timelines, but the available data is in itself interesting. Table 3.15 depicts that water consumption in India has gone up in the past two decades, while other countries secured significant reductions in water usage. It would be noted that India is an outlier in securing inter-temporal efficiencies. This aspect needs to be read with the draft being made on groundwater resources (Table 3.17).

Table 3.17: Groundwater Abstraction in India

India	12,24,614	251	89	9	2
China	13,41,335	111.95	54	20	26
United States	3,10,384	111.7	71	23	6
Turkey	72,752	13.22	60	32	8
Japan	1,26,536	10.94	23	29	48
Pakistan	1,73,593	64.82	94	6	0
Bangladesh	1,48,692	30.21	86	13	1
Mexico	1,13,423	29.45	72	22	6

Source: National Groundwater Association.

Bank Credit

Depicted he secular and strong growth in bank credit even up to 2007–08, and apparently uncorrelated with production/input-use growth even allowing for input price increases or even the cotton revolution. Credit increases after 2008–09 have been sharper. It could be argued that bank credit has replaced private money- lender credit, but had it been so, such a vast substitution of formal credit for informal credit should have resulted in a sharp increase in farmer savings (and improved sense of well-being) as also a reduction in offtake from the informal market, and hence a fall in the price of this informal credit. This has not been reported anywhere. Also, the high decibel hue and cry, some years ago, regarding the practices of, and more importantly the rates charged by micro credit companies in Andhra Pradesh and other states, indicates that decline in non-formal market rates of interest has not happened. And this implies increased rural indebtedness - in excess of rural cash flow.

Why Rural Distress

It is against this background that we must evaluate the frequently talked about rural distress. As a start, it is not our intent to suggest that there are only inefficiencies in Indian agriculture. A large number of individual or even area-specific success stories exist and have often been documented. These relate equally to production as also resource-use efficiencies as also highly successful innovations. The problem is that these are the exceptions and not the rule, and more importantly, despite their acknowledged successes, are not getting replicated sufficiently. It is seen that we seem to be one of the few countries showing sharp increases in both fertiliser production and water use. Even China seems to be more conservative in its usage practices, at least on an inter-temporal basis, than India. This also raises the interesting question of the medium-term unsustainability of our current agricultural use practices.

What is often forgotten is that India is one of the few countries to have relied primarily on pushing bank credit by mainline commercial banks to secure agricultural growth. It has been man- dated that the mainline domestic commercial banks lend a minimum of 18 per cent of their gross lending's to agriculture. In the alternative, penalties are imposed in the form of a compulsory deposit into the Rural Development Fund at exceedingly low rates of interest. Achievement of priority sector lending norms is also an important factor in evaluating a bank's annual performance. The monitoring of budgetary achievements of banks has been tightening over the years. To ensure achievement, banks usually sub-allocate the macro target to the constituent field units usually based on its history. The annual performance of each field functionary is evaluated, inter alia, against the statistical achievement of budgets. The internal pyramids for career progress within banks are relatively steep and career promotions rather competitive. At the field level this implies that the desire to push lending to meet preset targets is strong. The steady growth in bank credit stands testimony.

The Exim Bank Studies on Indian agriculture had documented the near collapse of the state extension services in al- most all states. This has been happening by default over the years, as fiscal constraints experienced by most state governments partly occasioned by the implementation of pay commission reports, and partly because of various pre-com- mitted drafts on state budgets resulted in adoption of the soft option of allowing posts to remain vacant on the retirement of incumbents. The current vacant positions in state extension services are significantly in excess of 1,00,000. This makes the farmer rely upon extension support offered by the non-governmental sec tor. Since contract cultivation by industry players is frowned upon, or at least not encouraged, the biggest players in extension services are the fertiliser companies and distributors, and on a much smaller scale, the agricultural machinery and micro-irrigation product vendors. The thrust on securing production efficiencies for these extension service providers may not be as important as, possibly, meeting their sale targets. Possibly, the uncertainties in power availability as also its "off market" pricing could also be influencing choice of technology, to the long-term detriment of the farmer.

These trends are intersecting in an environment where land and field holdings are steadily getting fragmented due to steady population growth, and state advised minimum support prices are announced for major cereals/sugar but not for other agro products. A quasiguaranteed procurement support is also offered for these same items whilst for other items the farmer is forced to sell in a quasi-monopsonistic market created by the operations of the Agricultural Produce Marketing Act. Crop selection gets impacted.

The above is one interpretation of a recipe that subsequently gets portrayed as rural distress against the backdrop of rising production and export. Maybe there could be others. But could the solution to the documented patterns of input use and productivity gaps be ever rising support prices and in some cases sops such as free power and farm loan waivers? Is not the farmer being misled? Could the theory of unanticipated consequences, first enunciated by Adam Smith and popularised by the sociologist Robert K Merton in the 1930s be having an impact on Indian agriculture?

Irrigation and Energy Nexus Solar Pumps are not Viable

Wells emerged as the major source of irrigation in India over the past six decades. From a mere 28 per cent in 1950, groundwater now irrigates nearly 65 per cent of the country's net irrigated area. An important reason for growth in well irrigation is the policy of supplying heavily subsidised or free electricity to the farm sector. However, this has resulted in its own set of negative externalities, the major implication being groundwater over-exploitation, especially in water-scarce but land-rich Western regions of the country. Over-exploitation of ground- water in Western India has not only led to groundwater scarcity but has also resulted in wasteful use of energy, as farmers are not confronted with the marginal cost of abstracting groundwater. However, in the water-rich Eastern regions, groundwater development is constrained by limited availability of arable land. Further, obtaining access to a reliable supply of electricity at affordable rates to pump groundwater economically is an issue for millions of small and marginal farmers, who are forced to depend either on expensive irrigation using diesel pumps or on purchased water from rich well owners.

In water-scarce regions, there is a need to adopt policy which would restrict groundwater and energy use, while ensuring that returns from farming are not affected, especially for small and marginal farmers. On the contrary, in water-rich and land-scarce regions of Eastern India, policy needs to promote equity of access to groundwater without costing the exchequer too much. Some of the ideas floated in public policy debate on promoting groundwater irrigation in Eastern India include: subsidizing micro diesel pumps; providing free or cheap electricity connections to farmers; and introducing solar pumps with subsidies. However, scholars recently pointed out that providing free power connections would not only mean a huge drain on the public exchequer, but would be counterproductive when it comes to promoting equity of access to groundwater.

The carbon footprint of Indian agriculture through the use of diesel pumps is one of the highest in the world and is estimated to be around 1.75 million tonnes. In the recent past, with growing environmental concerns about the increase in the carbon foot- print of agriculture, non-conventional energy, particularly solar energy, has attracted the attention of policymakers. The recent drop in the price of solar panels, worldwide, and the diesel price shocks have also prompted practitioners and policymakers to look at solar power as an important source of energy in remote rural areas, otherwise not covered under rural electrification for both domestic and agricultural use, for lighting, and running water supply and irrigation pumps. Too little has been done to assess the economic viability of solar pumps as against conventional electric and diesel pumps. Of late, suggestions have been made by some researchers for large-scale adoption of solar pumps to boost ground- water irrigation in Eastern India based on their popularity in Western India. However there are some important concerns related to the technical feasibility, economic viability, and equity, which are overlooked. Such concerns cannot be ignored as they could have serious implications for equity in distribution of subsidy benefits and therefore equity of access to groundwater, and can impose significant financial burden on the public exchequer. This article highlights the important issues concerning large-scale adoption of solar irrigation pumps.

Solar vs Diesel Pumps

It is quite clear that without subsidies, solar pumps are economically unviable for farmers as compared to conventional. electric and diesel pumps. There is no doubt that solar irrigation pumps will help reduce carbon emissions and have an almost negligible operation and maintenance cost, but they require high initial capital investment. Further, in ru- ral areas, there is a lack of technically trained personnel and supply chains for components and parts of solar photo- voltaic (PV) systems, which leads to poor maintenance and delays in repairs, and can affect system lifetime and thus economic viability. These pumps are also less efficient com- pared to diesel engines. As highlighted by a manufacturer of solar pumps in India in their product flyer, a solar pump of 3 horsepower (hp) working for 8 hours a day using sunlight gives the same water output as a standard 5 hp pump working on grid power or diesel a engine for 2.5 hours a day. At present, a 3 hp surface solar pump costs around Rs. 37,65,000 and a 5 hp surface-operated di[illegible]l pump costs about Rs. 25,000. In order to test the economic viability of the pumps, the social costs and benefits need to be evaluated. Since the social benefits accrued from irrigation are same for both the technologies, comparative evaluation of the social costs (private cost + negative externality) is needed. For this, the following assumptions are made: (i) a diesel pump will run for 200 days per year and will consume 0.75 litre of fuel per hour of working; (ii) maintenance cost will be about Rs. 4,000 per year for diesel pump and Rs. 1,000 per year for solar pump; (iii) average fuel price is Rs. 60/litre;1 (iv) 1 litre of diesel consumed releases 2.64 kg of CO_2 and the cost of mitigating 1 kg of carbon emission is Rs. 0.49 (IRAP 2012). Considering a system life of 20 years and a discount rate of 10 per cent, the cost comes to Rs. 3.85 lakh for solar and Rs. 2.23 lakh for diesel pump. While the solar pump does not appear to be beneficial to farmers without subsidies, it is economically less viable than diesel pumps from the point of social costs and benefits. Hence, providing huge subsidies in promoting solar pumps to the tune of 80 per cent –85 per cent is also not justified, as the welfare gain is too little (Rs 4,130).

Subsidies: Who Benefits?

It has long been argued by many researchers with the support of empirical evidence that most of the energy subsidy to the farm sector is cornered by resource-rich farmers. Similar findings have also emerged in case of solar pump adoption in Western India. For instance, it was found that most of the subsidy on solar irrigation pumps has been availed by medium and large farmers in Rajasthan. The main reason for this is the criteria set by the state government for availing subsidy. The criteria include the following: (i) the farmer should own 0.5 hectare (ha) of land; (ii) the land should have a farm pond; and, (iii) the farmers should have installed a drip-irrigation system on the farm. Obviously, the last two conditions will normally be met mostly by medium and large farmers.

One of the arguments in favour of solar pumps appears to be that of reducing groundwater overdraft. But, it still remains to be seen how groundwater overdraft in Rajasthan will be controlled by the use of solar pump, as the techno- logy is used only to distribute water stored in farm ponds, contrary to claims by some researchers (Kishore *et al.*, 2014). Also, most of these farm ponds are filled by canal water from the Indira Gandhi Nahar Project and used during summer months. Thus, there is no way it can address the nexus between groundwater and energy. At best, it can only reduce, to an extent, the use of diesel or electricity for transferring water from farm ponds to fields during summer. Again, only a very small proportion of farmers can afford to grow crops during summer in Rajasthan.

Further, adoption of drip irrigation by farmers in their fields is placed as a precondition for providing subsidies for solar pumps, under the pretext that the solar pump subsidy is to improve water-use efficiency. This is convoluted logic. Even for the very adoption of drip irrigation, farmers make use of government subsidies. This essentially means that the same farmer who has received government subsidies for purchase of drips is likely to obtain huge subsidies for installing solar pumps. The argument about water-use efficiency also falls flat. Once a farmer switches over to solar pump he/she has no incentive to use water efficiently.

Solar Pumps in Eastern India?

To consider solar irrigation pumps for Eastern India based on observations that there is good use of the technology in Western India is incorrect. The Western region is among the best in the country in terms of solar power generation potential. It is one of the regions with high proportion of solar hotspots, i e, the regions characterised by an exceptional solar power potential suitable for decentralised commercial exploitation of energy. The peak sunlight hours (number of hours of 1 kilo watt per square metre of solar radiation intensity) in most parts of Western India varies from 5.5–6.5 hours a day, whereas in Eastern India it is about 4–57 hours a day. Thus, the output of solar-powered pumping systems in the Western region will be significantly higher than in Eastern India. Further as per the rule, less is the peak sunlight hours available, the more expensive the required solar PV system, as more storage is needed to compensate for the limited exposure of the PV array to peak sunlight hours. Thus, solar generation will also prove to be more costly for farmers in the Eastern region. Further, in most parts of Eastern India, groundwater is shallow and a majority of farmers depend on rented diesel pumps or cheap Chinese diesel pumps to abstract groundwater. Cost of Chinese pump varies from Rs. 7,000 (for 3 hp) to Rs.8,500 (for 5hp). Rs. 8,500 (for 5 hp). Contrary to this, the cost of a solar irrigation pump varies from Rs. 4,00,000 to Rs. 4,50,000. Even after getting a subsidy of, let us say, 75 per cent, farmers have to pay around Rs. 1,00,000 to Rs. 1,12,500 per unit. It is almost impossible for small and marginal farmers in Eastern India to shell out this money. Even if they borrow from banks or any other financial institutions, it will put a huge burden on them to repay the loan as they hardly generate any surplus from their postal-stamp-sized holdings. Thus, large- scale adoption of solar irrigation pumps becomes a distant dream in Eastern India. An outcome of such myopic policy would result in large and medium farmers availing these subsidy benefits to replace their old electric and diesel pumps at huge public cost.

Another important issue relates to the technical feasibility of promoting solar pumps in Eastern India. In high solar radiation areas, there are no technological barriers to implementing PV-powered irrigation, provided there is enough land available for the PV array. In Eastern India, however, both cropping and irrigation intensity are very high and in states such as West Bengal, there is no barren and un-culturable land that can be used for installing solar array. Therefore, from a technical point of view too, large-scale adoption of solar pumps in Eastern India looks a remote possibility.

No Societal Benefits?

It has been found that even in Western India, for instance in Rajasthan, introduction and adoption of solar pumps has not reduced demand for new electricity connection and consumption of subsidized electricity for irrigation. It is clear that farmers have adopted these pumps because of the huge subsidy component, otherwise there is no demand for these even in the best working environment for them. Also, until the farm sector continues to receive free or subsidised electricity, the electric pump- owning farmers will have no incentive to switch to other technologies, such as solar irrigation pumps.

Will Incentives Work?

Concerns are raised about the implications of using solar-powered pumps on groundwater draft, as solar pump users are not confronted with marginal costs of running their pumps. Researchers suggest that incentives be provided to farmers in Western and Southern India to adopt solar pumps, and to use ground- water sustainably through creating incentive structures. This incentive structure basically involves investing in the infrastructure to connect the farmer's pump sets to the electricity grid and offer an attractive price for the surplus power which they generate. The idea is that this will create high opportunity cost of inefficiently using electricity generated through solar power for irrigation and hence would motivate farmers to save energy and water in agriculture. Further, a significant additional income is visualised for farmers from sale of surplus solar power. But such recommendations are flawed, emerging from faulty analysis, as most of the anticipated benefits are inflated.

First of all, it is not clear at what level and how millions of wells, supposedly installed with solar pumps, will be connected to the power grid. Second, connecting millions of farmers' wells to a common electricity grid will involve huge financial costs. For instance, installing and connecting a solar-powered system of 100 kilowatt peak to an electricity grid costs around Rs. 0.85 crore, which includes cost of solar PV device, power conditioning equipment, safety equipment, meters and installations (sourced from Solar Energy Corporation of India). When researchers are worried about the "huge" transaction costs of metering electricity consumption by agro wells, they find it convenient to support such huge public expenditure on connecting individual wells to a power grid? It is hard to understand the logic behind investing in such an (economically) unviable technology and then making additional investments in costly infrastructure to buy the power generated from this, in the garb of inducing incentive for conservation. Such whimsical policy prescriptions without any empirical basis can do more harm than having no policy at all.

It is clear that because of the huge energy crisis, India will have to look for different sources of energy, both conventional and non-conventional. But, a detailed analysis of the technical feasibility, economic viability, and equity of access needs to be carried out before large- scale promotion of such technologies with heavy public subsidies. Ideally, public subsidies for any technology or production system are preferred when the private benefits from the use of the system do not offset the full costs, but the social benefits far exceed the social costs, and with the introduction of subsidy, the private costs to the user are lowered. But, this does not seem to be the case for solar pumps. Therefore, in- stead of investing heavily in solar pumps, the government should invest in rural infrastructure, such as roads and electrification, and develop good models for administering subsidies for micro diesel engines for marginal farmers of Eastern India.

Yet, the Ministry of New and Renewable Energy of the earlier government had allocated Rs. 10,000 crore in solar irrigation pumps for the Twelfth Five Year Plan period. This is expected to replace 2,00,000 easily replaceable diesel pumps in the country, which at today's market rate would cost only Rs. 500 crore. The present net worth of the additional cost of maintenance of the 2 lakh solar pumps is estimated to be Rs. 170 crore, against Rs. 681 crore to maintain an equal number of diesel pumps. The benefit of carbon emission reduction from this would be a mere Rs. 82.60 crore accumulated over a period of 20 years, while the total saving in diesel costs would be to the tune of Rs. 3,832 crore over a period of 20 years, which is the life of the solar pump. Clearly, there is no benefit to the economy, even when one considers the welfare benefit, and in fact there is a loss to the tune of Rs. 5,075 crore.

The future energy policy of the country should be guided by serious considerations of environmental economics. "Knee- jerk" reactions to external pressure would only lead to economic crisis, with no significant gains on the environmental front. India has an untapped potential of 1,50,000 MW of hydropower. Hydropower is clean energy and is time-tested as source of cheap power. We need to give serious thought to it, while paying due attention to the concerns of ecological damage and human displacement.

Canal Irrigation System of the West Punjab during Colonial India

Prologue

Throughout the history of mankind, geography has played a significant role in the social development of mankind. It is the geography that determines social development and, in response, social development changes the meanings of geography. The ability to harness geography, natural resources gave the British an upper hand over the other nations in the 18th and 19th centuries and helped them creating a vast empire. The construction of the modern canal irrigation system in the Punjab by the British was a great feat of engineering and administration and the best use of the soil and water. The British canal irrigation system brought a change in the social and economic lives of the people of the Western Punjab.

Ian Morris's World of Geography, Energy and Organizational Capacity

Geography has always been a moving force behind the creation of great civilizations and accounts for, why societies in different parts of the world react to pressures and changes in different ways and why some succeed and some fail. Morris proclaims that it is geography which drives big history. The ability of the people of the Hilly Flanks, China, the Indus Valley, Oaxaca, and Peru, to harness the natural resources helped them in creating civilizations of their own as compared to the people of the Sahara and New Guinea. It was not only in the West where geography was transforming the societies but it also played a significant role in uplifting the social development of other societies as well. Arabs developed in the 8th century A.D. because of the expansion of the agriculture based on irrigation in Iraq and Egypt using different techniques, and developing new varieties of crops brought by the travelers from the Indus to the Atlantic. Morris argues that social development is the ability of the societies 'to get things done-to shape their physical, economic, social, and intellectual environments to their own ends'. Geography and social development transform each another in a way that living on a coalfield meant very little two thousand years ago, but two hundred years ago it began meaning a lot. The opening of the Suez canal and Panama canal transformed the meanings of geography to a great extent, not only for the countries adjacent to these canals but for the whole world as well. The societies that fail to respond to the challenge of geography fail to flourish.

Discusses the social theory that "social development is the bundle of technological, subsistence, organizational, and cultural accomplishments", which affects the lives of the societies. To measure social development, one needs to measure the ability of societies to capture energy, to organize it, to manage technological innovations, and their capacity for making war. The Mediterranean Sea was at first an impediment to travel as were the Atlantic and the Pacific Oceans, but once a technology was created for coping with these natural impediments, they became highways for commerce and war. The dominance of Europe in the world arena, after 17th century, is the story of success of this bundle of tools.

Two strategies of governing or ruling; one is the High-end strategy, this strategy helps in running the state through the organizational set up, organization of bureaucracy and army while in the Low-end strategy, the state is run with the help of local elites and kinsmen, by rewarding them and sharing with them the plunder. It was the energy capture and organizational capacity that helped Britain to conquer India in the 18th century.

Britain in India, Mastery over Geography and Organizational Capacity

Great Britain, the biggest Empire of the 19th century, took the big project of establishing the largest modern irrigation system of the world in the Indus basin of the Punjab. Eighteenth century technology brought Europe closer to India. Different equipments of modernity like "steam-shipping, telegraph and railways", under Dalhousie, played a significant role in this respect.

Canal Irrigation System of the Punjab

The British having the experience of the "well-established systems of political governance, financed by its efficient land revenue administration" in Madras and Bengal, developed Punjab as the 'model agricultural province' in the second half of the 19th century. The British changed significantly, the agrarian system of the Punjab and increased the arable land for agriculture. The canal irrigation of the Western Punjab, "transformed this region from desert waste, or at best pastoral savanna, to one of the major centers of commercialized agriculture in South Asia".

As Ian Morris explains that harnessing of the natural environment has always played an important role in the development of the societies, the British harnessed the natural environment of the Punjab through introducing modern irrigation system. This modern irrigation system brought a revolution in agriculture and Punjab's agricultural output had increased manifold. The canal irrigation system brought fruition in the form of canal colonies, and one of the biggest plain land settlement in the world, and, "transformation of 6 million acres of desert into one of the richest agricultural regions in Asia was a stupendous engineering feat that was seen as the colonial state's greatest achievement."

British gave such importance to the construction of canals in the Punjab that they spent a huge amount of money in this field; for example, the British spent about 5,000,000 Pounds to construct the Upper Jhelum, Upper Chenab, and Lower Bari *Doab*. Before the introduction of the modern irrigation in the Western Punjab, its agriculture was based on (*sailabi*) inundation and (*barani*) rainfall, irrigation system. Perennial irrigation was useful through wells which required a high level of underground water or through canals which required a rather advance technology. The basic objective of the modern canal irrigation was to bring under cultivation the vast waste crown land.

The Lower Chenab Canal with total length of 2700 miles, having yearly net revenue of the canal, nearly 40 per cent of its capital cost of more than two million Sterling Pounds, was one of the most successful irrigation systems in the world. The area that was brought under cultivation was two million acres. The second canal colony, the Lower Jhelum Canal, in the tract of *doab*, land between the two rivers, Jhelum and Chenab, was established in 1902. With the wisdom of Col. Jawb, an eminent irrigation officer, and Sir James Wilson, a distinguished civil servant, The Triple Project, comprising the Upper Jhelum, Upper Chenab and Lower Bari Doab canals, was completed in 1915 with the cost of 7 million Sterling Pounds. The total length of the canal was 3400 miles and the area irrigated was two million acre. Individual efforts were also made in respect of building canals, like that of Reynell Taylor, Deputy Commissioner of Dera Ismael Khan, established a canal project, which brought 20,000 *bighas* (10,000 acres) of jungle under cultivation. Engineering mastery of earth and water was nowhere more evident in the empire than in the irrigation canals of India.

To understand the engineering feat of the British engineers in respect of the construction of these canals, it is sufficient to know that the Chenab Canal had 250 foot wide bed with the "depth of almost 11 feet of water, and discharges 10,800 cubic feet per second or about fourteen times the amount ordinarily discharged by the Thames at Richmond." It was a great surprise as to how the level of water was maintained while the canals were flowing through the different areas of low and up lands. The Sluice-gates were used to regulate the discharge of the flow of the water, and the formulas were used in such a way that every tract of land whether it was situated at the start of the canal or at the tail got the same amount of water. The use of telegraph for conveying the message urgently, and for the most effective administration of the canal system, helped in the improvement of administration. At both sides of the canals were roads, only for the official use, then the trees and then a road for the common travelers. The trees were cared for so efficiently that they were marked with numbers. It was duty of the official to count the trees every day. The administration was very efficient and competitive.

Before the colonial rule in the Punjab, most of the population lived semi nomadic lives, in summer at the bank of rivers while in winter they migrated to the bar area, *rahna* (where there is water and pastoral land for the animals). It was a cycle of seasonal migration. After the canals construction, *chaks* were alsochalkedout with wide bazar, streets, hospitals, schools and pastoral land, totally different from the older villages, *mozas*, having narrow streets and no facilities that were available in the *chaks*.The British administration directly affected the rural life and helped People in living a settled life, with a sense of the rule of law.

There was no permanent solution to cop with the menace of famine before the establishment of canal irrigation system but when once established with modern communication system of railways and roads, and responsible government, the effects of famine got diminished to a great extent. When peasants were given rights of ownership, they cultivated the land with much labour, resulting in the permanent supply of food and the communication system. Railways, roads and telegraph greatly helped in controlling the effects of famine. Thus the modern canal irrigation system transformed the lives of the people of the Punjab from nomadic to the settled one and undoubtedly the manifestation of settled life in the form of increasing civility and polish and falling down rate of crimes did appear in canal irrigated Punjab.

Land revenue was one of the big source of income for all governments of India so the British too, gave much importance to the said department and established 'Land Revenue Systems'27 to extract more revenue from the agricultural land. This Land Tenure System helped a lot in modernizing the agriculture of the Province of Punjab where every inch of arable land was measured and classified. When agriculture flourished owing to the irrigation system of the Province, 'commercialization of produce' increased under the British Raj, the regional markets were not only linked with the all-India market but with the world market as well. The railway integrated India into the world economy, as an exporter of cash crops and as an importer of manufactured goods. Modern canal irrigation system improved communications. The agriculture output increased highly, 'Punjab produced a tenth of British India's total cotton crop and a third of its wheat'. After the completion of the railway line from the Punjab to Karachi, Multan was the largest city in the value of trade in the province, in fiscal year 1880- 81. Its value of export was (8,404,104) Rs, value of Imports Rs. 3,703,382, and the total value Rs. 12,107,486.

Canal irrigation project of British Punjab totally changed the demography of the province. Different canal colonies were established in the doabs, like Sidhnai Colony in the Multan District in1886 with an area of 57000 *bighas*, Sohag Para Colony was situated in Montgomery District in 1886-88, 90,000 acres was brought under cultivation, Chunian Colony in the southern part of the Lahore district was completed in 1906, with allotted area of 103000 acres, Lower Chenab Colony in the Rachna Doab, the largest of the canal colonieswith an allotted area of two million acres. Another colony, Lower Jhelum Colony was developed in Shahpur District, with its headquarters at the newly founded town of Sargodha. The next important project was lower Bari Doab Colony which was developed between 1910 and 1925. Nili Bar Colony was the last major project, which was done beautifully by the British, in the Montgomery and Multan Districts. It covered approximately an area of about 800,000 acres. Thus, an urbanization process started due to the agricultural developments in the Punjab. The British established colonies on such modern lines that land was fixed for forests, market and experimental farms for fruit, seed, and vegetable. These canal colonies proved 'extremely profitable to the government'.35 It is clear that before the modern canals, dams and bridges, the large tracts of land between the rivers were uninhabited. Thus, change in geography in the form of, land contourment, agriculture, population migration, and establishment of institutions, helped in changing the meaning of the lives of the people. Whenever there are ownership rights for the peasants, improvement comes.36The British government gave ownership rights to the peasants of Punjab and it enhanced the production of the land and brought socio-economic change in the lives of the people. The British constructed the perennial canal irrigation system in the Punjab that brought a 'major economic change in this part of the British India.'

Colonial Administration and Organization Capacity

The British invested their ideas of management and organization in India. When they came to India as rulers, the Indian society was devoid of such institutionalization, the former had introduced. It was a triumph of scientific vision of governance. The history of irrigation in India, where British built large new irrigation works, that increased colonial revenues and expanded commercial production, provides a dramatic illustration of this notion. The colonized irrigation was a sort of 'imperial science', their mastery over the natural environment. Colonial administration brought a huge change in the agricultural economy of the province of the British Punjab, the Western Punjab. Firstly, British administered the canals through 'highly personalized action', a kind of 'paternalistic administrative structure' that transformed into bureaucratic rules to administer the canals and the canal colonies. The British administration assured the supply of water. British administration was so kind and effective that the peasants considered the administrators their '*ma-bap*'(parents), and they believed that the availability of water was impossible without the 'aid of the rulers.'

The bureaucracy under the British Raj in the Punjab helped greatly in big projects like the "ambitious canal construction and land distribution projects", but the administration after independence failed to get such achievements. The British colonial agriculture was based on the canal irrigation which is the proof of not only the socio-economic development in the Punjab but it also confirms the performance of the colonial state.

Failure of the Advantage of Backwardness

Morris argues that advantage of backwardness and slothfulness helps in rising social development. But in the context of Pakistan, having advantage of backwardness, we failed miserably in getting benefits from the backwardness. Yes, it is right that the British developed canal system and the communication system for their own benefits but it also served a lot for the social development of the local people. When the British left India they did not ruin the canal irrigation system before their departure or had not packed the system for the Great Britain. They left the system in beautiful shape and good structure. After the departure of the British, we did not benefit from the canal system and ruined it owing to our mismanagement and poor rule of law. During the colonial rule, state performed in a better way in respect of the rule of law. After the emergence of Pakistan, both society and the state are in constant decline in respect of governance and morality, "in stark contrast to the British period and in marked resemblance to the pre British period."

The colonial canal irrigation system has had great importance for the people of the Western Punjab. It brought a visible change in the socio-economic lives of the people, and brought a sense of urban life with modern amenities for the people. The British got mastery over the geography of the West Punjab and transformed the meaning of the lives of the people of the region. It not only helped the British in their imperial gains but was a great blessing for the people of the region as well. The present study confirms that Ian Morris is quite right in respect of the importance of geography in the social development of the societies. The colonial agriculture in the Punjab based on canal irrigation contributed a lot in the field of ending of internal warfare among semi-nomadic tribes and banditry, and changed their life pattern.

4

Agriculture during British Period

Farming methods and practices in India have a long history, and have been evolved out of the experience of ages. They have well stood the test of time. The technological advances in the world, *e.g.* the use of fertilizers, insecticides, fungicides, irrigation pumps, machines, *etc.*, have mostly taken place in the present century and are hardly sixty years old. It should cause no surprise if a country of India's size with millions of small farmers has not yet been able to adopt them as fully as has been done by some countries of the West and Japan in the East.

Agriculture Department

Government took little interest in the problems of agriculture before the mid-seventies of the 19th century. They were pre-occupied almost exclusively with problems of revenue, police and judicial administration. It was the great Bengal and Asom Famine of 1866 and the experience of organizing relief that forced the attention of the Government to the need for the development of agriculture as a means of averting famine and mitigating the hardships of the people. The first proposal to constitute a special Department of Agriculture at the Centre was put forward by the commission appointed in 1866, following the severe famine of that year but was turned down as premature.

Guided by the needs of Lancashire textiles mills, the British Government undertook measures in 1869 to improve the production of long staple cotton in India. This could perhaps be cited as the initial measure towards some action in developing Indian agriculture. In the same year the proposal for a Special Department was repeated by the Manchester Cotton Organization which led to a serious consideration of the subject by the Government. However, this attempt also did not achieve much success.

It was only in 1871 that the Department of Revenue, Agriculture and Commerce was created. It was recognized that provincial Departments of Agriculture must form an essential part of any scheme of agricultural development. The Department has a very brief existence of about seven years and was merged in the Home Department in 1879 for reasons of financial stringency. Its only achievement was to evolve a system for the collection of agricultural statistics and other data.

Very useful suggestions were made by the successive Famine and Irrigation Commissions for the development of agriculture. But, as observed by the Royal Commission of Agriculture, "The Government took no immediate action and the next ten years were spent mainly in conferences." It was the result of the recommendations of the report of the Famine Commission, 1880-81 that a separate Department of Revenue and Agriculture was revived in 1881. It was fully realized even in the initial stages that without a provincial agency, no programme of agricultural improvement, which emanated from the Headquarters, could be productive of tangible results, that the main responsibility for agricultural research and experiment must fall on provincial governments and all that the Centre could do was to give a general lead. The new department was directed to take up immediately the question of the form to be given to provincial department.

Provincial Departments of Agriculture were established in UP (then known as the NWP), Bombay, Bengal, Madras and C.P. between 1875 and 1905. Their functions were clearly defined by the Government of India and consisted largely of collection of statistical and economic data. There was some research notably in Bombay, and a few sporadic experiments in the cultivation of American cotton, groundnut, potatoes and fruit in some other provinces. But not much progress could be made by such efforts in the absence of a sound policy and an efficient organization.

Voelcker Report

It was realized very early by the Secretariat that without technical advice no advance was possible. Dr. J.A. Voelcker, distinguished agricultural chemist, was brought over to India in 1889 to conduct an enquiry into the character of the soils and agricultural conditions. This is regarded the "first serious step to frame a policy of agricultural research suited to the conditions of India". This was followed by the appointment, in 1891, of an Inspector-General of Agriculture with advisory functions. His duties were specified as "the systematic study of Indian agriculture, its conditions and remediable defects; the supervision and development of provincial agricultural departments; the establishment of improved agricultural methods and new staples and generally the direction of the agricultural policy of the government". An imperial mycologist and an entomologist were added to the department in 1901-3. Among the developments of this period mention should be made of the establishment of the Imperial Institute of Veterinary Research in 1889 and the Botanical and Zoological Surveys in 1890.

The Famine Commission Report, 1901

The report of the Famine Commission of 1901 inaugurated the next significant stage in the growth of agricultural departments. The Commission recommended the strengthening of the scientific staff of agricultural departments and legislation to set up mutual credit societies. During the period of Lord Curzon these recommendations were implemented and agricultural departments were thoroughly reorganized. There was an expansion of the scientific staff of the departments and of their activities. With the setting up of the Imperial Agricultural Research Institute at Pusa and bringing together with it the scientific staff of the department, under a single roof, organized agricultural research may be said to have begun in India. Directors of Agriculture were appointed in all the provinces. Circles were marked out and placed under Deputy Directors who were responsible for experimental seed and demonstration farms within their jurisdiction. Agricultural colleges were reorganized and research stations and experimental farms were also set up. The Imperial agricultural Service was constituted in 1906. Altogether the years from 1905 to 1914 were years of the steady growth in agricultural research and education and other services. The substantial advances made during these years by provincial departments were largely due to the stimulus and guidance of the central government and grants made by it.

The 1919 Reforms

After the reforms of 1919, agriculture became a transferred subject. Under the new arrangements, the central government was to look after the central agencies and institutions for research. The provincial governments could not make much headway since the resources at their disposal were limited. During the period some irrigation works were completed. Under 1935 Reforms, the sphere of central government's responsibilities was further restricted. It was to look after the subjects of all India importance only.

Royal Commission on Agriculture

A Royal Commission on Agriculture was appointed in 1926. It was required to examine and report on the prevailing conditions of agriculture. The commission was asked to investigate, particularly, (1) the measures being taken for the promotion of agricultural and veterinary research, experiment, demonstration and education, the compilation of agricultural statistics, the introduction of new and better crops and improvement in agricultural practices, dairy farming, and the breeding of stock; (2) the existing methods of transport and marketing of agricultural produce and stock; (3) the methods by which agricultural operations were financed and credit afforded to agriculture; and (4) main factors affecting the rural prosperity and welfare of the agricultural population.

The recommendations of the Commission were made with a view to bringing about all round efficiency in the field of agricultural production. The aim was to make agriculture a profitable industry rather than a mode of living. The Commission covered such subjects as fragmentation and subdivision of holdings, irrigation, improvement of livestock, rural education, cooperation, marketing *etc.* One of the basic recommendations of the Commission was to change the outlook of the cultivator himself so that he may take an active interest in bringing about improvements. It suggested that "rural problem should be tackled a whole in all its various aspects simultaneously." It assigned and charged the government with specific responsibility for bringing improvements in rural life. The Commission's Report is still considered a reference publication on agricultural policy.

Great Depression

Worldwide economic downturn that began in 1929 and lasted until about 1939. It was the longest and most severe depression ever experienced by the industrialized Western world. Although the Depression originated in the United States, it resulted in drastic declines in output, severe unemployment, and acute deflation in almost every country of the globe. But its social and cultural effects were no less staggering, especially in the United States, where the Great Depression ranks second only to the Civil War as the gravest crisis in American history.

Economic History

The timing and severity of the Great Depression varied substantially across countries. The Depression was particularly long and severe in the United States and Europe; it was milder in Japan and much of Latin America. Perhaps not surprisingly, the worst depression ever experienced stemmed from a multitude of causes. Declines in consumer demand, financial panics, and misguided government policies caused economic output to fall in the United States. The gold standard, which linked nearly all the countries of the world in a network of fixed currency exchange rates, played a key role in transmitting the American downturn to other countries. The recovery from the Great Depression was spurred largely by the abandonment of the gold standard and the ensuing monetary expansion. The Great Depression brought about fundamental changes in economic institutions, macroeconomic policy, and economic theory.

Timing and Severity

In the United States, the Great Depression began in the summer of 1929. The downturn became markedly worse in late 1929 and continued until early 1933. Real output and prices fell precipitously. Between the peak and the trough of the downturn, industrial production in the United States declined 47 per cent and real GDP fell 30 per cent. The wholesale price index declined 33 per cent (such declines in the price level are referred to as "deflation"). Although there is some debate about the reliability of the statistics, it is widely agreed that the unemployment rate exceeded 20 per cent at its highest point. The severity of these declines becomes especially clear when they are compared with America's next worst recession of the 20th century, that of 1981–82, when real GDP declined just 2 per cent and the unemployment rate peaked at under 10 per cent. Moreover, during the 1981– 82 recession prices continued to rise, although the rate of price increase slowed substantially (a phenomenon known as "disinflation").

The timing and severity of the Great Depression varied substantially across countries. **Table 4.1** shows the dates of the downturn and upturn in economic activity in a number of countries. **Table 4.2** shows the peak-to-trough percentage decline in annual industrial production for countries for which such data are available. Great

Britain struggled with low growth and recession during most of the second half of the 1920s, due largely to its decision in 1925 to return to the gold standard with an overvalued pound. Britain did not slip into severe depression, however, until early 1930, and the peak-to-trough decline in industrial production was roughly one-third that of the United States. France also experienced a relatively short downturn in the early 1930s. The French recovery in 1932 and 1933, however, was short-lived. French industrial production and prices both fell substantially between 1933 and 1936. Germany's economy slipped into a downturn early in 1928 and then stabilized before turning down again in the third quarter of 1929. The decline in German industrial production was roughly equal to that in the United States. A number of countries in Latin America slipped into depression in late 1928 and early 1929, slightly before the U.S. decline in output. While some less developed countries experienced severe depressions, others, such as Argentina and Brazil, experienced comparatively mild downturns. The depression in Japan started relatively late (in early 1930) and was, by comparison, mild.

Table 4.1: Dates of the Great Depression in Various Countries

(In Quarters)

Country	*Depression Began*	*Recovery Began*
United States	1929:3	1933:2
Great Britain	1930:1	1932:4
Germany	1928:1	1932:3
France	1930:2	1932:3
Canada	1929:2	1933:2
Switzerland	1929:4	1933:1
Czechoslovakia	1929:4	1933:2
Italy	1929:3	1933:1
Belgium	1929:3	1932:4
Netherlands	1929:4	1933:2
Sweden	1930:2	1932:3
Denmark	1930:4	1933:2
Poland	1929:1	1933:2
Argentina	1929:2	1932:1
Brazil	1928:3	1931:4
Japan	1930:1	1932:3
India	1929:4	1931:4
South Africa	1930:1	1933:1

The general price deflation evident in the United States was also present in other countries. Virtually every industrialized country endured declines in wholesale prices of 30 per cent or more between 1929 and 1933. Because of the greater flexibility of the Japanese price structure, deflation in Japan was unusually rapid in 1930 and 1931. This rapid deflation may have helped to keep the decline in

Japanese production relatively mild. The prices of primary commodities traded in world markets declined even more dramatically during this period. For example, the prices of coffee, cotton, silk, and rubber were reduced by roughly half just between September 1929 and December 1930. As a result, the terms of trade declined precipitously for producers of primary commodities.

Table 4.2: Peak-to-Trough Decline in Industrial Production in Various Countries (Annual Data)

Country	*Decline*
Unites States	46.8 per cent
Great Britain	16.2 per cent
Germany	41.8 per cent
France	31.3 per cent
Canada	42.4 per cent
Czechoslovakia	40.4 per cent
Italy	33.0 per cent
Belgium	30.6 per cent
Netherlands	37.4 per cent
Sweden	10.3 per cent
Denmark	16.5 per cent
Poland	46.6 per cent
Argentina	17.0 per cent
Brazil	7.0 per cent
Japan	8.5 per cent

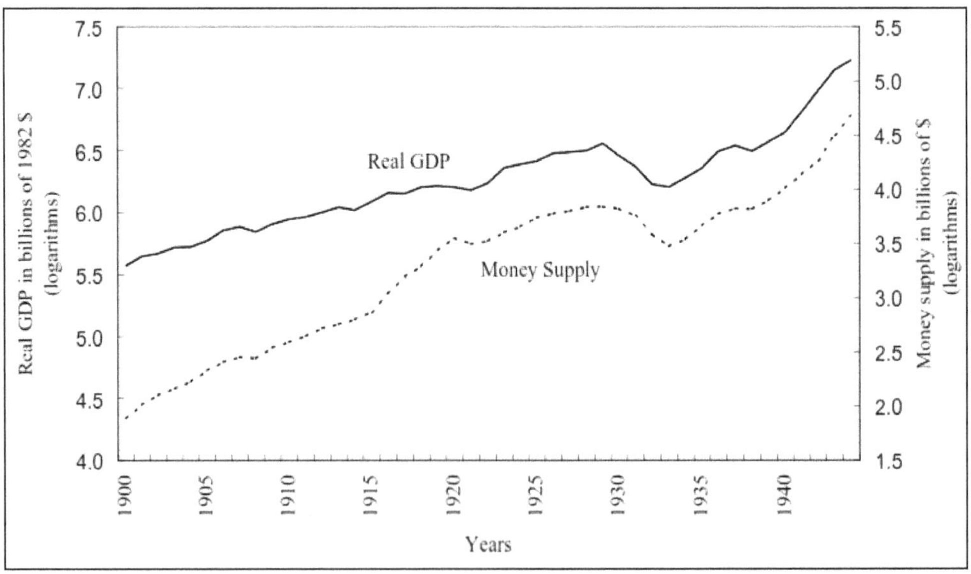

Figure 4.1: Money and Output in the United States.

The U.S. recovery began in the spring of 1933. Output grew rapidly in the mid-1930s: real GDP rose at an average rate of 9 per cent per year between 1933 and 1937. Output had fallen so deeply in the early years of the 1930s, however, that it remained substantially below its long- run trend level throughout this period. In 1937–38 the United States suffered another severe downturn, but after mid-1938 the American economy grew even more rapidly than in the mid-1930s. U.S. output finally returned to its long-run trend level in 1942.

Recovery in the rest of the world varied greatly. The British economy stopped declining soon after Britain's abandonment of the gold standard in September 1931, though genuine recovery did not begin until the end of 1932. The economies of a number of Latin American countries began to strengthen in late 1931 and early 1932. Germany and Japan both began to recover in the fall of 1932. Canada and many smaller European countries started to revive at about the same time as the United States, early in 1933. On the other hand, France, which experienced severe depression later than most countries, did not firmly enter the recovery phase until 1938.

Causes of the Great Depression

The fundamental cause of the Great Depression in the United States was a decline in spending (sometimes referred to as aggregate demand), which led to a decline in production as manufacturers and merchandisers noticed an unintended rise in inventories. The sources of the contraction in spending in the United States varied over the course of the Depression, but they cumulated into a monumental decline in aggregate demand. The American decline was transmitted to the rest of the world largely through the gold standard. However, a variety of other factors also influenced the downturn in various countries.

Sources of Recovery

Given the key roles of monetary contraction and the gold standard in causing the Great Depression, it is not surprising that currency devaluations and monetary expansion became the leading sources of recovery throughout the world. There is a notable correlation between the time countries abandoned the gold standard (or devalued their currencies substantially) and a renewed growth in their output. For example, Britain, which was forced off the gold standard in September 1931, recovered relatively early, while the United States, which did not effectively devalue its currency until 1933, recovered substantially later. Similarly, the Latin American countries of Argentina and Brazil, which began to devalue in 1929, had relatively mild downturns and were largely recovered by 1935. In contrast, the "Gold Bloc" countries of Belgium and France, which were particularly wedded to the gold standard and slow to devalue, still had industrial production in 1935 well below its 1929 level.

Devaluation, however, did not increase output directly. Rather, it allowed countries to expand their money supplies without concern about gold movements and exchange rates. Countries that took greater advantage of this freedom saw greater recovery. The monetary expansion that began in the United States in early 1933 was particularly dramatic. The American money supply increased nearly 42 per cent between 1933 and 1937. This monetary expansion stemmed largely from a substantial gold inflow to the United States, caused in part by the rising political tensions in Europe that eventually led to World War II. Worldwide monetary expansion stimulated spending by lowering interest rates and making credit more widely available. It also created expectations of inflation, rather than deflation, and so made potential borrowers more confident that their wages and profits would be sufficient to cover their loan payments if they chose to borrow. One sign that monetary expansion stimulated recovery in the United States by encouraging borrowing was that consumer and business spending on interest-sensitive items such as cars, trucks, and machinery rose well before consumer spending on services.

Fiscal policy played a relatively small role in stimulating recovery in the United States. Indeed, the Revenue Act of 1932 increased American tax rates greatly in an attempt to balance the federal budget, and by doing so dealt another contractionary blow to the economy by further discouraging spending. Franklin Roosevelt's New Deal, initiated in early 1933, did include a number of new federal programs aimed at generating recovery. For example, the Works Progress Administration (WPA) hired the unemployed to work on government building projects, and the Agricultural Adjustment Administration (AAA) gave large payments to farmers. However, the actual increases in government spending and the government budget deficit were small relative to the size of the economy. This is especially apparent when state government budget deficits are included, because those deficits actually declined at the same time that the federal deficit rose. As a result, the new spending programs initiated by the New Deal had little direct expansionary effect on the economy. Whether they may nevertheless have had positive effects on consumer and business sentiment remains an open question. United States military spending related to World War II was not large enough to appreciably affect total spending and output until 1941.

The role of fiscal policy in generating recovery varied substantially across other countries. Great Britain, like the United States, did not use fiscal expansion to a noticeable extent early in its recovery. It did, however, increase military spending substantially after 1937. France raised taxes in the mid-1930s in an effort to defend the gold standard, but then ran large budget deficits starting in 1936. The expansionary effect of these deficits, however, was counteracted somewhat by a legislated reduction in the French workweek from 46 to 40 hours–a change that raised costs and depressed production. Fiscal policy was used more successfully in Germany and Japan. The German budget deficit as a per cent of domestic product increased little early in the recovery, but grew substantially after 1934 as a result of spending on public works and rearmament. In Japan, government expenditures, particularly military spending, rose from 31 to 38 per cent of domestic product between 1932 and 1934, resulting in substantial budget deficits. This fiscal stimulus, combined with substantial monetary expansion and an undervalued yen, returned the Japanese economy to full employment relatively quickly.

Economic Impact

The most obvious economic impact of the Great Depression was human suffering. In a short period of time world output and standards of living dropped precipitously. As much as one- fourth of the labour force in industrialized countries was unable to find work in the early 1930s. While conditions began to improve by the mid-1930s, total recovery was not accomplished until the end of the decade. The Depression and the policy response also changed the world economy in crucial ways. The Great Depression hastened, if not caused, the end of the international gold standard. Although a system of fixed currency exchange rates was reinstated after World War II under the Bretton Woods system, the economies of the world never embraced that system with the conviction and fervour they had brought to the gold standard. By 1973, fixed exchange rates were abandoned in favour of floating rates.

Both labour unions and the welfare state expanded substantially during the 1930s. In the United States, union membership more than doubled between 1930 and 1940. This trend was stimulated both by the severe unemployment of the 1930s and the passage of the National Labor Relations (Wagner) Act (1935), which encouraged collective bargaining. The United States established unemployment compensation and old age and survivors' insurance through the Social Security Act (1935), which was passed in response to the hardships of the 1930s. It is uncertain whether these changes would have eventually occurred in the United States without the Depression. Many European countries had experienced significant increases in union membership and had established government pensions before the 1930s. Both of these trends, however, accelerated in Europe during the Depression.

In many countries, government regulation of the economy, especially of financial markets, increased substantially during the Great Depression. The United States, for example, established the Securities and Exchange Commission in 1934 to regulate new stock issues and stock market trading practices. The Banking Act of 1933 (also known as the Glass-Steagall Act) established deposit insurance in the United States and prohibited banks from underwriting or dealing in securities. Deposit insurance, which did not become common worldwide until after World War II, effectively eliminated banking panics as an exacerbating factor in recessions in the United States after 1933.

The Depression also played a crucial role in the development of macroeconomic policies intended to temper economic downturns and upturns. The central role of reduced spending and monetary contraction in the Depression led British economist John Maynard Keynes to develop the ideas in his *General Theory of Employment, Interest, and Money* (1936). Keynes's theory suggested that increases in government spending, tax cuts, and monetary expansion could be used to counteract depressions. This insight, combined with a growing consensus that government should try to stabilize employment, has led to much more activist policy since the 1930s. Legislatures and central banks throughout the world now routinely attempt to prevent or moderate recessions. Whether such a change would have occurred without the Depression is again a largely unanswerable question. What is clear is that this change has made it unlikely that a decline in spending will ever be allowed to multiply and spread throughout the world as it did during the Great Depression of the 1930s.

Russell Report

In accordance with the recommendations of the Royal Commission to review periodically the working of the Imperial Council of Agricultural Research which had been established on its recommendation. Sir John Russell, Director, Rothamsted Experimental Station, England, visited India in 1936. According to him, the Council had admirably discharged its primary duty of coordination and promotion of agricultural research in the field of crop production. He stressed the need for further use of existing knowledge, rather than the accumulation of more knowledge and stated that knowledge gained in the laboratory should be applied to cultivator's fields. He suggested that the burden of the Council should be lessened by asking universities to undertake all investigations of a scientific or non-technical nature.

He recommended that the results of research in agriculture should be made widely public. It had not been done in the past. He suggested that the Council should take the responsibility of putting the result of research into practice and act as a Development Commission. He further remarked that the Council's investigations should be directed to increasing the output per acre of food crops with a view to ensuring food supplies and liberating land for the growth of the supplementary and fodder crops for the production of milk.

Some other Important Reports

A number of committees have from time to time gone into the problem of agricultural administration and machinery. All of them have helped to expand agricultural organization. Mention may be made of the Patel Committee (1945 – 46) as a result of whose recommendations the Directorate of Economics and Statistics became a separate unit of the Ministry; the enquiry by N.Gopalaswami Ayyanger (1949), which recommended the creation of the Bureau of Natural Resources and Agriculture with a coordinating minister for the Ministries of Agriculture, Food and Works; the W.R. Natu Enquiry 1951; and more recently the enquiry Mr. Keifer (1965). Fullscale general enquiries have also been made of the machinery of the Government of India leading to proposals for its reorganization by Richard Tottenham in 1945 by Shri R.A. Gopalaswami in 1952 and Dr. Appleby in 1954. It may also be mentioned here that the Food Grains Policy Committee 1947, the Grow More Food Committee 1952 and the Balwantrai Mehta Committee 1957 also have made recommendations for the reorganization of the machinery of agricultural administration at different levels. The desire for administrative improvement has been continuous, the suggestions for reform numerous, but the practical results of the enquiries have been rather disappointing. Piecemeal changes and adjustments have been made from time to time. There has been some bureau shuffling off and on but no thorough reorganization that could assure the efficient functioning of the machinery for accelerating development.

Agricultural Production Board

In order to streamline agricultural production in the country, an Agricultural Board was constituted in 1963. The main function of the Board was to bring about expeditious implementation of the programme for the development of agriculture including animal husbandry forestry, and fisheries. The working of the board was recently reviewed and the Minister of Food, Agriculture, Community Development and Cooperation is its Chairman and the Minister of Irrigation and Power, the Minister of Planning, the Minister of State in the Ministry of Finance, and the Member Agriculture, Planning Commission, are its members. The function and powers of the board are: (i) to secure effective and continuous coordination between the concerned ministries and departments at the Centre and between the Centre and the states, in matters relating to the formulation and execution of development programmes; (ii) to review the progress of development programmes, ensure efficient arrangements for supplies and services, locate difficulties, if any, and endeavour to remove them; and (iii) to take such other measures as would ensure speedy and effective implementation of the development programmes at the Centre and in the states. The decisions of the Board will be treated as the decisions of the Government of India as long as they are within the plan provisions and within the framework of the approved budget provisions of the ministries concerned.

Agricultural Development: An Overview

Immediately after Independence, the Government realized that as the bed rock of India's economy, agriculture and allied activities have remained the focal point of the country's planned economic development. In fact, with around half of our rural workforce deriving their livelihood from agriculture and allied activities (agriculture being the largest private activity), the performance of this sector still holds the key to improvements in real incomes and living standards of the bulk of Indian population. Despite phenomenal diversification of the country's economy, this sector even now constitutes around 15 per cent to the Gross Domestic Product (GDP). It was thus realised, there can be no sustained growth of the Indian economy without broad-based progress of our agriculture. This is also the experience of most of the fast growing countries of East Asia.

The sector produces 51 major crops, provides raw material to the country's agro-based industries and fetches nearly one-sixth of the total export earnings. India has in addition the added advantage of a rich endowment of nature. Her most abundant natural resource is her diverse agro-climatic conditions, pernnial rivers, tropical location, diverse climate, large arable land (second largest) suitable fertile soil qualities, abundant sunlight, spread of monsoon rains, comparative advantage in the global markets and a receptive and resurgent farming community. As against world share of arable to total land area of only 11 per cent, India is bestowed with 51 per cent. Our climate, being moderate, enables us to grow two to three crops a year; whereas in most parts of the world, due to severe winters, only one crop can be grown a year. India also has more irrigated land than any other country.

The sector produces 51 major crops, provides raw material to the country's agro-based industries and fetches nearly one sixth of the total export earnings of the country. These resources if complemented with the suitable technology can create wonder for the Indian economy. Present need is to focus upon the quality of agricultural production so that a niche is created for Indian agricultural products in the domestic and foreign markets. Areas where climate is favourable and resources are available to produce exportable produces should be identified and encouraged to produce quality products. These pockets should be developed on the pattern of export promotion parks where production should be targeted at the overseas markets only. Basmati rice producing areas is one of such examples.

At the same time it has to be accepted that: India has only 2 per cent of the world's land, 4 per cent of its fresh water resources but carries 16 per cent of the world's population and 10 per cent of its cattle. India has geographical area of 329 million hectares of which 47 per cent cultivated, 23 per cent under forest, 7 per cent under non-agricultural uses, remaining 23 per cent falls under the categories of barren and un-culturable land, culturable wasteland, permanent pastures and fallow lands. An estimated 150 million hect of degraded land suffers from water and wind erosion. Per capita availability of land declined from 0.89 hectare in the early 1950s to 0.37 hectare in the mid 1990s and is expected to reduce further to 0.19 hectare by 2020. India produces more than 200 million tones of foodgrain from 142 million hectares of cultivated land. Irrigated areas account for 37 per cent of crop-land and contribute 55 per cent of foodgrain production whereas rainfed areas account for 63 per cent of cultivated land but contribute 45 per cent of foodgrain production.

Pre-independence Scenario

Agriculture in India till Independence was traditional with a very slow growth rate having little application of science and technology. In a study on "Growth and Instability in Indian Agriculture', during the first 24 years of the century foodgrains production increased at an average annual (linear) rate of 0.3 per cent. The next 24 years, however, presented a completely different picture. During this period, foodgrains production showed a declining trend of 0.02 per cent per annum (linear) on the average, in spite of the fact that droughts turned out to be relatively moderate and less frequent. The National Commission on Agriculture (1976) in their study on progress of agricultural development also observed that during the period 1931-47, production and productivity of both foodgrains and non foodgrains exhibited mostly a declining trend.

During the period 1931-47, besides irrigation, improved seeds and agricultural education were the other technological factors which had started making some impact on the agricultural economy of the country. Inspite of increased imports, fertilizer use remained limited to plantations and commercial crops which alone could fetch remunerative prices. Other factors like types of crops grown and manner of farming remained traditional. Agricultural implements used were largely conventional and whatever new types of iron ploughs were publicised were not widely accepted by the farmers. Apart from being expensive and strenuous to bullocks, these were unsuitable for small fragmented holdings and less effective in uprooting weeds and, at the same time, were prone to expose and dry out comparatively more of the under-soil moisture. Moreover, there was lack of repairing facilities.

The improved seed technology was mainly confined to cash crops though the Indian Institute of Agricultural Research had initiated work of multiplication and dissemination of improved varieties right from its establishment. Research was conducted with the object of introducing new crops, improving indigeneous types and producing new and better varieties with regard to rice, wheat, gram, sugarcane, cotton, jute, tobacco and oilseeds. Though it is now difficult to evaluate the impact of improved seeds on yield rates of individual crops, during the period substantial progress was achieved in bringing additional areas under improved varieties of commercial crops.

Grow More Food (GMF) Campaign

The Grow More Food (GMF) campaign launched in 1942 marked the beginning of a sustained national endeavour at increasing agricultural production and improving agriculture. Its main objective was to achieve immediate increase in food production through extension of area under cultivation, including diversion of acreage from cash crops like cotton, extension of irrigation and use of inputs like improved seeds, manures and fertilizers. From 1943, in the wake of increasing scarcities and the Bengal famine, the Government of India began assuming wider responsibilities in regard to food production, a field which for over two decades was the primary concern of provincial governments. It began to direct the GMF campaign and also sought to bring about greater involvement of provincial governments in the drive, by offering them special assistance. The Foodgrains Policy Committee and the Famine Enquiry Commission, appointed during this period, not only endorsed the campaign, but also gave suggestions for improving agriculture (NCA, 1976).

Statement of Agriculture and Food Policy in India

A major development in the immediate pre-Independence period was the first ever elaboration in January 1946, of an all India policy in agriculture known as 'Statement of Agriculture and Food Policy in India.' The ten objectives of the policy included: increase in production of foodgrains and protective foods; improvement in methods of agricultural production and marketing; stimulating production of raw materials for industry and exports, securing remunerative price for the producer and fair wages to the agricultural labour; ensuring fair distribution of the food produced and promoting nutritional research and education (NCA, 1976).

Post-Independence Policy Framework and Development-Initial Phase

Despite these developments, agricultural performance remained at a low key and together with the problems created by the Second World War, the food situation steadily deteriorated. The partition of the country in 1947 and the growing population aggravated the critical food situation. The government, therefore, felt necessary to give urgent attention to agriculture so as to alleviate widespread hunger and malnutrition and facilitate the nation's general economic progress. The Grow More Food (GMF)campaign, which was launched earlier in 1942 in order to accelerate food production, was, therefore, placed on a planned basis from 1947-48 for a period of five years.

The Foodgrains Policy Committee appointed in 1947 under the Chairmanship of Shri Purshotamdas Thakurdas recommended greater attention to minor irrigation, development of local manurial resources, distribution of improved seeds, and land reclamation. The Report also impressed the need for planning production of fertilisers. A greater reliance on the use of science and technology for faster growth in agriculture was emphasized. The impact of GMF was minimal as there was no real breakthrough in any crop and the increase in foodgrains production was very modest. In view of the persisting stagnation in agricultural production, a committee was appointed in 1952 (V.T. Krishnamachari Committee) to review the working of GMF campaign. The committee felt that the lack of success of the programme was mainly due to its failure to arouse enthusiasm in the countryside. Its important recommendations related to:

a) The setting up of countrywide extension service organisation;
b) The acceleration of minor irrigation programme; and
c) The provision of adequate credit and quality seed, and development of local manurial resources.

The basis of the policy framework in the initial years of planning was really based on the recommendations/reviews of all these committees.

Fifty Years of Indian Agriculture: An Appraisal

The quantum of plan outlay, its financing and the targets set for the agricultural sector were all decided through the planning process at the state and central levels. The first three Five-Year Plans concentrated on growth with some institutional changes. The intermediaries in agriculture, like Zamindars and Jagirdars, were abolished within a few years after independence and the actual tillers, accounting for about 40 per cent of the cultivated area, became the owners. This provided a major incentive to the growth of agriculture in large parts of the country. In the mid-sixties, a new technology in the form of high-yielding varieties (HYVs) was introduced for cereals. Apart from the new technology, public investment in agriculture particularly in irrigation, was stepped up significantly. The public sector played an important role in promoting agricultural research and education. Large investments were made for the development of research system under the aegis of the Indian Council of Agricultural Research and the State Agricultural Universities. Simultaneously, a well designed extension network was created for disseminating new technologies to the cultivators. The administered price policy has provided incentives to the farmers. In short, successive Five-year Plans have aimed at improving the infrastructure through irrigation, stepping up the use of fertilizers and improved varieties of seeds and effecting institutional changes, supply of credit through the banking system, ensuring remunerative prices for agricultural commodities, *etc.* There has been a significant increase in the use of modern inputs.

The agricultural development strategy followed since independence particularly since the mid-sixties, has paid rich dividends. In terms of growth, the performance has been quite impressive in the post-independence period as compared to the pre-independence period. The all crop growth of around 2.7 per cent per annum in the post-independence period (during 1949-98) was much higher than the negligible growth rate of around 0.3 per cent per annum in the first half of this century. It shows that the long-term growth rate of agriculture during the period 1949-50 to 1998-99 has been around 2.7 per cent per annum. The growth rates for sub-periods show that agriculture recorded much higher growth during 1980s and 1990s than in the earlier decades though there was, some deceleration in 1990's decade (**Tables 4.3).** The production of foodgrains increased from 51 m.t. in 1950-51 to a record level of about 209 m.t. in 1999-00. The country has been able to meet its food needs inspite of about 2 per cent per annum growth in population. Commendable progress has been made in increasing the production of commercial crops like cotton, oilseeds, sugarcane, besides livestock products, poultry, fisheries, fruits and vegetables. Within the agriculture sector, animal husbandry and fisheries have shown a much impressive growth (**Table 4.4**).

Table 4.3: Average Annual Growth Rates in Agriculture and Allied Sector's GDP

Period	*Agriculture*	*Animal Husbandry*	*Forestry*	*Fishing*	*Agriculture and Allied (Total)*
1950-51 to 1966-67	1.90	1.02	1.85	4.73	1.77
1966-67 to 1980-81	3.72	2.33	0.13	3.16	3.01
1980-81 to 1996-97	3.85	7.97	–0.67	5.67	3.91
1990-91 to 1996-97	2.16	7.32	–0.94	6.50	2.90

Source: Computed from CSO,"National Accounts Statistics", Various Issues.

Table 4.4: Production of Major Crops, Milk and Fish 1950-51 to 1998-99 (million tones)

Crops	*1950-51*	*1970-71*	*1980-81*	*1990-91*	*1998-99*
Foodgrains	50.82	108.42	129.59	179.39	203.00
Rice	20.58	42.22	53.63	74.29	85.99
Wheat	6.46	23.83	36.31	55.14	70.78
Coarse Cereals	15.38	30.55	29.02	32.70	31.46
Pulses	8.41	11.82	10.63	14.26	14.81
Sugarcane	57.05	126.37	154.25	241.05	295.73
Cotton (m.bales)	3.04	4.76	7.01	9.84	12.18
Nine Oilseeds	5.16	9.63	9.37	18.61	25.21
Milk	17.0	21.2*	31.6	53.9	74.70@
Fish	0.75	1.76	2.44	3.84	5.26@

@ Anticipated

Source: Agricultural Statistics at a Glance, 2000.

The green revolution has transformed agriculture from subsistence to production-intensive farming. Over the first five decades, the gross cropped area has gone up by almost a third, production has quadrupled, yield tripled and coverage under irrigation double. A point for elation for India has been the increase in food production by 4 times from 51million tones in 1950-51 to over 208 million tones during 1999-2000. Crop-wise, there were 17 times increase in potato production, 11 times in wheat, 5 times each in oilseed and sugarcane, 4 times each in rice and cotton. India today has become the largest producer of pulses, groundnut, Jute, milk and vegetables and the second largest producer of wheat, rice, sugarcane and fruits.

Since the driving force for agricultural development was the food crisis, the focus of all the efforts came to be placed more on the areas, which already had better infrastructure in terms of irrigation systems and roads *etc.* in order to overcome the crisis quickly. Therefore, it was no coincidence that the Green Revolution took its roots in Punjab first and then spread to Haryana and part of Western Uttar Pradesh. A few other districts and regions in the country, where the conditions were similarly conducive also took advantage of the Green Revolution. Thus, regions such as Punjab, Haryana, Western Uttar Pradesh *etc.* have benefitted more, during the initial phase of the Green Revolution, than others. This is partly explained by agroclimatic factors, and partly by earlier introduction of high yielding varieties and better irrigation.

Green Revolution thus helped in eradicating poverty in a way that no other development activity has done. Poverty in Punjab in 1993-94 was just 12 per cent while at all India level it was 36 per cent. This has however, resulted in accentuating regional disparities in the immediate post-Green Revolution period. A district level analysis of the growth rates between the early sixties and the early eighties shows that out of 281 districts covered, 17 reported an absolute decline in output; 82 districts recorded a growth rate of less than 1.5 per cent a year, while 23 had more than 5 per cent per annum. An important feature of the 1980s and the 1990s, however, is that there has been much more equitable spread of agricultural growth. After performing poorly during the early years of the Green Revolution, many of the States, where poverty is widespread namely Asom, Bihar, Odisha, Madhya Pradesh and West Bengal, have shown significant growth in the 1980s. Oilseeds have also gained in the dry belt of Rajasthan, Madhya Pradesh, Karnataka and Maharashtra. An interesting development during this period was that agricultural growth had spread to all the regions of India. Apart from the eastern region, States like Madhya Pradesh and Rajasthan also recorded higher growth during this period. Thus, there seems to have been a reduction in the regional disparities in output growth during the 1980s and 90s.

Indian Agriculture at Cross Roads

Malthusian predictions of 1798 appeared in its stark reality when India was declared as a 'basket case' and a sinking ship by the international community during mid sixties. Living from ship to mouth, the sustained cooperation between ICAR and CIMMYT supplemented by research of Indian scientists brought about a sea change in the Indian food economy and India today has emerged as the second largest wheat producer of the world. Indian agriculture has at the same time moved from food scarcity to food surpluses. As discussed earlier, during the last 50 years, agricultural production has increased manyfold. Agricultural research, supplemented with modern inputs carried out by the Indian Council of Agricultural Research and the state agricultural universities, has played a pivotal role, particularly after the Green Revolution, in increasing productivity of crops, which has so far kept pace, with the population growth.

This euphoria on the food front has, however not been without accompanying problems. We are acutely aware today, as never before in the past, that there are both costs and benefits attached to the intensification of agriculture to feed a population of over one billion and growing. The natural resource base, namely land and water, which sustains the production process, is under considerable stress. Per capita land availability has halved since independence and will decline further in the next quarter century. Of the total geographical area, about 50 per cent suffers from varying degrees of degradation. Unless great efforts are undertaken, there will be continuing decline in biodiversity, which is the source of genetic material needed to make further advances in plant and animal breeding. There could also be accelerated global climate change. The impact of all these factors on our land resources compels us to focus on the issue of sustainable production.

The green revolution areas of Punjab Haryana and Western U.P. have started showing signs of fatigue. We may go into the problems faced by Punjab agriculture. The water resource there is being over-exploited because of the adoption of high water requiring cropping sequences. Water table in Central districts of the State is declining at 23 cm per annum, whereas, the Western districts are facing an acute problem of water logging. The water table has risen by 4 to 5 metres within a decade's time. If the process of declining water table continues for the next 15 years, state would have to spend Rs. 2000 crore on submersible pumps for replacing 2.5 lakh centrifugal pumps. In view of such a constraint, there is a need to adopt technologies related to ground water recharge and crop related practices having no adverse effect on water resources.

Punjab soils are primarily light textured, poor in organic carbon and nutrient resources. The nutrient reserves of soil are being exploited at a much faster rate than its replenishment due to extremely high cropping intensity (almost 200 per cent) and extensive adoption of rice-wheat cropping system. The organic content of soil has gone down from 0.5 to 0.2 per cent. Micro-nutrient's deficiency on large scale is affecting the crop as well as animal health adversely. Accumulation of salt on the soil surface in the water-logged south-western zone where underground water is brackish is a matter of serious concern. Therefore, there is an urgent need to sustain the soil-health through integrated nutrient management (using chemical fertilizers in conjunction with organic manures and incorporation of crop residues). It has been estimated that 59 lakh tonnes of organic waste containing 2.6 lakh tonnes of nutrients are burnt with paddy straw every season (1997-98), in addition to creating pollution problems of the environment. Likewise, press mud- an industrial waste from sugar factories can be put to use for reclaiming/improving alkaline soils. Thus, for ecologically sustainable agriculture, there is an urgent need to substitute chemical fertilisers with biowastes to maintain productivity level in future.

After attaining an exemplary rise in production, Punjab agriculture has come to a stage beyond which sustaining the growth rate is an ardous task. Being mainly the agricultural economy, the overall development of the state has slowed down. During 1993-94, the per capita real income was Rs. 4,025 at 1980-81 prices which increased to Rs. 4,627 (Quick Estimate) in 1998-99 showing a growth of 15.0 per cent against a comparative figure of 26.51 per cent (at 1993-94 prices) for the country. The average annual growth rate in the state agriculture, including livestock, has declined from 5.44 per cent in the Sixth Plan to 3.32 per cent in Eighth Plan which is projected to fall further to 2.38 per cent in the Ninth Plan. This is much lower as compared to other sectors of the economy.

Decelerating Growth

Our soil resources in the country continue to undergo increasing stresses due to population pressure, loss of vegetative cover, soil degradation, pollution, and climatic changes. The vagaries of nature, with floods in one part of the country and drought in the other, and unseasonal and erratic rains, adversely affect crop production. Thus looking into our resource and targets to be achieved for crop production there needs to be a comprehensive policy on development and adoption of low input sustainable agriculture (LISA) technologies.

Growth of Indian agriculture (**Table 3.3**) is showing clear signs of deceleration. Having touched the highest levels (3.72 per cent) during the eighties, the nineties are likely to be around 2.3 even when the all time bumper crop of 1998-99 is taken into account. This is when we had good favourable monsoon over a decade in a row and vagaries of South-west monsoon did not have adverse impact on food production as they did during the eighties. What is most disturbing in this scenario is the declining trend in productivity. This is true not only for foodgrains, but also for all crops. Pulses and cotton have also shown a negative growth. The continued increase in the livestock and fisheries sub sectors is, however, quite encouraging.

Waste Land Problem

Our record in tackling problems of denudation and soil erosion has unfortunately been most unsatisfactory and wasteful. Although soil conservation schemes for agricultural lands have been on the scene for nearly half a century, they have failed to make any significant dent on the problem on account of disregarding the "complete mini-watershed" principle. As a result, excessive run-off from denuded forest lands that are almost invariably situated in the higher reaches of watersheds has caused great damage to the terraces and bunds on agricultural fields, particularly because such bunds are not correctly aligned along contour lines but built along field boundaries. This basic flaw is common to all the other schemes which aim essentially at soil and water conservation and explains why they too have not succeeded in their objectives. The position is not very much different with regard to water-logging and salinization. It appears that we have just not taken any notice of these issues. As the Eight Plan document admits, there has so far not been even a systematic survey of the extent and location of lands that are affected by this malady. The Ninth Plan also appreciates that the deterioration in land and water resources has been in the form of land degradation, water logging and decline in water table. There is a greater need to have an integrated approach in the management of agricultural nutrients, chemicals and in taking effective measures to deal with the overall pollution problems.

The present scenario in wastelands development is depressing in the extreme because while there is just no talk of wastelands that are suffering from waterlogging and salinisation, even the responsibility for attending to denuded wastelands has been split between two Ministries and there is yet no institutional arrangement for bringing them together so that they may follow the "complete mini-watershed" approach on the ground. It is necessary to remember in this connection that since forest wastelands and non-forest wastelands are inextricably juxtaposed in most situations, neither of these can be tackled alone in a cost-effective manner under the existing administrative arrangements.

This by no means is the end of the story. For even if we were to correct this anomaly by once again creating a unified agency for dealing with both forest and non-forest wastelands, and even if we were also to place responsibility for the reclamation of waterlogged and saline lands squarely on such an agency, the case for treating "wastelands development" as a separate subject by itself, would still be untenable. For such an approach would necessarily push the problem of over 96 odd million hectare of degraded agricultural lands further into the background, where indeed they are already languishing. This would be a great tragedy because contrary to popular belief schemes of the Department of Agriculture for the conservation of soil and water on eroding agricultural lands are not at all doing well because they are being implemented in violation of the "complete mini-watershed" principle.

Better Land Management

Per capita availability of land had declined from 0.89 hectares in the early fifties to 0.37 hectares in the mid nineties due to population increase. The shortage of land is compounded by the increase of waste land, which is estimated to be 63.8 million hectares. We are loosing around 15 million hectares of land due to spread of urbanization and the demand for more land for housing, industries and communication. There is then hardly any possibility, in such a situation to provide 0.5 hectares of arable land per capita which was the minimum required for ensuring food and other essential needs of an individual. Future of food security thus lies in conserving prime land for agriculture and restoring the biological potential of degraded lands. Emphasising on the need for effective community-led land-care movement, the people should be in the forefront to manage land and water resources and the scientists must provide leadership in the movement.

While doing so, we ought to be mindful of the soil health. An effective community-led land care movement, people should be in the forefront to manage land and water resources and scientists must provide leadership in the movement. At the same time, the vital role of women who are the worst victim of land degradation must be acknowledged. Green Revolution can be converted into an evergreen revolution if there is a paradigm shift in our research and development strategy leading to a change from a purely commodity-centered approach to an integrated natural resource management. Therefore, the scientific and agricultural experts will have to join with the farmers and the common people in a popular movement for land protection. While there is a global concern for land care, greater political and public support in this area is needed. Land resource management is a challenging task as it will have to take care of the twin needs of increasing food production and improving land resources. There is the need for according high priority for development of rainfed areas and a common approach to watershed development. The fragile eco-system poses serious implications for the mankind and there is a need for sound land and water management. It is heartening to note that India had demonstrated that it was possible to reverse the trend of land degradation through people centered efforts.

Our Land is one of our most valuable assets. It continues to feed our growing population. However, there is increasing evidence that the land's capacity to go on producing enough food for the nation may be under threat – rising biotic pressures, expansion of agriculture into fragile ecosystems, over-exploitation of groundwater, water logging resulting in the salinisation of soil. Overuse of chemicals and practice of slash-and-burn agriculture are leading to declining soil health. Farmlands are being made vulnerable due to loss of topsoil and small hills have been transformed into wide gullies and non-reclaimable ravines. Soil erosion due to wind and water, waterlogging, salinity, alkalinity, mining and improper farming practices are well known cause of soil degradation. It has been said about soils that "a handful of good soil is a miracle. Treated well by man it will support life almost indefinitely. Abuse it, the soil will collapse and die taking man with it".

According to available estimates, of the 175 mha of degraded and wastelands that the country possesses, around 150mha suffer from erosion of the topsoil and its attendant ills (such as floods and gully formation, *etc.*) and around 25mha from wind erosion. Over 5.3billion tones of top soil is lost every year due to soil erosion. Nature takes about a thousand years to form one cm. of soil. The average soil loss is estimated to be over 16 tonnes/hectares/year which translate into approximately 1mm each year or 1cm every decade. Soil loss varies from less than 1 tonne/hectare/year in dense forests to more than 80 tonnes/hectare/year in the Shivalik Hills. For every tone of foodgrain, about 105kg. of nutrients are removed from the soil, a cereal crop absorbs 35-46kg. of NPK to produce a tonne of grain, leguminous crop absorbs 35-46kg. of NPK to produce a tonne of grain, leguminous crop absorbs 210kg. NPK to produce 1 tonne of grain and legumes fix upto 450kg. of Nitrogen/hectare/year.

There is no doubt that lands subject to erosion constitute the biggest single threat to country's economy. For not only do such lands suffer an increasing loss of productivity because of the progressive loss of the fertile top-soil but they also contribute to the loss of a great deal of priceless sweet water by way excessive run-off along denuded slopes. This run-off loaded as it is with soil also causes a great deal of damage by contributing to the causation of floods and the premature siltation of river beds, tanks and reservoirs - before it reaches the sea. And since a large part of this water could have been, under better conditions of land management, retained either as soil moisture (so vital for rainfed agricultural lands) or as ground water (which is the mainstay of the country's irrigation infrastructure today), its loss is also a major reason for droughts. Floods and droughts are indeed two sides of the same coin of poor land management and both can be moderated very substantially by preventing excessive soil erosion.

Considering that prevention is better than cure, and that wastelands have gone out of production in any case, the protection of erosion-prone agricultural lands against further deterioration needs very high priority. This means that we should start worrying a little less about wastelands, but a little more about degraded lands, which if they are not saved in time may also get slowly converted into wastelands. We cannot also continue to be complacent with regard to the dangers of depletion and deterioration faced by lands that do not belong to either of these categories and are believed to be in good health. Such lands are around 89 mha (264 -175 mha) in extent and comprise of around 50 mha of good agricultural lands and around 39 of good forests. Finally together we can care for land:

- ✰ By educating, informing and sensitizing all stakeholders about various aspects of this precious resource and its sustainable use.
- ✰ By nourishing the land through conservation of rain water, recharging of ground water and judicious water use.
- ✰ By anchoring soil to the earth through tree plantation, afforestation and enhanced green cover.
- ✰ By enriching soil fertility through planting of leguminous crops and adoption of suitable crop rotations that support diversification.
- ✰ By nurturing soil health through recycling organic wastes and adoption of Integrated Plant Nutrient Management (IPNM) techniques.
- ✰ By protecting land from overuse of chemical pesticides through adoption of Integrated Pest Management (IPM) practices.
- ✰ By saving land from overgrazing through adoption of suitable animal husbandry and "Social" and other fencing.
- ✰ By adoption of resource conservation technologies such as zero-tillage, reduced tillage, surface seeding, bed-planting, use of drip and sprinkler irrigation and agronomic practices that promote precision agriculture, save water, energy and improve factor productivity.
- ✰ By creating suitable incentives to conserve land such as security of land tenure, productive land conservation techniques, people's participation, charges and sanctions.
- ✰ By empowering local communities to make and implement decisions in order to tap knowledge, enthusiasm and energy and creating forums for negotiation and conflict resolution.
- ✰ By adopting the integrated watershed approach for rainfed farming which promotes need- based soil conservation measures in arable and non-arable lands.

Rainfed Farming

Rainfed agriculture, a dominant sector of agricultural economy, is practiced over 64 per cent of crop land and contributes about 40 per cent of food grains production in the country. Bulk of pulses and oilseeds are produced under rainfed conditions. Rainfed farming accounts for 25.7 per cent of agricultural GDP and 9 per cent of total GDP in the country. This sector supports 42.2 per cent of the country's production including meat, milk, hides and skins, wool, *etc.* which are largely produced in rainfed areas. Rainfed areas contribute 36 per cent of the agricultural export and 10.26 per cent of the total exports of the country.

These areas are under heavy human and livestock pressure and subject to the process of ecological degradation. The restoration of ecological balance on such fragile ecological endowments holds the key to the environmental health of the country. Development of rainfed areas is thus important from economical, social and environmental points of view.

In the case of rainfed areas, hilly regions, *etc.* it has been accepted that a watershed approach, which is a natural hydrological unit, for planning for development is essential. Allied to this is need for adopting a Farming System approach to enhance incomes of farmers from the alternative use of land pattern. The process of rainfed area development on watershed basis has to be based on a holistic view for comprehensive and integrated development.

The earliest experiments in watershed management in India were conducted in the semi-arid tropics under the aegis of ICRISAT. Between 1975 and 1988, fourteen cropping systems were evaluated along with a range of management practices. The Indian Council for Agricultural Research also took up forty-six model watersheds in the dryland areas of the country in 1982. During the seventies and eighties the thrust of policy shifted to the improvement and stabilisation of productivity in rainfed areas. It was understood that, in these areas, the main focus of development strategy should be on minimizing risk through appropriate land management systems which ensured greater retention of soil moisture for increased production. The emphasis was therefore shifted to the area development approach taking the watershed[1] as the unit of development. The components include appropriate cropping systems, dryland horticulture, fodder and economically viable plantations in an integrated manner.

Development of rainfed agriculture figures as one of the high priority area in the Twenty point programme (T.P.P.) of 1982 and new T.P.P. of 1986. Four pilot projects were operationalised during 1984 with World Bank assistance to provide an integrated approach. A vigorous National Watershed Development Project for Rainfed Areas (NWDPRA) was launched in the Seventh Five Year Plan and restructured during Eighth Plan to achieve the twin objectives of sustainable production bio-mass and restroration of ecological balance in the vast rainfed areas in the country. The programme is being implemented in sixteen states and extends

---

1 GOI(1994) : Ministry of Rural Areas and Employment, Guidelines for Watershed Development which give details of various programmes related to watershed and soil conservation.

to ninety-nine districts. Around 2,328 watersheds are receiving treatment under this programme. The idea of an integrated treatment of the watershed was also behind the centrally sponsored schemes aimed at reducing the inflow of sediment into the river valley projects and flood prone rivers. Two such schemes are currently in operation, one in the catchments of River Valley Projects (RVP) and the second as an integrated watershed management scheme in the catchments of Flood Prone Rivers (FPR) in about 900 priority watershed of 35 catchments spread over in India.

In an integrated watershed development programme, all principal practices, *viz.* soil and water conservation, improved farming systems, dryland horticulture, pasture land development and forestry need to be taken up simultaneously. In the implementation of these practices, two phases can be identified. The first one involves fairly long-term public and private investment. This should be followed by a phase of efficient management of resources and technologies on a sustainable basis. Both these aspects of investment and management of watershed practices are influenced by several factors. Once these practices are identified, the impact of these watershed measures and the factors influencing the adoption are most significant in economic analysis. Therefore, analysis needs to be focussed on the impact of technologies, costs, benefits and distributional aspects.

Watershed Management Practices should aim at achieving the twin objectives of restoration of ecological balance and promotion of diversified sustainable farming systems through synergistic combination of annual crops, agro-forestry, dryland horticulture, livestock management, household production systems, *etc.* both in arable and non-arable areas. The holistic approach of treating arable-land, non-arable land and drainage-lines on the principles and practices of watershed management are to be pursued. Local initiative, self-help culture and participation of the watershed community as partner in the project planning and implementation and not merely as beneficiary, is the central theme of the strategy and approach.

Sustainable Development

Probably the first comprehensive definition of sustainable agriculture was put forth by the US Farm Bill, 1990. It has defined sustainable agriculture "as an integrated system of plant and animal production practices having a site specific application that will, over the long term, satisfy human food and fiber needs, enhance environmental quality and conservation of natural resources base, upon which the agricultural economy depends, make the most efficient use of non-renewable resources and on-farm resources; and integrate, where appropriate, natural biological cycles and controls, sustain the economic viability of farm operations; and enhance the quality of life of farmers and society as a whole".

Sustainable agriculture has been recognized as agriculture that combines modern technological innovations with proven resource conservation and food and fibre production practices to protect environmental quality, maintain and enhance profitability, preserve rural communities, and produce a safe and adequate supply of food for current and future generations. The sustainable agriculture entails management of agricultural activities to protect air, soil and water quality and to conserve biodiversity, thereby increasing long-term productivity and profitability, as well as enhancing human health and well being. Production of a safe, high quality and affordable supply of food and fibre in a manner that protects and conserves natural resources. Natural resource conservation, profitability, productivity and human health are all interrelated aspects of sustainable agriculture.

Of the total geographical area of 328.8 million hectares of the country, approximately 142 million hectares are arable lands of which about 55 million hectares are irrigated and the remaining rainfed. The management of these agricultural lands can affect the quality of ecosystems and the conditions of natural resources over a significant portion of the country's land base. The farmers should, therefore, be encouraged to control land degradation, use environmentally sound pest and nutrient management techniques and take other actions that would preserve the health of ecosystems and conserve natural resources, particularly soil and water. If these two resources are not properly used and conserved, we will be faced not in the too distant future with serious problems which may be difficult or very expensive to solve later. With the intensification of agriculture and an increased use of chemical fertilizers and pesticides, not only the ground water is polluted, the soil health is also affected, particularly due to depletion of organic matter, a main source of nutrients and fabric of soil structure. In canal-irrigated areas, the most pervasive damage has occurred from waterlogging and salinization.

Hence, natural resource management has become a critical issue, because sustainable management of water and soils, combined with sustainable agricultural development, is the only hope for providing food, feed, fuel and fibre for the present generation, while assuring that future generations have equal access to the same resources in order to ensure their own survival. A key consideration in approaches to achieve sustainability in agriculture is to give serious thought to possible impacts of technical advancements in short as well as long term perspective and their cross sectoral linkages. If the technology does not proceed or match the new situations, the consequences can be disastrous to the state's economy. The options for achieving sustainability are never confined to technological innovations. Revival or redevelopment of the old practices abandoned as a result of technological or socio-economic or political forces in the past, with appropriate scientific and institutional inputs, is also considered to be an equally effective option of mitigating many of the present problems.

For sustainability of agriculture, it is essential to achieve optimal production with minimal external inputs and enhanced use of on-farm resources, maintain productivity, reduce the level of production risk, protect the potential of natural resources, prevent degradation of soil and water quality and satisfy human needs. Besides it should give adequate economic returns and have minimal adverse environmental impacts. The Ninth Plan does talk about environmental sustainability, but the problem does not go beyond raising the issue, and lacks concrete steps.

Capital Formation in Indian Agriculture

There is continuing concern about the falling share of public investment in agriculture. This has happened mainly because a large proportion in agriculture. This has happened mainly because a large proportion of public expenditure has been going into current expenditure in the form f increased level of subsidy for food, irrigation, fertilizer, electricity, credit and other agricultural inputs rather than creation of assets. Annual investment in agriculture was Rs.13523 crore in 1993-94, one third of which was public investment. In 1999-2000 it rose to Rs.18656 crore (38 per cent increase over 1993-94) but the share of public investment decline to 25 per cent.

Investment in agriculture is estimated at 1.5 per cent of GDP in 1999-2000 as against the public sector stagnation (0.71 per cent rate of growth) in the nineties. It was the private sector (7.6 per cent annual) growth which has saved the situation. Official estimates put private sector investment in agriculture towards the end of the 1990s to be more than 50 per cent higher, as compared to the early 1990s. During the seven year period 1993-94 to 1999-2000 agriculture investment has ranged between a low of 1.4 per cent to a high of 1.6 per cent of GDP (**Table 4.5**)

Table 4.5: Gross Capital Formation in Agriculture (at 1993-94 prices)

(Rs. Crore)

Year	*Total*	*Public*	*Private*	*Per cent Share*		*Investment in Agriculture as*	
				Public	*Private*	*Per cent of GDP*	*Per cent of Current Ex.*
1993-94	13523	4467	9056	33.0	67.0	1.6	3.4
1994-95	14969	4947	10022	33.0	67.0	1.6	3.3
1995-96	15690	4848	10842	30.9	69.1	1.6	2.8
1996-97	16176	4668	11508	28.9	71.1	1.5	2.3
1997-98	15953	3979	11974	24.9	75.1	1.4	1.7
1998-99	16384	3846	12538	23.5	76.5	1.4	1.4
1999-2000*	18656	4668	13988	25.0	75.0	1.5	1.4

* Quick Estimates; ** This does not include forestry, logging and fishing..

Definitions/Concept of terms used in above Table.

(i) **Gross Fixed Capital Formation**: Gross fixed capital formation represents the gross value of fixed assets such as buildings roads, bridges and other construction works, machinery and equipment *etc.*, acquired or from own account production of such assets during the year. The type of assets in the public agriculture sector comprises of mainly buildings, construction works and machinery and equipment created for irrigation.

(ii) **Change in Inventories**: Change in inventories represents the value of net additions in raw materials, works in progress and finished products during the year.

(iii) **Gross Capital Formation**: Gross capital formation is the total of gross fixed capital formation and change in inventories.

(iv) **Departmental Commercial Undertaking (DCU):** Departmental enterprises are unincorporated enterprises owned, controlled and run directly by the

Government. These are mainly engaged in furnishing the kinds of goods and services which are often produced by business establishments but do not hold or manage financial assets and liabilities apart from their working balances and business accounts, payable and receivable.

(v) **Non-Departmental Commercial Undertaking (NDCU)**: Non-departmental enterprises comprise government companies registered under Companies Act 1956 and having 51 per cent or more share of government and public corporations set up in the Central/State Government Act wholly, or mainly, owned and/or controlled by the public authorities excluding housing boards, municipal corporations and improvement trusts.

Some of the factors for slower growth in the investment/capital formation in agriculture could perhaps the ascribed to some of the following factors:

(a) Large part of the allocation in public sector is going out in the form of subsidies,

(b) Growing recurring expenditure on maintenance of existing assets thus draining out the resources from investment;

(c) Rising time and cost over-runs of ongoing projects

(d) Relatively lower allocations for irrigation and Agricultural Research,

(e) Inadequate credit support and credit infrastructure in rural area.

With the burden on productivity – driven growth in the future, this must also be stimulated further through appropriate policies. Considering that nearly 70 per cent of India still lives in villages, agricultural growth will continue to be the engine of broad-based economic growth and development as well as of natural resource conservation, leave alone food security and poverty alleviation. Acceleration investment are needed to facilitate agricultural and rural development through:

- ✰ Productivity increasing varieties of crops, breeds of livestock, strains of microbes and efficient packages of technologies, particularly those for land and water management, for obviating biotic, socioeconomic and environmental constraints;
- ✰ Yield increasing and environmental-friendly production and post-harvest and value-addition technologies.
- ✰ Reliable and timely availability of quality inputs at reasonable prices, institutional and credit supports, especially for small and resource-poor farmers, and support to land and water resources development.
- ✰ Effective and credible technology, procurement, assessment and transfer and extension system involving appropriate linkages and partnerships; again with an emphasis on reaching the small farmers;
- ✰ Improved institutional and credit support and increased rural employment opportunities, including those through creating agriculture-based rural agro-processing and agro-industries, improved rural infrastructures,

including access to information, and effective markets, farm to market roads and related infrastructure;

- Particular attention to the needs and participation of women farmers; and
- Primary education, health care, clean drinking water, safe sanitation, adequate nutrition, particularly for children (including through midday meal at schools) and women.

This will need to be supported through appropriate policies that do not discriminate against agriculture and the rural poor. Given the increasing role of small farmers in food security and poverty alleviation, development efforts must be geared to meet the needs and potential of such farmers through their active participation in the growth process. There is the need to facilitate and support community level action by private voluntary organization, including farmers groups aimed at improving food security, reducing poverty, and assuring sustainability in the management of natural resources. In addition, government should enhance efforts to ensure good nutrition and access to sufficient food for all through primary health care and education for all.

Modern biotechnology tools, genetic engineering, as well as conventional breeding methods are all expected to play important roles in the generation of higher yielding, past and stress resistant varieties of rice, wheat, maize and other cereal crops. The availability of genetic innovations in developing countries will depend on continued high levels of investment in agricultural research, both at the international and the national levels. Free and unhindered access to germplasm to breeders is absolutely crucial to the rapid dissemination and adoption of improved germplasm. This free movement and the dissemination of modern biotechnology innovations to developing countries are hampered by increased patent protection and private sector investments. There is an urgent need to address this problem of free access to technology in the future.

Increased attention will also have to be given to development of sustainable system that protect the natural resource base. Recent evidence of resources degradation and declining productivity in some intensively cropped areas is of particular concern. Also population driven intensification of agriculture without the use of external inputs, is leading to a serious problem of maintaining soil fertility. In addition to investments in research, infrastructure, investments, particularly in irrigation, transport and market infrastructure development are equally important for sustaining the productivity and profitability of food crop production. It is also necessary to mobilize the best of science and development efforts (including traditional knowledge and modern scientific approach) through partnerships involving national and international research institutions, NGOs, farmer's organizations and private sector in order to tackle the present and future problems of food security and production.

5

Agribusiness and its SWOT Analysis

While conducting SWOT analysis of agricultural sector in India it shows that the abundant availability of raw material, easy availability of labour, growing capacity of food processing industry, by-default 'organic' production in many parts of the country and vast pool of scientific and technical manpower are the strengths of the sector. The agribusiness-led growth has good potential to contribute to sustained economic development of the country. Vast intapped resources of East India is nothing short of a sleeping tiger – waiting to be awakened. The increasing demand for processed and high- value food in domestic and international markets, globalization and integration on international consumer market offer further opportunities for development of agribusiness in the country. Small and fragmented holdings, lack of government support, low value-addition, poor infrastructural development, lack of awareness and consequently, poor management practices, *etc.* are some of the weaknesses of the sector. Increased competition and import threat, environmental threat, threat to unorganized retail players, thread from nexalites – some 125-131 districts are badly affected. Maintenance of health and safety standards are some of threats being faced by Indian agriculture. Investment is required in public infrastructure to facilitate market linkages and to induce private investment in cold chains, food processing and other market infrastructural development besides investment in R&D. Appropriate policy suppo9rt is also required to strengthen supply chain and value addition.

Agricultural Education

When India attained independence in 1947, there were 17 agricultural and veterinary colleges. While these colleges were largely responsible for training agricultural graduates, the State Departments of Agriculture and Community

Development executed the programmes of research and extension. The relationship between agricultural colleges and research departments was not close enough to ensure maximum utilization of findings of research. Annexure- A.

Although higher agricultural education was extremely costly and the resources available were limited, a large number of new agricultural colleges sprang up during early post-independence period under pressure of regional interests. The number of agriculture/veterinary colleges almost doubled during the period 1953 to 1960. There are at present 92 agricultural and animal science colleges in the country, 50 of which offer post-graduate courses. Their total enrollment is of the order of 13,500 per annum of which about 2,000 enter post graduate institutions each year. This rapid multiplication of agricultural colleges affiliated to traditional universities in spite of inadequate financial support led to a certain degree in the sliding down in standards of education and soon became a serious problem. Accordingly, the pace of progress remained slow and production technology developed at these institutions did not keep pace with the fast changing requirements of our agriculture. Moreover, this expansion took place along the traditional pattern. It was not oriented to meet the needs of economic progress and change that was overtaking the country. It was soon realized that both the system of education as well as the set up of the institutions in the field of agriculture and animal sciences needed to be reorganized if it was to serve as an effective vehicle of agricultural progress and development. This necessitated a re-examination of the existing system of agricultural education. Recognizing the weakness of the then existing educational system and the need for linking programmes of agricultural education closely with production programmes, the University Education Commission (1948) headed by Dr. S. Radhakrishnan suggested the establishment of 'Rural universities'.

The birth of the idea of Agricultural universities, however, can be traced to the First Joint Indo-American Team which was set up on the advice of Dr. Frank W. Parker, who was T.C.M. Advisor to the Ministry of Food and Agriculture, Government of India. The report submitted by the Team in September, 1955, laid the foundation of the scheme. The Second Joint Indo-American Team set up on 12 September, 1959 was headed by the Vice President of the Indian Council of Agricultural Research. It had three representatives of American Land-Grant Universities, *viz.* Dean Arthur D. Weber, Dean A. E. Darlowand Dean Arthur L. Deering, and Dr. Martin G. Weiss representing the U. S. Department of Agriculture. The Indian members of the Team were Dr. B. N. Uppal, Dr. L. Sahai, Mr. Lal Singh, Mr. P. D. Nair, Dr. M. D. Patel, Dr. K. C. Naik, (Secretary) Dr. J. S. Patel, and Mr. Ibne Ali. The Team submitted its report on July 11, 1960, and recorded that there was widespread demand from many states for the establishment of agricultural universities. The Team recommended that assistance in establishment of an agricultural university should only be granted when there was adherence to basic principles such as (1) autonomous status, (2) location of Agricultural, Veterinary Animal Husbandry, Home Science, Technological and Science Colleges on the same campus, (3) integration of teaching by offering courses in any of these institutions to provide a composite course, and (4) integration of education, research and extension.

About ICAR

The Indian Council of Agricultural Research (ICAR) is an autonomous organization under the Department of Agricultural Research and Education (DARE), Ministry of Agriculture, Government of India. Formerly known as Imperial Council of Agricultural Research, it was **established on 16 July 1929** as a registered society under the Societies Registration Act, 1860 in pursuance of the report of the Royal Commission on Agriculture. The ICAR has its headquarters at New Delhi.

The Council is the apex body for coordinating, guiding and managing research and education in agriculture including horticulture, fisheries and animal sciences in the entire country. With **100 ICAR institutes** and **71 agricultural universities** spread across the country this is one of the largest national agricultural systems in the world.

The ICAR has played a pioneering role in ushering Green Revolution and subsequent developments in agriculture in India through its research and technology development that has **enabled the country to increase the production of foodgrains by 5 times, horticultural crops by 9.5 times, fish by 12.5 times, milk 7.8 times** and **eggs 39 times** since 1951 to 2014, thus making a visible impact on the national food and nutritional security. It has played a major role in promoting excellence in higher education in agriculture. It is engaged in cutting edge areas of science and technology development and its scientists are internationally acknowledged in their fields.

The work of this Team was greatly facilitated by a blueprint on agricultural universities which was prepared by Dean H. W. Hannah in 1956. The contribution made by Dean Hannah to this scheme while he was living at Terai Farm, which had been reclaimed from a marsh and scrub forest, is basic. On the basis of this blueprint, the Uttar Pradesh Government submitted a proposal to the Government of India in September, 1956 to establish an agricultural university at Rudrapur in Terai, now called Pantnagar. The Government of India approached the problem in a cautious manner and agreed to the setting up of the agricultural university in Terai as an experimental measure in the Second Five-Year Plan. However, there was demand from many more states for such universities, and in 1961 the Government of India accepted the need for a few more such universities during the Third Plan period and suggested that the existing colleges or institutions which had strong departments for teaching and research should serve as nuclei for such universities.

Since 1960, nine Agricultural Universities have come into existence by 1968 in the States of Andhra Pradesh, Madhya Pradesh, Mysore, Odisha, Punjab, Rajasthan, U.P. and West Bengal, today there a total of nine Agricultural Universities. The development patterns as also the functions and responsibilities delegated to these Universities, however, have varied from state to State and in some cases do not strictly conform to the central concept behind this institution building process as laid down in the Model Act for Agricultural Universities. Transfer of Statewide agricultural research, education and extension to the Agricultural Universities is the first requisite if the objective of integration is to be achieved and these institutions are to make their contribution on the agricultural production of a state. It is our hope that this situation would be soon corrected in the interest of agricultural development and the reorganized system of agricultural education would meet with the same success in this country as the Land Grant Colleges system in the U.S.A. from which we have borrowed heavily in developing this new pattern.

Cummings' Committee Report

In 1960 the Government of India appointed a committee headed by Dr. Ralph W. Cummings to advise the State Governments on the legislation for the establishment of agricultural universities. The main idea was that the new agricultural universities should have the essential features that characterize the system and that they have a uniform base to carryover the functions with which they were charged. The report of the Committee published in 1962 spelled out the distinctive features of the agricultural universities as compared to the existing universities and provided guidelines for their development. On the recommendations made by this Committee, the Indian Council of Agricultural Research developed a model act which could be adopted with such changes as were necessary by the newly developing agricultural universities. This is an important milestone in the development of these universities for which Dr. Cummings and his colleagues deserve much credit.

Relationship of Indian Agricultural Universities with the U.S. Land Grant Universities

The T.C.M., now A.I.D., program that started in India in 1955, was on a regional basis for agricultural education. The trend to one Agricultural University for each state took shape in 1960 and all AID support has been on a state basis since 1963. There are at present eight Agricultural Universities being assisted by the Agency for International Development through six U. S. Land Grant Institutions. This university to university relationship is the bridge to scientific and cultural understanding between two great democracies. These relationships are: Uttar Pradesh Agricultural University and the University of Illinois, Punjab Agricultural University and the Ohio State University, Andhra Pradesh Agricultural University and Kansas State University, University of Udaipur and the Ohio State University, Jawaharlal Nehru Krishi Vishwa Vidyalaya and the University of Illinois, Odisha University of Agriculture and Technology and the University of Missouri, Mysore University of Agricultural Sciences and the University of Tennessee, and Maharashtra Agricultural University and Pennsylvania State University.

Through USAID and these U. S. Universities, assistance has been given for training Indian Agricultural University's faculties. At present there are 4,500 returned participants from AID programs and over one-fourth have been related to agriculture, veterinary medicine, engineering and home science. Specialists from the U. S. Universities serving with Indian counterparts have been and are serving in teaching, research and extension education. At present, the U. S. teams are nearly at full strength in India. Their impact is real and most welcome. Other AID assistance has included limited amounts of equipment for teaching and research. All of these inputs - participants, specialists and equipment have aided the Indian Agricultural Universities to become real centers of new ideas and practices for India's agriculture.

U.P. Agricultural University, Pantnagar

The first agricultural university was started at Pantnagar on a 12,000 acre farm in 1960. The University has four constituent colleges, namely those of agriculture, veterinary medicine, engineering and technology, and basic science and humanities. The college of home science is planned for inauguration next year.

Punjab Agricultural University, Ludhiana

The next university to follow was the Punjab Agricultural University at Ludhiana in October, 1962. As I have been associated with this institution since its very birth, it would be relevant to give a few details. On account of partition of India on 15 August, 1947, the famous Lyallpur Agricultural College fell to the share of Pakistan. The refugee farmers of Ludhiana, who mostly came from Lyallpur Canal Colony were conversant with the benefits of an agricultural college and they vacated a school building to house the agricultural college. I was then Director General of Rehabilitation, rehabilitating rural refugees from West Pakistan in India. I allotted about 1,000 acres of land in an evacuee village for the college. By 1955 a modern college building was already under construction for the college of agriculture. In 1962, largely due to the interest taken by the Chief Minister, Shri Pratap Singh Kairon, who was keenly interested in agricultural development, Ludhiana was selected for the second agricultural university. Due to the dedicated efforts of the first Vice Chancellor and the enthusiastic support of the Chief Minister, Mr. Kairon, the University made rapid progress. Apart from the college of agriculture, a college for basic sciences, a home science college and an engineering college have been started.

University of Udaipur, Udaipur

In 1962, an agricultural university was established in Udaipur. To this University was affiliated the college of agriculture at Udaipur and Jobner and the veterinary college at Bikaner. Subsequently, college of agricultural engineering and home science were added, and an existing college in basic science and humanities was added as a constituent unit. This University, however, will be entrusted with the responsibility for agricultural research in the state in the Fourth Plan period.

Odisha University of Agricultural Science and Technology, Bhubaneswar

The University began to operate from February 1963, having taken over the State colleges of agriculture and veterinary science. The college of basic sciences and humanities was added in 1964 and a college of agricultural engineering in 1966. With the addition of a college of home science within the next few years the complex of essential units of the agricultural universities would be complete. This University has partly taken up the responsibility of agricultural research from the state Department of Agriculture.

Andhra Pradesh Agricultural University, Hydrabad

The Andhra Pradesh Agricultural University was established in June 1964. It took over the three existing agricultural colleges, two veterinary colleges and a home science college. A new college of basic science and humanities has been added while a college of agricultural engineering is expected to be started. The state Government has transferred research responsibility to the University in 1966 along with staff and 6,000 acres of land and laboratories.

Mysore University of Agricultural Sciences, Bangalore

The Mysore University of Agricultural Sciences located at Bangalore commenced operating from October 1965 with the transfer of the two State agricultural colleges and a veterinary college. The long range plan of the university includes the establishment of new colleges of basic science and humanities, fisheries, agricultural engineering, forestry, home science, dairying and horticulture. Most of the research organization of the state Government has been transferred to the university along with 3,000 acres land in the experimental stations. As in several other universities, the extension activities are mainly the organization of refresher courses for state Government extension worker-s though direct extension work is also in progress around the campus.

The Unive.rsity of Kalyani, Kalyani

The University of Kalyani was established in 1960 with faculties of arts, sciences and education. It was recognized as an agricultural university in 1965. This University has still to assume the responsibility of agricultural research and extension education from the Department of Agriculture of the State Government. Maharashtra Agricultural University, Rahuri Maharashtra Agricultural University was inaugurated in March 1968. It consists of eight state agricultural colleges and two veterinary colleges in addition to two affiliated private agricultural colleges. The central university campus is proposed to be built at Rahuri. The University would soon have the statewide research responsibility.

Contributions of Agricultural Universities to Agricultural Production

The concept of integration of teaching, research and extension has already proved its worth through remarkable progress made in the field of agricultural education, research and extension by the new Agricultural Universities. There is perceptible improvement in the quality of education. There are more competent teachers, better equipped libraries, laboratories, and farm-B. The internal examination system is geared for continuous preparation on the part of both the students and the teachers.

These institutions are largely responsible for the development of the high-yielding varieties of wheat, maize, bajra and jowar'. Unprecedented high crop yields have been recorded. Agronomic and plant protection practices to exploit maximum yield potentials have been developed and effectively extended to the farms. These institutions today are serving as fountain heads of new knowledge earned through purposeful, problem-solving research and have become main centers of dissemination of useful knowledge to farming community. Some of the best training for farmers is offered by the Agricultural Universities. There are numerous functional specialists who have gained confidence through experience in successfully applying scientific knowledge to the solution of practical problems.

The working conditions and incentives that they offer to the faculty and the students are providing opportunities for productive work, and are fostering team spirit and a healthy change in the outlook of all the teachers and the researchers, and the government administration. They are winning the confidence of the farmers. They have assumed leadership in science, education and extension. Their direct contribution to programmes like pedigree seed production, fertilizer use and National Crop Demonstrations is highly impressive. In this connection, the inter-institutional collaboration within the country and the international collaboration with the U. S. Universities needs special mention. Agricultural Universities are participating most effectively in the execution of the various coordinated programmes of agricultural research initiated by the ICAR. Collaboration with U. S. Universities is aimed principally for advanced training of University personnel at institutions of repute in the U. S. and securing subject-matter specialists on long or short term basis for helping to raise the standards of teaching and research at these institutions.

Besides improvement in quality, the new system of education has reduced the 'wastage' in higher education. This not only saves cost but provides training opportunities for more students. It will not be out of place to quote some of the specific contributions made by the Agricultural Universities. The soybean projects operating at the U. P. Agricultural University, Pantnagar and Jawaharlal Nehru Agricultural University, Jabalpur, have demonstrated that a number of soybean varieties are highly adapted to these areas and the crop can be grown successfully. Introduction of sugar beets in the eastern region of the U. P. is another example of a new crop having been adapted successfully in the region. The U. P., Punjab and Andhra Pradesh Agricultural Universities have already made significant contributions in the cereal improvement programs coordinated by the Indian Council of Agricultural Research in cooperation with The Rockefeller Foundation. In the field of extension, there the Tungbhandra fertilizer use project executed by the University of Agricultural Sciences, Mysore, is an outstanding example of how new varieties of crops and agronomic practices can be speedily spread among the farming communities. The University took up the development of intensive farming in 10,000 acres of any irrigated land under the project with the assistance from OXFOM and the USAID. The project has been highly successful. The College of Engineering of the Punjab Agricultural University which has received considerable support from the Ford Foundation has provided much needed technical guidance to private companies which are now engaged in the manufacture of agricultural implements such as fertilizers- cum-seed drills, threshers, water-lift pumps, sprayers and dusters.

The Punjab State is known for its progressive farmers, An ambitious scheme of rural electrification has been taken up and private tube wells at the rate of 20,000 per year are being energized. In the utilization of ground water, this is a major development. No doubt, the drought of 1965-66 has been the major stimulus in ground water utilization. The spread of high-yielding varieties of wheat developed at the University has been rapid and there is now adequate seed to cover entire irrigated areas in this state.

In the field of animal sciences the technique of preserving semen in coconut milk medium has enabled long distance transport of semen in the states of Mysore, Uttar Pradesh and Andhra Pradesh. The poultry feed projects of the PAU and University of Udaipur as well as the animal cross-breeding project of the PAU are other noteworthy examples.

The Report of the Education Commission 1964 - 66

The impact which the scheme of agricultural universities has made on policy makers is evident from the report of the Education Commission (1964-66) headed by Dr. D. S. Kothari, the Chairman of the University Grants Commission. The Commission had recommended the establishment of at least one agricultural university in each state. They have further recommended that all aspects of research on agriculture should be the concern of the agricultural universities. Implementation of these recommendations would further enlarge the area under the control of these universities. It would lead to the integration of teaching, research, and extension education where it does not exist at present. That an agricultural university provides a better environment for research than a State Department of Agriculture, however, is still to be realized in a number of states, where vested interests which believe in holding on to what you possess are still strong.

Employment of Agricultural Scientists

The scheme of intensive agriculture provided employment to a large number of agricultural graduates. The fertilizer and agro-chemical industries which are now developing at a fast rate have also provided a new field of employment. However, there is need for making plans for the absorption of agricultural engineers by developing schemes for soil conservation and irrigation. There is already large scale unemployment among civil engineers in India and there is an apprehension that unless steps are taken now to find new avenues of employment for the agricultural engineers, a similar situation may arise in this field. So far as agricultural graduates are concerned, some of them will be returning to their own farms for work rather than relying on government jobs. This is mainly due to the fact that only in the last two or three years, due to assured remunerative prices for cereal crops, agriculture for the first time in the history of India has become a paying profession. With the acceleration of rural electrification schemes which are closely connected with tube well programs for utilization of ground water, there is every possibility of a number of young men returning to their farms, linked with it is the need for large scale programs for rural roads. In the State of Punjab an ambitious programme of rural link roads has already been taken up. Rural electrification coupled with assured remunerative prices for the crops would insure absorption of increasing number of young men on their own farms.

Examples of areas where the agricultural graduates have been leaders and where leadership will continue to expand are in seeds, pesticides, fertilizers, credit and other agro-business enterprises. For instance, the use of pesticides application have increased over seven times since 1961, nitrogen over four times and phosphorous by five times. Many other examples could be used but time does not permit.

Association of Indian Agricultural Universities

An Association of Agricultural Universities has been formed and registered and, it is hoped, will soon start functioning. It is expected to playa key role in the development of Agricultural Universities in India on the same lines as the National Association of state Universities and Land Grant Colleges of the U.S.A. We intend to draw liberally from your experience in this field and count on your guidance and support. Summing up, there are nine Agricultural Universities, out of which eight have sisterhood relationships with six U. S. Land Grant Institutions. There is likelihood of three to four more agricultural universities coming up in the next few years. The eight Agricultural Universities which have cooperation with the U. S. Universities, have 63 colleges in all. Out of these there are 25 colleges of agriculture, 11 of veterinary medicine, 8 of engineering, 8 of basic sciences, and 7 of home science. These institutions are in various stages of development. Transfer of agricultural research on statewide basis to the remaining agricultural universities is an urgent necessity. The necessity of establishing faculties of basic sciences and humanities as part of agricultural universities is yet to be realized. Wherever they have been established, they have strengthened programs of education in agricultural sciences and have stimulated basic research. That all this has happened during a period of eight years is a creditable achievement and attributable to Indo-American cooperation. In fact, it is in the field of agricultural education and research that the contribution of the U.S.A. has been most effective and useful. The U. S. Government and the Rockefeller and Ford Foundations have generously supported this new development by providing technical assistance in the form of specialists and equipment as well as training of faculty members of Indian Agricultural Universities at U. S. institutions, for which India is highly grateful.

Detailed Note on the Pantnagar University

Govind Ballabh Pant University of Agriculture and Technology (**GBPUA and T, Pantnagar University** or **Pantvarsity**) is the first agricultural university of India. It was inaugurated by Jawaharlal Nehru on 17 November 1960 as the "Uttar Pradesh Agricultural University" (UPAU). Later the name was changed to "Govind Ballabh Pant University of Agriculture and Technology" in 1972 in memory of the first Chief Minister of Uttar Pradesh, statesman and Bharat Ratna recipient Pandit Govind Ballabh Pant. The University was declared as the best among all the other State Universities in India, in 1997. The University lies in the campus-town of Pantnagar in the district of Udham Singh Nagar in the state of Uttarakhand. The university is regarded as the harbinger of Green Revolution in India.

History

Pre-establishment Groundwork (1949 -1959)

The first Education Commission of India (1949) headed by Sarvepalli Radhakrishnan recommended setting up rural universities in India on American land-grant model. Uttar Pradesh (UP), under chief - ministership of Govind Ballabh Pant, took the first step and in 1954 invited an Indo- American team headed by Dr. K. R. Damle, the Vice-President of ICAR, to consider an area around Tarai State farm in erstwhile Nainital district as a possible site for a rural university. This area was actually a dense forest near Himalayan foothills and the government was using this area to rehabilitate Hindu, Sikh and other migrants from West Pakistan, in the aftermath of the partition of 1947. Encouraged by favourable view of the Damle team, two senior government officials- H S Sandhu and A N Jha visited United States to look for collaborations with US Universities. In consultation with University of Illinois dean Dr H W Hannah, the government of state of Uttar Pradesh presented a proposal to the Government of India in 1956 for establishing a Land-grant style university. Thereafter, a contract between the Government of India, the Technical Cooperation Mission and few US land grant universities, was signed to promote agricultural education in India. The US universities included the University of Tennessee, the Ohio State University, the Kansas State University, the University of Illinois, the Pennsylvania State University and the University of Missouri. The task of mentoring the proposed university in UP was assigned to the University of Illinois which signed a contract in 1959.

Initial Years and Development (1960-1972)

The University was dedicated to the Nation by the first Prime Minister of India Jawaharlal Nehru, on 17 November 1960. In the early Days, Illinois faculty served the university in designing its education system and putting in place an effective research and extension system. Six to eight Illinois faculty used to stay in Pantnagar at a time serving between 2 and 4-year terms for a period of 12 years. Dr. William Thompson, a team member on site at Pantnagar, shared that it was unusual for the project to start a university in a place with nothing. All buildings and facilities had to be constructed in the jungle there. In 1965, drastic upheaval of the university board of directors, which was spurred by lack of state government support for the institution, caused removal of the entire administrative and governance team of the university. D P Singh was named vice chancellor of the university with complete control over its affairs until a new board of directors was chosen. Under Singh's leadership, many necessary upgrades took place, and the university flourished. The University of Illinois left Pantnagar in 1972, when president Richard Nixon ordered Americans out of the near east. The UP Act XI-V of 1958– the founding legislative act was later amended under UP Universities Re-enactment and Amendment Act 1972 and the University was rechristened as Govind Ballabh Pant University of Agriculture and Technology keeping in view the contributions of Pandit Govind Ballabh Pant. Pantnagar is a symbol of successful partnership between India and the United States. US greatly influenced the development of the university through its funding of several university programs. Many of the university's research activities were funded by the Rockefeller Foundation, the Ford Foundation, the Indo-U.S. Technical Cooperation Programme, and the United States Agency for International Development (program Public Law 480). The establishment of this university brought about a revolution in agricultural education, research and extension. It paved the way for setting up of 41 other state agricultural universities in the country.

Role in Ushering Green Revolution

Pantnagar University soon became a significant force in the development . d transfer of High Yielding Variety seeds and related technology. The Mexican wheat varieties, developed by Norman Borlaug were tested in Pantnagar and locally adaptable selections, like Pantnagar Kalyan Sona and other improved varieties were released for farmers. The university utilised its 16,000 acres (65 km^2) of land to launch one of the largest seed production programs at that time, under the brand name Pantnagar Seeds, which became a household name in rural India. The contribution of the university was recognized by Norman Borlaug, who described Pantnagar as "Harbinger of Green Revolution".

Colleges

Pantnagar is a non-affiliating university and its academic activities are organized through its nine constituent colleges (faculties),which run large undergraduate (UG)programs. All Masters and PhD students are part of College of Post Graduate Studies, however they have strong association with one of the other colleges as per their academic major. Prior to 2012, the university also had two off-campus colleges namely, 'College of Forestry and Hill Agriculture, Ranichauri' (founded 1991) and 'Veer Chandra Singh Garhwali College of Horticulture, Bharsar' (founded 2002). However, recently, through an act of Uttarakhand Legislative Assembly[36] these colleges have been transferred to newly incorporated Uttarakhand University of Horticulture and Forestry. Distributed among colleges and other units of the university, there are 763 teachers and officers, 59 technical personnel, 631 administrative personnel and 1425 class III employees, amounting to a total staff strength of 2878 against a students strength of 2800–3000, which makes teacher taught ratio of about 1:6 and student-staff ratio about 1:1.

Table 5.1: List of Colleges

Sl.No.	*College*	*Year Founded*	*Annual Student Intake*	*Faculty Strength*
1	College of Agriculture	1960	425	170
2	College of Veterinary and Animal Sciences	1960	255	81
3	College of Technology	1962	532	98
4	College of Basic Sciences and Humanities	1963	185	72
5	College of Post Graduate Studies	1963	*	**
6	College of Home Science	1971	125	25
7	College of Fisheries	1985	41	13
8	College of Agribusiness Management	1996	90	9
9	International School for Agriculture	2010	17	**

" Counted in associated College, ** Participating faculty from various departments

Education System

The university shifted from a trimester to a semester system in the 1980s to reduce the workload on students. There are 200–210 net instructional days per year, which is highest in India. The minimum duration of all UG programs of the university is 8 semesters (4 years), except BVSc and AH (10 semesters, 5-year). All the UG programs are infused with extensive practical training and include one semester of compulsory internship/work experience. All Masters programs are of 4 semester (2 years), except MCA (6 semester/3 years). Normal duration for PhD is 6 semesters. The university follows a credit based GPA system for evaluation. In 1998, it shifted from the US style 5-point and letter grade system to a 10-point GPA based system, for easy conversion to percentages, which are more commonly used in India.

The university follows 100 per cent internal evaluation on continuous basis. A course has typically 2 pre-final exams (popularly called as "hourlies"), 1 lab final exam and 1 semester-end final exam, totaling 4 evaluation cycles per semester. Typically, a UG student registers for 8 courses in a semester, leading to 8 x 4= 32 exams *i.e.* on an average basis, 1 exam on every fourth working day. Since its inception, Pantnagar University is known for its academic rigour for keeping its students always busy.

Research

Being an Agriculture and Technology university, the main focus of research is on agriculture and engineering. Research is coordinated by a Directorate of Experiment Station and carried out through ~70 subject-matter departments spread across colleges, 14 specialized research centres located in the campus, 7 dedicated off-campus research stations and 10 off-campus horticultural research-cum-extension centres. So far, the university has released 211 varieties, many of which played important role in Green Revolution. The work of the university in introducing soybean as a crop in India is well known. As Uttarakhand has been declared an 'Organic state', the present thrust of research is on Organic farming and Biological pest control. The university has developed a pregnancy diagnostic kit for cattle and its Salmonellosis vaccine for poultry is in advanced trials. In 1991, engineers of the university developed a 'Zero-till Ferti seed drill' for No-till farming along with National Agro Industries Ludhiana Punjab, which has been immensely popular in Haryana, Punjab and other areas of Indo-gangetic plains. CIMMYT/ CGIAR has described zero tillage technology based on the Pantnagar seed drill as the most widely adopted resource conserving technology in the Indo-Gangetic Plains, till date.

Extension

Extension is one of the three mandates of the university. University organizes informal continuing education programs for rural people for achieving better diffusion of innovation. The university reaches out to a wider clientele through short term training programs at its campuses, Farmer Field Schools, Participatory technology development programs, radio and television broadcast, community radio based narrowcasting, print media, Farmer's fairs, and other innovative communication technologies. The university maintains a State Agricultural Management and Extension Training Institute and a network of 11 Krishi Vigyan Kendras (Farm Science Centres, KVKs) located in various districts.

Residential Life

Pantnagar is a residential university. Hostel accommodation is compulsory, except for the wards of university employees who live in the campus. In the Pantnagar campus, there are 23 hostels (called bhawans), including seven for girls, two for married couples and one for working women. The older hostels are named after important personalities of India: Mohandas K. Gandhi, Subhas Chandra Bose, Jawaharlal Nehru, Lal Bahadur Shastri, Govind Ballabh Pant, Sardar Vallabhbhai Patel, Rabindranath Tagore, Sarojini Naidu, Kasturba Gandhi, Mokshagundam Visvesvarayya, Chittaranjan Das (I and II) and Vivekananda with the exception of a hostel named Silver Jublee, which commemorates 25 years of the university. The names of newly constructed bhawans are inspired from the various geographical regions of the Uttarakhand – Gangotri, Yamunotri, Shivalik, Mandakini, Bhagirathi, Alaknanda *etc.* – with an exception of a new hostel named 'Golden Jublee', commemorating 50 years. Most of older hostels are along the 'third arc' as per the original 'concentric semicircles' campus plan, while newer ones are on radial roads or other parts of the campus. There are two guest houses, Lambert's Square and International Guest House, which has a hostel wing for foreign students and an attached Farmer's Hostel. The hostels have student-managed or, in few cases, outsourced cafeteria-style food services. All the faculty members and employees stay in bungalows and apartments on campus. Two Market Namely Chhoti (small) and Bari (big) markets are also present in the campus.

Extra Curricular Activities

Most of the students' activities are organized through college-level professional societies, like the Agriculture Society, the Engineering Society, the Management Society, *etc.* These societies organize popular fests, fairs, cultural nights, treasure hunts and competitions year round. Apart from these, there are departmental societies, hostel societies and other organizations like Vivekanand Swadhyay Mandal, Sanskritic Chetna Parishad, *etc.* which organise popular student activities. The Pantnagar Chapter of SPIC MACAY regularly organizes classical music and dance concerts by renowned artistes. CHETNA Student Wing is an extension of CHETNA NGO (Delhi) working for upliftment of underprivileged and poor children in the campus and nearby areas.

The university National Cadet Corps (NCC) chapter organizes basic military training and adventure sports for student-cadets. The Remount and Veterinary (R and V) core is a Cavalry regiment of NCC, which trains students for professional equestrianism (horse riding). Pantnagar is among very few universities in India to have a full-fledged Air Squadron under NCC Air Wing, which promotes aviation and airmanship among students through its activities organized in the airport near the university campus. All undergraduate students of Pantnagar take part in community service, which fetches them two credits towards their degrees in form of a mandatory National Service Scheme course involving 240 hours/10 days of social service. Students also take part in liberal education courses offered by the university like dance, music, yoga, Indian cultural heritage, personality development, English vocabulary, Indian languages (Tamil, Punjabi), foreign languages (French, German), *etc.* Stevenson stadium is the hub of sports activities, which are organized through 17 sport clubs.

Forest Research Institute (India)

The **Forest Research Institute** is an institute of the Indian Council of Forestry Research and Education and is a premier institution in the field of forestry research in India. It is located at Dehradun in Uttarakhand, and is one of the oldest institutions of its kind. In 1991, it was declared a deemed university by the University Grants Commission.

History

Founded in 1878 British Imperial Forest School, by Dietrich Brandis. In 1906, reestablished as the 'Imperial Forest Research Institute', under the British **Imperial Forestry Service** (IFS). The 'Indian Forest College' was established in 1938; officers recruited to the Superior Forest Service by the states and provinces were trained there.

Architecture

Established as Imperial Forest Research Institute in 1906, Forest Research Institute (FRI) Dehradun, is a premier institution under the Indian Council of Forestry Research and Education (ICFRE). Styled in Greeko Roman Architecture by C.G. Blomfield, the main building is a National Heritage which was inaugurated in 1929. FRI Dehradun is one of the oldest institutions of its kind and acclaimed the world over. The Institute's history is virtually synonymous with the evolution and development of scientific forestry, not only in India, but over the entire sub-continent. Set in a lush green estate spread over 450 hectares, with the outer Himalaya forming its back drop, the institute's main building is an impressive edifice, marrying Greco-Roman and Colonial styles of architecture, with a plinth area of 2.5 hectares. The Institute has a developed infrastructure of all equipped laboratories, library, herbarium, arboreta, printing press and experimental field areas for conducting forestry research, quite in keeping with the best of its kind anywhere in the world. It is 7 km from Clock Tower, on the Dehradun - Chakrata motorable road and is of the biggest forest based training institute in India. Most of the forest officers are a product of this institute. The F.R.I. with its majestic building also houses a Botanical Museum for the attraction of tourists.

Location

FRI and College Area campus is a census town, between Kashmir in the north and the Indian Military Academy to the south. The Tons River forms its Western Boundary.

Training Facility

Within its campus it hosts IGNFA (Indira Gandhi National Forest Academy), which is a separate organization of the Indian Ministry of Environment and Forests (MoEF) for training officers selected for **Indian Forest Service** (IFS) by Union Public

Service Commission. It also hosts the Central Academy for State Forest Services (CASFOS). The Wildlife Institute of India previously was on campus but now has shifted to Chandrabani, and is now an independent organization under MoEF. IIFM is an independent autonomous organization of MoEF and has good liasoning with FRI for forestry research and related activities. The deemed university of FRI runs four MSc courses *viz.* Cellulose and Paper Technology, Environment Management, Forestry Management, Wood Science Technology and two P.G.Diploma courses in Natural Resource Management and Aroma Technology. It also enrolls large number of research scholars every year for PhD.

Museum

FRI also contains a museum on forestry. It is open from 9:30am to 5:00pm daily, which an entry fee of Rs 25 per person and a nominal entry fee for vehicles. There are six sections in the museum:

1. Pathology Museum
2. Social Forestry Museum
3. Silviculture Museum
4. Timber Museum
5. Non-Wood Forest Products Museum
6. Entomology Museum

Indian Agricultural Research Institute (IARI)

The **Indian Agricultural Research Institute (IARI)** commonly known as **Pusa Institute** is India's premier national Institute for agricultural research, education and extension. Situated in Delhi, it is financed and administered by the Indian Council of Agricultural Research (ICAR). The IARI was responsible for the research leading to the "Green Revolution in India" of the 1970s.

History

The institute was originally established in 1905 at Pusa, Bihar, as Agricultural Research Institute (ARI), which was changed to Imperial Institute of Agricultural Research in 1911 and then again in 1919 as the Imperial Agricultural Research Institute. With the financial assistance of an American Philanthropist, Henry Phipps. Jr. Phipps was a family friend of Lady Curzon, who was the daughter of American millionaire and wife of Lord Curzon, the Viceroy of India. Phipps stayed as a guest of Curzons during his visit to India. More importantly, Phipps left behind with them a donation of $30,000, which was utilized for the establishment of the institute. He laid the foundation stone of Agricultural Research Institute and college on 1 April 1905. However the institute was damaged during the devastating Bihar earthquake of 1934 which occurred on 15 January 1934. Thereafter the Secretary of State approved the transfer in July, 1943. The Standing Finance Committee of the Union Assembly finally announced on 25 August 1934 in Shimla, the decision to shift the institute to New Delhi at the approximate cost of Rs. 3.8 million (US$60,000). To a place that is now called Pusa in New Delhi. The new campus at New Delhi was inaugurated on 29 July 1936, while the new building of the Imperial Institute of Agricultural Research was inaugurated by then Viceroy of India, Lord Linlithgow on 7 November 1936.

Post-Independence: 1947 Present

Post-independence, the institute renamed Indian Agricultural Research Institute, and in 1950 the Shimla sub-station of institute developed Rust-resistant varieties of wheat, including Pusa 718, 737, 745, and 760. In 1958, it was recognized as a "deemed university" under the UGC act of 1956 of Parliament and since then it has awarded MSc and PhD degrees. What remained of the institute at the original location was downgraded to an agricultural research station until 1970, when the Government of Bihar established the Rajendra Agricultural University at the location.

Overview

It has served the cause of science and society with distinction through first rate research, generation of appropriate technologies and development of human resources. In fact, the Green Revolution was born in the fields of IARI and its graduates constitute the core of the quality human resource in India's agricultural research and education. The Institute has all along been adjusting and improving its policies, plans and programmes to effectively respond to the needs and opportunities of the nation. During the fifties, the advancement of scientific disciplines constituted the core programme and provided the base for its fast expansion in the 1960s and 1970s in all its three interactive areas, namely, research, education and extension. Besides basic research, applied and commodity research gained great importance resulting in the development of several popular high yielding varieties of almost all major crops and their associated management technologies, which brought about an unprecedented increase in the national food and agricultural production.

The campus is spread over 500 hectares (5.0 km^2), 8 km west of New Delhi Railway Station. This was initially outside Delhi, but over the decades the city has grown much beyond the campus. Indian Agricultural Statistics Research Institute is affiliated with and is located in the campus of the Indian Agricultural Research Institute.

Indian Council of Agricultural Research (ICAR)

Agriculture sector occupies the centre stage of India's social security and overall economic welfare. Since Independence, India has witnessed significant increase in foodgrain production (green revolution), oilseeds (yellow revolution), milk (white revolution), fish (blue revolution), and fruits and vegetables (golden revolution). All these became possible owing to the application of cutting edge of science coupled with the positive policy support, and hard work of Indian farmers. The Indian Council of Agricultural Research (ICAR), an apex organization for agricultural research, has been at the forefront to lead these agricultural revolutions in the country, making India not only self-sufficient in food but also with surplus. As a forward looking organization, fully realizing the emerging complex challenges,

ICAR has set a vision to attain **'Rainbow Revolution' covering the** entire spectrum of activities in agriculture which will make India a developed nation free of poverty, hunger, malnutrition, and environmental safety. Towards this goal, it is operating two prestigious and mega projects, *viz.* National Agricultural Technology Project with emphasis on production system research, organization and management reforms and innovations in technology dissemination and Agricultural Human Resource Development Project with emphasis on improving the quality of agricultural education. The ICAR began as the Imperial Council of Agricultural Research, an autonomous body (a registered society) in 1929. Presently the Union Minister of Agriculture is the President of the ICAR Society. The Director-General (DG) is the principal Executive Officer and is also the Secretary to the Department of Agricultural Research and Education (DARE). DARE is the nodal department for all related scientific and development activities and bilateral scientific collaborations with other countries. The Council has its Headquarters at New Delhi, and a vast network of institutes all over the country, consisting of 45 Institutes, four National Bureau, 30 National Research Centres, ten Project Directorates and 80 All India Coordinated Research Projects. The functions of ICAR are similar to those of University Grants Commission (UGC) in respect of agricultural education. The technical functions of the Head- quarters are grouped into eight subject-matter divisions, each headed by an eminent scientist as Deputy Director-General. Under the aegis of the ICAR there are 28 state agricultural universities, four deemed-to-be-universities and one central agricultural university, 261 *Krishi Vigyan Kendras* (KVKs: farm clinics) in the rural districts of the country for transfer of technology, and eight trainers training centres. For staff training and addressing research management issues, it has established an institute for human resource development called the National Academy of Agricultural Research Management (NAARM). ICAR is one of the leading agricultural research and development systems in the world, having 30,000 personnel, with more than 7000 engaged in active research and management.

The Mandate

The mandate of ICAR is:

- ☆ To plan, undertake, aid, promote and coordinate education, research and its application in agriculture, agro-forestry, animal husbandry, fisheries, home science and allied sciences.
- ☆ To act as clearing house of research and general information through publications and information system, and instituting and promoting transfer of technology programmes.
- ☆ To provide, undertake and promote consultancy services in the fields of education, research, training and dissemination of information.
- ☆ To look into problems relating to broader areas of rural development concerning agriculture, including post-harvest technology, by developing cooperative programmes with other organizations such as the Indian Council of Social Sciences Research, Council of Scientific and Industrial Research, Bhabha Atomic Research Centre and universities.
- ☆ To do other things considered necessary to attain the objectives of ICAR.

ICAR is engaged in developing new crop varieties with specific traits that improve yield and nutritional quality along with tolerance/resistance to various biotic and abiotic stresses. Besides, it matches crop production and protection technologies to target agro-ecologies. A total of 104 varieties of different crops were released for different agro-ecological niches. To ensure effective seed chain for making quality seed available to farmers, 11,835 tonnes of breeder seeds of recommended varieties of different field crops were developed. The adoption of improved varieties and crop management technologies has resulted in enhancement of production and productivity of cereals, pulses and other field crops.

Crop Science Division

The Crop Science Division has played a key role in ushering in the era of Green and Yellow Revolutions in the country. It is the largest Division of the Council with focus on the development of improved crop cultivars and appropriate crop production-protection technologies, and basic and strategic researches in crop science. The Division has ten institutes, one bureau, five project directorates, seven national research centres and 30 all India coordinated research projects. In addition, a large number of adhoc research projects and revolving schemes are also in operation.

Achievements

- ☆ Over 2,300 high-yielding verities and hybrids of field crops have been developed, released and notified for their commercial cultivation.
- ☆ First in the world to develop hybrid cultivars of cotton, grain pearl millet, pigeonpea, castor and safflower, and second to develop hybrid cultivars of rice and sorghum. Hybrid cotton is a landmark achievement in hybrid research.
- ☆ The varieties of foodgrains, particularly that ofwheat and rice, have been instrumental in ushering in the area of Green Revolution in mid-60s and sustaining the momentum of productivity enhancement in post-green revolution period.
- ☆ The improved varieties played a catalytic role in the adoption of other components of improved technology package, such as fertilizers, pesticides, irrigation water and transforming the mindset of farmers from conservative to technology-responsive ones.
- ☆ Incorporation of resistance to diseases and tolerance to abiotic stresses in high yielding background have enabled insulation of crop plants against these stresses and thus provided stability in food production and food security.
- ☆ Development of short duration varieties of rice, sorghum, cotton, pigeonpea, chickpea, greengram, blackgram *etc.* has opened up awareness for multiple cropping systems and helped in enhancing cropping intensity.
- ☆ Developed/deployed the concept of new plant type in various crop plants particularly wheat, rice, maize, sorghum *etc.* to upgrade genetic ceiling of yield potential.

- ☆ Spectacular success has been achieved in introduction and improvement of new crops, such as soybean and sunflower. India is now the fifth largest producer of soybean in the world.
- ☆ Improved varieties of sugarcane, wheat, rice, maize, sorghum, groundnut, mustard *etc.* developed in India have been used for commercial cultivation in many other countries.
- ☆ Developed new breeding methods, mating designs and analyses, and germplasm screening techniques for evaluation of resistance/tolerance to biotic and abiotic stresses.
- ☆ Developed experimental transgenics in cotton and rice by incorporating genes for insect resistance. Also developed protocols for micropropagation.
- ☆ The etiology, epidemology and the management of major diseases/insects pests have been worked out, facilitating the forecasting system and in developing location-specific integrated pest management (IPM) modules for sustainable crop production. Protocols have been developed for mass multiplication and release of biocontrol agents.
- ☆ Adoption of IMP modules has helped in lowering the quantum of pesticide requirements and promoting non-chemical eco-friendly approaches.
- ☆ Established the National Gene Bank at the National Bureau of Plant Genetic Resources, New Delhi, one of the World's leading gene banks, for long-term storage of seed and other planting materials. About 0.20 million accessions have been conserved.
- ☆ Steady increase in breeder seed production resulting in enhanced supply of quality seeds to the farmers. About 26,000 q. of breeder seed is being produced annually and supplied for production of foundation seed and in turn certified seed.
- ☆ Seed production technologies for various crops refined, particularly with reference to hybrid seed production.
- ☆ Developed a unique concept of multi-disciplinary, multi-locational approach in crop improvement in the form of All India Coordinated Research projects which led to synergistic cooperation for the development of widely adapted/location-specific cultivars and production technologies.
- ☆ Between 1950-51 and 1998-99, the production of foodgrains increased from 50.8 to 202.5 mt, oilseeds from 5.2 to 25.7 mt, cotton from 3.0 to 12.8 mt and sugarcane from 5.2 to 290.7 mt. The productivity of wheat increased four times and that of rice, maize and cotton three times.
- ☆ Development and adoption of new varieties of oilseeds and complementary technologies doubled oilseeds production in a decade (12.6 mt during 1987-88 to 24.4 mt during 1996-97), generally known as the Yellow Revolution.
- ☆ Increased food production has transformed the ship to mouth nation of early 1960s into a food secure one with exportable surplus of certain commodities.

- India has become the second largest producer of wheat and rice and is also amongst the top exporters of rice.

Milestones

- As of July, 2006 it has developed a vaccine against bird flu. The vaccine was developed at the High Security Animal Disease Laboratory, Bhopal, the only facility in the country to conduct tests for the H5N1 variant of bird flu. It was entrusted with the task of developing a vaccine by the ICAR after the Avian Influenza outbreak in February. The ICAR provided Rs. 8 crore for the purpose.
- 2009: In December 2009, it announced that it was considering a policy to provide open access to its research.
- ICAR scientists were the first in the world to sequence the pigeon pea genome. it was a purely indigenous effort by 31 scientists led by Nagendra Kumar Singh of NRCPB. The first draft of the sequence was published in J. Plant Biochem.

Horticulture Division

The Horticulture Division has nine research institutes, 11 national research centres and 15 coordinated research projects.

Achievements

- India has emerged as the second largest producer of fresh fruits and vegetables in the world.
- India is presently the largest exporter of spices and cashew.
- A total of 460 high-yielding varieties and hybrids of horticultural crops have been developed. As a result, productivity of banana and potato has gone up three-times each, and cassava two- times.
- Regular bearing mango hybrid, export quality grapes, multi-disease resistant vegetable hybrids, high value spices and tuber crops of industrial use have been developed.
- Substantial increase in production of banana plantlets through tissue culture, use of drip irrigation, chemical regulation of mango flowering and high-density orcharding in fruit crops have been obtained.
- True potato seed (TPS) technology is standardized.
- Low cost environment-friendly cool chambers for on-farm storage of fruits and vegetables have been developed.

National Bureau of Plant Genetic Resources (NBPGR)

National Bureau of Plant Genetic Resources (NBPGR) was established in 1976 by ICAR, at Pusa Campus, New Delhi. The institute has been vested with the responsibility to plan, undertake and coordinate activities and services related to plant genetic resources including collection, exchange, quarantine, evaluation, documentation, conservation and utilization of crops plants and horticultural plants, their land races and wild relatives. Besides its Headquarters at the Pusa Campus and Experimental Farm at the Issapur village near Delhi, the institute has 12 regional stations/exploration base centers/quarantine stations/satellite stations located in diverse agro-climatic zones of the country. NBPGR is the nodal organization for developing, operating and coordinating the Indian Plant Genetic Resource System. The system comprises base collections of germplasm of different crops kept under long term storage at NBPGR headquarters and a network of over 30 National Active Germplasm Sites located throughout the country. These sites are responsible for evaluation, multiplication and storage of germplasm. NBPGR has been able to build up 2,02,228 accessions from 183 species in its base collections in the National Gene bank (at -20°C) as of June 2001.

Animal Sciences Division

The Division has seven research institutes, one bureau, two project directorates, six national research centres, and 11 coordinated research projects.

Achievements

- ☆ A number of new genotypes in cattle - Karan Swiss, Karan-Fries and Frieswal developed for increased milk production. Murrah, Nili, Ravi and Surti buffaloes improved for milk production
- ☆ Three new high producing strains of sheep for fine wool, carpet wool and mutton evolved; and a strain of Mohair goat developed.
- ☆ Two hybrids of fast-growing poultry broilers and four high producing layer strains developed and released.
- ☆ Various immuno-biologicals, vaccines, immuno- diagnostics, indigenous drugs and medicines against various infectious and non-infectious diseases of livestock and poultry, developed.
- ☆ Monoclonals have been developed for the diagnosis of various diseases and reproductive disorders.
- ☆ ELISA-based diagnostic tests for various livestock and poultry diseases developed for precise and rapid diagnosis. Effective disease monitoring and surveillance have resulted in reduced morbidity and mortality.
- ☆ Process, techniques and equipment for the manufacture of quality milk and dairy products with reliable quality testing methods developed.
- ☆ Methods of preparation of different recipes and preservation of meat and eggs developed, and quality testing protocols for egg, meat and meat products standardized.
- ☆ Blending of camel hair and Angora rabbit wool with other fibres and converting those into yarn suitable for making finished goods achieved.
- ☆ Nutrient requirements of various categories of livestock for different production functions studied and standards set up.

- Technologies for utilization of cereal straws, agro-byproducts, conventional and non-conventional feeds developed area-wise and region- wise.
- Artificial insemination and embryo transfer technology used for improvement of native germplasm, and production and multiplication of elite germplasm.

Fisheries Division

Division has six institutes, one bureau and one research centre. In addition a good number of ad-hoc research projects and revolving fund schemes are also in operation.

Achievements

- Blue Revolution has been attained by enhancing fish production from 0.75 million mt in 1951 to 5.4 million mt in 1997.
- India has emerged as the second largest producing country in the world in freshwater aquaculture.
- Phenomenal growth of marine products export.
- Indigenous design of fishing craft and gears.
- Developed national standards for fish inspection and quality control.
- Nutritional evaluation of major fish species and fishery products.
- Created national collection centre for characterization and storage of important marine microorganisms.
- Developed value-added fishery products for export market.
- Technology development of Retortable Pouch Process as a substitute for canning fish.
- Commercial production of chitin and chitosan from shrimp head and shell.
- Hatchery technology for shrimp.
- Semi-intensive shrimp farming.
- Fattening of lobsters and crabs
- Artificial feed for shrimp farming.
- Technology package for broodstock management, production of fingerlings and grow-out systems for major finfish, shellfish and molluscs.
- Culture and utilization of sea weeds.
- Technology package for mass culture of 11 species of micro-algae.
- Production of ornamental fish under hatchery conditions.
- Breeding of seabass under controlled conditions.
- Commercial production of cultured pearls from pearl oysters.
- Induced breeding of major carps, catfishes and other finfishes.

- Production of freshwater pearls.
- Development of vaccines and formulation of drugs for fish diseases.
- Genetically improved, rohu, CIFA IR-I.
- Commercialization of fish feeds for inland aquaculture.
- Production of *mahseer* and snow-trout in hatcheries.
- Conservation of endangered species.
- Cryopreservation of milts of consumable important fish species.
- Enhanced fish productivity of reservoirs.

Agricultural Engineering Division

The Division provides engineering inputs for mechanization of conventional and protected agriculture, conservation of produce and by-products from quantitative and qualitative losses, and value addition and agro-processing enterprises for additional income and employment, energy management in agriculture and rural living for increasing production and productivity, and reducing drudgery.

Achievements

- Manual, animal and power-operated seeds drills, and planters, weeders; animal drawn multi-purpose tool frame; zero-till/strip-till- drills, till-planters; puddlers; manual and self- propelled rice transplanters; tall tree, orchard, and high clearance sprayers; self-propelled, walking/riding and tractor/power tiller mounted vertical conveyor reaper; and multi- crop threshers.
- 126 types of agricultural machines developed, of which more than 60 have been commercialized and 23 released for front-line demonstrations.
- Hand-operated groundnut-cum-castor decorticator, low-cost grains, pulse, oilseed mills; power puffing machine; crop residue fired and solar drier; cleaners, sifters and graders; low cost improved storage structures for foodgrains, evaporative cooled structures for fruits and vegetables.
- Machinery for soybean processing and solid, granular and liquid jaggery developed.
- Technologies for cotton and jute products developed.
- Process for the preparation of shellac, lac dye and wax from lac factory effluents; insulating varnishes, melfolac, primer and paint compositions, shellac bond powder, aleuritic and jalaric acids, perfumery compounds developed.
- Energy efficient equipment and packages; enhanced system efficiency in use of animate power; tractors and power tillers; low cost biogas plants; pyrolysed briquetted fuels; low cost solar cookers and water heaters, solar tracking device; high efficiency cooking stoves; gasifiers for process heat and mechanical power; charcoal briquetting machine; portable charring kiln developed.

- ☆ System improvement and safety in centrifugal pumps; solution to corrosion and incrustation in tube-wells; construction of wells in hard rock areas; surface, sub-surface and vertical drainage.
- ☆ Establishment of Agricultural Research Information System (ARIS), a WAN connecting 28 State agricultural universities and their 120 zonal agricultural stations, 49 ICAR institutes, 10 project directorates, 25 NRCs and its headquarters in *Krishi Bhavan*.

Natural Resources Management Division

The division has nine research institutes, one bureau, two project directorates and two national research institutes.

Achievements

- ☆ Soil map of the country on 1:7 million scale, state map on 1:250,000 scale and district soil maps on 1:50,000 scale have been prepared. Soil degradation map of the country on 1:4.4 million scale has also been prepared.
- ☆ Twenty agro-ecological zones and sixty agro-ecological sub-regions of the country based on physiography, soils, climate, length of growing period and available soil moisture are mapped on 1:4.4 million scale.
- ☆ Modification of land configuration on Alfisols, Vertisols and Inceptisols for on-farm rainwater management in rainfed areas.
- ☆ An integrated strategy of managing rainfed areas through watershed development projects in several parts of the country.
- ☆ Soil and water conservation treatments in mountainous watersheds to runoff and soil loss
- ☆ Off-season tillage to improve moisture conservation and weed control on Alfisols.
- ☆ Techniques of sand dune stabilization and shelter belt plantation were developed for arresting the movement of sand dunes in the arid zone.
- ☆ Cropping sequences and intercrop combinations developed for irrigated and rainfed areas.
- ☆ Water-use efficient irrigation schedules for major crops were evolved resulting in saving of irrigation water.
- ☆ Water-use efficient, micro-irrigation methods and technologies for utilization of available water in scarce areas were developed for irrigation, resulting in considerable saving of water and significant increase in crop yields.
- ☆ Technology for reclamation of alkali soils has been adopted in 1.0 M ha in Haryana, Punjab and Uttar Pradesh.

- ☆ Sub-surface drainage technology developed for waterlogged saline soils in Punjab, Haryana, Rajasthan, Gujarat, Karnataka, and Andhra Pradesh.
- ☆ Critical growth stages of various crops and cropping systems with respect to water stress and water requirement have been identified.
- ☆ Methodology of artificial groundwater recharge developed for excess groundwater utilization by the use of cavity wells, irrigation-cum-recharge well and by percolation tank in hard rock areas.
- ☆ Relay, parallel, multiple and multi-storey cropping systems resulted in improved cropping intensity and productivity.
- ☆ Agri-silviculture, agri-silvi-horticulture, agro- horticulture, silvi-pasture systems have been developed and evaluated for different agro- ecological regions.
- ☆ Agriculture production strategies developed based on weather forecast.

Agricultural Education Division

The Education Division provides administrative support to the Central (CAU) and the National Academy of Agricultural Research Management and development grants to State Agricultural Universities, CAU, and Deemed- to-be Universities in ICAR.

Achievements

- ☆ Establishment of Accreditation Board for quality assurance to clientele.
- ☆ Grants of over Rs.760 millions sanctioned to SAUs, Central University and DUs to strengthen and update infrastructure and faculty improvement during VIII Plan.
- ☆ Revision and updating of course curricula of all 11 undergraduate (UG) courses completed and adopted by all SAUs.
- ☆ Qualifying National Eligibility Test (NET) made compulsory for recruitment at Assistant Professor/Lecturer level.
- ☆ To reduce inbreeding and promote cultural exchange, 300 National Talent Scholarships in UG, 437 Junior Research Fellowships (JRF) and 200 Senior Research Fellowships (SRF) are awarded to meritorious students in every academic session.
- ☆ 200 foreign students from Iran, Ethiopia, Nepal, Bhutan, Bangladesh, Eritrea, Mauritius, Uganda and Yemen were admitted in UG and PG courses in 1998-99 academic session.
- ☆ Organized summer/winter schools in different disciplines to train scientists and faculty members.
- ☆ Arranged training programmes under Centres of Advance Studies to train scientists/teachers in current advancement in several areas.
- ☆ A workshop was held for writing university level books and 35 titles were finalized.

- ☆ National Information System on Agricultural Education (NISAGE) developed.
- ☆ Nine National Professors and 25 National Fellows are in position.

Agricultural Extension Division

This is the backbone of the ICAR System for technology assessment, refinement and transfer to the farmers.

Achievements

- ☆ 25,000 farmers and farm women are trained every year in agriculture and allied fields such as crop production, plant protection, livestock production and management, soil and water management, farm machinery and tools, and home science.
- ☆ Vocational training is imparted to 46,000 rural youth in poultry, dairying, piggery, beekeeping, fisheries, fruit and vegetable preservation, maintenance and repairing of farm machinery and tools, and hybrid seed production.
- ☆ In-service training programme to upgrade the knowledge and skills in transfer of technologies in agriculture and allied areas for 2,200 extension functionaries in KVKs and Trainers' Training Centres (TTCs).
- ☆ Front-line Demonstration (FLD) on oilseed and pulse crops benefitting 11,000 farmers.
- ☆ Training of trainers in frontier areas such as dryland agriculture, animal production, horticulture, freshwater, aquaculture, marine fisheries, hill agriculture, agricultural engineering and women in agriculture in TTCs.
- ☆ Multi-locational advance varietal trials of 17 selected crops in 60 centres under irrigated and rainfed conditions under separate cropping sequences in a joint collaborative Technology Evaluation and Impact Assessment Project with Crop Science Division.
- ☆ Identification, assessment and development of gender-specific technologies at the National Research Centre for Women in Agriculture, Bhubaneshwar (Odisha). Training of farm women in farm implements and tools at the sub-centre of NRCWA at the Central Institute of Agricultural Engineering, Bhopal (Madhya Pradesh).
- ☆ Establishment of 40 Agricultural Technology Information Centres in ICAR Institutes and SAUs. These centres will provide a 'Single Window' delivery system for technology products, services and information available in the institutions to the farmers.

Annexure A: List of Agricultural Universities in India

Telangana

1. Professor Jayashankar Telangana State Agricultural University, Hyderabad
2. Sri Konda Laxman Telangana State Horticultural University, Hyderabad

Andhra Pradesh

1. Dr. Y.S.R. Horticultural University, Venkataramannagudem, Tadepalli Gudem Mandal, West Godavari District
2. Sri Venkateswara Veterinary University, Tirupati
3. Acharya NG Ranga Agricultural University, Bapatla,,

Asom

1. Asom Agricultural University, Jorhat

Bihar

1. Bihar Agricultural University, Bhagalpur
2. Rajendra Agricultural University, Samastipur

Chhattisgarh

1. Indira Gandhi Agricultural University, Raipur

New Delhi

1. Indian Agricultural Research Institute, New Delhi

Gujarat

1. Anand Agricultural University, Anand
2. Junagadh Agricultural University, Junagadh
3. Navsari Agricultural University, Navsari
4. Sardarkrushinagar Dantiwada Agricultural University, Banaskantha

Haryana

1. Chaudhary Charan Singh Haryana Agricultural University, Hisar
2. Lala Lajpat Rai University of Veterinary and Animal Sciences, Hisar
3. National Dairy Research Institute, Karnal

Himachal Pradesh

1. Chaudhary Sarwan Kumar Himachal Pradesh Krishi Vishvavidyalaya, Palampur
2. Dr. Yashwant Singh Parmar University of Horticulture and Forestry, Solan

Jammu and Kashmir

1. Sher-e-Kashmir University of Agricultural Sciences and Technology of Jammu, Jammu
2. Sher-e-Kashmir University of Agricultural Sciences and Technology of Kashmir, Srinagar

Jharkhand

1. Birsa Agricultural University, Kanke

Karnataka

1. Kerala Agricultural University library
2. University of Agricultural Sciences and technology, Bangalore
3. University of Agricultural Sciences, Dharwad
4. University of Agricultural Sciences, Raichur
5. University of Horticultural Sciences, Bagalkot
6. Karnataka Veterinary, Animal and Fisheries Sciences University, Bidar
7. University of Agricultural and horticultural sciences, shimoga

Kerala

1. Kerala Agricultural University, Vellanikkara, Thrissur
2. Kerala University of Fisheries and Ocean Studies, Kochi
3. Kerala Veterinary and Animal Sciences University, Wayanad

Madhya Pradesh

1. Jawaharlal Nehru Agricultural University, Jabalpur
2. Rajmata Vijayaraje Scindia Krishi Vishwa Vidyalaya, Gwalior
3. PDKV Akola Agricultural University

Maharashtra

1. Dr. Balasaheb Sawant Konkan Krishi Vidyapeeth, Dapoli
2. Dr. Panjabrao Deshmukh Krishi Vidyapeeth, Akola
3. Maharashtra Animal and Fishery Sciences University
4. Mahatma Phule Krishi Vidyapeeth, Rahuri
5. Vasantrao Naik Marathwada Agricultural University, Parbhani
6. Central Institute of Fisheries Education, Mumbai

Manipur

1. Central Agricultural University, Iroisemba

Nagaland

1. Nagaland University

Odisha

1. Odisha University of Agriculture and Technology, Bhubaneswar

Punjab

1. Punjab Agricultural University, Ludhiana
2. Guru Angad Dev Veterinary and Animal Sciences University, Ludhiana
3. Desh Bhagat School of Agriculture Sciences Desh Bhagat University, Mandi Gobindgarh

Rajasthan

1. Maharana Pratap University of Agriculture and Technology, Udaipur
2. Swami Keshwanand Rajasthan Agricultural University, Bikaner
3. Shri Karan Narendra Agriculture University, Jobner
4. Agriculture University, Jodhpur

Tamil Nadu

1. Tamil Nadu Agricultural University, Coimbatore, Tamil Nadu, India
2. Tamil Nadu Veterinary and Animal Science University, Madhavaram, Chennai, India
3. Tamil Nadu Fisheries University, Nagapattinam, Tamil Nadu, India
4. Annamalai University, Chidambaram, Tamil Nadu, India

Uttar Pradesh

1. Chandra Shekhar Azad University of Agriculture and Technology, Kanpur
2. Banda University of Agriculture and Technology, Banda
3. Narendra Dev University of Agriculture and Technology, Faizabad
4. Sam Higginbottom Institute of Agriculture, Technology and Sciences, Allahabad
5. Sardar Vallabhbhai Patel University of Agriculture and Technology, Meerut
6. Banaras Hindu University, Varanasi
7. Aligarh Muslim University, Aligarh
8. Indian Veterinary Research Institute, Bareilly
9. Uttar Pradesh Pandit Deen Dayal Upadhyaya Pashu Chikitsa Vigyan Vishwavidyalaya Evam Go Anusandhan Sansthan,Mathura
10. Mahatma Gandhi Kashi Vidyapith University, Varanasi
11. Chaudhary Charan Singh University, Meerut

Uttarakhand

1. G.B. Pant University of Agriculture and Technology, Pantnagar

West Bengal

1. Bidhan Chandra Krishi Viswavidyalaya
2. Uttar Banga Krishi Viswavidyalaya, Cooch Behar
3. West Bengal University of Animal and Fishery Sciences, Kolkata
4. Visva-Bharati University, Santiniketan

6

Grow More Food

Failed to advice its objectives, various steps taken were:

1. Increasing area irrigated
2. Launching of Community Development Programme, the idea was involve the people at large in the community
3. Intensive Agriculture District Development Programme was launched in 1960
4. Since the programme was restricted to most progressive District, its impact was net visible.
5. Intensive Area Programme was launched in 1964 to have a higher coverage.
6. Japanese method of rice cultivation as also tried.

The next effect of all these activities that India imported 20 million tonnes of food grain from 1966-67. The international community declared India as a sinking ship not possible to recover.

Meanwhile Mexico had been tying high yielding varieties. Taking a.............. the Mexican Expert.

1. Increasing Area Irrigated

Irrigation and Power Projects in the First and Second Five-Year Plans

At the time of Independence in 1947, India was facing a severe food shortage. On account of the economic slump of the 1930s, followed by World War II, the development of irrigation had been slowed down since 1930, but the population had increased by 70 million. With the partition of the country, a large part of the irrigated area of the country went to Pakistan. The development of new irrigation facilities, thus, became an urgent necessity. A large number of new projects had already been investigated as part of the post-war reconstruction plans. Some of these were immediately taken in hand. The projects were in Odisha, Bihar, West Bengal, the United Provinces of Agra and Oudh (U.P.), Karnataka (Mysore), Maharashtra and Gujarat (Bombay Presidency).

During 1950-51 was serious (**Table 6.1**) with a population of 361.1 million and a food production of just 50.8 million tones. Urban population in this was just 17.3 per cent during 1951, per capita availability of foodgrains was just 140kg. The national Government was ceased with the problem and took positive steps to increase production. The first step was to take up various irrigation projects (**Table 6.2**). Besides these some other major projects like Rihand Project (UP) 1952-62, Chambal Project -1954-67, Nagarjuna Sagar Project, Andhra 1955-74, The Rajasthan Council Project 1956 were initiated. Net irrigated area increased from 20.8 to 31.1 million hect. during 1950-51 to 1970-71. The share of Government canals increased from 34.3 per cent to 38.5 per cent. Besides irrigation which was a major programme.

Table 6.1: Food Production -1950-51

Items	*Area*	*Production*	*Yield*
Cereals	78.2	42.2	668
Pulses	19.0	8.4	442
Total Food Grain	97.3	50.8	522

Irrigation from 1948 to 1963

The total are under irrigation has increased since Independence from 18.7 million hectares (46.7million acres) in 1947-48 to 25.6millin hectares (63.4million acres) in 1962-63; an increase of about 36 per cent in 15 years. By sources, the distribution of this irrigated area was as follows.

	Million hectares	*Million acres*
Government canals	9.69	23.95
Private canals	1.17	2.90
Tanks	4.73	11.70
Wells	7.66	18.95
Other sources	2.39	5.90
Total	25.64	63.40

Expenditure on Irrigation

The total capital outlay on State-owned irrigation projects in the whole of India during the one-and-a-half centuries from 1800 to 1950 was little under 1,000 million rupees, whereas the corresponding expenditure during the short period of ten years of the first two Five-Year Plans, from April 1951 to March 1961, was about 7,000

Table 6.2: Salient Features of Major Irrigation Projects Costing 100 Million Rupees and above During 1951-1961

Project	*Source of Water*	*Storage/ Non-storage*	*Capital Cost (million rupees)*	*Year of Completion*	*Gross Area to be Irrigated*		*Cost per Area Irrigated (Rs.)*
					Acres	*Ha.*	
			Andhra Pradesh				
1. The Tungabhadra Project	The Tungabhadra River	Storage	335.600	1956 (dam)	672,345	27,089	504/ha; 204/acre
			Bihar				
2. Maithon	The Brakar River	Storage	138,900	1957	270,000 (under irrigation barrage)	109,265	–
3. Konar	The Konar River	Storage	99,400	1955	104,000 (under irrigation barrage)	42,087	–
4. The Panchet Hill	The Damodar River	Storage	148,600	1959	680,000 (under irrigation barrage)	55,390	–
5. The Durgapur Barrage	The Damodar River	Non-storage	1,313.00	1955	134,400	55,390	–
			Gujarat				
6. The Kakrapara Project	The Tapi River	Non-storage	116.507	1957	651,662	263,718	442/ha; 179/acre
			Madhya Pradesh				
7. The Gandhi Sagar	The Chambal River	Storage	–	1960	–		–
			Chennai				
8. The Lower Bhawani Project	The Bhawani River	Storage	120.000	1955	207,000	83,770	1,218/ha; 493/acre
Mysore							
9. The Ghataprabha Left Bank (Stage I)	The Ghataprabha	Non-storage	54,500	1956	1,000,000	40,469	1,346/ha; 545/acre

Contd...

Table 6.2–*Contd...*

Project	*Source of Water*	*Storage/ Non-storage*	*Capital Cost (million rupees)*	*Year of Completion*	*Gross Area to be Irrigated*		*Cost per Area Irrigated (Rs.)*
					Acres	*Ha.*	
			Odisha				
10. The Hirakud Dam Project	The Mahanadi River	Storage	122.164 (irrigation)	1956	600,000	242,811	504/ha; 204/acre
			Punjab				
11. Harike	The Sutlej River	Non-storage	872.91	1952-53	–	–	–
12. Bhakra-Nangal	The Sutlej River	Storage	988.302 (irrigation)	–	3,604,000	1,458,489	680/ha; 275/acre
			West Bengal				
13. The Mayurakshi (dam at Masanjore and barrage at Tilpara)	The Mayurakshi River	Storage	204.609	1955-56	560,000 (kharif) 50,000(rabi)	226,624(k) 20,234 (r)	–

K: Kharif, R: Rabi.

Source: Irrigation Development in India.

million rupees. During the five years of the Third Plan, another 6,000 million rupees was provided for irrigation projects, excluding flood control and drainage works. Thus, the expenditure on irrigation in one single year now is more than what was spent in a century in the past.

Comparing with the latest position, while population increased from 361.1millions to 1210.8 million during 1950-51 to 2011. The corresponding production of foodgrains was 244.5 million. It is interesting to observe that while population increased 3.35 times, production increased 5.2 times during this period.

Rihand Project (UP) 1952-62

The concentration of a gravity-type dam was started in 1952 near the village of Pipri in the Mirzapur District of U.P. The dam is 271ft (82.6) high and has a reservoir with a storage capacity of 8.6 million acre-feet (39,095.3 million m^3) and a water spread of 180 square miles ($466m^2$) completed in 1962, it has five generating units of 50,000 kw each and provides electric power for the eastern parts of U.P. The water stored at the reservoir also makes available increased steady supplies on the zone for extending irrigation on the existing Zone Canal System of Bihar.

2. Launching of Community Development Programme in India

Background

After 53 years of independence and a population of one billion, India is the largest democracy in the world. To sustain this democracy and freedom, it is necessary to ensure economic empowerment and better quality of life for all the citizens. Unfortunately, the annual per capita Gross National Product of USD 370 in India is one of the lowest as compared to USD 860 in China, USD 2450 in Asia, USD 5170 in the world and over USD 20,000 in the developed countries. Despite significant advancement in industrial development, Indian economy is dependent on agrobased activities. Over 65 per cent of the population living in rural areas are primarily dependent on agriculture for their livelihood. A majority of these families spend over 90 per cent of their earnings on basic needs such as food, fuel and health care. Over 35 per cent of the rural families being poor, tackling their problems is the national priority to ensure social justice and better quality of life.

Problems of the Poor

Poverty in India is judged by the income generated by the family. A family of 4-5 members need at least Rs. 11,000 (USD 275) per annum, to meet their minimum basic needs. Those who cannot earn this income are classified as poor. Unemployment and underemployment are the major causes of rural poverty. The other factors are small land holdings, harsh agro-climatic conditions, poor infrastructure and limited opportunities for rural industrial development, poor health care, illiteracy, social suppression, addiction to alcohol and exploitation by vested interests.

Land is the major resource in India, which provides livelihood to the rural population. According to a survey conducted in 1991-92, about 11.25 per cent of the rural families are landless and among the land holders, over 69.35 per cent own less than 1 ha (marginal farmers) and 21.25 per cent own between 1 and 2 ha (small farmers). Water is another critical resource required to enhance the agricultural production. Out of the total arable area of 169 million ha, only 28 per cent area is under irrigation and the rest is dependent on rains, where hardly one crop can be grown in a year. About 40 per cent of the cropping area is located in low rainfall regions where the employment opportunity is hardly for 40 to 50 days in a year and crops generally fail twice in five years. Decline in many traditional occupations and poor institutional infrastructures have further reduced job opportunities. Thus about 90 per cent of the rural population, who are deprived of adequate land holding have to look for other means of livelihood for their survival. Livestock is another important natural resource owned by the poor. Generally the rural families keep 1-2 cows or buffaloes for milk, a pair of bullocks for farming and a few sheep, goats and poultry for supplementary income. Although fodder is in scarcity, most of the poor families maintain large herds and let them graze on common lands. Dairy husbandry has good scope, as there is good demand for milk. However, most of the important breeds of cattle and buffaloes are genetically eroded and about 80-85 per cent of the livestock are unproductive and cause undue pressure on forest resources while the rural poor are unable to take advantage of these animals. It is feasible to make sustainable use of these natural resources through application of appropriate technologies and use of idle labour.

Apart from inadequate earning for livelihood, the rural people also suffer from poor health arising from starvation, lack of immunization, hygiene and sanitation. Over 25 per cent villages do not have year-round supply of drinking water and about 75 per cent of the potable water sources are polluted. Consumption of polluted water is the main source of illness particularly during the rainy season. Most of the villages have no drainage and sanitation and less than 10 per cent of the rural people use latrines. Medical and immunization services do not reach remote villages. Illiteracy in rural areas is high particularly among women, varying from 55 per cent to 75 per cent. This has a direct bearing on the population growth, child health and education. With the annual birth rate of 28 per 1000, as compared to 23 in Asia, the family size has been growing without any addition to the income. Unable to meet the growing needs, the rural poor have to depend on money lenders, to meet their emergencies and fall into the debt trap. They often try to forget their problems by consuming alcohol. While some migrate to cities, others live in chronic poverty. They lose confidence in others as well in their ability to live a decent life. This is a vicious cycle.

Community Development Programme

The concept of community development in India was initiated well before independence. Even during the struggle for independence, under the leadership of Mahatma Gandhi - the Father of the Nation, considerable attention was given to rural upliftment and reconstruction. He emphasized on a nineteen point constructive programme for complete independence by truthful and non-violent means. He often warned the leaders that true independence could be enjoyed only when the rural economy was strengthened and poverty eradicated. He advocated communal harmony, economic equity, social equality, dead diction from alcohol and narcotics, promotion of 'khadi' (hand spun and hand woven cloth) and village industries, sanitation, health care, education and empowerment of women. The aim was to generate gainful employment in rural areas and to improve the quality of life.

The Government of India Act, 1935 under the British Rule, while confirming autonomy on the provinces, included rural development as an important programme to be initiated for the welfare of the people. With the outbreak of World War II and Bengal Famine of 1943, food supply was a critical problem in most parts of the country. Emphasis on food production was provided through 'Grow More Food Campaign' which included the supply of free seeds, subsidies for construction of wells and embankments, supply of manure, fixing a minimum price for grain, *etc.* However, there was limited scope under the 'Zamindari' system where a small number of rich farmers owned the agriculture lands and the rest of the villagers worked as labourers.

After the independence of India in 1947, community development assumed high priority. In 1948, a pilot community development project was launched through the Etawah Project. Later in 1952, the Government of India launched 55 Community Development Projects, each covering about 300 villages or a population of 30,000. This programme was multi-dimensional but the major emphasis was placed on agricultural production, as the areas selected for launching the project were located in irrigated areas or where the rainfall was assured. In 1953, the National Extension Service Project was launched with similar objectives to cover larger areas, including the dry regions. This project of three-year duration, demarcated the blocks of 150-300 villages as manageable units for initiating community development programmes. The objectives and activities of the Project were modified from time to time and continued as a permanent multi-function extension agency in each block. These community development blocks were treated as normal administrative units for planning and development with regular budgetary allocations.

By the end of the First Five Year Plan (1952-57), 1114 blocks covering 163,000 villages were in operation and by sixties, the community development programme covered the entire country. The programme was characterised by the following features: comprehensive in content; economic progress as the core objective; flexible programmes and posting of a multipurpose worker at the village level. The plan defined the central objective of planning as creation of conditions in which living standards are reasonably high and the citizens have full and equal opportunity for growth and justice (Sachidananda, 1988). The programme aimed at upliftment of the rural poor, covered agriculture, animal husbandry, roads, health, education, housing, employment, social and cultural activities. While aiming at economic development through agriculture and cottage industries, efforts were made to improve literacy, health, sanitation, housing, transport and communication. To implement the multi-facet programme, an extension organisation, headed by a Block Development Officer (BDO) was established at each block or the revenue tehsil, with a team of subject specialists and village level workers (VLW). Each VLW covered a population of 5000-6000, spread over 5-10 villages to implement various development programmes launched by different departments. The VLWs were expected to meet the farmers and persuade them to take part in various development schemes. The BDO was assisted by eight Extension Officers, one each for agriculture, animal husbandry, panchayat, cooperation rural industries, rural engineering, social education and women and child welfare. In addition, a medical officer with support staff was posted in each block to provide medical assistance. The extension officers reported primarily to the BDO and to their senior as well, in the respective department, based at the district headquarters. The BDOs reported to the District Collector, who is the administrator of the district. The Development Commissioner, at the state level was responsible for coordinating community development through the District Collectors. At the National level, the administration of the community development programme was entrusted to the Planning Commission (Figure 6.1).

Level		*Administrative Head (Government)*	*Non-Official Head*
Nation	↓	Planning Commission	Prime Minister/Chairman Planning Commission
State	↓	Development Commissioner	Consultative Committee
District	↓	District Collector/ Chief Executive Officer, Z.P.	Chairman District Development Committee
Block	↓	Block Development Officer	PresidentPanchayat Samiti
Village	↓	Village Level Worker	ChairmanGram Panchayat

Figure 6.1: Multi-level Planning Framework for Community Development in India.

During the initial phase of community development, the government officials prepared the plan under the guidance of the Planning Commission. There was no opportunity for the community to demand any facilities to solve their problems. As the intention was to ensure people's participation, advisory committees were subsequently established at various levels. A state level consultative committee consisting of state legislators was constituted for advising on community development. District development or planning committee consisting of officials and non-officials was formed under the chairmanship of the collector. A block level advisory committee was constituted with block level officers and non-officials, member of the parliament and state legislators, heads of educational institutions, progressive agriculturists and representatives of voluntary organizations. As the non-officials were generally interested in matters of personal concern, they did not look at their roles in the proper perspective. Furthermore, as the development programmes approved at the central and state level were rigid, the consultation process remained merely an academic exercise.

From 1950s till mid 1970s, there was no significant achievement and poverty was on the verge of increasing. The Indian economy had become slower compared to those in the East and South East Asia over the post independence period. The levels of living were unacceptably low for a large section of the population. The Land Reforms Act of 1956 did remove vestiges of the feudal-colonial rule from the scene, but the enactment of the laws did not help the poor and landless to gain control over the land, particularly in North India. There were fluctuations in the poverty status but the most prominent increase in poverty was observed in the late sixties and early seventies (55 per cent - 69 per cent), when rainfall levels were less than normal and monsoon failures prolonged beyond a year. Even the core programme of agricultural development failed to enhance food production. In early seventies, India was compelled to import foodgrains. Trickling benefits of development were wiped out by the growing population. Improvement in agricultural productivity improvement in the absence of education and infrastructural development was not sustainable. The resources allotted for community development during 1952-67 were also so low that it worked out to hardly Rs.10 per head over this period.

Panchayati Raj Institutions

In 1957, five years after launching the community development programme, the Government appointed the Balvantrai Mehta Committee to suggest measures to remove obstacles in implementing the programme. The Committee recommended the formation of a three-tier-system of rural local Government, to be called 'Panchayati Raj' (Rule by Local Councils). These were Gram Panchayat (Village level), Panchayat Samiti (Block level) and Zilla Parishad (District level). The aim was to decentralize the process of decision making and to shift the decision making centre closer to the people, encourage their participation and place the bureaucracy under the local people's control. However, the Panchayati Raj was not able to fulfill all the expectations of the people and planners. A major reason was the domination of socially and economically privileged sections of the local community, who ignored the welfare of the weaker sections. The other reasons were lack of harmony among the elected members due to political fractions and prevalence of corruption and inefficiency.

In mid sixties, the national priority was shifted to agricultural production and a distinct technological orientation was given to agriculture. Under the Fourth and Fifth Five Year Plans (1969-74 and 1974-79), the central government introduced independent administrative hierarchies to carry out special programmes, bypassing the Panchayati Raj institutions. Special programmes like Small Farmers' Development Agency (SFDA), Intensive Agricultural Areas Programme (IAAP), Intensive Agricultural District Programme (IADP), Tribal Development Agency (TDA), Marginal, Small Farmers and Agricultural Labourers Development Agency (MFAL) and area development agencies such as Command Area Development, Drought Prone Area and Hill area Development Programmes were financed and operated directly by the Central Government.

Integrated Rural Development Programme

The special programmes to boost the agricultural production in late sixties certainly helped to raise the food production but the benefits were largely reaped by those who had necessary resources. Small and marginal farmers trailed behind, as they were not directly benefited by the Green Revolution. On the contrary, their economic position worsened as the rich farmers declined to offer their lands for shared cropping. Thus to tackle the problems of the rural poor, the Integrated Rural Development Programme (IRDP) was introduced in 1979, with specific focus on the weaker sections of the society, particularly those living in poverty and to involve them in programme implementation. While the earlier programmes emphasized on the delivery systems which suppressed self-reliance, there was good scope for people's initiatives to build up their economy with dignity. Heeding to the criticism, further changes were made in the programme during the Seventh and Eighth Plans (1985-1990 and 1992-1997 respectively). These included the linkage between infrastructure and employment schemes, designing of the programme as a credit based self-employment activity, rather than a subsidy distribution and decentralisation of programme implementation through District Rural Development Agency and block authorities. Several sub-schemes such as Development of Women and Children in Rural Areas (DWCRA), Training of Rural Youth for Self-employment (TRYSEM), National Rural Employment Programme (NREP), Jawahar Rojgar Yojana (JRY) were also launched to target the weaker section of the society.

By mid eighties, the Government was able to meet the minimum needs of the poor, which included elementary education, health, water supply, roads, electrification, housing and nutrition. In 1993-94, about 32.37 per cent of the population in India was poor. The percentage of population living in poverty was high by about 17-22 per cent among the Scheduled Castes and Scheduled Tribes, compared to general categories. This was mainly due to small land holdings, landlessness or illiteracy.

Decentralized Planning for Rural Development

Based on the Sivaraman Committee report, the Planning Commission issued guidelines to all the State Governments in 1987 to consider the block as the unit for planning. The task of planning at the district level was entrusted to the District Planning and Development Council or District Planning Board which had wider representation of the society. This body consisted of elected as well as nominated representatives headed by a Minister or District Collector or a non-official. It was responsible for setting up policy guidelines apart from coordination, monitoring, review, finalization of annual plans and Five Year Plans and collection of data. At the block level, the officers of different departments prepared the plan as per the guidelines received from their district heads.

This process needed further articulation to prepare the plans in consultation with the Gram Panchayats for implementing the development programmes more effectively and economically. Planning at the village level can be very effective as it can address the problems directly and facilitate development at the grassroot level.

Emphasis on People's Participation

In spite of many drawbacks in launching community development programmes, the Panchayati Raj has made significant contribution to the development of the country by creating awareness among the public and by developing political leadership. The system has also helped in reducing the gap between the bureaucracy and the people. Centralisation of power and non-involvement of people in the process of development have been the major concerns ever since the introduction of Panchayati Raj. Although, it was widely acknowledged as the only hope for activating people's participation which is the soul of a democratic system, the system had generated tension and factions. This necessitated Panchayati Raj reforms through the 73rd Constitutional Amendment in 1992, which empowered the PR institutions to shoulder the responsibility of development and decentralised planning. Till then, all the functions were carried out by the government machinery and there was no scope for participation by the villagers. This had created a dependency syndrome and enabled the government officials to dictate terms to the people.

Under this constitutional amendment, 29 items of development were transferred to PRIs. These can be grouped under the following sectors: Agriculture; Forestry and Environment; Industries; Infrastructure, minimum needs; Social welfare; Poverty Alleviation and Maintenance of community assets. Considering the weak status of the Gram Panchayats to facilitate village level micro-planning for development, the District Planning Committee has been strengthened with members representing various government and non-government organizations.

To facilitate the planning at micro-level, it is proposed to strengthen the Gram Sabha (village assembly). The Gram Panchayat can use the Gram Sabha as a forum for discussion and finalization of annual plans. Such a forum can also set the priority for implementing various development programmes. Simultaneously, a suitable mechanism should be developed to sustain the interest of the villagers in Gram Sabha activities. In the absence of adequate participation, vested interests may influence the proceedings for their own benefits. Initiatives from farmers' organizations, self help groups, educational institutions and other voluntary organizations to nominate their representatives on the Gram Sabha can ensure their participation in the proceedings and safeguard the interest of the common people.

Presently, through the democratic process, over 3.3 million elected leaders have assumed positions at different levels of the Panchayati Raj administration covering 227,678 village Panchayats, 5906 Panchayat Samitis and 474 Zilla Parishads. The mandatory provision of 33 per cent reservation for women and SC/STs have resulted in active participation of these groups. The women could place some priority issues like safe drinking water, girls' education and basic health services on the Panchayati Raj agenda.

Role of Non-Government Organizations

The major factor influencing the successful implementation of rural development in India is motivation of the poor families to ensure their active participation. Development of suitable people's organizations is also necessary to avail the benefits of various schemes, particularly by the illiterate poor. To ensure people's involvement in the development process and to take them into confidence, they should be involved in the programme, right from the stage of planning. Several innovative programmes implemented on a pilot scale have shown that many rational suggestions made by even the poor people can be easily implemented to produce good results. Such an approach demands flexibility in the programme. The target families should be motivated to assume the responsibility of implementing the programme, while the implementing agencies should play the role of a catalyst. In this task of integrated rural development, voluntary agencies can play an important role.

The voluntary organizations or non-government organizations (NGOs) in general focus their voluntary action and service to tackle the problems of the common people or to help them face their challenges. The main characteristic of voluntary agencies is human-touch. Organizations with strong voluntarism and professionalism can implement the programme efficiently. Dedication of the volunteers and staff, their relationship with the community, flexibility in the programme, innovative approaches to solve the problems can result in greater success. Agencies having deep roots in rural areas can serve the community more effectively (Paul Choudhary, 1990).

It was only during the seventies that the government recognized the role of voluntary agencies in supplementing government's effort in rural development. Prior to this, the NGOs formed by religious institutions and enlightened public were involved in operating hospitals and educational institutions. With assistance from international donor agencies, the Ministry of Agriculture formed an independent organization called 'Freedom from Hunger Campaign' to support the voluntary organizations involved in rural development. This Organization was re-organized and renamed as 'People's Action for Development-India (PAD-I). PAD-I was merged with Council for Advancement of Rural Technology in 1986 and renamed as Council for Advancement of People's Action and Rural Technology (CAPART). Since then, CAPART has been providing financial assistance to voluntary agencies engaged in rural development. In the last two decades, several initiatives of the non-government organizations have had significant impact on the development. Widespread success of these initiatives have now encouraged many state governments to launch schemes to promote people's participation and several centrally sponsored schemes have stipulated the development of community based organizations to plan and implement the programme.

With better opportunities for promoting self-employment through investment in agriculture and micro-enterprise, facilities were created for availing soft loans from banks and other financial institutions. The Reserve Bank of India issued guidelines to all the bankers not to insist on collateral security up to Rs.25,000 drawn by the poor for investment in development activities. As a result, over Rs.250 billion was distributed as rural credit in 1995-96, of which 50 per cent amount was from the cooperatives and the rest from other banking institutions. Nevertheless, as the formal banking operations were not convenient for many villagers because of their rigidity, distance and high cost of operation, several innovative banking institutions have been established by the NGOs.

3. Intensive Agriculture District Programme - (1960)

The Intensive Agriculture District Programme (IADP) was launched as the first major experiment in intensive agriculture development in the year 1960. The inadequate rate of growth of agriculture during 1950's was a matter of serious concern to the government. The steep fall in crop productions in the drought year 1957-58 focused attention on the seriousness of the food situation up substantially. In these circumstances, the government of India invited a team of agriculture experts sponsored by the Ford Foundation to make a careful study of Indian agriculture and make recommendations future actions. The team visited India early in 1959 and submitted its report entitled "India's food crisis and steps to meet it in April that year. The team observed that "India is facing a crisis in food production. The crux of the problem is food enough for the rapid increasing population. The team went on to say that "this target (of 110 million tones) can be achieved if an all out emergency food production programme is undertaken. The best in Indian agriculture is comparable to the best in other countries, but the average level is unduly low. The task before the country is to develop ways of raising the low average to the higher levels that many Indian cultivators have achieved.

The Government of India accepted in general, the recommendations made by this team. The IADP popularly known as package programme was launched in three districts in 1960 - 61 and four districts in 1962 - 63. The seven districts were:

i. Thanjavur (Tamil Nadu)
ii. West Godavari (Andhra Pradesh)
iii. Shahabad (Bihar)
iv. Raipur (Madhya Pradesh)
v. Aligarh (Uttar Pradesh)
vi. Ludhiana (Punjab)
vii. Pali (Rajasthan)

In the second phase the programme was extended to nine more districts by 1963 - 64. In addition to these, this progrmme has now been extended to 12 more districts in the country and is in operation in 28 districts in all.

Objective

The major objectives of the programme were to demonstrate the most effective ways of achieving rapid and significant increase in agriculture production through adoption of a package of improved practices and to suggest successful and adoptable innovations, approaches and procedures which could be adopted in similar areas of the country. The programme envisaged the selection of favourable areas maximum irrigation facilities and minimum of natural hazards, providing simultaneously all the essential elements of production such as adequate supply of fertilizers credit *etc.* It aims at an integrated and intensified approach to the problem of agricultural production in areas which are more responsive to such production efforts. As its immediate goal, the programme sought to achieve rapid increase in the level of agricultural production through a concentration of financial technical, extension and administrative resources. In the long run it aimed at a self - generating break through in productivity and raising the production potential by stimulating the human and physical process of change.

As this programme demanded more intensive work, the number of village level extension workers was doubled in the community development blocks in the IADP districts. Along with the number of extension officers, the cooperative staff were also increased at the block and district levels. IADP was a pace - sitting demonstrational programme which showed how to increase production more rapidly on one hand and a path finding experimental programme which facilitated the development of new innovations, new ideas and procedures for wider adoption in agriculture development.

It had covered all the important cash crops grown in the districts although emphasis was to be laid on the major food - grain crops such as paddy, wheat and millets. While the role of the central government was to provide overall direction, guidance and coordination, the programme was of the state government financed by the ford foundation through the central government now the programmed is financed by state government in 28 districts where it is in operation.

4. Since the Programme was Restricted to most Progressive District, its Impact was Net Visible

The objective of the Intensive Area Development Programme (IADP) or "Package Programme" is to promote the use of inputs such as fertilizers, improved seeds and insecticides, etc, in their most profitable combinations in areas possessing the maximum of advantages in regard to physical, climatic and institutional factors wit h a view to ensuring maximum returns on the package of these scarce factors in the shortest possible time. The district s selected therefore, are those which enjoy the maximum of irrigation facilities, where the natural hazards like soil-erosion, floods and water-logging etc, are the minimum, and where service cooperatives are well-developed and the cultivators are receptive to improved practices. The basic infrastructure is thus given and growth is to proceed through the intensification of current inputs in their effective combinations. The Programme, therefore, does not require a long gestation period and increases in yields can be expected even within a single production year, though a longer period would be required to produce striking improvements in yields through the increased level of input use.

The Report

The period covered by the Report is three years, *viz.*, 1960-61, 1961-62 and 1962-63. In some districts, however, the Programme was started in 1961-62. The achievements in regard to input use are impressive. According to the Report, the consumption of nitrogenous fertilizers in these districts, on an average, doubled and that of phosphatic fertilizers trebled between 1960-61 and 1962-63. The quantity of seeds treated against seed-borne diseases increased by about 50 per cent and the area treated against pests and diseases increased by about 36 per cent between 1961-62 and 1962-63. Credit under the Package Programme is production oriented and conventional rules regarding credit worthiness are relaxed somewhat wit h a view to enabling the tenants to participate in the production plans. The loans advanced by Central Cooperative Banks in these districts increased by 50 per cent between 1959-60 and 1961-62. Short- term loans advanced by Primary Credit Societies increased by 37 per cent while the amount of medium-term loans registered an increase of over 300 per cent.

To assess the changes in the yields of principal crops in the package areas, crop-cutting experiments are conducted under the Bench-Mark Survey. Crop- cutting experiments are also conducted in the "controlled" areas for each of these districts. Controlled areas are selected in such a way that they are as similar as possible to the Programme district in respect of factors like irrigation, rainfall, soil types, crop-pattern, credit and marketing facilities and agricultural practices. The difference in production trends between Programme districts and Controlled areas could move appropriately be ascribed to the Programme.

A Comparison

It is necessary to examine whether the development efforts in the Package areas contributed to the improvement in yields relative to those in Controlled areas. Such an improvement can be expected even in the period experiencing bad weather because areas where Intensive Programmes of development are in progress should be relatively less vulnerable to the vagaries of climate. An attempt is made to examine this by comparing the ratios of Package yields to Controlled yields in the two years (**Table 6.3**). The ratio has improved for rice in West Godavari and Raipur districts, while it showed a deterioration in Thanjavur. The ratio increased for many important crops in Ludhiana. In Aligarh certain principal crops such as wheat and gram showed an improvement in their yields relative to those in Controlled areas, although bajra and maize showed a decline. The ratios for Pali indicate a decline in the relative yields in the case of many crops. Thus two districts, viz, Thanjavur and Pali show a clear deterioration in crop yields relative to Controlled areas. Considering the fact that the proportion of area irrigated is quite low (20 per cent) in Pali, Thanjavur is the only district showing deterioration in the relative yields despite favourable physical conditions (about 84 per cent of cultivated area is irrigated in this district).

But these inferences rest *on* the assumption that weather has affected crop yields in the same direction and wit h equal intensity in Package and Controlled areas. However, weather is known to behave differently even between adjoining districts. To the extent this phenomenon holds true for the period under study, the comparisons made would yield a distorted picture. It is essential, therefore, to collect data on principal weather indicators such as rainfall for Package as well as Controlled areas without which meaningful comparison could be attempted. Another factor to be kept in view while making such comparisons is that crop-cutting experiments are extended to new areas every year as the coverage under Package Programme increases. In Thanjavur the number of crop-cutting experiments as well as area covered increased substantially between the two years. The comparison will be distorted if the yield levels in the *newly* covered areas are different from those in older areas due either to the differences in soils or to the level of input use being relatively higher in areas already under the Programme. It will be useful, therefore, if, in addition to the average yield rates, separate figures of yields are also given for older areas and for those newly covered. There is yet another factor initiating

Table 6.3: Yields In IADP and Controlled Areas

(Quintals per hectare)

	1960-61	1961-62	1962-63	Per cent Change Between Col. (1) and (2)	Col. (2) and (3)
	1	2	3	4	5
1. Thanjaur (Madras)					
Rice (First) I	15.2	17	15.4	11.8	-9.4
C	-	17.3	17.7	-	2.3
(Second) I	14.4	17	14	18.1	-17.6
C	-	16.1	15.6	-	-3.1
2. West Godavari (Andhra)					
Rice: First I	14.6	16.9	13.8	15.8	-18.3
C	-	17.9	14.5	-	-19
Second I	10.9	15.2	18.2	39.4	19.7
C	-	13.6	15.9	-	16.9
3. Shahbad (Bihar)					
Rice I	10.8	12.9	11.8	19.4	-8.5
Wheat I	8	6.8	7	-15	2.9
Gram I	5.6	5.3	5.4	-5.4	1.9
4. Raipur (M.P.)					
Rice I	-	9.8	7.9	-	-19.4
C	-	9.4	7.2	-	-23.4
5. Aligarh (U.P.)					
Rice I	2.7	5.2	7.9	92.6	51.9
C	-	5.5	9.9	-	80
Maize I	2.4	5	13.4	108.3	168
C	-	4.6	15.8	-	243.5
Wheat I	11.5	12	11.9	4.3	-0.8
C	-	15	13.5	-	-10
Barley I	9.1	11.3	11.3	24.2	N.C.
C	-	15.8	12.3	-	-22.2
Gram I	9.4	7.8	9.5	17	21.8
C	-	12.4	8.6	-	-30.6
Pea I	11.4	10.5	9.6	-7.9	-8.9
C	-	15.4	14.7	-	-4.5

Contd...

Table 2.1–*Contd...*

	1960-61	1961-62	1962-63	Per cent Change Between Col. (1) and (2)	Col. (2) and (3)
	1	2	3	4	5
6. Ludhiana (Punjab)					
Maize I	-	20.1	11.4	-	-43.3
C	-	20.9	12.7	-	-39.2
Cotton I	-	3	1.6	-	-46.7
C	-	3	1.4	-	-53.3
Wheat I	15.6	17.6	17.5	12.8	-0.1
C	15.9	17.6	1534	10.7	-12.5
Gram I	9.4	10.5	10.5	11.7	N.C
C	9.5	11.2	9.6	17.9	-14.3
7. Pali (Rajasthan)					
Maize I	9.9	9.7	9.1	-2	-6.2
C	-	7.9	10.5	-	32.9
Bajra I	1.7	3.2	2.5	88.2	-21.9
C	-	2.7	2.5	-	-7.4
Jowar I	1.2	2.8	1.3	133.3	-53.6
C	-	2.8	1.2	-	-57.1
Wheat I	9.5	10.4	9.4	9.5	-9.6
C	-	12.4	12.8	-	-3.2
Barley I	10.3	11.8	10.8	14.6	-8.5
C	-	14.9	11.2	-	-24.8

I: IADP Area; C: Controlled Area; NC: No Change.

comparison of yields between Package and Controlled areas. Part of the improved inputs, particularly fertilizers, supplied in the Package areas may be finding their way into the Controlled areas and raising crop yields there. This is quite likely in view of the fact that these inputs are supplied liberally and Controlled areas are generally contiguous to Package areas and are, by definition, most suited to input-intensification. The comparison under such circumstances would tend to underestimate the progress in the Package areas.

In view of the difficulties mentioned above, the comparison between Controlled and Package areas cannot, by itself, provide a reliable method for evaluating the progress. It is important, therefore, to evolve some additional criteria for measuring the progress under the Programme. The present Bench-Mark Surveys could be utilized to measure the changes in yields for the same plots or for relatively homogenous areas and to work out net returns at different levels of input use under varying climatic conditions. Cost-benefit analyses made for the sample farms are far more useful for evaluation purposes than the analyses available for demonstrations laid out on cultivators' plots not only because the former would be relatively free from reporting basses but also because they help in explaining the lag between the actual and expected growth and in assessing the growth possibilities as distinct from growth potentialities indicated from the demonstrations laid out.

Table 6.4: Percentage difference in Yields per acre between Fields using Fertilizers and those Not Using Fertilizers

	1961-62		*1962-63*	
	M	*O*	*M*	*O*
	(1)	*(2)*	*(3)*	*(4)*
1. Thanjaur (Madras)				
Rice: Kuruvai	7.1	-	4.2	-
Samba	6.1	N.C.	13.3	7.1
Thaladi	39.4	20.8	10.2	7.6
2. West Godavari (Andhra)				
Rice: First crop	29.9	38.9	28.9	-15.5
Second Crop	28.5	43.8	-1.5	32.8
3. Shahbad (Bihar)				
Rice: Irrigated	7.1	50	6.7	12.7
Unirrigated	-	31.4	-	21.4
4. Raipur (M.P.)				
Rice: Irrigated	19.1	41.7	34.6	23.9
Unirrigated	7	53.4	10.8	15.5
5. Aligarh (U.P.)				
Wheat : Irrigated	58.3	31	4	12.2
6. Ludhiana (Punjab)				
Maize : Irrigated	8.6	1	59.3	-11
7. Pali (Rajasthan)				
Maize : Irrigated	2.7	-	8.7	-

M: Fields using Farm Yard Manure; O: Fields not using farm yard manure, NC: No change.

5. Intensive Area Programme was Launched in 1964 to have a Higher Coverage

During the third five year plan 30 per cent increase in food grain production was achieved through IADP. The intensive promotion of agriculture was very popular among policy- makers and administrators. As a result of this a revised version of IADP with less intensive and therefore less costly programme was formulated and launched in selected blocks of 150 districts. It was named as IAAP. The selected blocks were to have the same physical conditions as in the case of selection of

Table 6.5: Ratios of Yields per acre in IADP Areas to those in Controlled Areas

	Difference Between		
	1961-62	*1962-63*	*Col (2) & (1)*
1. Thanjaur(Madras)			
Rice (First)	0.98	0.87	-0.11
(Second)	1.06	0.90	-0.16
2. West Godavari (Andhra)			
Rice: (First)	0.94	0.95	0.01
(Second)	1.12	1.14	0.02
3. Raipur (M.P.)			
Rice	1.04	1.10	0.06
4. Aligarh (U.P.)			
Bajra	0.95	0.80	-0.15
Maize	1.09	0.85	-0.24
Wheat	0.80	0.88	0.08
Barley	0.72	0.92	0.20
Gram	0.63	1.10	0.47
Pea	0.68	0.65	-0.03
5. Ludhiana (Punjab)			
Maize	0.96	0.90	-0.06
Cotton	1.00	1.14	0.14
Wheat (0.98)*	1.00	1.14	0.14
Gram (0.99)	0.94	1.09	0.15
6. Pali (Rajasthan)			
Maize	1.23	0.87	-0.36
Bajra	1.19	1.00	-0.19
Jowar	1.00	1.08	0.08
Wheat	0.84	0.73	-0.11
Barley	0.79	0.96	0.17

* Figures in brackets are ratios of yields in the IADP areas to those in Controlled area for the year 1960-61.

districts for IADP. Under this programme 20 to 25 per cent of the cultivated area of the country was brought ujnder the intensive agricultural development.

Implementation of IAAP was accepted by Agricultural Production Board and came into operation in March 1964. This programme also followed the package approaches of use of improved methods. The uses of interrelated factors of physical, social and institutional were also followed in a strategic combination mainly to produce an impact on agricultural production. The management of these programmes did not function as envisaged. There were many weaknesses of deficiency in inter-agency and inter-personal coordination, inadequate staff motivation, malpractices, non-formulation of local production plans on proper lines and delay in delivery of inputs to farmers. However, the production and productivity were modest. The highly adverse conditions (droughts) during 1966-68 served as a big blow. The foodgrains output was still insufficient to meet the rising domestic demands. Imports were also continued to supplement the local production.

Institutional Village Linkage Programme (IVLP)

IVLP is an innovative program developed by the ICAR to help scientists to have direct interaction with the farming community so that appropriate technologies are developed for farmers. Here research, extension and farmers establish firm links by carrying together the assessment and refinement functions in the technology development and dissemination process. This helps the research system to generate a cafeteria of technologies, which are more productive in small production system, more profitable in commercial production system and gender sensitive for removal of drudgery of farmwomen.

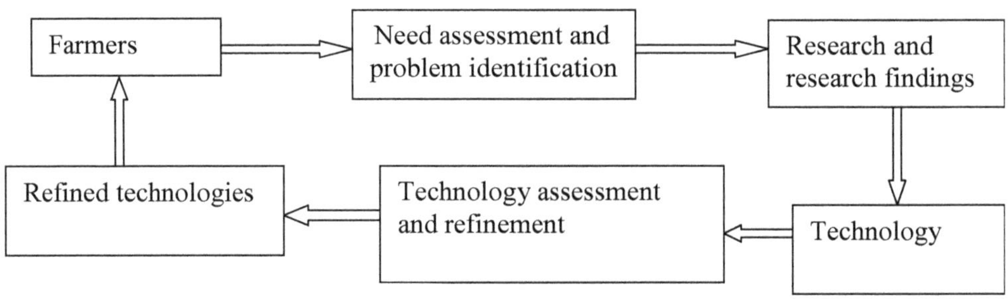

Research and Extension as an Integral Part of Technology Development

Objectives of IVLP as per ICAR guidelines are as follows

1. To introduce technological intervention with emphasis on stability and sustainability along with productivity of small production systems.
2. To introduce and integrate the appropriate technologies to sustain technological interventions and their integration to maintain productivity and profitability taking environmental issues into consideration in a comparatively well defined production system.
3. To introduce and integrate the appropriate technologies to increase the agricultural productivity with marketable surplus in commercial on and off-farm production systems.
4. To facilitate adoption of appropriate post-harvest technologies for conservation and on farm value addition to agricultural products, by-products and waste for grater economic dividend and national priorities.
5. To facilitate adoption of appropriate technologies for removal of drudgery, increase efficiency and higher income of farm women.
6. To monitor socio-economic impact of the technological/technology modules based on environmental at meso and mega levels.

Methodology of implementation of IVLP as per ICAR Guidelines

1. Selection of participating Institutions
   a. ICAR and ICAR Institutes
   b. SAUs and their Regional Resarch Statins/Zonal Research Stations. c. Krishi Vigyan Kendras

   Based on availability of multi-disciplinary team of scientists, laboratory facilities and transport *etc.*
2. Selection of village

   One village or a cluster of villages to cover about 1000 farm families.
   a. The selected village should not be far away from the research station,
   b. Should have asses through road
   c. Should be a relatively poorly developed in agriculture
3. Agro-Eco-System analysis

   Using Participatory Rural Appraisal Methods to gather information about
   a. The resource availability with the farmer's present production practices
   b. The extent of impact of the previous programs in the area
4. Constitution of multi-disciplinary team of scientists
   a. Core team : Those scientists whose disciplines are essentially needed
   b. Optional team : The other disciplines depending upon the needs of the area.
5. Selection of team leader
6. Provide training to the multi-disciplinary team
7. Plan for Technology Assessment and Refinement
   a. Intervention points based on problem – cause analysis
   b. Basket approach considering the problems
   c. Weightage to indigenous knowledge
8. Implementation of action plan
   a. On-farm research
   b. Demonstration
   c. On-farm trials
9. Monitoring and Evaluation
   a. Regular visit of team members
   b. Technical staff posted for the village
   c. IVLP card separately devised for the purpose.

High Yielding Variety Programme (HYVP-1966)

HYVP is launched in 1966, which helped the country in attaining self-sufficiency in food. The technological development did not remain confined to the introduction of high yielding crop varieties alone. These were combined with the application of high analysis and balanced fertilizer, irrigation, plant protection, improved implements etc, which made a 'green revolution' possible in the country. The pervasive influence of high yielding technology spread to other areas of farm production such as animal production, such as animal production, fishery, sericulture, social forestry *etc.*

Punjab, Haryana and Western parts of UP were initially selected for the phased launching of this strategy. The cultivation of HYV since 1966-67 had resulted in a substantial increase in foodgrains production. Wheat production was doubled. Rice production also had a substantial increase, though not as much as in the case of wheat. The target of coverage of 2.5 crore hectares of area under HYVs of cereals and millets under fourth five year plan was exceeded. The coverage was more than four crore hectares.

6. Japan Method of Rice Cultivation

Yayoi period storehouse Many archeologist looked upon the introduction of wet land rice farming techniques as the technological advancement that marked the beginning of the Yayoi period (400 B.C."A.D. 300) and the end of the Jomon period. In Kyushu people at red-kerneled rice. For a long time the earliest evidence of rice farming was dated to around 300 B.C. which worked nicely into models that it was introduced when the Koreans, forced to migrate by upheaval in China n the Warring States Period (403-221 B.C.", arrived around the same time. Later a number of Korean objects, dated between 800 and 600 B.C." were found. These discoveries upset the neatness of the model. Then in the early 2000s, grains of wetland rice were found in pottery from northern Kyushu dated to 1000 B.C. This called into question the dating of the entire Yayoi period and caused some archeologist to speculate that maybe wet-land rice farming was introduced directly from China. This assertion is backed up somewhat by similarity in skeletal remains of 3000-year-old skeletons found in Quinghai province in China and Yayoi bodies unearthed in northern Kyushu and Yamaguchi prefecture. Japanese immigrants introduced superior rice varieties to the United States. By the early 1920s, 85 per cent of the rice in California was of Japanese origin.

Rice Agriculture

Rice is one of the world's most labor intensive foods. In Japan the planting and harvesting is done mostly with machines, but in much of the world these chores - along with weeding, and maintaining the paddies and irrigation canals - are still largely done by hand, with water buffalo helping with the plowing and preparing of the fields. Between 1000 to 2000 man or women hours are required to raise a crop on 2.5 acres of land. This fact that rice is so labor intensive tends to keep a lot of the population on the land.

Rice is also a water thirsty crop, requiring lots of rain or irrigation water The wet rice grown in most Asia, needs hot weather after a period of rain, conditions provided by the monsoons that affected many of the places where rice is grown.

Rice farmers can often produce multiple crops a year often by adding no or little fertilizer. Water provides a home for the nutrients and bacteria that enrich the soil. Often the remains or previous crops or the burned the remains or previous crops are added to the soil to increase its fertility. Kevin Short wrote in the Daily Yomiuri, "Much of the most rice produced in Japan is still grown using commercial chemicals such as herbicides and pesticides. Organic farming requires a heavy input in terms of weeding and other labor. Many farms are now owned and managed only by an elderly couple, and there is just not enough labor available to farm organically. Fortunately, however, the chemicals used, as well as the delivery systems, have improved greatly over the past several decades. Pesticides, for example, are far milder, and are usually sprayed directly onto the rice plants by hand, or from low-flying radio-controlled helicopters. In the past, powerful chemicals were dropped from a higher altitude, and wound up falling heavily on the surrounding lands as well."

Growing and Harvesting Rice

The water depth in the paddy is increased as the rice seedlings grow and then gradually lowered in increments until the field is dry when the rice is ready to be harvested. Sometimes the water is drained during the growing season so the field can be weeded and the soil aerated and then water is put back in. Rice is harvested when it is a golden-yellow color several weeks after water has been completely drained from the paddy and the soil around the rice is dry. In many places rice is still harvested with a sickle and bundled into sheaves and then threshed by cutting the top inch or so of the stalks with a knife and removing the grains by slapping the stalks over propped up boards. The rice is placed on large sheets and left to dry on the ground for a couple of days before being taken to the mill to be processed. In many villages around the world, farmers usually help each other to harvest their crops.

There are also harvesting machines. Some diesel-powered rototiller-tractors and mechanical rice transplanters are available with harvesting attachments. Large machines are not used to harvest rice because they can not maneuver around the paddy without messing them up. Plus, most rice paddies are small and divided by dikes. Large machines need long tracts of uniform land to do their job efficiently. Kevin Short wrote in the Daily Yomiuri, "The tractors used in the harvest are tiny, but nonetheless very well-designed. A typical ride-on-top machine cuts several rows of rice at a time. The rice grains are automatically separated from the stalks, which can be either tied into bundles or chopped into pieces and scattered back into the paddy. On some models the rice grains are automatically loaded into bags, while on others they are temporarily stored in an onboard bin, then transferred to a waiting truck via a suction-powered boom."

After the rice harvest the stubble is often burned aling with waste products from the harvest and the ashes are plowed back into the field to fertilize it. Hot summers often translate to meager rice harvests and lower quality rice. Shortages of high-quality rices often result in bags of blended rice in which it isn't always clear what is in the mix. Some of the blends are created by "rice masters" who are skilled at getting the best taste at the lowest cost from their blends.

Problems with Rice Cultivation

Nets protect fields from birds Bacterial leaf blight, plant hoppers, rodents and stem borders are the major rice destroying pests. These days the biggest threat to the world rice crops is the leaf blight, a disease which wipes out as much as half the rice crop in some parts of Africa and Asia, and annually destroys between 5 and 10 per cent of the world's total rice harvest. In 1995, scientist cloned a gene that protects rice plants from leaf blight and developed a genetically-engineered and cloned rice plant that resists the disease. The trend towards reliance on only a few strains of highly-productive rice plants worldwide has the potential of causing a disaster. If these strains suddenly become vulnerable to a disease or pests, huge amounts of crops could be destroyed, causing severe food shortages or even a famine. If many strains are used and some of them are destroyed by disease or pests, there are still many remaining stains producing rice and overall food supply is not jeopardized.

While demands for food increase, land used to grow rice is being lost to urbanization and industry and the demands of a growing population. Demographers estimate that production of rice must increase by 70 per cent over the next 30 years to keep up with a population that is supposed to grow by 58 per cent before the year 2025. Much of the rice grown in coastal plains and river deltas is vulnerable to sea-level rises caused by global warming. Sometimes fertilizers and pesticides leak out of the paddies and damage the environment.

Rice Production in Japan

Washing rice in the 19th century Most rice farming is done by "Sunday farmers" who make most of their income in some other way. There is little profit in raising rice and many do it "only because we inherited the land." A typical farmer with a hectare of rice fields produces about a ton of rice. Thus is about enough to feed his extended family and make generous gifts to friends. When the cost of machines, tools and supplies is factored the family loses money, but say the effort is worthwhile because the rice tastes good and there is sense of satisfaction in producing it oneself.

Rice production has been hurt by declining rice consumption which in turn is caused in part by declining birthrate and the aging society. Annual rice consumption per capita has fallen from 67.3 kilograms in 1996 to 61.4 kilograms in 2005. Farmer are paid about $150 for a 60 kilograms of rice. These 60 kilograms in turn are sold for about $300 at stores. Rice prices fell sharply in 2007 as a result of a supply glut caused by too many farmers growing rice over other crops because it was seen as more profitable.

The government keeps a stockpile of rice that ideally is around 1 million tons. It often buys rice from farmers during periods of oversupply. The Japanese government is trying to encourage the use of rice flour to make bread and noodles to stimulate demand and encourage more people to grow rice. These days more farmers are using rice as feed for chickens and pigs. One of the affects of this is that rice-fed chickens produce yokes that are nearly white (corn feed is one reason why egg yokes are usually yellow). Pork from rice-fed pigs is said to be quite tasty. Using rice in this way helps rice farmers find markets for their crops and increases Japan's efforts to be more self-sufficient in food production.

Rice Farming in Japan

A few weeks before the rice is planted the paddies are plowed and filled with water. The transplanting is done with machines. Spraying fertilizers and insecticides takes only a couple of hours. The most time-consuming task is planting seeds to produce the seedlings. Rice seedlings are grown in separate beds. First the seeds are soaked in water, then sewn in dense mats designed to fit into special devices mounted on rice planters. The seedling are sold in palettes that sell for about ¥750 a piece. It takes about a dozen or so of these pallets to plant a small field. The planting is done with a machine that looks like jet ski with robot wheels. There is an angled splay on the back that feeds the palettes to small pincers that whirl around and pluck off a couple of shoots at a time and stick them in the muck. Two-wheeled machines are used for small fields. Four-wheel machines are much faster and used in bigger fields. Hard to reach areas are don by hand. With these machines planting a field takes just a few hours. Sometimes learning to drive the machines takes longer than the field work.

The planters have incredibly dexterous mechanical fingers that pluck patches of seedlings from the mat and plant them in neat. The harvesting is also done with machines that look minaiture versions of the large harvesters that you see in the United States harvesting wheat.

Study on Evaluation of Consolidation of Holdings Programme - 1969

The consolidation of holdings started in almost all the States during the First Plan Period. Consolidation measures were designed to remedy two distinct conditions prevailing in Indian agriculture, firstly, the division of land into under-sized units not suitable for proper cultivation and secondly, the excessive dispersal of parts of a single farm. An evaluation of the Consolidation of Holdings Programme. was taken up by the PEO at the instance of the Planning Commission in eight States where more than 4.05 lakh hectares of land were consolidated.

The report on the study was brought out into two parts. Part-I of the report highlighted some of the important aspects of the programme of consolidation of holdings by comparing the situation prevailing in the selected States, Part-II of the report brought out the special problems of consolidation in the States of Maharashtra, Punjab, Haryana and Uttar Pradesh.

Objectives

The main objective of the evaluation study were:

1. To find out the extent to which the measures relating to the consolidation of holdings programme have had the desired effect and
2. To identify the directions in which improvements can be made.

Sample Size/Sample Selection

The sample consists of 18 districts 36 tehsils/talukas, 106 villages and about 1100 cultivators in 8 selected States where more than 4.05 hectares of land were consolidated. These states were Gujarat, Madhya Pradesh, Maharashtra, Mysore, Punjab, Haryana, Rajasthan and Uttar Pradesh.

Districts had been taken as the primary sampling units. The requisite number of districts were selected in each State systematically with probability proportional to the area consolidated from a stratified list of districts arranged in the descending order of the area consolidated. From each selected district, two tehsils were selected with probability proportional to the area consolidated. From each of the selected tehsils, a sample of 3 villages was drawn with probability proportion to the area consolidated from each selected village, a sample of about 10 or 11 cultivators was drawn by the method of Systematic Sampling with equal probability.

Period of Study

The field study was conducted in 1967-68.

Main Findings

1. The amount of work expected of the staff was more than they could do. The staff, therefore, resorted to short cut methods in order to achieve targets in terms of time. The detailed attention that the work required could not be given.
2. Not only the staff in position was found inadequate but also no arrangements for training of the staff for consolidation existed in some States like Madhya Pradesh, Mysore and Gujarat.
3. The actual expenditure on administration was fairly below the amount allotted. Since in some states no arrangements or no adequate arrangements had been made; the shortfall in utilization of finance to some extent impeded the progress of consolidation work.
4. In the Bijapur district of Mysore and the Ahmedabad district of Gujarat, 95 per cent and 80 per cent of the selected respondents did not feel the need for consolidation. In all other selected districts, there was an appreciable number of respondents who did not feel the need for consolidation.
5. Attitudes to consolidation were unfavourable and their existence was a reflection of the weak publicitywork done before and during the consolidation operations.
6. In Gujarat and Maharashtra, the criterion of backwardness for selecting areas for consolidation had an adverse effect on the progress of the programme. This criterion was, therefore, revised in favour of more developed areas, areas under the watershed of majorand medium irrigation works and areas covered by the intensive agricultural programmes.
7. In majority of selected areas, the record of rights was not uptodate and the work involved *in* bringing them uptodate was tremendous. The cultivators, in many cases being illiterate, were unable to **understand** the entries in the record of rights without assistance. As a result, in many cases the cultivators were *unable to* lodge their objections in time.

8. Future developmental requirements of the villages have not always been kept in mind while drawing up the scheme for consolidation. Besides, correct valuation of the land was an extremely *Intricate* problem because in certain cases, inferior lands were deliberately valued at higher rates so **that there was** a gain in the exchange of land. Another factor which led to incorrect valuation of the land was that of productivity of the soil taken into consideration while fixing the value of the land was the same that had been entered into records years earlier. No consideration was given to possible improvement that had increased productive capacity of the land.

Major Suggestions

1. It is not enough to provide for prevention of fragmentation in the Prevention of Fragmentation and consolidation of Holdings Acts of different States. A careful study has to be made of the relevant provisions of the consolidation of Holdings Acts and the Acts relating to Succession and the suitable amendments made wherever necessary to prevent the re-emergence of fragmentation.
2. The publicity work for consolidation should be directed at showing the advantages of consolidation and at pointing out the irrationality of attachment to a particular piece of. land. For demonstrating the advantages of consolidation, organised publicity at official level is necessary and the organisational and administrative arrangements as well as the publicity media have to be carefully worked out.
3. In order to ensure the smooth and satisfactory progress of consolidation programme, the valuation of the land should be done by a high powered committee consisting of retired senior official of the Revenue department and two men of public importance Whatever be the agency to value the land or to hear the appeals, the whole procedure should be streamlined and followed expeditiously.

The East Punjab Holdings (Consolidation and Prevention of Fragmentation) Act, 1948

Short Title, Extent and Commencement

It is hereby enacted as follows:

1. This Act may be called the East Punjab Holdings (Consolidation and Prevention of Fragmentation) Act, 1948.
2. It extends to the whole of the State of Haryana.
3. This section shall come into force in the principal territories at once and the remaining provisions of the Act shall come in such area and from such date as the State Government may by notification appoint in this behalf, and different dates may be appointed for the coming into force

of different provisions of the Act and in the transferred territories on the 3rd April, 1957.

Interpretations

In this Act, unless there is anything repugnant in the subject or context:

(a) "Consolidation Officer" means an officer appointed as such under section 14 by the State Government and includes any person authorised by the State Government to perform all or any of the functions of the Consolidation Officer under this Act;

(b) "Consolidation of Holdings" means the amalgamation and the redistribution of all or any of the lands in an estate or sub-division of an estate so as to reduce the number of plots in the holdings;

(c) "common purpose" means any purpose in relation to any common need, convenience or benefit of the village and include the following purposes:

(i) extension of the village Abadi;

(ii) providing income for the Panchayat of the village concerned for the benefit of the village community.

(iii) village roads and paths; village drains, village wells, ponds or tanks; village water-courses or water-channels; village bus stands and waiting places; manure pits; hada rori; public latrines; cremation and burial grounds; Panchayat Ghar; Janj Ghar; grazing grounds; tanning places; mela grounds; public places of religious or charitable nature; and

(iv) schools and playgrounds, dispensaries, hospitals and institutions of like nature, waterworks or tube-wells, whether such schools, playgrounds, dispensaries, hospitals, institutions, waterworks or tube-wells may be managed and controlled by the State Government or not.

(v) "fragment" means a plot of land of less extent than the appropriate standard area determined under this Act: Provided that no plot of land shall be deemed to be a fragment by reason of any diminution in its area by diliuvion;

(vi) "land" means land which is not occupied as the site of any building in a town or village and is occupied or let for agricultural purposes or for purposes subservient to agriculture or for pasture, and includes the sites of buildings and other structures on such land;

(vii) "notified area" means any area notified as such under section 3;

(vii) "owner" means in the case of unalienated land the lawful occupant and when such land has been mortgaged, owner means the mortgagor; in the case of alienated land, owner means the superior holder;

(viii) "prescribed" means prescribed by rules made under this Act;

(ix) "Settlement Officer (Consolidation)" means an officer appointed as such under section 20 by the State Government and includes any person authorised by the [State] Government to perform all or any of the functions of the Settlement Officer (Consolidation) under this Act;

(x) "standard area" in respect of any class of land means the area which the State Government may from time to time determine under section 5 as the minimum area necessary for profitable cultivation in any particular notified area and includes a standard area revised under the said section;

(xi) "sub-division" means a part of an estate recorded as a sub-division, *pati, tarf or pana* in a record-of-rights prepared under section 31 of the Punjab Land Revenue Act, 1887, provided it forms a compact block; and

(xii) words and expression used in this Act but not defined have the meanings assigned to them in the Punjab Land Revenue Act, 1887.

Determination of Standard Areas and Treatment of Fragments

Determination of Notified Area

The State Government may, after such inquiry as it deems fit, specify any estate or sub-division of an estate as a notified area for the purposes of this Chapter of this Act.

Settlement of Standard

1. The State Government may, after such inquiry as it deems fit provisionally settle for any class of land in any notified area the minimum area that can be cultivated profitably as a separate plot.
2. The State Government shall by notification and in such other manner as may be prescribed publish the minimum areas provisionally settled by it under sub-section and invite objections thereto.

Determination and Revision of Standard Areas

1. The State Government shall, after considering the objections, if any, received within three months of the date of publication of the notification under sub-section of section 4 in the estate concerned and making such further inquiry as it may deem fit, determine the standard area for each class of land in such notified area.
2. The State Government may, at any time, if it deems fit expedient so, to do, revise a standard area determined under sub-section such revision shall be made in the manner laid down in section 4 and sub-section (1) of Section 5.
3. The State Government shall, by notification and in such other manners may be prescribed, give public notice of any standard area determined under sub-section (1) or revised under sub-section (2).

Entry in the Record of Rights

1. On notification of standard area under sub-section (3) of section 5 for a local area all fragments in the local area shall be entered as such in the record-of-rights.
2. Notice of every entry made under sub-section (1) shall be given in the prescribed manner.

Transfer and Lease of Fragments

1. No person shall transfer any fragment in respect of which a notice has been given under sub-section (2) of section 6 unless thereby the fragment becomes merged in a contiguous survey number or recognized sub-division of a survey number.
2. Notwithstanding anything contained in the Punjab Tenancy Act, 1887, no such fragment shall be leased to any person other than a person cultivating any land which is contiguous to the fragment.

Fragmentation Prohibited

No land in any notified area shall be transferred or partitioned so as to create a fragment.

Penalty for Transfer or Partition Contrary to Provisions of Act

The transfer or partition of any land contrary to the provisions of this Act, shall be void.

Valuation of Fragment

Any owner of a fragment who intends to sell it shall make an application in this behalf to the Collector for determination of its market price and the Collector shall, after hearing the applicant and owners of the contiguous survey numbers or recognised sub-divisions of survey numbers determine the market price, and such determination shall be final and conclusive for the purposes of this chapter.

Transfer of Fragment

The owner referred to in the preceding section shall in the first instance offer the fragment for sale to the owners of contiguous survey numbers or recognized sub-divisions of survey numbers, and in case of their refusal to preceding section, may transfer it to the Government for the purpose of State on payment by the Government of such price as aforesaid to persons possessing interest therein as the Collector may determine, and thereupon the fragment shall vest absolutely in the Government for the purpose of the State free from all encumbrances.

Partition of Estate Assessed to Payment of Act

When a decree is transferred to the Collector under section 54 of the Code of Civil Procedure, 1908, for the partition of an undivided estate assessed to payment of revenue to the Government in any notified area for which standard areas have been fixed, or for the separate possession of a share of such an estate, no such partition or separation shall be made so as to create a fragment.

State Government or Local Authority not to Acquire Land so as to leave Fragments

1. Notwithstanding anything contained in any law for the time being in force no land shall be acquired by the State Government or any local authority or sold at any sale held under the orders of any court so as to leave a fragment.
2. If any land acquired by the State Government or any local authority is in excess of its requirements, it shall be offered for sale in the first instance to the owners of contiguous survey numbers or recognised sub-divisions of survey numbers at price at which it was acquired under sub-section (1).

Consolidation of Holdings

Scheme to Provide Compensation

The scheme prepared by the Consolidation Officer shall provide for the payment of compensation to any owner who is allotted a holding of less market value than that of his original holding for the recovery of compensation from any owner who is allotted a holding of greater market value than that of his original holding.

Occupancy Tenancies

The scheme prepared by the Consolidation Officer may provide for the distribution of land held under occupancy tenure between the tenants holding a right of occupancy and his landlord in such proportion as may be agreed upon between the parties.

When the scheme is confirmed under section 20, the land so allotted to the occupancy tenant and the landlord shall, notwithstanding anything to the contrary contained in the Punjab Tenancy Act, 1887, or in any other law for the time being in force, be held by each of them respectively in full right of ownership, and the right of occupancy in the land allotted to the landlord shall be deemed to be extinguished.

Power to make Provision in the Scheme to Partition Joint Lands and Joint Occupancy Tenancies

1. Notwithstanding anything contained in Chapter 9 of the Punjab Land Revenue Act, 1887, except section 17, thereof, the scheme prepared by the Consolidation Officer may provide for the partition of land between joint-owners of land, or between, joint-tenants of a tenancy in which a right of occupancy subsists, in accordance with the share of each owner or tenant in the land or tenancy, as the case may be, if :

   (a) such share is recorded under Chapter 4 of that Act as belonging to him, or

(b) the right of such owner or tenant to such share has been established by a decree which is still subsisting at the time of preparing the scheme, or

(c) a written acknowledgment of such right has been executed by all persons interested in the admission or denial thereof.

2. When the scheme is confirmed under section 20, the land so partitioned shall, notwithstanding anything to the contrary contained in any law for the time being in force, be held by each such owner or tenant in full right of ownership or tenancy, as the case may be and the rights of other joint-owners or joint tenants, in the land shall be deemed to be extinguished.

Amalgamation of Public Roads *etc.* within Scheme for Consolidation of Holdings

1. Whenever in preparing scheme for the consolidation of holdings, it appears to the Consolidation Officer that it is necessary to amalgamate any road, street, lane, path, channel, drain, tank, pasture or other land reserved for common purposes with any holding in the scheme he shall make a declaration to that effect stating in such declaration that it is proposed that the right of the public as well as of all individuals in over the said road, street, lane, path, channel, drain, tank, pasture or other land reserved for common purposes, shall be extinguished or, as the case may be, transferred to a new road, street, lane, path, channel, drain, tank, pasture or other land reserved for common purposes laid out in the scheme of consolidation.
2. The declaration in sub-section (1) shall be published in the estate concerned in prescribed manner along with the draft scheme referred to in section 19.
3. Any member of the public or any person having any interest or right, in addition to the right of public highway, in or over the said road, street, lane, path, channel, drain, tank, pasture or other land reserved for common purposes or having any other interest or right which is likely to be adversely affected by the proposal may within thirty days after the publication of the declaration under sub-section (1) state to the Consolidation Officer in writing his objection to the proposal, the nature of such interest or right and the manner in which it is likely to be adversely affected and the amount and particular of his claim to compensation for such interest or right:

   Provided that no claim for compensation on account of the extinction or diminution of the right of public highway over such road, street, lane, path, channel, drain, tank, pasture or other land reserved for common purposes, shall be entertained.
4. The Consolidation Officer shall, after considering the objection, if any, made to the proposal, submit it with such amendments, if any, as he may consider necessary to the Settlement Officer (Consolidation), together with

the objections received, his recommendations thereon and a statement of the amounts of compensation, if any, which in his opinion are payable, and of the persons by whom and of the persons to whom such compensation is payable. The decision of the Settlement Officer (Consolidation), on the proposal and regarding the amount of compensation and the person by whom such compensation, if any, is payable, shall be final.

Preparation of Record-of Rights

a. The Consolidation Officer shall cause to be prepared a new record-of-right in accordance with the provisions of the Punjab Land Revenue Act, 1887, in so far as these provisions may be applicable, for the area under consolidation, giving effect to the repartition and order in respect thereof made under the preceding section.

b. Such record-of-right shall be deemed to have been prepared under section 32 of the Punjab Land Revenue Act, 1887

Right to Possession of New Holdings

1. If all the owners and tenants affected by the repartition, as carried out under sub-section (1) of section 21 agree to enter into possession of the holdings allotted to them there under, the Consolidation Officer may allow them to enter into such possession forthwith or from such date as may be specified by him.
2. If all the owners and tenants as aforesaid do not agree to enter into possession under sub-section (1), they shall be entitled to possession of the holdings and tenancies allotted to them from such date as may be determined by the Consolidation Officer and published in the prescribed manner in the estates concerned; and the Consolidation Officer shall, if necessary, put them in physical possession of the holding to which they are so entitled, including standing crops, if any, and for doing so may exercise the powers of Revenue Officer under the Punjab Land Revenue Act, 1887.
3. If there are standing crops on any holding of which possession has been given under sub-section (2), the Consolidation Officer shall determine in the manner provided under this Act the compensation payable in respect of such crops by the person put in possession, who shall, within six months of the date of possession, pay such compensation to the person or persons from whom possession was transferred, and in case of default such compensation shall be recoverable from him as an arrear of land revenue.
4. If any person from whom compensation is recoverable under the scheme fails within 15 days of the date referred to in sub-section (2) to deposit such compensation in the prescribed manner, it shall be recoverable from him as an arrear of land revenue, and in such case the amount realised after

deducting the expenses, shall be paid to any person having the interest in the holding.

Management and Control of Lands for Common Purposes to Vest in Panchayats or State Government

As soon as a scheme comes into force, the management and control of all lands assigned or reserved for common purposes of the village under section 18:

(a) in the case of common purposes specified in sub-clause (iv) of clause (bb) of section 2 in respect of which the management and control are to be exercised by the State Government, shall vest in the State Government; and

(b) in the case of any other common purpose, shall vest in the panchayat of that village;

and the State Government or the Panchayat, as the case may be, shall be entitled to appropriate the income accruing there from for the benefit of the village community, and the rights and interests of the owners of such lands shall stand modified and extinguished accordingly.

Provided that in case of land assigned or reserved for the extension of village abadi or manure pits for the proprietors and non- proprietors of the village, such land shall vest in the proprietors and non-proprietors to whom it is given under the scheme of consolidation.

Coming into Force of Scheme

1. As soon as the persons entitled to possession of holdings under this Act have entered into possession of the holdings, respectively allotted to them the scheme shall be deemed to have come into force and the possession of the allottee affected by the scheme of consolidation, or as the case may be, by repartition, shall remain undisturbed until a fresh scheme is brought into force or a change is ordered in pursuance of provisions of sub-sections (2), (3) and (4) of section 21 or an order passed under section 36 or 42 of this Act.
2. A Consolidation Officer shall be competent to exercise all or any of the powers of a Revenue Officer under the Punjab Land Revenue Act, 1887 (Act 17 of 1887), for purposes of compliance with the provisions of sub-section (1).

Rights of Land-Owners and Tenants after Consolidation same as before

A land-owner or a tenant shall subject to the provisions of [section 16 and 16-A] have the same right in the land allotted to him in pursuance of the scheme of consolidation as he had in his original holding or tenancy, as the case may be.

Effect of Consolidation of Holdings on Evacuee Property

1. If in pursuance of a scheme for consolidation of holdings any land which is evacuee property within the meaning of the Administration of Evacuee

Property Act, 1950 (31 of 1950), is or has been exchanged for any other land which is not evacuee property, then such other land shall, as from the date of the coming into force of the scheme, be deemed to be evacuee property declared as such within the meaning of the said Act and the original evacuee land shall, as from such date, be deemed to have ceased to be evacuee property.

2. If a displaced person to whom any land has been leased or allotted by the Custodian under the conditions published by the notification of the Government of Punjab in the Department of Rehabilitation No. 4891-S or 4892-S, dated the 8th July, 1949, has been allotted some other land in lieu thereof in pursuance of any scheme, for consolidation of holdings, then such other land shall be deemed to have been leased or allotted, as the case may be, to such displaced person under the aforesaid conditions within the meaning of the Displaced Persons (Compensation and Rehabilitation) Act, 1954 (44 of 1954).
3. In this section, the expressions 'Custodian' and 'displaced persons', have the meaning, respectively assigned to them in the Administration of Evacuee Property Act, 1950 (31 of 1950) and the Displaced Persons (Compensation and Rehabilitation) Act, 1954 (44 of 1954).

Power to Vary or Revoke Scheme

A scheme for the consolidation of holdings confirmed under this Act may, at any time, be varied or revoked by the authority which confirms it subject to any order of the State Government that may be made in relation thereto and a subsequent scheme may be prepared, published and confirmed in accordance with the provisions of the East Punjab Holding Act.

Planning Commission and NITI Aayog

Functions of Planning Commission

The 1950 resolution setting up the Planning Commission outlined its functions as to:

a. Make an assessment of the material, capital and human resources of the country, including technical personnel, and investigate the possibilities of augmenting such of these resources as are found to be deficient in relation to the nation's requirement;
b. Formulate a Plan for the most effective and balanced utilisation of country's resources;
c. On a determination of priorities, define the stages in which the Plan should be carried out and propose the allocation of resources for the due completion of each stage;
d. Indicate the factors which are tending to retard economic development, and determine the conditions which, in view of the current social and political situation, should be established for the successful execution of the Plan;

e. Determine the nature of the machinery which will be necessary for securing the successful implementation of each stage of the Plan in all its aspects;

f. Appraise from time to time the progress achieved in the execution of each stage of the Plan and recommend the adjustments of policy and measures that such appraisal may show to be necessary; and

g. Make such interim or ancillary recommendations as appear to it to be appropriate either for facilitating the discharge of the duties assigned to it, or on a consideration of prevailing economic conditions, current policies, measures and development programmes or on an examination of such specific problems as may be referred to it for advice by Central or State Governments.

Evolving Functions

From a highly centralised planning system, the Indian economy is gradually moving towards indicative planning where Planning Commission concerns itself with the building of a long term strategic vision of the future and decide on priorities of nation. It works out sectoral targets and provides promotional stimulus to the economy to grow in the desired direction.

Planning Commission plays an integrative role in the development of a holistic approach to the policy formulation in critical areas of human and economic development. In the social sector, schemes which require coordination and synthesis like rural health, drinking water, rural energy needs, literacy and environment protection have yet to be subjected to coordinated policy formulation. It has led to multiplicity of agencies. An integrated approach can lead to better results at much lower costs.

The emphasis of the Commission is on maximising the output by using our limited resources optimally. Instead of looking for mere increase in the plan outlays, the effort is to look for increases in the efficiency of utilisation of the allocations being made.

With the emergence of severe constraints on available budgetary resources, the resource allocation system between the States and Ministries of the Central Government is under strain. This requires the Planning Commission to play a mediatory and facilitating role, keeping in view the best interest of all concerned. It has to ensure smooth management of the change and help in creating a culture of high productivity and efficiency in the Government.

The key to efficient utilisation of resources lies in the creation of appropriate self-managed organisations at all levels. In this area, Planning Commission attempts to play a systems change role and provide consultancy within the Government for developing better systems. In order to spread the gains of experience more widely, Planning Commission also plays an information dissemination role.

The Planning Commission was set up in 1950, as an advisory body, the Prime Minister being its ex-officio Chairman. The Planning Commission is concerned broadly with economic planning and organizing the planning, but not implementing it.

The chief objective of Planning was defined as initiating "a process of development which will raise living standards and open out to the people new opportunities for a richer and more varied life". Another objectives of the commission was to stabilize the growth of population over a reasonable period, to ensure that the economy, becomes self reliant in years to come. Planning in India derives its objectives from Directive Principles of State Policy.

Advantages of Economic Planning

(i) It is only through economic planning that a nation can hope to improve the living standards of its people and to use its resources to optimum capacity;

(ii) Economic planning prevents lop-sided growth to national economy by selecting the most productive combination of projects and by judicious allocation of available resources;

(iii) It helps to correct and eliminate regional imbalances by focusing attention on the needs of backward areas;

(iv) It promotes co-ordination between the different sectors of the economy;

(v) It is the only reliable way in which the State can try to remove economic disparities and establish an egalitarian society free of social and political tensions;

(vi) It provides the people with the best criterion by which to judge the performance of their governments which, in a planned economy, have to work for achieving pre-determined targets and rates of growth.

Imperative and Indicative Planning

In case of imperative planning economic activities belong to public sector, while in the other type they belong to the private sector. It is believed that the Five-year Plans provide a long-term perspective to the economy in market-friendly fashion.

National Development Council (NDC)

NDC helps in the implementation of the Plan and helps in the resources of the country being pooled together for uniform development of all regions. The National Development Council has as Members the Chief Ministers of all States as also Members of the Planning Commission. It is presided over by the Prime Minister of India as Chairman. It helps in working out common policies in the important spheres of the country's economy and thus ensures balanced economic development of the country as a whole. The Five-year Plans are finally approved by the National Development Council.

Growth Performance in the Five-Year Plans (per cent per annum)

Sl.No.	Plan	Target	Actual
1.	First Plan (1951-56)	2.1	3.60
2.	Second Plan (1956-61)	4.5	4.21
3.	Third Plan (1961-66)	5.6	2.72
4.	Fourth Plan (1969-74)	5.7	2.05
5.	Fifth Plan (1974-79)	4.4	4.83
6.	Sixth Plan (1980-85)	5.2	5.54
7.	Seventh Plan (1985-90)	5.0	6.02
8.	Eighth Plan (1992-97)	5.6	6.68
9.	Ninth Plan (1997-2002)	6.5	5.5

The Planning Commission and its Future

Last week in his Independence Day speech, the Prime Minister announced the intention of the government to replace the Planning Commission. Furthermore, he has asked people to give their suggestions on the next steps in replacing the Commission with another suitable institution. Some of the reactions to this announcement were that this decision was welcome. We do not need state directed planning in the era of liberalization; this was a common refrain heard last week.

However, we do not subscribe to that sentiment that planning itself is unnecessary. While the nature of planning in the form of obsessively restrictive and centralized budget controls is clearly outdated, a liberalized economy requires as much of planning as before, but of a completely different nature. Such planning would focus more on strategy and less on interference with budgetary allocations, which can be left to individual ministries, state governments and local governments to do, within their functional responsibilities.

The suggestion to do away with the Planning Commission in its present form is a welcome one, for the reason that it is a paradoxical institution with no constitutional basis for its existence. However, if the replacement institution is going to dish out more of the same thing; centralized budgeting, then the announcement of the Prime Minister would be counter-productive. It will completely defeat the purpose if the doing away of the Planning Commission results in the setting up of a similar extra-constitutional body.

While steps are taken to the Planning Commission, do hope that the history of the Planning Commission will be studied to understand its purpose, evaluate whether that purpose has been well served and then take a decision on the next steps. It is probably not well known that the Planning Commission is actually designed to function as the Secretariat of another overarching body, the National Development Council. This was not the original intention, though. Prime Minister Jawaharlal Nehru established the Planning Commission in 1950, the year that the Republic was established. He was inspired by the centralized Soviet model and entrusted the task of formulating the national Five-Year plans to it. However, by

1952, he realized that for a national plan to be implemented in a federal structure, there was a need to obtain a centre-state consensus on approving the plan. Thus, in 1952, a resolution was passed at a meeting of all chief ministers of states convened by the union government, to establish a National Development Council (NDC) chaired by the Prime Minister and comprising of all the cabinet ministers and chief ministers of states, in order to guide the preparation of India's Five-Year Plans.

Thus, the NDC was positioned as a general body to the Planning Commission and was also entrusted the task of considering and approving the national plan prepared by the Planning Commission and to review its implementation. In doing so, it could also deliberate on important national social and economic policy issues. In 1983, following the recommendations of the Sarkaria Commission on centre-state relations, the Constitution was amended to provide for an inter-state council (ISC) chaired by the Prime Minister and comprising of the chief ministers of all states and union territories and administrators of UTs without legislatures and six central cabinet ministers. In addition, four more central cabinet ministers could be inducted as permanent invitees to ISC. The council has been positioned as a recommendatory body with the duty of investigating and discussing subjects in which some or all of the states or the union and one or more of the states have a common interest, as may be brought up before it, making recommendations on any such subject, in particular recommendations for better coordination of policy and action with respect to that subject; and deliberating on such other matters of general interest to the states as may be referred by the chairman to the council.

As can be seen, there is a remarkable similarity between the composition of NDC and ISC. Furthermore, the ISC's constitutionally mandated scope of responsibility is wider than that of the Planning Commission. These observations lead to a solution that satisfies the Constitutional mandate, while still preserving the need for planning. Quite easily, NDC could have morphed itself into ISC and the Planning Commission could have been positioned to become ISC's secretariat. This would have ensured a natural evolution of the idea of having a deliberative body in our federal set-up. However, the opportunity to do this in the post-Sarkaria Commission era was lost. Therefore, the NDC continues to exist and function as a separate body; in 2011, it held its fifty fifth meeting.

It is time to take this idea off the shelf and implement it now. The Inter-State-Council currently lives in a gilded cage, with little of its staff to do. Though it has a permanent secretariat, with a senior officer heading it, all it does is hold seminars and discussions. Yet, it must not be lost on us that the ISC is has been given a constitutional status, whereas both the NDC and the Planning Commission are founded on the basis of executive orders. So here are a couple of suggestions to take forward the intention of the government announced by the Prime Minister. First, the task performed by the NDC can be taken over by the Inter-State Council. It therefore becomes the final arbiter of India's planning exercises, whatever might be their nature in the future, in the same manner that the NDC approved of national plans in the past.

Second, the successor to the Planning Commission, by whatever name called, becomes a planning focused Secretariat and decision support system to the Inter-

State Council. Once these institutional relationships are rejigged, one needs to examine what exactly would be the role of the new institution that replaces the Planning Commission. More of that in my next blog.

Summary: Five Year Plans in India

Plan	*Notes*
First Plan (1951-56)	It was based on Harrod-Domar Model.
	Community Development Program launched in 1952
	Focus on agriculture, price stability, power and transport
	It was a successful plan primarily because of good harvests in the last two years of the plan
Second Plan (1956-61)	Also called Mahalanobis Plan named after the well known economist
Target Growth: 4.5%	Focus - rapid industrialization
Actual Growth: 4.27%	Advocated huge imports through foreign loans.
	Shifted basic emphasis from agriculture to industry far too soon.
	During this plan, prices increased by 30 per cent, against a decline of 13 per cent during the First Plan
Third Plan (1961-66) Target Growth: 5.6%	At its conception, it was felt that Indian economy has entered a take-off stage. Therefore, its aim was to make India a 'self-reliant' and 'self-generating' economy.
Actual Growth: 2.84%	Based on the experience of first two plans, agriculture was given top priority to support the exports and industry.
	Complete failure in reaching the targets due to unforeseen events - Chinese aggression (1962), Indo-Pak war (1965), severe drought 1965-66
Three Annual Plans (1966-69) Plan holiday	Prevailing crisis in agriculture and serious food shortage necessitated the emphasis on agriculture during the Annual Plans
for 3 years.	During these plans a whole new agricultural strategy was implemented. It involving wide-spread distribution of high-yielding varieties of seeds, extensive use of fertilizers, exploitation of irrigation potential and soil conservation.
	During the Annual Plans, the economy absorbed the shocks generated during the Third Plan
	It paved the path for the planned growth ahead.
Fourth Plan (1969-74) Target Growth: 5.7%	Main emphasis was on growth rate of agriculture to enable other sectors to move forward
Actual Growth: 3.30%	First two years of the plan saw record production. The last three years did not measure up due to poor monsoon.
	Influx of Bangladeshi refugees before and after 1971 Indo-Pak war was an important issue
Fifth Plan (1974-79)	The fifth plan was prepared and launched by D.D. Dhar.
Target Growth: 4.4% Actual Growth: 3.8%	It proposed to achieve two main objectives: 'removal of poverty' (Garibi Hatao) and 'attainment of self reliance'
	Promotion of high rate of growth, better distribution of income and significant growth in the domestic rate of savings were seen as key instruments
	The plan was terminated in 1978 (instead of 1979) when Janta Party Govt. rose to power.

Contd...

Contd...

Plan	Notes
Rolling Plan (1978-80)	There were 2 Sixth Plans. Janta Govt. put forward a plan for 1978-1983. However, the government lasted for only 2 years. Congress Govt. returned to power in 1980 and launched a different plan.
Sixth Plan (1980-85) Target Growth: 5.2% Actual Growth: 5.66%	Focus - Increase in national income, modernization of technology, ensuring continuous decrease in poverty and unemployment, population control through family planning, *etc.*
Seventh Plan (1985-90) Target Growth: 5.0%	Focus - rapid growth in food-grains production, increased employment opportunities and productivity within the framework of basic tenants of planning.
Actual Growth: 6.01%	The plan was very successful, the economy recorded 6 per cent growth rate against the targeted 5 per cent.
Eighth Plan (1992-97)	The eighth plan was postponed by two years because of political uncertainty at the Centre Worsening Balance of Payment position and inflation during 1990-91 were the key issues during the launch of the plan. The plan undertook drastic policy measures to combat the bad economic situation and to undertake an annual average growth of 5.6 per cent Some of the main economic outcomes during eighth plan period were rapid economic growth, high growth of agriculture and allied sector, and manufacturing sector, growth in exports and imports, improvement in trade and current account deficit.
Ninth Plan (1997-2002) Target Growth: 6.5% Actual Growth: 5.35%	It was developed in the context of four important dimensions: Quality of life, generation of productive employment, regional balance and self-reliance.
Tenth Plan (2002-2007)	Goals: To achieve 8 per cent GDP growth rate Reduction of poverty ratio by 5 percentage points by 2007. Providing gainful high quality employment to the addition to the labour force over the tenth plan period. Universal access to primary education by 2007. Reduction in gender gaps in literacy and wage rates by atleast 50 per cent by 2007. Reduction in decadal rate of population growth between 2001 and 2011 to 16.2 per cent. Increase in literacy rate to 72 per cent within the plan period and to 80 per cent by 2012. Reduction of Infant Mortality Rate (IMR) to 45 per 1000 live births by 2007 and to 28 by 2012. Increase in forest and tree cover to 25 per cent by 2007 and 33 per cent by 2012. All villages to have sustained access to potable drinking water by 2012. Cleaning of all major polluted rivers by 2007 and other notified stretches by 2012.

Contd...

Contd...

Plan	Notes
Eleventh Plan	Goals:
(2007-2012)	Accelerate GDP growth from 8 per cent to 10 per cent. Increase agricultural GDP growth rate to 4 per cent per year.
	Create 70 million new work opportunities and reduce educated unemployment to below 5 per cent.
	Raise real wage rate of unskilled workers by 20 per cent.
	Reduce dropout rates of children from elementary school from 52.2 per cent in 2003-04 to 20 per cent by 2011-12. Increase literacy rate for persons of age 7 years or above to 85 per cent.
	Lower gender gap in literacy to 10 percentage point. Increase the percentage of each cohort going to higher education from the present 10 per cent to 15 per cent.
	Reduce infant mortality rate to 28 and maternal mortality ratio to 1 per 1000 live births
	Reduce Total Fertility Rate to 2.1
	Provide clean drinking water for all by 2009. Reduce malnutrition among children between 0-3 years to half its present level. Reduce anaemia among women and girls by 50 per cent.
	Raise the sex ratio for age group 0-6 to 935 by 2011-12 and to 950 by 2016-17
	Ensure that at least 33 per cent of the direct and indirect beneficiaries of all government schemes are women and girl children
	Ensure all-weather road connection to all habitation with population 1000 and above (500 in hilly and tribal areas) by 2009, and ensure coverage of all significant habitation by 2015
	Connect every village by telephone by November 2007 and provide broadband connectivity to all villages by 2012
	Increase forest and tree cover by 5 percentage points.
	Attain WHO standards of air quality in all major cities by 2011-12.
	Treat all urban waste water by 2011-12 to clean river waters.
	Increase energy efficiency by 20 percentage points by 2016-17.

Centre Embraces Consultative Policy-Making

The government on Thursday replaced the 65year-old Planning Commission -a relic of the Socialist era -with a Niti Aayog or National Institution for Transforming India, marking a major shift in policy making by involving states.

To be headed by PM Narendra Modi, the new body will have a governing council comprising CMs of all states and Lt Governors of Union Territories and will work towards fostering "co-operative federalism" for providing a "national agenda" to the Centre and states. The membership of states has rendered the National Development Council, a body of CMs that approves five-year Plans, defunct.

The new body, which would remain at its predecessor's address in Lutyens' Delhi, will have a vice-chairperson and CEO to be appointed by the PM.

Columbia University professor Arvind Panagariya was under consideration for vicechairperson, sources said. The Niti Aayog, which will serve as a policy think tank for the central and state governments, will have 7-8 full-time members and two part-time members, while four Union ministers would serve as ex-officio members. The two part-time members would be from leading universities and research organizations; the number of full-time members has not been specified as yet.

Experts, specialists and practitioners with relevant domain knowledge would also be part of Niti Aayog as "special invitees nominated by the PM". Modi, who announced the dismantling of Planning Commission in his maiden Independence Day speech last year, hailed the replacement in a series of tweets, saying the inclusion of CMs and LGs will "foster a spirit of cooperative federalism". "Through Niti Aayog, we bid farewell to a `one size fits all' approach towards development. The body celebrates India's diversity and plurality," Modi tweeted.

The scrapping of the Planning Commission addresses the long-held view that the body, which was conceived at a time when the Centre was supposed to occupy "commanding heights of economy", had outlived its relevance because of diminished role of the state and the emergence of the private sector and also due to the foci of economic activities shifting to states. Beginning with former Prime Minister Rajiv Gandhi's dismissing the body as a "bunch of jokers", the Commission's utility had

been increasingly called into question, with state governments protesting against their limited role in the planning process despite Congress's protests against what it has called the assault on Nehruvian policies.

The Modi government focused on the lament of the states and an important evolutionary change from the past will replace a centre-to-state one-way flow of policy by a genuine and continuing partnership with the states. The declared intent to empower states has played well as nonNDA chief ministers have also supported the dissolution of Planning Commission. Significantly, Modi and the official statement also stressed that the Niti Aayog will pay special attention to the poor who have not benefited from development process. "A pro-people, pro-active and participatory development agenda stressing on empowerment and equality is the guiding principle behind NITI Aayog", Modi said in one of his tweets. It also said that the new institution "has to provide at the Centre and state levels with relevant strategic and technical advice" on key policy issues.

Landmark Change in Policy Making - Modi Sarkar Puts Bharat First with PM-Led NITI Aayog to have all CMs and Lieutenant-Governors as its Key Members

The NITI Aayog, which will replace the Planning Commission, will be responsible for formulating what the government described as "a Bharatiya approach to development". The new organisation -NITI stands for National Institution for Transforming India -is being seen as a think tank that will foster "cooperative federalism" rather than take a top-down approach.

"An important evolutionary change from the past will be replacing a Centre-to-state one way flow of policy by a genuine and continuing partnership with the states," the government said in the statement unveiling the body on New Year's Day. "The institution must have the necessary resources, knowledge, skills and ability to act with speed to provide the strategic policy vision for the government as well as deal with contingent issues."

The Planning Commission, which was set up in 1950, was among the principal targets of the Narendra Modi government that came to power in May, seeing it as symptomatic of a way of thinking that was obsolete. "In the context of governance structures, the changed requirements of our country point to the need for setting up an institution that serves as a think tank of the government -a directional and policy dynamo," the government said. "The proposed institution has to provide governments at the central and state levels with relevant strategic and technical advice across the spectrum of key elements of policy."

The body will be headed by the prime minister and comprise a vice chairperson, besides full-time and part-time members. Aside from that, it will have up to four members of the council of ministers and a chief executive officer of secretary rank. No names have been announced yet but the rooms of the deputy chairman and members of the erstwhile Planning Commission at what used to be its headquarters at Yojana Bhawan in New Delhi are being readied for the team that is likely to assume charge shortly.

Arvind Panagariya, professor of economics at Columbia University, is among those being tipped as vice chairperson. He couldn't be immediately reached for comment. "NITI Aayog will seek to provide a critical directional and strategic input into the development process," the government said, pitching it as an institution that will create a greater stake for states in development. "The Centre-to-state one-way flow of policy, that was the hallmark of the Planning Commission era, is now sought to be replaced by a genuine and con tinuing partnership of states."

This is sought to be achieved through a governing council consisting of all chief ministers and lieuten ant-governors of union territories besides regional councils set up for a specified tenure and experts, specialists and practitioners as special invitees nominated by the prime minis ter. The decision comes nearly three weeks after Modi held consultations, with chief ministers at a meeting where most favoured restructuring of the Socialist-era Planning Commission while some belonging to t Congress opposed disbanding the existing setup.

Modi had announced in his Independence Day speech that the Planning Commission would be replaced by a new body that was in sync with the global economy. The government said on Thursday that the Aayog will provide a "nation al agenda" for the prime minister and chief ministers while recognising that "strong states make a strong nation". It will also interact with other national and international think tanks, as also with educational and policy research institutions.

According to Arun Maira, former member of the Planning Commission, the structure of NITI Aayog may not be entirely new but getting the chief ministers involved in formulating the country's development vision was the right way to go. "The idea to create an institution where states' leaders will be part and parcel of the collective thinking with the Centre and other stakeholders in formulating a vision for the development of the country is right on as compared with the previous structure, where a handful of people formulated the vision and then presented it to the National Development Council (NDC)," he said. "This was not entirely absorbed and adopted by the latter." NDC comprises chief ministers of all states.

The announcement drew criticism from Congress, which cited the move as being against the tenets of Jawaharlal Nehru, the country's first prime minister and a driving force behind the establishment of the Planning Commission. "Constructive reform of Planning Commission vs alteration of identity and basic structure cos of anti nehruvianism and congism: support former not latter," party spokesperson Abhishek Singhvi tweeted.

While the Planning Commission went through numerous operational makeovers over the years -ranging from being a simple planning body to a powerful 'control commission' to a fiscal decentralisation instrument to an official think tank -voices had begun to grow louder for an overhaul even before the new government took charge.

Subsequently, a consultation process was launched for suggestions on the structure and role of the new body and there was a strong view that this should play the role of a catalyst and provide a platform to the Centre, states and experts to discuss issues and arrive at the best solutions. NITI Aayog will create a knowledge, innovation and entrepreneurial support system through a collaborative community of national and international experts, practitioners and partners, the government said. It will offer a platform for resolution of inter-sectoral and inter-departmental issues in order to accelerate the implementation of the development agenda. In addition, it will monitor and evaluate the implementation of programmes and focus on technology upgradation and capacity building.

The objective of the new body is to evolve a shared vision of national development priorities, sectors and strategies with the active involvement of states. It will develop mechanisms to formulate credible plans at the village level and aggregate these progressively at higher levels of government. Aayog will design strategic and long-term policy and programme frameworks and initiatives, and monitor their progress and their efficacy.

Aims and Objectives of NITI Aayog

NITI Aayog will seek to provide a critical directional and strategic input into the development process. The centre-to-state one-way flow of policy, that was the hallmark of the Planning Commission era, is now sought to be replaced by a genuine and continuing partnership of states.

NITI Aayog will emerge as a "think-tank" that will provide Governments at the central and state levels with relevant strategic and technical advice across the spectrum of key elements of policy.

The NITI Aayog will also seek to put an end to slow and tardy implementation of policy, by fostering better Inter-Ministry coordination and better Centre-State coordination. It will help evolve a shared vision of national development priorities, and foster cooperative federalism, recognizing that strong states make a strong nation.

The NITI Aayog will develop mechanisms to formulate credible plans to the village level and aggregate these progressively at higher levels of government. It will ensure special attention to the sections of society that may be at risk of not benefitting adequately from economic progress.

The NITI Aayog will create a knowledge, innovation and entrepreneurial support system through a collaborative community of national and international experts, practitioners and partners. It will offer a platform for resolution of inter-sectoral and inter-departmental issues in order to accelerate the implementation of the development agenda.

In addition, the NITI Aayog will monitor and evaluate the implementation of programmes, and focus on technology upgradation and capacity building.

Through the above, the NITI Aayog will aim to accomplish the following objectives and opportunities:

- An administration paradigm in which the Government is an "enabler" rather than a "provider of first and last resort."
- Progress from "food security" to focus on a mix of agricultural production, as well as actual returns that farmers get from their produce.

- ☆ Ensure that India is an active player in the debates and deliberations on the global commons.
- ☆ Ensure that the economically vibrant middle-class remains engaged, and its potential is fully realized.
- ☆ Leverage India's pool of entrepreneurial, scientific and intellectual human capital.
- ☆ Incorporate the significant geo-economic and geo-political strength of the Non-Resident Indian Community.
- ☆ Use urbanization as an opportunity to create a wholesome and secure habitat through the use of modern technology.
- ☆ Use technology to reduce opacity and potential for misadventures in governance.
- ☆ The NITI Aayog aims to enable India to better face complex challenges, through the following:
- ☆ Leveraging of India's demographic dividend, and realization of the potential of youth, men and women, through education, skill development, elimination of gender bias, and employment
- ☆ Elimination of poverty, and the chance for every Indian to live a life of dignity and self-respect
- ☆ Reddressal of inequalities based on gender bias, caste and economic disparities
- ☆ Integrate villages institutionally into the development process
- ☆ Policy support to more than 50 million small businesses, which are a major source of employment creation
- ☆ Safeguarding of our environmental and ecological assets.

Major Highlights

1. The new National Institution for Transforming India (NITI) will act more like a think tank or forum and execute programs by taking the States along with them. This is in sharp contrast with the defunct Planning Commission which imposed five-year-plans and allocated resources while running roughshod over the requests of the various States.
2. NITI will include leaders of India's 29 states and seven union territories. But its full-time staff – a deputy chairman, Chief Executive Officer and experts – will answer directly to the Prime Minister of India, who will be chairman.
3. The opposition Congress mocked the launch as a cosmetic relabelling exercise – the new body's acronym-based name means 'Policy Commission' in Hindi, suggesting a less bold departure than the English version does. Several believe that is consistent with the negativism that has become the hallmark of the Congress.

4. Despite being blamed by critics for the slow growth that long plagued India, the Commission survived the market reforms of the early 1990s, riling Mr Modi with its interventions when he was Chief minister of industry and investor friendly Gujarat.
5. Mr Modi, elected by a landslide last year on a promise to revive flagging growth and create jobs, had vowed to do away with the Planning Commission that was set up in 1950 by Congressman and Prime Minister Jawaharlal Nehru.
6. But his plans been derided by the Congress party, which wants to defend the Nehru legacy and describes Mr Modi's vision of "cooperative federalism" as cover for a veiled power grab.
7. India's first Prime Minister Jawaharlal Nehru, a socialist who admired Joseph Stalin's drive to industrialize the Soviet Union, set up and chaired the Commission to map out a development path for India's agrarian economy.
8. In 2012, the Planning Commission was pilloried for spending some Rs. 35 lakh to renovate two office toilets, and then it was lampooned for suggesting that citizens who spent Rs. 27 or more a day were not poor.
9. The commission had remained powerful over the decades because it had emerged as a sort of parallel cabinet with the Prime Minister as its head.
10. The Commission's power in allocating central funds to states and sanctioning capital spending of the central government was deeply resented by states and various government departments.
11. The NITI Aayog will also seek to put an end to slow and tardy implementation of policy, by fostering better Inter-Ministry coordination and better Centre-State coordination. It will help evolve a shared vision of national development priorities, and foster cooperative federalism, recognizing that strong states make a strong Nation.

How will NITI Aayog differ from Planning Commission?

Unlike the Planning Commission, the new body is for governance across the public and private domains, the Cabinet resolution says. "Everyone has a stake in ensuring good governance and effective delivery of services. Creating Jan Chetna, therefore, becomes crucial for people's initiative. In the past, governance may have been rather narrowly construed as public governance. In today's changed dynamics – with 'public' services often being delivered by 'private' entities, and the greater scope for 'participative citizenry', governance encompasses and involves everyone." The NITI Aayog will also seek to put an end to slow and tardy implementation of policy, by fostering better inter-ministry and better Centre-State coordination.

Will it be Modi's economic bureau? Yes. Modi took the chief ministers into confidence on December 7 to go ahead with the new institution to strengthen "cooperative federalism" by promoting the concept of "Team India," noting that even then prime minister Manmohan Singh said on April 30 last year that the current structure of the Planning Commission has "no futuristic vision in the post-reform period".

What Else Does the Resolution Say?

The Cabinet resolution, which provides the ideological framework, invokes Mahatma Gandhi on constant development in life, BJP ideologue late Deen Dayal Upadhyaya for uplifting the downtrodden, Swami Vivekananda, Constitution's architect Dr. Ambedkar and Tamil poet and philosopher Thiruvalluvar, to let the NITI Aayog act as the "pillars" that provide a "Bharatiya approach to development." It prescribes 13 objectives on which the NITI Aayog has to remain focussed.

Will NITI devolve more powers to states? The new body will allow state governments to determine the architecture of economic growth and development, the resolution says: "The one-size-fits-all approach, often inherent in central planning, has the potential to create needless tensions and undermining the harmony needed for national effort. The resolution quotes Dr. Ambedkar that it is "unreasonable to centralise powers where central control and uniformity is not clearly essential or is impracticable."The Planning Commission was unpopular with chief ministers as it engaged in promotion of Centre-to-state one-way flow of policy."

7

Green Revolution in Mexico, 1943 – 1965

In 1942, agriculture in Mexico was in a deplorable condition. Mexico was importing more than 50 per cent of the wheat that it consumed, as well as a considerable percentage of its maize. Wheat yields were low and static, with a national average yield of 750 kilograms per hectare, even though most of the wheat was grown on irrigated land. The use of chemical fertilizers was almost unknown.

In 1943, a co-operative agricultural research and training programme was launched in Mexico. This was a pioneer co-operative project between the Mexican Ministry of Agriculture and the Rockefeller Foundation, initiated at the request of the Mexican Government, for assistance in increasing the production of maize, and beans. The scientist chosen by the Rockefeller Foundation for the Mexican programme in 1944 was Dr. Norman E.Borlaug, who later on became the Director of the Wheat Department of the International Maize and Wheat Improvement Centre (CIMMYT), Mexico. After preliminary work in 1943, plant breeders, soil scientists, plant pathologists and entomologists, working as a team, began a concentrated attack on the various aspects of wheat production in 1944. An in-service (intern) training component was added to the research programme to train a new generation of Mexican scientists – while they were assisting with the development of the research programme. Provision was also made for fellowships to enable the most promising of these young scientists to study abroad for advanced degrees, hopefully in preparation for positions of leadership in Mexican agriculture.

Research from the outset was production-oriented and restricted to that which was relevant to increasing wheat production. Researches in pursuit of irrelevant academic butterflies were discouraged, both because of the acute shortage of scientific manpower and because of the need to have data and materials available as soon as possible for use in the production programme. To accelerate progress in varietal development, two generations of all segregating materials were grown each year. One generation was sown close to seal level in Sonora at 28 degrees north latitude in the fall when the days were progressively shorter; the second was sown near Toluca, at 18 degrees latitude and 2500 meters above sea-level during the summer when days were progressively longer. Through the use of this technique we developed high-yielding, day-length-insensitive varieties with a wide range of ecologic adaptation and a broad spectrum of disease resistance – a new combination of uniquely valuable characters in wheat varieties.

The wheat breeders at CIMMYT, in addition to collecting samples of Mexican varieties, imported seeds from the major wheat-producing countries of the world, including the United States, Canada, Argentina, Australia, and several African countries. They made selections, crossed them and had the first improved varieties available for distribution by the autumn of 1948. By 1955, 75 per cent of the Mexican wheat crop was derived from them.

'Norin 10', Dwarf Wheats from Japan

Borlaug produced high-yielding wheat varieties by adopting some new concepts in plant breeding. The most important concept adopted by him is that of an efficient plant type capable of responding to very high doses of fertilizers and irrigation. He has able to do so by incorporating dwarfing genes into these varieties. The short-strawed germplasm from 'Norin -10' crosses developed by Dr. O.A. Vogel, of the Washington State University, was obtained by Borlaug in 1953; it was crossed with various Mexican and Colombian wheats, and dwarf lines were released in 1961. "These wheats", wrote Albert Moseman in 1964, "yielded up to 120 bushels per acre in the Yaqui Valley of Mexico and now occupy more than 85 per cent of Mexico's wheat acreage." By 1965, the Mexican wheat production showed a pre-acre yield as much as 400 per cent as compared with that in 1950 and was becoming well known internationally. Borlaug is the first wheat breeder to succeed in the use of dwarf parents in the evolution of commercial wheat varieties. Further, he bred with a view to producing lines which were insensitive to day length. The result is that the varieties produced by him have unprecedented adaptability and are fit to be grown over a wide area in the world. He also achieved stability in resistance to diseases by incorporating greater variability through multi-lineal approach. The wheat varieties produced by him are remarkably resistant to common wheat diseases.

From 1945 to 1965, the production of maize, the basic food crop of Mexico, increased four times. The production of cotton jumped from 107,360 tonnes to 605,000 and became Mexico's major export commodity, providing funds for the industrialization of the country. Wheat production increased six time. In 1945, Mexico imported 10,175,000 bushels of wheat, by 1965 she was exporting it, yields per acre having tripled.

Nine Factors in the Green Revolution in Mexico

Though crop improvement was the key factor in the Mexican Green Revolution, there were also other factors which contributed to it. These spectacular increases in productivity, according to Dunbar, may be accounted for by a least nine events: the Mexican Revolution, land reform, expansion of rural education, extension of roads, extension of irrigated area, provision for agricultural credit, improvement of crop varieties, use of farm machinery and use of fertilizers.

The Mexican Government encouraged the agricultural revolution by building roads and irrigation facilities. The highway system, including paved, surfaced and unsurfaced roads, increased from 6,138 to 34,967 miles (3,814 to 21,727km) from 1929 to 1964. In 1926, the Calles administration created a National Irrigation Commission which, with the assistance of American irrigation engineers, began the reclamation by the Government of Mexican arid land, Between 1926 and 1958, this agency and its successor, the Ministry of Hydraulic Resources, brought under irrigation 5,554,561 acres (2247,853 hectares), most of it in the north and north-west. On this land today is grown 80 per cent of Mexico's wheat and 95 per cent of its cotton.

Another factor in the Mexican agricultural revolution as in the American has been the provision for agricultural credit. Again, President Calls was a pioneer. His government in 1926 created the National Bank of Agricultural Credit to provide loans to farmers through local co-operative associations. Since the credit facilities offered by this bank were unsuited to the needs of the ejido farmers, Cardenas created the National Bank of Ejidal credit. Closely allied to the increase in productivity was the increased use of power-driven machinery and the consumption of fertilizer. "By the end of the 1950, mexico had more tractors than any other Latin American country. The use of fertilizers was five times as much in 1958 as it was eight years earlier.

8

Green Revolution in India

In the 20th century, massive public investments in modern scientific research for agriculture led to dramatic yield breakthroughs in the industrial countries. The story of English wheat is typical. It took nearly 1,000 years for wheat yields to increase from 0.5 to 2 metric tons per hectare, but only 40 years to climb from 2 to 6 metric tons per hectare. Modern plant breeding, improved agronomy, and the development of inorganic fertilizers and modern pesticides fueled these advances. Most industrial countries achieved sustained food surpluses by the second half of the 20th century, and eliminated the threat of starvation. These advances were much slower in reaching developing countries. The colonial powers invested little in the food production systems of these countries, and by independence, their populations were growing at historically high rates. By the mid-1960s, hunger and malnutrition were widespread, especially in Asia, which increasingly depended on food aid from rich countries. Back-to-back droughts in India during the mid-1960s made the already precarious situation worse, and a 1967 report of the U.S. President's Science Advisory Committee concluded that "the scale, severity and duration of the world food problem are so great that a massive, long-range, innovative effort unprecedented in human history will be required to master it."

In response, the Rockefeller and Ford foundations took the lead in establishing an international agricultural research system to help transfer and adapt scientific advances to the conditions in developing countries. The first investments were in research on rice and wheat, two of the most important food crops for developing countries. The breeding of improved varieties, combined with the expanded use of fertilizers, other chemical inputs, and irrigation, led to dramatic yield increases in Asia and Latin America, beginning in the late 1960s. In 1968, U.S. Agency for International Development (USAID) Administrator William S. Gaud coined the term

"Green Revolution" to describe this phenomenal growth in agriculture. To achieve higher yields for rice and wheat, scientists needed to develop plants that were more responsive to plant nutrients and that had shorter, stiffer straw to support the weight of heavier heads of grain. They also needed to develop varieties that could mature quicker and grow at any time of the year, thereby permitting farmers to grow more crops each year on the same land. New varieties also needed to be resistant to major pests and diseases, which flourish under intensive farming conditions, and to retain desirable cooking and consumption traits.

Borrowing from rice-breeding work undertaken in China, Japan, and Taiwan, the International Rice Research Institute (IRRI) in the Philippines developed semi-dwarf varieties that met most of these requirements. Similar achievements were made for wheat after Norman Borlaug (later awarded the Nobel Peace Prize for his work) crossed Japanese semi-dwarf varieties with Mexican wheats at what is now known as the International Center for Maize and Wheat Improvement (CIMMYT) in Mexico. Although the term Green Revolution originally described developments for rice and wheat, high-yielding varieties (HYVs) have since been developed for other major food crops important to developing countries, including sorghum, millet, maize, cassava, and beans. Moreover, a full-fledged system of international agricultural research centers now works on many aspects of developing-country agriculture (the Future Harvest Centers that make up the Consultative Group on International Agricultural Research).

Impacts on Agricultural Production

The adoption of HYVs occurred quickly. By 1970, about 20per cent of the wheat area and 30 per cent of the rice area in developing countries were planted to HYVs, and by 1990, the share had increased to about 70 per cent for both crops. Yields of rice and wheat virtually doubled. Higher yields and profitability also led farmers to increase the area of rice and wheat they grew at the expense of other crops. And with faster-growing varieties and irrigation, they grew more crops on their land each year. These changes more than doubled cereal production in Asia between 1970 and 1995, while population increased by 60 per cent. Instead of a widespread famine, cereal and calorie availability per person increased by nearly 30 per cent, and wheat and rice became cheaper.

Latin America experienced significant gains as well, but the impact in Sub-Saharan Africa was much more modest. Poor infrastructure, high transport costs, limited investment in irrigation, and pricing and marketing policies that penalized farmers made the Green Revolution technologies too expensive or inappropriate for much of Africa.

Social Impacts

The Green Revolution led to sizable increases in returns to land, and hence raised farmers' incomes. Moreover, with greater income to spend, new needs for farm inputs, and milling and marketing services, farm families led a general increase in demand for goods and services. This stimulated the rural nonfarm economy, which in turn grew and generated significant new income and employment of its own. Real per capita incomes almost doubled in Asia between 1970 and 1995, and poverty declined from nearly three out of every five Asians in 1975 to less than one in three by 1995.The absolute number of poor people fell from 1.15 billion in 1975 to 825 million in 1995 despite a 60 per cent increase in population. In India, the percentage of the rural population living below the poverty line fluctuated between 50 and 65 per cent before the mid-1960s but then declined steadily to about one-third of the rural population by 1993. Research studies show that much of this steady decline in poverty is attributable to agricultural growth and associated declines in food prices.

The Green Revolution also contributed to better nutrition by raising incomes and reducing prices, which permitted people to consume more calories and a more diversified diet. Big increases occurred in per capita consumption of vegetable oils, fruits, vegetables, and livestock products in Asia.

Problems with the Green Revolution

A revolution of this magnitude was bound to create some problems of its own. Critics charged that the Green Revolution resulted in environmental degradation and increased income inequality, inequitable asset distribution, and worsened absolute poverty. Some of these criticisms are valid and have been or still need to be addressed. But there is a tendency today to overstate the problems and to ignore the appropriate counterfactual situation: what would have been the magnitude of hunger and poverty without the yield increases of the Green Revolution and with the same population growth?

The Green Revolution in Asia stimulated a large body of empirical literature on how agricultural technological change affects poor farmers. Critics of the Green Revolution argued that owners of large farms were the main adopters of the new technologies because of their better access to irrigation water, fertilizers, seeds, and credit. Small farmers were either unaffected or harmed because the Green Revolution resulted in lower product prices, higher input prices, and efforts by landlords to increase rents or force tenants off the land. Critics also argued that the Green Revolution encouraged unnecessary mechanization, thereby pushing down rural wages and employment. Although a number of village and household studies conducted soon after the release of Green Revolution technologies lent some support to early critics, more recent evidence shows mixed outcomes. Small farmers did lag behind large farmers in adopting Green Revolution technologies, yet many of them eventually did so. Many of these small-farm adopters benefited from increased production, greater employment opportunities, and higher wages in the agricultural and nonfarm sectors. Moreover, most smallholders were able to keep their land and experienced significant increases in total production. In some cases, small farmers and landless laborers actually ended up gaining proportionally more income than larger farmers, resulting in a net improvement in the distribution of village income.

Development practitioners now have a better understanding of the conditions under which the Green Revolution and similar yield-enhancing technologies are likely to have equitable benefits among farmers. These conditions include: (1) a scale- neutral technology package that can be profitably adopted on farms of all sizes; (2) an equitable distribution of land with secure ownership or tenancy rights; (3) efficient input, credit, and product markets so that farms of all sizes have access to modern farm inputs and information and are able to receive similar prices for their products; and (4) policies that do not discriminate against small farms and landless laborers (for instance, no subsidies on mechanization and no scale biases in agricultural research and extension).These conditions are not easy to meet. Typically, governments must make a concerted effort to ensure that small farmers have fair access to land, knowledge, and modern inputs.

Another shortcoming of the Green Revolution was that it spread only in irrigated and high-potential rainfed areas, and many villages or regions without access to sufficient water were left out. Although evidence suggests that even in these cases villagers obtained important indirect benefits through increased employment and migration opportunities and cheaper food, the benefits were rarely sufficient to prevent further widening of income gaps. In India, for example, poverty in many low-potential rainfed areas has improved little even while irrigated and high-potential rainfed areas have progressed. Regional inequalities have worsened in China as well.

The Green Revolution has also been widely criticized for causing environmental damage. Excessive and inappropriate use of fertilizers and pesticides has polluted waterways, poisoned agricultural workers, and killed beneficial insects and other wildlife. Irrigation practices have led to salt build-up and eventual abandonment of some of the best farming lands. Groundwater levels are retreating in areas where more water is being pumped for irrigation than can be replenished by the rains. And heavy dependence on a few major cereal varieties has led to loss of biodiversity on farms. Some of these outcomes were inevitable as millions of largely illiterate farmers began to use modern inputs for the first time, but inadequate extension and training, an absence of effective regulation of water quality, and input pricing and subsidy policies that made modern inputs too cheap and encouraged excessive use also created negative environmental impacts. These problems are slowly being rectified without yield loss, and sometimes with yield increases, thanks to policy reforms and improved technologies and management practices, such as pest-resistant varieties, biological pest control, precision farming, and crop diversification.

Often ignored, however, is the positive impact of higher yields in saving huge areas of forest and other environmentally fragile lands that would otherwise have been needed for farming. In Asia cereal production doubled between 1970 and 1975, yet the total land area cultivated with cereals increased by only 4 per cent.

Overall, the Green Revolution was a major achievement for many developing countries and gave them an unprecedented level of national food security. It represented the successful adaptation and transfer of the same scientific revolution in agriculture that the industrial countries had already appropriated for themselves. The Green Revolution also lifted large numbers of poor people out of poverty and helped many nonpoor people avoid the poverty and hunger they would have experienced had the Green Revolution not occurred. The largest benefits to the poor were mostly indirect, in the form of lower food prices, increased migration opportunities, and greater employment in the rural nonfarm economy. The direct benefits to the poor through their own on-farm adoption, greater agricultural employment, and empowerment have been more mixed and depend heavily on local socioeconomic conditions. In many cases inequalities between regions and communities that adopted Green Revolution technologies and those that did not also worsened. At the same time, the Green Revolution had many negative environmental impacts that have still to be adequately redressed.

Agricultural research remains a potent force for good in the developing world and is the key to increasing yields further to meet the continuing growth of food needs in developing countries. This need is especially urgent in Sub-Saharan Africa, which has yet to experience an agricultural revolution of its own. But simply adding to the pile of food will not be enough. The indirect benefits for the poor are likely to be weaker in the future as globalization and trade make food prices less responsive to local production and as agriculture becomes less important to the livelihoods of the poor. Policymakers will need to target the poor more precisely to ensure that poor people receive greater direct benefits from new technologies. New technologies will also need to be more environmentally sustainable. By building on the strengths of the Green Revolution while seeking to avoid its weaknesses, scientists and policymakers can take significant steps toward achieving sustainable food security for all the world's people.

New Green Revolution: Producer Cos Help Farmers Reap Profits

Farmers are joining India Inc in mind, body and spirit. In a quiet revolution underway across the countryside, growers are setting up companies, replete with balance sheets, professional CEOs, board of directors, and income tax returns. By pooling together the land and produce of their shareholders, these companies are signing lucrative deals with large retail chains, food companies and exporters keen to establish reliable supply chains. As many as 200 companies have been formed by farmers from different parts of the country and another 100 are likely to be set up by March next year. With an average 1,000 members, more than 2 lakh farmers are now shareholders in what are known as 'producer companies' under the Companies Act. "It was a job offer like any other in the market," said Sanjeev Kumar Gupta, managing director, Devbhumi Natural Products Ltd., based in Uttrakhand, with 2,200 shareholders. The company exports certified organic honey and also supplies silk and other produce to Fab India, the clothing and home furnishing chain.

"Agri-business enterprises are increasingly looking for direct tie-up with farmers to source agricultural produce. Producer companies are an important solution for aggregation and quality management," said Bhaskar Reddy, director, FICCI. "The trigger was an amendment to the Companies Act in 2004 to allow producer companies," said Pravesh Sharma, Managing Director, Small Farmers Agri-Business Consortium, under the ministry of agriculture, which has given seed capital to 400 producer groups and helps negotiate bank finance. But registrations gathered steam only recently when farmers realized the value of economies of scale and collective negotiation in a more market-driven agriculture. Most companies are located in Maharashtra, Karnataka, Andhra Pradesh and Tamil Nadu.

"Organized retail is doubling its share every three years or so and is likely to play an increasingly important role in influencing the nature of agricultural markets in the coming decade. Farmer-owned companies demonstrate the benefits of aggregation of production base and produce at a time when 83 per cent of the land holdings are owned by small and marginal farmers". Producer companies offer farmers several benefits. They allow growers to formulate a collective production and marketing strategy in line with customer requirement, negotiate better terms with buyers, buy inputs such as seeds and chemicals in bulk, and receive technical guidance at their doorstep. Some companies also undertake premium distribution and help in receiving subsidy from agriculture departments. "In rural areas where producers are dispersed and scattered and communications are not optimal, the importance of these organizations is even greater," said VV Sadamate, advisor, Planning Commission. An added attraction of the corporate structure is that it protects farmers from the influence of local politicians and the state government, which has been the bane of farmer cooperatives.

"The law does not allow producer companies to get non-producer shareholders or members. This means private equity funds, other corporates or politicians cannot acquire stake or hold a post," said Sharma. Aslew of NGOs are acting as sponsor institutions and consultants for these companies. From identifying the opportunity, mobilising potential farmer shareholders, exploring the market potential in the product value chain, writing the business plan, and getting the company registered, appointing staff, to filing annual returns with the registrar of companies, these consultants are helping farmers learn the ropes, said N Madhu Murthy, project director at Hyderabad-based Access Livelihoods Consulting India.

With barely a dozen consultants operating in this sector, demand for their services is growing leaps and bounds. "Our client base has grown from 35 to 150 in last three years. Now we are adding 10 new clients a year," he added. Farmer companies are dealing in a wide variety of crops, including cereals, horticultural produce, spices, oilseeds and pulses. Madhya Pradesh alone has 17 producer companies, of which 15 are crop based, one is engaged in milk production and one in poultry farming. With 41,560 shareholders, they have a combined annual turnover of Rs. 38 crore. "Some of the large companies have annual turnover of almost Rs. 20 crore," said Murthy.

In most producer-owned companies, 10 farmers, usually active members of the community, are appointed as chief promoters. The company floats shares that are picked up by other farmers. These shares can be transferred but can't be listed on the stock exchange. "Our company shares were priced at Rs. 10 and at least 400 shareholders bought 100 shares each. That is no small thing, given that they are marginal farmers living off subsistence agriculture," said Gupta. Like the rest of India Inc, access to credit and working capital is the chief hurdle in the growth of producer companies.

Commercial banks are reluctant to lend without substantial collateral, which becomes a challenge for companies that don't have factories and other large physical assets. "Sometimes the chief promoters offer their land as collateral," said Murthy.

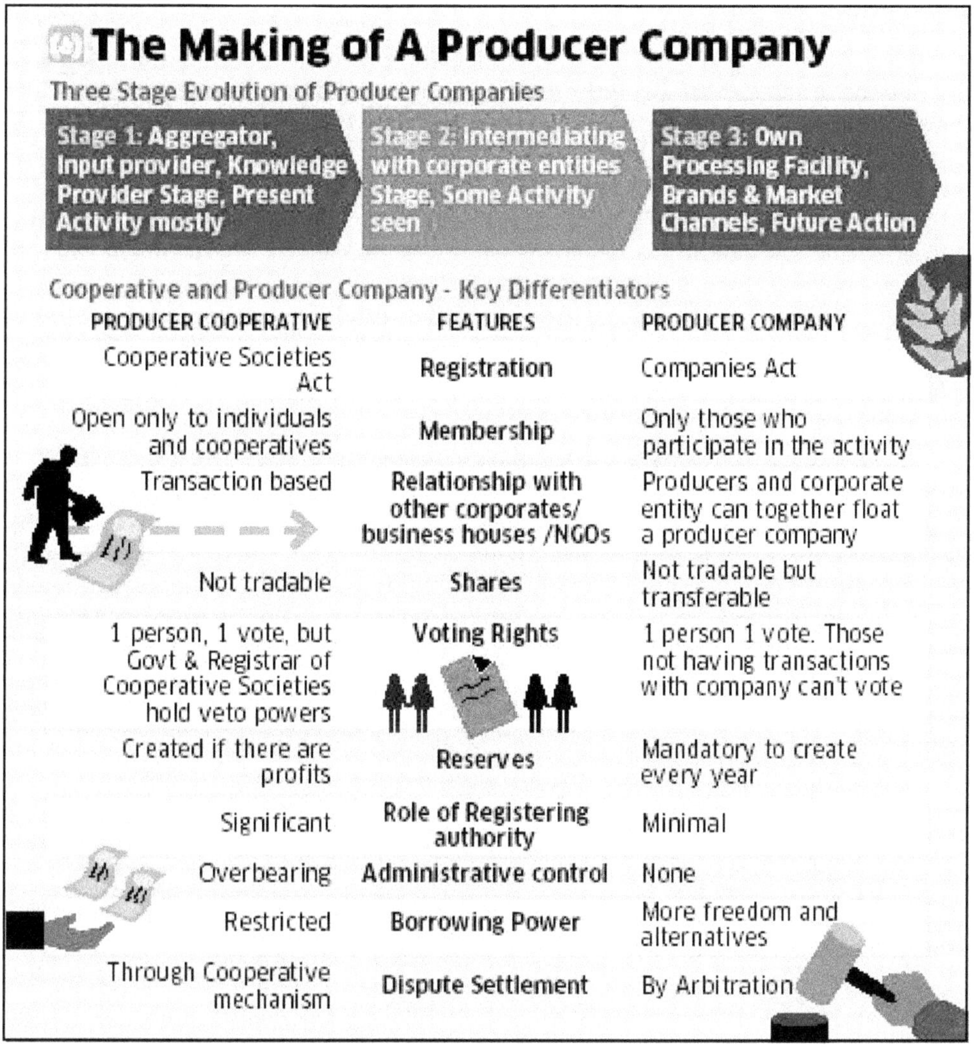

The Making of A Producer Company

Three Stage Evolution of Producer Companies

Stage 1: Aggregator, Input provider, Knowledge Provider Stage, Present Activity mostly → Stage 2: Intermediating with corporate entities Stage, Some Activity seen → Stage 3: Own Processing Facility, Brands & Market Channels, Future Action

Cooperative and Producer Company - Key Differentiators

PRODUCER COOPERATIVE	FEATURES	PRODUCER COMPANY
Cooperative Societies Act	Registration	Companies Act
Open only to individuals and cooperatives	Membership	Only those who participate in the activity
Transaction based	Relationship with other corporates/ business houses /NGOs	Producers and corporate entity can together float a producer company
Not tradable	Shares	Not tradable but transferable
1 person, 1 vote, but Govt & Registrar of Cooperative Societies hold veto powers	Voting Rights	1 person 1 vote. Those not having transactions with company can't vote
Created if there are profits	Reserves	Mandatory to create every year
Significant	Role of Registering authority	Minimal
Overbearing	Administrative control	None
Restricted	Borrowing Power	More freedom and alternatives
Through Cooperative mechanism	Dispute Settlement	By Arbitration

With few options to raise money except through share capital, it becomes tough to scale up. "We have so many plans for expanding our supply chain, brand our products, and attract good professionals to our company. But it needs money," says Gupta. Luckily, the government is now more than willing to lend a hand "We spend Rs. 2,000 per farmer over two years," said SFAC's Sharma. SFAC has now tied up with Nabard for loans to producer companies.

The Coming Crisis

Rain-dependent areas, which account for 60 per cent of India's agricultural land, are in the throes of a growing manmade crisis. Fix it and India's rural poverty problem will be sizeably addressed. Ignore it and face severe farmer distress and food shortages.

The Green Revolution

This dilemma is acutest in regions that don't receive adequate rainfall. Rain-dependent areas can be broadly split into two: 'drylands', which receive less than 750 mm of rain a year; and rainfed areas, which receive more than 750 mm. Comprising arid and semi-arid ecosystems, drylands stretch from Gujarat in the west till Eastern Madhya Pradesh; and from Rajasthan till the southern tip of India. To reduce their vulnerability to rains, farmers here used to grow jowar, bajra and pulses. While lowyielding, such crops were less affected by variations in rainfall. It was a 'safe' existence. Farmers invested little in their fields. Why invest if returns are uncertain? In the same field, they planted multiple crops. For instance, Jowar or pulses, both drought-resistant, would be planted alongside wheat, which gave high yields in normal rains. They also maintained livestock or, if forests were in the vicinity, gathered minor forest produce.

The green revolution came in the sixties. Tasked with ensuring food security, it pushed high-yielding varieties (HYVs) of wheat and rice over jowar, bajra *et al.*, It began in the floodplains of the north. Where, as canals came up, farmers, realising rainfall risk was a thing of the past, switched to HYVs. In the drylands, the story evolved differently. The green revolution came here in bits and pieces. The seeds and fertilisers reached. So did the exhortations to farmers to adopt 'modern' farming. What did not reach was water. Predictable water supply is something the farmers created for themselves. When electricity came, they invested in groundwater pumps. What followed was transformative. In Malwa (MP), for instance, till the early-1970s, farmers grew jowar during the rains and Malwi Ghehu, a local wheat variety, after that. Once the pumps came in, farming became a yearlong activity. Cash crops like soya displaced jowar. HYVs of wheat displaced Malwi Ghehu. Below the Malwa plateau, the same set of changes played out more recently, as groundwater pumps came just eight years ago.

This is the story across India. Groundwater (tube wells) has been the mainstay of addition to irrigation resources. According to a Planning Commission report, titled 'Synopsis of Groundwater Resources in India', by Himanshu Kulkarni, P.S. Vijay Shankar and Sunderajan Krishnan, in 1960-61, canals and tanks accounted for 61 per cent of non-rain water for irrigation, compared to 0.6 per cent for tube wells. In 2002-03, the share of canals and tanks was down to 33 per cent, while tube wells had increased to 39 per cent. "Of the addition to irrigated area of 25.7 million hectares between 1970 and 2000, groundwater accounted for over 85 per cent. At present, about 61 per cent of the irrigation is from groundwater," says the report. That story is now beginning to wind down, jeopardising the shift made by farmers in dryland areas and raising serious livelihood and food-security issues.

Is Winding Down

In parts of Malwa, for example, groundwater levels have plunged from 50 ft in the 1970s to 700 ft now. The Planning Commission study, which is an unnerving read, says the crisis is acute in six states. As per the study, if the ratio of groundwater extraction to groundwater recharge is less than 70 per cent, it is considered safe; 70-90 per cent, semicritical; 90-100 per cent, critical; and more than 100 per cent, overexploited. Between 1995 and 2004, the proportion of districts in semi-critical, critical and over-exploited has grown from 5 per cent of the agricultural area and 7 per cent of the population to 33 per cent and 35 per cent respectively. The report says that the "level of groundwater extraction is unsustainable" in Punjab, Rajasthan and Haryana; and that Tamil Nadu, Gujarat and UP are "fast approaching that stage". These six states accounted for half the foodgrain production in 2008-09. There are other problems. Soils are weakening. Badhia Naval Singh farms 8 bighas at Mansinghpura village in Bagli tehsil. Earlier, he says, if he pulled out a tuft of grass, he would see earthworms in the soil. "Dikhaye dena ab band ho gaye hain (They do not show up anymore)," he says. Imbalanced fertiliser use is skewing the chemical composition of soils, which are hardening and retaining less moisture. Says the 11th five-year plan document: "Nearly two-thirds of our farmlands are either degraded or sick."

Rainfall itself is becoming more and more unpredictable. While talking to Shagan Sanaf Chotu, a larger farmer in Mansinghpura, it suddenly starts raining, quite heavily. It is the middle of May. "But it never rains in May," says Chotu. Even as all this changes, hedging across livelihoods is less feasible now. The green revolution increased yields partly by shortening plants' height. This reduced fodder for livestock. Encroachments into village commons have reduced grazing spaces for cattle. As forests retreated, villagers traditionally dependent on minor forest produce have one less fallback option. In all, farmers have no option but to invest in agriculture to keep their heads above water – even though it now yields diminishing returns. Two years ago, Singh borrowed. 1 lakh for a tube well. Since the local bank doesn't lend for tube wells, he borrowed from fellow villagers, at 36 per cent interest. "It will take us five years to repay this loan," his wife says acerbically. "Till then, we will live on mirchi roti." What will exacerbate matters is climate change, whose worst sufferers are expected to be India's drylands and sub-Saharan Africa. The question is: how does one equip dryland farmers to deal with heightened weather and rainfall risk without pushing them back to the low risk, low output subsistence trap?

Vacuum in Research

Badhia says the HYVs are "nazuk" (delicate). They give a good yield only if all the conditions are favourable. He says older varieties were more resilient but also that he can't go back to them. "Mandis want cotton with long staples – about 32-38 mm," he says. "The old varieties were 14-15 mm." B Venkateswarlu, director of the Central Research Institute for Dryland Areas (CRIDA), says short-stapled cotton was used to spin dhotis, but this is the age of readymades. "The spinning mills have shifted to long-staple cotton, which is easy to spin and more durable. The farmer has no choice." That is one problem. Farmers' need for cash, with the need to buy farm inputs, send kids to school, etc, is greater than before. But the market wants attributes that traditional crops lack. Which raises another obvious question: why aren't better varieties available? There hasn't been enough research. The central body for coordinating agricultural research in India is the Indian Council for Agricultural Research (ICAR). For the longest time, most of its research was on rice and wheat – the focus of the green revolution. Yields of rice and wheat increased even as those of coarse cereals stagnated. "In Punjab, Haryana and UP, R&D into two crops will take care of 80 per cent of the land and farmers," says Ramesh Chand, director, National Centre for Agricultural Economics and Policy (NCAP). "In dryland areas, you need R&D in 10-12 crops to reach out to a similar proportion of farmers and land."

Only in the nineties did yields of jowar, bajra and maize begin to climb again. But, this has not been accompanied by consumer demand. Diets have changed, partly due to affluence, partly due to the PDS changing people's tastes. Pulses and oilseeds, stapes of dryland agriculture, continue to stagnate. Worse, state agricultural universities (SAUs), which are best placed to undertake locally relevant research, face a cash crunch. "ICAR has 4,000 scientists, the SAUs have 20,000," says Chand. "But the money that comes in goes entirely into salaries. Money for research has almost vanished." Says the 11th Plan: "SAUs are so poorly funded by their state governments that some are in chronic overdraft, only a quarter are accredited, and almost all rely on ICAR for research funding."

Green Revolution Reloaded

Increasing number of agri-entrepreneurs in Punjab are helping farmers become job creators from job seekers.

Drifting away from traditional farming of paddy and wheat never came that easy for the agri-entrepreneurs of Punjab. There were few takers of their ideas when they set off on a new journey. Not only providing employment to hundreds of youth but also inspiring others to become 'job providers' from 'job seekers'. These out of the box thinkers are heralding a second wave of green revolution in Punjab.

On his visits to his native village Latala in Punjab Jassi Khangura noticed that many people were leaving for cities to seek employment, because opportunities for work in villages were limited even for those who owned land. He also noted that the social equations in the villages were heavily tilted towards men. And, women from the poorest sectors had no real opportunity at all. These were the issues he set out to address by forming the Macro Dairy Ventures (MDVL). A social enterprise dairy project, MDVL is located in villages near Ludhiana, providing highest quality community dairy infrastructure across villages and partners women entrepreneurs and contract farmers around its farms. Khangura, who is also the sitting MLA of Qila Raipur, designed and implemented the project to provide socially acceptable vocational opportunity for people from the poorest rural communities.

With an initial investment of around 100 crore the project started in 2008 with the first community dairy unit at village Phallewal and the central state-of-the-art processing unit at Majara. Khangura's MDVL project also provides market linkage for women entrepreneurs through its brand TruMilk, a highend 100 per cent pure cows' milk product, across Punjab and NCR. With a turnover in 2010 of around 50 crore, "It is amazing to see a positive impact of the project on the lives of the rural poor. When I see these women entrepreneurs ensuring their children are educated and are able to pay for medicines because of their partnership with MDVL, it's definitely all worthwhile. We currently have almost 1,000 such women. And, soon this figure will catapult to 3,200," Khangura said.

Before becoming an aquaculturist, Rajvinder Pal Singh Rana, 38, from Mullanpur in Ludhiana, served in the Indian army and later worked for a MNC. He got into the fisheries business in 2001, without any financial help from any institution and started with only 2.5 acre on leased land. Rana now employs 14 people who remain at the farm round the clock. How did he do it? "I take training from various agencies on fish breeding, business development, FCR ration, fish processing and fish and water management," Rana said, who is also working on 'fish skin' and have applied for a patent. Apart from manufacturing a Tag machine which he sells at 250, Rana also produces fish oil at comparatively lower price than others in the market. Vikram Aditya Ahuja, 44, is currently an executive director of Zamindara Farmsolutions (ZFS), a Fazilka-based farm equipment bank. Vikram's philosophy is simple; a second green revolution will happen when farmers' debt burden is eased.

ZFS helps small and marginal farmers improve the economics and sustainability of their work. It offers a combination of methods–farm equipment on pay per use basis, extension services, educating farmers etc to achieve their goals. The debt burden in Punjab outstrips the national average. With nearly 13 lakh farmers, Punjab has a per capita debt of 50,000 compared with the national figure of 15,000. On March 31, 2009, Punjab-based farmers owed banks a whopping 35,000 crore. Each acre in Punjab is under a debt of 15,000. Farmer suicides have been on the rise throughout India from 1997. "Farmers in North India have traditionally taken expensive bank loans to buy expensive farm equipment (tractors). This equipment have been by and large underutilised due to the small size of the farm holdings (4.2 million hectares are cultivated by 1.5 million families) and the seasonality of use," Ahuja of ZFS said.

Punjab grows rice on 70 lakh acre (out of 4.2 million hectares). Farmers burn nearly 140 lakh tonne of crop residue (rice straw or naad). This causes damage to the environment, affects public health, dries up soil moisture, burns down the good top layer of soil and kills friendly insects. "Farmers follow this practice (of burning straw) due to lack of awareness, nonavailability of new technologies. We feel that this threat can become a major opportunity by creating awareness and making relevant machines available on affordable rates," Ahuja said. ZFS is associated with 6,000 farmer families and has trained 250 village youths in farm mechanisation. It also trained more than 4,000 farmers through campaign 'Earn–Don't burn'. It claims to have dissuaded 800 families from taking expensive farm loans (approx 30 crore) to buy farm machinery by offering machines on 'Pay for use' basis. Saving an yearly interest of 300 lakh in total.

A large number of younger farmers from the region are exploring options in Africa nations like Ethiopia. Recently world's largest grower of roses, Karuturi Global Ltd invited the farmers from Punjab to invest in Ethiopia. Karuturi Global Ltd has evolved a revenue-sharing model for Indian farmers to share their expertise and lead the world towards self-sufficiency in food, said Sai Ramakrishna Karuturi, who is the founder and managing director of Karuturi Global. Puneet Singh, a progressive young farmer from Chandigarh region, has signed a contract with Karuturi. "Our expertise would add value there business," Singh said. "It gives Indian farmers an opportunity to enjoy the economies of scale that comes along with large scale commercial farming seen in the United States and Australia."

Some Comments

In the 1960s, the Green Revolution allowed less developed countries, such as India, to overcome chronic food deficits. Basically, the Green Revolution stands for producing more food and other agricultural products from less land. Modernization is one of the main concepts in the Green Revolution. The practices were made up of using high-yielding varieties of seeds, modifying farm equipment, and substantially increasing chemical fertilizers. This allowed growth and sustainability. At the beginning of the Green Revolution, there was a large growth in Indian agriculture however, instability arose and the Green Revolution was on a rapid decline. In the end, it caused a shortage of water. When water is the primary source of survival, life seems difficult when there is a large shortage of the one thing that can make assurance of life.

Before the Green Revolution was introduced prior to the 1960s, farmer's main goal was to produce wheat and rice. These varieties had a low yield per hectare, which means that these crops took one year to produce and in order for farmers to increase production, there would have to be a change. The change would have to consist of irrigation facilities, fertilizers, and pesticides. In order for these changes to work properly, there would have to be a sufficient quantity of water and fertilizers. At the beginning, many farmers thought if they could double their production of crops in one season, that they would do whatever they could do increase crop production. Improving high yielding varieties of wheat was a major factor, which finally led to the Green Revolution.

The introduction of high-yielding varieties of seeds and the increased use of chemical fertilizers provided the agriculture industry in Indian increase in production. The Green Revolution was thought to pave the way for rapid industrial growth, but in the end it did exactly the opposite. It created a shortage. At the time, when the Green Revolution first began, it was considered one of the most significant technological achievements in the agricultural industry. The Green Revolution dramatically increased global food production over the next two decades, particularly in India.

Green Revolution Leader

Mankombu Sambasivan Swaminathan is an Indian genetics and international administrator who took the leading role in India's Green Revolution. His main goal was to eliminate world hunger and poverty, especially in India by using environmentally sustainable agriculture, sustainable food security and the preservation of biodiversity. He called the preservation of biodiversity the evergreen revolution. Swaminathan was apart of the Indian Council of Agricultural Research and he was the administrator of Agriculture from 1979-1980, which is when his early work was started in the involvement of the Green Revolution. M.S. Swaminathan was hailed as the father of the Green Revolution in India.

Impacts

The increase production of wheat fueled a self-sufficiency of food for India. The high yielding seeds and irrigation facilities, brought enthusiasm to many farmers in India. By seeing better profits from the Green Revolution, the farmers began to enjoy life more with better earnings, knowing that they had the ability to provide for their family. Unfortunately, with the rise in use of chemical pesticides and fertilizers, there were many negative effects on the soil and land. Despite the early benefits, it became apparent that there were many negative impacts from the green revolution. After the Green Revolution began in India, there was a change in the land use patterns, known as the degradation of land. Whereas primarily there were only one crop planted per year, before the Green Revolution. There were two to three crop rotations every year, the land quality diminished and the land quality had suffered. Due to the impute of heavy chemical fertilizers, a strain on the carbon material within the soils were created. There also has been a loss in bio diversity in farm lands because since there has been an increase in chemical pesticides and fertilizers, many insects have been killed and the birds that create homes in rural areas have found new areas to live because of the negative side effects that the pesticides give off to the environment.

As a result of chemical use in the land, contamination of ground water affected the health of the people who are consuming the agricultural goods that now contained pesticides and chemicals. This directly affected the health of Indians, who were not used to putting such chemicals in their bodies.

It was not only the environmental and health problems that arose; there were also many social problems that occurred in the late 20th century because of the green revolution. Since the farming industry was booming and farmers had higher incomes, there was an increase demand for more land which created more demand for farm families who took on more farming land that they could manage.

Critics believe that the Green Revolution resulted in environmental degradation, increased income inequality, inequitable asset distribution, and worsened poverty levels in India. The majority of the large farmers were able to adapt to the new technologies because they had better irrigation, fertilizers, and seeds. However, the Green Revolution affected just as many of the smaller farmers during the Green Revolution. It is believed that the Green Revolution encouraged mechanization, which pushed down rural wages and employment, and increasingly impoverished small farmers. When the Green Revolution occurred, there was a spread only in irrigated and high-potential rainfed areas, which created many villages without sufficient water. Even though there were more employment opportunities and cheaper food, not having a sufficient amount of water in a village is crucial for survival. There was some retreating of water from natural watersheds; this was replenishing water, which was pumped from areas that can be quickly replenished by the rainwater, however, there was not a sufficient amount of water for survival.

Double Cropping

Double cropping was a primary feature of the Green Revolution. The idea was for farmers to have two crop seasons within one year. This meant that the crop production would double within one year. This was based primarily on the natural one monsoon per calendar year. For farmers to have to double the crops per season there would have to be one artificial monsoon. These were created from a large irrigation facility. Dams were built in rural areas to collect large volumes of monsoon rainwater. Before the Green Revolution, this water was wasted. This was a simple irrigation technique that rural farmers adopted. However, in a year without monsoon, this led to failure.

Consequences, Damage and Results

The consequences of the Green Revolution to the rural farming areas, dating significantly from the 'year of the drought'. There were flaws in the technological achievements, which created this drought in India. The new technologies relied on the monsoon seasons every year for the water supply in reassuring to keep up with increased land use and production. When the monsoon failed in 1972-73, the 'year of drought' occurred; this had a major effect on the Green Revolution. There was much damage caused by the Green Revolution. There was an excessive amount of fertilizers and pesticides used which polluted waterways and poisoned agricultural workers. The beneficial wildlife and insects that lived in the farming areas were also killed which made the land not as healthy for farmers to work with.

There are four main economic results of the Green Revolution. The crops were under high-yield varieties and needed more water, fertilizers, pesticides, and other chemicals. This stimulated growth of the local manufacturing sector. For example, the industrial growth created new jobs and contributed to the country's GCP. There was an increase in irrigation, which created a need for new dams to collect the monsoon water. The water was stored and used to create hydro-electrical power. This boosted the industrial growth, created new jobs, and improved the quality of life within the people living in the villages. India paid back all of their loans from the World Bank to create the dams in rural areas. By an outside view, this looks good on behalf of India for paying back such a large loan and it looks good in the eyes of other lending agencies.

Economy

The irrigation and growth of wheat and rice are very important to the economy and production of livelihood in India. These are the main products that most families can afford and live off on a daily basis. Both wheat and rice varieties require carefully controlled irrigation. About 75 per cent of cropped land is devoted to food grains such as rice, wheat, maize, and barley. Rice and wheat contribute to approximately 70 per cent -90 per cent of the food requirements for the people of India. Because of the high yielding varieties, the soils are too dry when the monsoon season fails when the agricultural industry needs an adequate amount of water for the rice and wheat to grow properly, but that water is not available, it poses a great problem to the success of the Green Revolution. Countries involved in the Green Revolution have subsidies in part with the prices of the fertilizers, pesticides and the production of wheat and rice. At the beginning, there were no problems with the Green Revolution until the farmers and government started to see problems arising.

The sudden change in agricultural techniques developed caused a rapid change creating economic imbalance among farmers, which contributed to large interregional agricultural disparities. It has been brought to the attention of many farmers that crop productivity needs to increase in order to keep up with the competitive agricultural market. In order to survive the market, may need to increase productivity in order to keep an income coming in for a family. This requires new technology being brought into their farms to succeed in the agricultural industry in India, and yet not all farms can afford this new technology. There has been an ongoing debate in India surrounding the survival of farmers: whether to increase agricultural productivity or improve food quality. Ever since the Agreement on Agriculture was created in 1999 from the World Trade Organization this has been a topic surrounding not only farmers but this problem affects mostly everyone living in India. This idea was created to replace agriculture price support with direct payments to farmers from production dates.

Protests

One would think that protests to get peoples attention about the growing problem from the green revolution would not be wasting food, however, that is not the case. In the district in Andhra Pradesh, farmers had dumped cartloads of tomatoes on the streets to express their struggles as farmers. With the price of tomatoes increasing for a normal family, it was almost impossible for farmers to keep producing these goods and selling them at a reasonable price. In another instance near Punjab, potato farmers demonstrated their negative views towards the Green Revolution by throwing potatoes onto highways to get people in the community to start talking about the agricultural problem arising. It is believed that the productivity will bring more income to farmers, which will no doubly be true, however, the factors of the environmental issues because of the over productivity does not come into place.

Farmer Suicide

Suicide has become common in rural farms on the outskirts of major Indian cities. The pressure farmers have to produce a certain amount of crops yearly puts many stresses on top of their busy lives. Farmers who cannot produce a certain amount of crops, they believe that the stress about not being able to provide for their families, which they believe life not worth living any longer. In 2002, there were over 1000 suicides reported from 12 districts of Maharashtra, India. The rising amount of debt that farmers collect from having to purchase fertilizers and pesticides puts a large strain on themselves. Within the process of producing the second round of crops within one year, when the second round does not produce as quickly or they do not have the same quality, farmers get very distressed that they will not be able to sell these crops. The government in India has been in denial about the number of suicides that have occurred since 1987, which has arisen to over 10,000 suicides committed by farmers. This is within the same time that the Green Revolution began to tumble and not be as successful.

Food Security

At this time in India, during the food crisis, farmers cannot pay the wages. There is a shortage of farm laborers because of the higher prices that the farmers need to pay the laborers. The increasing price from seeds, fertilizers and water, cause the wages to increase and it will make growing food more expensive. Small farmers cannot afford this especially since they need to provide for their family.

After over forty years since the green revolution, the soil and water in India has been poisoned and infected with pesticides it has caused a major problems to the food security in India. The World Bank supports the green revolution because they wanted to focus on the environment, sustain economic growth, which was not possible without sustainable environment. India borrowed money from the World Bank in support of the green revolution. In this end, we see that the Green Revolution is not enough. If the population keeps growing in the developing world, it is gong to be close to impossible for everyone to get enough nutrients and food for a country such as India. The cost of production is becoming very expensive and as we can see with the Green Revolution, when putting growth hormones in the agricultural industry, there are more political and environmental problems that arise.

Spread of Green Revolution: Role of Policies, Technologies and Institutions

The GR was seen as a means of achieving selfsufficiency in food production, infusing dynamism to the economy and propelling rural transformation. All five countries made significant investments in research and development (R&D), extension, technology and infrastructure development to achieve selfsufficiency in the production of food crops, mainly rice, wheat and maize. They also provided substantial input subsidies (for fertilizers, irrigation and energy) and credit to enable farmers to make the most of innovative practices in farming. Farmers were quickly able to utilize technologies developed at international agriculture research centres such as the International Rice Research Institute (IRRI) in the Philippines, the International Maize and Wheat Research Centre (CIMMYT) in Mexico and the AVRDC: The World Vegetable Centre in Taiwan. A closer look at the policies as well as technological and institutional interventions in the five countries reveals some striking variations. Bangladesh and India chose a deliberate strategy of intensive agriculture. Bangladesh, which now has one of the most mechanized and labourintensive agricultural sectors in south Asia, experienced significant changes, particularly in rice cultivation with the widespread adoption of modern rice varieties, mechanization of tilling and intensive use of inputs. These changes enabled it to achieve food security to a large extent, though at the cost of a diminution of its genetic resources. Through the GR, India eradicated famines and persistent food shortages and attained food selfsufficiency in a short span of 10 to 15 years. Thailand's agriculture underwent significant developments in the years following the First National Economic and Social Development Plan in 1961. Substantial investments were made in infrastructure, technology, irrigation, research and extension, besides establishing rural credit support and institutional systems for facilitating market incentives and trade flows. The adoption of GR technologies, including high yielding variety (HYV) seeds, significantly boosted the production of irrigated rice in the central plains and maize in the rainfed uplands.

Vietnam initiated economic reforms (Doi Moi) in 1986 with the goal of creating a "socialistoriented market economy". The major reform in the agricultural sector was Resolution No 10 of the Politburo (1988), which abolished the collective farming sys tem and allocated land to individual households on a longterm basis. The price reforms of 1989 liberalised all conventional price controls, including interest and foreign exchange rates. The new exchange rate policy dramatically altered the income of farm households by giving them full control over production and free access to markets. The reforms also greatly contributed to improving the terms of trade in favour of agriculture.

In sharp contrast, development policies in Korea were guided by rapid growth through industrial expansion and urbanisation, which were considered essential for alleviating poverty and setting right the socioeconomic disruptions caused by the Korean War (195053). While pursuing such policies, Korea heavily taxed its agriculture sector by maintaining low grain prices (using PL 480 imports from the US) during the 1950s and 1960s. However, there was a reversal in this policy in the 1970s and Korea began subsidising agriculture to achieve selfsufficiency. To support its large number of small and poor farmers, productivity enhancing policies and infrastructure improvement programmes were implemented, which included enlarging farms, improving drainage and developing water resources. Korean agriculture greatly changed after the late 1980s when import barriers were removed under strong international pressure, forcing it to be sensitive to changing global markets. Since joining the World Trade Organisation (WTO) in 1995, Korea has made efforts to further strengthen its agricultural sector while sticking to its WTO commitments. Korea has also introduced direct payment programmes, though this is not in full conformity with the Green Box measures of the Agreement on Agriculture (AoA).

Food Self-Sufficiency and Commercialisation of Agriculture

The GR significantly increased the production of foodgrains in all the five countries, as is evident from Table 8.1. All the countries, barring Korea, achieved a consistent increase in food production, as indicated by the average annual production in three sub periods. In Korea, annual average food production considerably declined from 8.98 million tonnes during 197690 to 7.21 million tonnes during 19912009.

Spectacular growth in the production of wheat and rice made India almost self sufficient in foodgrains in less than a decade of the GR. Vietnam attained food self sufficiency by 1989 and has been a leading rice exporter ever since. Korea witnessed a deceleration and even negative growth in food production during the entire period. Average annual per capita food production in Korea declined from 249 kg during 196175 to 228 kg during 197690 and further to 156 kg during 19912008, mainly due to policies that favoured the conversion of land under staple crops (paddy and barley) to highvalue crops (vegetables and fruits). Though Bangladesh experienced a drop in per capita food production from 247 kg (196175) to 232 kg (197690), the situation improved with an increase to 252 kg during 19912008. Though India's trend growth rates show a significant growth in food production, the growth in percapita food production was negative (0.41 per cent) during 19912008, indicating worsening food availability.

Table 8.1: Trends in Production of Foodgrains (1961 to 2009)

Period	*Bangladesh*	*India*	*Korea*	*Vietnam*	*Thailand*
			Average annual production (million tonnes)		
1961-75	16.27	100.85	7.59	10.02	14.31
1976-90	22.80	156.57	8.98	14.83	21.74
1991-2009	36.17	228.46	7.21	33.02	29.91
			Average annual simple growth in food production (per cent)*		
1961-75	2.54 (0.24)	3.18 (2.80)	2.57 (0.16)	1.15 (0.12)	4.58 (0.44)
1976-90	2.57 (0.61)	3.24 (4.77)	0.38 (-0.07)	4.62 (0.65)	1.71 (0.50)
1991-2009	3.21 (1.25)	1.49 (3.12)	–0.44 (-0.04)	4.22 (1.29)	2.90 (0.75)
			Average annual per capita food production (kg)*		
1961-75	247 (-2.59)	192 (0.86)	249 (-0.20)	244 (-2.47)	405 (0.98)
1976-90	232 (0.39)	212 (1.86)	228 (-4.89)	257 (5.99)	432 (1.80)
1991-2008	252 (4.12)	226 (-0.41)	156 (-2.25)	415 (11.48)	468 (7.50)

* Figures in parentheses indicate linear trends growth rates.

Source: Estimated from FAOSTAT (www.faostat.org).

EPW Issue No.4, 2012.

Much of the contribution to the growth in food production came from a single crop, rice, which also was the major benefactor of the GR in all the countries, except India. The contribution of rice to total food production was as high as 96 per cent in Korea, 95 per cent in Bangladesh, 89 per cent in Vietnam and 88 per cent in Thailand. In India, rice contributed only 56 per cent to total foodgrains production, followed by wheat at 30 per cent. The yield impact of GR technologies was quite significant in the case of rice. Among the five countries, Korea achieved the highest yield levels of 4.0 to 4.8 tonnes per ha over 1961 to 2008. In Vietnam, the rice yield varied between 1.92 and 4.8 tonnes per ha and in Bangladesh, it varied from 1.68 to 3.71 tonnes per ha during the same period. India (1.48 to 3.1 tonnes per ha) and Thailand (1.78 to 2.82 tonnes per ha) had relatively lower levels of rice productivity.

Besides benefiting from the GR, the five countries also made healthy investments in R&D, infrastructure development and extension programmes to increase the production of commercial crops for exports. Vietnam and India were growing tropical cash crops such as rubber, tea and coffee even earlier to boost export earnings and the GR period coincided with the dynamic growth of commercial agriculture in these countries. Nevertheless, the promotion of cash crops did not result in large scale diversification of agriculture in Bangladesh and Vietnam with food crops (mainly rice and wheat) continuing to dominate the gross cropped area in these countries (86 per cent and 72 per cent, respectively). But India, Korea and

Thailand were more successful in crop diversification and the area under nonfood/commercial crops was about 44 per cent to 47 per cent of the gross cropped area in 2007. Apart from Bangladesh, the growth in commercial agriculture enabled the other countries to gain significantly from increased exports over a period of time, as Table 8.2 shows (see also Figure 8.1).

Table 8.2: Trends in Agricultural Exports, 1989 to 2009 (in million dollars)

Period	*Bangladesh*	*India*	*Korea*	*Thailand*	*Vietnam*
1989-91	157	2,843	1,125	5,760	667
1992-94	122	3,181	1,204	6,594	976
1995-97	130	5,667	1,741	9,077	1,984
1998-2000	125	4,939	1,628	7,176	2,384
2001-03	99	5,753	1,726	8,617	2,210
2004-06	192	9,112	2,258	13,126	3,743
2007-09	270	16,559	2,896	20,716	6,349

Source: FAOSTAT (estimated).

EPW Issue No.4, 2012.

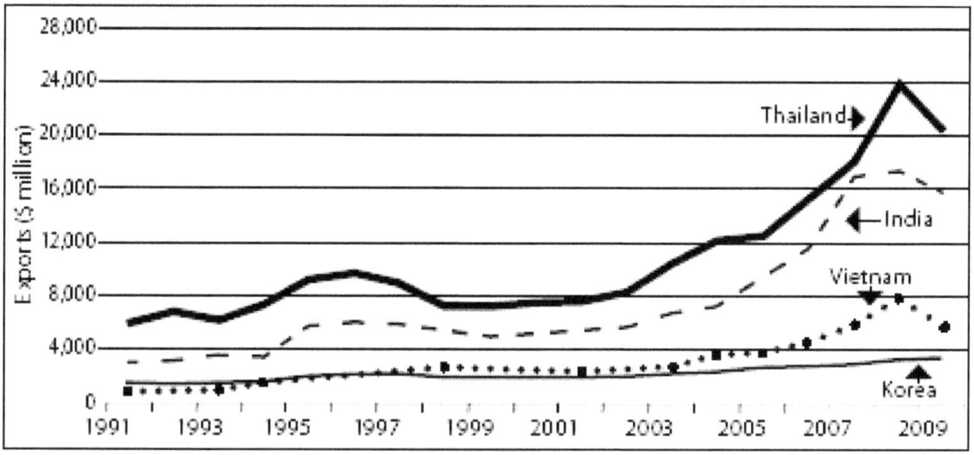

Figure 8.1: Trends in Agricultural Exports of Major Asian Countries.

Thailand had the highest gains between 200406 and 200708 with its agricultural exports jumping from $13,126 million to $20,716 million. In relative terms, India's agricultural exports increased by 82 per cent, followed by Vietnam (69 per cent), Thailand (58 per cent), Bangladesh (40 per cent) and Korea (28 per cent) during the period 200409. With the exception of Vietnam, the other four countries experienced a drop in exports between 199597 and 19982000 due to the 1997 financial crisis. By and large, it may be observed that trade liberalization policies adopted by these countries significantly helped them diversify exports through improved market access and increased engagement in free trade agreements (FTAs). However, much of the increase in exports came from the nonfood crop sector. All five countries, especially Bangladesh and Korea, experienced a decline in exports of food and food products, with an unfavourable balance of trade in the food sector.

Structural Changes and Contraction of Agriculture Sector

A closer look at the sectoral composition of GDP is appropriate to see if the structural transformation in the five countries resulted in a progressive shift from agriculture to other sectors. Broad trends suggest that the economies witnessed a phenomenal increase in the value of agricultural GDP, though with a drastic fall in the relative share of the agricultural sector in GDP (Table 8.3). India's agricultural GDP increased from $80,351 million (198589) to $94,445 million (199099) and further to $1,46,735 million in the last decade (200008). The period 19962000 witnessed a significant reduction in agricultural GDP in Thailand and Korea due to the Asian financial crisis. Notably, all countries (except Korea) experienced a significant rise in agricultural GDP in the last decade with the maximum growth realised by Vietnam (113 per cent), followed by India (55 per cent), Thailand (38 per cent) and Bangladesh (22 per cent). Despite the significant increase in the value of agricultural GDP and

Table 8.3: Trends in Agricultural GDP and the Sectoral Composition of GDP (1980 to 2010)

Period	*Bangladesh*	*India*	*Korea*	*Vietnam*	*Thailand*
		Agricultural GDP (current $ million)			
1985-89	7,531	80,351	16,899	8,344	8,994
1990-99	9,974 (32.4)	94,445 (17.5)	26,236 (55.2)	13,183 (58.0)	5,186 (-42.3)
2000-08	12,213 (22.4)	146,735 (55.4)	25,844 (-1.5)	18,184 (37.9)	11,029 (112.7)
		Agriculture value added (per cent)			
1980-89	31.6	32.0	13.4	17.4	41.6
1990-99	27.1	27.6	6.6	10.4	30.2
2000-10	21.0	19.5	3.5	10.5	21.8
Per cent change	–33.5	–39.1	–73.9	–39.7	–47.6
		Manufacturing value added (per cent)			
1980-89	21.2	26.0	39.5	32.5	25.9
1990-99	24.1	26.5	41.4	40.0	29.3
2000-10	27.2	27.7	37.1	43.5	39.7
Per cent change	28.3	6.5	–6.1	33.8	53.3
		Services value added (per cent)			
1980-89	47.2	42.0	47.1	50.1	32.5
1990-99	48.8	45.9	52.0	49.6	40.5
2000-10	51.8	52.8	59.4	46.0	38.5
Per cent change	9.7	25.7	26.1	–8.2	18.5

Figures in parentheses indicate the percentage change between periods.

Source: World Bank, World Development Indicators, 2010 (compiled

EPW Issue No.4, 2012.

the rise in agricultural exports, the relative importance of agriculture declined in all countries, which was clearly an indication of sectoral transformation (Table 8.3). The decline in the share of agricultural GDP was more pronounced in Korea (74 per cent) and Vietnam (48 per cent). In Korea, the share of agricultural GDP declined from 13.4 per cent (1980s) to 3.6 per cent (2000s). If this trend continues, Korea may soon join Japan and the US, where the share of agriculture in the national GDP is hardly 2 per cent. Thailand also experienced a decline in the share of agriculture from 17.4 per cent to 10.2 per cent.

The pattern of sectoral transformation looks somewhat different across the five countries. For instance, the services sector grew dynamically in all the countries except Vietnam, where manufacturing growth was more pronounced.

Corporate Farming by the State

As part of a diversification plan, four special purpose vehicles (SPVs) were created in the state – the Council for Citrus and Agri Juicing, the Organic Farming Council of Punjab (OFCP), the Council for Value Added Horticulture and the Viticulture Council of Punjab.

Set up in 2006 as a joint venture by Pepsico and the state government's Punjab Agro Foodgrains Corporation (PAFC), the Council for Citrus and Agro Juicing leased land and undertook corporate farming to promote horticulture. It had 75 acres owned by the PAFC in three locations in the state for growing citrus saplings, an operation managed by Pepsi. The council also leased 2,000 acres at Rs. 8,000 to Rs. 12,000 an acre for 12 years from farmers and aimed to enhance this to 21,000 acres by 2007 to grow citrus fruit varieties imported by Pepsico. Pepsi sold saplings at Rs. 50 each to the PAFC and the planted farms were managed by the council. However, even as trials at the Punjab Agriculture University (PAU) and the horticulture department were on, the council invited farmers to offer their lands on lease under two schemes spanning six and 12 years. In 2007, reports of poor fruit retention, incompatibility of rootstocks and a few varieties yielding no juice began trickling in from the trial sites. Following media reports, the council was asked to reduce its land target to 5,000 acres. The government stopped funding it and refused to stand guarantee for loans it planned to raise. It was instead asked to hunt for a venture capitalist. Even Pepsico, which had by this time grown thousands of plants at its nursery in Jallowal near Jalandhar, brought in a few new varieties to replace the non-performing ones. However, it was too late for farmers who had bit the diversification bait in 2006 and early 2007. They alleged breach of contract by the council and demanded their land back so that they could return to growing wheat and paddy. While no citrus plantations have come up on the lands of most farmers, the varieties grown on the lands of others have failed. To make matters worse, the farmers have neither been paid their rent in a timely fashion nor has the council been monitoring the orchards.

There have also been allegations of funds being misappropriated and an enquiry into this was initiated by the state government recently. In addition, two units set up by the council for multipurpose juicing at Hoshiarpur and Abohar have been idle for two years and plans are on to privatise them. They failed partly because Tropicana stopping buying juice from them when it found a cheaper alternative.

Role of the State

The state and its agencies affect agricultural development in many ways by incentivizing or dis-incentivizing some activities over others. In the context of contract farming, it is not just direct incentives to agencies and farmers that count, but also the agricultural policy. For example, contract farming influences the direction of ecological change through two actors. One, the contracting agency lays down a production schedule for farmers at the farm level. By determining the crop to be grown and the husbandry practices farmers have to follow, it influences the effect such farming will have on the environment. The government, as the main source of conservation measures – advisory, financial and material – is the second actor. Farmers' access to these measures is, to a large extent, determined by government policy. Thus the contracting agency and the government have a larger role to play in environmental or ecological change than farmers because they occupy "privileged" positions in chain of decision-making chain.

State-Sponsored Contract Farming

A state-sponsored programme for contract farming in Punjab, which was to divert one million hectares from wheat and paddy over five years (2002-07), failed to deliver. Most of the problems farmers faced related to production and quality (such as that of seeds and extension) and not being able to market produce (other than peas) in the open market (Dhaliwal *et al.*, 2003). There were also instances of corruption and malpractice in the programme due to conflicts of interest among implementing agencies and a lack of monitoring.

For the contract-farming programme to be successful there should have been greater attention paid to the selection of crops, a quick and effective contract enforcement and dispute resolution system, and mechanisms to limit the number of parties in a contractual arrangement and the fiscal risks to the state government. In addition, farmers' organisations capable of contracting with sponsors with a view to reduce transaction costs, better information flow and improving farmers' negotiating positions should have been in place (World Bank 2003, 2004).

More importantly, the programme excluded small farmers in most cases as firms preferred to deal with larger farmer for reasons of transaction costs, eligibility conditions like landholding, irrigation and education and absence of farmers' collectives. Table 7.4, which gives the average profile of contract growers across crops and agencies, shows that the contracted farmers had average holdings much larger than the average size of a holding in the state (9.5 acres). For example, none of the sample contract growers with Frito-Lay had less than 10 acres of land, in spite of 70 per cent of the holdings in the state being less than 10 acres (Singh, S 2008a). Another study on the programme found "no marginal farmer (in the size group of below one hectare) operating under contract farming. A handful of small farmers (in the size group of one to two hectares) were operating". Yet another study noted "the majority of the acreage registered in the project (contract farming by PAFC) is held by larger farmers, who tended to receive greater benefits from participation".

Table 7.4: Crop and Agency-wise Average Size of Holding of Contract Growers in Punjab

Study and Year	*Contracting Agency*	*Crop under Contract*	*Average Size of Contract Grower Holding (Acres)*	*Average Area Under Contract Crop (per cent)*
Singh, S 2002	Frito Lay (Pepsi)	Potato	53	8
Singh, S 2002	Nijjer Agro	Tomato	22	23
Singh, S 2002	HUL	Tomato	78	33
Singh, S 2002	Frito Lay (Pepsi)	Chilly	90	4.5
Asokan and Singh 2006	A M Todd	Mint	57	
Kumar 2006	Many MNCs and	Many crops	37	12
local firms	together			
Singh, S 2008a	Frito Lay (Pepsi)	Potato	63	53
Singh, S 2008b	A M Todd	Mint	40	27
Roopam Singh 2008	Frito Lay (Pepsi)	Potato	75	–

Markfed, a parastatal agency, placed advertisements in local newspapers a few years ago asking those willing to grow at least three acres of basmati paddy under contract farming to contact its district managers. The questions which arise from this type of offer are how many small farmers can spare three acres for basmati paddy? How many can spare it for contract farming? And how many would like to spare it for contract farming with Markfed?

From Trickle Down to Leapfrog

How to Go Beyond the Green Revolution?

The diversification of rural and agricultural production could bring some hope to the Indian countryside, but only the most integrated social groups and areas will benefit from it, making the others wait for an elusive trickle down. Perhaps a more fruitful process could be the growth of organic farming and the payment by the government for the ecosystem services that farmers do produce - often without being aware of it. Through a historic leapfrogging this could perhaps bring to the fore the disadvantaged areas and groups.

Most of the recent studies on the Indian countryside underline the prevailing high level of poverty, persistent social inequities, slow diversication and limited employment growth. While there exists a general consensus on the symptoms of the disease, experts disagree on the remedy to cure it. Various options can be chosen that can be clustered into two groups of policies. One is inspired by free trade and the belief that economic liberalization shall allow the taking off of some social groups and some regions. In this perspective, concentrating investments and measures on these groups and areas will engender a growth that little by little shall "trickle down" to groups and areas set aside. In the long term, capitalist farming and integration of agriculture into agribusiness are the solution to rural problems of India. Another policy, inspired by planning, is based on the belief that action by the State is necessary so that an excessive polarization does not prevent equitable economic development and social justice. Its advocates have no faith in the "trickle down" formula; they argue that rural development programmes and public investment in remote areas remain necessary, not only on behalf of equity but also for efficiency. Too large and many social, sectoral and spatial inequities may prevent sustainable growth and long-term development.

The history of independent India gives no reason for hope of success of either of these two policies. Homogenisation policies by the Nehruvian state have not succeeded in bringing the whole nation into the "mainstream", while the present liberalization policies clearly enlarge social and spatial gaps between the rich and the poor. The trickle-down effect does exist, but not to such an extent that it could eventually encompass the whole of India. Is it then really impossible to maintain both growth and equity, efciency and justice? How to reach this "leapfrog" effect enabling multiple growth poles to simultaneously emerge without having to wait for an elusive trickle down to happen? Here again there is no consensus on the ways to achieve it, but it is the goal of this paper to briey beat some tracks leading to new paradigms. We argue in the rst section of this paper that even if the gloomy appraisal on rural situation must be nuanced, it remains that the dominant paradigm since the green revolution must be replaced by new processes. The second section is devoted to an important hope, namely, diversication of rural and agricultural production, but it is feared that only the most integrated social groups and areas shall benet from it, letting the others wait for an elusive trickle down. Perhaps a more fruitful process could be the growth of organic farming and the payment by the government of the environmental services that farmers do produce – without being aware of it. According to some optimistic minds, this, as addressed in the third and last section, could perhaps bring to the fore the disadvantaged areas and groups through a historic leapfrogging. Gloomy Appraisal of the Rural Situation The declining share of agriculture in the Indian gross domestic product (GDP) corresponds to a long-run trend. If only the postreform is considered, according to the Economic Survey, the share of agriculture and allied sectors in the GDP was 26.9 per cent in 1999-2000, but only 14.5 per cent in 2010-11. Considering the tremendous growth in services, however, this could be seen as a positive proof of the "emergence" of India in its march towards development. The fact remains that the share of agriculturists in the working population is not decreasing at the same speed as the share in the GDP, implying that the number of marginal farmers and landless is increasing. Moreover, the star crops of the green revolution, wheat and rice, now show stagnating yields, and in many places it seems that the law of diminishing returns cannot be avoided. Higher land productivity is reached only by a disproportionate quantity of investment and inputs. The prevalent pessimism about this matter may be misplaced, however, the example of Egypt shows that an important margin of progress still exists without new technological breakthrough, only by a better use of resources.

Thanks to a better management of irrigation (and of drainage), in spite of rural poverty and lack of capital similar to India, with about 65 quintal of wheat per hectare and 90 quintal of paddy, Egypt has yields greater than Punjab by 55 per cent and 70 per cent, respectively. Rural poverty remains pervasive in India. "Longitudinal" case studies over a long period do not bring out a homogeneous picture of social changes and mobility. Many of them highlight a diversication of livelihoods across classes thanks to education, non-farm economy, and for poorer groups and areas, more migration that often brings back "new condence and skills as well as higher income". The evolution, however, strongly varies according to places and social groups; and poor people in an advanced area are usually less poor than poor people in an underdeveloped area. In northern Tamil Nadu, a region that is far from destitute, agricultural productivity has stagnated, social gaps are maintained in spite of real income growth, rural proletariat got feminized, economic diversication is progressing to the main advantage of big farms, and the proliferation of tubewells has had negative environmental impacts. While it is necessary to analyse how to get out of poverty, it remains urgent to understand the factors that explain how households can slide into it. Factors of mobility, downward and upward, are very different and asymmetrical. "Downward spirals into poverty" are triggered by social expenses for wedding or dowry, by the death or sickness of household members (not to speak of alcoholism), or by the loss of assets after some drought or taking over by some moneylender. Upward mobility is rather engendered by off-farm jobs and migration.

"A silver lining in this otherwise dismal picture" is that wages are increasing in the rural areas. The growth rate for male agricultural workers is roughly the same as the one for non-agricultural workers (only the wages of female agricultural workers are growing more slowly than the wages of female non-agricultural workers). These data t with the ofcial decline of the rural poverty rate. Remember, however, that the annual income of workers is more crucial than the level of daily wages: real wages may increase while the number of employment days is declining. Now this is the case: according to the National Sample Survey Office (NSSO), agricultural wage employment has decreased yearly by 0.9 per cent between 1993-94 and 2004-05. The main factor explaining the wage increase is probably the decit, at least seasonal, of labour. This decit has at least two reasons – not to speak of the impact of the recent National Rural Employment Guarantee Scheme. One is the growing disdain of agriculture: educated people are more and more refusing to take on farming. The other reason is migration, or more precisely, mobility: people migrate to cities, to the Gulf countries, to irrigated areas, to coffee plantations, or simply commute daily to a non-agricultural job nearby. The rate of urban growth remains low in India and for each of the last three decades (1981-2011) net rural-urban migration ("rural exodus") never explained more than 21 per cent of the urban population growth. This low rate, however, hides migrations to peri-urban fringes as well as a high mobility (daily commuting, seasonal migration, return after some years in city, etc). The winds are slowly changing. The 2011 Census showed that urban India's increase was greater than that of rural India's by nearly half a million people.

Water shortage might well accelerate this mobility. The most suitable places for dams have been since long equipped (and face many problems, notably salinization of soils, poor drainage, silting of reservoirs). The new projects create controversies that are technical and ecological as well as due to the greater political awareness of the people likely to be displaced. Hence, it is very doubtful that "nearly 40 per cent of the available irrigation potential of 58.5 million hectares from major and medium irrigation projects in the country still remains to be exploited". Such an estimate is based on very theoretical calculations limited in time (was any long-term cost-benet analysis made, including silting and salinisation?), in space (have all the impacts on downstream and upstream areas been evaluated?) and in the gamut of issues (ecological, social and political outcomes) which must be taken into account, not only the watering potential.

As for groundwater, in most of India the tables are declining, and only in the eastern regions untapped groundwater (and surface water) resources remain in quantity. For sure, it is a problem of collective action: groundwater is considered a private good instead of a common resource, and it is wise that the new Groundwater Model Bill, 2011, recognizes groundwater as a public trust. But scarcity is not only a social construct, it is also a natural given. In semi-arid climates water is anyway a scarce resource. Watershed development, check dams, water harvesting and the like cannot lead to more optimism. Not to speak of the dubious political economy of many projects, including alleged success stories, the impacts of watershed development cannot be very positively assessed when regional and national levels are considered. If in the upstream part of a river basin all the local watersheds are equipped with check dams and water harvesting public works, no water will be left downstream: irrigated areas in deltas, lagoons and mangroves will be severely affected.

As a conclusion to this section, there is a necessity of changing the paradigm of rural development that was dominant up to now. It was based on maximization of land productivity; farms had to be as autonomous as possible from agro-processing industries considered as a potential urban and corporate threat; and cereals (or more exactly wheat and rice) were supported by public policies on behalf of national self-sufciency. India should go beyond the legacy of the green revolution and accept a new paradigm that is elaborated below, based on new agricultural and rural activities; on more integration of farms into marketing chains; and on new visions of agriculture as not only an economic sector of production but also a key social and environmental service provider. The conditions necessary for such a positive change, however, are far from granted.

Diversification: Still the Good Old Trickle Down?

Rural policies in India are made difficult to design because of at least three factors: (1) vested interests, sometimes masked by social reasons (e g, reservations of oilseed agro-processing to small-scale industries); (2) agriculture is a state subject, hence the central government cannot directly enact laws on agricultural labour, taxes, and land price issues; (3) general poverty in the country prevents any large margin of manoeuvre on wages or prices. On the one hand, it is difficult for the government to have minimum legal wages enforced, because many small farmers cannot afford engaging expensive labourers during harvest, and labour-demanding crops such as paddy should not be jeopardised. On the other hand, agricultural prices cannot be supported beyond a oor level lest they become out of reach for consumers with limited purchasing power. Both employers and consumers are too poor. Since the taxpayers are few (for political rather than economic reasons), the government is in a quandary for funding support to the agricultural sphere.

If little action is possible regarding wages and prices, at least two solutions remain: First, increase productivity, both of soil (yields per hectare) and of investment (*e.g.* by simplifying the marketing chain). The former has become a difficult challenge not only because of agronomic difficulties (pest resistance and the like), but also for cultural reasons: nowadays agriculture in India is increasingly considered a painful, non-protable, risky and derogatory activity; educated sons refuse to take over the ancestral farm; many well-off farmers lead a process of "extensication", with lower investments, lower yields per hectare but more net income. Education is a way of "leapfrogging", and many agriculturists prefer to invest in their children's education rather than in their farm. But this leapfrog may be social much more than spatial; emigration of educated people for better jobs cannot develop the area of origin if little remittance is sent back. On the contrary, this brain drain will make the area poorer than before. Second, develop new rural activities inside the agricultural sector (new crops) or outside it (economical diversication).

This "diversication" should not be confused with the already present "diversity" of productions that exists in many regions, especially in rain-fed agriculture. In south Karnataka, in order to reduce risks and come as close as possible to self-reliance, farmers grow one crop of nger millet followed by pulses, with some patches of oilseeds. They own some cattle, sheep, goats, hens; in short, diversity is an old key characteristic of Indian rain-fed farming systems. Diversication for high-value crops (vegetables, owers) or poultry or shrimps is a different matter. Diversication is partly demand-driven by the domestic market. This is an important point, since erce criticism is heard against diversication policies, arguing that they are relying on exports at the expense of the Indian consumer. The current changes in the domestic market allow diversication to be domestic-based to a large extent. The following three long-term factors make it an important process to eventually concern the whole India:

(a) Change in Consumption Pattern: This "food transition" is explained by Bennet's and Engel's laws. Due to higher purchasing power, the rise of middle classes, urbanisation and general development, the share of meat, fruit and vegetables will increase in the Indian diet. Note, however, that some food and taste habits are specic to India for cultural reasons (little vegetable consumed; little "hot" meat because of tridosha clasication; religious values favouring vegetarianism). What I called the "Hindu food transition", in a reference to the famous "Hindu rate of growth", argues that unlike northern countries and south-east Asia, pork and vegetable consumption will not grow rapidly unlike other white meat such as poultry. In particular, the decrease in cereal consumption per capita is not fully compensated by the rise in consumption of meat, vegetables, *etc.* Hence, on the whole calorie availability per capita is declining in India. The average calorie intake per person per day has fallen over time, a unique phenomenon that goes contrary to the general theory of food transition (Landy 2009b) and that cannot but hinder the growth speed of non-cereal demand.

(b) Urbanisation: Urbanisation, for various reasons (working women, cost opportunity of cooking time, smaller households, cultural change, etc), accelerates this food transition. The urbanisation rate of India is still ofcially low, but food transition is also accelerated by the "cultural urbanisation" and pseudowesternisation that are occurring in many rural areas, especially among well-off families.

(c) Changing Relative Prices between Cereals and Highvalue Agriculture: During the 1980s, dairy, fruits and vegetables, as well as poultry or sh were made cheaper thanks to technological and marketing boosts. Once again, however, some nuances have to be brought out. Just consider that maize is too expensive for many Indian animal-rearers to buy – but cheap enough for one-sixth of the Indian production to be currently exported, creating a vicious circle since these exports make domestic prices higher. That imports are controlled by the Food Corporation of India shows that maize is officially still considered as human food rather than animal feed, indirectly preventing meat prices from falling enough to be within the reach of poor consumers. Beside these demand-driven factors, there exist supply-side factors, such as "improvements in infrastructure (specically roads and markets) and technology (relative protability and risk in different commodities)". Liberalisation is a priori a positive factor for agricultural diversication. Less protectionism brings cheaper inputs (notably feed, for poultry) and easier exports (aquaculture) and encourages greater market orientation of farm production. It frees private capital that is allowed to be invested in new sectors (oilseed processing), and it makes new technology available. It remains to be seen, however, whether demand factors always t with supply factors, and whether the change in consumption always corresponds to a change in production. Is there not a gap, at least in time? Cheaper imports of subsidised poultry (say, from Europe) could compete with Indian producers for feeding new consumers.

Moreover, it cannot be denied that the external markets will represent a good part of the outlet, which will also create risk and uncertainty. Global phyto-sanitary standards prevent the easy access of Indian products to markets abroad since these are sometimes of poor quality. Diversiûcation also includes sectoral change outside agriculture. India remains far from the successful rural diversication of many south-east Asian countries, even though the trend is positive. More worrying, the absolute number of workers employed in agriculture does not decline: according to the NSSO, the labour force in agriculture, which was 242.5 million people in 1993-94, had gone down to 237.6 million in 1999-2000 but rose again to 259.1 mil.

in 2004. Whatever the validity of these data, the demographic pressure on land is still important, preventing any consolidation of farms and decline in the number of agricultural labourers. It is hardly surprising, consequently, that the map of off-farm male rural workers looks like the map of poverty.

Between 1993-94 and 2004-05, however, rural non-agricultural employment increased by 3.5 per cent yearly, whereas total agricultural employment increased by 0.4 per cent only. If we consider that meanwhile some "traditional" artisans (weavers, potters, *etc.*) have disappeared, whilst "modern" rural industry and services as well as commuters having an urban job were growing, this data shows that diversication is slowly taking shape. What remains to be seen is where, for what and for whom. In particular, it is feared that a part of these non-agricultural jobs is "distress diversication" born out of rural unemployment, with low income and low productivity, rather than "willful diversication".

Does this double (agricultural and non-agricultural) diversication represent a big hope for small farmers? Non-cereal agriculture brings more income per hectare and needs more labour (or as much, in the case of rice). "Smallholders can leapfrog from their foodgrain-based systems to high-value agriculture to augment their incomes. They have certain advantages, primarily abundant family labour, as most of the high-value commodities are labour-intensive", it is often argued. But they also have many disadvantages, as most of the high-value commodities are capital-intensive and market dependent, fragile and risky (Shiva and Bedi 2002). If you add another key constraint, credit, not to speak of the low level of education of many petty farmers, it is highly debatable that "the evidence is that the primary production centres of highvalue commodities are largely concentrated with smallholders". That diversication will take off is certain, but that this process will bring benet mostly to the small farmers is much less sure. Not to speak of economic constraints, indebtedness, patronage, asymmetries of information and interlocked markets of products, credit and labour "remind us that the study of livelihood diversication is more than simply about multiple income sources; it relates to the current economic and political transformations whether originating from global, national or local levels".

If we consider only the local level, we nd "social structures of accumulation" that segment the rural (and urban) society to the benet of "intermediate classes" and at the expense of dominated groups. Hence, diversication may not only benet mainly the already well-off groups and areas, but, on the reverse, polarisation and excessive social and spatial hierarchies may prevent diversication to occur. Inequality can often inhibit "the potential of 'growth linkages' from agriculture for bringing about rural economic diversication" pass the constraints faced by small farmers, at least two (non-exclusive) solutions exist: (a) Contract Farming: In the case of contract farming, "the rm controls the production process without owning or operating the farms but ensures assured procurement of output and remunerative prices". It allows for compressing the supply chain by reducing the number of intermediaries, so that allegedly the farmer receives a higher share of the nal consumer price (presently one-fth, in the case of fruits and vegetables). It make input supply easier for the farmer and guarantee an outlet at a xed price, thus overcoming the shortage of capital and information that many smallholders suffer from. Contract farming is expanding in India due to liberalization, but in the Punjab it is accelerated by indirect contracts with agro-processing rms through the state, namely, the Punjab Agro Foodgrains Corporation – with limited success. Often the smallholders are excluded from the retail chain.

Producer Groups and Cooperatives: They can allow economies of scale and prevent smallholders to be alone while bargaining an agreement with a wholesaler or a corporate company. Already, dairy and vegetable rms are today contracting through farmers' associations rather than with individuals. But how to make real cooperatives (not the present type of "government cooperatives" that exist in most villages) ourish in such a hierarchical and segmented rural society? Some success stories exist in India – and not only milk cooperatives in Gujarat. Generally speaking, villages with a large proportion of small farmers belonging to the same caste without excessive agrarian hierarchy or political bias could be pioneers in cooperative building. All in all, production per se should not be the only link of the chain focused on. Both ends of the chain should be tackled. Upstream: Strict terms of reference by the government should be made compulsory in contract farming, in order to "frame" agreements and prevent abandoning small farmers alone in front of agribusiness rms. (Reforms in India should not mean less state, but a better state.)

Downstream: Processing the agricultural output allows better price, thanks to added value and less market uctuation. In India, the ratio of produced fruit and vegetable that are processed is very low (even if the commonly published percentage of 2 per cent is probably an underestimate because production is overestimated). Increasing this ratio could also create rural off-farm employment if factories are set up near the production areas. It is also necessary to improve transport and marketing infrastructures (roads, cold chains, etc). The risk of bias is not only social, but also spatial. A study of the International Crops Research Institute for the SemiArid-Tropics (ICRISAT) mapped the share of high-value commodities in the total value of agricultural output (Rao and Kar 2005). Central India is little diversied, and with the notable exception of Bihar-Jharkhand, the BIMARU states – Bihar, Madhya Pradesh, Rajasthan and Uttar Pradesh – belong mostly to the low diversication group. In short, diversication is spatially elitist. Could this hurdle be overcome by targeted public investments and programmes? Fan *et al.* (1999) have argued that marginal returns to public investment in weakly integrated areas are high. Others like Ravallion (2005) and Farrington *et al.* (2006: 451) claim that the combination of exclusion from some markets, interlocking of others, and generally high levels of risk and vulnerability severely constrain how and how far [the poor farmers] can engage in markets, and therefore, limit their opportunities to take up new options and activities in new locations or emerging sectors.

What is true for poor social classes is true for poor areas, and it is feared that spatial exclusion of some regions can prevent them to benet from such opportunities Hence, the better-endowed groups will be the rst to benet. From big to small farmers, from peri-urban areas endowed with market and good infrastructures to the countryside far from cities and transportation axes, the usual social and spatial trickle down must be expected, as weak and slow as it can be, rather than some illusory drastic change that would miraculously balance the existing inequities between regions and between classes. Some surveys, true, bring out success stories in contractual arrangements between small dairy farmers and multinational companies such as Nestlé in Punjab: "the vertical coordination in high-value food segment helps in lowering the transaction costs and market risks of smallholders" (Joshi *et al.*, 2007: 407), but some others are much less optimistic.

How to avoid that contract farming means "enslaving agriculture to corporations" (Shiva and Bedi 2002: 65) if contract statements are too harsh for the farmer? Many cases in developed countries (milk or chicken producers in France, for instance) do not hold out optimism for the fate of small farmers in developing countries. An important bias in studies underlining the better situation of smallholders in contract farming compared with smallholders without contract is that they hardly take into account the fact that processing companies select ex ante the farmers to deal with: consequently it is not astonishing that, ex post, contract farmers are better off than others. Environmental Services and Organic Farming Environmental services and organic farming are issues that might bring more optimism regarding a possible leapfrogging. Organic Farming: Can the Last Be First? Like diversication, the trend in favour of organic farming is both demand- and supply-driven. On the demand side, food safety has become a goal as important as food security in a country where greater health awareness and higher purchasing power for the middle and upper classes make the consumers check more carefully the quality of foodstuffs before buying them. On the supply side, loss of soil fertility due to improper irrigation or chemical over-fertilisation, health hazard of pesticides and the cost of industrial inputs, are some factors that push for an environmentally sustainable agriculture.

In this approach, the "progressive farmers" of the green revolution will become outmoded if they do not renounce, at least partly, the most unsustainable traits of articialised agriculture. On the opposite, the presently so-called "backward farmers" who in rain-fed poor areas (69 million hectares) hardly use chemical fertilisers, no pesticides, and have stuck to traditional crops and local desi varieties have a new card to play. As the website of the Uttarakhand Organic Commodity Board put it, they are "organic by default". Is it a renewed version of the tale, "The Tortoise and the Hare"? Can the slowest eventually win the race? If the small poor farmers and the rain-fed poor regions turned to organic farming, the last social and spatial groups could become the rst. At least this is what could be dreamed of. Many hurdles exist, however.

On the supply side, fertilisers are already commonly used in most rain-fed areas, local varieties have become rather rare even in these parts of India, and it remains to be proved that the present "tradition" is always "organic". Moreover, rain-fed areas are often out of the main marketing networks and selling organic products, for the time being, often needs the support of non-governmental organisations (NGOs) for reaching the few consumers interested in it. Lastly, on the demand side, consumers remain fussy regarding the taste and the look of organic foods. If they are ready to pay for more expensive products, they are not ready to accept smeared fruits or vegetables with a bizarre shape. Moreover, organic marketing chains must adhere to strict standards, national or international. Certiûcation of production systems can raise the incomes of farmers and give an impulse to the diversication processes. But it is a difficult constraint, very demanding in time and expertise. It is remarkable, therefore, that certied areas have seen a dramatic growth in recent years.

Organic farming needs integration in agro-processing marketing chains, the constraints and drawbacks of which have been addressed above. Even though some processing should be done at the farm level for bringing added value to the producer, the integration into a marketing chain will bring him or her into some forms of dependence. A solution could be the building of chains of "fair trade", where the end product is sold at a somewhat higher price than a normal product, but the consumer is aware that this selling price includes not only an economic added value but also a social and environmental added value. In India some tea and coffee are already "fair traded", but in this country, like anywhere, some frauds and misuse exist. Today, globally, fair trade is developing itself between southern producers and northern markets, but south-south fair trade is also slowly taking root (Alter Eco in Brazil) and such marketing chains are also slowly growing in India, in particular in hilly areas. In short, the building of organic farming chains, from small farmers to rich consumers, from rain-fed poor areas to metropolitan cities, cannot but be a very long-term process. Probably big farmers and corporate farms, as well as irrigated areas will enter such chains before "the last". There will be no large-scale leapfrogging through organic farming.

Environmental Services – The 'Multifunctionality' of Agriculture: In 2007 in Coorg (Kodagu district, Karnataka) I visited coffee estates that have undertaken homestay activities. In most cases, the planter's wife is putting up the tourists: accommodation, food, camping re amid coffee plants under evergreen forest cover. Such a new activity is based on a double need from the estate perspective: the nancial need to have an economic "cushion" in case of lower coffee prices; the social need for the planter's wife to break a monotonous life. This activity is also demand-driven, linked to the growth of the middle classes who see so-called "nature" as a resting environment rather than a space of production. For them it does not matter whether the forest has been partly cleared for letting space to coffee plants fed with chemical fertilisers and pesticides. Seeing "greenery" is sufficient to feel relaxed, far away from the congested life of Bangalore or Mumbai.

In these home stays, about one-third of the tourists ask questions regarding coffee cultivation, ora and fauna. This is not negligible. But many of them take the landscape as granted and are not interested in the way it has been constructed by farmers. A large proportion of the middle class has lost any contact with the agricultural sphere, and exactly like in the northern countries, considers an agricultural landscape as "natural" because it is vegetal. Or, if they are aware that it is an arte fact, they consider it as a "heritage", i e, ancient, unchanging and destined for eternity. This view may appear naïve, but there is some wisdom in it, that goes much beyond the specic case of Coorgi home stays and is valid for the whole of India. That agriculture is part of "nature" may sound foolish if deforestation, a depleting water table, use of chemicals, etc, are considered. It may not sound so if positive environmental impacts are highlighted. Capitalisation through the delivery of global public goods preventing a greenhouse effect is promoted by the Kyoto Protocol. Recent coffee planting on degraded forests in Western Ghats has increased carbon sequestration, both in the soil and above ground. Paddy elds in lowlands prevent ooding and allow water table recharge. Well-managed agriculture prevents erosion. It also maintains, as the Coorgi lady said, "nice-looking greenery", which is attractive for tourists. In France, ve-year-long "sustainable agriculture agreements" (contrats d'agriculture durable) can be signed between farmers and the State for the grant of nancial public support in exchange for the farmer's commitment regarding "agro-environmental actions" (management of animal excreta, non-pollution of water table, biodiversity conservation, etc, as well as organic farming chain or activities favouring employment growth). India too must go beyond the productivist vision of agriculture and recognize ecosystem services. Without forgetting its production goal that remains essential as highlighted since the national and global food crisis in 2006-08, other functions must be taken into account:

- Economical Goals at the Level of the Production Unit: Unless this goal is not considered, if a farmer cannot make enough money, he will sell his farm, at least in peri-urban areas where land is coveted for other uses – or none of his sons will be willing to take over it. "The future emphasis of agricultural policy ought to be on maximising farm household incomes rather than generating food surpluses".
- Social Function of Agriculture: Often considered as a "parking sector", a "residual" activity, agriculture must remain vibrant even in the poor countryside, unless rural exodus increases in scale and makes urban management and slum problems even more difcult. This social utility deserves some support.
- Ecological Function of Agriculture: Its role in biodiversity conservation, erosion prevention, etc, cannot be under emphasised, at least in the case of sustainable agriculture.
- Cultural Function: What would south Indian culture be like if paddy elds were to disappear to the advantage of rubber, banana or casuarina? What would be the food of rural south Karnataka people without mudde, nger millet bowls?

Not only coconut trees are necessary to attract tourists in Kerala (economic function), they are a key ingredient of traditional food, of local medicine – of Malayali identity. The agricultural landscape is much more than a space of production: it is a patrimony, often built generation after generation, carrying social, political and cultural identities. It is like a palimpsest where the legacy of former values, passed utilities, history of societies and cultures as well as present functions can be read. Lastly, one more reason makes it urgent to acknowledge what is called in the European Union's jargon of the "multifunctionality" of agriculture: since this multifunctionality is acknowledged in Europe, and to a large extent, at the World Trade Organization, the Indian farmers cannot be on the same level playing field, if they do not benefit from the same support as their French or United States counterparts. Multi-functionality is the reason – not (only) a pretext – for keeping some subsidies in the "green box" of the permitted "programmes that are not targeted at particular products and include direct income supports for farmers that are not related to (are "decoupled" from) current production levels or prices. They also include environmental protection and regional development programmes".

They are allowed without limits, provided they cause "minimal distortion" to trade. Even China is now granting some small direct subsidy to farmers, acre-wise, that is decoupled from actual production. Using these supports is efficient and politically painless, unlike prohibitions that are not easy to implement. Though in Kerala conversion of paddy fields is banned, many farmers are replacing paddy by banana or rubber. Conversion would slow down if support to paddy growers, decoupled from production but based on environmental services, was granted by the government. Cynically speaking, one could say that the main outcome of prohibition is to increase bribes and corruption. Hence it is suggested to use the carrot rather than the stick, and to provide strong incentives to grow environmentally sound crops.

The role of the demand side appears essential for driving positive changes in Indian agriculture (diversification, organic farming, green tourism, etc). It is hoped that the "emergence" of India and the growth of the middle classes, that up to now concern mainly the urban sphere, shall also indirectly touch the countryside through demand by these consumers. Another important actor remains in place, however. The role of the government is not to be reduced in spite of liberalization. As seen above, the growth of private agribusiness role in agriculture could be a parallel process, instead of an alternative to the improvement of state action in matters of public investment and of rules for contract farming. As in other matters, liberalisation is not characterised by the substitution of one actor by another, but by the addition of new actors without the former, here namely the state, necessarily disappearing. "Liberalisation" is nothing but a multiplication of actors (Ruet *et al.*, 2009). Similarly, the "participation" hype in natural resource management (irrigation, forest, etc) cannot make us forget tutions cannot be democratic, effective and sustainable without a proper framing by the state. Now, the government is not an isolated actor, and many of its decisions are responses to citizens' demands. Farmers' power in India has been analysed as an exception by Varshney (1994: 3), since, in most of the developed countries, the historical paradox of rural power can be stated as follows: although in the process of economic development the populous countryside loses power, the combination of a democratic polity and an industrialized economy later seems to empower it.

In India, rural power rose in the 1980s in spite of under-industrialization – just because, unlike developed countries where universal franchise had been introduced during the industrial revolution, "independent India was born agrarian as well as democratic". In the future, the Indian situation will be probably less paradoxical: being smaller than other economic sectors, agriculture will be supported by the government through less of a fiscal strain. Richer consumers shall accept higher farm prices, at least for some specic foodstuffs. India will come closer to the situation of rich countries, all the more so if the following statement is accepted by all the decision-makers: except for a few rare types of farms such as in Argentina or Australia, agriculture in the world needs public-support to be sustainable.

Even the European or American farmers get subsidies (or payment for social and ecological services), so it should be intellectually and politically accepted that the Indian farmers should also get subsidies. Each country has the right to reach food security with a sustainable agriculture that provides environmental and cultural services beside productive and social functions. Accepting this right and trying to achieve it shall force to reconsider the relationships between agriculture and state under a new light.

White Revolution

"Operation Flood started the White Revolution in India and made our country self-sufficient in milk and this was achieved entirely through the cooperative structure. Today around 12 million farmers in more than 22 states across the country own around 250 dairy plants handling around 20 million litres of milk a day. This is a remarkable achievement. While we in India tend to take our achievement for granted, this feat elicited high praise and admiration throughout the world." "In 1955 our butter imports were 500 tons per year; today our cooperatives alone produce more than 12,000 tons of butter. Similarly, we imported 3000 tons of baby food in 1955; today our cooperatives alone produce 38,000 tons of baby food. By 1975 all imports of milk and milk products stopped. The import permitted was that of food aid under Operation Flood."

"A separate agency called as the Indian Dairy Corporation (IDC) was created to receive grants of food aid and use it in the Operation Flood, this agency was also headed by the marvelous Dr. Verghese Kurien." "Milk powder production increased from 22,000 tons in the pre-Operational Flood year to 1,40,000 tons by 1989." "In 1998 the World Bank published a report on the impact of dairy development in India and looked at its own contribution to this. The audit revealed that of the Rs. 200 crore the world bank invested in Operation Flood, the net return on India's rural economy was a massive Rs. 24,000 crore each year over a period of 10 years to which no other dairy programme has ever matched." "Moreover when Amul started there were so many competitors in the market like Polson and other foreign players but Amul stood the test of time and over the years emerged as India's favorite."

Milk production in India increased from 17 million tons in 1950-51 to 84.6 million tons in 2001-02 and 137.69MT in 2013-14. Therefore, from being a recipient of massive material support from the World Food Program and European Community in the 1960s, India has rapidly positioned itself as the world's largest producer of milk. Milk production in India during the last five decades is shown in Figure 8.2 and Tables 8.5 and 8.6.

Milk production in the country was stagnant during the 1950s and 1960s, and annual production growth was negative in many years. The annual compound growth rate in milk production during the first decade after independence was about 1.64 per cent; during the 1960s, this growth rate declined to 1.15 per cent. During the late 1960s, the Government of India initiated major policy changes in the dairy sector to achieve self-sufficiency in milk production. Producing milk in rural areas through producer cooperatives and moving processed milk to urban demand centers became the cornerstone of government dairy development policy.

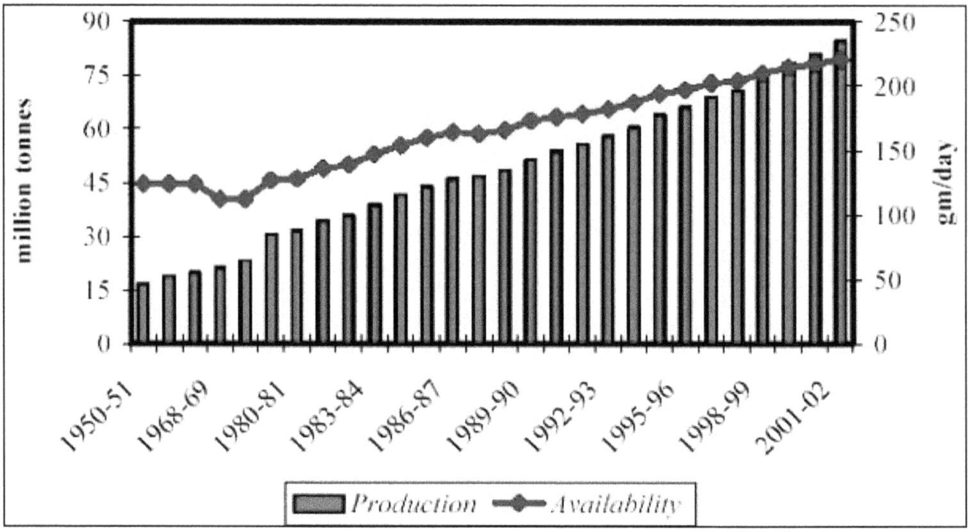

Figure 8.2: Milk Production and Consumption Trends in India: 1950-51 to 2001-02.
***Source*: GoI, 2003.**

This policy initiative gave a boost to dairy development and initiated the process of establishing the much-needed linkages between rural producers and urban consumers.

India ranks first in milk production accounting for 17 per cent of world production. During 2013-14, milk production peaked at 137.69 MT, thus becoming an important secondary source of income for 70 million rural households engaged in dairying. The average year-on-year growth rate of milk, at 4.18 per cent vis-à-vis the world average of 22 per cent, shows sustained growth in availability of milk and milk products for the growing population.

Table 8.5: Annual Growth Rate (per cent) of Production of Major Livestock Products in India

Period	*Milk*	*Eggs*	*Wool*
1950-51 to 1960-61	1.64	4.63	0.38
1960-61 to 1973-74	1.15	7.91	0.34
1973-74 to 1980-81	4.51	3.79	0.77
1980-81 to 1990-91	5.68	7.80	2.32
1990-91 to 2000-01	4.21	4.46	2.01

Source: GOI, 2003.

The performance of the Indian dairy sector during the past three decades has been very impressive. Milk production grew at an average annual rate of 4.57 per cent during the 1970s, 5.68 per cent during the 1980s, and 4.21 per cent during the 1990s. The country's milk production reached 139.69MT in 2013-14.

Table 8.6: Annual Growth Rate (per cent) of Milk, Eggs, and Wool in India: 1975-76 through 2001-02, by plan

Plan	Year	Milk	Eggs	Wool
5th Five Year Plan	1975-76 to 1979-80	2.91	3.5	1.49
6th Five Year Plan	1980-81 to 1984-85	6.42	8.4	2.67
7th Five Year Plan	1985-86 to 1989-90	4.37	7.23	1.88
8th Five Year Plan	1992-93 to 1996-97	4.41	4.58	0.80
9th Five Year Plan	1997-98 to 2001-02	4.13	4.34	2.14

Source: GOI, Basic Animal Husbandry Statistics 2002, Department of Animal Husbandry and Dairying, Ministry of Agriculture.

This growth was achieved through extensive intervention by the Indian government, as well as through increased demand driven by population growth, higher incomes, and urbanization. Until 1991, the Indian dairy industry was highly regulated and protected. Milk processing and product manufacturing were mainly restricted to small firms and cooperatives. High import duties, non-tariff barriers, restrictions on imports and exports, and stringent licensing provisions provided incentives to Indian-owned small enterprises and cooperatives to expand production in a protected market. Indian policy makers saw the development of the dairy sector as a measure to create supplementary employment and income among the small and marginal farming households and landless wage earners, as milk production takes place in millions of rural households scattered across the country.

Despite its being the largest milk producer in the world, India's per capita availability of milk is one of the lowest in the world, although it is high by developing country standards. The per capita availability of milk, which declined during the 1950s and 1960s (from 124 gm per day in 1950-51 to 121 gm in 1973-74) expanded substantially during the 1980s and 1990s and reached about 226 gm per day in 2001-02 (Figure 8.2). The per capita consumption of milk and milk products in India is among the highest in Asia. However, it is still below the world average of 285 gm per day and the minimum nutritional requirement of 280 gm per day as recommended by the Indian Council of Medical Research (ICMR).

Several factors have contributed to the increased milk production in the country. First, milk and dairy products have cultural significance in the Indian diet. A large portion of the population is lacto-vegetarian, so milk and dairy products are an important source of protein in the diet. The demand for milk and dairy products is income elastic, and growth in per capita income is expected to increase demand for milk and milk products. Empirical evidence has shown that the composition of an average Indian's food basket is gradually shifting toward value-added products, including milk and dairy products. The proportion of income spent on milk and milk products increased from 11.7 per cent in rural areas and 14.7 per cent in urban areas in 1970-71 to 21.6 and 16.7 per cent in 1999-00, respectively (Annex -1). Other socioeconomic and demographic factors, such as urbanization and changing food habits and lifestyles, have also reinforced growth in demand for dairy products. On the supply side, technological progress in the production and processing sectors, institutional factors, and infrastructure play an equally important role. The linking of rural small producers with urban consumers through producers' cooperatives was a true institutional innovation in the Indian dairy sector.

Given its high income elasticity, the demand for milk and dairy products is expected to grow rapidly. A study conducted by Saxena (2000) using National Sample Survey (NSS) data for 1993-94 showed that income elasticity of demand for milk and milk products is higher (1.96 national level) in rural areas (ranging from 1.24 in Punjab to 2.92 in Odisha) than in urban areas (ranging from 0.99 in Punjab to 1.78 in Bihar). The northern region in general and Gujarat in the western region show low income elasticity of demand for milk and milk products. The high values of income elasticity for different states in the eastern region-varying from 2.5 to 2.9 in rural areas and from 1.5 to 1.8 in urban areas-show a very strong preference for milk and milk products with an increase in income. Further increases in per capita income and changing consumption patterns would lead to acceleration in demand for milk and other livestock products in India and thus would give a boost to this sector. Radhakrishna and Ravi (1994), Gandhi and Mani (1995), Kumar (1998), Dastagiri (2001), and others have estimated demand and income elasticity of demand for milk and milk products, and show similar trends (Table 8.7).

Table 8.7: Income/Expenditure Elasticity of Demand and Estimates of Demand for Milk in India

	Rural	*Urban*	*Demand for Milk by 2020 (million tons)*
Radhakrishna and Ravi (1992)	1.15	0.99	–
Gandhi and Mani (1995)	1.70	1.06	–
Kumar (1998)	–	–	126.0-182.8@
Saxena (2000)	1.96	1.32	85.7#
Delgado, *et al.* (2001)	–	–	132.0
Dastagiri (2001)	1.36	1.07	147.21

Notes: @: Estimates based on 4 per cent growth in GDP (126.0), 5 per cent growth (142.7), and 7 per cent growth (182.8); #: Estimates for 2007-08.

Delgado *et al.* (2001) have estimated per capita consumption of milk products in developing countries to be about one-third that of developed countries in 2020; however, in aggregate terms, 60 per cent of world milk consumption will take place in developing countries, which is a major shift from the early 1990s, when the developed countries consumed 59 per cent of world milk production. The projected growth rate for milk is expected to be around 4.3 per cent during 1993-2020. Kumar (1998) projected demand for milk at 142.7 million tons by 2020 at 5 per cent growth in GDP (182.8 million tons at 7 per cent growth in GDP). The estimates given by Saxena (2000) are different than other estimates and project demand for milk to reach its peak at 85.7 million tons in 2007-08 and decline thereafter. Saxena argued that the domestic market may expand if a rise in per capita income is more in favor of lower income groups and regions, as the income elasticity of demand for such groups and regions (eastern) is much higher. The wide variations in demand estimates are mainly due to different assumptions of elasticity, population projections, and other parameters.

Regional Patterns of Growth

There are large interregional and interstate variations in milk production as well as in per capita availability in India. About two-thirds of national milk production comes from Uttar Pradesh, Punjab, Rajasthan, Madhya Pradesh, Maharashtra, Gujarat, Andhra Pradesh, and Haryana. However, there have been some shifts in milk production shares of different states. In 2001-02, Uttar Pradesh was the largest milk producer in the country with about 16.5 million tons of milk, followed by Punjab (8.4 million tons), Rajasthan (6.3 million tons), Madhya Pradesh (6.1 million tones), Maharashtra (6 million tons), and Gujarat (5.6 million tons). During 1982-83 triennium ending (TE), the top five milk-producing states were Uttar Pradesh (18.5 per cent), Punjab (10.1 per cent), Rajasthan (9.8 per cent), Gujarat (6.8 per cent), and Haryana (6.6 per cent). During TE 2001-02, Uttar Pradesh (19.5 per cent), Punjab (9.9 per cent), Rajasthan (7.5 per cent), Maharashtra (7.3 per cent), Madhya Pradesh (7.2 per cent), and Gujarat (6.6 per cent) were the largest producers. The share of Andhra Pradesh, Gujarat, Karnataka, Kerala, Maharashtra, Punjab, Uttar Pradesh, and Odisha increased between 1991 and 1999-01, while the share of Bihar, Haryana, Madhya Pradesh, Rajasthan, Tamil Nadu, and West Bengal declined. The regional shares of milk production are presented in Figure 8.2. Major milk-producing regions in the country have good resource endowment and infrastructure. The eastern region is lagging behind in terms of dairy development. The government has initiated various dairy development programs, especially for the eastern and hilly regions.

There are also wide variations in per capita availability of milk in the country. The average per capita availability is lowest in the eastern region and highest in the northern region. The average per capita availability of milk during 2000-01 was highest in Punjab (997 gm/day), followed by Haryana (645 gm), Himachal Pradesh (354 gm), Rajasthan (300 gm), and Gujarat (296 gram). Only 10 states had higher than the national average per capita availability of milk (220 gram/day). The per capita availability is low in the eastern and northeastern states. Milk production and per capita availability in major states during TE 1998-00 are presented. The average per capita consumption of milk and dairy products is lower in rural areas than in urban areas, even though milk is produced in rural areas.

Policies Influencing the Dairy Sector

Agriculture, including the dairy sector, is state controlled, and state governments are primarily responsible for development of the sector. The central government supplements the efforts of the state governments through various schemes for achieving accelerated growth of the sector. Despite the importance of dairying in the Indian economy, especially for the livelihoods of resource-poor farmers and landless laborers, government policy toward this sector has suffered from the lack of a clear and strong thrust and focus. The first attempt to conceive a set of policies for livestock development in India was the Royal Commission on Agriculture (1928). We can divide the government policies into three distinct phases; pre-Operation Flood, post-Operation Flood, and post-reform period.

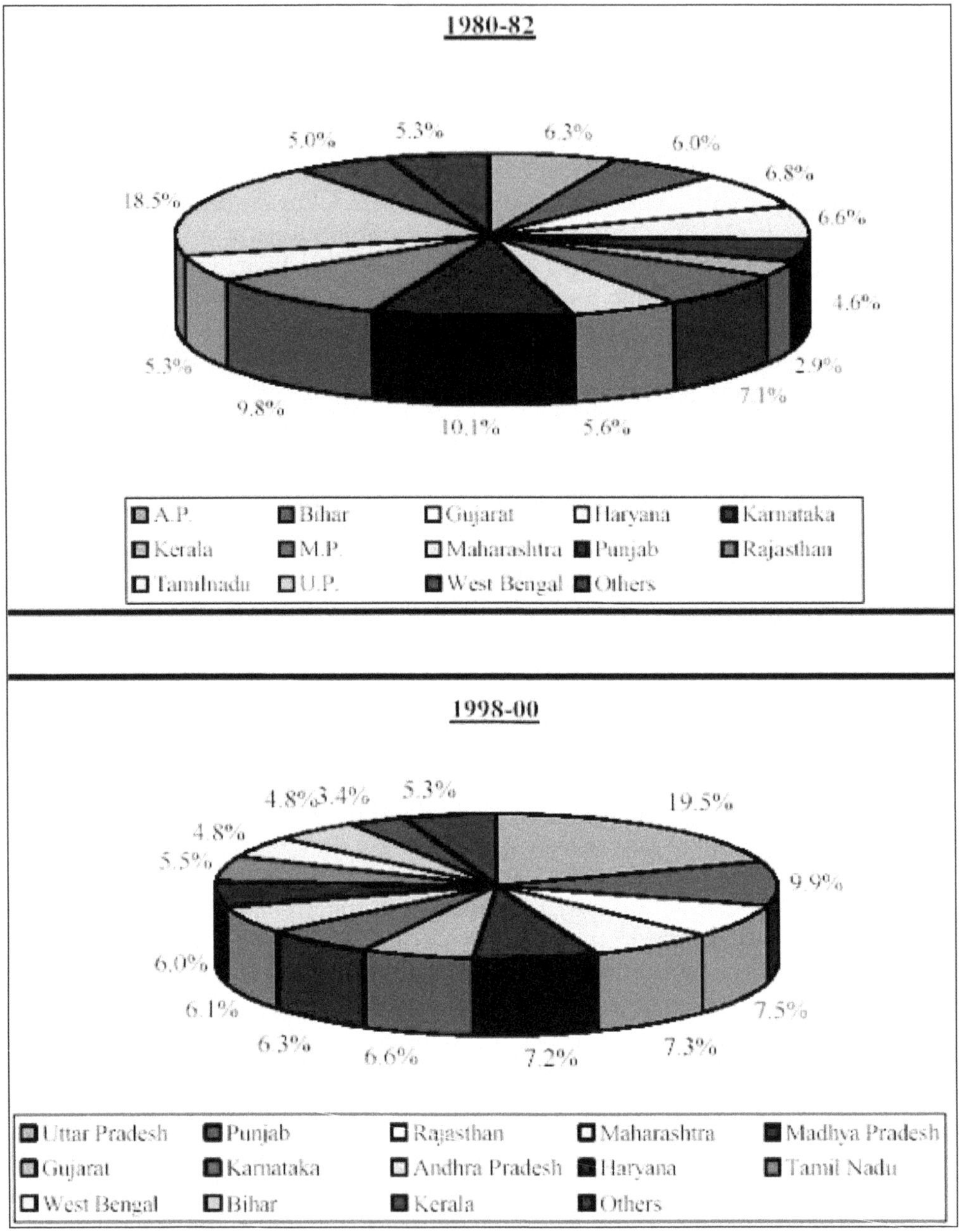

Figure 8.3: Share of Milk Production in India by State: 1980-82 and 1998-00.
***Source*: NDDB, 2003a.**

Operation Flood- A White Revolution

Introduction

Operation flood (also known as the White revolution of India) was a rural development programme started by India's National Dairy Development Board (NDDB) in 1970. One of the largest of its kind, the programme objective was to create a nationwide milk grid. It resulted in making India the largest producer of mil and milk products, and hence is also called the white revolution of India. It also helped reduce malpractices by milk traders and merchants. This revolution followed the Indian green revolution and helped in alleviating poverty and famine levels from dangerous proportions in India during the era.

Earlier, Milk was always available in plenty, but the sad truth was that the middleman was in control of marketing and siphoned away the major share of the farmer's profits. During seasons of plenty, the farmers were forced to drastically cut down their prices to sell off their surplus supplies. And during lean periods, milk production fell considerably, resulting in a shortage, unable to meet the market demand. To deal with this problem, 'Operation Flood' was launched in 1969-70 to initiate and organize cooperatives that would be wholly responsible for the procurement, storage and marketing of milk, thereby eliminating the need and role of the middleman completely. These cooperatives would also convert surplus milk into a range of dairy products so farmers were assured a regular income and fair share of profits, around the 'Operation Flood' heralded the beginning of the White Revolution that brought tangible profits to milk farmers and initiated a dramatic change in their lives.

Operation Flood

Operation Flood has helped farmers, direct their own development, placing control of the resources they create in their own hands. A 'National Milk Grid', links milk producers throughout India with consumers in over 700 towns and cities, reducing seasonal and regional price variations while ensuring that the producer gets a major share of the price consumers pay. The bedrock of Operation Flood has been village milk producers' cooperatives, which procure milk and provide inputs and services, making modern management and technology available to members. Operation Flood's objectives included:

- Increase milk production ("a flood of milk")
- Augment rural incomes
- Fair prices for consumers

Programme Implementation

Operation Flood was implemented in three phases

Phase I

Phase I (1970–1980) was financed by the sale of skimmed milk powder and butter oil donated by the European Union (then the European Economic

Community) through the World Food Programme. NDDB planned the programme and negotiated the details of EEC assistance. During its first phase, Operation Flood linked 18 of India's premier milk sheds with consumers in India's major metropolitan cities: Delhi, Mumbai, Kolkata and Chennai; thus establishing mother dairies in four metros. Operation flood, also referred to as "White Revolution" was a gigantic project propounded by Government of India for developing dairy industry in the country. The Operation Flood – 1 originally meant to be completed in 1975, actually the period of about nine years from 1970-79, at a total cost of Rs.116 corers.

As start of operation Flood-1 in 1970 certain set of aims were kept in view for the implementation of the programmers. Improvement by milk marketing was made by organizing dairy sector in the metropolitan cities including Bombay, Calcutta, Madras, and Delhi. The objectives of commanding share of milk market and speed up development of dairy animals respectively hinter lands of rural areas with a view to increase both production and procurement.

Phase II

Operation Flood Phase II (1981–1985) increased the milk sheds from 18 to 136; 290 urban markets expanded the outlets for milk. By the end of 1985, a self-sustaining system of 43,000 village cooperatives with 42.5 lakh milk producers were covered. Domestic milk powder production increased from 22,000 tons in the pre-project year to 1, 40,000 tons by 1989, all of the increase coming from dairies set up under Operation Flood. In this way EEC gifts and World Bank loan helped promote self-reliance. Direct marketing of milk by producers' cooperatives increased by several million liters a day.

Phase III

Phase III (1985–1996) enabled dairy cooperatives to expand and strengthen the infrastructure required to procure and market increasing volumes of milk. Veterinary first-aid health care services, feed and artificial insemination services for cooperative members were extended, along with intensified member education. Operation Flood's Phase III consolidated India's dairy cooperative movement, adding 30,000 new dairy cooperatives to the 42,000 existing societies organized during Phase II. Milk sheds peaked to 173 in 1988-89 with the numbers of women members and Women's Dairy Cooperative Societies increasing significantly. Phase III gave increased emphasis to research and development in animal health and animal nutrition. Innovations like vaccine for Theileriosis, bypassing protein feed and urea molasses mineral blocks, all contributed to the enhanced productivity of milk animals.

Effects of the White Revolution

The white revolution gave a major boost to the dairy sector in India in the late 1960s by producing milk in rural areas through smallholder producer cooperatives and moving industrially- processed milk from these small holder sources to the urban demand centers, thus, establishing a much needed linkage between the rural producers and urban consumers. The United Nations has commended India's "White Revolution," saying a sharp increase in the production of milk has achieved twin goals of raising incomes of rural poor families and nutrition status of the people. India's milk production rose from around 30 million tonnes in 1980 to an estimated 87 million tonnes by 2003 and despite increasing population, availability per person rose from less than 50 kilo calories per day in 1980 to 80 kilo calories per day in 2000, a report on hunger by the Food and Agriculture Organization (FAO) said. The report forecasts that India's dairy production will triple by 2020.

"With government policies that facilitate rural credit and provide essential support services to promote milk production, the White Revolution will continue to play a significant role in reducing poverty and hunger." FAO estimates that increasing milk production has boosted the incomes of 80 to 100 million families, the vast majority of whom are marginal or small farmers whose plots are often too small to support their families and landless laborers who depend on common grazing lands and forests for fodder.

Matching Demand and Supply

The white revolution carried on the extensive dairy development programs on the supply side, and this coupled with the increasing demand for value added milk products on the consumer side along with the country's population growth, increased urbanization and higher income, led to increased demand for milk. The various other factors contributing to the growth in milk production are given below:

- In the India diet, milk and milk products play a significant role.
- With a large lacto- vegetarian population, milk and dairy products are an important source of protein in diet.
- There is a perceptible shift towards the value- added food habits in which milk products form a larg part.
- On the supply side, technological progress in the production and processing sectors, institutional factors and infrastructure played an important role.

Protection to Milk industry

After the White revolution, smaller enterprises and cooperatives, with whom, milk processing and product manufacturing were mainly restricted, were given protection by means of high import duties, non- tariff barriers, restrictions on import and export and stringent licensing provisions. This also provided supplementary employment and income among the small and marginal farming households and landless wage earners.

Before the White revolution started, the farmer's income was derived almost entirely from seasonal crops. The income from milch buffaloes was undependable. Only private traders and middlemen gained as the marketing and distribution system was controlled by them. But currently, the scheme has ensured average returns of Rs.1600 to Rs.1700 per month, with the beneficiaries being mostly women. The Amul project seeks to bring poor households into the mainstream economy by transforming dairying into active income generating enterprises. As many as 7500 below poverty line beneficiaries were targeted. The number has already been surpassed, and the project has already covered over 8000 beneficiaries. All aspects of dairying have been exhaustively covered under the programme, in which innovations and modern technologies have been put to the best use for the rural population.

White Revolution and Gender Revolution

The White revolution did not only affect the Indian Economy but the Indian social scenario as well. Traditionally, India's dairy cooperative societies were run by men, but with time, this has gradually changed. Today, about 18 per cent of cooperative members are women and nearly 2 all- women cooperatives are functioning in the country. Over 1000 women in Gujarat's Sundarnagar district have proved that modern technologies can ameliorate the worries of poor rural women. Gujarat in fact, spearheaded the 'Operation Flood' movement in the 1970s, which sought to maximize milk production and profits through scientific techniques and attempted to free milk producers from the tyranny of the middlemen. Today, several NGOs are working to change this scenario to be more inclusive of women. The Self Employed Women's Association (SEWA), Ahmadabad, Gujarat was the first one to start the concept of all- women dairy cooperatives in Banaskantha district.

Initially, women were hesitant about depositing milk with the dairy. But through intensive training sessions, they began to see the benefits of associating with the dairy. On an average, a woman is able to deposit one to two litres of milk per day. The women associated with the dairy say that it is not the higher price they get for their milk that makes them keep coming back. There are other benefits, like bonus, payment every 10 days, information about fodder, mineral mixture and cattle health training. Bank accounts and transactions, as well as the day- to- day functioning of the cooperatives are operated entirely by women.

The women of these villages have developed an unshakeable faith in the machines. They are especially pleased with the measurement of fat content in milk. This helps them fix the price of the milk depending on its fat content, instead of the flat rate that the middlepersons gave them. They have also learnt techniques to ensure better health for their cattle, through de- worming and mineral mixture. This, in turn, leads to increased fat content in milk. Further, artificial insemination - which the women can afford because of the profits that the cooperatives earn - results in a better breed of cattle. From the above, we reach the conclusion that the White revolution has had a tremendous impact on the economic as well as the social life of the rural sector of India and has helped in the development of rural sector an established a linkage between the rural and the urban sector of India.

Success of White Revolution

As we discussed earlier, white revolution gave a major boost to the dairy sector in India in the late 1960s by producing milk in rural areas through smallholder producer cooperatives and moving industrially- processed milk from these small holder sources to the urban demand centers, thus, establishing a much needed linkage between the rural producers and urban consumers. The White Revolution started in Gujarat and with its success over the years, other states have started implementing the concept of the White Revolution with a view to simultaneously improve and develop the economic and social parts of their rural sectors and linking them with their urban sectors.

Emulating the concept of Amul, Uttarakhand is preparing to bring in the white revolution through a new 10- point programme prepared by the agricultural minister, Mr. Trivendra Singh Rawat. The first target of the government is to add 200 dairies with an addition of 40,000 liters of milk in order to make the hill state surplus. Under this programme, the government has tied up with the Bank of Baroda to provide easy loans for buying cows and buffaloes and set up dairies. Once these dairies have been setup, the government will start buying milk and produce various dairy products through value- addition.

The state of Punjab is ushering a second white revolution by adopting modern dairy technology which is seen as a variable and profitable alternative. With the aim of providing the latest information about commercial dairy farming, the Punjab government has also started a new dairy technology course at five of its old centers and three new centers. The state dairy development department also provides new techniques, better sheds, cattle breed and artificial insemination to the milk producers to increase milk production in the state. Financial help at very low rate of interest is being provided to candidates who successfully undertake training. Punjab is the second largest milk producing state in India, producing eight million tonnes annually, which is around 10 per cent of the country's milk production.

Second White Revolution

After starting a White Revolution in the country and changing the way India consumes milk and its products, Amul is ready to start a second revolution. The cooperative milk giant is looking to double its milk production capacity by 2020. Years ago, the unique cooperative movement transformed India from an imports-dependent country to a self-sufficient one. Now, the demand for milk has increased with people's purchasing power going up and Amul is hoping to meet this demand with its capacity expansion. GCMMF, which markets all Amul products, expects its sales to shoot up to Rs. 27,000 crore in 2020 from an estimated Rs. 6,700 crore in 2008-09. To achieve the 2020 target, Amul will increase the capacities of the existing 30 dairy plants and add some new ones.

Criticisms

Some critics of the project argue that the emphasis on foreign cow breeds has been instrumental in the decimation of Indian breeds. Foreign breeds give higher yields, but require more feed and are not suited to Indian conditions. Critics also argue that the focus on the dairy sector during this period came at the expense of development, research, and extension work in other areas of Indian agriculture. There is also the criticism that the product from the White Revolution, namely milk and dairy products (like food grains from the harvests using Green Revolution methods and practices) is qualitatively, not exactly 'technically', inferior to the product obtained employing traditional methods and practices geared to smaller population levels which had only to be 'scaled up' for larger populations.

Moreover, the developed countries' heavy subsidies on dairy products pose an obstruction to the Indian Dairy growth. The distortion of global dairy prices due to heavy subsidies by the European Union and the U.S.A, besides other developed countries, has rendered Indian Dairy products non- competitive in the international market. There has been a consistent rise in India's share in world milk production – from 9.9 per cent in 1990 to 12.3 per cent in 1996 and further to 14.5 per cent in 2003. India is now the world's largest milk producer. However, the opening up of the Indian market to foreign goods has raised much concern about the status of the Indian dairy industry in the post-WTO era.

Conclusion

In the end, we may conclude by saying that though every concept has its loopholes and drawbacks, the concept of the White Revolution is relatively much more beneficial. 60 years ago when the concept of White Revolution first came into existence, prior to that, the Indian dairy sector had a very low productivity rate. But after this concept was introduced, it brought about a revolution in the dairy sector. The productivity rate increased and it opened new employment avenues for the rural sector. It further contributed to the empowerment of women. It helped in removing the middlemen, thus, establishing a link between the rural and the urban sector. It came as a relief to the landless farmers and farmers with smaller lands which were inadequate for farming. Thus, it may be said that the white revolution has contributed a lot to the Indian economy and more of such innovative ideas must be brought in light so as to ensure the quick development of the rural and the urban sector and furthermore, of the country as a whole.

One of the indicators of a sector's importance is the budget allocation to that sector. The investment pattern in animal husbandry and dairying during various plan periods is given in Table 8.8. The plan outlay (at current prices) of central and centrally sponsored schemes under animal husbandry and dairying has increased from Rs. 22 crore in the First Plan to Rs. 1,545.64 crore in the Ninth Plan and Rs. 2500 crore in the Tenth Plan. The outlay for dairying increased from Rs. 781 crore in the First Plan to Rs. 900 crore in the Eighth Plan and then declined in the Ninth Plan to Rs. 469.5 crore (all figures are at current prices). The allocation to animal husbandry and dairying as a percentage of total plan outlay varied from 0.98 per cent during the Fourth Plan to about 0.18 per cent during the Ninth Plan (Figure 8.4). However, in most cases the bulk of the budget is eaten up by wages and other administrative costs of the government departments. Although the dairy sector occupies a pivotal position and its contribution to the agricultural sector is the highest, the plan investment made so far does not appear commensurate with its contribution and future potential for growth and development.

The low productivity of Indian cattle has been the central concern of livestock policy throughout the last century. In the First Five Year Plan, the Key Village Scheme (KVS) was launched to improve breeding, feed and fodder availability, disease control, and milk production. To meet urban areas' need for milk, the government promoted state-owned dairy plants to handle milk procurement, processing, and marketing. In 1959, the government Delhi Milk Scheme (DMS) was set up to

Table 8.8: Growth Pattern of Livestock Population in India: 1951-1992 (millions)

Species	*1951*	*1956*	*1961*	*1966*	*1972*	*1977*	*1982*	*1987*	*1992*	*1997**
Cattle	155.3	158.7	175.6	176.2	178.3	180.0	192.5	199.7	204.6	175.0
Adult female cattle	54.4	47.3	51.0	51.8	53.4	54.6	59.21	62.07	64.36	–
Buffalo	43.4	44.9	51.2	53.0	57.4	62.0	69.78	75.97	84.21	84.03
Adult female buffalo	21.0	21.7	24.3	25.4	28.6	31.3	32.5	39.13	43.81	–
Total bovines	198.7	203.6	226.8	229.2	235.7	242.0	262.4	257.8	289.0	259.0
Total livestock	292.8	306.6	335.4	344.1	353.4	369.0	419.6	445.3	470.9	452.5
Annual Growth Rates (per cent)		*1951-56*	*1956-61*	*1961-66*	*1966-72*	*1972-77*	*1977-82*	*1982-87*	*1987-92*	*1992-97*
Cattle		0.43	2.04	0.07	0.24	0.19	1.35	0.74	0.48	–
Adult female cattle		–2.76	1.52	0.31	0.61	0.45	1.63	0.95	0.73	–
Buffalo		0.68	2.66	0.69	1.61	1.55	2.39	1.71	2.08	–
Adult female buffalo		0.66	2.29	0.89	2.40	1.82	0.76	3.78	2.28	–
Total bovines		0.49	2.18	0.21	0.56	0.53	1.63	1.01	0.94	–
Total livestock		0.93	1.81	0.51	0.53	0.87	2.60	1.20	1.12	–

Note: *: Excludes the data for Bihar, Dadra Nagar, and Haveli.

Source: GOI, Basic Animal Husbandry Statistics 2002, Department of Animal Husbandry and Dairying, Ministry of Agriculture, Government of India.

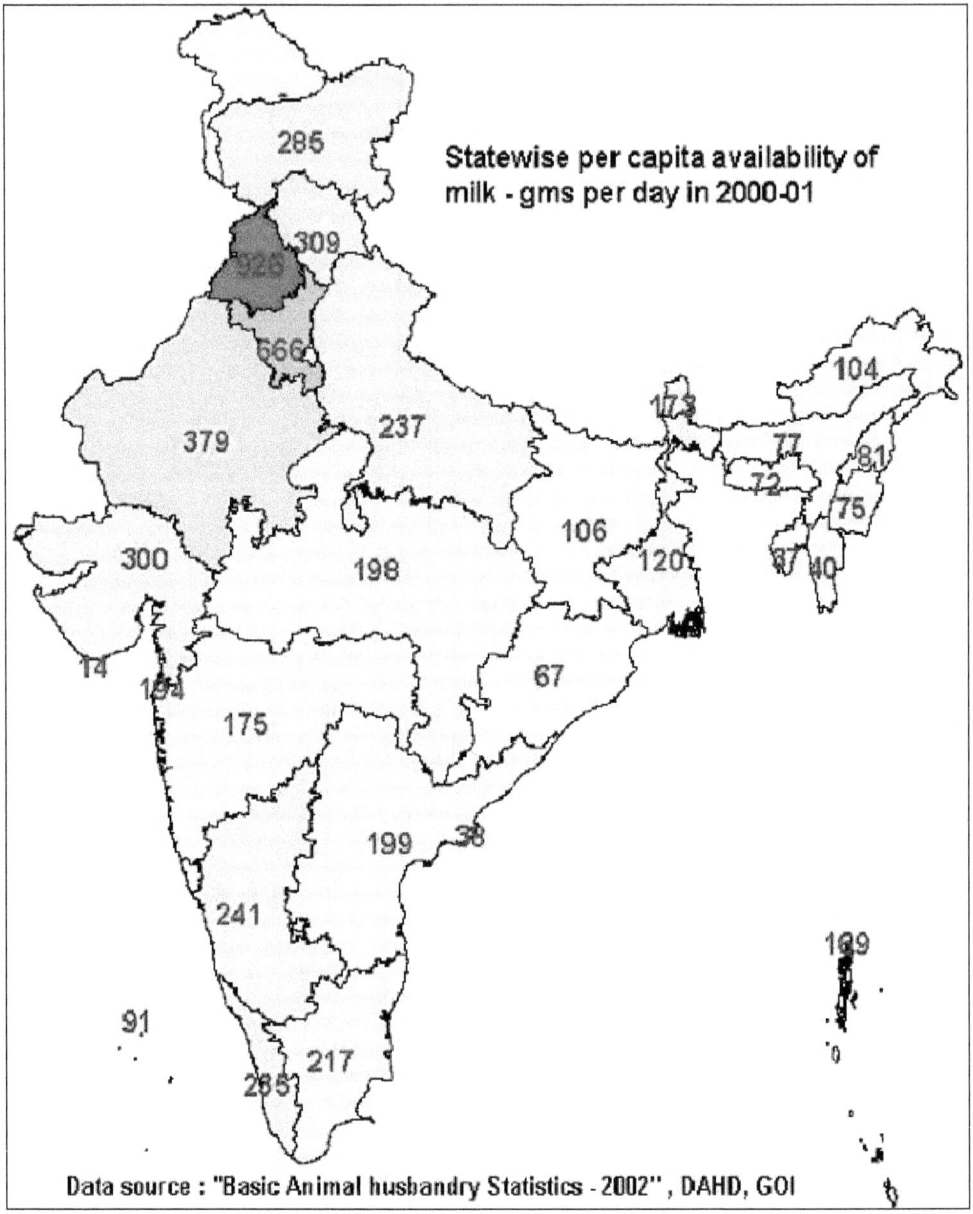

Figure 8.4

supply milk to the urban population of Delhi. This scheme adopted the method of departmental milk procurement from the milk-producing areas around Delhi by setting up its own milk collection and chilling centers. Though the collection was started from small milk vendors initially, it ultimately ended up creating big contractors who purchased milk from the small vendors and supplied it in bulk to the milk scheme. The same policies and strategies continued in the Second Five-Year

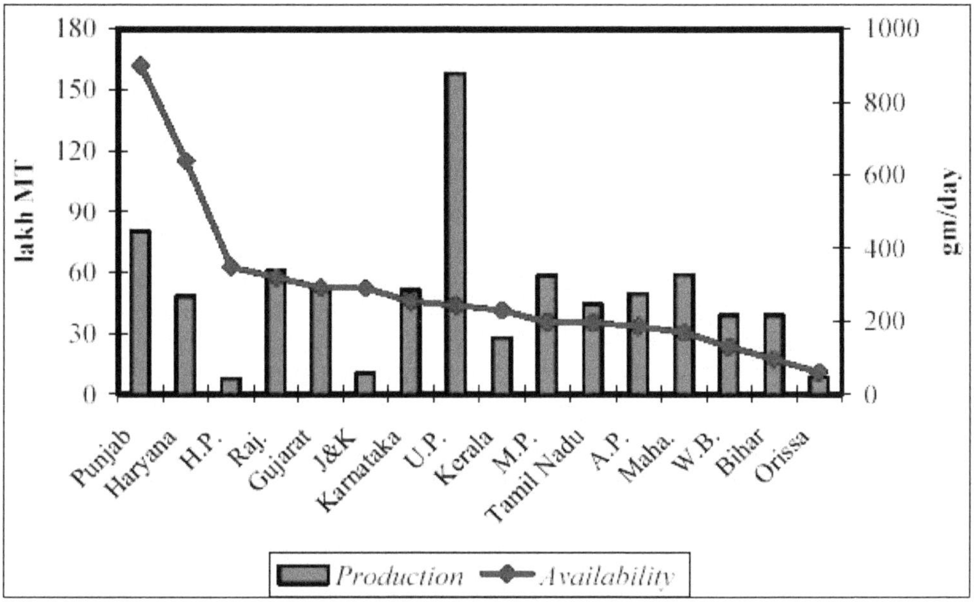

Figure 8.5: Total Milk Production and per Capita Availability of Milk in Major States in India: TE 2000-01.
***Source*: GoI, 2003.**

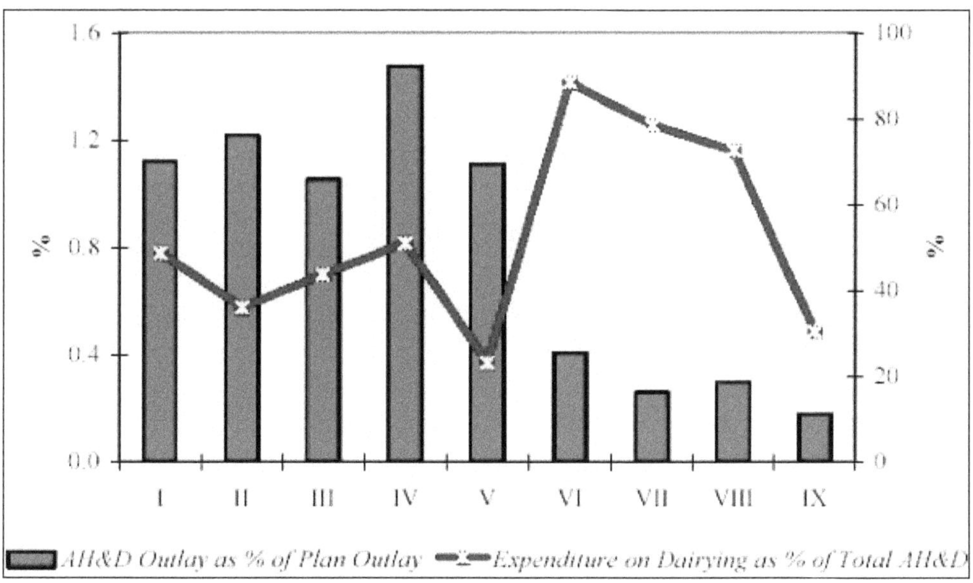

Figure 8.6: Share of Animal Husbandry and Dairying Outlay in Total Plan Outlay during different Plan Periods.
***Source*: GoI, 1999.**

Plan. In 1976, the National Commission on Agriculture concluded that the KVS could not meet its objectives because, due to a shortage of funds, it did not stress feed and fodder development and marketing of milk. The Third Plan emphasized the need to develop dual-purpose animals for milk as well as draft use; crossbreeding of nondescript indigenous cattle was introduced during this plan. The Intensive Cattle Development Programme (ICDP) was launched in areas with high milk potential.

The disappointing performance of the dairy sector during the 1950s and 1960s concerned policy makers, and the Government of India undertook a far-reaching policy initiative. Dairy development through producers' cooperatives and milk production based on milk sheds in the rural areas, modeled on the successful experience of dairy cooperatives in Gujarat, became the cornerstone of the new dairy sector policy. This policy initiative turned the Indian dairy sector around and led to all-around growth with several unarticulated spread effects.

The Government of India launched a massive dairy development program popularly known as Operation Flood (OF) from 1971 to 1996. The program was initially started with the help of the World Food Program (WFP) and later continued with dairy commodity assistance from the European Economic Community (EEC) and a soft loan/credit from the World Bank. Under this program, rural producers were organized into cooperatives so they would have an assured market, remunerative prices, and inputs and services for milk production enhancement, such as better feed and fodder, breed improvement through artificial insemination, and disease control measures. The program was unique in its approach inasmuch as the gift dairy commodities received by India under the program were not consumed by free distribution but were used to manufacture liquid milk, and funds thus generated were reinvested in rural areas in milk production enhancement activities. This coordinated and innovative effort has greatly increased milk production and ushered in a "White Revolution," making India the world's largest milk producer.

However, concerns were raised about these government controls and licensing requirements for restricting large Indian and multinational players from making significant investments in this sector. The government has amended the MMPO from time to time; the major amendment was made in March 2002, when restrictions on setting up milk processing and milk product manufacturing plants were removed and the concept of milkshed was also abolished. This amendment is expected to facilitate the entry of large companies, which would definitely increase competition in the domestic markets.

The second major development in Indian dairy sector policy came when India signed the Uruguay Round Agreement on Agriculture (URAA) in 1994 and became a member of the World Trade Organization (WTO), which made India open up its dairy sector to world markets. The import and export of dairy products was delicensed and decanalized, and trade in dairy products was allowed freely, with certain inspection requirements. The first major step was taken in 1994-95, when the import of skim milk powder (SMP) and butter oil was decanalized; restrictions on the remaining products were removed in April 2002. Moreover, there was a significant reduction in the import tariffs on dairy products after trade liberalization.

However, India had bound its import tariffs for dairy products at low levels in the Uruguay Round schedules.

Livestock Population Trends

India has one of the largest livestock populations in the world, accounting for about 57 per cent of the world buffalo population and 16 per cent of the cattle population (GOI, 2002). The growth pattern of the livestock population during 1951 and 1997 is given in Table 4. Between 1951 and 2002, the cattle population increased from 155.3 million to 175.1 million. The cattle population grew by less than 1 per cent per year between 1951 and 1997, while the buffalo population almost doubled (2.24 per cent per year) during the same period. The cattle and buffalo stocks witnessed a significant acceleration in growth during 1977 to 1982 compared to the previous five years. The rate of increase in the cattle (2.04 per cent/year) and buffalo (2.66 per cent/year) populations was highest between 1956 and 1961 among all the periods considered. The turning point in the composition of the draft animal population was 1977; male cattle population declined from 73.22 million to 61.14 million between 1977 and 1982, and the corresponding decline among male buffalo population was over 1.96 million (GOI, 1999). This declining trend, however, is not uniform across the states. Agriculturally advanced states such as Punjab, Haryana, Andhra Pradesh, Kerala, and Tamil Nadu witnessed a sharp decline in the male draft animal population due to farm mechanization, while the less progressive and hilly states such as Asom, Bihar, Madhya Pradesh, Odisha, and West Bengal showed increasing dependence on work animals.

The crossbreeding program, after a slow start during the 1960s, spread very fast, and the successive rounds of livestock census clearly established the speed with which crossbreeding spread in different parts of the country. In 1992, crossbred cattle constituted about 4.5 per cent of total cattle and about 9.5 per cent of total cows in the country. In states such as Punjab and Kerala, the proportion of crossbred cows is substantially higher than in other states. The proportion of crossbred cows increased while that of indigenous cows declined, indicating the increasing importance of crossbred cows over indigenous cows. The proportion of female buffaloes also increased significantly, from 30.2 per cent to 36 per cent between 1982 and 1992.

In the global context, the performance of the Indian dairy sector appears impressive in terms of livestock population and total milk production but extremely poor in terms of productivity. The main reasons for low yields are inadequate availability of feeds and fodder in all seasons, non-availability of timely and good animal health care and breeding services, and lack of credit. The average milk productivity per year per cow increased from 731 kg in 1989-91 to about 1,014 kg in 1999. Although average annual milk production per animal has improved substantially, it is far below the world average (2,071 kg/year) and that of countries such as Israel (8,785 kg), the United States (8,043 kg), and Denmark (6,565 kg). The available data on milk yield indicate that average productivity went up substantially in the case of cows during the 1970s and 1980s. There is an increase in the yield of buffaloes also, but it is less sharp than that of cows. A key factor accounting for the sharper increase in cow milk yield is the increasing proportion of crossbred cows.

As in milk production and availability, there are wide interstate variations in milk yields. In general, buffaloes have higher yields than indigenous cows, but crossbred cows are more productive than either indigenous cows or buffaloes. The average productivity of local cows is highest in Haryana (4.11 kg/day), followed by Punjab (2.88 kg/day) and Gujarat (2.84 kg/day). For crossbred cows it is highest in Punjab (8.36 kg/day), followed by Gujarat (7.96 kg/day) and West Bengal (7.82 kg/day). The average productivity of buffaloes is highest in West Bengal (6.26 kg/ day), followed by Haryana (5.64 kg/day) and Punjab (5.62 kg/day).

Blue Revolution at a Crossroads

Introduction

There is now little doubt that the world's fisheries are in crisis. Mounting scientific evidence points to dramatic declines in global catches. Increasingly, many are making the case that farming fish offers a solution to meeting the growing demand for seafood that catching fish cannot provide. Aquaculture now accounts for roughly one-third of the world's total supply of food fish and undoubtedly the contribution of aquaculture to seafood supplies will increase in the future. Aquaculture has the potential to become a sustainable practice that can supplement capture fisheries and significantly contribute to feeding the world's growing population. However, instead of helping to ease the crisis in wild fisheries, unsustainable aquaculture development could exacerbate the problems and create new ones, damaging our important and already-stressed coastal areas.

The vast majority of aquaculture takes place in Asia. In 2002, over 70 per cent of worldwide aquaculture production was in China alone. Most farmed fish and shellfish are grown in traditional small-scale systems that benefit local communities and minimize the environmental impact. Utilizing simple culture technologies and minimal inputs, these systems have been used for centuries. The net contribution of these traditional aquaculture systems can be great as they offer many benefits, including food security in developing nations.

However, as happened with the "green revolution" of agriculture in the last century, the current "blue revolution" of aquaculture is becoming an industrial mode of food production. An emerging trend is toward the increased development of farming high-value carnivorous fish species using environmentally and socially damaging systems. Farming fish on an industrial scale, especially carnivorous fish is rapidly expanding; the number of different species farmed and geographic regions where they are farmed increases continually. Largely controlled by multinational corporations, industrialized farming of carnivorous fish such as salmon requires the intensive use of resources and exports problems to the surrounding environment, often resulting in environmental impacts and social conflicts.

Some segments of the aquaculture industry are long overdue for reform. What is required is a paradigm shift in how we think about aquaculture, particularly its interaction with natural and social systems. This new paradigm should be based on sustainable development-"the management and conservation of the natural resource base, and the orientation of technological and institutional change in such a manner to ensure the attainment and continued satisfaction of human needs for present and future generations. Such development conserves land, water, plant and genetic resources, is environmentally non-degrading, technologically appropriate, economically viable and socially acceptable." Sustainable aquaculture must consider the ecological, social, and economic aspects of development (Figure 8.7).

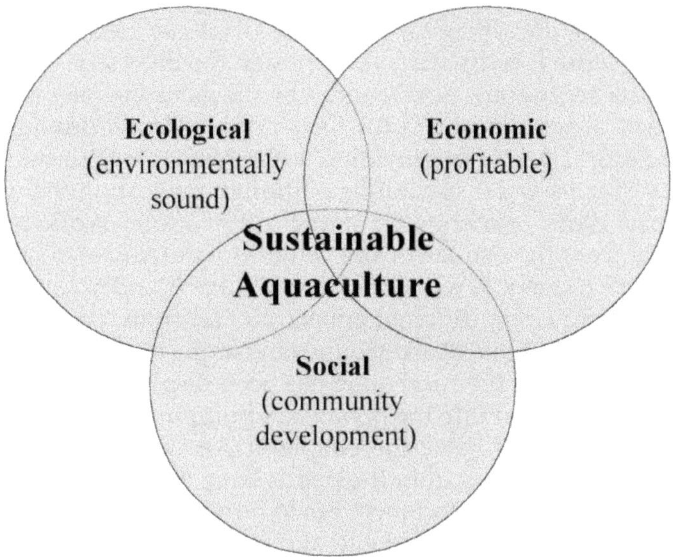

Figure 8.7.

Background

Historical Overview

The town of Turners Falls, Massachusetts, on the Connecticut River about 150 miles west of Boston, was one of the many small New England towns that thrived during the Industrial Revolution at the turn of the twentieth century. The textile factories and pulp mills of that era are long gone, but today the town is part of another development that could have global reverberations. This time, it is part of what might someday be called an *integrated* industrial revolution. One of its newer businesses, housed in a corrugated steel building not far from the river, produces fish. Aquafutures, as the name implies, is not a traditional New England fishing enterprise of the kind that launched its boats from the Massachusetts or Rhode Island coast during the country's early history. Through the use of water recirculation technology, the company one of many in its fast-growing business is helping to radically change the way fish are produced, and ultimately to determine what role aquaculture will play in supplying food to the world.

Inside this one-acre facility, amidst fish tanks, processing facilities, and support systems, Aquafutures' employees raise hybrid striped bass a cross between white and striped bass. Shipped in as fingerlings that weigh 2 grams, the bass are kept in this secure environment for about 10 months during which they are given high-protein feeds. Once they reach market- size of about 800 grams, they are shipped to restaurants and supermarkets. While this facility may seem far removed from nature, it is in fact fairly well designed to mesh with the resource constraints of natural systems.

By using state-of-the-art re-circulating technology, with computers monitoring temperature, oxygen, and chemical levels, the plant can re-circulate each liter of water 250 times before it is discharged. Although the process is energy-intensive, the electrical costs are to some degree offset by the gains in water efficiency. High water demand is a major problem for food production worldwide, whether in irrigating agricultural land or in providing future habitat for fish. Aquafutures has minimized not only its water use but its pollution, replicating the waste manage practices of local dairy farmers who storehouse manure. Workers periodically scoop out waste from the fish tanks and sell it as a nutrient-rich manure to local corn and vegetable farmers. A world away, on Phuket Island in southern Thailand, aquaculture is having a very different impact. Not far from four-star resort hotels, shrimp- breeding ponds have blanketed the landscape. In the village of Ao Goong (meaning Bay of Prawns), the rural quiet has been displaced by a constant drone of generators circulating air into the ponds and pumping out wastes. Unlike their contemporaries in Turners Falls, the people of Ao Goong have not welcomed this new business. The wastes from the ponds have been simply dumped on the ground, causing the villagers' coconut trees to turn brown and their well water to go bad. They can no longer use the wells, and have had to haul in drinking water from somewhere else. When the wastes from the cultivated shrimp were redirected into the bay, wild shrimp were killed - destroying the traditional local economy. For generations, the villagers have made their living by fishing shrimp from the sea, but the arrival of commercial prawn farmers is putting them out of business.

As the people of Phuket Island have learned, the environmental and social costs of shrimp farming can be enormous. Shrimp is the most valuable commodity being produced by the booming aquaculture industry, and arguably the most destructive. But it is not alone: a wide range of indiscriminate and poorly planned fish farming operations are leaving their heavy footprints- destabilizing coastal ecology, degrading farmland, and polluting drinking water supplies. The conundrum is that fish-farming, if properly done, offers great advantages in resource-use efficiency over both traditional fishing and conventional agriculture - advantages the world can ill afford to squander in an era of impending food scarcity. Indeed, integrated fish and rice farming has formed the backbone of traditional agriculture in Asia for centuries. As the world's expanding human population drives up global demand for new food supplies, the great potential of aquaculture, combined with some of the horrendous ecological and human impacts it has had so far, have created seemingly intractable tensions between its promoters and its critics. How that tension is resolved will go a long way toward determining how-or with what limitations - we will be able to eat in the future.

The Blue Revolution

As far as is known, the earliest fish farming (or aquaculture) took place in China, where the common carp has been a staple food for millennia. A document called "Fish Culture Classic," written by Fan Li nearly 25 centuries ago (in 460 B.C.), details numerous experiments on the sizes of fish ponds, stocking rates, and harvest yields. In rural inland areas, often on marginal lands, small carp ponds supplemented the output from farm crops for centuries; the ponds were supplied with nutrients from an adjoining pig pen, or from other animals, ducks, food preparation, or night-soil. Several times a year, the ponds were cleaned out and the resulting rich bottom soil was applied to neighboring fields. When Emperor Lee came to power during the Tang Dynasty (618–917 A.D.), growers shifted from cultivating carp to other types of fish. The shift was an accident of history: the word for carp sounds like the Emperor's name "Lee" in Chinese, and farmers were apprehensive about the connotations of any talk about "eating carp." Other species were put in the ponds. And, because supplies of newly hatched fish varied from season to season, farmers had to make do with whatever they could get, sometimes raising different species together in the same pond. Gradually, a stable and highly productive system of poly-culture raising many fish together–evolved.

Chinese pond-culture systems continue to be among the most productive freshwater fisheries in the world today. One of the keys to their continued success is that nearly all the wastes are put to productive use, so the systems are largely self-sufficient and closed: the waste of one species becomes food for another. In more open systems, unused nutrients attract infestations of bacteria, insects, birds, and other organisms, which upset the balance and deplete the food supply needed by the fish.

In the polyculture pond, available nutrients are matched closely to the feeding habits of various fish. Silver carp and tilapia feed on phytoplankton (microscopic plants), while bighead carp graze on zoo- plankton (microscopic animals). Grass carp, wuchang fish, and common carp eat green fodder, while common carp, black carp, and mud carp forage in sediments at the bottom of the pond. Species are thus combined to maximize yields and minimize inputs, providing a valuable source of food and protein for local consumption. In a pond about two meters deep, for example, an experienced farmer may raise up to 10 tons of fish per hectare each year. This sort of small-scale aquaculture has also been long practiced by rural villages in Indonesia and India, using local fish varieties. During the early twentieth century, Chinese immigrants brought their experience with poly-culture to Thailand. From there, small-scale "integrated" fish farming that which efficiently combined the feeding of fish with the raising of plants–spread throughout Southeast Asia. But it was not until well past mid-century that commercial aquaculture began its boom in earnest.

During the late 1960s and 1970s, fish farming was touted by international development agencies as a new source of protein that would sidestep the problems of traditional agriculture–which was facing a gamut of problems with soil erosion, increasing fertilizer needs, and scarcity of fertile land and irrigation water. Fish farming was viewed as a simple yet elegant means of providing food security for growing populations in developing countries. Simply by flooding marginal lands and releasing easy-to-grow fish that most people could afford to buy, social planners predicted, farmers in those countries could create a glob- al revolution in food production. Indeed, in many ways, aquaculture is a more reliable source of protein than either traditional fishing or farming. Compared to ocean fishing, which can be disrupted when storms or territorial disputes turn boats back, fish farming is less dependent on the weather, less dangerous, and less vulnerable to seasonal ebbs. And while commercial livestock operations depend on a few dozen species of animals, the number of species of fish that can be successfully farmed runs to the thousands.

As a source of animal protein, farmed fish are–in a grain-limited world–a godsend. The estimated "grain feed conversion rates"–the amounts of rice or wheat needed to produce a ton of product–are far more economical for fish than for pork or beef and on par with that of chicken. On average, whereas four kilograms of grain are required for each kilogram of pork, and seven for each kilogram of beef, only two kilograms of grain are needed to produce one kilo- gram of either fish or chicken. In terms of how much of the animal is actually consumed by people, fish are also ideal: approximately 65 per cent of the raw weight of finfish is eaten, compared with 50 per cent of raw weight of chicken and pigs and 40 per cent of sheep. (Because fish are supported by water, they don't have to put as much of their growth energy into bones, so more of their weight is edible.) Fish are also low in fat and cholesterol, and a substance found in fish oil, omega-3 fatty acids, helps reduce blood clotting and in turn the risk of heart attacks– all notable advantages over other meats.

With these and other benefits in mind, in the early 1980s, international aid agencies and resource planners called for a "Blue Revolution"–mirroring the high hopes of the agricultural Green Revolution that had begun nearly half a century earlier, but somehow not anticipating that the Blue Revolution would also replay many of the most vexing problems the Green Revolution had eventually encountered. Aid agencies funded research aimed at genetically modifying fish to resist disease, grow faster, taste better, and thrive in highly controlled environments. And, as had occurred with such developments in agriculture, yields increased dramatically. In 1993, researchers in the Philippines announced that they had successfully cultivated a strain of tilapia that grows 60 per cent faster than its wild cousin. World Bank officials called it the "aquatic chicken." And in a 1995 report, the World Bank referred to fish farming the "next great leap on food production," thanks to genetically improved fish, increased inputs, and technological manipulation of the growth cycle.

Today, aquaculture constitutes one of the fastest growing sectors in world food production. From 12.4 million tons in 1990, farmed fish production nearly doubled to 23 million tons by 1996. For every 5 kilo- grams of beef produced worldwide, there are now 2 kilograms of farm-raised fish. Clearly, aquaculture will play an increasingly prominent role in the world food supplies, meat markets, and seafood production. By no small coincidence, the explosion in aqua- culture production is occurring at the same that the world's fishing fleets have suffered catastrophic losses and many ocean fish stocks have dropped to all-time lows. Marine harvests still account for 80 per cent of world seafood supplies, but fish farmers are quickly altering the balance. Nearly 40 per cent of the salmon consumed today have lived in captivity for most of their lives, compared to just 6 per cent a decade ago. Likewise, 40 per cent of the mollusks–which include oysters, clams, and mussels–and 65 per cent of fresh- water fish have lived mostly in farm environments. The fact that world seafood supplies continue to increase at all is due almost entirely to the phenomenal growth in aquaculture.

Parallel with the rise in output, aquaculture increased from a $9.5 billion industry in 1984 to $36.2 billion in 1995. Its most profitable commodity is shrimp, the frenetic explosion of which has been called an "aquatic gold rush." In the past decade, for example, annual production of giant tiger prawns in Thailand has exploded from 900 tons to 277,000 tons. The world total has jumped more than five- fold, to 500,000 tons. Shrimp farming is now a $6.3 billion industry that spans 50 countries. Driven by industrial countries' growing appetites for seafood, shrimp farming has become a major export industry in Bangladesh, Indonesia, India, Thailand, and China. Yet, all of those countries have problems producing enough food for their own populations. Although high-value carnivorous species dominate the net worth of the industry, production in volume is dominated by freshwater- predominantly vegetarian -fish. In 1995, four of the top five cultured species by weight were species of carp. Worldwide, 10.4 million tons of carp were farmed in 1995–nearly half of all aquaculture production. Other non-carnivorous species, such as catfish, tilapia, milkfish, and shellfish account for an additional 7 million tons, with relatively benign environmental impacts. Almost 85 per cent of aquaculture now occurs in low-income food deficit countries, with China leading the way. Between 1990 and 1995, Chinese out- put increased from 5.8 to 12.8 million tons accounting for 62 per cent of the world's total aqua- culture production. By 2010, it is planning to double the area under fish cultivation, to 7 million hectares. (By comparison, China's total grain area covered 5.7 million hectares in 1994.)

While harvests of wild fish appear to have topped out at close to the maximum level the oceans can sustain, the promise of aquaculture is that it can provide a valuable supplement to traditional fish harvests. Under favorable conditions, estimates the U.N. Food and Agriculture Agency (FAO), aquaculture could supply up to 39 million tons of fish by 2010–about 70 per cent more than is produced today. As defined by the United Nations Food and Agriculture Organization (FAO), aquaculture is the "farming of aquatic organisms including fish, molluscs, crustaceans and aquatic plants. Farming implies some sort of intervention in the rearing process to enhance production, such as regular stocking, feeding, protection from predators, *etc.* Farming also implies individual or corporate ownership of the stock being cultivated' Aquaculture is the fastest growing sector of the world food economy, increasing by more than 10 per cent per year and currently accounts for more than 30 per cent of all fish consumed.

While the world community has only recently viewed aquaculture as a potential solution to the dilemma of depleted oceans, it is by no means a new practice. In fact, the advent of aquaculture dates back millennia, though its exact origins are unknown. It most likely grew out of necessity – foraging and hunting were not sufficient to provide a stable source of food to local communities. While there are many parallels to agriculture, the development of aquaculture has progressed more slowly than terrestrial farming because of the unfamiliar nature of the ocean terrain and characteristics of aquatic organisms. A large proportion of organisms that humans rely on for protein and sustenance come from the sea. Currently, approximately 16 per cent of animal protein consumed by the world's population is derived from fish, and over one billion people worldwide depend on fish as their main source of animal protein. Worldwide consumption of fish as food has risen from 40 million tons in 1970 to 86 million tons in 1998.

Once thought of as an abundant, inexhaustible resource, the world ocean faces a significant loss of essential diversity. This loss is occurring at an alarmingly rapid rate, due to the combined effects of overfishing, habitat destruction, pollution, and profound ecological and biotic change caused by global warming as well as the human-mediated transfer of marine organisms. According to the FAO "About 47 per cent of the main stocks or species groups are fully exploited and are therefore producing catches that have reached, or are very close to, their maximum sustainable limits." Clearly, additional means of producing fish must be developed in order to maintain a sufficient supply of food for an ever-growing population. Aquaculture offers one way to supplement the production of wild capture fisheries and it will continue to increase in importance as demand increases in the future.

It was not until after World War II that aquaculture gained much attention as a potentially large- scale industry. A shift in economic conditions in developed nations of the world led to an increase in the demand for fish such as salmon, shrimp, eels, and sea basses, all of which can be produced profitably through aquaculture. In the 1960's, aquaculture became a significant commercial practice in Asia where it had mainly been used as a small-scale means of local community food production for thousands of years. In the last few decades, worldwide aquaculture production has increased significantly. In 1970 aquaculture operations composed 3.9 per cent of all fish production, compared to 27.3 per cent in 2000. Worldwide, total fish production from aquaculture operations has increased steadily at a rate of 9.2 per cent per year. But, aquaculture has not yet become the large-scale global food replacement for the numerous food-poor areas of the world, as many thought it would be.

The Problems

As stated earlier about fish potential, such plans are ambitious, and rather than sticking with the Chinese model of complex, closed-loop production, development aid agencies and lucrative export markets have encouraged farmers to increase output with fewer species and intensified production. The industry has placed more reliance on artificial inputs such as fishmeal and high-protein feed, chemicals, and antibiotics. In recent years, diverse pond systems have been replaced with a range of high-tech, high-volume systems: floating cages in marine water; tanks with mechanical controls on feed, light, and growth stimulation; and structured raceways that provide artificial flows of water and diversion channels to simul te the natural environment. In short, recent forms of aquaculture have developed into the equivalent of raising fish in marine feedlots–or factories.

Along with these new systems have come new trade-offs, many of which parallel the problems of the Green Revolution that fish farmers were trying to avoid: high volumes of waste, pollution, poor land- use practices, disease outbreaks, risks of genetic inbreeding, and inequitable use of resources. In terms of environmental impacts, shrimp and salmon farming, which together accounted for 6.3 per cent of world output in 1995, are the worst offenders. Indeed, shrimp and salmon farming constitute two of the most resource- intensive food production systems in the world. But the same practices that produce such disastrous results in intensive shrimp ponds and salmon netcages, are being adopted, to varying degrees, in all types of modern fish farming. To raise fish, large quantities of water are needed as a medium for growth and to replenish oxygen and remove wastes. Exact needs vary according to the type of fish raised, size of farm, rate of ground seepage and evaporation, and production rate. In general, intensive production of common carp and tilapia requires the roughly the same amount of fresh water, per ton of meat, as the not-insignificant amounts needed to raise grain-fed cattle or pigs. (To raise one ton of fish requires 8 tons of water, compared to 5 tons of water for pork and 8.5 tons for grainfed beef.) Intensive shrimp production requires up to 10 times more water. These high demands can place enormous pressures on local water supplies. In the Ranot region of Thailand, for example, development of shrimp ponds has been so rapid that the average groundwater level in the area fell by four meters between 1989 and 1991.

One of the key benefits of aquaculture is its potential for reducing pressure on marine fisheries. Yet, for carnivorous species such as, shrimp, salmon, trout, bass, and yellowtail, aquaculture actually increases demands on marine production in order to provide feed for the farmed fish. The carnivores require amino acids from other fish for growth, which are provided in the form of high-protein feed pellets made from wild fish. An estimated five kilo- grams of oceanic fish reduced into fishmeal are required to raise one kilogram of farmed ocean fish or shrimp, representing a large net protein loss. Intensive aquaculture's high demand for fishmeal is now a leading factor contributing to marine over- harvesting. Worldwide, about a third of the global fish harvest, or 30 million tons, goes to nonfood uses, primarily animal feed, fish meal, and oils. Of this, 17 per cent goes to feed fish, while the balance is used to feed cattle and poultry. It has been estimated that by 2010, carnivorous fish on farms could be taking all of the world's fishmeal, using up needed protein that could otherwise be used for direct human consumption - a redistribution of marine biological wealth from the poor to the rich.

Further pressure is put on marine ecosystems by aquaculture's reliance on wild stocks to supply seed. Fish culture in inland lakes and reservoirs of Cambodia, Thailand, and Vietnam rely heavily on small, wild-caught fish for fingerlings, while industrial shrimp farmers divert estuarine waters into large ponds and essentially cherry-pick the best larvae for cultivation, leaving the remains behind for wild harvesting. Added to these high levels of water and food inputs, is the need for land. Under pressure from Western and Asian business investors and multinational corporations, local governments and authorities from countries such as El Salvador and Thailand have been coerced into granting access to prime agricultural and coastal lands to aquaculture development. Inland fields are flooded for catfish ponds, mangroves are cut down to make way for shrimp farms, and coastal waters are fenced in to raise salmon, mollusks, seaweed, and even cod, often with little regard for previous uses of the land. Whereas traditional forms of aquaculture are designed to be compatible with surrounding resource uses, intensive aquaculture is often imposed on local ecosystems, with significant environmental consequences that extend far beyond the farm.

One of the biggest problems is water pollution. Because fish are raised directly in the aquatic environment, fish waste, uneaten feed, and fecal matter accumulate in the receiving waters and can float directly downstream and into water supplies. Scottish researchers have estimated that between 300 and 1,000 kilograms of solid wastes are produced from each ton of fish raised–up to a ton of waste for each ton of fish. Most of the waste is in the form of organic solids, dissolved nutrients, and inorganic nutrients such as nitrogen and phosphorus. Though they are not toxic, these wastes can cause over nutrification and the spread of algal "blooms"–massive blankets of green slime on the water's surface that precipitate bacteria growth, deplete oxygen, and kill much of the life in the water below. Because of high stocking rates, salmon netcage farms in British Columbia, on the Pacific coast of Canada, were found to be producing the waste equivalent of a city of half a million people.

Raising large volumes of fish in controlled environments also puts abnormal stress on the fish themselves, which increases vulnerability to outbreaks of disease both on the farm and in surrounding waters. Fish biologists have identified more than 50 bacterial diseases and host of other health problems–threats that are prompting aquafarmers to inject antibiotic, vaccines and chemicals to rid their crops of illness. Scottish authorities report, for example, that a newly approved organophosphate chemical called azamethiphos is being used to control the sea lice that infest the flesh of salmon and other fish, but that this chemical also has a tendency to kill young lobsters and herring. Escapes of domesticated, farmed fish into rivers, lakes, and coastal areas have become a growing risk to already decimated stocks of wild fish. Yet, such escapes are not uncommon. In tropical freshwater systems, two-thirds of escaped fish species have become established. In Europe, 30 per cent of all exotic aquatic species came originally from fish-farm escapes. Salmon are especially vulnerable to displacement by farmed stock, as they have specific territorial behavior that is essential for survival in the wild. An accident with a ripped net in 1997 allowed nearly 300,000 farmed Atlantic salmon in Puget Sound to escape into the Pacific. These non-native fish could undermine efforts by federal and state officials to restore endangered wild populations in the Columbia River and elsewhere. restore endangered wild populations in the Columbia River and elsewhere.

While mismanaged fish farms can "leak" into surrounding areas, they can also exhaust the areas set aside for them. Over time, their high demands for water, feed and sometimes for electric power and other artificial inputs, combined with high levels of pollution and disease, can lead to declining yields and, eventually, pond abandonment. Again, shrimp farming provides the most dramatic illustration: Between 1965 and 1985, shrimp farmers in Taiwan intensified production from 1.4 tons per hectare to 3.6 tons per hectare. But by the late 1980s, the industry in Taiwan had completely collapsed due to a series of disease outbreaks and financial disasters– leading many operators to abandon their ponds. This practice is not unlike the slash-and-burn farming of tropical forests, which become quickly depleted and are often abandoned after a few years. Between 1985 and 1995, approximately 150,000 hectares of shrimp ponds fell into disuse worldwide, according to estimates by the World Wildlife Fund International. And the rate of abandonment–like that of slash-and-burn deforestation–appears to be increasing. While the average fish pond is good for about 10 years, for high-density shrimp ponds, the typical lifespan is even shorter–about 5 years.

Such unsustainable practices have exacted a heavy cost both to local peoples and to the environment. Once limited to ponds in Taiwan and Thailand, shrimp farming -reinforced by armed members of the international "prawn mafia" - has swelled into a wave of development spreading across South Asia and Latin America, and threatens to overtake shoreline areas in Africa as well. Ironically, the people who suffer the most from such development are often those who have been the most dependent on fishing for their livelihoods. The traditional fishers are not the ones profiting from the new mass-production operations. What happened in the Thai village of Ao Goong, for example, is happening now in Bangladesh. Since the 1980s, villagers in the district of Khulna, about 350 kilometers southwest of Dhaka, have been protesting the incursions of commercial shrimp cultivation. In the late 1980s, the area was declared a shrimp-free zone at the villagers' insistence. However, influential shrimp traders persuaded local authorities not only to throw out the zoning but to physically break down protective embankments around agricultural land, flood the fields with saltwater to ruin the crops, and open the way for more shrimp development. In recent years, village women have been physically and sexually abused by shrimp farmers and their guards, and at least 100 villagers have been killed in clashes over land acquisition disagreements.

Environmental Costs

Like other forms of intensive food production, industrial-scale fish farming generates significant environmental costs. These depend on a number of factors and are rarely evaluated before farming begins or before operations are expanded. In presiding over the rules and regulations that govern the establishment of aquaculture ventures, many governments have largely overlooked the environmental problems that may arrive with the advent of large-scale fish production.

The expansion of some forms of aquaculture, particularly salmon and shrimp farms, has proven to be destructive to the natural environment and populations of aquatic animals. Industrial scale farming of salmon in netpens and shrimp in coastal ponds are the most problematic because they require the intensive use of resources and export problems to the surrounding environment. Unless significant changes are made, the increasing production of high value carnivorous fish promises to be just as unsustainable.

There are environmental concerns with other systems, such as inland ponds, flow-through systems, and closed systems, but they generally have fewer environmental problems overall, especially when properly managed. For a brief overview of the major environmental implications of the different systems, as well as the species commonly cultured in each, please see Appendix-I.

Social Conflict

The advent of industrial aquaculture has not only been a source of environmental concern, it has also been a source of social concern. Intensive practices used by some in the aquaculture industry pose significant threats to traditional, community-centered societies. Shrimp, for example, is a highly valued species in such regions as the United States, Japan, and Europe. However, it is mostly farmed in tropical areas, with production controlled by multinational corporations. Because shrimp production yields such a high profit, foreign or non-local individuals tend to exploit the areas where they can best be grown, without regard for the environmental or social implications, and then market them elsewhere, taking the profits as well as the product out of the region. The ecological impacts of large-scale shrimp production can leave the farming area unsuitable for small-scale production, damage local fisheries, and as a result damage the local economy.

Finfish farming, especially salmon, has been promoted as a means of generating socioeconomic benefits in rural coastal areas. As farming methods have become more intensive, however, employment opportunities have declined. For example, according to the Norwegian government, between 1994 and 2000 the number of people employed raising salmon from hatcheries to harvest declined by 18 per cent while production more than doubled. Additionally, as the industry attempts to further reduce costs and increase profits, it likely will become more sophisticated and employment will continue to decline. The industry has also impacted coastal communities where salmon are fished commercially. This became evident in the 1990's as salmon farms in Chile took advantage of cheap labor to grow and export large quantities of farmed salmon, depressing the prices paid to commercial salmon fishermen in the United States and Canada. Partly as a result, many fishermen along the Pacific Coast of North America lost significant sums of money or went out of business. Between 1990 and 2002, the number of commercial salmon fishermen in Alaska declined from 10,487 to 6,567 and the value of the salmon harvest fell from $559 million to $130 million in 2002.

Making Improvements

Recently, governments and industry have made efforts to curb unsustainable aquaculture activities. Many have realized that the focus of aquaculture must evolve into the development of an industry that is both environmentally and socially sustainable in the long term. Positive changes are being made with this growing interest and awareness of improving the sustainability of aquaculture. With the expected expansion of aquaculture in the coming years it will be vital that sustainable practices be implemented and further developed to avoid environmental and social problems.

Sustainable Alternatives: A Paradigm Shift

Many have looked to aquaculture to fill in the gaps left by dwindling fisheries. Undoubtedly, this is a genuine possibility. Its realization, however, will depend on how the aquaculture industry develops in the future. At this point in time, the development of aquaculture is at a crossroads. Unsustainable development will only generate short and medium-term profits for multinational corporations at the expense of long-term ecological balance and social stability. More sustainable development alternatives are needed to ensure that in the future aquaculture can contribute to the growing need for seafood products.

There are a number of alternative ways forward in the development of aquaculture, which can offer more sustainable solutions. In some cases these methods have been around for centuries, but they have rarely been adopted in the modern aquaculture industry, and in other cases they are innovative practices that can be explored by aquaculture proponents. Alternatives include ecological aquaculture, organic aquaculture, polyculture, mollusc farming, and closed and low discharge systems. These alternative practices have been successfully implemented in different areas of the world; however, they must be examined for their application on a wider scale. While each of the potentially sustainable practices mentioned in the following discussion does have some environmental impacts, they can be greatly minimized if the systems are managed well. In addition to environmental considerations, social and economic aspects must be considered when assessing each of these practices.

Sustainable Aquaculture

The most destructive forms of aquaculture still comprise a small share of global output. But they are disproportionately threatening to the future viability of aquaculture as a whole, and to the level of confidence we are likely to have in what could turn out to be a critical food source in confronting future food scarcity. The problem is that while shrimp and salmon are produced in smaller volumes -so far -than carp or mollusks, they are more profitable and more in demand by the expanding international market. The ripple effects of the land grab being felt around the rims of Asia, South America, and the Caribbean can be seen in the rivers and ponds farther inland as well: more pressure is being put on farmers to shift to rapid, mass-production monoculture. Both natural systems and the local human communities they sup- port are being elbowed out.

A critical need, then, is to promote commercially attractive alternatives to slash-and-burn-style aqua- culture. One of the advantages of aquaculture is the great variety of available species and methods of cultivation that can be adapted to fit local resources and local needs, whether for supplemental food, income, or jobs. In this respect, though the scale and volume of the small family fish farm may hold little interest in the fast-track commercial world, the model of Chinese polyculture offers a key benefit. By using household and livestock effluent as a nutrient, the traditional system neatly solves two problems that in the modern world tend to be treated separately. In recent decades, however, the increased volumes of waste produced–driven by rising population and rising per-capita consumption–has turned waste disposal into a high-profile problem. The business of waste disposal has been catapulted from a marginal trade for urban waste disposal can be developed that allow the nutrient or energy value of waste to be turned from a net cost to a net value, those strategies could offer attractive opportunities for investment.

One such strategy is to recycle waste nutrients into agriculture (see Gary Gardner, "Human Waste: Pollutant or Resource?" January/February). A parallel one is to use the waste to feed fish–and having the fish serve as a means of wastewater treatment. Some studies indicate that fish can reduce water treatment costs by 30 to 90 per cent. In parts of India, this principle has been practiced for decades. Since 1850, the wetlands east of Calcutta have been used for brackish-water aquaculture. Until the 1920s, the water supply came from the River Bidyadhari. When it ran dry in 1928, fish farmers began experimenting with sewage, dividing the wet- lands into a series of ponds for aerating sewage, growing fish and vegetables, and diverting semi-treated water to agricultural areas. Today, Calcutta has one of the largest integrated resource recovery systems in the world. More than 12,000 hectares of wetlands are divided into four integrated recover y schemes: garbage-composed vegetable farms, sewage-fed brackish-water aquaculture, wastewater-fed freshwater fishponds, and paddy fields using fishpond effluent. Smaller systems of this kind are also operating in China, Germany, Thailand, and Vietnam.

The human health risks associated with sewage-fed fish have been addressed to some extent through low- tech filtration and maturation ponds, careful monitoring, and public education campaigns about the risks of eating raw fish. Thorough cooking can destroy most pathogenic bacteria, although the people handling the waste and raising the fish remain at risk of illness. A 1990 report by the Asian Institute of Technology, ICLARM, the UN Development Program, and the World Bank Water and Sanitation Program found "no evidence that consumption of fish from sewage-fed ponds increases the risk of excreta-related disease." An alternative application of this principle is to take it one step further, using the waste generated by aquaculture as an input to another industry–as in the case of the Massachusetts-based Aquafutures company selling its fish waste to local vegetable farmers. Many facilities in the United States and Europe are now cultivating hydroponic vegetables, fruits, and herbs along with fish. Plants float directly on fish pond water with their roots suspended, or in a connected side-channel or pond. Facilities in Missouri, for example, raise tilapia in a tank and then release the fish effluent water into gravel beds, where it feeds vegetables and herbs. The water can then be recirculated back to the fish tank. And, what works for fish waste could eventually work for other waste streams as well: hydroponic systems can be as effective at removing high-tech methods available, according to a recent report by the New York-based Environmental Defense Fund.

The technical capability to recirculate water may be a key to the success of such systems in places where fresh water is in short supply. Although complex in design and expensive to build, recirculating systems have been widely adopted by large cities in such diverse countries as China, India, Germany, Israel, and the United States. Used in fish production, they can reduce total water use to one-twenty-fifth of that required by a traditional flow-through system–while giving the farmer bet-

control of other inputs and ultimately more efficient production. Israel leads the world in the development of water recirculating systems that combine aquaculture and agriculture. In a typical Israeli system, fish farming is the first in a chain of users, in which each link of the chain produces a product of high economic value by allowing several users to benefit from the same water. On the Israeli-Egyptian border, closed pond systems appear from an airplane as large bubbles on the desert -aquatic greenhouses teeming with tens of thousands of tilapia and carp. These systems have advanced from flow-through methods -in which water is continually flushed through to remove wastes-to biofiltration methods with compartments for removing wastes and recirculating water. (Bio-filtration is essentially a natural cleansing process that uses bacteria to degrade the organic waste from fish.) These bubble environments are highly con- trolled, with paddle wheel-like devices stirring the water to keep it oxygenated, and pH levels, temperature, and nutritional content monitored and adjusted by computer. The bubbles are made of a special plastic that absorbs heat energy during the day and releases it at night. Under these optimized conditions, tilapia grow to full size in about 12 months, thus producing much larger yields than in traditional open ponds where they take 17 months.

What is Necessary to Make Fish Farming Sustainable	**And How it Can Be Done**
Stop the decimation of sensitive ecosystems	Restrict locations of fish farms to assure no net loss of mangrove forests or other threatened environments ☆ Use biofiltration to degrade fish waste ☆ Maintain buffers between ponds, water sources, and filtration systems
Protect local peoples' traditional livelihoods	Locate farms so there is no displacing of local fishing and spawning grounds ☆ Guarantee local people access rights to fishing grounds, forests, and agricultural land
Reverse net protein loss	Raise fish that require little or no fish-meal in their diets ☆ Promote consumption of herbivorous species such as catfish, crayfish, tilapia, carp, and mollusks
Stop Escapes	Ban net-cages from coastal waters ☆ Monitor farms for escape routes, and keep cages and holding facilities in good condition ☆ Raise native strains rather than introduced species ☆ Employ conventional breeding programs rather than transgenic technologies
Halt Pond Abandonment	Reduce stocking rates of fish ☆ Restrict inputs of chemicals, antibiotics, and pesticides ☆ Allow ponds to lie fallow ☆ Remove wastes regularly ☆ Provide regular maintenance ☆ Require monocultures to pay for equivalent area of land rehabilitation ☆ Rotate crop production and maintain a variety of species

Recirculate Water	Charge fish farms for water use and develop market incentives for recirculating technologies ✰ Invest in R&D to improve recirculating technologies.
Integrate with Other Industries	Use household wastewater to feed fish ✰ Use locally available ingredients to make feed ✰ Raise hydroponic vegetables with fish ✰ Apply fish-farm manure to agricultural land ✰ Reuse water from fish farms for other industries ✰ Provide tax incentives for closed-loop production and resource efficiency ✰ Certify and label aquaculture products produced sustainably

A drawback of these systems is that they are expensive to set up and consume prodigious amounts of energy. The investment costs for recirculating systems run between $4 and $8 per kilogram of annual output, compared to $2.20 to $3.30 for a pond or raceway system. On the other hand, if all aquaculture operations were charged for their discharges, these systems would be more competitive. In another kind of industrial synergy, U.S. electric utilities are now promoting aquaculture as a way to recycle warm water discharge from power plants. Leasing utility land and selling power plant waste heat to farmers growing fish in warehouses can make good use of this untapped resource, which would otherwise be a pollutant. The partnership also provides the utility with a new high-demand customer. Near Bismarck, North Dakota, the Golden Fisheries company operates a year-round aquaculture facility for warm water species such as tilapia, using hot waste- water from a coal-fired power plant across the street. In addition to the wastewater, however, the 2-acre dome structure consumes 4.1 million kilowatt-hours of electricity per year, as much as a small town. If energy efficient methods are not adopted, the high energy demand of these intensive systems could make them part of the problem rather than a potential solution.

As with traditional agriculture, technological advances can help address some of the problems associated with aquaculture. Norwegian researchers, by designing fish food that is more easily digested, and by targeting it to the fish more precisely, have reduced the amount of feed needed to produce salmon by 50 per cent, resulting in an 80-per cent reduction in discharges of solids. They have also found a way to feed salmon successfully without adding fish oil, though it could be a decade or so before the new formula is commercially viable. Soon they expect to eliminate the need for fishmeal altogether, by genetically altering the salmon's nutritional requirements. And new efficiencies are being gained in cleanup, as well as in production. Trout farmers in Idaho, for example, have made some progress in reducing phosphorus discharge by switching from sinking feed pellets to ones that float, thereby reducing the loss of uneaten food.

Ecological Aquaculture

Ecological aquaculture has been defined as "an alternative model of aquaculture research and development that brings the technical aspects of ecological principles and ecosystems thinking to aquaculture, and incorporates – at the outset – principles of natural and social ecology, planning for community development, and concerns for the wider social, economic, and environmental contexts of aquaculture." There are six main principles of ecological aquaculture: to preserve the form and function of natural resources; to ensure trophic level efficiency (using animal wastes and plants, rather than fishmeal as sustenance); to ensure that chemicals and nutrients from the system are not discharged as pollutants; to use native species so as not to contribute to "biological pollution"; to ensure that the system is integrated into the local economy and community in terms of food production and employment; and to share the practices and information on a global scale. Ecological aquaculture focuses on the development of farming systems that preserve the environments in which they are situated and enhances the quality of these environments while at the same time maintaining a productive culture system. All aspects of the operation are interconnected in order to minimize negative impacts on the community, both natural and social. Ecological aquaculture can also be incorporated into sustainable fisheries management and coastal zone management.

Organic Aquaculture

Sustainability is one of the main goals of organic food production. Aquaculturists who aim for organic farms wish to "manage food production as an integrated, whole system that is an 'organism' whose individual parts mesh together into one whole production system." In organic food production, all parts of the operations are connected and integrated with each other, such as the nutrient inputs, the animals, the environment, and the wastes being produced.

Organic fish producers must comply with all of the same regulations that other organic certified producers do. Some substances or practices are prohibited from organic operations. For example, the addition of antibiotics to the fish feed is tightly regulated and the inclusion of genetically modified organisms is strictly forbidden in organic production. Rather than rely on the use of chemicals and drugs to improve the production of their fish, farmers instead optimize the living conditions, through lower stocking densities and cleaner, healthier water.

Organic aquaculture standards have been developed in many nations around the world and they are in the final stages of development in the United States. Some of the basic principles of organic aquaculture according to the International Federation of Organic Agriculture Movements are as follows: to encourage natural biological cycles in the production of aquatic organisms; using feed that is not intended or appropriate for human consumption; using various methods of disease control; not using synthetic fertilizer or other chemicals in production; and using polyculture techniques whenever possible. There are several obstacles to the implementation of organic aquaculture, including: farming carnivorous fish with a diet of wild (non-organic) fish, management and recycling of wastes, escapes of fish, and controlling diseases and parasites.

Similar to organic certification, other certification and labeling programs also are under consideration. These "ecolabels", together with organically farmed seafood, have great potential for allowing consumers to know that the fish they purchase were farmed in environmentally friendly ways. Additionally, they could provide responsible farmers with a way to take advantage of the increasing consumer demand for food grown with environmentally friendly practices as well as receive a premium price for their product.

Polyculture and Integrated Aquaculture

Polyculture and integrated aquaculture are methods of raising diverse organisms within the same farming system, where each species utilizes a distinct niche and distinct resources within the farming complex. This may involve the rearing of several aquatic organisms together or it could involve raising aquatic organisms in conjunction with terrestrial plants and/or animals. In either case, the wastes from one organism are used as inputs to another, resulting in the optimal use of resources and less pollution overall.

Polyculture systems can provide mutual benefits to the organisms reared by creating symbiotic relationships while allowing for a balanced use of the available aquatic resources, whereas intensive monoculture systems extract resources from the system and place more stress on the surrounding environment. In addition, integrated systems can increase the economic efficiency of fish farms through improved conversion rates of input materials. Polyculture and integrated aquaculture have the potential to address some of the problems that arise from the intensive rearing of single finfish species. For example, the integration of fish culture with the culture of algal and/or shellfish species shows potential for reducing the risks of eutrophication and also for exploitation of the large amounts of wastes produced by fish farms. Further research is needed however, to determine the effectiveness of such systems, especially in open marine environments.

Polyculture systems are not a new concept; on the contrary, they have been used for centuries. For over one thousand years fish farmers in China have produced four of the most widely cultivated fish species together in the same pond: silver carp (a phytoplankton filter feeder), grass carp (a herbivorous plant feeder), common carp (an omnivorous feeder), and bighead carp (a zooplankton filter feeder). This type of system utilizes available food and water resources, with the effect of reducing costs and increasing efficiency and production. Although still experimental, other systems, such as the integration of seaweed, fish, and abalone culture, and the polyculture of shrimp and tilapia, have proved to be ecologically efficient methods for growing a variety of organisms and may increase profits at fish farms. It should also be noted that although polyculture systems based on netpens may prove beneficial for waste reduction, they fail to eliminate other problems associated with netpen aquaculture, specifically escape of fish, disease transfer, and discharge of chemicals.

Mollusc Farming

While not a new activity, mollusc farming offers an excellent alternative to finfish farming. With proper siting and planning, mollusc farms can have minimal impact upon coastal ecosystems and communities. In some cases the presence of mollusc farms can actually improve the water quality of the existing environment. Oysters, mussels, clams, and scallops are some typically farmed mollusc species.

In most cases mollusc farming does not have the same impacts upon the aquatic environment that finfish culture does, such as nutrient loading and eutrophication. Unlike other farmed species, shellfish remove organic particulates, suspended matter, and nutrients from the water and tend to improve the overall quality of the water. Because shellfish are filter feeders they act as natural biofilters in the water, removing phytoplankton, sediments, and organic particles. For example, it is estimated that every kilogram of shellfish meat harvested results in a removal of 16.8 grams of nitrogen. While relatively benign when properly planned and managed, in some instances, mollusc farming has been responsible for the introduction of non-native species and diseases as well as conflicts with other user groups.

Closed and Low Discharge Systems

Recirculating Systems

Concerns for water conservation and reduced waste discharges have prompted the increased use of closed recirculating aquaculture systems. Recirculating systems generally consist of land- based tanks with constantly flowing water. The systems are made up of three basic components: culture chamber, settling chamber, and biological filter. Water enters the culture chamber, flows through the settling chamber and then moves through the biological filter to remove additional particulate matter. The water is then circulated back through the systems' culture chambers. Recirculating systems conserve water and allow producers to control all of the environmental factors that might affect their plants and animals. For example, aquaculturists have complete control over temperature, salinity, oxygen, predators and introduction and transfer of diseases.

Recirculating systems, however, can be costly to operate, as they are highly dependent on electricity or other power sources. Pumps must be used in order to maintain the constant flow of water and often water must be heated or cooled to the desired temperature. Backup systems must be in place in case of a power failure. A less expensive and more environmentally friendly option would be to take advantage of alternative energy and heating sources. Solar, wind and geothermal power are being considered as is heated water obtained from the waste products of manufacturing, electricity production, and composting. For example, tilapia farms can use the cooling water from power plants as a low cost warm water source. The warm water, which is necessary for growing tilapia, would otherwise be wasted.

Recirculating systems have less of an impact upon the environment because of their closed nature – wastes and uneaten feed are not simply released into the ambient environment in the manner that they are with netpens and exotic species and diseases are not introduced into the environment. In recirculating systems, wastes are filtered out of the culture system and disposed of in a responsible manner. Recirculating systems can be built just about anywhere, including in urban settings where they can use existing structures and be placed close to markets, thereby reducing transportation costs. Recirculating systems can be used to grow a wide variety of fish species year-round in controlled environments. Species commonly grown in recirculating systems include hybrid striped bass and tilapia. Additionally, much research has been dedicated to developing recirculating systems for marine species of fish and this technology holds much promise.

Inland Pond Culture

Pond aquaculture was the earliest type of aquaculture to be practiced, dating back several thousand years. It remains one of the most common systems used to farm fish. Over 75 per cent of the farmed freshwater fish in China are produced in constructed ponds and in the U.S. nearly all of the farmed catfish are raised in ponds. Other species commonly farmed in ponds include tilapia, hybrid striped bass, and crayfish. When properly managed with little or no water exchange, wastes are not continuously released from the culture system as they are in netpens or flow through systems. Wastes, including solids, organic matter, and nutrients that are discharged can be released into settling ponds or constructed wetlands to further reduce their impact. In low or zero discharge pond systems, the pond itself develops a sort of ecosystem. Natural biological, chemical, and physical processes in ponds remove much of the wastes from the water. Waste accumulation and subsequent discharge from ponds is minimal as long as production is maintained within the limits of the natural processes to remove or assimilate wastes.

Blueprint for a Blue Revolution

If the fish-farming industry succeeds in trans- forming itself so as to become both highly productive and sustainable, what will it look like? Here's what the research suggests:

First, fish will be more popular than ever. In the past half-century fish consumption has boomed, but as the industry takes better advantage of its inherent resource efficiencies, it will boom even more. The most ubiquitous food of urbanized societies in the future may not be the hamburger, but the fishburger.

Second, there will be a shift in what kinds of fish people like. In recent years, growing numbers of people have shifted from meat-eating to vegetarian diets. For some of the same reasons, in future years many more could shift from eating carnivorous fish to eating varieties that are lower on the food chain. Fewer may choose to eat shrimp, salmon, or shark when they eat out, opting instead to eat plant-fed varieties such as tilapia, catfish, carp, or crayfish. Many menus will indicate that the fish are certified to have been sustainably produced. If the menus don't specify, many customers will ask.

Third, more fish will be produced locally, and the fish served in restaurants will be more likely to be local–and therefore more likely to be fresh. The impetus toward local production will come from growing recognition that sustainable fish farming is best done using readily available resources rather than relying on feeds hauled at high energy cost half-way around the world.

Fourth, there will be a great variety of fish-farming methods used–some based in outdoor ponds, some in fenced-off areas of rivers or bays, some in buildings. But virtually all would, in one way or another, integrate fish production with other industrial or ecological activities: growing hydroponic vegetables, recycling urban waste, purifying fresh water, keeping rivers or wetlands healthy. This will happen because integrated businesses will use resources more efficiently, and therefore be more profitable.

Fifth, consumers of fish will be more likely to believe that how the fish are raised is an important ecological and ethical issue. It is consumer awareness, more than government regulation, that originally gave impetus to the development of such vital movements as the booming market for organic food, the demand for dolphin-free tuna, and the stigmatizing of fur coats or ivory bracelets. Such awareness could eventually put a damper on the huge profits now flowing into unsustainable shrimp- or salmon-producing operations.

Finally, if all goes well, the farmed fish of the future will be both a thriving local industry in every area of the world and a major opportunity for business investment. Because of its nutritional benefits, as well as its inherent resource-efficiency advantages when properly designed (with recirculating water, *etc.*), fish farming will move from a boom-town industry still dominated in many places by get-rich-quick entrepreneurs to a more stable and less exploitative sector of the mainstream economy. The shrimp mafias will be pushed aside in favor of people who are willing to be more accountable to their communities. How will all this happen? Such changes rarely occur without help from committed organizers–but that help is coming, as efforts to protect mangrove forests and local fishers take their place alongside other worthy causes. Organizers of sustainable fish-farming initiatives could engage the services of well known chefs and food editors, for example, to show how carp can be made every bit as delicious as salmon. A strategy that has already proven successful in other areas is to distinguish sustainable from unsustainable production via certification programs. Successful precedents can be found in at least three industries: timber, paper, and tunafish. Timber producers can apply for special labels that certify their wood has been grown by sustainable forestry methods (no clearcutting and no decimation of endangered tropical hardwoods, for instance), and the growing market for labeled wood is providing an incentive for growers to change their practices. Paper producers, similarly, can reach a growing market for products containing post-consumer recycled fiber, or made without ecologically damaging chlorine bleaching, by qualifying for the appropriate labels. And canned tuna can't carry the now familiar "Dolphin-free" labeling if it has been caught by purse seine nets which encircle schooling dolphins that swim with tuna.

The certification effort for fish so far has focused on wild oceanic fish. The World Wildlife Fund is helping to set up a Marine Stewardship Council to conduct such screening. One of the obstacles it faces, however, is the difficulty of monitoring underwater activities in regions that extend over vast areas of the Earth's surface. Aquaculture, in contrast, should be much easier to observe and certify. Of course, the criteria for certification may not be easy to establish, because the very nature of sustainable aquaculture is its relative ecological complexity and variety (as opposed to the high-volume monocultures favored by the shrimp mafia). The kinds of criteria worthy of certification have been studied closely by the FAO, which has issued a set of guidelines for sustainable aquaculture development. Some of its key recommendations are included in the accompanying table, which summarizes the key objectives to be achieved by redesigning the fish- farming industry.

To make fish farming truly sustainable will probably require a carrot-and-stick combination of policies. The carrot will be found in the emergence of robust new markets, much like the booming markets for organic foods, natural fibers, and products made with- out animal testing or ingredients, that have become multi-billion dollar sectors of the economy in the United States and Europe. The stick will be wielded at least in part by the concern of governments that now recognize the huge damage being done to their coasts, peoples, and economies by loot-and-run commercial aquaculture operations. Several governments have clamped down on such operations as at least a temporary means of stemming the damage until less damaging alternatives can be established. Water scarce Egypt has banned the diversion of water for fish farming, and China now prohibits converting arable land to aquaculture ponds. Honduras has instituted a moratorium on new shrimp farms; Norway and Scotland have stopped building salmon netcages in coastal waters; and India has banned shrimp farms within 500 meters of the High-tide zone.

As these restrictions are put in place, however, no government is likely to lose sight of the immense potential that properly managed fish-farming represents for their future food security and economic vitality. The temptations of this industry are not unlike those that accompanied the green revolution half a century ago. The difference is that while the green revolution did indeed produce huge advances in food production (ironically allowing our sense of crisis about expanding population to be deferred another generation), it also brought devastating problems of land erosion, water depletion, pesticide poisoning, and other problems of high-intensity monoculture. But as the Environmental Defense Fund recently concluded, aquaculture does not have to be an inefficient or polluting industry. And indeed, governments may be more savvy about those pitfalls this time around. The ultimate success of aquaculture may lie in the ability of its developers to leapfrog the mistakes of agriculture–to resist putting chemicals in the water and hormones in the fry, and consuming vast quantities of resources to get their product out. The fish farmers' ticket to the future is to align their business with the growing movement toward integrated, closed-loop production that is already making waves in the agricultural and timber industries.

Future Directions and Recommendations

It is clear that there are ways to approach aquaculture that allow for the production of healthy seafood while avoiding the numerous pitfalls of some current aquaculture practices. It is also apparent that despite some progress, much needs to be done to reach a point where aquaculture can become a sustainable method of providing food for the world's growing human populations. Unfortunately, some of the efforts to further develop aquaculture are currently being focused on practices that are most damaging, such as salmon netpen culture and shrimp culture in coastal ponds.

In order for aquaculture to develop into an environmentally and socially responsible food production endeavor, we recommend the following:

Implement more Ecologically Sustainable Practices

- Possibilities include:
  - Raising different species together in polyculture and integrated systems. For example, finfish with molluscs, vegetables, or seaweeds.
  - Treating wastewater by using settling ponds, constructed wetlands, filters, or other methods.
  - Raising molluscs and other low trophic level species.

Transition to the use of more closed systems and low discharge systems, especially those that provide total containment of fish and recovery/reuse of wastes.

- More efforts should be made to improve recirculating aquaculture systems and other closed and low discharge systems.
- Funding agencies should place greater emphasis on further developing closed systems and other environmentally friendly aquaculture techniques.
- The use of renewable energy sources should be further explored.

Significantly Decrease or Eliminate the Dependence on Wild Fisheries

- Research should continue into the replacement of fisheries products with plant-based ingredients in the feeds for farmed animals.
- Government policy should foster reduced use of fishmeal and fish oil in feeds for farmed animals.
- Greater emphasis should be placed on growing molluscs and other low trophic level species, which do not require fishmeal or fish oil in their diets.

Establish Organic Standards and other Eco-labels for Aquaculture Products

- Standards should be developed that address each of the issues associated with fish farming without simply watering down the concept of "organic" production for the sake of applying it to aquatic organisms.
- Organic standards should start with low trophic level fish and shellfish since standards for carnivorous fish will prove to be problematic because of issues with feeds.
- Organic aquaculture systems for some species may have to be land-based to be consistent with organic principles, such as the management and recycling of wastes.
- For public acceptance, labels and certification systems must have independent 3rd party involvement at all stages of development, implementation, assessment, and control.
- Criteria for certification should be objective and verifiable; and information about the certification process and the product should be available to consumers.

Annexure 1: Aquaculture Systems and their Environmental Impacts

System Type	*Common Species*	*Environmental issues*	*Overall Impact*
Mollusc culture	Oysters, clams, mussels, scallops, abalone	Introduction of exotic species (target and non- target) and diseases.	Minimal, especially with proper planning and management. Sometimes positive asfarmed animals canactually improve water quality.
Recirculating/closed	Tilapia, hybrid stripedbass, sturgeon	Small amount of waste discharge. Energy intensive.	Minimal. Can beimproved with greater focus on alternative energy sources.
Inland pond	Catfish, carp, yellowperch, tilapia, hybrid striped bass	Some waste discharge. Some use of exotic species.	Moderate. Can beminimized with proper management.
Flow-through/raceway	Hybrid striped bass, trout	Some waste discharge.	Moderate. Can beminimized with best management practices.
Coastal pond	Shrimp	Discharge of wastes. Destruction of mangroves.	High environmental impact.
Netpen and/or cage	Salmon, sea bass, seabream, grouper, tuna, snapper	Discharge of biological and chemical wastes directly to the natural environment. Escapes of fish. Introduction of exotic species and transfer of diseases.	High environmental impact.

Develop Sustainable Aquaculture Operations that Provide Long-term Social and Economic Benefits to Communities

- ☆ Traditional farmers should be encouraged to diversify their sources of income by raising fish in addition to terrestrial crops.
- ☆ Recirculating aquaculture systems should be constructed in urban areas. Government should promote efforts to renovate or retrofit old factories or other existing buildings for aquaculture purposes.
- ☆ The social impacts of an expanded aquaculture industry should be considered so as not to negatively impact communities through the loss of jobs in other sectors.

Ensuring Food Security through Golden Revolution: Prospects, Achievements, and Bottlenecks

Introduction

India is bestowed with varied soil and agro-climatic conditions, which is highly favourable for growing a large number of horticultural crops from tropical to temperate throughout the year and often termed as natural reservoir of horticultural biodiversity. Besides, existence of such a favourable biodiversity, ensuring the food security for around 120 crores of population is the major challenge for the agricultural sector of the country. The agrarian community of the country has efficiently exploited this natural endowment and achieved a significant leap in horticultural production after post independence with seven times jump.

Presently horticultural crops occupy around 10 per cent of the gross cropped area in India and contributing 12.5 per cent and 14.0 per cent in world's fruit and vegetable production respectively (NHB Database-2011). Within the last 19 years (from 1991- 92 to 2010-11), horticultural cropped area in India has increased from 12.8 million ha to 21.83 million ha and production from 96.6 million tonnes to 240.53 million tonnes, recorded sharp increase of 70.6 per cent and 149 per cent, respectively (NHB Database-2010- 11). Presently, the horticulture sector of country has recorded significant growth rate of 5-6 per cent during last decade (Singh, 2007) and contributing more than 29 per cent in agricultural GDP (Gross Domestic Product). The credit for outstanding leap in horticulture sector goes to the multi-dimensional but united approach of policy planners and researchers associated with various organizations especially National Horticultural Board (NHB), Indian Council of Agricultural Research (ICAR), Agricultural and Processed Food Products Export Development Authority (APEDA) and Agricultural Universities under the aegis of Ministry of Agriculture, Government of India (GOI) and marked as "Golden Revolution" for the country.

However, fast increasing population has dimmed the impact of revolution in mitigating the projected problems *i.e.* starvation, malnutrition and poverty. The past efforts in the investment have been rewarding and India is fast emerging as a horticultural hub and reckoned as a leader in the global arena. The projected research and development has made quantum jump in production, productivity, availability, and export. Several nutritionally potent horticultural crops like kiwi, olive, gherkins, kinnow, oil palm *etc.* were also introduced for commercial cultivation. Besides, "*Golden Revolution*" several other revolutions are also going on in India and other countries related to specific fields (Table 8.9)

Table 8.9: Major Revolutions in India and other Countries

Revolution	*Associated to*
Black Revolution	Petroleum Products
Blue Revolution	Fish Products
Brown Revolution	Leather/Non-conventional (India)/cocoa production
Golden fiber revolution	Jute Production
Green Revolution	Foodgrain (Cereals, wheat and leguminous plant) production
Grey Revolution	Fertilizer Production
Pink Revolution	Onion Production/Prawn Production/Pharmaceutical (India)
Red Revolution	Tomato/Meat Production
Round Revolution	Potato Production
Silver Fiber Revolution	Cotton Production
Silver Revolution	Egg/Poultry Production
White Revolution	Milk/Dairy production (in India-operation flood)
Yellow Revolution	Oilseeds production
Evergreen Revolution	Overall Agricultural Development

The world's populace is projected at 9 billion by the year 2030 and majority of inflation will be confined in the developing countries, where chronic food shortages and malnutrition persist (Chopra, 2010). This projected inflation of populace will certainly squeezed the per capita availability of natural resources, and may hinder the equilibrium and sustainability of agricultural systems due to over exploitation of natural resources, which will ultimately aggravate the menace like poverty, starvation, malnutrition and increase in food prices. Regrettably, unlike other horticulture rich countries, average Indians do not get the basic daily requirement of fruits and vegetables and our Human Development Index is very low. This is because a substantial amount of this valuable produce is lost due to improper post harvest management and lack of processing facility. To achieve our target of a hunger free country and reduce poverty, prominence needs to be given to loss reduction (post-harvest technology) that is essential to make available more fruits and vegetables and give better economic return from the existing level of production.

Potential of Horticultural Crops for Ensuring Food and Nutritional Security

Horticultural crops form an important part of the daily diet as effective nutritional supplements and the most prominent source for nutritional security, despite their well-known significance from the economic viewpoint. Its significance further amplified in the countries like India where majority of population is vegetarian and undernourished. The fruit and vegetable sectors of horticulture are most potent to fulfil food requirements of the country and elevate the nutritional and economic status of people. In general, antioxidants, vitamins, minerals, and fibres are considered as main nutrients contributed by horticultural crops to a balanced diet, and thus special attention should be addressed to this group of nutrients. Higher productivity, wider adoptability towards soil and climatic conditions and lower cost of cultivation makes horticultural crops as an ultimate choice for sustaining livelihood and nutritional security to a large section of people in India having very meagre share of land holding. Productivity of fruits like banana (37 t/ha), grapes (23.6 t/ha), papaya (37.1 t/ha), pineapple (16.0 t/ha), and vegetables like potato (18.8 t/ha), onion (16.3 t/ha), cabbage (22.1 t/ha) are much higher than the wheat paddy rotation *i.e.* only 8 t/ha. These productivities have ample potential to ensure food security under ever squeezing per capita availability of natural resources (Table 8.10).

Table 8.10: Productivity (t/ha) Gap of Important Horticultural Crops (2010-11)

Crops	*Productivity (T/Ha.)*			
	India	*World*	*Higher Potential (Country)*	*Cap (t/ha.) Potential for improvement*
Fruits	11.7	10.9	24.4 (USA)	–12.7
Mango	6.6	7.7 (mango + guava)	15. 8(Brazil)	–9.2
Guava	12.0	14.9	42.8 (France)	–32.8
Apple	10.0	20.8	59.3 (Indonesia)	–23.4
Banana	35.9	17.6 (orange)	35.0 (Indonesia)	–26.2
Citrus	8.8	9.3	16.3 (USA)	–5.2
Grapes	11.1	NA	NA	NA
Pomegranate	6.9	24.7	85.8 (Indonesia)	–46.2
Papaya	39.6	22.3	69.5 (Indonesia)	–53.6
Pineapple	15.9			
Vegetables	17.3	18.8	37.2 (Spain)	–18.6
Potato	22.7	17.7	44.3 (USA)	–21.6
Onion	14.2	19.1	30.3 (Turkey)	–16.1
Tomato	19.1	32.8	81.0 (USA)	–61.9
Brinjal	17.5	25.0	49.2 (Egypt)	–31.7
Cabbage	21.5	27.7	66.1 (Japan)	–44.6
Cauliflower	18.3	16.9 (Cauliflower + broccoli)	24.8 (Pakistan)	–6.5
Spices	1.8			
Plantation Crops	3.6			
Flowers loose	5.4			
Productivity of Horticulture crops	11.0			

Source: National Horticulture Board Database–India for the years (2010-11).

The horticultural crops offer unique advantage to nutritional security as well. Fruits and vegetables provide substantial amount of nutrients, important for human health and can play significant role for improving the nutritional intake especially of pre- dominantly vegetarian population. They are important source of micronutrients, vitamins, minerals and folic aci. Fruits have proved to be essential for a balanced diet as a rich source of vitamins, protein, minerals, organic acids, carbohydrates, and antioxidants. Whereas, the nutritional value of vegetables as a vital source of essential minerals, vitamins and dietary fibres has well recognized. Vegetables play an important role in human nutrition as a neutralizing agent for acid substances produces during digestion of high-energy foods. Due to their high water content and fibre, leafy vegetables and roots probably aid in digestion and utilization of more concentrated food in human diet. Man needs a wide range of nutrients to perform different metabolic functions and to lead a healthy life and their deficiency leads to a number of diseases in children and adults. Fruits and vegetables contain nutraceutical substances that provides medical and health benefits and their daily consumption have been strongly associated with reduced risk for some forms of cancer, heart disease, stroke, and other chronic diseases. Some components of fruits and vegetables (phytochemicals) are strong antioxidants and function to modify the metabolic activation and detoxification/disposition of carcinogens, or even influence processes that alter the course of the tumour cell. It is better to consume a variety of commodities rather than limiting consumption to a few with the highest antioxidant capacity as it varies greatly among fruits and vegetables. Nutritive constituents of fruits and vegetables that have a positive impact on human health and their sources are present in Table 8.11.

The Fruit Crops

The fruit plants have immense ability to withstand various soils and climatic stresses, which make them suitable to rehabilitate degraded and wastelands, which are abandoned due to unfit for cultivation of cereal and food grains. India is the second largest producer of fruits after China and contributes 12.5 per cent in global fruit production (Table 8.11) (NHB Database-2011). The country has wide ranges and varieties of fruits in its basket namely mango, banana, citrus, pineapple, papaya, guava, sapota, jackfruit, litchi, and grape, among the tropical and sub-tropical fruits. Similarly, apple, pear, peach, plum, apricot, almond and walnut count among the temperate fruits. The aonla, ber, pomegranate, annona, fig, phalsa are among the arid zone fruits are the important fruit species. India is the world's largest producer of mango, banana, sapota and acid lime. About 39.5 per cent of the world's mangos and 23 per cent of world's banana are produced in India. Country has achieved laurels as highest productivity in the world for banana, grapes, citrus fruits, litchi, papaya and acid lime.

Besides all above achievements in fruit production, the actual potential of the country has yet not fully exploited. The major fruit growing belts are confined in subtropical and tropical parts of the country and a limited area has been harnessed in the temperate region. Productivity and fruit quality of several fruit crops in the country are much below to their actual potential and the losses during marketing are very high, sometimes more than 30 per cent. The poor post harvest handling and lack of adequate infrastructures has significantly reduced the per capita availability of fruits, which is merely 70 g/day against the dietary recommendations of WHO, which is 120 g/day (Dhillon, 2005). Very negligible proportion of fruit production is being exported from the country in form of fresh as well as processed fruit products.

Table 8.11: A Glance of all India Area and Production of Horticultural Crops (Area -000ha., Production-000 tonnes)

Crop	*2008-09*		*2009-10*		*2010-11*	
	Area	*Production*	*Area*	*Production*	*Area*	*Production*
Fruits	6100.90	68465.53	6477.83	73525.51	6383	74878
Vegetables	7980.71	129076.78	8214.58	136189.53	8495	146554
Spices	2629.40	4144.90	2466.8	4018.7	2940	5350
Plantation	3217.3	11336.4	3258.7	11249.3	3306	12007
Flowers						
Loose	166.5					1031
Cut		113618.0		50759.0		69027
Total Horticultural crops	20661.6	214715.9	21155.7	226872.1	21825	240531

Table 8.12: Bioactive Substances Present in some Horticultural Crops

Commodity	*Bioactive Substances*	*Commodity*	*Bioactive Substances*
Aonla	Vitamins, Polyphenols	Beans	Genistein, isoflavonoid
Apple	Flavonoid, polyphenols	Brocoli	Glucosinolate,selenium
Citrus	Vit. C, Bioflavonoids, chalcone	Carrot	a-carotene, lycopene
Grapes	Flavonoid, phenols	Chili	Capsaicin, capsinthin
Starwberry	Flavonoid, phenols	Ginger	Gingerol
Tomato	Lycopene	Turmeric	Curcumin
Onion	Quercetin	Radish	Iso-thiocyanate

There is ample possibility to cover-up this gap through integrated approach projected to harness maximum productivity potential of fruit crops through efficient and sustainable utilization of natural resources under precision farming practices. The Government of India has introduced several schemes in form of techno-financial assistance for the development of tissue culture units, vegetable seed production farms, protected cultivation, Hi-tech nurseries, rejuvenation of senile orchards, promotion of INM and precision farming, organic farming, bee keeping, mushroom production *etc.* under National Horticultural Mission. All these programmes are projected to increase production of quality fruits and reduction in processing and transportation losses. This effort have rewarded in terms of increase in fruit production of the country, however the achievement is much below than the actual potential. The harnessed productivity of major fruit crops is much below to their actual potential, which can be further improved through scientific orchard management.

The Vegetables Crops

Vegetables play a vital role in the nutritional security of Indian populace that is dominantly vegetarian. Its role becomes more pertinent considering the medicinal and protective attributes for various diseases. Higher productivity and economic return in vegetable cultivation attracted attention of farmers' specially small and marginal farmers who are having very small land holdings. India is a treasure house of wide vegetable biodiversity, which inhabited 175 types of vegetables including 82 leafy vegetables, and 41 root (tuber and bulb) crops cultivated in varied agro climate ranging from tropical to temperate (Singh *et al.*, 2006). However, vegetable sector of the country is mainly concentrated in the cultivation of potato, onion, tomato, brinjal, okra, peas, cabbage and cauliflower, which jointly contributes 66.3 per cent to total vegetable production. Potato alone contributed 28.9 per cent in total vegetable production during year 2010-11 (Table 8.11) (NHB Database-2011). India is the second largest producer of vegetable and contributed 14 per cent to the world's vegetable production during year 2010-11 (NHB Database-2011) from 8.5 million ha of cultivated area. However, per capita availability at the farm-level is around 210 g/day, compared to the recommended level of 280 g/day (Singh *et al.*, 2006). Data in respect of the highest productivity potential and actual productivity of the country reveal ample potential for improvement of productivity through use of quality hybrid seeds under scientific cultivation practices like precision farming, integrated nutrient, and pest management and post harvest value addition to reduce various losses (Table 8.10).

Spices and Plantation Crops

Spices are high value and low volume export oriented commodities, which yield aromatic and pungent principles commonly used for flavouring and seasoning of food and beverages and also used in Ayurvedic medicines. India is the centre of origin and diversity for major spices like black pepper and cardamom and possibly for ginger and turmeric and occupies a pre-eminent position in production and global trade of spices and foreign exchange earnings. About 60 different kinds of spices can successfully be grown in India but spice sector of the country is mainly concentrated on the cultivation of chillies, garlic, turmeric, ginger, coriander, cumin, and pepper, which covers about 89 per cent of the total cultivated area under spice and jointly contributed to 91 per cent of total spice production (Table 8.11) (NHB Database-2011). Productivity level of the India in respect of spices is also considerably low in comparison to other countries. For example average yield black pepper in India is only around 290 kg/ha, which is much below to the Malaysia average yield which is 2925 kg/ha (more than 10 times higher). Similarly, In case of cardamom, India's average yield is 120 kg/ha while in Guatemala it is 200 kg/ha. In India, even the gap between national average and the realizable yield is very wide. Realizable yield for pepper and cardamom is around 2445 kg/ha, and 1625 kg/ha, respectively, whereas, the national productivity of pepper and cardamom is merely 290 kg/ha and 120 kg/ha, respectively. The focal cause of poor crop productivity is unscientific crop management practices and lack of knowledge of precision farming. Bridging this gap through effective implementation of various extension tools like site specific crop demonstration, precision farming practices, information and communication technology (ICT) *etc.* is the thrust need and having potential to increase country's production many fold (Vision-2020, IISR).

Plantation crops constitute a large group of crops including coconut, areca nut, oil palm, cashew, tea, coffee, rubber, and cocoa. The major plantation crop of the country is coconut, which covers 57.4 per cent of area and share more than 90 per cent in total production (Table 8.11) (NHB Database-2011). India is the largest producer, processor, consumer and exporter of cashew in the world. India is also the leading producer of areca nut in the world with 53 per cent share of global output. Coconut, cashew nut, areca nut and cocoa are the major plantation crops of India that jointly cover 15.5 per cent of total cropped area under horticulture and contribute only 5.3 per cent to the total horticultural production of the country. Traditional cultivation practices and small farm holdings are the major causes of poor productivity of plantation crops. However, they play an important role in view of their export potential as well as domestic requirements and in employment generation and poverty alleviation programs particularly in rural sector.

Postharvest Management of Fresh Produce as a Backbone for Developing Countries

The ultimate goal of crop production is to provide quality produce to the consumers at reasonable rate. Most of the horticultural crops are highly perishable and post harvest losses are very significant under present scenario of post harvest management, which is insufficient to meet the requirement of the country. The break-up of these losses, which occur due to poor post harvest management facilities and practices are poor handling (30 per cent), poor storage (30 per cent), poor transportation (30 per cent), presence of large number of intermediaries (5 per cent) and the lack of knowledge about better preservation techniques (5 per cent) (Maheshwar and Chanakya, 2006).

The Golden revolution was mainly focused on the production of horticultural crops, which was achieved significantly. However, the development of needful storage infrastructure for post harvest storage was neglected in revolution programmes. This infrastructural lack imposes various post harvest losses and causes a significant gap between the gross production and net availability. Along with the increasing production of horticultural crops particularly fruits and vegetables there post harvest loss is increasing counting about 30- 35 per cent of total production in the absence of proper post-harvest handling/storage, preservation, and cold chain infrastructure (Kitinoja *et al.,* 2011). There is an urgent need for food processing technologies and packaging equipment, especially in areas of value- added fruit and vegetable processed products. Ready- to-eat/serve snacks and convenience foods segment, which is growing at the rate of 20 per cent, also offers good potential. Business in handling, packaging, and storage equipment sector for Horticultural-products also offers good prospects in India. The existing storage and handling system for fresh horticultural produce is traditional and there is an urgent need for modernization and creation of infrastructure to reduce large-scale wastage. A study jointly conducted by the management consultancy firm, McKinsey and Co. and the Confederation of Indian Industry (CII), determined that at least 50 per cent of the production of fruits and vegetables in the country is lost due to wastage and value destruction. The cost of wastage is estimated at Rs. 23, 000 crores (US$4.6 billion) on an annual basis (Rolle, 2006). This urgently necessitates developing an integrated approach right from production to post harvest management and processing in order to ensure increased income and employment for all sections of the rural people be economically sustainable and contribution towards food and nutrition security of the people. The growth potential for food business in India, however, is very promising. In recent times, factors such as changing food consumption patterns, increased spending on value-added products, increase in income levels, rapid urbanization, increasing number of women joining the workforce and changing lifestyles has brought about a marked shift among consumers for processed food items.

Amenities that could be created to all the important production centres as well as in large markets include post-harvest handling infrastructure *viz.*, humidified storage apparatus for fresh produce; sorting and grading equipment; technology to reduce microbial contamination; cooling systems, cargo transport air-conditioning and refrigeration systems for railways; refrigerated containers/trailers; and vans for transport of perishable products. Reduction of loss would mean increasing food availability in fresh or processed form. There is a need for adequate support for establishing Cold Chain facilities and modern markets, including setting up of Terminal Markets and creating modern humidified storage facilities at the Ports and the Airports also. Development of cold chain infrastructure is essential to exploit the potential of the horticulture sector in distant both global and domestic markets. Support to the growing organized retail sector through efficient supply chain systems could also boost domestic consumption, encourage value addition, and reduce loss (Maheshwar and Chanakya, 2004; Kitinoja *et al.*, 2011).

A further dilemma is that fruits and vegetables are not uniformly available throughout the country and some areas suffer from scarce supply even when there is a glut in other parts. Marketing, storage and packaging of both fresh and processed products would play a vital role in price permanence of delicate horticultural produce. An immense force to food processing and storage will add value to the product. It will increase the income of farmers, create employment, and promote rural industrialization. Infrastructure for handling of perishable commodities has been inadequate. Recently, initiative has been made to develop the requisite infrastructure, improvement in the Port facilities, setting up of Quality Testing Labs, Cold Chain facilities *etc.* Reduction of moisture loss from the product is the major quality of packaging materials. A solution to moisture loss problems from produce appeared with the development and wide distribution of semi permeable plastic films. Airflow through the ventilation holes allows hot fruit or vegetable to slowly cool and avoid the build-up of heat produced by the commodity in respiration. Holes are also important in cooling the fruit when the packages are placed in a cold storage, especially with forced air-cooling. Ventilation holes improve the dispersal of ethylene produced. Research and development effort is required to develop edible films and coatings that have good packaging performance besides being economical. Improved packaging will become more essential as International trade expands after globalization.Better packaging should be of immediate value in reducing waste. Much background research on packaging of perishable products and flowers is needed simulating the actual handling conditions expected during marketing (Murthy *et al.*, 2009).

Losses of fresh produce take place at the levels 1/3 between the farm and road head, another 1/3 between road head and wholesale markets and the remaining 1/3 between wholesale market and retailer. There is need to train the self-help groups, unemployed/under employed youth, and growers including house wives on scientific pre-harvest and post-harvest management including processing and preservation. The introduction of training young boys and girls in the field of "Preservation of fruits and vegetables" is to yoke the widely expanding horizon of job opportunities in the area. Extension personnel engaged in horticultural developmental activities either in government sector, agricultural universities or in food science and technology centres should be delegated the duty to provide the required training to growers/producers and training the trainers for thorough orientation in maturity determination, grading and packing standards for all commodities. There is need to improve packing containers, method of handling during transport and marketing. Application of correct techniques can reduce the losses by about 70 to 80 per cent. There is also need to improve the skill in handling of preservation equipment for large scale, working in hygienic manner, maintaining sanitation, preparation of preserved products, and identification of spoilage and application of remedial measure. Besides, the concept of supply chain management should be followed in this sector for over all development because supply chain management (SCM) represents the management of the entire set of production, manufacturing/transformations, distribution, and marketing activities by which a consumer is supplied with a desired product.

At last, investments are needed to develop necessary infrastructure in this sector including setting up of research and development laboratories and pilot scale plant in the agricultural universities and grant for setting up/expansion/ modernization of existing food processing industries in the state. Appropriate policies and schemes for development and transfer of new technologies are urgently needed to propagate non-polluting, cheap technologies *e.g.*, irradiation and biotechnology for preservation of horticultural produce. The efforts are needed to capitalize our strengths like long growing-season, diverse soil and climatic conditions and removing gaps/constrains to meet the goal of moving towards an alarming horticultural growth in India.

Conclusion and Future Directions

The investments especially in the post-harvest horticultural technology can have a major impact on reducing post harvest losses and increasing the food security, ultimately, improvement in the incomes with the use of existing production techniques without increasing production. The formulation of Postharvest Working Group in each states of the country would be very useful in providing a forum for communications among all those concerned with postharvest biology and technology research and extension. The future research and extension activities must be toward the maintaining quality especially flavour and nutritional content and ensuring safety (avoiding chemical and microbial contamination). A major attention and funding for postharvest research and extension efforts should be given to different governmental as well as private organizations. The fruitful postharvest research may be realized via internships, faculty exchanges, and human resource development for staff in university laboratories and research centres, laboratory upgrades, improved access to web-based information.

9

Food Situation Immediately after Independence

From 1935 to 1942, food production was going down. It was this situation that alarmed the British and, along with the rise of the Freedom Movement, expedited their exit from India. There was little use of chemical fertilizers for food crops, as there was no price support. The only hope lay in irrigating more of the rainfed land.

Research Organizations

Major research organizations established during the British period were (a) Forest Research Institute. It was originally founded in 1878 as British Imperial Forest School In 1906, it was re-established as Imperial Forest Research Institute. (b) Imperial Agricultural Research Institute was originally established in 1905 at Pusa Bihar. (c) Imperial (now Indian) Council of Agricultural Research was established in 1929. These three organizations, having been renamed as Indian' continued after Independence and are playing a central role in the development of agriculture. They have been discussed in detail in this study.

Contract Farming Act

The state government is ready with the draft of the Contract Farming Act. The legislation envisages a four-member Contract Farming Regulatory Commission, to be headed by a Chief Secretary-level officer, with two legal luminaries and an agricultural expert as members. It would have the authority to hear complaints with regard to violation of the Act. The legislation would enable farmers to enter into a contract with sugar mills, rice mills and the food packaging industry to sell their produce. If a sugar mill owner signs a pact with farmers that he would purchase the entire cane stock produced by them, it would become legally binding on him to do so. In case he violated the pact, he would have to pay a penalty besides the production cost. On the other hand, the farmers would have to fulfill the commitment made by them to the mill owner. In the absence of any legislation, many companies have duped farmers in the past. Repeatedly let down by cooperative sugar mills, farmers in the state had stopped growing cane because of which many cooperative mills had to be shut down.

DECODING THE PROPOSED LEGISLATION

The legislation would benefit farmers by ensuring a market for their produce.

It would guarantee availability of raw material to agro-based industries

It would provide an additional channel for selling and buying agricultural produce

The Centre has been pressing state governments to promote direct marketing and eliminate middlemen.

Whereas the legislation would benefit the farmers and ensure a market for their produce, it would guarantee availability of raw material to the agro-based industry. As of now, the entire agricultural produce is sold under the Punjab Agriculture Produce Marketing Act. Private firms buying farm produce have to abide by the Act. The proposed Contract Farming Act would now provide an additional marketing channel for selling and buying agriculture produce. The Centre has been pressing various state governments to promote direct marketing and eliminate the middlemen.

Why Not Farmers?

The Director of Agriculture and Industries, Punjab has issued instructions to zamindars on what they should and on what they should not grow this year. All cultivators would do well to restrict their sowings of toria and sarson this year. Owing to the war the exports of sarson to Europe are likely to be small. On the other hand the prices of wheat, barley and gram are sure to be high. The Russian and Canadian crops are short and Argentine sowings are a million acres down, so the English and European demand for wheat will be large. In 1911 when the Russian barley crop was short large exports of barely from the Punjab went to England for brewing; the same thing is likely to happen next spring. Gram is sure to be wanted to Europe for horses and cattle." – The following excerpt was published in The Tribune on September 9, 1914.

Assured Market

The farmer's greatest concerns have always been reasonable price and assured market. Given this, the Indian farmers can rise to the occasion to meet the country's requirements. The Government of India set up a Commission for Agricultural Costs and Prices (CAPC). The Commission announces a Minimum Support Price (MSP) for specified crops and the government or its agencies are charged with the responsibility to lift the produce if the market price falls below the MSP. The Commission at present announces these prices for more than 20 commodities. However, the arrangements for lifting of produce have worked satisfactorily only for wheat and rice (paddy).

The food situation in the country in the 1960s and early 1970s is well known. We were under great pressure to increase domestic production. The Government of India, state governments and their various corporations were charged with the responsibility to lift market arrivals of wheat and paddy if prices went below the MSP. Feeling assured, the farmers also rose to the occasion and made their contribution by adopting new seeds and technologies to make country self-sufficient in food grains.

Private Sector Participation

The question, however, is whether it is the government only which should be responsible for this. Should not the private sector and other organisations concerned with agriculture produce also take part in this process of assuring reasonable price and assured lifting? Apart from wheat and rice, there are other important crops like cotton, sugarcane, oil seeds *etc.* and fruits and vegetables, particularly potatoes, oil seeds, pulses and the like. Arrangements regarding these are nowhere as good as for wheat and rice. In case of cotton, there is the Cotton Corporation of India (CCI) which mainly carries out buying and selling of cotton on commercial basis. Whenever there is a glut, the government does direct the CCI to step in and effect purchases to avoid undue fall in price. The CCI does enter the market to support the price but it is limited intervention and not like the one for wheat and rice.

Farmers and Sugar Mills

In case of sugarcane, the prices are announced by the government. However, these mostly remain advisory in nature. Ultimately, it is a matter between the farmers and the sugar mills. When the sugar prices fall, the mills generally withhold sugar from the market and wait for better times. This delays the payments to the farmers. This delay can sometime be in years. Recently, in one of the states, the government had to give Rs. 10,000 crore to enable the mills to clear the farmers' dues. The sugar industry is in the hands of large cooperative or industrial houses. They are sometimes referred to as "sugar barons". Local sales and import/export takes place according to the interest of these large players which may not always be in the best interest of the farmers.

Advice on Horticulture

In case of horticulture, there is hardly any national-level organisation which advises the farmers in advance regarding weather, likelihood of pests/disease attack, expected market conditions *etc.* The government of Himachal Pradesh has done well in this regard. The Himachal Pradesh Horticultural Produce Marketing and Processing Corporation Limited has done very good work in marketing and processing but not so much in horticultural operations in advance. In case of vegetable, particularly potatoes, the situation is far from satisfactory. Every four to five years, there is a glut of potatoes in the market. There is no advance warning to farmers. When the situation has already occurred, the Government asks Nafed as also state organisations like Markfed, Hafed etc to enter the market. This does provide some relief but seldom gives adequate return to the farmers. There have been years when in case of potatoes, the cost of the empty bori was more than the price of a bori full of potatoes. The market prices had fallen so low that farmers could not recover even transport and marketing costs.

In UK, there is a Potato Board. Every year, they assesses the demand for table potatoes, for processing industry as also for horses and cattle feed and the like. They then decide how much area should go under potato. Farmers have to take permit from the board to cultivate potato. The board organises marketing and ensures lifting. In case of pulses again, there is no assured lifting of the produce at appropriate price. In a prominently vegetarian society, pulses are an important source of protein. There have been number of 'missions' yet the situation is not satisfactory. Similarly in edible oils, no such system of assured marketing and reasonable prices has come up as there is for wheat and rice.

Again, in December 6, 2014 (1914) the following appeared in The Tribune. "Forecast of Spring Harvest – CROP telegrams for the week ending November 28 indicate prices as fluctuating with an upward tendency. The United Provinces have practically recovered from the effects of the last famine, and standing crops are doing well and prospects are favourable. The Punjab Government reports: "The weather remained generally dry. The condition of standing sugarcane, cotton and other crops is average to good. Picking of cotton and harvesting of other autumn crops continue. Sowing of wheat is in progress and is normal or very normal. Prospects of spring crops are good. Cattle are generally healthy. Prices of food grains range between warning and scarcity rates".

Concern for Farmer

Nearly 100 years ago, when governments were primarily responsible for law and order, collection of the revenue and the like, yet farmers remained their important concern. Now one seldom comes across such a news item in the papers where farmer is kept advised about sowing as well as the conditions of the crop and market possibilities. Today the government is not the only player in this field. The question therefore is should this remain only the government's responsibility? Initially in the field of industry, there was lot of emphasis on the public sector but progressively it is the private sector that is becoming the larger player. How far and how effectively are we using the present IT and other technologies available in the field of agriculture? Market analysis both at the national and international level are available aplenty. Weather forecasting is much better but still farmers are left to themselves and seldom receive timely advice and help.

World View

Today we have organisations like International Chicago Board of Trade which gives a world view of important agricultural commodities, how the demand and supply position etc is. We have strong cooperative organisations for sugar, cotton as also other private sector associations of these. However, these organisations seldom shoulder the responsibility of either advising the farmer about better agricultural practices or properly forecasting the market. They are far from organizing lifting of produce. This year it was well known in advance that China is drastically going to cut their cotton import. Seventy per cent of our cotton exports go to China, yet nobody advised the farmers to reduce the area under cotton.

Food Security

Considering the importance of food security and the experience through which country has gone in 1960-70s, arrangements for wheat and rice should continue as at present with the government remaining fully responsible for fixing the price as well as assuring lifting of the produce. However, for other crops, whether agricultural or horticultural, Central and state governments should encourage private sector organizations to step in and help in maintaining reasonable price as well as lifting. There are many big organizations in sugar, cotton, oil seeds *etc.*

Their roles need to be properly defined and some responsibility cast on them. They should be asked, if necessary even by legislation, to discharge their responsibility towards the farmers. We could also consider boards like UK Potato Board for selected agricultural and horticultural crops. Today, in addition to the government, we have many cooperative and other such organizations in the agricultural sector. How one wishes that the farmer is kept advised as good as 100 years ago.

Prices Agriculture and Food Management

After remaining high for a prolonged period, inflation is finally trending down. Average Wholesale Price Index inflation declined to 3.4per cent in 2014-15 (April – December) as compared to an average of 6 per cent during 2013-14. The WPI inflation even breached the psychological level of 0 per cent in November 2014 and January 2015. Consumer price inflation released by the Central Statistics Office reached 5.1 per cent in January 2015. This is lower than the targets of 8 per cent set for January 2015 and 6 per cent for January 2016 given by the Reserve Bank of India in its report on the new monetary policy framework. Prices of the major commodity groups contributing to high inflation, namely 'eggs, meat, and fish', fruits and vegetables, and fuel, have all softened. The major developments driving the stubborn inflation down were falling global commodity prices, especially of crude oil, decline in the growth rate of rural wages, moderation in the increase in minimum support prices as also slack in economic activity. In so far as high food inflation contributed to elevated headline inflation, for sustainability of low inflation the policy focus should be on enhancing the resilience of the agriculture sector and eliminating leakages, inclusion and exclusion errors, and various distortions created by the present food policy. Growth in agriculture has now to increasingly come from non-price factors. Markets for agricultural commodities have to be made more competitive in the interests of both producers and consumers. The High Level Committee headed by Shri Shanta Kumar has given useful recommendations on proposed changes in the food policy.

The upside risk to inflation outlook also emanates from uncertainties surrounding the monsoon, international crude oil prices, and the stability in the value of the rupee, particularly in the event of monetary tightening by the US Federation.

Trends in WPI and CPI Inflation

Wholesale Price Index

Headline inflation measured in terms of the Wholesale Price Index (WPI) (base year 2004-05-100) which remained persistently high at 6-9 per cent during 2011-13 moderated to a low of 3.4 per cent in 2014-15 (April –December) on the back of lower food and fuel prices. During the first quarter of 2014-15, WPI headline inflation was at 5.8 per cent as mainly food and fuel prices were high. In second and third quarters of 2014-15, WPI inflation declined to 3.9 per cent and 0.5 per cent respectively Table 9.1. WPI food inflation, which remained high at 9.4 per cent during 2013-14 moderated to 4.8 per cent during April-December 2014 following sharp correction in vegetables prices since December 2013 (except March 2014) and moderation in prices of cereals and eggs, meat, and fish. A fuel has larger weight in the WPI, the decline in fuel prices led to a sharper fall in the WPI as compared to the Consumer Price Index (CPI) (base year 2010=100). Inflation in manufactured products has remained within a narrow range since 2013-14. The WPI headline inflation (provisional) in January 2015 stood at -0.4 per cent. The build up inflation rate in the financial year till January 2015 was -1.1 per cent compared to a build up rate of 5.2 per cent in the corresponding period of the previous year.

Table 9.1: Quarter-wise Inflation in WPI Broad Groups (in per cent)

	Weights	*2013-14*				*2014-15*		
		Q1	*Q2*	*Q3*	*Q4*	*Q1*	*Q2*	*Q3 (P)*
All Commodities	100.0	4.8	6.6	7.1	5.4	5.8	3.9	0.5
I. Primary Articles	20.1	6.5	12.4	13.6	6.8	7.5	4.1	0.4
II. Fuel and Power	14.9	7.7	11.9	10.8	10.1	9.6	4.4	-4.0
III. Manufactured products	65.0	3.3	2.4	2.9	3.3	3.8	3.6	2.0
All Food	24.3	7.7	11.8	11.9	6.2	6.9	5.0	2.5
Core Inflation	55.0	2.6	2.4	3.1	3.7	4.0	3.6	2.0

Source: Office of Eco. Adviser, Deptt. of Industrial Policy and Promotion (DIPP) P: Provisional.

Consumer Price Index

The Central Statistics Office (CSO) has started releasing state-wise and all-India rural, urban, and combined CPIs since January 2011. Retail inflation as measured by the CPI (combined) (base year 2010=100) remained stubbornly sticky around 9-10 per cent for the last two years. Like WPI inflation, CPI inflation has also moderated significantly since the second quarter of 2014-15. It declined to an all-time low of 5 per cent in Q3 of 2014-15 (Table 2). The Reserve Bank of India (RBI) had announced its intent to anchor its monetary policy stance to headline CPI (combined) inflation from April 2014.Taking note of the sustained moderation in retail prices, it has signaled easing of the monetary stance by reducing policy repo rates by 25 basis points from 8 per cent to 7.75 per cent on 15 January 2015. The CSO has revised the base year from 2010 to 2012 (Box 1) and released the revised series on 12th February, 2015 along with inflation data for January, 2015. CPI inflation in terms of the revised series stood at 5.1 per cent in January, 2015.

BOX - 1: Changes in CPI New Series

The CSO, Ministry of Statistics and Programme Implementation (MOSPI) has been, since January 2011, releasing separate rural, urban, and combined CPIs on monthly basis with base year (2010=100) for all-India and states/UTs. In addition to this, separate rural, urban, and combined Consumer Food Price Indices (CFPI) for all India were released from May 2014. The weighting diagram for the new CPI series was derived on the basis of average monthly consumer expenditure of an urban/ rural household obtained from the Consumer Expenditure Survey data (2004-05) of 61st Round of the National Sample Survey (NSS). The CSO has revised the base year of the Consumer Price Index from 2010=100 to 2012=100 and the revised index numbers were released on 12 February 2015. The basket of items and weighting diagrams for the revised series have been prepared by using the Modified Mixed Reference period (MMRP) data of the Consumption Expenditure Survey, 2011-12 of the 68th Round of NSS.

Table 9.2: Quarter-wise Inflation in CPI (base 2010=100) Broad Groups (in per cent)

	Weights	*2013-14*				*2014-15*		
		Q1	*Q2*	*Q3*	*Q4*	*Q1*	*Q2*	*Q3 (P)*
General	**100.0**	**9.5**	**9.7**	**10.4**	**8.4**	**8.1**	**7.4**	**5.0**
I. Food, beverages and tobacco	49.7	11.0	11.1	12.9	9.2	8.9	8.6	4.8
II. Fuel and light	9.5	8.4	7.9	7.0	6.3	5.2	4.0	3.4
III. Others	40.8	7.9	8.2	8.0	7.9	7.6	6.7	5.5
Food (CFPI)	42.7	11.1	11.4	13.6	9.3	9.1	8.8	4.5
Core inflation (Non-food non-fuel)	42.9	8.0	8.2	8.1	8.0	7.7	6.8	5.7

Source: CSO, P- Provisional.

Persistence of food inflation in recent years has been the major contributing factor in high headline inflation. There have been wide variations in inflation in commodities within food sub-groups across states, across commodities, and across seasons indicating supply constraints. Demand pressures exerted by high rates of growth of rural wages were mostly reflected in high prices of protein items like, milk, eggs, meat, and fish and also fruits and vegetables. During 2014-15, particularly in the third quarter, CPI food inflation declined considerably as compared to the previous year, partly on account of base effect, but also due to the seasonal softening of fruit and vegetable prices. Late arrival of the monsoon exerted some pressure on vegetable prices during June-August 2014, but the prices came down subsequently which helped significantly in the moderation of overall CPI inflation. CPI inflation in the fuel and light group registered consistent decline during 2014-15, touching 3.4 per cent in the third quarter following the sharp decline in international crude oil prices. Core inflation (non-food non-fuel) declined to 5.7 per cent in the third quarter of 2014-15 as against 8.1 per cent in the corresponding quarter of the previous year largely on account of the slack in economic activity. Housing and transport contributed to the significant decline in core inflation. Inflation in housing declined to 8 per cent in the third quarter of 2014-15, after remaining in double digits during 2012 and 2013. Inflation in the transport and communication sub-group under the miscellaneous category registered a significant decline of 1.8 per cent during the third quarter of 2014-15, in line with the continued easing of global crude oil prices. In the sub-category others, which largely includes services, inflation dropped to 8.5 per cent during the same period after experiencing double-digit inflation through 2012 and 2013. The overall trends in WPI and CPI inflation are shown in Figure 9.1.

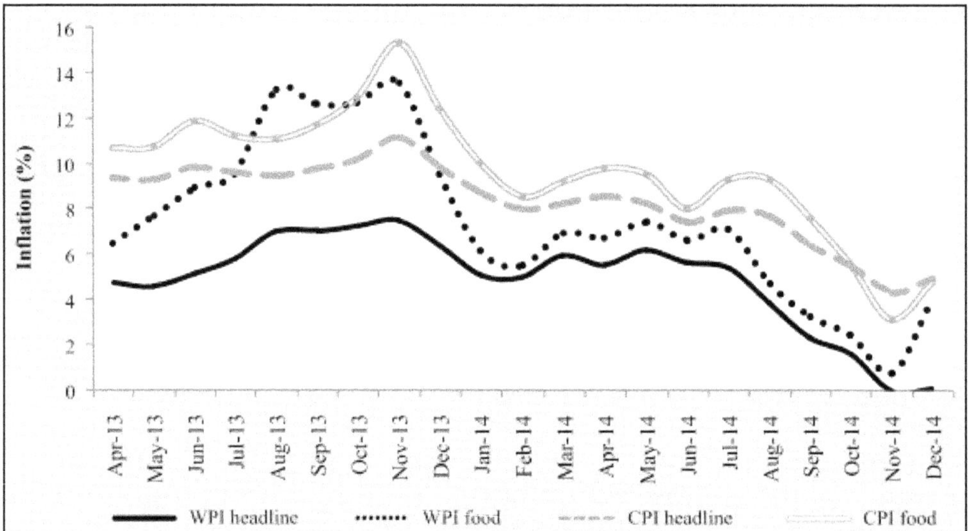

Figure 9.1: Inflation in WPI and CPI.

Soil Health Card Explained

Prime Minister Narendra Modi launched a nationwide 'Soil Health Card' scheme from Suratgarh in Rajasthan. Harish Damodaran explains what the scheme entails and its potential benefits for Indian agriculture.

What Exactly is this New Scheme?

It aims to provide a soil health card to every farmer in India over the next three years. Taking the country's total cultivable area of around 141 million hectares and an average holding of one hectare, it means distributing 141 million cards. The plan is to cover 30 million farmers in 2015 and 55 million each in the following two years.

Isn't that too Ambitious?

On the face of it, yes. Issuing the cards requires collecting soil samples from individual fields and testing these in laboratories. A soil testing laboratory can analyse 10,000-odd samples on an average annually. If one-third of 141 million fields are to be covered every year, we would need 4,700 laboratories, as against only about 1,250 –1,050 stationary and 200 mobile – today. How this capacity gap is going to be bridged isn't clear. But given that the government wants soil testing to be done once in three years in every field, it opens up opportunities for the setting up of laboratories "by a new class of entrepreneurs even in small towns", according to Modi.

If I am a Farmer, what will my Soil Health Card Show?

It will tell me the quality and fertility status of the soil in my field. Such an analysis is based on quantitative parameters, both physical and chemical. The former relate to soil depth, colour, texture, surface/subsurface hardness, compaction, *etc.* The chemical parameters include soil pH (to indicate acidity or alkalinity), electrical conductivity (a measure of salinity) and the levels of available nutrients: primary (nitrogen, phosphorous, potash), secondary (sulphur, calcium, magnesium) and micro (iron, zinc, copper, manganese, boron, molybdenum).

What Use is Such Information to me as a Farmer?

The quality and yields of the crops I grow are significantly influenced by the capacity of my soil to supply plant nutrients in the desired amounts to facilitate optimal growth. It is more than useful to know what is good or bad about my soil. If it is acidic, I need to apply lime, just as soil alkalinity and salinity can be substantially corrected through gypsum treatment and ensuring no water-logging.

Will this Help Reduce my Fertilizer Consumption?

Fertilizer consumption depends not just on the available plant nutrients in my soil, but also what crops I am growing. Wheat or rice, for example, require more nitrogen, while phosphorous is the most important nutrient in pulses and potassium is indispensable for tomato, banana or pineapple. Information on soil nutrient status is not meant to reduce fertilizer consumption as much as helping choose the right fertilizers in the right combination. Currently, I may be applying two bags of urea, which only contains 46 per cent nitrogen. If my field is deficient in sulphur, it might make sense to replace one bag of urea with ammonium sulphate that has 20.6 per cent nitrogen and also 24 per cent sulphur. Over time, as soil testing becomes an established practice, the market could shift increasingly to customised fertilisers. Instead of selling plain urea or di-ammonium phosphate, companies may start offering the same fertilisers fortified with zinc or boron. Further, they could introduce customised products, tailored for particular crops in specific regions (say, potato in Farrukhabad or groundnut in Anantapur).

Why is the Government Pushing the Soil Health Card Scheme?

The scheme is intended to promote balanced fertilisation, leading to improved soil health. But no less important a consideration is containing the mounting fertiliser subsidy bill, which, if fully provided for, comes to over Rs. 100,000 crore annually. Two-thirds of this is on account of urea alone. If the scheme helps in weaning away farmers from overuse of urea, both they as well as the government stand to gain. Besides, it could make urea price decontrol politically more acceptable.

How much will Soil Testing Cost?

Soil testing laboratories are now typically charging Rs. 150 or so per sample for basic parameters: pH, electrical conductivity, and primary and secondary nutrient content. For micronutrients, the cost can be Rs. 100 or more per parameters. This is because their testing requires use of costly reagents and equipment such as atomic absorption spectrophotometer (AAS). An AAS costs anywhere between Rs. 15 and Rs. 25 lakh, while other equipment (pH and conductivity meters, shaking apparatus, electronic balance, drying oven, etc) come for another Rs. 10 lakh. If one were to also add chemicals, glassware, standby generator and other materials, the capital cost for a testing laboratory may be upwards of Rs. 75 lakh. All these costs – plus salaries to lab technicians and utility charges – have to be recouped from farmers or the government in case it chooses to subsidise soil testing.

India's Soil Crisis

It is projected that India will need to grow its agricultural output 12 per cent to feed its people in 2020. Much of this has to come by growing more on the same amount of land. This land, however, is weakening and withering

The changing nature of soils - for the worse - is a refrain with farmers in these parts, even across the country. There are dramatic changes in Indian soils in about 40 years. They mirror similar developments elsewhere in the world. In China, for instance, a two year study by its Ministry of Water Resources, concluded that wind and water erosion threaten 17 per cent of its land, portending a 40 per cent drop in its food production in the next 50 years.

According to "Degraded and Waste Lands of India", a report by the Indian Council for Agricultural Research (ICAR) and the National Academy for Agricultural Sciences, about 141 million hectares of our total geographical area of about 328.2 million hectares is under cultivation. Of this, about 100 million hectares - or 70 per cent - is heading down a path where it will be incapable of supporting farming. Not only are these excesses and imbalances reducing the productivity and life of soils, they are now starting to show up in our bodies. For, if the soils are deficient in some nutrients, so are the foodcrops grown on them. Pharma companies, have consequently started adding Zinc, Copper, Selenium, Chromium, etc, to fortify their vitamin tablets. IISS uploaded maps of 18 states that show nutrient deficiency district by district. The maps paint a scary picture. Large parts of India are deficient in two or more critical nutrients. Areas like the Indo-Gangetic plains – Punjab, Haryana, Uttar Pradesh and Bihar - which produce nearly 50 per cent of our grains and feed about 40 per cent of our population - are seeing multiple deficiencies.

Why is this happening? In the past, farmers used to plough the stalks left standing on the field after the harvest, cowdung, etc, back into the soil. This ensured that nutrients taken out of the soil were replenished. The green revolution, which started in the sixties, changed all that. Highyielding crop varieties need more water and nutrients – which span from macronutrients like Nitrogen and Phosphorus to micro-nutrients like Copper and Boron. Apart from retarding growth in yields, this unbalanced use has also damaged soils. The ICAR report estimates that 6.98 million hectares, or 2 per cent of India's total geographical area, have acidic soils. These are mostly in North East India, south Chhattisgarh and Kerala. Another 6.7 million hectares are salt-affected. In the absence of historical data on nutrients, the degree of decline cannot be ascertained. The irrefutable point is that these soils are increasingly incapable of supporting agriculture.

Indeed Field studies at IISS and elsewhere suggest earthworm populations, organic carbon levels and yields are the highest when a combination of balanced fertilization and farmyard manure is used. But India doesn't have enough cow dung. Organic waste can be used instead of cow dung, but we need to figure out better ways to segregate organic and inorganic waste, and to get it across to fields. Here, Andhra might have an answer. As a part of its organic farming drive, the state is helping farmers set up composting pits where agricultural and household wastes can decompose into manure that is then ploughed back into the soil. Apart from that, it is handing out soil health cards to its farmers. Such solutions need to be encouraged. Or the country will find itself contemplating Malthus.

- ☆ Nitrogen: Deficient in Western Punjab, Haryana, U.P., Rajasthan, Gujarat, Maharashtra, parts of Bihar, Jharkhand, Madhya Pradesh, Andhra Pradesh and Tamil Nadu
- ☆ Zinc: (Per cent of districts low or very low in Zinc)
  - ❐ Over 50 per cent in Uttar Pradesh, Bihar, Madhya Pradesh and Tamil Nadu
  - ❐ 20 per cent to 50 per cent in Haryana Gujarat and Andhra Pradesh
  - ❐ Less than 20 per cent in Punjab and Odisha

NITROGEN is an essential part of all proteins. Its deficiency results in slow, stunted plant growth. As for **ZINC**, it is needed for building enzymes and for DNA transcription

- Phosphorus : Deficient in parts of Himachal, Haryana, UP, Rajasthan, Gujarat, Maharashtra, parts of Bihar, Jharkhand, Madhya Pradesh, Andhra Pradesh, Tamil Nadu, Bengal and Asom.
- Iron (Per cent of districts low or very low in Iron)
    - Less than 20 per cent in Punjab, Haryana, Bihar, Gujarat, Madhya Pradesh

PHOSPHORUS helps convert light energy into chemical energy. It is also important for plant growth and seed formation. **IRON** is necessary for building Chlorophyll

Manganese : less than 20 per cent in Uttar Pradesh

Boron : its deficiency may is still being prepared. But shortages are widespread in Odisha, Andhra, Tamil Nadu, Bengal, Asom, Punjab, Western UP and Haryana.

POTASSIUM reduces water loss. **SULPHUR** helps make amino acids and vitamins. **COPPER** builds cell walls and is needed for photosynthesis. **MANGANESE** is necessary for building chloroplasts, which carry out photosynthesis

The changing chemical composition of soils

- Acidic Soils: Change in the Chemical composition of the soil
- Salt Affected Soil: Deposits of fertilizer salts on top of the soil.

State of Agriculture in India

Agriculture in India is the means of livelihood of almost two-thirds of the workforce in the country. It employees around 48 per cent of total population and 42 per cent of total geographical area. It is therefore considered to be the most important sector of India Economy.

Population

Total population of India by May 2013 was 1252million, of which rural population was 68.8 per cent.

A Saga of Success

From a nation dependent on food imports to feed its population. India today is not only self- sufficient in grain production but also has a substantial reserve. The progress made by agriculture in the last four decades has been one of the biggest success stories of free India. Agriculture and allied activities constitute the single largest contributor to the Gross Domestic Product, almost 33 per cent of it. Agriculture is the means of livelihood of about two-third so the workforce in the country. This increase in agricultural production has been brought about by bringing additional area under cultivation, extension of irrigation facilities, the use of improved high-yielding variety of seeds, better techniques evolved through agricultural research, water management, and plant protection through judicious use of fertilizers, pesticides and cropping practices.

Area, Production and Yield

Table 9.3 gives area, production, and yield figures for different crops in 2013-14. In 2013-14, total foodgrain production has been estimated at 265.6 million tonnes as per the second Advance Estimates (AE), which is higher by 8.5 million tonnes than the 2012-13 production and 22.1 million tonnes than average foodgrain production during the last five years. As per the 2nd Advance Estimates for 2014-15, total foodgrains production in the country is estimated at 257.07 million tonnes which is the fourth highest quantity of annual foodgrains production in the country. It may be noted that despite deficiency of 12 per cent in the monsoon rainfall during the year, the loss in production has been restricted to just around 3 per cent over the previous year and has exceeded the average production during the last five years by 8.15 million tonnes.

As compared to last year's production of 265.57 million tonnes, current year's production of foodgrains is lower by 8.5 million tonnes. This decline has occurred on account of lower production of rice, coarse, cereals and pulses due to erratic rainfall conditions during the monsoon season-2014. To improve resilience of the agricultural sector and bolster food security including availability and affordable access our strategy for agriculture has to focus on improving yield and productivity. Though yield/productivity in foodgrains and pulses has increased post-2000, the yield gaps vis-à-vis other countries are wide and even within different states yields vary widely, showing that there are possibilities of raising production by increasing yield of most of the crops without necessarily increasing prices (**Table 9.4**).

An inverse relationship is noticed between increase in yield over time and the average cost of production of various crops in real terms. For example, for rabi crops a 10 per cent increase in yield resulted in a 2.1 per cent to 8.1 per cent decline in the average cost of production of various crops in real terms. (Price Policy for Kharif Crops, February 2014). This clearly points towards the fact that productivity increases, especially in low productivity states/regions, can significantly contribute towards reducing cost-push food inflation. Yield is contingent upon several factors like variety and quality of seeds, soil quality, irrigation – including quality of water –fertilizers - including their proportion pesticides, labour, and extension services. Prices received by farmers and the certainty or assurance of getting a particular price also incentivize farmers to take to a particular crop and use quality inputs in its cultivation.

Table 9.3: Area, Production and Yield 2013-14*

(Area: million ha; Prod.: million tonnes; Yield: kg/ha)

Group/Commodity	*Area*	*Per cent Change in Area*	*Production*	*Per cent Change in Production*	*Yield*	*Per cent Change in Yield*
Foodgrains	126.0	4.3	264.8	3.0	2101	–1.3
Rice	43.9	2.7	106.5	1.3	2424	–1.5
Wheat	31.2	4.0	95.9	2.6	3075	–1.3
Jowar	5.8	–6.1	5.4	1.7	850	–8.2
Maize	9.4	8.3	24.4	9.2	2566	–0.7
Bajra	7.9	8.0	9.2	5.5	1198	2.9
Pulses	25.2	8.3	19.3	5.3	764	–3.2
Gram	10.2	20.3	9.9	12.3	967	–6.7
Tur	3.9	0.0	3.3	9.7	848	9.2
Oilseeds	28.5	7.6	32.9	6.4	1153	–1.3
Groundnut	5.5	17.6	9.7	105.8	1750	75.9
Rapeseed and mustard	6.7	4.7	8.0	–0.5	1188	–5.9
Cotton[b]	11.7	–2.3	36.7	7.2	532	9.4
Sugarcane	5.0	0.0	350.0	2.6	70	0.0

Source : Directorate of Economics and Statistics, Department of Agriculture and Cooperation.

Notes : *Fourth AE., A Includes cereals, coarse cereals, and pulses., b Bales of 170 kg.

Table 9.4: Average, Maximum and Minimum Yield of Major Crops 2013-14

(Yield: kg/ha)

Crops	*All-India Average*	*Maximum*	*Minimum*
Rice	2416	Punjab (3952)	Madhya Pradesh (1474)
Wheat	3145	Punjab (5017)	Andhra Pradesh (500)
Maize	2676	Tamil Nadu (5372)	Asom (898)
Jowar	957	Andhra Pradesh (1661)	West Bengal (280)
Gram	960	Andhra Pradesh (1439)	Tamil Nadu (653)
Tur	813	Bihar (1667)	Andhra Pradesh (542)
Groundnut	1764	Gujarat (2668)	Himachal Pradesh (600)
Rapeseed and Mustard	1185	Gujarat (1723)	Tamil Nadu (241)
Soyabean	1012	Andhra Pradesh (1612)	Uttar Pradesh (577)
Sugarcane	70522	West Bengal (114273)	Jammu and Kashmir (1000)
Cotton#	510	Punjab (750)	Maharashtra (358)

Source : Directorate of Economics and Statistics, Department of Agriculture and Cooperation

Note : # Thousand bales of 170 kg each.

Crops

The 1970s saw a multi-fold increase in wheat production that heralded the Green Revolution. In the next decade rice production rose significantly; in 1995-96, rice production was 79.6 million tones. The total grain production crossed 211 million tones in 2001-02, and touched 264 million tones in 2013-14 a big leap from 51 million tonnes in 1905-51. To carry improved technologies to farmers, a National Pulse Development Programme, covering 13 states, was launched in 1986. Efforts to boost pulse production were augmented further by the Special Food Production Programme. In 2001-02, pulse production was 13.52 million tones and 19.57million tones in 2013-14.

With some states offering more than the statutory minimum price, sugarcane production also received a boost, and in 2013-14 a record 383.8 million tones was registered.

Irrigation

The central government initiated the Accelerated Irrigation Benefit Programme (AIBP) in 1996-97 for the completion of incomplete irrigation schemes. Under the AIBP, Rs. 67,195.47 crore of central loan assistance (CLA)/grant was released up to 31 December 2014. An irrigation potential of 85.03 lakh ha is reported to have been created under the AIBP by states from major/medium/minor irrigation projects till March 2013. The Command Area Development Programme has also been amalgamated with the AIBP to reduce the gap between irrigation potential that has been created and that is utilized. Suggestions for a National Water Grid for transferring water from water surplus to water deficit areas have been made from time to time. In spite of these schemes, Indian agriculture is still heavily rainfall dependent with just 35 per cent of total arable area being irrigated, and distribution of irrigation across states is highly skewed. Focus on micro-irrigation systems like drips and sprinklers would significantly increase water-use efficiency and productivity. The wide gap between gross cropped area and gross irrigated area which has not improved much since the First Five Year Plan period needs to be bridged for increasing productivity, production, and resilience (Figure 9.2).

Over 500 hydrological stations collect and transmit data through 400 wireless stations for issuing forecasts for 157 sites. About 5,000 forecasts are issued in a year with 94 per cent accuracy. The country also receives international support, with the World Bank as a primary source, for developing its water resources. International cooperation is also envisaged in setting up a National Centre for information on water and power. As there is a broad seismic belt in the country, particularly along the Himalayan, and the Kutch region and parts of Maharashtra, scheme is being evolved to collect all data on seismic activity at various dam sites.

Fertilizers

The fertilizer industry in India has grown tremendously in the last 30 years. The Government is keen to see that fertiliser reaches the farmers in the remote and hilly areas. It has been decided to decontrol the prices, distribution and movement of phosphatic and potassic fertilisers. Steps have been taken to ensure an increase

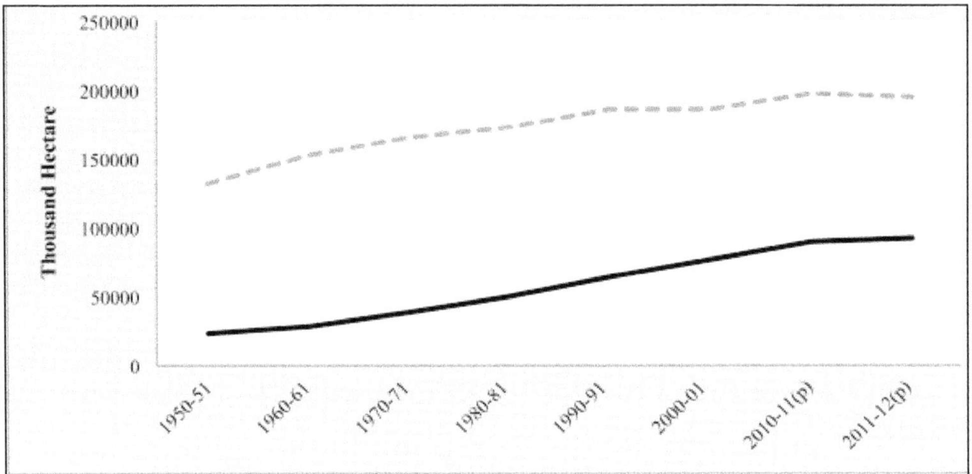

Source: Department of Agriculture & Cooperation (DAC).

Figure 9.2: Gross Irrigated Area vis-à-vis Gross Cropped Area.

in the supply of non-chemical fertilisers at reasonable prices. There are 66 fertiliser quality control laboratories in the country. Since bio-fertilisers are regarded as an effective, cheap and renewable supplement to chemical fertilisers, the Government is implementing a National Project on Development and Use of Bio-fertilisers. Under this scheme, one national and six regional centers for organising training, demonstrating programmes and quality testing of bio-fertilizers has been taken up.

It was a challenging decision of the Government to take Bombay High gas through a 1,700 km pipeline to feed fertiliser plants located in the consumption centers of North India. However, the major policy which has ensured the growth of the fertiliser industry is the thrust on accelerating fertiliser consumption by fixing, on the one hand, low and uniform price for fertilisers, and on the other hand providing the manufactures adequate compensation through the retention price and subsidy scheme.

The following major initiatives were taken in the fertilizer policy of the government in 2014- 15: (i) Notification of the Modified New Pricing Scheme (NPS-III) for existing urea units on 2 April 2014 in order to address the issue of under- recoveries of the existing urea units on account of freezing of fixed cost at the 2002-03 level. The modified policy has been implemented for a period of one year from the date of notification. (ii) Further, the government had notified the New Investment Policy 2012 on 2 January 2013 to facilitate fresh investment in the urea sector to make India self-sufficient. The amendment to New Investment Policy – 2012 has been notified by the Department of Fertilizers on 7 October 2014. As against the targets for domestic production of 89.68 lakh tonnes and 33.51 lakh tonnes for nitrogen and phosphate for April-November 2014, actual production was 82.86 lakh tonnes and 25.05 lakh tonnes respectively.

Animal Husbandry, Dairying, and Fisheries

India ranks first in milk production, accounting for 17 per cent of world production. During 2013-14, milk production peaked at 137.69 MT, thus becoming an important secondary source of income for 70 million rural households engaged in dairying. The average year- on-year growth rate of milk, at 4.18 per cent vis-à-vis the world average of 2.2 per cent, shows sustained growth in availability of milk and milk products for the growing population. In the poultry segment, the government's focus, besides framing suitable policies for enhancing commercial poultry production, is on strengthening the family poultry system, which addresses livelihood issues. Egg production was around 73.89 billion in 2013-14, while poultry meat production was estimated at 2.68 MT.

Fisheries constitute about 1 per cent of the GDP of the country and 4.75 per cent of agriculture GDP. The total fish production during 2013-14 was 9.58 MT, an increase of 5.96 per cent over 2012-13. Fish production during the first two quarters of 2014-15 has also shown an increasing trend and is estimated at 4.37 MT (Provisional). For sustainable and continuous growth of the livestock sector by emulating the success achieved in the dairy and poultry sectors, across species and regions, the National Livestock Mission has been launched in 2014-15 with an approved outlay of Rs.2,800 crore during the Twelfth Plan. This Mission is formulated with the objective of sustainable development of the livestock sector, focusing on improving availability of quality feed and fodder, risk coverage, effective extension, improved flow of credit, and organization of livestock farmers/rearers. Given the high contribution of protein items in inflation, the growth rate of this sector has to match the rising demand reflected in increasing share of these items in consumption expenditure.

Food Processing

The Ministry of Food Processing Industries, set up in July 1988, is the central agency of the Government responsible for developing a strong and vibrant food-processing sector with a view to create increased job opportunities in rural areas, enable the farmers to reap benefit from modern technology, create surplus for exports and stimulating demand for processed food. A new seeds policy has been adopted to provide access to high-quality seeds and plant material for vegetables, fruit, flowers, oilseeds and pulses, without in any way compromising quarantine conditions. Initiatives have been taken to encourage private sector investment in the food processing industry.

Agricultural Research

The apex body for education, research, training and transfer of technology in the field of agriculture is the Indian Council of Agricultural Research (ICAR), established in 1929. India's transformation from a food deficit to a food surplus country is largely due to ICAR's smooth and rapid transfer of farm technology from the laboratory to the land. ICAR discharges its responsibilities through 43 research institutes, four national research bureaus, 20 national research centers, nine project directorates, 70 all-India coordinated research projects, and 109 Krishi Vigyan

Kendra's (farm science centers). Besides, the programme of Agricultural Education is coordinated by ICAR with the curricula and other normative guidance given to the 26 agricultural universities and four national research institutes.

Oilseeds Production

A Technology Mission on Oilseeds was launched in 1986 to increase production of oilseeds in the country and attain self-sufficiency. Pulses were brought under the Technology Mission in 1990. After the setting up of the Technology Mission, there has been consistent improvement in the production of oilseeds. The oilseed production, which was 108.3 lakh tonnes in 1985-86, has increased to 247.30 lakh tonnes in 1998-99, the highest production so far. However, due to unfavourable weather conditions prevailing in the major oilseeds-growing states, the oilseed production during 1999-2000 and 2000-01 suffered some decline. It has now recovered to 329 lakh tones in 2013-14.

The increase in production was largely contributed by soyabean, rapeseed and mustard. Production of pulses has seen many ups and downs, which is expected to be checked under the mission. The country grows mainly nine oilseeds, with groundnut, rapeseed and mustard accounting for 62 per cent of total production. Lately, soyabean and sunflower have shown major growth potential.

Drinking Water

A Technology Mission on Drinking Water and Related Water Management has been constituted to cover the residual problem villages and provide potable water at 40 liters per capita per day, and 70 liters per capita per day in desert areas inclusive of 30 liters for cattle. The mission is tackling the problem through 55 mini-missions in project districts and countrywide problem-oriented sub-missions A Village Level Operation and Maintenance (VLOM) pump called India Mark-II has been developed and is being exported to 40 countries.

Monsoon Season Rainfall

Major part of Indian agriculture depends upon monsoon. The monsoon rainfall from June 1, 2005 to July 20, 2005 was normal/excess in 29 meteorological subdivisions (81 per cent) and deficient/scanty in the remaining 7 (19 per cent). A better monsoon season rainfall from June 1, 2005 to July 20, 2005 in comparison to annual average for 2000-04.

Procurement

To enhance efficiency of procurement and public distribution and to extend the benefits of MSP to local farmers, the Decentralized Procurement (DCP) scheme has been adopted by some state governments. The central government is urging all state governments to adopt the DCP scheme so that costs of distribution can be saved and outreach of price support mechanism to the farmers in hitherto weaker areas can be improved. To overcome the problem of gaps in the flow of information about procurement operations on day-to-day basis, an Online Procurement Monitoring System (OPMS) has been evolved for reporting and monitoring on a daily basis, procurement operations for wheat, paddy, and coarse grains in the country. Two decisions that will impact procurement and stocks of rice and wheat from kharif marketing season (KMS) 2014-15 and rabi marketing season (RMS) 2015-16 are: (a) To limit procurement from states that are declaring bonus over and above the MSP to the extent of targeted PDS (TPDS)/other welfare schemes (OWS) requirements (In the case of non-DCP states declaring bonus, the FCI will not take part in MSP operations in those states.) and (b) To cap the percentage of levy on rice at 25 per cent.

This decision has successfully led to dropping of the practice of giving bonus over and above MSP for paddy in states like Chhattisgarh and Madhya Pradesh in KMS 2014-15 and it is expected that the state governments of Madhya Pradesh and Rajasthan will avoid giving bonus for wheat also in RMS 2015-16 in view of this policy. The procurement levels in KMS 2014-15 are lower in both Chhattisgarh and Madhya Pradesh as compared to the previous year and there is re- emergence of competition in the market. **Table 9.5** gives procurement, off-stake and stock figures from 2003.

Table 9.5: Public Distribution System: Procurement, Off-Take, and Stocks (million tones)

Year	*Procurement*			*Off-take*			*Stocks*		
	Rice	*Wheat*	*Total*	*Rice*	*Wheat*	*Total*	*Rice*	*Wheat*	*Total*
2003-04	22.9	15.8	38.7	25.0	24.3	49.3	13.1	6.9	20.7
2004-05	24.7	16.8	41.5	23.2	18.3	41.5	13.3	4.1	18.0
2005-06	27.6	14.8	42.4	25.1	17.2	42.3	13.7	2.0	16.6
2006-07	25.1	9.2	34.3	25.1	11.7	36.8	13.2	4.7	17.9
2007-08	28.7	11.1	39.9	25.2	12.2	37.4	13.8	5.8	19.8
2008-09	34.1	22.7	56.8	24.6	14.9	39.5	21.6	13.4	35.6
2009-10	32.0	25.4	57.4	27.4	22.4	49.7	26.7	16.1	43.3
2010-11	34.2	22.5	56.7	29.9	23.1	53.0	28.8	15.4	44.3
2011-12	35.0	28.3	63.4	32.1	24.2	56.3	33.4	20.0	53.4
2012-13	34.0	38.2	72.2	32.6	30.1	62.8	35.5	24.2	59.8
2013-14	31.3	25.1	56.4	29.2	28.2	57.4	30.6	17.8	49.5
2014-15*	16.2	28.0	44.2	4.5	3.8	8.3	23.5	37.3	61.6

Source: Ministry of Food, Consumer Affairs and Public Distribution, Government of India.
Note: * as on 9.1.2015.

10

Agricultural Development in India Since Independence

Agriculture plays an essential role in the process of economic development of less developed countries like India. Besides providing food to nation, agriculture releases labour, provides saving, contributes to market of industrial goods and earns foreign exchange. Agricultural development is an integral part of overall economic development. In India, agriculture was the main source of national income and occupation at the time of Independence. Agriculture and allied activities contributed nearly 50 per cent to India's national income. Around 72 per cent of total working population was engaged in agriculture. These confirm that Indian economy was a backward and agricultural based economy at the time of Independence. After 61 year of Independence, the share of agriculture in total national income declined from 50 per cent in 1950-66, 18 per cent in 2013-14. But even today more than 45 per cent of workforce is engaged in agriculture. In spite of this, it is also an important feature of agriculture that is to be noted that growth of other sectors and overall economy depends on the performance of agriculture to a considerable extent. Because of these reasons agriculture continues to be the dominant sector in Indian Economy.

Since independence India has made much progress in agriculture. Indian agriculture, which grew at the rate of about 1 per cent per annum during the fifty years before Independence, has grown at the rate of about 2.6 per cent per annum in the post-Independence era. Expansion of area was the main source of growth in the period of fifties and sixties after that the contribution of increased land area under agricultural production has declined over time and increase in productivity became the main source of growth in agricultural production. Another important facet of progress in agriculture is its success in eradicating of its dependence on imported foodgrains. Indian agriculture has progressed not only in output and yield terms but the structural changes have also contributed. All these developments in Indian agriculture are contributed by a series of steps initiated by Indian Government. Land reforms, inauguration of Agricultural Price Commission with objective to ensure remunerative prices to producers, new agricultural strategy, investment in research and extension services, provision of credit facilities, and improving rural infrastructure are some of these steps.

In order to achieve the goal of self sufficiency in agriculture, new agricultural strategy was initiated in 1966-67. The fundamental of this strategy is the application of science and technology for increasing yield per hectare. This strategy, known as New Green Revolution, is based on the extension of high yielding varieties responsive to heavy doses of fertilizers and the package of improved practices in selected areas with assured rainfall or irrigation facilities. The programmes included under the new strategy are: (1) the high yielding varieties programme, (2) multiple cropping programme, (3) integrated development of dry areas, (4) plant protection measures, (5) increased use of fertilizers, and (6) new irrigation concept.

Notwithstanding these progresses, the situation of agriculture turned adverse during post- WTO period and this covered all the sub sectors of agriculture. The growth rates in output of all crops decelerated from 2.93 per cent to 1.57 per cent. The livestock declined from 4.21 per cent to 3.40 per cent. The fisheries declined from 7.48 per cent to 3.25 per cent. Only, forestry witnessed a sharp increase from 0.09 per cent to 1.82 per cent.

The crop sector, which forms largest segment of agriculture, showed poorest growth during post-WTO period in comparison to all other periods. Further, within crop sector, all crops except sugar showed declining trend between initial years of reforms and post-WTO period. This deceleration is very high in Cereals, Corse Cereals, Pulses, Oilseeds, and Drugs and Narcotics. The growth rate turned negative in the case of pulses. Both dominant nature of agriculture and decelerating growth trend in agriculture attracts attention of policymakers, researchers and economists. The main cause of failure of all development policy for agriculture is that there is no availability of any separate development strategy for Indian agriculture. Agriculture production is a biological process, agriculture is diminishing returns activity because land is ultimately a fixed factor of production and the demand for agricultural commodities is income inelastic. These characteristics make different to agriculture from other sector. Therefore, a separate policy for agricultural development is must This is due to the fact that we had not available necessary data to study the characteristics of Indian agriculture. But presently we have come a long way from Independence and now we have long-terms data pertaining to Indian agriculture. So, the present study makes attempt to fill this gap.

In this context, the present paper extensively evaluates performance and progress of Indian agriculture since Independence. Besides comparing facts and figures, we also examined sources of agricultural growth and instability of Indian agriculture for evaluating performance and progress of Indian agriculture. The paper also finds out determinates of agricultural production by using production function approach and verifies the results of decomposition of agricultural growth.

This paper covered period from 1950-51 to 2005-06. We have chosen 1950-51 as starting period because all required data is not available (some data are available but sources are not authentic) for period before 1950-51 and 2006-07 as end period because the most recent data pertained to 2006-07 at the time of finalising the paper.

The rest of the paper is organized as follows. Section (ii) reviews previous agricultural policies. Section (iii) evaluates changes and performance of Indian agriculture since Independence. In this section, we discussed change in whole scenario of Indian agriculture over period of time. Section (iv) guesstimate sources of agricultural growth by decomposition analysis. Section (v) calculates instability in production, area, and yield of principal crops. Section (vi) estimates determinants of agricultural production by production function approach. Concluding remarks are presented in the final section.

Agricultural Policy: A Review

In this section, we try to trace out the principle government policies for promoting agricultural development. For the overall development of Indian agriculture, many institutional and infrastructural changes have been introduced since Independence. Broadly, agricultural policy followed during this period can be distinguished in four phases: first phase considered from 1947 to mid sixties, second phase considered period from mid sixties to 1980, third phase included period from 1980 to 1991, and forth phase includes period from 1991/92 onwards. The first phase of agricultural policy witnessed tremendous agrarian reforms, institutional changes, development of major irrigation project and strengthens of cooperative credit institution. The most important contribution of land reforms was abolition of intermediaries and giving land titles to the actual cultivators. This released productive forces and the owner cultivators put in their best to augment production on their holdings. Land reforms were important in increasing agricultural production during this phase. The Community Development Programme, decentralised planning and the Intensive Area Development Programmes were also initiated for regenerating Indian agriculture that had stagnated during the British period. In order to encourage the farmers to adopt better technology, incentive price policy was adopted in 1964 and the Agricultural Price Commission was set up to advice the Government on the fixation of support prices of agricultural crops. Despite the institutional changes and development programmes introduced by the Government during this phase, India remained dependent upon foreign countries for food to feed the rising population.

The second phase in Indian agriculture started in mid 1960s with adoption of new agricultural strategy (Green Revolution). The new agricultural strategy relies on high-yielding varieties of crops, multiple cropping, the package approach, modern farm practices and spread of irrigation facilities. The biggest achievement of this strategy has been attainment of self sufficiency in foodgrains. Agrarian reforms during this period took back seat while research, extension, input supply, credit, marketing, price support and spread of technology were the prime concern of policy makers. The next phase in Indian agriculture began in early 1980s. This period started witnessing process of diversification which resulted into fast growth in non-foodgrains output like milk, fishery, poultry, vegetables, fruits etc which accelerated growth in agricultural GDP during the 1980s. There has been a considerable increase in subsidies and support to agriculture sector during this period while public sector spending in agriculture for infrastructure development started showing decline in real term but investment by farmers kept on moving on a rising trend.

The fourth phase of agricultural policy started after initiation of economic reform process in 1991. Economic reforms process involved deregulation, reduced government participation in economic activities, and liberalization. Although there is no any direct reforms for agriculture but the sector was affected indirectly by devaluation of exchange rate, liberalization of external trade and disprotection to industry. During this period opening up of domestic market due to new international trade accord and WTO was another change that affected agriculture. This raised new challenges among policymakers. Because of this, a New Agricultural Policy was launched by Indian Government in July 2000. This aims to attain output growth rate of 4 per cent per annum in agriculture sector based on efficient use of resources. It seeks to achieve this objective in a sustainable manner and with equity. This was first time when government released a national agriculture policy. The policy document discusses what ought to be done in agriculture but the subsequent step, how and when policy goals and objective would be achieved is not discussed (Chand, 2003). Therefore, it is highly desirable to prepare action plans at both centre and state level in quantity terms to implement the new policy agenda in a time bound framework.

Changing Agrarian Economy since Independence

In this section we focused on how agrarian economy has changed since Independence. Keeping this view in mind this section follows land use pattern, population and agricultural workers, distribution of operational holding, and cropping pattern.

Land Use Pattern

The basic factor in agriculture is land. A knowledge about land use pattern is vital to understand whether the utilisation of land in India is at its full potential or far from its full potential. In India the classification of land has had its roots in agricultural statistics. Till 1950, the land in India was broadly classified into five categories: (i) Area under forests; (ii) Area not available for cultivation; (iii) Uncultivated lands including current fallows; (iv) Area under current fallows; and (v) Net area sown. But then it was realised that such a classification did not give a clear picture of the actual area under different categories of land use required for agricultural planning. Hence, a reclassification was adopted from March 1950. Under it, land in India now classified under nine different categories. These are as: (i) forests; (ii) barren and uncultivable lands; (iii) land put to non-agricultural uses; (iv) cultivable wastes; (v) permanent pastures and other grazing lands; (vi) miscellaneous tree crops and groves not included in the net area sown; (vii) current fallows; (viii) other fallows; and (ix) net sown area.

Table 10.1 shows changes in land use pattern in India since 1951/52.The total geographical area of the country is 328726 thousand hectares in which 93 per cent area is reporting area which means that the area for which record is available. It was 88 per cent in 1950/51. The net sown area has risen by 18.44 per cent from 1950/51 to 2000/01. The net sown area is only 46 per cent of total reporting area that was 41 per cent of total reporting area in 1950/51. The area under non agricultural use has increased from 12690 thousand hectares to 24070 thousand hectares since 1950/51. But barren and uncultivable land has fallen from 37484 thousand hectares to 17709 thousand hectares. Both the cultivable waste land and fallow land have also decreased during this period. But even today 4.4 per cent of total reporting area is available as a cultivable waste land and 4.8 per cent of total reporting area is fallow land. This indicates that there is scope to increase the net sown area by at least 5 to 10 per cent by improving both cultivable waste land and fallow. Gross sown area was 131893 thousand hectares in 1951/52 and it has increased to 185704 thousand hectares in 2001/02. This shows that only 11 per cent of net sown area was used for more than one crop in 1951/52 and this figure increased to 31 per cent in 2001/02. This point out that gross sown area can be increased by 70 per cent of net sown area through intensive cropping.

Table 10.1: Change in Land Use Pattern in India from 1950-51 to 2001-02

Category	*1951-52*	*1961-62*	*1971-72*	*1981-82*	*1991-92*	*2001-02*
Geographical Area	328726	328726	328726	328726	328726	328726
Reporting Area forland use	287827 (87.56)	299151 (91)	304141 (92.52)	304272 (92.56)	304900 (92.75)	305014 (92.79)
Area under non-agricultural use	12690 (4.41)	14795 (4.95)	16972 (5.58)	19686 (6.47)	21465 (7.04)	24070 (7.89)
Barren anduncultivable land	37484 (13.02)	35921 (12.01)	27996 (9.20)	20010 (6.58)	19270 (6.32)	17709 (5.81)
Net sown area	119400 (41.48)	135399 (45.26)	139721 (45.94)	141928 (46.64)	141632 (46.45)	141416 (46.36)
Gross sown area	131893	152772	165791	172630	185742	185704
Cropping Intensity	111	115	118	123	130	131
Forest land under good tree cover	48889 (16.98)	54189 (18.11)	63771 (20.97)	67422 (22.15)	67866 (22.25)	69511 (22.79)
Misc. tree crops and groves	7881 (2.73)	4500 (1.50)	4284 (1.41)	3715 (1.22)	3761 (1.23)	3370 (1.10)
Cultivable Wastelands	23929 (8.31)	18632 (6.23)	17456 (5.74)	16475 (5.41)	14994 (4.92)	13405 (4.39)
Current fallows	13808 (4.80)	11155 (3.73)	12669 (4.16)	13173 (4.33)	14672 (4.81)	14643 (4.80)
Old fallows	15154 (5.26)	10478 (3.50)	8312 (2.73)	9862 (3.24)	9941 (3.26)	10304 (3.38)
Permanent pastures and grazing lands	8592 (2.98)	14082 (4.70)	12960 (4.26)	12007 (3.95)	11299 (3.71)	10586 (3.47)

Note: Figure in parentheses indicate percentage to Reported Area.

Source: Agricultural Statistics at a Glance (2008).

Changing Agricultural Structure

We look at changing structure of Indian agriculture in terms of employment and land holding. The share of agriculture in employment declined from about 82 per cent in 1950/51 to about 72 per cent by 2001. During the same duration, the share of agriculture in total GDP also declined from 54.66 per cent in 1950/51 to 24 per cent by 2001.Among agricultural workforce about 45.6 per cent are registered as agricultural labour and the rest, *i.e.*, 54.4 per cent as cultivators while 28.1 per cent was registered as agriculture labour and the rest as cultivators in 1950/51. This indicates that agricultural workforce shifted from cultivators to agricultural labours (**see Table 10.2**).

Table 10.2: Population and Agricultural Workers in India since 1950-51 (In Millions)

Year	*Total Population*	*Average Annual Exponential Growth Rate (per cent)*	*Rural Population*	*Cultivators*	*Agricultural Labourers*	*Total*	*Share of Agriculture to GDP (Per cent)*
1951	361.1	1.25	298.6 (82.7)	69.9 (71.9)	27.3 (28.1)	140.0 (100.0)	54.66
1961	439.2	1.96	360.3 (82.0)	99.6 (76.0)	31.5 (24.0)	188.7 (100.0)	49.14
1971	548.2	2.22	439.0 (80.1)	78.2 (62.2)	47.5 (37.8)	180.4 (100.0)	43
1981	683.3	2.20	523.9 (76.7)	92.5 (62.5)	55.5 (37.5)	244.6 (100.0)	37.5
1991	846.4	2.14	628.9 (74.3)	110.7 (59.7)	74.6 (40.3)	185.3 (100.0)	30.32
2001	1028.7	1.95	742.6 (72.2)	127.3 (54.4)	106.3 (45.6)	234.1 (100.0)	24

Note: Figure in parentheses indicate percentage to Reported Area.

Source: Agricultural Statistics at a Glance (2008).

There were 48900 million operational holding in 1960/61 and covered area of 131400 million hectares. These operational holding increased to 115580 million in 2000/01and covered area of 163357 million hectares. It shows that the number of operational holding has increased by about 66680 million units but the area covered by then has not significantly increased. It implies that size of operational holding has been reducing. **Table 10.3** shows that the number of marginal and small holdings and the area under such holdings has increased while the number of semi-medium, medium, and large holdings and the area under such holdings has reduced. It reveals that the inequalities in the distribution of land among the cultivators has reducing trend but the number of uneconomic holdings has an increasing trend, *i.e.*, small and marginal holdings are increasing in both number and percentage.

Table 10.3: The Percentage Distribution of Operational Holding by Size Class,1960-61 to 2000-01

Category of Holdings	*1960-61*	*1970-71*	*1980-81*	*1990-91*	*2000-01*
Marginal	40.69	50.6	56.4	59.4	63.0
Small	22.29	19.1	18.1	18.8	18.80
Semi-Medium	18.8	15.2	14.0	13.1	11.7
Medium	13.4	11.3	9.1	7.1	5.4
Large	4.9	3.9	2.4	1.6	1.02
(ii) Share in Operated Area (In Percentage)					
Marginal	6.6	9.0	12.0	15.1	18.82
Small	12.17	11.9	14.1	17.4	20.18
Semi-Medium	19.93	18.4	21.2	23.2	23.96
Medium	30.51	29.8	19.6	27.0	23.84
Large	30.74	30.9	23.0	17.3	13.21
(iii) Average Size (In Hectares)					
Marginal		0.41	0.39	0.39	0.40
Small		1.44	1.44	1.43	1.41
Semi-Medium		2.81	2.78	2.76	2.72
Medium		6.08	6.02	5.9	5.80
Large		18.1	17.41	17.33	17.18

Note: 1. Marginal – 0 to 1 hectare; Small – 1 to 2 hectare; Semi-medium – 2 to 4 hectare; Medium – 4 to 10 hectare; and Large – 10 and above hectare. Source: Agricultural Census Division, Ministry of Agriculture, New Delhi.

Changes in Cropping Pattern

Cropping pattern means the proportion of area under different crops at a particular period of time. A change in cropping pattern means a change in the proportion under different crops. **Table 10.4** indicates that the area under non-food crops as a proportion of the total cropped area is increasing but still there is dominance of food crops. At the beginning of the economic planning in India, 76.7 per cent land was put under food crops and about 23.3 per cent on non-food crops. By 2001, area under food crops had come down to 65.83 per cent and under non-food crops has increased to 34.17 per cent. This shift in the allocation of area from food crops to non-food crops reflect a change from subsistence cropping to commercial cropping. This shifting of land from food crops to non-food crops was mainly influenced by the prevailing price in market and profitability per hectare.

Similarly, here it can also be concluded that, there is preponderance of cereals, about 54.43 per cent of the area is devoted to the production of cereals, while only 11.4 per cent is devoted to pulses. Though, the area under both cereals and pulses is increasing but the rate of increase in area under cereals is greater than that of pulses. It means whatever cropped area increased as a result of irrigation facilities, chemical fertiliser, and high yielding varieties of seeds, a greater part of it is devoted to foodgrains. Within cereals, area under coarse cereals is gradually declining since 1950/51. This is due to fact that coarse cereals are inferior goods.

Table 10.4: Changes in Cropping Pattern in India since 1950-51

Crops	*1950-51*	*1960-61*	*1970-71*	*1980-81*	*1990-91*	*2000-01*
Rice	23.5	22.3	22.6	23.3	23	24.03
Wheat	7.6	8.5	11	12.8	12.9	13.84
Corse Cereals	29.9	29.4	27.8	24.6	19.6	16.55
Total Cereals	61.1	60.2	61.4	60.8	55.5	54.43
Total Pulses	15.6	15.5	14	13.2	13.5	11.4
Total Foodgrains	76.7	75.7	75.4	73.9	68.9	65.83
Sugarcane	1.3	1.6	1.6	1.7	2.1	2.49
Condiments andSpices	0.9	1	1.1	1.2	1.3	1.5
Fruits and Vegetable	1.7	1.9	2	1.7	3.6	4.39
Total Oilseeds	8.3	8.3	8.9	9.1	13.5	13.56
Total Fibres	5.1	5.7	5.5	5.3	4.7	5.22
Tobacco	0.3	0.3	0.2	0.3	0.2	0.16

Source: Agricultural Statistics at a Glance (2008).

Furthermore, **Table 10.4** also shows that area under fruits and vegetables and oilseeds is gradually increasing since 1950/51. This is because the consumption pattern is shifting from cereals to non-cereals. This phenomenon can be seen in **Table 10.5**.

Following inferences can be drawn in these two tables.

1. The share of food in the total budget expenditure of a household has been showing a decline over the years in both rural and urban sector. In rural all India, the share of food declined from 72.9 per cent in 1972-73 to 55 per cent in 2004-05 and in urban all India it declined from 64.5 per cent in 1972-73 to 42.5 per cent in 2004-05.
2. Although cereal continues to be the important constituent of a household's food bosket, its share in total budget is declining.
3. The share of pulses is also showing declining trend.
4. The consumption of vegetables and fruits, milk, meat, egg, and fish, and edible oil has shown an increase.

The structural shift in consumption pattern is on account of the consumption diversification effect because of easy access to supply, changed tastes and preferences, and change in relative prices (Radhakrishna and Ravi 1992; Kumar 1998; Murthy 2000). Increasing urbanization and economic growth reduces per capita demand for cereals and increases the demand for non-cereals food items. Modernisation of agricultural also bears a similar negative relationship with the per capita consumption of cereals. Mechanization of agricultural activities and

Table 10.5: Trends in Food Consumption Pattern from 1972-73 to 2004-05, All India (Rural and Urban)

Sector	*Per cent Share of Major Food Groups in Total Expenditure*										
	Year	*All Food*	*Cereals*	*Pulses*	*Milk and Milk Products*	*Edible Oil*	*Egg, Fish and Meat*	*Vegetables*	*Fruits and Nuts*	*Sugar*	*Beverages etc.*
Rural	72-73	72.9	40.6	4.3	7.3	3.5	2.5	3.6	1.1	3.8	2.4
	87-88	64.0	26.3	4.0	8.6	5.0	3.3	5.2	1.6	2.9	3.9
	93-94	63.2	24.2	3.8	9.5	4.4	3.3	6.0	1.7	3.1	4.2
	99-00	59.4	22.2	3.8	8.8	3.7	3.3	6.2	1.7	2.4	4.2
	2004-05	55.0	18.0	3.1	8.5	4.6	3.3	6.1	1.9	2.4	4.5
Urban	72-73	64.5	23.3	3.4	9.3	4.9	3.3	4.4	2.0	3.6	7.6
	87-88	56.4	15.0	3.4	9.5	5.3	3.6	5.3	2.5	2.4	6.8
	93-94	54.7	14.0	3.0	9.8	4.4	3.4	5.5	2.7	2.4	7.2
	99-00	48.1	12.4	2.8	8.7	3.1	3.1	5.1	2.4	1.6	6.4
	2004-05	42.5	10.1	2.1	7.9	3.5	2.7	4.5	2.2	1.5	6.2

Source: 61st Round Report of NSS, Ministry of Statistics and Programme implementation, New Delhi.

improvement in infrastructure also contribute to reduction in energy requirement and thus less cereal consumption (Rao 2000).

Performance of Indian Agriculture

Output Growth

Agricultural growth is one of the main facets of India's economic development and national food sufficiency policies. Tables 10.6-10.7 show the growth rate of agriculture by sector and different crop wise. The aggregate agricultural output increased annually at 2.6 per cent during period from 1950/51-2006/07. Disaggregating of aggregate agricultural output growth into sub periods shows that annual growth rate of agriculture was the highest during the period 1981/82-1990/91 and the lowest during period 1950/51-1965/66. Further disaggregating of agriculture into sub sectors; crop, livestock, forestry, and fishing, shows that fisheries and livestock were the main sources of the acceleration in growth rate of agricultural

Table 10.6: Average Annual Compound Growth in Value of Output (Group Wise) (in per cent)

Group	*1950/51–1965/66*	*1966/67–1980-81*	*1981/82–1990/91*	*1991/92–2006/07*	*1950/51–2006/07*
Crop	2.4	2.6	2.4	2.2	2.5
Livestock	1.3	3.2	4.4	3.4	3.1
Forestry	1.3	0	0	1.5	0.6
Fishing	4.5	3.1	5.8	4.0	4.2
Aggregate	2.1	2.5	2.8	2.6	2.6

Note: Annual compound growth rates have been calculated by using log linear model.

Table 10.7: Average Annual Compound Growth in Value of Output (Crop Wise)

Group	*1950/51–1965/66*	*1966/67–1980-81*	*1981/82–1990/91*	*1991/92–2006/07*	*1950/51–2006/07*
Rice	3.6	2.6	4.0	0.9	2.5
Wheat	3.4	6.4	3.2	1.4	4.7
Coarse Cereals	1.6	0.9	0.7	0.5	0.9
Pulses	1.1	0.5	1.6	0.4	0.6
Oilseeds	2.5	1.4	5.4	0.7	2.6
Sugar	4.4	2.5	2.6	4.1	2.9
Fibres	3.1	2.1	2.6	1.6	2.2
Drugs and Narcotise	2.1	3.2	2.2	2.6	2.7
Condiments and Spices	2.5	3.2	4.5	4.9	3.3
Fruits and Vegetables	1.8	4.3	2.1	4.3	3.8
Others	00	0.4	0	4.6	0.8

Note: Annual compound growth rates have been calculated by using log linear model.

output in 1980s. The growth rate of aggregate agricultural output turned up 3.29 per cent during the initial years of reforms, which was 0.43 percentage point higher than the previous period. However, the situation of agriculture turned adverse during post- WTO period and this covered all the sub sectors of agriculture. The growth rates in output of all crops decelerated from 2.93 per cent to 1.57 per cent. The livestock declined from 4.21 per cent to 3.40 per cent. The fisheries declined from 7.48 per cent to 3.25 per cent. Only, forestry witnessed a sharp increase from 0.09 per cent to 1.82 per cent.

The crop sector, which forms the largest segment of agriculture, grew annually at 2.5 per cent since 1950/51. The acceleration rate of crop sectors fluctuated around 2.5 per cent during all sub periods. It grew at the lowest rate during post-WTO period in comparison to all other periods. Further, within crop sector, all crops except sugar, condiment, spices, fruits and vegetables showed declining trend between 1950/51-1965/66 and 1991/92- 2006/07. This deceleration is very high in Cereals, Corse Cereals, Pulses, Oilseeds, and Drugs and Narcotics. Similar declining trend in growth rate of all crops is also confirmed by **Table 10.8** that shows average annual compound growth rate of output in physical term for all major crops.

Table 10.8: Average Annual Compound Growth in Real Term (Crop Wise)

Group	*1950/51–1965/66*	*1966/67–1980-81*	*1981/82–1990/91*	*1991/92–2006/07*	*1950/51–2006/07*
Rice	3.6	2.6	4.1	1.0	2.5
Wheat	3.7	6.3	3.3	1.4	4.7
Coarse Cereals	2.2	0.9	0.7	0.7	0.9
Pulses	1.6	0.2	1.4	0.2	0.5
Oilseeds	2.8	2.1	5.6	0.9	2.9
Cotton	3.6	2.5	3.2	2.7	2.4
Jute and Mesta	4.4	2.1	0.7	1.5	1.5
Sugarcane	6.2	3.0	2.9	0.8	3.1
Tobacco	2.4	2.1	0	–0.1	1.5
Potato	5.4	7.2	4.9	3.1	5.4

Note: Annual compound growth rates have been calculated by using log linear model

These growth rates are lower than the growth rate of rural population. Thus, the clear implication of this growth trends is that the per capita output in agriculture is declining. This seems to be one of the causes for rising disparity between rural and urban areas in India.

Net Availability of Foodgrains

Table 10.9 shows net availability of foodgrains for per capita per day. An average availability of foodgrains per capita per day was 429.8 gram in 1950s and increased to 475.5 gram during 1990s. Further, it decreased to 446.6 gram in the first decade of 21st century. Within foodgrains, all food crops reveals similar trend except coarse cereals. The availability of coarse cereals is continuously declining.

Table 10.9: Net Availability of Foodgrains (per day) India (in grams per capita per day)

Decades	*Rice*	*Wheat*	*Other Cereals*	*Cereals*	*Gram*	*Pulses*	*Foodgrains*
1951-60	178.1	65.9	119.4	363.4	27.5	66.4	429.8
1961-70	188.3	91.1	113.5	392.9	22.6	54.7	447.5
1971-80	183.0	114.8	100.9	398.7	17.0	43.5	442.2
1981-90	198.1	143.3	83.2	424.6	13.2	39.6	464.2
1991-2000	209.3	162.7	67.7	439.7	12.3	35.8	475.5
2001-2005	194.7	159.9	59.0	414.2	9.8	32.4	446.6

Source: Agricultural Statistics at a Glance (2008).

Input Use Pattern

Agricultural production and efficiency largely depend upon the inputs applied and the methods adopted. In India, "while population grows, the land surface is fixed and of this only a certain proportion is available for cultivation" (Planning commission, 1961). Further scope for bringing extra land under the plough is limited. If more production is to be got out of this existing area, the problem has to be tackled on a wide front. This can be done by applying inputs in a more intensive way and by adopting modern methods of production through use of improved technology, besides making an adequate provision for institutional financing, better methods of marketing, *etc.* Technical factors, *i.e.*, technology have received increasing emphasis and the recent breakthrough in agriculture is the outcome of these factors. These technological factors comprise (i) irrigation; (ii) Consumption of fertilisers and manure; (iii) Improved seed, and (iv) agricultural implements.

Water is another basic factor in agriculture next only to land. Only rainfall is the natural source of water in agriculture. But rainfall is the most unreliable and is marked by wide variations in different parts and also variation from year to year in its quantity, incidence, and duration. Therefore, only artificial supply of water through irrigation is the way to overcome the problem of deficiency of water. Irrigation water comes from two sources: surface water and ground water. Surface water is provided by the flowing water of rivers or the still water of tanks, ponds, lakes, and artificial reservoirs. The surface water is carried to the filed by canals, distributaries, and channels. Ground water is tapped by sinking wells where drought animals, diesel or electric power is utilized to take out water. In India canals, tanks, wells including tubewells are the principal sources of irrigation. Since 1950-51, considerable importance had been attached to the provision of canal irrigation and well irrigation. Even though 40 per cent of irrigation is supplied by canals, now well irrigation has caught up rapidly irrigation by tubwells has been expanded considerably. In the meantime, tanks and other source of irrigation are declining in importance (see **Table 10.10**).

Table 10.10: Sources of Irrigation (million.ha.)

Sources	*1950/51*		*1970/71*		*2000/01*	
	Area	*Per cent*	*Area*	*Per cent*	*Area*	*Per cent*
Canals	8.29	44.0	12.80	40.5	17.1	31.3
Wells includingTubewells	5.98	31.7	12.10	38.3	30.9	56.6
Tanks	3.61	19.1	4.10	13.0	3.1	5.7
Others	0.97	5.2	2.60	8.2	3.5	6.4
Total	18.85	100	31.60	100	54.6	100

Source: Indian Agriculture in brief, 21st Edn, 1986 CMIE, Statistical Abstract, India, 2004

Table 10.11 shows that the net irrigated area has risen by 163 per cent from 1950/51 to 2000/01. This increment in irrigated area is very nominal and only 39 per cent of net sown area is irrigated area. This figure is very unsatisfactory and it is matter of concern that why only 39 per cent of net sown area is irrigated area. It is also concerning matter that growth in gross irrigated area is also very nominal. Thus, there is scope to increase agricultural production by increasing both net and gross irrigated area.

Table 10.11: Changes in Irrigated Area in India from 1950-51 to 2000-01 (In thousand hectares)

Category	*1950-51*	*1960-61*	*1970-71*	*1980-81*	*1990-91*	*2000-01*
Net irrigated area	20853	24661	31103	38720	48023	54836
Gross irrigated area	22563	27980	38195	49775	63204	75821
Irrigation intensity (in per cent)	108	113	123	128	132	138

Irrigation intensity is calculated by author.

Source: Agricultural Statistics at a Glance (2008).

In any scheme for boosting agricultural output, the use of chemical fertiliser has an important role. India's soil though varied and rich in deficient in nitrogen and phosphorus- two plants nutrients which together with organic manure influence crop return, which population rising at a first rate, the use of larger and larger doses of chemical fertiliser is the only way to augment our foodgrains production. The new agricultural strategy was based on increased use of fertiliser. Since adoption of the new agricultural strategy in the sixties, the consumption of chemical fertiliser has been growing rapidly. The Government is also encouraging the use of fertiliser through heavy subsidies. That is why the consumption has gone up abnormally high from 70000 tonnes in 1950-51 to 2,177,000 tonnes in 1970-71, 12,546,000 tonnes in 1990-91 and 19,145,000 tonnes in 1999-2000 (**see Table 10.12**). The fertiliser consumption per hectare of gross cropped area has also gone up steadily, from 0.50 kg in 1950-51 to 74 kg in 1995-96. The corresponding figures for developed countries are much higher than the Indian Agriculture (see **Table 10.13**).

Table 10.12: Production, import, and Consumption of Fertilizer

	1950-51	1960-61	1970-71	1980-81	1990-91	2000-01
A. Nitrogen						
Production	09	98	830	2164	6993	10961
Imports		399	477	1510	414	154
Consumption		210	1487	3678	7997	10920
B. Phosphatic						
Production	09	52	229	841	2052	3743
Imports			32	452	1016	396
Consumption		53	462	1214	3221	4215
C. Postassic						
Production						
Imports		20	120	797	1328	1541
Consumption		29	228	624	1328	1567
D. All fertilizers						
Production	39	150	1059	3005	9045	14704
Imports	52	419	629	2759	2758	2090
Consumption	70	292	2177	5516	12546	16702

Source: Indian Economic Survey, 2002-03.

Table 10.13: Fertiliser Consumption per Hectare of Agricultural Land in Selected Countries in 2004-05

Countries	*N*	*P*	*K*	*Total*
Egypt	476.2	67.6	11.3	555.1
Bangladesh	128.6	37.4	18.5	184.5
India	65.0	25.7	11.4	102.1
Japan	116.0	137.4	109.5	362.9
Korea Rep.	208.3	82.9	103.5	394.7
Pakistan	92.3	26.6	NA	119.2
Sri Lanka	69.3	14.8	25.8	109.9
Belarus	47.7	12.1	55.6	115.4
Denmark	80.3	13.0	36.7	130.0
France	80.1	26.0	33.1	139.2
Germany	104.5	21.1	28.1	153.7
Netherlands	142.4	29.6	77.4	249.4
Norway	101.4	12.4	44.7	158.5
Poland	53.4	24.8	37.9	116.1
United Kingdom	66.8	16.0	22.3	105.1

Source: Agricultural Statistics at a Glance (2008).

The low consumption may be due to the poor economic condition of farmers, lack of assured irrigation (58.5 per cent of the cropped area lack irrigation facilities), inadequate demonstration and promotion for the use of fertilisers, insufficient supply at the proper time, high price of fertiliser, absence of soil testing facilities so as to recommended the precise deficiencies in the soil and recommended proper dose of fertiliser, and wrong notion among some conservative farmers regarding the use of chemical fertilisers.

Improved seeds have played vital role in augmenting agricultural production in developing countries like India. These seeds not only help in increasing in agricultural production by 10 to 20 per cent but introducing new characteristics in the biological structure of the plant. For, example researches have made it possible to develop such seeds which are quick maturing, provide higher agricultural yield and are resistant to insects, diseases and droughts. In India the success of Green Revolution is partly associated with the use of high yield variety (HYV) seeds. The HYV programme was started in 1966. Between 1967-68 and 1996-97 the area under HYVP has witnessed 12.6 times increase (from 6.07 million ha to 76 million ha). The success of the programme remains most marked in the case of wheat and rice. The HYV programme has led to 4.84 times increase in the output of wheat from 1966-67 (11.39 million tonnes) to 1990-91 (51.1 million tonnes) and 1.78 times increase in the production of rice from 1967-68 (37.6 million tonnes) to 1990-91 (74.3 million tonnes).

The implements and tools used by the Indian farmers are primitive, crude, and obsolete which impede the development of modern agriculture. New farm machineries not only save time, reduce cost of production but also increase agricultural production. These machineries replace the animal and human power and perform various works of agriculture ranging from ploughing, showing, and harvesting to the marketing of the produce. There is difference of opinion amongst scholar regarding the mechanisation of agriculture. In fact small and scatter land holding, cheap and abundant human labour and poverty amongst farmers go against total mechanisation of Indian agriculture but the possibility of limited mechanisation is not ruled out. In many cases where the use of animal and human power has become costlier, mechanization is proving to be boon for agriculture. Even small farmers prefer to use these machineries to save the time and money.

Studies show that sufficient progress has been made in respect of farm mechanisation in India. For example, the number of tractors which was less than 10000 in 1950-51 increased to 1 lakh in 1970-71 and 14.5 lakhs in 1990-91. Similarly, the number of diesel pump sets increased from 80,000 in 1950-51 to 48.5 lakhs in 1990-91 and electric irrigation pump sets from 26,000 in 1950-51 to 91 lakhs in 1990-91. But most of the mechanisation has largely been confirmed to the rich farmers belonging to the developed areas of the country.

Sources of Growth in Indian Agriculture

Any change in the output of a crop in physical term depends fundamentally on the changes in the area under the crop and its average yield. To measure the effect of area, productivity and their interaction in increasing the crop output, differential equation given by Sharma (1977) was used:

$$\Delta P = A\,\Delta Y + Y\,\Delta A + \Delta A\,\Delta Y \qquad (1)$$

The first term on the right hand side is considered as yield effect, second term as the area effect and the third as the interaction effect. Thus, total change in output can be decomposed into three effects; yield effect, area effect, and the interaction effect due to change in yield and area.

During 1950/51-1965/66, area and yield both almost equally contributed in growth of rice, wheat, and coarse cereals. But for non-food crops, expansion area was dominant source of output growth **(see Table 10.14)**. During 1965/66-1980/81, increase in yield was comparatively more contributed in output growth of all major crops except Wheat and Jute and Mesta. For Wheat, 40.49 per cent of output growth was contributed by expansion in area and 34.46 per cent was contributed by increase in yield. Remaining part (25 per cent) of output growth of Wheat was contributed by interaction of area and yield. During 1981/82-1990/91, an increase in yield was more contributed in output growth of all major crops except oilseeds and sugarcane.

Table 10.14: Sources of Agricultural Growth in Period (1950/51–2006/07)

Crop	*Change in output*	*Area Effect*	*Yield Effect*	*Interaction*
		1950/51–1965/66		
Rice	10010	3112.7 (31.1)	5991.1 (59.85)	906.16 (9.05)
Wheat	3940	1868.43 (47.42)	1606.83 (40.78)	464.74 (11.79)
CoarseCereals	6040	2723.24 (45.1)	2817.82 (46.65)	498.93 (8.26)
Pulses	1530	1600.83 (104.63)	–57.27 (-3.74)	–10.89 (-0.7)
Oil seeds	1240	2174.12 (175.33)	–665.26 (-53.65)	–280.24 (-22.6)
Cotton	307.7	183.04 (59.49)	94.08 (30.57)	33.28 (10.82)
Jute and Mesta	444.6	563.22 (126.68)	–60.99 (-13.72)	–57.78 (-13)
Sugarcane	66940	37767 (56.42)	17604 (26.30)	11633 (17.38)
Tobaco	30	14.62 (48.73)	16.92 (56.4)	0.94 (3.13)
Potato	2420	1660.1 (68.60)	383.04 (15.83)	383.04 (15.83)
		1966/67–1980/81		
Rice	11890	3598.7 (30.3)	7431.4 (62.5)	879.12 (7.4)
Wheat	20440	8276.38 (40.49)	7044.7 (34.46)	5118.93 (25.04)
CoarseCereals	4970	–1765.48 (-35.52)	7269.1 (146.26)	–533.61 (10.74)
Pulses	2280	128.18 (5.62)	2123.52 (93.14)	32.64 (1.43)
Oil seeds	2940	1112.8 (37.85)	1560 (53.06)	270.4 (9.2)
Cotton	295.8	–2.28 (-0.77)	297.92 (100.72)	–0.76 (-0.26)
Jute and Mesta	284.4	190.44 (66.96)	80.64 (28.35)	12.96 (4.56)
Sugarcane	61420	14924 (24.3)	40268 (65.56)	6478 (10.55)
Tobaco	130	25.02 (19.25)	97.02 (74.63)	6.93 (5.33)
Potato	6150	1934.4 (31.45)	2734.5 (44.46)	1512.7 (24.60)

Contd...

Table 10.14–*Contd...*

Crop	Change in output	Area Effect	Yield Effect	Interaction
		1981/82–1990/91		
Rice	20660	3392.8 (16.4)	16240 (78.6)	1027.4 (5)
Wheat	17690	3432.73 (19.40)	13070.14 (73.9)	1198.4 (6.8)
CoarseCereals	1610	–4493.29 (-279.1)	7103.175 (441.12)	–1025.74 (63.7)
Pulses	2750	396.06 (14.40)	2264.8 (82.36)	77.9 (2.83)
Oil seeds	6530	3348.36 (51.28)	2496.12 (38.22)	691.68 (10.59)
Cotton	333.2	–102.9 (-30.89)	475.54 (142.72)	–36.58 (-10.98)
Jute and Mesta	154.8	–170.4 (-110.1)	371.45 (239.95)	–41.99 (-27.13)
Sugarcane	54690	29180 (53.35)	22445 (41.04)	3518 (6.43)
Tobaco	40	–35.16 (-87.9)	79.64 (199.1)	–5.43 (-13.58)
Potato	5300	2339.28(44.14)	2476.1(46.72)	586.44(11.06)

During 1991/92-2005/06, again yield's contribution in output growth of all major crops except Wheat, Sugarcane, and Potato was greater than expansion in area. It is a matter of great concern as to why the productivity of wheat and sugarcane remains stagnant **(see Table 10.15)**.

Table 10.15: Sources of Agricultural Growth in Period (1991/92- 2006/07)

Crop	Change in output	Area Effect	Yield Effect	Interaction
Rice	16370	1838.5 (11.23)	14203 (86.8)	349.65 (2.01)
Wheat	18010	11755 (65.27)	5181.42 (28.77)	1093.76 (6.07)
CoarseCereals	6930	–3872.84 (-55.88)	12710.36 (183.41)	–1894 (-27.33)
Pulses	2090	650.26 (31.11)	1374.94 (65.79)	74.42 (3.56)
Oil seeds	4660	71.9 (1.54)	4556.64 (97.78)	17.6 (0.38)
Cotton	1926.1	317.52 (16.48)	1348.2 (70)	258.72 (13.43)
Jute and Mesta	189	–265.9 (-140.7)	546.12 (288.95)	–78.72 (-41.65)
Sugarcane	68940	65408 (94.88)	2933.8 (4.25)	756.36 (1.1)
Tobaco	–30	–2.34 (7.8)	48.256 (-160.9)	–6.73 (22.44)
Potato	7520	5884.48 (78.25)	1188.8 (15.81)	427.04 (5.68)

Instability in Indian Agriculture

Instability is one of the important decision parameters in development dynamics and more so in the context of agricultural production. An analysis of fluctuations in crop output, apart from growth, is of importance for understanding the nature of food security and income stability. Wide fluctuations in crop output not only affect prices and bring about sharp fluctuation in them but also results in wide variations in disposable income of the farmers. The magnitude of fluctuations depends on the nature of crop production technology, its sensitivity to weather, economic environment, availability of material inputs and many other factors. High growth in production accompanied by low level of instability for any crop is desired for sustainable development of agriculture.

In this section we examined extent of production, area, and yield instability for major crops. The instability in area, production and yield of major crops is measured in relative terms by the Cuddy-Della Valle index which is used in recent years by a number of researchers as a measure of variability in time series data. The simple coefficient of variation overestimates the level of instability in time-series data characterized by long term trends whereas the Cuddy-Della Valle index corrects the coefficient of variation.

Overall analysis indicates that production and yield instability for almost crops declined in post-reform period. But further it also indicates that area instability increased in the same period. Therefore, it can be concludes that reduction in production instability is mainly due to reduction in instability of yield and present instability in production is mainly because of increasing instability in area.

Concluding Remarks

Here we evaluates performance and progress of Indian Agriculture since Independence. In addition, we also analyzes sources of agricultural growth and determinants of agricultural production. We use the decomposition test to analyze sources of agricultural growth and the production function approach to analyze determinants of agricultural production over the period 1950/51 through 2005/06. The study indicates that there is scope to increase both net sown area and gross sown area. The study also highlights that only 39 per cent of net sown area is irrigated area. After evaluating the changes in agrarian structure, input use pattern and growth trend of agriculture, this paper point outs some points. These are: agricultural workforce shifted from cultivators to agricultural labours, the number of uneconomic holdings has an increasing trend, area under food crops shifted towards non food crops, and within food crops area under cereals has been shifting towards non cereals, growth trend of aggregate agriculture as well as all sub sector of agriculture except forestry is showing declining trend during post-WTO period. It is also observed in this study that production and yield instability declined for almost crop during post reform period while, area instability increased in the same period. This further indicates that instability in area became major responsible factor for production instability.

The decomposition analysis indicates that rising output per hectare is the predominant source of agricultural growth for most of the crops and crop groups. Disaggregating of reference period in four sub periods shows that expansion of agricultural land was the main source of agricultural growth during the period before 1965/66 after that the contribution of increased land area under agricultural production has declined over time and increase in productivity became the main source of growth in agricultural production.

The estimation of aggregate agricultural production function with both intercept and slope dummy indicates that land significantly affected the agricultural output growth during 1950/51-1964/65 and after that land became less significant and now labour and capital are significantly affecting the agricultural output growth. Thus, the result of the aggregate agricultural production function verifies the results of decomposition analysis.

11

Indian Seed System Development: Policy and Institutional Options*

Developments in the Indian seed industry and their impact on access and use of commercial seed by farmers have been examined. Various types of seed systems such as hybrids, self-pollinated crops, vegetatively propagated crops, crops with high seed volume, etc have been analysed. It has been shown that the commercial seed markets for hybrids are well developed, but these need improving flow of information to farmers and effective regulation of unscrupulous traders, *etc.* There are significant changes in terms of seed regulations, management of GM crops and protection of intellectual property. Since all these regulations are mutually enforcing, there is a need for developing institutional capacity for their enforcement, as well as flexibility to learn from the experience for future adaptation. There is a lot of scope for strengthening the seed system of 'orphan crops', where there is no participation of the private sector, and the public seed system is facing several resource and institutional constraints. In particular, there is a need for technological backstopping, developing partnerships with private and civil society organizations, and developing capacity at the local level. The results of farm surveys have shown that increasing proportion of farmers use commercial seed for quality considerations. The study has argued that there is a problem with variety selection, particularly of proprietary hybrids, due to lack of information, which has resulted into poor crop performance on several occasions.

---

* Harbir Singh, Prasoon Mathur and Suresh Pal.

Introduction

The Indian seed system has undergone a tremendous change during the past two decades. The system started with public seed corporations in the 1960s, matured over a period of time, ushering the green revolution in the country. These seed corporations ably met the twin objectives of efficiency (delivering seed to commercial farmers) and equity (catering to seed needs of small farmers and marginal areas). However, the situation changed significantly with the expansion of agricultural research system and entry of private seed companies. The State Agricultural Universities (SAUs) and their regional research stations meet the seed needs of a large proportion of farmers, particularly for self- pollinated crops. Private seed companies, on the other hand, are supplying increasing proportion of hybrid seeds. This trend of privatization in the seed system got steamed with the economy-wide reforms introduced by the government during the early-1990s and implementation of the New Seed Policy in the late-1980s. In particular, there is an increasing participation of transnational seed companies. These developments have, no doubt, increased the availability of improved seeds to commercial farmers, but their implications for seed quality and ability to meet seed requirement of small farmers or households in marginal areas are seen with skepticism.

Another significant development has been the shift to a regime of protection of intellectual property. It is feared that the protection of plant varieties would further accentuate private activities in the seed system, and transnational seed companies will dominate and exploit the system for their quest for profit. These companies will not only exploit local genetic material for their advantage, but would also charge high prices for seed, restricting the access of resource-poor farmers to improved seed. The problem is further compounded by the introduction of genetically-modified seeds (GMS). There are a number of health and environmental risks associated with the use of GMS. Although one can argue for the nature and extent of these risks, these apprehensions can't be ruled out completely. The question now arises how the seed system can be made more competitive and effective. What should be the mechanism for the regulation of GMS? Who will serve resource-poor farmers in the marginal areas? This paper has addressed all these questions; specifically, the study has suggested policy and institutional (organizational and regulatory) options for developing a sustainable seed system.

Rationale and Methodology

The importance of developing a competitive seed system can't be over-emphasized. The seed turnover rate is very low, and crop production losses are 15- 20 per cent due to poor seed health. A host of factors (technical, institutional and policy) constrain performance of the seed system. These factors have been analyzed in a number of studies in the recent past (Jaffee and Srivastava, 1994; Morris, 1998). However, these studies have mainly focused on the activities of private seed companies which primarily deal with hybrid seeds. The results of these crop- specific studies have not been able to present a complete picture of the seed system. Therefore, a systematic study analysing the system for various types of crops and regions was required. Moreover, there was a dearth of studies analysing structure of the system in terms of interface between public- public, public-private and private-private organizations. How these linkages affect the performance of seed system, and what are the conditions promoting these interfaces? How these linkages will change with the protection of plant varieties and use of GMS? What are the institutional mechanisms to ensure seed quality, safeguard interests of farmers and meet seed requirement of marginal areas? The present study has bridged this information gap.

The institutional economics framework was applied to assess the structure and performance of the seed system. This framework examines the organizations, activities, regulations and contracting arrangements for all elements of the seed system, *viz.* plant breeding, source seed, seed multiplication and conditioning, marketing, quality control, and consumer protection. In the first phase, structure of seed system, seed policies and regulatory issues were studied. It was followed by the case studies of seed system for commercial crops in developed region (cotton), high-value crops in marginal or hill region (vegetables), vegetatively propagated crop (potato), crop with high seed rate (groundnut), self-pollinated crops in developed region (paddy), *etc.* In terms of research issues, the case studies focused on the diversity of the seed system, institutional linkages, seed quality, information flow, *etc.* Farm surveys were conducted in 2003-04 and 2004-05 to assess the performance of the seed system in terms of delivery of improved seed, varieties and information to farmers. The details of farm surveys have been given in Table 11.1.

Table 11.1: Details of Farm Survey Conducted

Crop	*State*	*Districts*	*Number of Farmers*
Paddy	Haryana	Karnal	60
Cotton	Maharashtra	Yavatmal and Jalgaon	96
Potato	Uttar Pradesh	Firozabad and Barabanki	96
Vegetables	Himachal Pradesh	Solan and Mandi	96
Groundnut	Andhra Pradesh	Anantapur	72

Seed Industry: Evaluation and Contemporary Developments

The origin of Indian seed industry could be traced back to the establishment of national and state seed corporations during the 1960s, which continued to expand both in terms of their number and business during the 1970s. The industry underwent structural changes with the entry of private seed companies, mostly family-owned, during the 1980s and this trend continued in the 1990s also. The private seed companies focused mainly on hybrid seeds and a few large companies diversified into research and development (R&D) to increase their share in the seed market. The new seed policy of 1988 and the economy-wide reforms of 1991 attracted the multinational companies (MNCs) to India in a major way. Most of them entered through partnership with the national companies, and only a few established their independent seed business in the country. The industry was mainly governed by the Seeds Act (1966) and thrived largely on the material bred by the public plant breeding programmes of the institutes under the Indian Council of Agricultural

Research (ICAR) and SAUs. This is a typical development path followed by seed industries even in the developed countries (Morris, 1998).

The private seed sector has witnessed tremendous growth and now it supplies most of the hybrid seeds in the country. For example, the private sector is a dominant player in major cotton-growing states and public sector accounts only for 10-30 per cent of commercial seed supply. Even in self-pollinated crops like paddy, the share of public sector is nominal and the private sector supplies 60-80 per cent of commercial seed in the states of Haryana and Andhra Pradesh. Low marginal cost and risk in producing paddy seed and potential lucrative market for hybrid rice could explain greater private sector's participation in rice seed in the favourable production environments. In the case of inaccessible hilly areas also, the private sector supplies a significant proportion of commercial maize and vegetable seeds. All these signs point towards consolidation of this trend in future. Only in the case of high-volume seed crops like potato and groundnut, there is less participation of the private sector and the aggregate statistics do not reveal private sector's participation, if any. Entire potato seed in Uttar Pradesh (UP) and Himachal Pradesh (HP) is supplied by the government, although there are some private seed producers mostly in Punjab and Western UP, but there share is negligible. In the case of groundnut, besides state seeds corporations and government department, there are some producers' organisations and oil trading public agencies who supply seed to farmers and their role becomes significant in the year of seed scarcity. Wherever there is support of research stations and opportunities for seed supply, there are local seed producers, often progressive farmers, and their role is likely to increase whenever there are efforts to strengthen decentralized seed system. Thus, there are increasing trends towards the use of commercial seed and the private sector is playing a larger role in the seed supply.

Today, the Indian seed industry is heading towards a maturity phase with three major undergoing changes. First, private seed companies consider research and development (R&D) as an important mechanism to differentiate their product and enhance their market power. This tendency is likely to intensify further. The second major change is arising from the process of globalization and liberalization. The resource-rich MNCs with well- established R&D programmes overseas are expanding their activities through mergers and acquisitions and the national companies may find it difficult to compete with them. Third, the industry is going to be governed by multiple regulations, and protection of intellectual property rights (IPRs) is emerging as an important factor to shape its growth and performance. India has put in place all the necessary legislations and institutions to strengthen the IPRs regime to comply with the World Trade Organization (WTO). Concomitantly, other regulations like those dealing with development and commercialization of genetically-engineered (GE) crop varieties and access to and use of genetic resources have become important for the seed industry. How will these mutually-reinforcing developments affect the structure and performance of the seed industry? The following section deals with these issues.

Policy Issues

The contemporary issues in public plant breeding mainly relate to sustainability of funding and management of intellectual property in the context of modern science. The sustainability of funding requires attracting additional resources from both public and private sectors. The public resources could be mobilised through improving effectiveness and efficiency of the programmes with greater accountability in utilization of these resources. This requires a number of organizational and management reforms to rationalize allocation of resources, facilitation of knowledge-flow in the system, management of intellectual property, regulation of private R&D, human resource development, *etc.* Private research resources would largely depend on the research infrastructure, including strength of public breeding programmes, research cost, size of the market for new technologies (*e.g.* commercial seed), protection of intellectual property for appropriation of research benefits, regulatory policies, *etc.* (Byerlee and Echeverria, 2002; Pray and Umali-Deininger, 1998). India is well placed in terms of availability of research infrastructure, scientific manpower and a vast market for improved technologies. Fiscal incentives like tax concessions for R&D expenditure, lower duty rates on import of R&D equipment, sales tax exemption (on seed), *etc.* are also conducive for attracting private investment. However, protection of intellectual property and regulatory policies for private research, especially on development and use of genetically modified organisms (GMOs) are evolving, and therefore these have been discussed in this paper.

The Protection of Plant Varieties and Farmers' Rights (PPVFR) Act, 2001 and the amendments of the Patents Act provide incentives to the private sector by bringing the domestic IPRs regime at par with the international regime envisaged under the WTO. The evidence indicates that there is increasing tendency to protect intellectual property and the number of patents filed by research organizations is increasing rapidly. The Seeds Act, 1966 is also under revision to provide a greater operational flexibility to seed industry and entrust it with the responsibility of ensuring seed quality. These developments are expected to further accelerate privatization of seed research in the country. In fact, private seed companies are finding it more attractive to develop and sell proprietary material to capture a significant proportion of the seed market. The effective implementation of the PPVFR Act is expected to promote private plant breeding in the country in the long-run. The immediate effect could be in terms of increased access to seeds developed by transnational seed companies. These companies may sell seed on their own or tie up with the national companies for multiplication and marketing of their material. It is also likely that transnational seed companies establish joint research ventures with the national companies, such as that between Monsanto and Mahyco. Whatever may be the path, Indian farmers may have multiple choice and access to improved seed, which can have positive effect on crop productivity. At the same time, this could create some degree of concentration in seed market because of substantial investments made by some of the transnational seed companies.

The provision of compulsory licensing and presence of a strong public breeding programme for developing varieties which can be delivered by public and private seed agencies, are useful options to control the monopolistic tendencies. The provisions of mandatory registration of plant varieties, farmers' right to sell unbranded seed, and disclosure of information on parents of hybrids are being discussed by the private seed industry and these may significantly influence relations among seed entities. Nevertheless, the issues like protection of genetic resources - a provision under the Convention on Biological Diversity - and to encourage free flow of seed among farmers are quite important from the system perspective. It is extremely important for the crops where traditional seed systems are dominant. Public ownership of genetic resources could also be used to bargain for access to proprietary technology to promote a competitive seed industry.

Another fear is that a private-sector dominated seed industry may not serve resource-poor farmers in marginal areas, and may also raise seed prices beyond the reach of small farmers. The available evidences support this trend. Therefore, the government should closely monitor the seed sector and intervene using an appropriate mix of the measures like laws, fiscal incentives and direct seed supply to farmers. It may be noted here that seed prices of hybrids have risen much faster because of greater participation of private seed companies and their quest to make profit. It is desirable that seed prices should be kept within the reach of farmers, particularly small farmers, and these should not rise beyond a limit not justified by the yield or economic advantage.

As regards the policy on biotechnology, India has approved commercial cultivation of first GM crop (*Bt* cotton) and taken a number of other initiatives. There are still a number of challenges to be addressed - development of research capacity, biosafety and management of public dialogue on controversial issues. Establishment of biotechnology capacity is relatively capital- and human resource-intensive. Both the public and private sectors will have to play an important role, and there is much potential for forging public-private linkages to enhance overall impact. These linkages could be further useful as advances in biotechnology have blurred the differences between pure science and agricultural science, requiring close linkages with general science and technology providers. It is more so when a major responsibility for promotion of biotechnology in India rests with the Department of Biotechnology in the Ministry of Science and Technology. These public-private linkages can be fostered by setting appropriate mechanisms for the sharing of cost and benefits, establishing joint ventures, and management and ownership of intellectual property.

Given the current debate on biotechnology in India and elsewhere, effective biosafety regulations must be in place that are credible, cost-effective and properly coordinated. There is no easy solution to resolve these issues. Ethical issues relating to alteration of biological system, protection of genetic resources, and biosafety and health effects are really critical. We need to take a collective judgement on these issues, which is possible only when adequate scientific evidence is passed on to all the concerned and a healthy debate is encouraged. Lack of information and debate breeds confusion, and delay in decision making.

A dimension often neglected in the regulation is coherence between various acts governing an industry or sector. For instance, biotech research in India is governed by a number of acts, notably the Seeds Act, Environment (Protection) Act, PVPFR Act, and Biological Diversity Act. It is important that there is coherence between these acts; otherwise some of the positive aspects of these provisions could be neutralized, hampering the growth of private sector research. Finally, there is inadequate flow of information about new technologies to farmers. Since much of this information is a public good, public institutions and government will have to take the major responsibility of disseminating information and educating farmer consumers.

Public Private Partnership

A number of theoretical concepts are applied to study the public-private partnerships. The new institutional economics literature sees the partnership as a strategy to minimize transaction costs associated with developing and enforcing contractual relations in provision of a good or service. The transaction costs are mainly determined by the frequency and uncertainty of a transaction, limit to rational behaviour of economic agents, and asset specificity of the transaction (Williamson, 1975). For example, a private seed company has to transact with public plant breeding programs for new varieties and source seed. A high transaction cost with high asset- specificity of establishing a plant breeding program may help develop partnership with public plant breeding programs (see ICRISAT model). On the other hand, a low transaction cost will favour marke - based transactions, while low asset-specificity can lead to vertical integration, bringing seed production and plant breeding under a hierarchical structure. The second important conceptual framework used is the recent developments in the theory of organisational behaviour. The analysis blurs the classical difference between public and private sectors, and underlines the need for partnerships for efficient provision of a good or service with equitable social benefits, whilst maintaining higher flexibility, and accountability of the private sector and social interest of the public sector. Other approaches focus on the traditional welfare analysis in use of scarce resources, development of networks of innovations for the given social and economic institutions, and incentives and relationships that shape the flow of knowledge and information.

One may infer that incentives, problems and risks associated with incentives, contextual realities and nature of goods or services are important for developing and enforcing partnerships. Since plant breeding is a risky activity with high asset-specificity, contractual relations that shape the flow of knowledge and material are critical for establishing research partnerships. Macro-economic policies and social and economic institutions further influence the attitudes and pace of research partnerships. For example, a greater reliance on market forces and the enabling institutions like IPRs may facilitate research partnerships, while public and private sector will continue to maintain a negative perception in an inward looking economic environment.

The ICAR, as an apex agricultural R&D organisation of the country, has initiated dialogues with private R&D organisations, NGOs and other stakeholders to develop partnerships. A number of policy decisions have been taken through a consultative process. These decisions underscored continuity of dialogue, sharing of resources, expertise and cost and benefits of technologies in a transparent manner, capacity building, and developing a culture of mutual confidence and trust. Although these initiatives are quite comprehensive and path breaking in several ways, there are only a few examples of successful partnerships. In the case of hybrid rice, ICAR, SAUs, IRRI and national private seed companies collaborated for development of male sterile lines, development of hybrids and refinement of seed multiplication technologies. The partnership upscaled the hybrid rice technology and intensified plant breeding and seed multiplication activities in the private sector. The technology has been commercialized and being adopted even in marginal areas of eastern India because of significant yield advantage.

In spite of the above-mentioned successes, there are certain constraints and rigidities which need to be addressed. The private sector would appreciate and value a timely response from the public organisations which is possible only in a decentralised system. Delay and uncertainty in establishing a partnership may enhance transaction cost. In order to avoid this, the private sector would prefer to enter into some kind of agreement with other private companies, or an international organisation. Perhaps the ICRISAT's consortium of private seed companies reflects this concern. A group of private seed companies, both national and international, has formed a consortium to fund the plant breeding program of ICRISAT for pearlmillet and sorghum. The member companies pay annual fee and have access to advanced breeding material. The material is available to the public plant breeding programs, but not to non-member seed companies. Private seed companies benefit from advanced breeding material and minimize their research cost, while ICRISAT is able to generate resources to fund its breeding programs for the crops. The National Botanical Research Institute, Lucknow, has taken lead in this direction for *Bt* cotton, but their initiative is constrained by the lack of freedom to operate. In addition to addressing technical issues like freedom to operate in an era of IPRs, development of mutual confidence and trust among the partners is a major issue which shall build over a period of time. However, it requires transparent procedures and commitment to honour the contract and confidentiality of information. While all these issues are important for working in a partnership mode, there are other modes of partnerships which are largely governed by markets. The markets for technologies provide ample opportunities to work independently, but complement each other's activities through market-based contracts. For example, public sector can develop inbred lines or semi-finished breeding material, which can be taken by the private seed companies on agreed conditions for development of a finished product. This can be further licensed to another private seed company for multiplication and distribution of seed. The government should play a facilitating role in such cases.

The examples discussed above demonstrate that some forms of public-private partnerships can be of mutual benefit and can serve the farming community in a more effective manner. There is a need to promote such partnerships in provision of improved seed to marginal and isolated agricultural production environments. One successful example of this has been the public delivery of private seed, mostly of vegetables, in inaccessible regions of Himachal Pradesh, where the government procures seeds from the private seed companies through open tender and the line department distributes them to farmers.

Seed Acquisition and Management by Farmers

Positive developments in the seed system should be reflected in the acquisition, management and replacement of seed by farmers. It is likely that easy access to quality seed and information would help farmers make informed-decisions regarding selection of variety, procurement of seed from a reliable source and replacement of seed frequently. This section deals with these issues. The discussion is based on the farm survey data conducted in the sample states. Here, it may be noted that these surveys were designed to capture different farm situations representing degree of commercialisation of agriculture and seed system, and therefore, may not necessarily be used for a wider generalisation of the results.

Use of Commercial Seeds

Of the total seeds of cotton, tomato and pea used in the case study states, more than 90 per cent were procured from the commercial sources. Even in the case of paddy, 60 per cent of the seeds used were commercial. Only in potato, farmers mostly used farm-saved seed and in a few cases, the seed was procured from other farmers. The proportion of commercial seed was 20 per cent (Table 11.2). In the case of groundnut, although the share of seed procured from commercial sources was very high, it was largely produced by progressive farmers and therefore, it was difficult to compare with other crops in terms of quality. This was primarily because of difficulty in supply of source seed caused due to seed- multiplication rate. This problem was solved in potato to some extent because of large-scale source seed production by research institutions and state government departments and seed multiplication done by small seed producers.

The trends in procurement of commercial seed were echoed by the area planted with different types of seeds. Almost entire area under cotton, tomato and pea was planted with commercial seeds. In case of paddy, about half of the area was sown with commercial seed. However, potato growers continued to be served largely by the traditional seed system, where seed saving and exchange was predominant **(Table 11.2).** These trends in the use of seed for various case study crops revealed that development of commercial seed markets in India has increased use of quality seed for most of the crops. This was expected for hybrids where farmers need to replace seed every year. But, increasing trends in the use of quality seeds of pea and paddy were quite encouraging. This showed that wherever there was economic advantage in the use of commercial seed, farmers did acquire fresh seeds from the market. And since there was increasing demand for quality seed, private sector did participate in the provision of seed. This situation exited in groundnut and potato, but low seed multiplication rate and high seed handling costs restricted the private sector's participation. In both the crops, there was no problem of genetic impurity of seeds, but there were other quality considerations like disease control, which deserved immediate attention. Technologies to manage diseases, handling and storage of seed and increasing supply of source seed would go a long way to strengthen the decentralized seed systems for these crops.

Table 11.2: Proportion of Seed Quantity, Source of Seed and Area Planted with Commercial Seed by Sample Households: 2003-04 and 2004-05

Sl.No.	*Particulars*	*Crops*					
		Cotton	*Tomato*	*Pea*	*Paddy*	*Potato*	*Groundnut*
1.	**(a) Production of seed quality procured (per cent)**						
	Commercial	91	99	98	60	21	35
	Other farmers	8	1	0	12	19	23
	Farm saved	1	0	2	28	60	42
	(b) Proportion of farmers procuring seed from different sources (per cent)						
	Commercial	98	98	98	61	37	41
	Other farmers	1	2	0	14	21	23
	Farm saved	1	0	2	25	42	36
2.	**Proportion of area planted with commercial seed (per cent)**						
		96	99	97	60	24	37

Source: NCAP farm survey.

Source of Commercial Seeds

What are the sources and reasons for procuring commercial seeds by farmers? Almost all farmers bought the seed of cotton, tomato and pea from the commercial sources, *i.e.* a private seed dealer or sale counter of the seed corporation or government department. In a majority of the cases, the seed was bought from private seed dealers, as almost the entire private seed and a significant proportion of public seed was being sold by them. Furthermore, all of these market transactions were in cash and hardly a few farmers bought seed on credit basis even when they were buying from a dealer known to them. The cases of farmers buying seed from other farmers were quite low both for paddy and potato grown in the favourable production environments (Table 11.3).

It was interesting in the sense that as the supply of commercial seed improved and farmers had a better access to seed markets, importance of the traditional seed system diminished. However, traditional seed systems were found to be very effective in terms of popularisation and seed provision of new varieties. Once a variety was accepted by the farmers, formal seed systems faced little risk in the production and delivery of new variety on a large scale. The reasons for farmers acquiring fresh seed were quite revealing. A majority of the farmers bought fresh seed for quality considerations. They felt that fresh seed available in the market was of better quality, and would give a distinct yield advantage. The proportion of such farmers was as high as 90 per cent for tomato and pea, and about 60 per cent for cotton and paddy. In cotton and paddy, a significant proportion of farmers (more than 30 per cent) bought fresh seed to change variety. Only in the case of potato and groundnut, where seed storage and management was rather difficult, farmers also bought seed as they had sold all the stock. It reiterated that there was seed demand for these crops but there were not adequate incentives for the private sector to participate. Perhaps fiscal incentives, coupled with technological developments to reduce the costs and risks would go a long way in attracting the private sector to the seed business of these crops.

Table 11.3: Sources of Information about Varieties

Crop	*Source of Information (Per cent of cases)*			
	Fellow Farmers	*Seed Dealers*	*Extension Agents*	*Others*
Cotton	70	17	–	13
Paddy	76	14	2	8
Potato	87	7	2	4
Pea	42	28	26	2
Tomato	49	42	3	6
Groundnut	80	–	16	4

The above results have clearly shown that Indian farmers replace seed frequently for quality considerations. Increased availability of seed and wider seed distribution networks have improved farmers' access to commercial seed. Discussion with farmers during farm surveys revealed that higher seed prices were not a problem for them if they could realize higher yield or cost reductions due to use of commercial seed. In case seed prices were high and there was germination failure due to adverse weather conditions, farmers found it difficult to arrange additional resources to procure fresh seed. This problem was indicated by some farmers growing cotton and vegetables. It is therefore important that farmers are either compensated by seed companies in case of inferior seed, or crop is insured against crop failures due to adverse climatic conditions.

Sources of Information

Delivery of information about seed of crop varieties on sale, seed agencies, seed and crop management practices are as important as delivery of seeds. Farmer-to-farmer dissemination of information on crop varieties was still dominant for all crops (Table 11.3). This was not surprising when the present extension methodology under the Training and Visit system emphasized working with a few contact farmers, which in turn, spread messages among their fellow farmers. Seed dealers turned out to be the second important source of information about crop varieties. This source was particularly important for vegetables, where almost entire seed is sold by the private dealers, and for crops like tomato, cauliflower and cabbage where most of the seed is imported. Only pea growers received information from diversified sources (fellow farmers, seed dealers and extension agents) because most of the seed sold in HP was of public varieties and the government department and private seed dealers had significant role in seed sale. Some seeds like *Bt* cotton required additional information to be delivered to farmers, and during the filed visits it was also found that the concerned seed company printed extension material for distribution along with *Bt* cotton seed. In some cases, seed dealers were well informed by the seed company about the new seed, who in turn educated the farmers. However, it was rather difficult to establish that to what extent the required guidelines were followed by *Bt* cotton growers, as there were very few growers in the sample area. Nevertheless, this indicated that seed companies and dealers will have to play a much larger role in educating farmers in the adoption of information-intensive crop varieties and associated management practices.

Conclusions

The study has concluded that the commercial seed markets for hybrids are well developed, but these need improvement in the flow of information to farmers and effective regulation of unscrupulous traders, *etc.* There are significant changes in terms of seed regulations, management of GM crops and protection of intellectual property. Since all these regulations interact closely, there is a need for developing institutional capacity for their enforcement, as well as flexibility to learn from the experience for future adaptations. There is a lot of scope for strengthening the seed system of 'orphan crops', where there is no participation of the private sector, and public seed system is facing several resource and institutional constraints. In particular, technological backstopping, developing partnerships with the private and civil society organisations, and developing capacity at the local level deserve special attention. The study has shown that increasing proportion of farmers use commercial seed for quality considerations. However, there are problems with the selection of crop varieties, particularly of proprietary hybrids, due to lack of information. This has resulted into poor crop performance on several occasions. Thus, empowerment of farmers with information about commercial seed market and new varieties and strengthening of the system to protect farmer seed users would go a long way in developing the seed system. In the case of high volume crops, technological empowerment of local seed agencies and farmers should be accorded high priority.

Biodynamic Farming Beckons

Organic farming as a concept has not attracted much the imagination of farmers in this part of the country. Of course, some farmers in this region have switched over to organic farming, but their number is negligible. Reason: there is no encouragement or incentive either from the Punjab Government or the Centre to promote organic farming in a big way. However, the Uttarakhand and Maharashtra

Governments are engaged in encouraging farmers to take to this type of farming.

There are certain NGOs, which have been working hard to promote organic farming. Obviously, farmers who believe in traditional agriculture have certain fears about organic farming. That is why they have not been showing the desired interest in it. The other reason is that there is no proper dissemination of information to farmers regarding organic farming. There is extensive lack of technical knowhow in this regard. The key fear of farmers is that it entails a fall in the production. Farmers have been made to believe that there is less production in case chemical fertilisers and pesticides are not used.

However, there are farmers such as Jaspal Singh Chatha, a public school-educated farmer, who have developed a lasting interest in organic farming. In fact, Jaspal Singh has moved ahead of the organic farming concept. He has been doing biodynamic farming, which is an advanced stage of organic farming. "There is a certain school of thought which believes that heavenly bodies such as the moon and other planets also play a role in the growth of various crops. I am a votary of that school of thought", says Jaspal Singh. "In earlier times, our elders used to seek the help of astrologers to find out the location of various planets before the sowing of any crop, even during harvesting time. I feel that the elders were aware about the effect of planets on the growth of various crops", he says.

"For instance, the positioning of the moon at the time of sowing, growing and maturing of some of the crops plays a role in their yield level", claims Jaspal Singh, who, after passing out from Yadavindra Public School, Patiala, and then graduating from DAV College, Chandigarh, started farming at his Mussewal village, near Nakodar, in Jalandhar district. He owns 30 acres and grows wheat, basmati, pulses such as *mahan* and moong and turmeric, mustard and linseed. "My entire produce of wheat and other crops is certified by the Institute of Marketecology (IMO)", says Jaspal Singh.

Initially, Jaspal faced a lot of problems. "It was not a smooth sailing in the beginning. However, now I am comfortable and also encouraging other farmers to adopt this technique of farming", he says. His elders shifted from Kenya to India in 1964. After graduation, he took the training in Uttarakhand in organic farming in 2004. "The first time I heard about this type of farming was in 2002. One woman Vanita Shah was doing organic farming in Uttarakhand. She told me that the training was imparted in her native state. After a formal training of a few weeks, I made up my mind to do this type of farming only and to bid goodbye to traditional farming forever.

"I started using bio-waste, especially the recycling of the crop residue for the purpose of compost, to nurse various crops and completely do away with the use of pesticides and other chemicals on my fields", says Jaspal Singh. "This year, the yield of turmeric on my farm was about 100 quintals per acre and I believe that is a very good level of yield", he adds. "I have sold the turmeric seed at Rs. 50 kg. That means I have earned Rs. 5 lakh from one acre. Obviously, I am very satisfied in this regard. The processed turmeric sells at Rs. 250 per kg. My wheat has a ready market now. It sells in the range of Rs. 2500 per quintal whereas at the initial stage, I faced problems to dispose of my foodgrains". There is a fall in the input cost for growing various crops.

The heavy use of chemicals has spoiled the texture of the soil. The soil health has deteriorated to the extent that it has become useless at some places. Chemicals are seeping in various foodgrains, leading to a spread of diseases such as cancer. There is high incidence of heart diseases, kidney problems and liver ailments. "All that is because of the irrational and injudicious use of chemical fertilisers and pesticides", says Jaspal Singh,who is also associated with various experts of biodynamic farming in other countries such as New Zealand and Switzerland.

Instead of chemicals and pesticides, he uses the CPP (cow pat pit) technique for the treatment of seed, soil and for folio spray. As much as 30 kg of CPP is enough for one acre of land. There is now Biodynamic Association of India, which is engaged in educating farmers with regard to this new technique of farming. "People in this part of the country should learn from our experience and shift to organic farming".

Seeds of Wellness

Seeds may be tiny but are big storehouses of many vital nutrients which boost immunity and prevent numerous diseases

Seeds are the embryo and the food supply of plants. Strictly speaking most cereals, pulses and nuts are also seeds but here we refer mostly to oilseeds like sunflower, pumpkin, sesame, flax, melon *etc.* Seeds are excellent sources of good fat (40-50 per cent), primarily poly and mono - unsaturated fats (PUFA and MUFA), the kind that lower bad cholesterol. Besides having a profound impact on our cholesterol levels, these boost immunity and prevent diseases. Research shows that frequent seed consumption is associated with lower levels of inflammatory markers (C-reactive protein, interleukin-6 and fibrinogen) which increase the risk to several diseases, including cardiovascular disease. These are cholesterol free and provide significant amounts of plant protein (20-30 per cent).

Seeds are low in starch and carbohydrates. These are high in fiber and have a low glycemic index (ability to raise blood glucose). This means that these do not cause spikes or steep increase in blood glucose levels. Since these seeds keep blood glucose levels more stable these keep one feeling fuller longer and prevent cravings. This makes them a good snack food compared to breads, biscuits, cakes, rusks, chips and other high glycemic food. This property also makes them useful in diabetes to improve glycemic control. Being the "future" plants, seeds are also loaded with micro-nutrients. These include B vitamins - B1, B2 and minerals like zinc, calcium, magnesium, phosphorous, potassium and iron. Seeds are a rich source of vitamin E and other natural antioxidants which have several health benefits. These are perfect cocktail for radiant skin and hair.

If snacking on them, seeds should be lightly roasted or heated gently to destroy unwanted components (allergens, contaminants, toxins) that may be present. Seeds can be sprouted which makes them richer in vitamins B and C, amino acids, and enzymes. Sprouts can be used in salads or sandwiches or mixed with curd/other sprouts. Sprinkle toasted seeds over curd or snacks. Poppy seeds are especially attractive as toppers or flavourings for breads, rolls, salads, and stir-fries. Combined with salt, these can be used to make "reduced salt" preparations.

Some Commonly Eaten Seeds

Flax Seeds

Traditionally known as linseed or alsi (in Hindi), these seeds are packed with huge amounts of omega-3 fats, beneficial fats similar to those found in fish. Flaxseeds owe their benefits to their high content of alpha linolenic acid (ALA), an essential fatty acid, a precursor to omega-3 fats, phytoestrogens (lignans) as well as their high-fibre content. Flax seeds are essential for normal growth and development. These are also known to play a role in the treatment of heart disease, high blood pressure, diabetes, arthritis, asthma, inflammatory, auto-immune disorders and cancer. Besides, good fat and lignans, flaxseeds are also a good source of vitamin B complex, vitamin E, soluble and insoluble fibre. Soluble fibre helps lower cholesterol levels and consequently lower risk of heart disease, while the insoluble fibre helps prevent constipation without worrying about dependency. Soluble and insoluble fibre aids regular bowel movement without abdominal cramping or bloating. Ensure it is taken with adequate fluids. These are also gluten free and ideal for those with sensitive intestinal tracts.

These also have mild estrogenic properties due to their lignan content. This makes them useful to treat menopausal problems, including brittle bones and hot flashes. As far back as Hippocrates, flaxseed tea has been used to treat sore throats, hoarseness and bronchial spasms. A tablespoon or two, powdered, taken with water, cereal, mil or curd is a good way to add them to your diet. Although there is no recommended daily amount, but many studies use 1-2 tablespoons of ground flaxseeds daily. Start with a small amount, such as one teaspoon, and slowly build up to two tablespoons in 7-10 days.

Eating the whole seeds will not provide most of the benefits. Ground seeds are preferable for deriving maximum benefit. Flaxseed oil is cholesterol free and provides omega-3 fats. However, it does not provide fibre or lignans present in the seeds. It is also very unstable and has a very limited shelf life. It breaks down with heat and must not be used for cooking.

Caution

Patients on tamoxifen should not consume large amounts and must check with their physician before adding in their diets. Pregnant women or those considering pregnancy consult with their healthcare providers.

Sunflower Seeds

Sunflower seeds are rich in vitamin B-complex, vitamin E, magnesium, calcium and proteins. These are useful for cholesterol reduction, inflammatory conditions, osteoporosis, to boost immunity and particularly for post-menopausal women. Toasted sunflower seeds can be added to stir fries and make excellent snacks.

Sesame Seeds

A good source of protein, vitamin E and calcium, these also provide iron, niacin, riboflavin and thiamin. Sesame seeds contain 50 per cent oil by weight which is essentially a good balance of unsaturated fats. Research shows these are a rich source of sesamin, which may have protective effects against hormone-related diseases such as breast cancer. Moreover, being a rich source of vitamin E, these also help decrease lipid per-oxidation thereby reducing the risk of cardiovascular diseases.

Sesame seeds are especially attractive as toppers or flavourings for breads, rolls, salads, and stir-fries.

Other Seeds

Seeds such as pumpkin, cucumber, melon and water-melon, are good sources of protein and fibre. These are valuable sources of vitamin E and B- complex. Pumpkin seeds are rich in Panagamic acid, an antioxidant. These four seeds are useful in stimulating "anti-stress" hormones and fighting fatigue. These also boost immunity, lower cholesterol, prevent arthritis and reduce food cravings. Pumpkin seeds are believed to positively influence prostrate health. Toasted seeds can be added to stir fries and make good snacks.

Alfalfa Seeds

Best taken as sprouts, alfalfa is useful in liver disease, detoxification of blood, anaemia, ulcers, colitis, arthritis and fluid retention. These are particularly useful for reducing blood cholesterol and blood sugar levels. These are also known to be useful in cancer prevention and have anti-fungal properties. Never eat alfalfa seeds without sprouting because these contain high levels of the toxic amino acid (canavanine). Over time, eating these seeds could result in impaired functioning of the platelets and white blood cells. Pregnant and nursing women should consult their physician before use.

Chia Seeds

Chia seeds are originally, from Southern Mexico. These seeds come from the desert plant Salvia hispanica, a member of the mint family. Chia seeds owe their benefits to their high content of alpha linolenic acid (ALA), phytoestrogens (lignans), high-dietary fibre content, protein as well as vitamins, minerals, including calcium, potassium and magnesium and certain antioxidants. Considering their good nutritional profile and health benefits, in 2009, the European Union approved chia seeds as a novel food.

These oval shaped, nutty flavoured seeds are packed with huge amounts of beneficial omega- 3 fats. The concentration of omega-3 fatty acids in Chia seeds is even more than flax seeds. Antioxidants present in Chia seeds help in their longer shelf life without these getting rancid. Antioxidants are also known to promote good health and offer protection against certain diseases, including diabetes, heart disease and certain cancers. Like flaxseeds, chia seeds are also a good source of soluble and insoluble fibre.

These seeds do not have to be ground to make their nutrients available to the body. Chia seeds may be eaten raw as whole seeds, ground or as sprouts. Although there is no recommended daily amount, but we may start with a small amount, such as a teaspoon, and slowly build up to 2 tablespoons in 7-10 days. Pregnant women or those considering pregnancy must consult with their healthcare providers. Seeds are versatile and convenient to consume. Seeds can be ground in your coffee grinder and once ground store in an air tight, opaque container in a refrigerator or freezer. Considering their nutritional value, these may be added to almost all sorts of dishes. These little powerhouses make excellent, inexpensive snacks. In fact, these make great substitutes for high-priced nuts in a poor man's diet. No wonder our grannies took the pains of removing, drying and deseeding seeds from water-melons and melons to be able to give their family their daily vitamins.

Seeds of Change

Seed is the only part of the agricultural chain that is growing at 20 per cent a year and is insulated from both government subsidies and weather vagaries. And, in the clamour for higher yields and greater farmer acceptance, seed companies are hitting a sweet spot in innovation.

There's an unmistakable buzz in and around the $1.5 billion Indian seed industry. "It has the potential to grow to $2.5 billion in the next five years." The last two-and-a-half years have seen six equity deals worth $153 million being signed. "For us, it's a play for the next 20 years, not the usual five," adds Manish Jain, whose private equity fund Axis Holdings has invested $5 million in Kaveri Seeds and is preparing to sew up two more deals. Multinationals are hunting for Indian players with research skills in cash crops like cotton, corn and vegetables. As are cash-rich homegrown companies like the Rs. 900 crore Nuziveedu, India's largest seed company. Also, fertiliser and agro-chemical companies, like Rallis India of the Tata Group, are diversifying here. "Acquirers have paid high premiums," says Kumar. Investors and companies are buying the macro story: the growing demand to feed a billion mouths, millions of whom transit to a superior diet every year; rising food prices; the world's second-largest arable area; and the $150 billion Indian agriculture sector. But that's not all they are buying into. If so, any agricultural business would pass muster. But most in India don't because they are either out of bounds, or are plagued by government subsidies or weather vagaries. By comparison, seeds is a free market for all practical purposes; it is breaking new ground in adoption and innovation; and, no matter what the weather, a farmer will sow.

Seed of Life

A pivotal change is in the offing in how farmers sow. About 70 per cent of the seeds sown in India are those saved from the previous year. Compared to new seeds, they yield less and are more disease-prone. The government has set a seed replacement rate–percentage of new seeds in a year–target of 40 per cent. "If that happens, the industry will grow three times". While the government is using policy to increase consumption of new seeds, companies are improving product quality. The industry is, today, at a point where cuttingedge technology meets high finance and industrial farming. The main challenge is to increase yields. Agricultural land is finite. So, to feed a population that is growing and eating more, that means drawing more output from the same land. That means losing less to the three giant killers: heat, thirst and pests. That means withstanding floods. That means lowering farming costs.

A tiny seed, which has greater potential for innovation than anything else in the agri chain, can deliver all these. "Seed is the agent for change," says Dr Gyanendra Shukla, director at Mahyco Monsanto Biotech India. "From farmers and industry to scientists and government, everybody understands that change is faster and cheaper by leveraging seeds."

Hybrid Play

Broadly speaking, there are two types of seeds. The first are varietals, which pollinate by themselves; they can be reused, but with diminishing returns. The second are hybrids, which are developed through controlled pollination and cannot be reused. Barring wheat and soybean, other crops have both types of seeds. In 2010-11, hybrids accounted for Rs. 5,000 crore of the industry's Rs. 8,000 crore turnover.

"Farmers have understood that varietals appears cheap, but are more expensive in the long run," says Reddy of Ganga Kaveri Seeds. "Seeds become tired, and lose yield potential and quality from one generation to the next." Amitabh Jaipuria, managing director of Monsanto India: "Conversion of area from varieties to hybrids will be the next big thing. As we saw in cotton, the availability of technology can lead to faster adoption." Monsanto transformed the Indian cotton market with *Bt* cotton, a hybrid introduced in 2002-03. Between that year and 2010-11, when 90 per cent of cotton farms had *Bt* seeds, yields increased from 302 kg a hectare to 503 kg a hectare. And production from 13.6 million bales to 32.5 million bales, turning India from cotton importer to the world's second-largest producer and exporter. Today, cotton accounts for half the hybrid seeds sold in India. Seed companies are looking to replicate that story in more crops. For them, hybrids promise growth and value, which varietals don't. "Seed accounts for 5-15 per cent of the cost of production". "A farmer is willing to pay for value, performance and consistent quality. If you deliver this, you can recover value."

At present, hybrids are available for cotton, corn, pearl millet, sorghum, sunflower and vegetables. Greater adoption of hybrids means more seeds sold, which will mostly benefit the private sector. There are two central government companies and 13 state seed corporations. But their efforts are inadequate, "Up to labs, government is good. But state seed corporations have become defunct." Companies are reinvesting profits in research, expansion and asset creation, in that order. "R&D is the only differentiator to increase value for a farmer," says Reddy. "With stronger IPR (intellectual property rights), companies are competing to show farmers a 15-20 per cent increase in yields."

R&D Push

Corn hybrids are a Rs. 700 crore market, with Rs. 70 crore being invested in R&D by Indian companies alone; investment in rice hybrids is a steady Rs. 15-20 crore a year. The DuPont Knowledge Centre at the Genome Valley near Hyderabad has grown to 400 scientists in two-and-a-half years. "We were attracted by the quality of people available in India". Companies need a large bank of germ plasm–the building block of a seed that houses traits–and capital. "It (germ plasm) is scarce, expensive and must be refined properly to meet breeding objectives".

Biotechnology is taking hybrids to another level, by tinkering with genes to alter traits of seeds. But launching a commercially viable hybrid is a slog, with a high risk of failure, which discourages smaller players. "The discovery of a new trait can take 2-10 years. From idea to dealer, it takes another 10 years, if all goes well," says Subbarao. "It can cost up to $120 million in the US to develop a transgenic product with a good chance of becoming an industrial success." Farm biotech is the fastest-growing segment of India's biotechnology industry, says Dr KK Narayanan, co-founder and managing director of Metahelix, which developed an alternative technology for *Bt* cotton. Biotechnology brought with it a new business model. Monsanto India, through its joint venture Mahyco Monsanto, has licensed its basic *Bt* cotton technology, based on genetic modification, to 35 companies. Such tie-ups allow Indian companies to leapfrog several years of research, and enter the market cheaply and quickly.

New Frontiers

Companies are making variants to be relevant in all eight agro-climatic zones of India. They are setting up shop in Africa and South-east Asia, where the climate is similar to India. They are venturing into new crops with untapped market potential. Monsanto, for instance, is investing in vegetable seeds, a Rs. 800 crore market. It has to start from scratch because vegetables that are popular abroad, like broccoli and lettuce, don't fly in India yet. Challenges will increase. The pool of scientists and researchers is limited. And expanding marketing and distribution calls for more spending. "Seed is a show-and-tell business," says Vidya Sagar Parchuri, CMD of the Rs. 650 crore Vibha Seeds. "Each year, we hold 10,000 demonstrations for 100,000 farmers. Village-night meetings, posters, hoardings and advertising are extra." All this eats into the 50 per cent gross margin that good companies typically earn.

Technology has increased risk by shortening the product shelf life from 10-12 years to 7-8 years. "After introducing a new seed, it takes two years to scale it up. It sells for four years and then declines because new rivals enter the market," says Subbarao, making a case for a strong pipeline of hybrids. As R&D costs increase, companies are expected to seek alliances with government research institutes and state agricultural universities, which can develop technology but need help to commercialise it. "The Green Revolution was driven by the public sector, the *Bt* cotton revolution by the private sector," says Shukla. "The next push will come from public-private partnerships." Companies say if policies encourage competition and innovation, promote a science-based regulatory regime and protect intellectual rights, investment and growth will continue. It's why they decry the proposed Seed Bill 2008, which penalises companies for crop failure. "It can't prevent the problem of counterfeit seeds," says Rao of Nuziveedu. It's why instead of subsidised seeds–which some states, including UP, Rajasthan, Gujarat and Bihar, offer–they suggest a cash subsidy. "Government subsidy takes away merit," says Jaipuria. Adds Rao: "Give a cash subsidy against a bill to give farmer the freedom to buy." Those are small issues, but the big story has been planted.

FARMING

Out of bounds for cos. They can, at best, do contract framing with many small farmers, but aggregation is painful. And, it also means dealing with weather risk

SEEDS

Only 25% of seeds in India are bought afresh annually. Companies have pricing power and intellectual protection, and are innovating to solve problems and increase yields

YIELDS (Kg per hectare)

INDIA

	2009-10	2022*
Rice	2,130	2,651
Wheat	2,830	3,380
Coarse Cereals	1,046	1,222
Cereals	2,077	2,364

*Projected

GLOBAL HIGH

RICE 9,731

WHEAT 9,036

COARSE CEREALS 9,079

CEREALS 8,886

SOURCE: NCAP, FAO Statistics, Standard Chartered Research

IRRIGATION

Drawn water does not reach many parts. Even where it does, the govt subsidy for micro-irrigation kits is 50%; many farmers can't afford even the balance

FERTILISERS

There's only so much fertiliser one can apply to a patch of land, which caps growth. Also, the vagaries of government control influence prices and company operations

LOGISTICS

Govt wants the private sector to set up storage facilities and cold chains, but players say it is not feasible without generous sops

FINANCING

Agricultural credit is the domain of banks. Outside of banks, there is microfinance, but the Andhra experience shows that politics can erase years of good work

FARM MACHINERY

About 82% of land is tilled by farmers whose holding is below two hectares, which dilutes the case for large-scale mechanisation

CROP PROTECTION

A bad crop rules out the need for herbicides and pesticides. And if he wants to save cash, this is the first cut back. Also, copying and commoditsation is rampant in this segment

THE SEEDS ACT, 1966

Definitions

In this Act, unless the context otherwise requires,

1. "Agriculture" includes horticulture;
2. "Central Seed Laboratory" means the Central Seed Laboratory established or declared as such under sub-section (1) of section 4;
3. "Certification agency" means the certification agency established
4. "Committee" means the Central Seed Committee constituted under sub-section of Section 3;
5. "Container" means a box, bottle, casket, tin, barrel, case, receptacle, sack, bag, wrapper or other thing in which any article or thing is placed or packed;
6. "Export" means taking out of India to a place outside India;
7. "Import" means bringing into India from a place outside India;
8. "Kind" means one or more related species or sub-species of crop plants each individually or collectively known by one common name such as cabbage, maize, paddy and wheat;
9. "notified kind or variety", in relation to any seed, means any kind or variety thereof notified under Section 5;
10. "Prescribed" means prescribed by rules made under this act;
11. "seed" means any of the following classes of seeds used for sowing or planting-

    (i) seeds of food crops including edible oil seeds and seeds of fruits and vegetables;

    (ii) cotton seeds;

    (iii) seeds of cattle fodder;

    and includes seedlings, and tubers, bulbs, rhizomes, roots, cuttings, all types of grafts and other vegetatively propagated material, of food crops or cattle fodder;
12. "Seed Analyst" means a Seed Analyst appointed under section 12;
13. "Seed Inspector" means a Seed Inspector appointed under section 13;
14. "State Government", in relation to a Union territory, means the administrator thereof;
15. "State Seed Laboratory", in relation to any State, means the State Seed Laboratory established or declared as such under sub-section (2) of section 4 for that State; and
16. "variety" means a sub-division of a kind identifiable by growth, yield, plant, fruit, seed, or other characteristic.

Central Seed Committee

(1) The Central Government shall, as soon as may be after the commencement of this Act, constitute a Committee called the Central Seed Committee to advise the Central Government and the State Governments on matters arising out of the administration of this Act and to carry out the other functions assigned to it by or under this Act.

(2) The Committee shall consist of the following members, namely:- (i) a Chairman to be nominated by the Central Government; (ii) eight persons to be nominated by the Central Government to represent such interests that Government thinks fit, of whom not less than two persons shall be representatives of growers of seed; (iii) one person to be nominated by the Government of each of the States.

(3) The members of the Committee shall, unless their seats become vacant earlier by resignation, death or otherwise, be entitled to hold office for two years and shall be eligible for renomination.

(4) The Committee may, subject to the previous approval of the Central Government, make bye-laws fixing the quorum and regulating its own procedure and the conduct of all business to be transacted by it.

(5) The Committee may appoint one or more sub-committees, consisting wholly of members of the Committee or wholly of other persons or partly of members of the Committee and partly of other persons, as it thinks fit, for the purpose of discharging such of its functions as may be delegated to such sub-committee or sub-committees by the Committee.

(6) The functions of the Committee or any sub-committee thereof may be exercised notwithstanding any vacancy therein.

(7) The Central Government shall appoint a person to be the secretary of the Committee and shall provide the Committee with such clerical and other staff as the Central Government considers necessary.

Quality Measurement

Measurement of seed quality can be made in the seed testing laboratory and in the field. There are certain quality attributes which could only be measured more precisely in laboratory. Accordingly. seed testing has been recognized as an essential aspect of quality seed production programme. The importance of the seed testing was realised in the country even before the enactment of the Seeds Act. In the late fifties. it was decided to have at least one seed testing laboratory in each State/Union territory. The Seeds Act was passed in 1966 and Seed Analyst and Laboratories have been assigned important role to ensure the quality of planting material for sale.

The main provisions deal with setting up of the seed laboratory, seed sampling, seed analysis and report/results. Thus all essential aspects of the seed testing have been taken care of in the Seeds Act and Seeds Rules.

Seeds Testing Laboratory

There is a provision to set up a Central Seed Laboratory and States Seed Laboratory to discharge various functions enshrined under the Seeds Act. :The Section 2(2) and Section 2(15) define these laboratories as under:

2(2): "Central Seed Laboratory means the Central Seed Laboratory established or declared as such under sub-section (1) of Section 4;

2(15): " State Seed Laboratory". in relation to any state. means the State Seed Laboratory established or declared as such under sub-section (2) of Section 4 for that State. the Section 4(I) and 4(2) of the Seeds Act specify that Central Government and State Governments could set up the Central Seed Laboratory and State Seed Laboratory respectively. The Section 4(1) and 4(2) read as under:

4(1): The Central Government may. by notification in the Official Gazette. establish a Central Seed Laboratory or declare any Seed Laboratory as the Central 271 Seed Laboratory to carry out the functions entrusted to the Central Seed Laboratory by or under this Act.

4(2): The State Government may. by notification in the Official Gazette. Establish one or more State Seed Laboratories or declare any Seed Laboratory as a State Seed Laboratory where analysis of seeds of any notified kind or variety shall be carried out by Seed Analysts under this Act in the prescribed manner.

Functions or the Seeds Laboratories

Central Seed Laboratory has been assigned important functions under the Seeds Act and Seeds Rules. The detail of the same is given as under:

Analysis of the sample received from accused vendor or complainant. As specified under Section 16(2). any accused vendor or complainant could request the Central Seed Laboratory through Court to Analyse the sample. The Section 16(2) reads as under:

16(2): "After the institution of a prosecution under this Act. the accused vendor or the complainant may. on payment of the prescribed fee. make an application to the court for sending any of the samples mentioned in clause (a) or clause (c) of sub- section (2) of Section 15 to the Central Seed Laboratory for its report and on receipt of the application. the court shall first ascertain that the mark and the seal or fastening as provided in clause (b) of sub-section (1) of Section 15are intact and may then despatch the sample under its own seal to the Central Seed Laboratory which shall thereupon send its revert to the court in the prescribed form within one month from the date of receipt of the sample, specifying the result of the analysis."

Analysis or the Service Sample

The Central Seed Laboratory could also analyse the service sample of which results to be used as information for seeding, selling or labelling purposes (Rule 2(i) of Seeds Rules: 1968).

Apart from above, the Central Seed Laboratory has been assigned to perform the following functions as indicated in Rule 5 of the Seeds Rules, 1968.

In addition to the function entrusted to the Central Seed Laboratory by the Act, the Laboratory shall carry out the following functions, namely:

(a) Initiate testing programmes in collaboration with the State Seed Laboratories designed to promote uniformity in test results between all seed laboratories in India;

(b) Collect data continually on the quality of seeds found in the market and make this data available to the Committee; and

(c) Carry out such other functions "as may be assigned to it by the Central Government from time to time.

Functions of the State Seed Laboratories

State Seed laboratories are meant to analyse the seed samples of any notifiedkind or variety in the prescribed manner. The State Seed Laboratory is expected to analyse the samples received from various sources for the following different purposes.

(i) Analysis of the samples received from Seed Certification Agencies set up " under Section 8 of the Seeds Act.

(ii) Analysis of the service samples: Seed users and seed producers could get seed sample tested to obtain the result to be used as information for seeding. Selling or labelling purpose.

(iii) Analysis of the samples received from Seed Inspector to determine the compliance of labelling requirements under Section 7 of the Seeds Act.

Seed Analyst

As per Section 2(12) "Seed Analyst" means a seed analyst appointed under Section 12.

Section 12 reads as under

"The State Government may. by notification in the Official Gazette. Appoint such persons as it thinks fit. having the prescribed qualifications. to be Seed Analysts and define the areas within which they shall exercise jurisdiction.

" The perusal of the Section 12 would indicate that State Government could. appoint the Seed Analyst through notification in the Official Gazette. A copy of the draft notification to be issued under Section 12 is given at page 278. The copy has been given to provide an idea about the draft of the notification and, infact, draft notification has to be prepared by the concerned Department and is to be vetted by the Department of Law of the concerned State.

The Seed Analyst should possess contain minimum qualifications as prescribed in Rule 20 of the Seeds Rules, 1968.111e same arc reproduced below:

Qualifications of Seed Analyst

A person shall not be qualified for appointment as Seed Analyst unless he –

(i) possesses a Master's or equivalent degree in Agriculture or Agronomy or Botany or Horticulture of a University recognized for this purpose by the Government and has had not less than one year's experience in seed technology; or

(ii) posses.<;c.a." Bachelor's degree in Agriculture or Botany of a University recognised for this prpose by the Government and has had not less than three year's experience in seed technology.

Duties of a Seed Analyst

1. On receipt of a sample for a analysis the Seed Analyst shall first ascertain that the mark and the seal or fastening as provided in clause (b) of the subsection (1) of section 15 are intact and shall not the condition of the seals thereon.
2. The Seed analyst shall analyze the samples in accordance with the procedures laid down in the Seed Testing Manual published by the Indian Council of Agriculture Research as amended from time to time. (Amendment No.7(17)/69-Seeds-Dev., dated 30-6-1973)
3. The Seed Analyst shall deliver in Form VII, a copy of report report of the result of the analysis to the persons specified in sub-section (1) of section 16, as soon as,may be but not later than 30 days from the date of receipt of samples sent by the Seed Inspector under Sub-Section (2) of the Section 15.(Amendment No.7(17)/69-Seeds-Dev., dated 30-6-1973)
4. The Seed Analyst shall from time to time forward to the State Government the reports giving the result of analytical work done by him.

Power to Notify Kinds or Varieties of Seeds

If the Central Government, after consultation with the Committee, is of opinion that it is necessary or expedient to regulate the quality of seed of any kind or variety to be sold for purposes of agriculture, it may, by notification in the Official Gazette, declare such kind or variety to be a notified kind or variety for the purposes of this Act and different kinds or varieties may be notified for different States or for different areas thereof.

Power to Specify Minimum Limits of Germination and Purity, *etc.*

The Central Government may, after consultation with the Committee and by notification in the Official Gazette, specify-

a. the minimum limits of germination and purity with respect to any seed of any notified kind or variety;
b. the mark or label to indicate that such seed conforms to the minimum limits of germination and purity specified under clause (a) and the particulars which such mark or label may contain.

Farmers Right Act

Farmers Right Act 2001 has been effectively Implemented to check the unfair practice of unauthorized persons/companies selling the seeds prepared by some other person/institution.

The enactment of this legislation would also facilitate internationally renowned companies to sell their quality seeds in the country. As per the provisions of the Act, the plant breeders/farmers can get their crop varieties protected, provided these are novel, distinct, stable and uniform. The breeders are entitled to save, use, sow re-sow, share exchange and even sell their seed, but they cannot sell branded seeds of a protected variety.

The farmers can claim compensation if a protected variety does not perform as had been claimed by the breeder. The union government has set aside a sum of Rs.2 crore to be utilized by the National Gene Fund to ensure that the due benefit of a protected variety goes to the breeder.

SEED BILL 2010

An Analytic View

Introduction

Seed is the most unassuming potent gift in the life of the farmer to make his life fruitful. Right to good food and right to safe food are the slogans of the day and the same can be achieved by attaining food sustenance and food security. Seed is the kernel of the life itself, the source of our food, when contaminated have an adverse effect on our health and also health of our planet. India is a country mainly relied upon agriculture and for boosting up agricultural output availability of good quality of seeds to farmers is inevitable. To regulate the seed sector the Parliament has enacted the Seed Act 1966. Due to recent innovations in the seed sector, entry of private industry and introduction of varieties of seeds and its importation in India, makes the existing Seed Act as redundant since it fails to serve useful purpose. Hence, it paves the way to Government to introduce a new Bill in the Parliament legislation to deal with several issues in the seed sector. Accordingly, the Seed Bill 2004 has been prepared and introduced in the Rajya Sabha on 9th December, 2004 and thereupon referred to Parliamentary Standing Committee on agricultural chaired by Prof. Ramgopal Yadav to study and submit its report. The Seed Bill 2004 introduced radical changes in the seed sector.

The Seed Bill 2004 was introduced in the Rajya Sabha on 9th December, 2004, as stated supra and the same has been referred to Parliamentary Standing Committee on Agriculture for submission of report.

Key Features of Seed Bill, 2004

1. The Seed Bill 2004 aims at promoting the production and supply of quality of seeds and also to regulate its quality for sale, export and import.

2. The Seed Bill 2004 seeks to repeal and replace the existing Seed Act, 1966.
3. To comply with W.T.O. obligations and to protect the farmers and small scale agriculturist from monopolistic activities of commercial seed producers and seed suppliers, the Parliament proposed to go ahead for a new legislation on seeds and made an attempt to bring home several radical changes in the seed sector.
4. According to the Bill all varieties of seeds have to be registered and certified. The Bill proposed to be established a Central Seed Committee (CSC) for which the Secretary to the Government of India, Department of Agricultural and Co-operation, Ministry of Agriculture is the ex-officio Chairman.
5. The CSC is a prime controlling and regulating body under the Bill with several Government representatives as its members in the said body apart from two representatives from the farmers' sector and two from the seed industry. The Bill proposes for constitution of a sub-committee called Registration Sub-Committee which shall maintain a National Register of Seeds for all varieties of seeds. Any type of seed for sale must be registered with the Registration Sub-Committee and its nomenclature must be entered in the National Register of Seeds.
6. At State level, State Seed Committees and State Seed Certification Agencies will be established which are advisory in nature.
7. The Bill does not restrict the farmers right to use or sell his farm seeds and planting material provided that he does not propose to sell the same under a brand name.
8. The provisions for compensation were introduced in the Bill. All the registered varieties and seed producers, distributors and vendors have to disclose the expected performance of the seeds under certain given conditions. In case of failure to perform the expected standards the farmer can claim compensation from the dealer, distributor or vendor under the Consumer Protection Act, 1986.
9. As a part of the regulatory method, the Bill also introduced Central and State Seed Testing Laboratories with required number of seed analysts.
10. Import of seeds would be subject to the Plant Quarantine (Regulation of Import into India) Order 2003 or any corresponding Order under the Destructive Insects and Pests Act, 1914. The Central Government can restrict export of seeds if it is deemed to affect the food security of the country.
11. The Bill allows self certification of seeds by accredited agencies and also allows the Central Government to recognize certification by foreign seed certification agency.
12. Every seed producer and dealer and horticulture nursery must be registered with the concerned State Government.

13. The transgenic variety of seed must get clearance as per the provisions of Environmental Protection Act, 1986 for obtaining its registration.
14. The Bill proposes to increase the penalties to regulate the seed business. The Bill has enhanced the penalties that were prescribed under the existing Seed Act 4 ½ decades ago.

Seed Bill 2004 Changes Proposed by Standing Committee on Agriculture

After placing the Bill before Rajya Sabha, the Standing Committee on Agriculture appointed a Parliamentary Committee to study the Bill under the Chairmanship of Prof. Ramgopal Yadav and he submitted his report on 20.11.2006 and recommends for certain changes in the original Bill.

a) The Committee recommends that Plant Variety and Protection of Farmers' Right Act 2001 (hereinafter called as PVPFR Act) be made fully operative before the Seed Bill 2004 is passed.
b) The Bill should not impose a condition that the farmers seeds also must conform to the minimum standards required to be maintained by the commercial producers for registering their seeds. The Committee feels that such a restriction imposed in the Bill will seriously effect the rights of the farers and recommended for deletion of such restrictions.
c) The Committee suggested for expansion of the definition of the "farmers".
d) The Committee discourages for private participation in the Seed Certification since it seriously conflict the interest of the farmers who practice traditional system of exchange and sale of seeds. Thus it strongly recommends for deletion of the self certification provisions.
e) Another notable feature recommended by the Committee is for introduction of price regulatory mechanism in the Bill to ensure that the farmers should not be charged with arbitrary prices by the seed producers and sellers.
f) The Committee recommends for enhancement of the penalties mentioned in the Bill. The Committee suggests to increase the penalty for contravention of provisions of law as a fine of Rs.50,000/- which may be extended up to Rs.2,00,000/- and imprisonment which may be extended up to 3 months. Likely, the penalty for supply of spurious and mis-branded seeds must be an amount of Rs. 2,00,000/- which may be extended up to Rs.10,00,000/- and imprisonment of three months which may be extended up to one year.
g) The Committee also recommends for introduction of provisions banning the print misleading pictures on seed packages which may inspire and attract the innocent farmers to buy the said seeds and ultimately land in losses due to nil production.
h) The Committee also recommends for introduction of machinery to deal with the compensation matters in the Bill itself. The Committee found

fault with for entrusting the task of dealing with compensation matters to the Consumer Forums constituted under the Consumer Protection Act.

i) The Committee recommends for introduction of Seed Crop Insurance for providing compensation to the farmers whose seeds do not give the desired yield. The Committee further says that the compensation to the farmers should be based on the expected performance as mentioned by the seed producers on the label of the seed package and that the said certification agency should also make liable in the compensation process, in case the seeds failure to give the promised results.

j) The Committee opined that there are chances for misuse of powers by the seed inspectors. The Committee suggest for effective regulatory machinery over the powers of the seed inspector. The search and seizure provisions may be exercised by the seed inspector with the prior permission of the District Collector or/and Magistrate. The Committee also recommends for exemption of farmers from the provisions of search and seizure since they are not selling branded seeds except the farm seeds saved from their own production.

k) The Committee further recommends that there should be atleast one state representative from each of agro climatic zones instead of geographical zones on rotation basis and that number of farmers' representatives must be increased to five and each one from different geographical zones.

l) The Committee further recommends that for reduction of the registration period 10 and 12 years instead of 15 and 18 years. The Committee suggests for deletion of the provisions of re-registration since it will tend to monopoly among certain seed producers.

m) The Committee recommends that seed testing includes seed germination and yield testing and there should be a pre-registration requirement. The Committee further recommends that the Bill should include provision for declaring the origin of the variety and its parental details by the person sought for registration to ensure that farmers exempted variety should not be misused by the said companies.

n) The Committee also recommends that the foreign seed certification agency should be recognized only if the seeds certified by the said agency are tested on Indian soil to confirm to the minimum requirements.

o) The Committee further recommends for deletion of restriction of period of two years mentioned in Sec. 45 of the Bill which is introduced for removal of difficulties.

Seed Bill-2010 (Amendments Proposed to Seed Bill 2004)

After the Parliamentary Standing Committee on Agriculture has submitted its report, the Government of India accepted some of the amendments to the Seed Bill 2004 and subsequently on 13.04.2010 notified the Seed Bill 2010 draft. The new Draft Seed Bill 2010 was introduced in the Rajya Sabha on 23.04.2010.

The issue of seed price regulation was not agreed upon. After 2002 with the introduction of *Bt* cotton the issues of royalties over seed technologies like Genetic Engineering etc have come into forefront.

In 2005 faced with the problem of price hike AP government sought the intervention of Monopoly and Restrictive Trade Practices Commission. In the last five years, even the AP, MP and Gujarat State Governments have faced serious problems with the Seed Companies in regulating the Seed sale in the state in fulfilling its objective of helping farmers to have timely access to good quality seed in adequate quantities.

1. In 2005, after establishing the large scale cotton crop loss was due to seed failure in Warangal dist, the State government asked Mahyco to pay compensation. This company refused to pay and moved to AP high court on paying compensation saying state govt. is harassing them. AP High court orders also were in favour of Mahyco and till date the company has not paid the compensation Mahyco refused to the compensation.
2. In 2006, after MRTP commission's ruling to reduce the bt cotton seed price, AP government reduced the cotton seed prices to Rs. 650 and Rs. 750 for bollgard I and II. Challenging this, MMB moved to Delhi high court on this issue.
3. In 2007, when Agriculture officers in Warangal district found that Mahyco *Bt* hybrids were being sold in Warangal market with being approved by the state government. they raided and seized the shop. Mahyco challenged that cotton seed was removed from Essential Commodities Act, hence Seed control order which draws powers from EC Act does not apply to cotton. At this juncture, AP government made a new act to regulate 'Transgenic cotton seed in AP'. However, all these Acts, including Seed control order, will be repealed once the 2010 bill is passed, there by taking away the rights of the farmers and also the powers of the State government.
4. In 2010, Monsanto filed case in AP High Court requesting to stop state govt. from reducing the royalty arguing that it does not have any power to do so. The case is still pending in the court.

Changes Proposed by Government of A.P.

The State of Andhra Pradesh is the seed hub of our country and most of the varieties of seeds are being produced in the State of Andhra Pradesh. Even though, the State Government has introduced a Draft A.P. Seed Bill 2004, the same has not been approved by the Centre since the Central Government proposes to introduce Seed Bill 2004 in the Parliament. Since the Government of A.P. felt that some of the provisions of the Seed Bill 2004 are not farmer friendly, made certain proposals for amendment to Seed Bill 2010. In fact, the Bill does not provide any independent powers to the State Governments to deal with or regulate the seed producers and also with regards to registration of seeds. Since no control is provided to the State Governments over the seed producers, the objectives of the Act cannot be achieved. The Government of A.P. also felt that there must be definite provisions for price regulation in the Bill and without that no purpose would be achieved under the Act to protect the farmers. The Government of A.P. also suggested for fixation of maximum retail price and introduction of provisions to that extent in the Bill.

Views of Centre for Sustainable Agriculture and its Suggestions on the Seed Bill 2004

National Seed Act should not be looked at in isolation and it should be looked in conjunction with other legislations like PVPFR Act, 2001, Biological Diversity Act, 2001, Environmental Protection Act, 1986 *etc.* Any legislation on seeds should first and foremost look at rights of the farmers and they should be protected. Certain provisions of Seed Bill 2004 are quite contrary to provisions of PVPFR Act 2001. Certain rights granted under the PVPFR Act are taken away and abridged under the present Seed Bill. Discourage the private participation in seed testing and certification since it leads to collusion and there by encouraging spurious seed in the market. Facilitative climate for growth should not be at the expense of the farmers. Facilitative growth cannot be interpreted as lax regulation.

Without resolving issues related to ownership of seed resources, boosting exports would only facilitate and legitimize bio-piracy. The in its current form the Bill seeks to violate the rights of the farmers in many ways. The Bill demands the farmer to compulsorily register themselves as seed traders if they are want to sell seeds. This is violation of farmers rights which was protected in PVPFR Act. Apart from that there is serious incompatibility among the Seed Bill 2004 and other related legislations also. The Bill further demands that the farmers' seed must conform to the minimum standards laid down in the Bill and the same is infringement of traditional rights of the farmer in dealing with the seeds.

There are several contradictions between the 2004 Bill and PVPFR Act, 2001. The 2004 Bill is an attempt to take away certain protections and rights provided to the farmers under PVPFR Act. The Bill failed to introduce an effective mechanism to deal with the compensation cases of the farmers in case of failure and proposed that appropriate mechanisms to estimate and deliver adequate compensation should be defined.

The provisions for two years provisional registration for transgenic variety of seeds since is a clear violation of provisions of Environmental Protection Act, 1986 and the Rules made there under. We suggest for appropriate changes for the provisions of registration of seeds and varieties. It recommends the system of licensing instead of registration. It also recommends for decentralization of powers between the Centre and the States for effective implementation of the provisions of the Bill. The system of centralizing seeds regulatory mechanism excludes the authority of States in approving the varieties that are suitable for their locations. License should be granted for each variety between 3 to 5 years and the same is renewable basing on the performance.

We Suggest Introduction of Required Provisions for Seed Insurance

The penalty clauses mentioned in the Bill do not create any deterrent or fear in the minds of the spurious producers and seed companies and the said provision cannot control the erring companies and producers unless the penalties are enhanced to maximum extent. We also recommend for increase of criminal liability, compensation and for introduction of provisions for price control *etc.* It also suggests for effective powers to the State Governments for proper implementation of the provisions of Seed Bill.

A delegation from Andhra Pradesh including the Minister for Agricultural of the State, all party MPs, farmers organizations and Civil Society representatives has represented for certain changes in the proposed bill. Taking into consideration all these things the Union Government organized a meeting at National level on 28.06.2010 at New Delhi providing opportunity to all the farmer unions, women groups, NGOs *etc.*

Then the Agricultural Minister Sri Sharad Pawar also suggested several amendments to the Seed Bill 2004. Later on 18.10.2010 the Union Cabinet approved the Draft Seed Bill 2010 and in November, 2010 the said Bill with some amendments was slated for reintroduced in the Rajya Sabha. Apparently the Government has not accepted most of the principal amendments proposed by many of the Members of the Parliament and also not taken into consideration the suggestions made at national consultation held at New Delhi. The principal amendments suggested by the Union Minister of Agriculture includes appointment of compensation committee, reduction of tenure of registration, amendment to the **"farmers"** definition, excluding the farmers from registering their varieties in the register, recognition of seed certification agencies situated in out side India *etc.*

In these circumstances for regulating the seed price and providing the seeds at reasonable prices to the farmers the following steps may be taken to include in the Seed Bill 2010:

a) The State Committee must be provided with powers to collect the data and review the prices of the seeds registered and sold after their approval.

b) The Central Government should frame seed price control order dealing with several aspects for regulation of prices by fixing a maximum retail price.

c) The State should have an apex body to regulate the price of the seeds

d) The royalty/trait fee charged by the companies over the patents/proprietary technologies should be regulated. Any such royalty/trait fee on all traits/patents should not exceed 5 per cent of the total cost of the seeds.

e) Seed bill should have a liability clause in case of failure and define a simple redressal mechanism for farmers.

f) The role of the State Government in implementing the Seed Act is not defined or specified in the Bill. The State Governments are almost power less to implement the provisions. Its role is only an advisory.

g) Further, there is no proper representation in the Central Seed Committee from the States. Even though, the agriculture is included in the State list of the VII Schedule of the Constitution, no effective regulatory powers are entrusted with the States for effective implementation of the Seed Act.

Seeds Bill to be Opposed in Rajya Sabha

Having failed to reach an agreement with the government on the price control mechanism, Opposition parties, including the Left parties, Samajwadi Party and Telugu Desam Party, plan to oppose the contentious Seeds Bill, 2004 that the government is going to move in the Rajya Sabha on Wednesday for consideration and passing. Agriculture Minister Sharad Pawar has given notice for at least 30 amendments to the Bill on which there are major differences between the government and the Opposition members on the price control mechanism and the penalty (for failure of seed) clause. The Bill seeks to regulate the production, distribution and sale of seeds and requires every seller of seeds, including farmers, to meet a certain minimum standard.

Farmers' organisations have opposed the Bill, which they say, has opposed the substantive amendments proposed by the Parliamentary Standing Committee. Mr. Pawar's meetings in July with MPs failed to reach an agreement on price regulation. "The amendments proposed by the government seriously compromise the rights of farmers and will lead to unrestricted commercialisation of varieties in the public domain including farmers' varieties," the All-India Kisan Sabha said here.

UPA Ministers Back Nitish Opposition to Seed Bill

A section of Congress is set to back the demand of chief ministers of BJP and other Opposition-ruled states to reject the Seeds Bill, 2010, in its present form. Bihar Chief Minister Nitish Kumar, spearheading the campaign against the bill, has told Prime Minister Manmohan Singh and Agriculture Minister Sharad Pawar that the bill, whose avowed purpose is to facilitate production and supply of seeds of quality, will put the peasantry at the "mercy of agri-businesses." This stand of Kumar found backing of a section within the Government and members of the National Advisory Council. Ministers like Jairam Ramesh are against rushing into the use of GM crops without adequate safeguards.

In separate letters to the Prime Minister and Pawar, Nitish Kumar said the act should not be passed by Parliament without according the Seed Committee the power to decide on the price of the seed. "Out experience with private companies producing hybrid maize seeds underscores this. While public sector companies like NSC sell seeds at less than Rs. 100 per kg, the cost of private seeds run into hundreds and thousands of rupees." Kumar said the Centre's attempts to intervene in an area that falls under state's domain – agriculture is a state subject – is unacceptable. "State governments should have the authority to fix the retail seed price and royalty charges. State governments should also be empowered to register seeds that are locally suitable and appropriate".

"In addition, authorisation of seed production, processing, storage, distribution and sale should be with the state government. If seeds have to be registered with the National Register of Seeds, it is imperative that state governments must be given the authority to decide on which of these registered seeds can be licensed to be used in their state," Kumar said. The support for the tough stand against is certain to make its passage difficult. "We cannot back this bill as it would damage agriculture and the goal of food security," said a Congress leader.

Farmers Advised to Prepare Wheat Seed at Home

The Agriculture Department has appealed to the farmers to prepare their own wheat seed for the next season instead of shelling out money to obtain it from outside agencies.

Harvesting was crucial for producing quality seeds. The crop should be harvested at optimum maturity with less humidity content. Early harvesting with more humidity content robbed the seeds of its germination ability while late harvesting raised the chances of fungi attack. Therefore, harvesting the crop for seed purpose should be done at the right time.

Early harvesting, when the grain has a lot of moisture content, has an adverse impact on the germination ability in seeds while late harvesting increase the chances of fungi attack. Therefore, harvesting the crop for seed purposes should be done at the right time.

The farmers must choose the best quality of crop and remove weeds and diseased plants. On storage, harvested seeds should be cleaned and dried in sunlight to bring down the moisture content below the 12 per cent mark to avoid a fungus attack.

The farmers should also remove the broken and damaged seeds before storage. They should treat the seeds with an insecticide and store it in airtight containers or new gunny bags. The seeds should be kept in a cool place, away from direct light and humid conditions.

Important Tips

- The crop should be harvested at optimum maturity when the grains have the least humidity content.
- The harvested seeds should be cleaned and dried in sunlight to bring down the moisture content below the 12 per cent mark avoid to prevent a fungus attack.
- Farmers should remove broken/damaged seeds before storage, treat these with insecticide and store these in alright containers or new gunny bags.
- The seeds should be kept in a cool place, away from direct light and humidity.

Some Important Seed Companies

JK Agri Genetics Ltd. (Hyderabad), Part of JK Organization

Date of MoU : August 17

Crops: Maize, Bajra, sorghum, cotton, fodder beet and vegetables

Main Feature: Training in DNA fingerprinting for testing varietal purity: company to provide cash funding and access to certain resources and technology.

Krishidhan Seeds Ltd. (Pune) part of Karwa group of Industries

Date of MoU: September 1

Crop: Maize, wheat, bajra, castor, mustard, legunes, soybean and sorghum

Main features: company to provide cash funding and access to certain resources and technology

Mossanto India Ltd. and Monsanto Holdings Pvt. Ltd. Subsidiary and Majority Owned Firms of Monsanto USA

Date of MoU : July 27

Crops: Hybrid maize, cotton, vegetables

Main features: production of hybrid cotton seeds on farmers' fields and scale up as feasible; fair chance for partnership in government's seed distribution schemes, evaluate feasibility of company's protected culture seeds, participation in state's centre of excellence at Bassi, to get land and facilities from government as per activities

PHI Seeds Ltd. (Hyderabad) part of DuPont business

Date of MoU:

Crops: Bajra, mustard, soybean and sorghum

Main features: Establish DNA fingerprinting facilities for testing varietal identify: postharvest training in fodder banking, company in provide cash funding and access to certain resources and technology

Advanta India Ltd (Hyderabad) associate company of UPL

Date of MoU: September 23

Crop: Mustard, maize and bajra: fodder crops, vegetables

Main features: Breeding, development of hybrids with improved agronomic parameters; company to provide cash funding and access to certain resources, technology

R&D in Agriculture

India needs to further liberalize its rules for research in agriculture. Any research in agriculture will need land. But it is still very difficult in India to get licences for research plots to breed seeds. There are also several restrictions on the material that can be tested here which need to be streamlined. Though a lot of areas have been opened up, improvements can be made. DuPont has aggressively made research in agriculture its top priority. In 2009, it ploughed half the total annual R&D budget of $1.4 billion into increasing global food production. Seeds and crop inputs are DuPont's top money spinners and generate a third of its $700 million annual sales. The company is already expanding its two-year-old research hub in Hyderabad that focuses on new products for use in agriculture and food, infrastructure, renewable energy and automotive industries. Subsidiary company Pioneer Seeds is in the second year of testing new corn seed that has higher resistance to pests and herbicides. DuPont is open to collaborating with local Indian companies that are developing cuttingedge technology and is always scouting for likely targets. The company aims to become a $1-billion player in the Indian market in the next two years even though it faces tough competition and margin pressure from other multinational giants such as Monsanto, Bayer and Syngenta that licence their technology to local players. India is currently among the top 10 global markets for the diversified American multinational.

"DuPont has identified three mega trends that contain the biggest opportunities and challenges of the future. These are increasing food production; decreasing dependence on fossil fuels, and protecting people and the environment. By sustainability we mean products that are good for the environment and also greater longevity of the solutions that we offer our customers. Chemicals are now a smaller and smaller part of their business. Agriculture also came to the rescue of the company that mostly sells to other companies and enterprises, during the global economic recession last year. "Unlike downturns in the past, this time it was a precipitous change over barely 12 weeks and you couldn't balance anything off each other.

The Seeds Bill, 2011: Some Reflections

Seed is a carrier of technology and the medium for translating scientif-ic achievement to the field. Often, Indian farmers do not distinguish be-tween grain and seed (Chand 2007), and use common farm produce as seed. The reasons for this are: (1) lack of awareness about the potential of quality seed, (2) the non-availability of good quality seed, and (3) high seed price. To a greater extent, this also explains the large gap between attain-able levels of productivity achieved in front line demonstration plots and the actual productivity at farm levels. This article examines the trends in sup-ply of good quality seed and hybrids, esti-mates the seed demand and supply gap and discusses the government response ad-dressing issues of seed production and availability through legislative measures, namely, the new seeds bill. It is noteworthy here that the draft of the seeds bill has been revised three times in last seven years (2004, 2008, 2010) mainly to accommodate the concerns expressed by farmers bodies, civil society and parliamentarians, and the latest 2011 version is still pending in the Upper House of Parliament. The article discusses the latest amendments proposed in the bill and their implications for improving farmers' access to quality seed.

1. Access to Quality Seed

The seed scenario in India is undergoing rapid change. During 2003-04, the private sector produced 62.8 lakh quintal and public sector 69.5 lakh quintal of seed. Seed production by the private and public sectors in the next seven years increased to 109 and 171 lakh quintal, respectively. However, it is important to point out here that the share of the private sector in total seed production does not portray the correct picture of market share. Its share in the market value of seed is much higher than the share in volumes as the major focus of the private sector has been on selling seeds for vegetables and hybrids. These fetch higher prices (Ramaswami 2002). Further, the private sector has a strong interest and dominance in the sup-ply of proprietary seed-like hybrids, as can be seen from the information presented in Table 1. Though accurate and detailed data is not available, interaction with the National Seed Association of India and information available in the annual reports and websites of the major seed companies together indicate that the private sector has developed three to 22 times the number of crop hybrids marketed by the public sector till 2001-02.

Table 10.4: Number of Hybrids in Major Field Crops Developed by the Private and Public Sector in India

Crop	*Till 2001-02*		*2002-03 to 2009-10*		*Total*		*Share of Private Sector Hybrids in Total Hybrids (Per cent)*
	Private Sector	*Public Sector*	*Private Sector*	*Public Sector*	*Private Sector*	*Public Sector*	
Cotton	150	15	43	10	193	25	88.5
Maize	67	3	36	25	103	28	78.6
Paddy	12	4	11	15	23	19	54.8
Wheat			3	0	3	0	100.0
Pearl millet	60	6	22	7	82	13	86.3
Sorghum	41	5	12	8	53	13	80.3
Pigeon pea			1	2	1	2	33.3
Soybean			2		2	0	100.0
Sunflower	35	6	13	10	48	16	75.0
Jute				23	0	23	0.0
Mesta				11	0	11	0.0
Castor			4	9	4	9	30.8
Green gram			1		1	0	100.0
Mustard			11	1	11	1	91.7
Safflower				2	0	2	0.0
Total	365	39	159	123	524	162	76.4

Source: Seed Association of India, 2002; ICAR research institutes' websites; annual reports and websites of major seed companies (JK Agri-Genetics, Nath Seeds, Monsanto India, Syngenta, Mahyco, Adventa, Nuzhiveedu, Indo-American, Ankur, Rasi and Ganga Kaveri Seeds).

There is discernible change in the role of the public sector in the development of hybrids after 2001-02 in all types of crops **(Table 10.4).** Till 2001-02, the private sector had developed 150 hybrids of cotton com-pared to 15 by the public sector. Similarly, the numbers of maize hybrids developed by the private and public sectors were 67 and three, respectively. In the next eight years (2002-03 to 2009-10), however, the share of the public sector increased from 8 per cent to 19 per cent in cotton, 4 per cent to 40 per cent in maize and 25 per cent to 58 per cent in rice. Similar changes are observed in the case of other crops also. The sum total of hybrids developed so far by the public and private sectors (with the limited information that is available for the latter) and also the share of the private sector in hybrids of different crops is pre-sented in the last three columns of Table 10.1. Cotton and maize have been the favourite crops for development of hybrids by both the public and private sectors. The private sector has also shown a strong interest in pearl millet, sunflower and sorghum. Con-sidering all crops together, the private sec-tor accounts for three-fourths of the total hybrids developed so far in the country.

The data for the last eight years, *i.e.*, after 2001-02, shows that the gap between the private sector and public sector in the development of hybrids narrowed down considerably compared to earlier, even though the private sector continued its dominance in crops like cotton, maize, pearl millet and sorghum. These are either cross-pollinated or amenable to economic production of hybrid seed. It is interesting to point out that even in the case of self-pollinated crops like paddy and wheat where the development of hybrids is quite cumbersome and costly, the private sector has ventured to develop and supply improved and hybrid seeds wherever it found commercial markets for these crops (like paddy in Andhra Pradesh and Haryana)

In order to meet farmers demand for good quality seed and to maintain competition in the seed and to maintain competition in the seed market, it is essential to maintain the same growth tempo witnessed during last five to six years in the public sector.

2. Demand for Quality Seed

To understand the size of the domestic market for quality seed, the total seed requirement for various crops was esti-mated by multiplying per hectare seed used by farmers by area under the crop in the country. Per hectare seed requirement for the selected crops was estimated by taking the weighted average of seed applied by sample farmers included in the Ministry of Agriculture's Cost of Cultiva-tion Scheme in various states, using area under the crop in that state as weight. This was then compared with total quan-tity of seed supplied by the private sector (other than farmers themselves) and the public sector to get an idea about the adequacy of quality seed. The ratio of quality seed distributed vis-à-vis seed requirement at the national level is pre-sented in **Table 10.5**. In the case of self-pollinated crops, farm-saved seed is supposed to last for three to four generations. However, the yield potential cannot be realised if due care is not taken regarding varietal and physical purity, seed health and vigour. Farm-level surveys and interactions with farmers reveal that farmers prefer to buy fresh seed every year, even of the self-pollinated crops, provided it is of high quality (Singh *et al.*, 2008). Thus, to make the analysis more realistic, the supply gap for quality seed for cereals, pulses, oilseeds and fibres in the country is esti-mated under three scenarios corresponding to 25 per cent, 33 per cent, and 100 per cent seed replacement rates (SRR). These estimates show that even for achieving the minimum SRR of 25 per cent at the national level, there is a deficit of quality seed to the tune of 36.4 lakh quintal for cereals, 7.8 lakh quintal for pulses and 12.7 lakh quintal for fibre crops. However, a closer look at seed demand for individual crops shows that we have adequate quantity of quality seed to achieve more than 25 per cent SRR for maize, pearl millet, rapeseed/mustard, sunflower, soybean and cotton (Table 10.2). When fresh seed is used every third year (SRR=33

Table 10.5: Seed Requirement and Gaps in Demand for Quality Seed in India (2008-09)

Crop	*Seed Requirement (Lakh quintals)*	*Quality seed Distributed (Lakh quintals)*	*Ratio of Seed Distributed/ Requirement (col.2/col.1)*	*Gaps in Demand for Quality Seed*		
				Scenario-I (SRB= 25 per cent)	*Scenario-II (SRB= 33 per cent)*	*Scenario-III (SRB= 100 per cent)*
Paddy	323.79	58.18	0.18	22.77	49.75	265.61
Wheat	367.71	74.83	0.20	17.10	47.74	292.88
Maize	18.77	7.94	0.42	–3.25	–1.68	10.83
Sorghum	10.20	2.41	0.24	0.14	0.99	7.79
Pearl millet	4.93	2.20	0.45	–0.97	–0.56	2.73
Barley	8.02	1.62	0.20	0.38	1.05	6.40
Ragi	1.90	0.25	0.13	0.22	0.38	1.65
Cereals	*735.32*	*147.43*	*0.20*	*36.40*	*97.68*	*587.89*
Gram	56.39	8.60	0.15	5.50	10.20	47.79
Pigeon pea	5.43	1.09	0.20	0.27	0.72	4.34
Green gram	6.37	1.23	0.19	0.36	0.89	5.14
Black gram	7.22	1.37	0.19	0.43	1.04	5.85
Lentil	7.25	0.59	0.08	1.22	1.83	6.66
Pulses	*82.67*	*12.88*	*0.16*	*7.79*	*14.68*	*69.79*
Rapeseed/mustard	3.81	1.63	0.43	–0.68	–0.36	2.18
Groundnut	64.02	15.90	0.25	0.10	5.44	48.12
Sunflower	1.12	0.80	0.71	–0.52	–0.43	0.32
Soybean	71.29	20.89	0.29	–3.07	2.87	50.40
Safflower	0.41	0.09	0.22	0.01	0.05	0.32
Sesame	1.06	0.18	0.17	0.09	0.17	0.88
Niger seed na		0.37	Na	na	na	na
Oilseeds	*142.07*	*39.49*	*0.28*	*–3.97*	*7.87*	*102.58*
Cotton	3.65	2.27	0.62	–1.36	–1.05	1.38
Jute	57.17	0.28	0.005	14.01	18.78	56.89
Fibres	*60.83*	*2.55*	0.04	*12.66*	*17.73*	*58.28*

na: Not available.

Source: Authors' calculations based on Commission for Agricultural Costs and Prices (2010) and Ministry of Agriculture (2010: 296).

per cent), the gap between seed requirements and the present level of availability increases to 112.4 lakh quintal for foodgrains and 17.7 lakh quintal for fibre crops. Additional seed requirement in the third scenario, that is when fresh seed is used every time, rises to 587.9 lakh quintal for cereals, 69.8 lakh quintal for pulses, 102.6 lakh quintal for oilseeds, 56.9 lakh quintal for jute and 1.4 lakh quintal for cotton.

In this scenario, seed requirement for paddy and wheat is estimated to be 265.6 and 292.9 lakh quintals, respectively. These estimates show that there is huge market demand for good quality seed, not only for hybrids but also self-pollinated crops. Except for crops like maize, pearl millet, sunflower, cotton, hardly one-third of demand for quality seed is met by present availability of seed. It needs no emphasis that use of low quality seed adversely affects crop productivity, and farmers would prefer 100 per cent seed replacement if quality seed is available at reasonable prices. Since the recent Seeds Bill 2011 aims at ensuring access to quality seed, the next section assesses its salient features and implications for the farm sector

3. Seeds Bill 2011

The very first draft of the Seeds Bill in 2004 came under a lot of criticism from farmers, civil society and parliamentarians. Subsequently, the Government of India referred the Seeds Bill 2004 to the Parliamentary Standing Committee on Agriculture (PSCA).4 Incorporating many of the PSCA recommendations, a Seeds Bill 2010 was again introduced in the Rajya Sabha in April 2010.5 Meanwhile, many state governments (particularly Andhra Pradesh, Kerala, Madhya Pradesh and Odisha) sought further amendments in the bill. Some of these suggestions from the states have been incorporated in the revised Seeds Bill 2011, introduced in the Rajya Sabha in February 2011. The revised bill is still awaiting discussion and approval by Parliament.

4. Areas of Concern

Though the proposed bill with the latest amendments addresses most stakeholder concerns, there are still some grey areas which need closer examination.

4.1 Quality Seed

Although the central aim of the proposed bill is to regulate the quality of seed for sale, import and export and to facilitate production and supply of quality seeds, nowhere does the bill define what quality seed is. Even if it can be discerned that quality seed will have to meet all the prescribed criteria of germination, genetic and physical purity and seed health, it would be more prudent if standards for quality seed are specifically defined in the proposed bill.

4.2 Seed Price Regulation

There are lots of apprehensions on how the proposed bill will ensure supply of quality seed at reasonable prices. In the absence of adequate competition in the seed sector, a few seed producers may be tempted to raise seed prices in order to realize their research and development (R&D) investments in the shortest possible time. Increased privatisation of the seed industry is found to be associated with enhancement in seed prices, which is likely to continue. The cost of establishing and maintaining plant variety rights will further add to seed cost. The moot question, therefore, is whether the proposed bill should have provision for regulating seed price? Are there other effective ways of ensuring supply of quality seed at reasonable prices?

Before jumping to conclusions on this issue, one should remember that the private sector will invest in seed development and delivery only when it finds a favourable institutional and policy environment for expected returns on its R&D investment. If the public sector alone can meet farmers' demand for quality seed, price regulation would not be a serious concern. But both the public and private sectors have a role, in some cases quite distinct, in meeting future requirement of seed. Past experience shows that the public sector has an edge over the private sector in terms of seed R&D, particularly for self-pollinated crops, while the private sector has been more proactive in meeting the demand for hybrid seeds. While there are institutions like the Competition Commission of India to deal with situations of monopoly control in seed industry, one of the effective ways of ensuring a reasonable seed price would be to create conditions for competition among private seed providers, and in some cases, even with public seed agencies. If a competitive environment alone does not serve the purpose of keeping a check on seed prices, and the market fails to meet farmers' demand for seed at reasonable price, the legislation could be amended in the future to provide for a seed price regulator.

4.3 Penal Clauses

Initially, the penalties for contravening the provisions of seed bill were very low and it was dubbed a toothless law with an unwritten licence for violation, malpractice and cheating (Bala Ravi 2010). The latest version stipulates that any person who sells seed which does not conform to the prescribed standards will be punishable with a fine ranging from Rs. 25,000 to one lakh. Further, the penalty for furnishing false information relating to seed standards, misbranding the seed, supplying spurious seed or selling non-registered seed, will attract a prison term of up to one year or fine up to Rs. 5 lakh or both. Although raising the penalties would keep a check on fly-by-night operators looking for windfall gains, particularly in the event of seed scarcity, the amount of penalty can be an effective deterrent only when it is linked to the volume of (seed) business.

4.4 Relevance of Other Legislations

The proposed bill does not seem to be in harmony with the Protection of Plant Varieties and Farmers' Rights Act (PPVFRA), 2001 which provides for compulsory licensing (Section 41) if reasonable requirements of the public for seeds or other propagating material of a particular variety have not been satisfied. Therefore, it would be logical if the registration of seed is linked to varieties protected under the PPVFRA. In the absence of such harmonisation, there is a room for unscrupulous seed producers/dealers to get useful seeds registered through the route of the Seeds Bill 2011, without sharing the benefits, as provided under the PPVFRA, with those who make efforts in evolving and conserving plant genetic resources.

5. Conclusions

It is widely accepted that the public sector alone cannot meet all the seed demand of our farmers. The private sector has managed to harness the benefits of technological innovation in some segments of the seed sector, as is evident particularly in Bt-cotton, maize and pearl millet hybrids. Further, the private sector has an edge over the public sector in capturing market signals and delivering the desired (seed) product in time. However, there have been instances of fleecing of farmers through supply of spurious and low-quality seeds and charging of exorbitant prices by the private sector. Thus, there is a strong need to promote competition in the seed sector to fill the gap between demand and supply of quality seed and to ensure supply reasonable prices. The proposed Seeds Bill 2011, with its central aim of enhancing production and supply of quality seed to meet future requirements for sustainable agricultural production, is an attempt in this direction. It addresses most concerns of stakeholders, except some grey areas which need closer examination.

The seed legislation should avoid excessive curbs on private sector activities. Through competition, effective monitoring and strong regulation, it should check the tendency to acquire control of the market through monopoly and cartelisation. If a competitive environment alone does not serve the purpose of keeping a check on seed prices, the seed legislation may be amended later to incorporate a provision for establishing a seed price regulator. It seems that the proposed legislation has got stuck in the never-ending process of consultation and accommodation of various (often conflicting) stakeholder concerns. As a result, the larger issue of ensuring farmers' access to improved agricultural (seed varieties and hybrids) technology, for raising and sustaining agricultural productivity in the country, is getting sidetracked.

As of now, the unreasonable delay and indecisiveness on these important issues is doing more harm than good. Therefore, the Seeds Bill 2011 should become an act at the earliest to avoid more delay in meeting the requirement for quality seed and realising the yield potential of improved agricultural technologies.

Seed Bill 2014 approved by National Assembly Panel

The National Assembly's Standing Committee on National Food, Security and Research Wednesday approved Seed Amendments Bill 2014 to improve the existing law, enabling it to meet the requirements of modern seed industry. The committee that met with Malik Shakir Bashir Awan in the chair here decided to increase the proposed penalty to Rs. 200,000 on sale of sub-standard seed on first offence and the amount would go up to Rs. 600,000 on the second offence.

Secretary Ministry of National Food Seerat Asghar told the meeting that the existing Seed Act 1976 does not commensurate with the requirements of modern seed industry. Thus there was a need to make some amendments to the law to make it suitable for the modern industry. Secretary Ministry of National Food, Security and Research Seerat Asghar told the National Assembly's Standing Committee that the Bill has been pending since 2007. Lately, it was introduced in the National Assembly on August 8, 2014 after the approval of the Federal Cabinet on May 15, 2014. Subsequently, it was referred to the National Assembly Standing Committee on NFS and R.

The meeting was further informed that the Bill clearly delineates the role of private sector (seed companies, dealers, processing units and fruit plant nurseries) and allows the private sector to produce basic seed for its multiplication and certification. Additionally, it allows the private sector to establish accredited seed testing laboratories. The bill authorizes the registration of Genetically Modified crop varieties (GM) subject to law.

The meeting was informed that Seed Act, 1976 was enacted by the Parliament to control and regulate the quality of seeds of various varieties of crops. The Act envisages public sector as the sole player in seed development and registration process with no role assigned to the private sector. The Seed Act 1976 does not fulfill the requirements of "Modern Seed Industry". Over the years, the capacity of the public sector has deteriorated and private sector is playing stronger and more vibrant role in the development of seed industry, the world over. The new innovations in hybrid technology and Genetically Modified crops, have transformed the seed industry. The new draft "Seed Amendment Bill, 2014" has been drafted considering the emerging realities of the modern times and will provide level playing field to both, public as well as the private sector.

Bihar's Concerns over Seeds Bill

According to our view the bill in its present form would lead to unrestricted commercialisation of varieties in the public domain, including farmers' varieties. "The proposed bill is not only anti-farmer but also brazenly favours multinationals in the garb of higher productivity. Any attempt to pass the bill in its present form will irrevocably damage Indian agriculture and make the goal of food security a distant dream".

Since the objective of the bill was to provide quality seeds, quality could not be divorced from cost. "The bill may, in fact, lose meaning if seeds are not available to farmers at affordable prices. With private companies, particularly multinationals, joining the field, cost becomes relevant. Our experience with private seed companies producing hybrid maize seeds underscores this".

Last year, the Bihar government had to give about Rs.61 crore as compensation to maize farmers when the private hybrid seed failed to form grain. "The problem of the non-setting of grains was not observed in public sector hybrids. This led to great distress among farmers, but they disowned their responsibility". "Our concern about rushing into the use of genetically modified [GM] crops without adequate safeguards is well-known. The proposed bill should lay down these safeguards explicitly. GM seeds should be registered only after extensive research to address concerns like adverse effects on biodiversity, ecology and human health, with dissemination of findings in the public domain. Seed imports should be allowed only after pest risk analysis and local adaptability assessment. There is a need for a Liability Clause to be introduced that makes seed exporters responsible for any pest outbreak and clean-up operations".

While public sector seeds were affordable, "the cost of private seeds runs into hundreds of rupees or even several thousands, in case of hybrid vegetable seeds." Besides, the provision of re-registration would increase the monopoly of seed companies for at least 20 years, he said, adding that the government should take responsibility for seed certification.

Compensation

Lamenting that no criterion had been laid for determining compensation to farmers for seed failure, it should be fixed at the difference between the value of expected produce as per the seed producer's claim and the actual produce. He said there should be a time-limit for payment of compensation, and that the definition of farmers – as described in the National Commission on Farmers – should include landless farmers cultivating leasehold lands.

"The onus of paying compensation should be determined clearly as stated in the bill," the penalties proposed in the bill were "trivial." Raising the issue of agriculture being a State subject, a wider role for States in regulation is required.

"State governments should have the authority through the bill to fix the retail seed price and royalty charges. This is the in light of the fact that apart from spurious seeds, it is also the high market price of seeds that contributes to the complex situation of the current agrarian distress, which results in farmers' suicides," it is desired that wanted all the powers of the Seeds Control Order of 1983– which will get nullified once the Seeds Bill is passed – to be incorporated in the proposed bill.

World's Top Seed Giants: Proprietary seeds account for 82 per cent of the commercial seed market worldwide

Monsanto (US)	*$4.9bln	23 per cent
DUPont(US)	$3.3bln	15 per cent
Syngenta (Switzerland)	$2 bln	9 per cent
Groupe Limagrain (France)	$1.2 bln	6 per cent
Land O'Lakes (US)	$917mln	4 per cent
KWS AG (Germany)	$702mln	3 per cent
Bayer Crop Science (Germany)	$524mln	2 per cent
Sakata (Japan)	$396mln	<2 per cent
DLF-Trifolium (Denmark)	$391mln	<2 per cent
Takii(Japan)	$347mln	<2 per cent
Total of Top 10	$14,785mln	67 per cent

* 2007 sales (share of global proprietary market).

Source: ETC Group.

Is Punjab on its Way to Wasteland?

The water table in Punjab has been dropping by nearly 65 cm every year. Unless the state implements major initiatives on water usage and fixes its canals, it would be facing the reality of barren lands.

Weeks ago, the Punjab agriculture department issued a notification, altering the schedule regarding the transplant of non-basmati and basmati varieties of paddy in the state. It fixed June 15 (substituting June 10) as a cutoff date to begin the transplantation of non-basmati varieties of paddy and July 5 for basmati. However, only a few days later, the notification was put in abeyance by the Chief Minister following pressure from some quarters. The seriousness of the problem regarding Punjab's subsoil and surface water is yet to sink in the minds of decision makers. The objective behind altering the schedule was to further rationalise the water usage but it now stands defeated.

The Green Revolution made the country self-sufficient in food but it also played havoc with subsoil and water. The state's political leadership has been fighting legal battles against its neighbouring states and the Centre for alleged discrimination in the allocation of river waters. The number of blocks where a bore well cannot be sunk without permission has gone up to 45. Such blocks are notified. Of the 138-odd blocks, 110 are overexploited as extraction far exceeds replenishment. Two of these blocks are critical while two are semi-critical. Of all the blocks, only 23 are "safe" as the subsoil water is not worth using for irrigation and drinking purposes due to the presence of excess alkaline in it.

Before the Green Revolution, there were only 26,000 tube wells in Punjab (1965-66). The figure is now 13 lakh. Reckless mining of subsoil water has caused the water table to drop by 65 cm per annum on an average during the past several years. Central Punjab is the worst affected. Over the past four years, there has been some improvement, though negligible, due to the fixing of the paddy transplantation period. The agriculture department says the water table is declining, but its "proportion" compared to the earlier period is down by about 15 cm.

Poor Rainfall

Punjab has not been receiving adequate rainfall over the years. Rainfall was been less than over 20 per cent than the average. Since 1901, there have been 32 deficient monsoon years in the state. Normal rainfall figure in Punjab is pegged at 502 mm. If rainfall is 19 per cent above or less than 502 mm, it is known as normal rainfall and if it is above 20 per cent or more, it is excess rainfall. About 20 per cent below normal is deficient rainfall.

However, there is no serious effort to harvest rainwater. Recently, the Chief Minister had proposed that rain harvesting in 10 marla and above houses should be made mandatory. But it is not clear when this proposal would be implemented. The runoff during monsoon is so rapid that its seepage is minimal. This adversely affects the recharging of the groundwater.

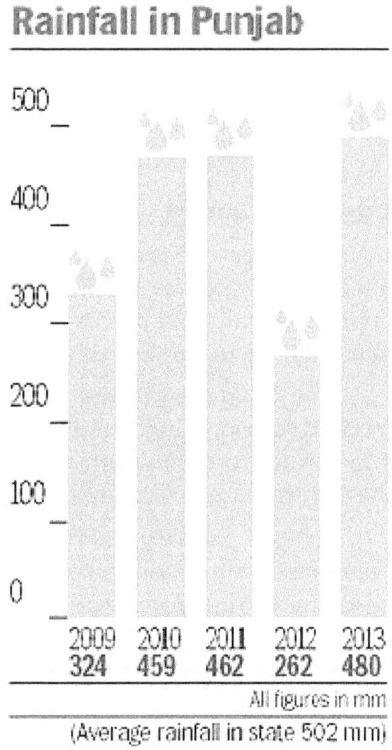

Tube Wells Rule

The state's agricultural economy, particularly the existing cropping pattern of paddy-wheat rotation, largely depends on subsoil water in spite of a good canal network. Of the total 42.90 lakh hectares under agriculture in the state, canal water is accessible to 30.28 lakh hectares. Canal water is available to 24.71 lakh hectares, but to a limited extent as water allowance is not enough to meet the cropping intensity needs (close to three crops) round the year.

Paddy requires at least 22 water doses from transplantation to harvesting stage. It needs watering at least once a week. Also, the water-carrying capacity of various canal systems has come down due to lack of funds for their upkeep. "We need at least 42 million acre feet (MAF) canal water annually to meet our agricultural, industrial, domestic and other needs against the allocated availability of 14.22 MAF from the three eastern river systems," says a senior official of the Punjab irrigation department. Punjab was allocated 14.22 MAF water from 34.34 MAF assessed in three eastern rivers on the basis of the 1921-60 flow series.

As of now, canal water meets less than 30 per cent of the demand. "In real terms, only 23 per cent of the total requirement for the agriculture sector is met from canal water, 72 per cent from tube wells and 5 per cent from rain in the rain-fed belt such as the Kandi area," the official says."A matter of grave concern for the future generation is that we have already started mining water from the second aquifer in the subsoil. We have already approached the dead-end. We are heading towards a barren state because accessing the last aquifer would not be possible for all. Our political leadership should take decisive steps for the conservation of water. Water harvesting, checking runoffs, judicious use of water and recycling of sewer water, *etc.* must be taken on priority," he adds.

Upgrade Canals

The over 100-year-old canal system is in a bad shape. The lining of brick-lined canals such as the Bhakra Mainline Canal and its offshoots has developed depressions at most places and is in a dilapidated condition. Lack of funds for the maintenance of canals has affected their water-carrying capacity.

"Our canal systems require rehabilitation, renovation, extension and modernisation," says a senior official. The state government had sought about Rs. 4,200 crore from the Centre for this purpose and a detailed report has been sent to the Central Water Commission for technical evaluation of the projects.

Being a granary of the country, Punjab needs a robust canal system. The Planning Commission can provide the state with 75 per cent funds of the total project cost. Under the Accelerated Irrigation Benefit Programme, Punjab spent Rs. 120 crore on the rehabilitation of the first Patiala Feeder and Kotla branch. This canal has been lined with bricks. For the relining of the Rajasthan and Sirhind feeders, a project worth Rs. 1,442 crore has been submitted to the Centre. A sum of Rs. 106 crore has been released so far.

Besides this, a project worth Rs. 3,500 crore has been submitted to the Central Ground Water Board, Ministry of Water Resources, for water table management.

Tapping it Right

Solutions are available to check the depleting subsoil water that has gone down beyond 20-30 metres at several places in the state. However, lack of political is not helping. Israel is the best example of how to best use limited water resources.

Micro irrigation: There are no takers for this programme at a big scale even though there is 50 per cent subsidy on the installation of drip, micro sprinkler and sprinkler irrigation systems. In case of small farmers, the subsidy is 60 per cent. But to date, only 28,000 hectares (far less than 1 per cent of the total land under agriculture in the state) have been covered.

Underground pipeline: This scheme is for individual farmers as well as group of farmers. Under it, 90 per cent assistance is provided to a group of farmers under the Rashtriya Krishi Vikas Yojna. To individual farmers, the assistance is 50 per cent. Under the group scheme, 1,27,000 hectares have been covered while individual farmers account for only 7,483 hectares.

Rainwater harvest: The watershed management programme was launched long back to harvest rainwater and recharge water levels in the Kandi area. Over 1,000 rainwater harvesting and recharge structures have been set up in the belt. Farmers use water from these structures to irrigate about 80,000 hectares. Levelling, cleaning: There is a programme to level land, clean ponds and clear drainage as part of water conservation measures. However, these measures have not taken off beyond the initial stage. Sewer water should be recycled and the canals should be maintained.

Regulate usage: The Water Usage Regulatory Authority, which has been set up by most states, is needed. It should regulate water use by various stakeholders. Farmers, who are given free power to operate tube wells, should be brought under the authority to monitor the optimum utilisation of water. It should have the powers to prosecute and fine people.

Diversification: The programme to diversify from water-guzzling crops like paddy should be implemented.

Ensuring Food Safety: Can we Learn from Others?

The mechanisms to manage food safety must be upgraded and expanded. Consumers should be regularly informed about the activities and conclusive actions taken by the regulator.

Food safety is a public good and the unalloyed right of consumers. It is unfortunate that a country that hosts one fifth of the world's consumers offers infinitesimal level of food safety to its citizens. It takes sporadic incidences such as the present Maggi case and the old episode of pesticides in PepsiCo's products to bring food safety to limelight.

But how much attention do consumer rights and concerns for food safety get on a routine basis? Apart from Maggi, there are innumerable packaged food brands which demand attention and check by the enforcement authorities. At present, packaged foods are attracting the curiosity of the policy makers, politicians and the consumers. However, a critical question is that do packaged foods represent the main problem or is there the need to focus the spotlight on the regulation of fruit and vegetable vendors, restaurants, hotels, fast-food chains and above all the large number of street food vendors serving a zillion Indian consumers daily?

The regulatory authorities overseeing food safety face a colossal task of managing the various touch-points along the food supply chain. Ensuring safe food is the responsibility of all including the producers, manufacturers, distributors as well as the retailers. However, the stakeholders in the food supply chain have little incentive to offer safe food because of informational asymmetries. The association of clear responsibility for unsafe food cannot be easily attributed to a single stakeholder or a specific food product because the outcome is observable after the good has been consumed. Also, random monitoring before consumption cannot assure that all the risks are identified and controlled. Therefore, the role of regulatory authorities is primal.

Consumers often want to know how regulatory and enforcement authorities judge food safety, the main factor signifying good food quality. To ensure food safety, the features of the food products are examined against established quality standards. These food safety standards are established by public and private agencies worldwide. The private sector is also proactive in many countries and large retailers like Tesco, Walmart, Carrefour, *etc.* enforce their own safety standards.

It is crucial that we in India learn quick lessons from the efforts of the public and private agencies operating worldwide. USA has the highest per capita consumption especially of packaged food products and the work done by its Food and Drug Administration (FDA) is noteworthy. It systematically examines the domestic and foreign foods in the US market and informs consumers about warnings, recalls as well as refusals in case of imported food products. Apart from this, FDA also provides food safety alerts, warning letters and information related to outbreaks. Further the agency disseminates general guidance to highly vulnerable age groups like children, pregnant women and seniors.

The European Union, the largest exporter and importer of food in the world, tries to ensure regulatory coherence among member countries. Further, a tool called RAFFS (the Rapid Alert System for Food and Feed) is employed to allow rapid sharing of information regarding risks to food safety. It was started as early as 1979 and offers round-the-clock notifications, alerts and border rejections (of imported food). Similarly, Japan, Australia, China and Russia have active networks of food safety organisations backed by their respective governments.

Many would argue that these are developed nations with established systems in place. However, these systems and policy initiatives shaped up after years of public demand as well as supportive governments. Considering the increasing demand of packaged and high value foods in India, the market size for these products is exploding. The mechanisms to manage food safety thus have to be upgraded as well as expanded. The consumers have to be regularly informed about the activities of the regulator as well as the conclusive actions taken by them. A strategic vision of policy makers as well as regulatory agencies is desirable to ensure at-least a minimum level of food safety to our consumers. Finally, the role of Indian consumers, who are the ultimate beneficiaries as well as the ones most affected, will be critical in bargaining for desirable levels of food safety and quality from the producers, manufacturers and the government bodies.

Development Yet to Rural India

The Socio-Economic and Caste Census (SECC) 2011 for rural India belies the claims of India's high growth and its underlying philosophy that the benefits of high growth automatically percolate down. The data also casts aspersions on claims that rural poverty has declined over time. The grim reality painted by the SECC 2011 makes a clear disconnect between high growth and its benefits reaching rural India. In spite of emphasis on inclusive development, the exclusion has been the norm for a majority of rural society.

The SECC, covering 640 districts, reveals that 56 per cent of the 17.91 crore rural households possess no land or property. In 75 per cent households, the monthly income of the highest earning member is less than Rs. 5,000. Less than 10 per cent households have a member earning above Rs. 10,000 per month. More than 90 per cent rural people are without a salaried job. Only 17 per cent households own two-wheelers and 2.46 per cent own four-wheelers.

One-fourth of the households are phoneless. The knowledge base of the rural society is abysmally low. Around 24 per cent households do not have a literate adult above the age of 25. The overall rural illiteracy rate is 36 per cent, which follows a skewed distribution across gender and caste. Only 3.45 per cent rural households have graduate degree holders. Another revelation is that 60 per cent households are deprived households, out of which 21.5 per cent belong to the SC/ST category. The deprivation index is computed on the basis of seven indicators comprising households with only one room, kutcha walls and kutcha roof; no adult member between the age of 16 and 59; female headed households with no adult male member aged between 16 and 59; households with disabled member and no able-bodied adult member; SC/ST households; households with no literate adult aged above 25; and landless households deriving a major part of their income from manual casual labour.

The gulf between rural and urban India widened during the post-economic liberalisation period. The epicenters of the liberalisation are industry, trade and investment. The agricultural sector has not only been bypassed by liberalisation, but also has become the victim due to a decline in public investment in agriculture, terms of trade turning against agriculture, rationalisation and partial withdrawal of subsidies to the sector.

However, the rural-urban divide is not unique to India. The annals of economic history of the developed world, particularly Japan during Meiji Restoration (1868-1912), clearly reveal that for industrialisation, agriculture has paid a heavy cost in the form of taxes. Michael Lipton, a distinguished economist and author of Why Poor Stay Poor: A Story of Urban Bias in Development (1977), has propounded the Urban Bias Thesis (UBT). Lipton argues that urban bias "involves (a) an allocation to persons or organisations located in towns, of shares of resources so large as to be inefficient and inequitable, or (b) a disposition among powerful (urban classes) to allocate resources in this way". Price distortion is at the centre of Lipton's thesis stating agricultural goods are underpriced whereas industrial goods are overpriced and thus the rural sector finances the development of industrialisation and urbanisation. A lively debate on UBT is continuing. Lipton and Robert Eastwood argue that UBT exists in post-liberalism period as better educated and better placed urban people have exploited most of the economic opportunities in the wake of liberalisation.

The opponents of UBT opine that mega projects, like international airports, can only be located in cities. The benefits of cities and mega industrial clusters do not confine only to urban areas. These benefits also trickle down to their rural hinterlands. They also argue that parallel to rural poverty and backwardness, urban poverty and slums are also a reality. Notwithstanding the criticism of UBT, it has made inroads into the arena of public policies, including at the World Bank level.

The impact of price distortions and economic liberalism is partly offset by the rural development policies. In India, the central and state governments have implemented more than 100 programmes for rural development. Unfortunately, the benefits of these schemes could not fully reach the target population mainly due to (i) wrong identification of the beneficiaries; (ii) corrupt practices associated with financial transactions; (iii) diversion of funds by the poor; (iv) non contribution of state share to funds; and (v) lack of political and administrative accountability. These policies, barring the Green Revolution, have also not widened the productive base of the rural economy, and as a consequence, the rural-urban divide has increased.

Thanks to SECC, the writing on the wall is clear. The census has built a strong case for introspection of policies from the rural perspective. For delivering their due to the rural people, a paradigm shift in economic policy-making and budget allocation is needed.

The Central government has replaced the Planning Commission by the NITI Aayog. The state governments should also constitute state NITI Aayogs, electing heads of the local governments on it. This decentralised process of policy-making would incorporate the priorities of people at the grass-roots level and enable fruits of policies reaching the target population. Economist Wallace Oats has built a case for decentralisation by developing a 'Decentralisation Theorem' that states, "for a public good – the consumption of which is defined over geographical subsets of the total population, and for which the costs of providing each level of output of the good in each jurisdiction are the same for the central or for the respective local government – it will always be more efficient (or at least as efficient) for local government to provide the pareto-efficient level of output for their respective jurisdictions than for the central government to provide any specified and uniform level of output across all jurisdictions".

Along with the decentralised policy-making process, the governments should also make liberal investment in rural education, health and infrastructure. Only this shift in budgetary allocations can help the rural sector to reap benefits of the flagship programmes, including Make in India, Skill India, Digital India, Swachh Bharat and Smart Cities. The diversification of the rural economy by promoting non-farm activities, specially rural industries, can also help in generating gainful employment for rural youth. Lipton opines that a given amount of public investment in a small agricultural sector or rural off-farm activities will yield higher returns vis-à-vis same investment in cities or large urban-based industries. Thus the existing big industry model be supplemented by the industry dispersal model, incentivising small and medium enterprises to shift to rural areas. Generation of rural employment and income at this juncture would be a boon to the Indian economy in the form of expanding market. Plugging leakages, promoting transparency and accountability in rural development programmes by making social audit an integral part can improve the outcomes of the programmes.

India Lost Original Forests 70 Times Delhi's Area in 14 Years

Records Show Loss of 10.6 m Hectares between 1999 and 2013

Is India's Forest Cover Increasing? Or is it Shrinking?

The government claims a steady increase in the country's forest turf since 1999. But official data indicate it may be missing the woods for the trees. Records show the country may have lost close to 10.6 million hectares, or about 70 times Delhi's geographical area, of original forests between that year and 2013.

This is more than three times the reported increase in the nation's forest cover in this period and 15 per cent of India's existing green cover. According to the biennial state of forest reports (SFRs) from the Ministry of Environment, Forest and Climate Change (MoEFCC), forest cover in the country went up by 3.3 million hectares between 1999 and 2013.

The Centre and states have been using the numbers from these reports, the latest released last year, to pat themselves on the back for "stabilising" the country's forest cover despite huge demand for land from industry and growing need for food, fodder and timber. The stated spread of forest cover, however, is mostly due to proliferation of plantations outside forest areas, masking a massive deforestation in the natural forests. The disparity came to light from the analyses of the data and the methodology adopted in the past few years by the Forest Survey of India (FSI), the nodal agency of the government that maps forest cover.

Due to technical limitations till 1999, the FSI calculated forest cover only over the traditional or legally notified forest areas. Because there are no digital maps of notified forests in the country, FSI used the regions marked as green patches, also known as "green wash areas", in the government's Survey of India (SOI) topographic sheets as a proxy to such forests.

Green wash areas represent the forest territories in the country at the time of the SOI studies carried out till the 1980s. FSI started digital interpretation of satellite images with improved resolution from 2001. With this, it began estimating forest cover for the entire country without differentiating between the notified forests and the plantations outside.

It now counts all land, irrespective of its ownership, with more than 10 per cent canopy cover over an area of more than one hectare as "forest cover", which now

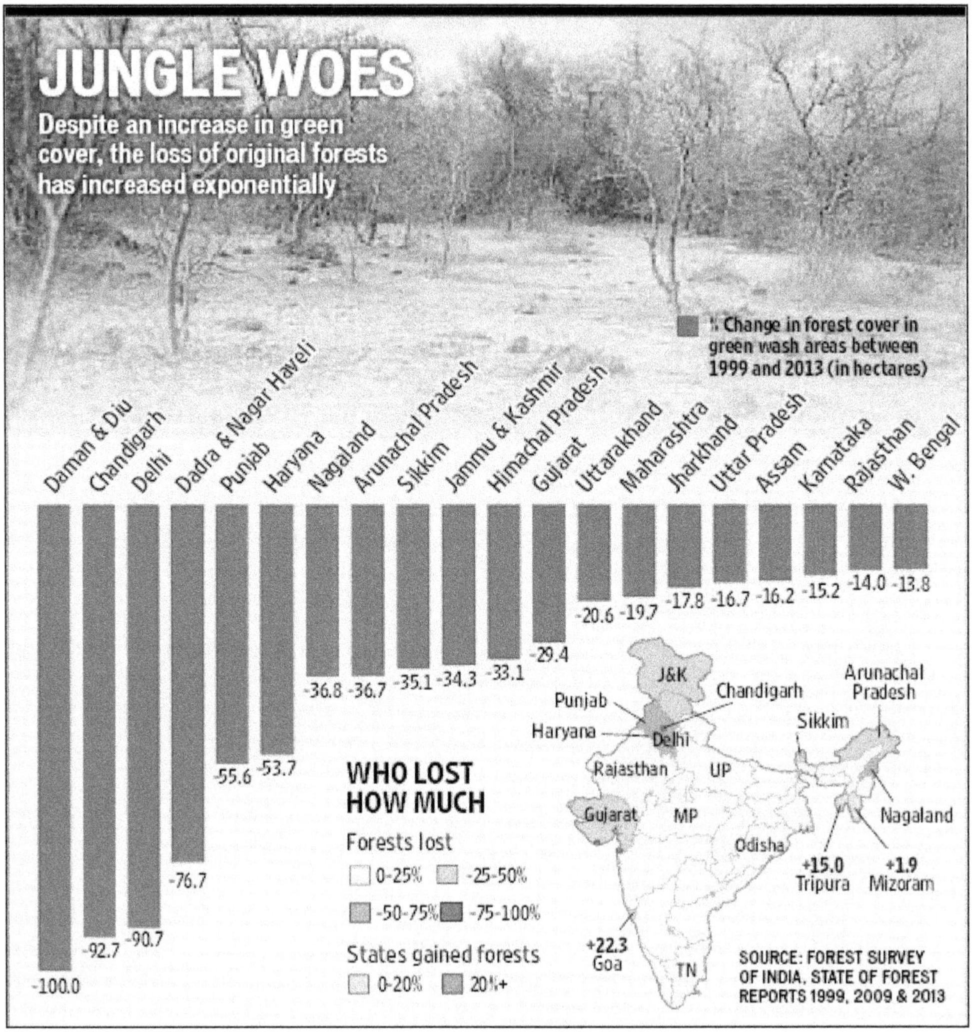

includes all plantations like tea and coffee gardens as well as orchards apart from traditional forests. Estimated through this method, India's forest cover showed a substantial growth over the years. But forestry analysts criticised the methodology as it did not give the idea of how our natural forests were faring.

Responding to this criticism, the FSI digitised the Survey of India topo-sheets and estimated the forest cover in green wash areas afresh in the SFR 2013. As per the report, out of the total 69.8-million-hectare forest cover in the country, 53.08 million hectares lie in green wash areas.

While the 2013 report does not make any reference to the pre-2001 data on forest cover in green wash regions, SFR 2009 contains the forest cover data for the period till 1999 revised for the technological advancements in methodology to make it comparable with the latest figures. According to the 2009 report, forest cover in green wash areas in 1999 was about 63.66 million hectares. This indicates a deforestation of at least 10.58 million hectares in original forests over 14 years. Experts say while any increase in green cover is welcome, the massive loss of original forests in the country should be a cause for concern.

12

Land Acquisition, Rehabilitation and Resettlement Bill, 2013

The Monsoon session of the Parliament saw the passing of the Right to Fair Compensation and Transparency in Land Acquisition, Rehabilitation and Resettlement Bill, 2013 that shall set to be repealing and replacing the century old Land Acquisition Act, 1896 that had been governing Land acquisitions in India. This Bill of 2013 is set to establish a fair and a strong structure for land acquisition in India and lays down a list of legal prerequisites that are mandatorily required to be first discharged before setting in motion the land acquisition process. This Bill aims to provide for fair compensation to the affected land owners and provide for transparency in the process of land acquisition. The primary highlight of this Bill is that it provides for rehabilitation and resettlement of all those persons who have been displaced from their land by such compulsory acquisition and have been adversely affected and impacted from the loss of their only source of livelihood. The present article is aimed to briefly describe some of the key features of the Bill.

Need of passing of Right to Fair Compensation and Transparency in Land Acquisition, Rehabilitation and Resettlement Bill, 2013

The Right to Fair Compensation and Transparency in Land Acquisition, Rehabilitation and Resettlement Bill, 2013 is an important legislation to be enacted by the Parliament this year. This legislation has come up in wake of the recent disputed and contested land acquisitions processes undertaken around the country. This legislation aims to address these issues and bring clarity on the subject of land acquisition in India.

The erstwhile Land Acquisition Act, 1896 was of the nature that mandated land acquisition without providing for adequate safeguard to the displaced land owners. The Act did not provided for any form of proper hearing or even adequate compensation to the land owners who were losing their land to acquisition. The Act of 1896 was also silent on the resettlement and rehabilitation part for displaced landowners and mandated land acquisition under the urgency clause without taking into consideration the interest of farmers and important village communities that were solely depending upon the land for their livelihood.

This bill in contrast introduces participation of government bodies along with panchayats, gram sabha and landowners in the process of land acquisition thereby adopting inclusive approach and involving all concerned people in the matters affecting land acquisition and thereby negating confusion and chaos over land acquisition and encouraging and protecting legitimate acquisitions and interest of farmers and land owners across board.

Some of the Highlights of the Bill 2013

- This Bill provides for a fair and transparent process of land acquisition for public purpose. The Bill clearly defines public purpose and provides for Social Impact Assessment (SIA) to be conducted before issuance of notification regarding acquisition of land. This SIA report is to be examined by an expert body that is to include besides other members, 2 persons representing Panchayati raj institutions. This expert committee shall be deciding as to the legitimacy of the project and thereby keeping the interest and concerns of land owners on board in the process of land acquisition. This report shall be in public domain and shall be available for scrutiny of the general public. The Bill also provides that consent of the affected parties or persons from land acquisition is bound to be taken before setting in motion the process of acquisition. The Bill also provides for public hearing after release of notification notifying land acquisition and provides for hearing in respect of finalization of scheme of resettlement and rehabilitation.
- The most unique feature of the bill is that it provides for compensation to affected parties along with provision for resettlement and rehabilitation of the affected parties. The bill proposes clear guidelines for payment of compensation. This bill provides that in case of rural properties and urban properties, compensation of 4 times the market rate of land and 2 times of market rate of land shall be provided to the affected parties for releasing their land in favour of land acquisition by government. The Bill provides that affected parties shall be rehabilitated and resettled by providing allowance for resettlement and employment and thereby takes into consideration the interest of land owners and endeavours to mitigate the harmful impact of land acquisition.
- This Bill is retrospective in operation and is aimed as the title of the bill suggests for providing fair compensation to affected parties and provides for transparent process of land acquisition. This Bill shall be applicable

to all those land acquisitions where compensation has not been provided or where award has not been passed or where possession of land has not been taken. Fresh land acquisition proceedings shall be initiated under this Bill 2013.

- ☆ The bill also endeavours to safeguard agricultural land and food security and thereby direct the respective state governments to impose limits on the acquisition of land under agricultural cultivation.
- ☆ The Bill also provides that in case where the acquired land is not utilised for the public purpose for which it had been undertaken then, it shall be either returned back to the respective land owner or shall go into the State Land Bank.
- ☆ The Bill also provides that in case where there is sale of the land acquired through land acquisition to the third party and this is sold at 40 per cent of the appreciated value, than the profits are to be shared with the land owners.
- ☆ The bill also provides that rights of Scheduled tribes and scheduled caste under Forest Rights Act shall not be affected under the land acquisition bill and they shall be continued to be provided with their age-old traditional dwellers rights of fishing *etc.*
- ☆ The bill also provides that no stamp duty or income tax shall be paid in respect of released land under the bill or compensation received under the Bill.
- ☆ The Bill also provides that no scheduled land shall be acquired for land acquisition under the Bill and in case of acquisition; consent of Gram sabha shall be mandatorily required to be taken.

Conclusion

This Bill of Right to Fair Compensation and Transparency in Land Acquisition, Rehabilitation and Resettlement is a path defining law on land acquisition that has served the dual purpose of providing transparent land acquisition process and also providing for fair compensation to the affected parties. This Bill truly from its provisions illustrates that processes of land acquisition can be undertaken with consideration of interest of all classes involved and can emerge as a winner for the cause of public interest for which it is undertaken and serve the interest of parties deeply affected be land acquisition. This Bill serves as a great model for other countries to follow in regard to land acquisition.

Love Affairs, Impotency behind Farmer Suicides in Country!

Union Agriculture Minister Radha Mohan Singh today sparked a major controversy by listing love affairs, impotency and dowry among reasons behind suicide by farmers across the country. The minister, replying to a question by CPM leader CP Narayanan in the Rajya Sabha, quoted National Crime Records Bureau (NCRB) data for the over 1,400 suicides in 2014 and 263 to date this year.

The reply did not rule out debt as one of the reasons, but those cited above were enough to trigger a storm with the opposition parties accusing the Narendra Modi government of being insensitive towards farm suicides. The ruling BJP tried to downplay the issue and distanced itself from the "ill-drafted written reply". Union Minister Prakash Javadekar said the "minister concerned would answer any queries on the statement".

Grim scenario

Suicides*	2012	2013	2014
Nationwide	13,754	11,772	5,650
Haryana	276	374	14
Punjab	75	83	24
HP	29	33	32
J&K	10	18	12
U'khand	14	15	nil

*Source: National Crime Records Bureau

Radha Mohan and his ministry officials maintained they had only put forth data sourced from an official agency (NCRB). Narayanan had asked "whether it was a fact that lack of support to poor and small farmers by way of greater inputs and better support price for their produces is the main reason for burgeoning suicides".

The Agriculture Minister, in a written reply, said: "According to NCRB, causes of suicides include family problems, illness, drug abuse or addiction, unemployment, property dispute, professional or career problems, love affairs, barrenness or impotency, cancellation or non-settlement of marriage, dowry dispute, fall in social reputation and unknown causes." The reply mentioned that suicides were also caused by agrarian reasons such as indebtedness, crop failure and drought.

The minister said the suicide figures (as reported by state governments due to agrarian distress) stood at 1,066, 890 and 1,400 in 2012, 2013 and 2014. The maximum deaths were reported from Maharashtra and MP. Punjab reported five deaths in three years.

Numbers Tell Tale of Farmers' Travails

The NCRB recently released the figures of farmers' sucides for 2014. The data officially proves the point that debt-stressed farmers and agriculture workers are taking their own lives.

The agrarian sector is in dire straits. Reports of farmers' suicides from the countryside, especially in predominantly agricultural states such as Punjab, have increased. Bad crop seasons in a row have hit farmers hard but the Government's healing touch is yet to be felt.

The National Crime Records Bureau (NCRB), released the figures of farmers' suicide for year 2014 recently. It has maintained data since 1995. The cumulative data indicates that on an average 42 farmers and agriculture workers commit suicide every day in the country. The NCRB has tabulated a figure of 3,094,26 suicides in the farm sector since 1995.

Obviously, data maintained by the Bureau does not project the real picture because most of the states fudge figures on technical grounds regarding the number of suicides. Surprisingly, the ruling dispensation of the country appears unconcerned. The apparent reason for such insensitivity is that the corporate sector has overwhelmed the mindspace of the ruling establishment. There are amazing similarities in what is happening in India now and what happened in America, in the second half of the 19th century, when the farm sector there faced a crisis due to a number of reasons, including the ripping-off of farmers by rail-road companies and banks. The major reason for the persistence of the crisis for decades was the American ruling establishment's tilt towards the industrialists, which is the case now in our own country. Agitated farmers formed alliances to fight an unfair system. Their persistent protest movements from time to time, made the American government enact laws to put a check on rip-offs by rail-road companies and banks charging exorbitant interest rates. We face a similar situation more than 100 years later in India. Because of the domination of the big corporate houses on the psyche of the Indian ruling establishment, there is no apetite to deal with the problems faced by the rural economy. Almost every week, agitated farmers either march to Delhi or to their respective state capitals to seek redressal of their grievances, the most pronounced of which is a high incidence of indebtedness. The large network of public and private sector banks has failed to free farmers from the stranglehold of private money lenders, though the institutional short-term and long- term credit touched a figure of Rs. 6,073,75 crore in 2012-13. It was just Rs. 4,6268 crore in 1999-00. The last NSS report says that in Andhra Pradesh, 93 per cent small and marginal farmers are under debt and in Telangana, 89 per cent. In Tamil Nadu, 82.5 per cent and in Punjab 53 per cent are under debt. The average level of debt in Telangana is Rs. 2.40 lakh per house hold. In all, as many as 52 per cent farmer households are under debt. Though the average amount of outstanding loan per agricultural household has been Rs. 47,000 but in states like Kerala the average amount of outstanding loan is Rs. 2,13600 per household. A recorded fact is that of the 5,650 farmers besides 6,710 agriculture workers, who committed suicide in 2014, the indebtedness and bankruptcy was the reason in 22.8 per cent cases and farm-related issues in 19 per cent of cases. Of all these farmers, about 72 per cent were small and marginal farmers, owning land up to two hectares (five acres). The issue was forced into the national collective consciousness after Gajendra Singh's suicide in full public glare at a Kisan rally in New Delhi. The problem of suicides in farm sector is not confined to India only. Compared to other professions, the suicide rate in the farm sector is said to be highest across the globe. Developed countries such as America, UK, France and Australia are also hit. It has been reported that the suicide is a third highest cause of death among farmers in France. Why such a high incidence of suicide cases in India compared to other countries? The Socio-Economic and Caste Census released few weeks ago by the Union Government tells a dismal story. Of the total 17,91,647,59 households, 74.49 per cent are such in which highest earning member receives less than Rs. 5000. Add to this a figure of 17.18 per cent of households where highest earning member takes more than Rs. 5,000 but less than Rs. 10,000 per month, it becomes more than 91 per cent. There are only 8.29 per cent households in which the highest earning member gets above Rs. 10,000 per month in the rural areas. As against this, in elite cities such as Chandigarh, there are families who pay more than Rs. 10,000 alone as electricity bill of their house what to say about other committed expenditure such as on grocery, milk and so on. Another aggravating fact. Only 9.68 per cent households have salaried jobs in the rural areas. Of it, 5.02 per cent salaried jobs are in the government sector, 1.12 per cent in the public sector and 3.58 per cent in private sector. In fact, these are the perhaps only households where earning is more than Rs. 10,000 per month. The source of income of 5.39 crore (30.10 per cent) households is cultivation and of 9.16 crore (51.14 per cent) is manual casual labour. In rural India, 56 per cent people own no land and the figure of landless people in states like Punjab is 65 per cent. In such an overall scenario of poor subsistence level in rural India, one can imagine the nature of deprivation and desperation. Obviously, the Green Revolution has proved to be a boon for the country which has become self sufficient in foodgrains. The foodgrain production reached near 270 million tonne last fiscal year, from 51 million tonne in 1950-51. However, Green Revolution's motto to grow more has landed farmers in trouble. From traditional farming, they moved to highly mechanised modern farming and took huge loans to introduce machinery to do farming. Rise in input cost burdened them with formal and informal loans, thus making farming unviable for most of them. Flourishing consumer culture and rising cost of living further added to their woes. There is no national policy in place to tackle the crisis in farm sector. Instead, respective governments have been announcing packages for the farmers in areas more prone to suicides especially cotton belts of Maharashtra, Andhra Pradesh, Karnataka and adjoining areas.

No package as such has been announced for farmers of Punjab. Even now states such as Madhya Pradesh and Chhattisgarh figure on the top of the suicide chart. No workable insurance policy has been crafted for the farm sector. No efforts have been made to educate farmers by deputing counsellors to the countryside. The political leadership has never strived to give a psychological boost to the sagging spirit of the farming community.

As farming has become unviable for about 83 per cent (65 per cent marginal owning less than one hectare and 18 per cent small owning up to 2 hectares), no mechanism has been developed to supplement farmers'income by introducing allied jobs for them or launching beneficial schemes to ensure flow of liquidity for them. There is no effort or will to restructure the rural economy. Farmers have been left to fend for themselves. Tragic results are for all to see.

The National Food Security Bill, 2013

A Bill to *provide for food and nutritional security in human life cycle approach, by ensuring access to adequate quantity of quality food at affordable prices to people to live a life with dignity andfor matters connected therewith or incidental thereto.*

BE it enacted by Parliament in the Sixty-fourth Year of the Republic of India as follows:

1. This Act may be called the National Food Security Act, 2013.
2. It extends to the whole of India.
3. It shall come into force on such date as the Central Government may, by notification in the Official Gazette appoint, and different dates may be appointed for different States and different provisions of this Act.

In this Act, unless the context otherwise requires,-

1. *"anganwadi"* means a child care and development centre set up under the Integrated Child Development Services Scheme of the Central Government to render services covered under section 4, sub-section (1) of section 5 and section 6;
2. "central pool" means the stock of foodgrains which is,-
   (*i*) Procured by the Central Government and the State Governments through minimum support price operations;
   (*ii*) maintained for allocations under the Targeted Public Distribution System, other welfare schemes, including calamity relief and such other schemes;
   (*iii*) kept as reserved for schemes referred to in sub-clause (*ii*);
3. "eligible households" means households covered under the priority households and the Antyodaya Anna Y oj ana referred to in sub-section (l) of section 3;
4. "fair price shop" means a shop which has been licensed to distribute essential commodities by an order issued under section 3 of the Essential Commodities Act, 1955, to the ration card holders under the Targeted Public Distribution System;
5. "foodgrains" means rice, wheat or coarse grains or any combination thereof conforming to such quality norms as may be determined, by order, by the Central Government from time to time;
6. "food security" means the supply of the entitled quantity of foodgrains and meal specified under Chapter II;
7. "food security allowance" means the amount of money to be paid by the concerned State Government to the entitled persons under section 13;
8. "local authority" includes Panchayat, municipality, district board, cantonment board, town planning authority and in the States of Asom, Manipur, Meghalaya, Mizoram, Nagaland and Tripura where Panchayats do not exist, the village councilor committee or any other body, by whatever name caned, which is authorised under the Constitution or any law for the time being in force for self-governance or any other authority or body vested with the control and management of civic services, within a specified local area;
9. "meal" means hot cooked meal or ready to eat meal or take home ration, as may be prescribed by the Central Government;
10. "minimum support price" means the assured price announced by the Central Government at which foodgrains are procured from farmers by the Central Government and the State Governments and their agencies, for the central pool;
11. "notification" means a notification issued LInderthis Act and published in the Official Gazette;

12. "other welfare schemes" means such Government schemes, in addition to the Targeted Public Distribution System, under which foodgrains or meals are supplied as part of the schemes;
13. "person with disability" means a person defined as such in clause (t) of section 2 of the Persons with Disabilities (Equal Opportunities, Protection of Rights and Full Participation) Act, 1995;
14. "priority households" means households identified as such under section 15;
15. "prescribed" means prescribed by rules made under this Act;
16. "ration card" means a document issued under an order or authority of the State Government for the purchase of essential commodities from the fair price shops under the Targeted Public Distribution System;
17. "rural area" means any area in a State except those areas covered by any urban local body or a cantonment board established or constituted under any law for the time being in force;
18. "Schedule" means a Schedule appended to this Act;
19. "senior citizen" means a person defined as such under clause (*h*) of section 2 of the Maintenance and Welfare of Parents and Senior Citizens Act, 2007;
20. "social audit" means the process in which people collectively monitor and evaluate the and of a programme or
21. "State Commission" means the State Food Commission constituted under section 22;
22. "State Government", in relation to a Union territory, means the Administrator thereofappointed under article 239 of the Constitution;
23. "Targeted Public Distribution System" means the system for distribution of essential commodities to the ration card holders through fair price shops;
24. "Vigilance Committee" means a committee constituted under section 37 to supervise the implementation of all schemes under this Act;
25. the words and expressions not defined here but defined in the Essential Commodities Act, 1955, or any other relevant Act shall have the meaning respectively assigned to them in those Acts.

Provisions for Food Security

1. Every person belonging to priority households, identified under sub-section (1) of section 15, shall be entitled to receive five kilograms of foodgrains per person per month at subsidized prices specified in Schedule I from the State Government under the Targeted Public Distribution System: Provided that the households covered under Antyodaya Anna Yojana shall, to such extent as may be specified by the Central Government

for each State in the said scheme, be entitled to thirty-five kilograms of foodgrains per household per month at the prices specified in Schedule 1.

Explanation.-For the purpose of this section, the "Antyodaya Anna Yojana" means, the scheme by the said name launched by the Central Government on the 25th day of December, 2000; and as modified from time to time.

2. The entitlements of the persons belonging to the eligible households referred to in sub-section (1) at subsidised prices shall extend up to seventy-five per cent. of the rural population and up to fifty per cent. of the urban population.
3. Subject to sub-section (1), the State Government may provide to the persons belonging to eligible households, wheat flour in lieu of the entitled quantity of foodgrains in accordance with such guidelines as may be specified by the Central Government."
4. Subject to such schemes as may be framed by the Central Government, every pregnant woman and lactating mother shall be entitled to- (*a*) meal, free of charge, during pregnancy and six months after the child birth, through the local *anganwadi,* so as to meet the nutritional standards specified in Schedule II; and (b) maternity benefit of not less than rupees six thousand, in such instalments as may be prescribed by the Central Government:

   Provided that all pregnant women and lactating mothers in regular employment with the Central Government or State Governments or Public Sector Undertakings or those who are in receipt of similar benefits under any law for the time being in force shall 110t be entitled to benefits specified in clause (b).
5. (1) Subject to the provisions contained in clause (*b*), every child up to the age of fourteen years shall have the following entitlements for his nutritional needs, namely:- (*a*) in the case of children in the age group of six months to six years, age appropriate meal, free of charge, through the local *anganwadi* so as to meet the nutritional standards specified in Schedule II: Provided that for children below the age of six months, exclusive breast feeding shall be promoted; (*b*) in the case of children, upto class VIII or within the age group of six to fourteen years, whichever is applicable, one mid-day meal, free of charge, everyday, except on school holidays, in all schools run by local bodies, Government and Government aided schools, so as to meet the nutritional standards specified in Schedule H.

   (2) Every school, referred to in clause (*b*) of sub-section (*I*), and *anganwadi* shall have facilities for cooking meals, drinking water and sanitation: Provided that in urban areas facilities of centralised kitchens for cooking meals may be used, wherever required, as per the guidelines issued by the Central Government.

6. The State Government shall, through the local *anganwadi*, identify and provide meals, free of charge, to children who suffer from malnutrition, so as to meet the nutritional standards specified in Schedule II.
7. The State Governments shall implement schemes covering entitlements under sections 4, 5 and section 6 in accordance with the guidelines, including cost sharing, between the Central Government and the State Governments in such manner as may be prescribed by the Central Government.

Food Security Allowance

In case of non-supply of the entitled quantities of foodgrains or meals to entitled persons (see details above), such persons shall be entitled to receive such food security allowance from the concerned State Government to be paid to each person, within such time and manner as may be prescribed by the Central Government.

Identification of Eligible Households

The percentage coverage under the Targeted Public Distribution System in rural and urban areas for each State shall, subject to sub-section (2) of section 3, be determined by the Central Government and the total number of persons to be covered in such rural and urban areas of the State shall be calculated on the basis of the population estimates as per the census of which the relevant figures have been published. (1) The State Government shall, within the number of persons determined under section 14 for the rural and urban areas, identify (a) the households to be covered under the Antyodaya Anna Yojana to the extent specified under sub-section (1) of section 3, in accordance with the guidelines applicable to the said scheme; (b) the remaining households as priority households to be covered under the Targeted Public Distribution System, in accordance with such guidelines as the State Government may specify. (2) The State Government shall update the list of eligible households, within the number of persons determined under section 14 for the rural and urban areas, in accordance with the guidelines framed under sub-section (1). The State Government shall place the list of the identified eligible households in the public domain and display it prominently.

Eforms in Targeted Public Distribution System

(1 The Central and State Governments shall endeavour to progressively undertake necessary reforms in the Targeted Public Distribution System in consonance with the role envisaged for them in this Act.

(2) The reforms shall, *inter alia*, include- (*a*) doorstep delivery of food grains to the Targeted Public Distribution System outlets; (*b*) application of information and communication technology tools including end-to-end computerization in order to ensure transparent recording of transactions at all levels, and to prevent diversion; (*c*) leveraging "aadhaar" for unique identification, with biometric information of entitled beneficiaries for proper targeting of benefits under this Act; (*d*) full transparency of records; (*e*) preference to public institutions or public bodies such as

Panchayats, self help groups, co-operatives, in licensing of fair price shops and management of fair price shops by women or their collectives; (*f*) diversification of commodities distributed under the Public Distribution System over a period of time; (*g*) support to local public distribution models and grains banks; (*h*) introducing schemes, such as, cash transfer, food coupons, or other schemes, to the targeted beneficiaries in lieu of their foodgrain entitlements specified in Chapter II, in such area and manner as may be prescribed by the Central Government.

Women Empowerment

(1) The eldest woman who is not less than eighteen years of age, in every eligible household, shall be head of the household for the purpose of issue of ration cards. (2) Where a household at any time does not have a woman or a woman of eighteen years of age or above, but has a female member below the age of eighteen years, then, the eldest male member of the household shall be the head of the household for the purpose of issue of ration card and the female member, on attaining the age of eighteen years, shall become the head of the household for such ration cards in place of such male member.

Grievance Redressal Mechanism

Every State Government shall put in place an internal grievance redressal mechanism which may include call centres, help lines, designation of nodal officers, or such other mechanism as may be prescribed. (1) The State Government shall appoint or designate, for each district, an officer to be the District Grievance Redressal Officer for expeditious and effective redressal of grievances of the aggrieved persons in matters relating to distribution of entitled foodgrains or meals under Chapter II, and to enforce the entitlements under this Act. (2) The qualifications for appointment as District Grievance Redressal Officer and its powers shall be such as may be prescribed by the State Government. (3) The method and terms and conditions of appointment of the District Grievance Redressal Officer shall be such as may be prescribed by the State Government. (4) The State Government shall provide for the salary and allowances of the District Grievance Redressal Officer and other staff appointed under sub-section (1), and such other expenditure as may be considered necessary for their proper functioning.

(5) The officer referred to in sub-section (1) shall hear complaints regarding non-distribution of entitled food grains or meals, and matters relating thereto, and take necessary action for their redressal in such manner and within such time as may be prescribed by the State Government. (6) Any complainant or the officer or authority against whom any order has been passed by officer referred to in sub-section (1), who is not satisfied with the redressal of grievance may file an appeal against such order before the State Commission. (7) Every appeal under sub-section (6) shall be filed in such manner and within such time as may be prescribed by the State Government. (1) Every State Government may, by notification, constitute a State Food Commission for the purpose of monitoring and review of implementation of this Act.(2) The State Commission shall consist of- (*a*) a Chairperson; (*b*) five other Members; and

(*c*) a Member-Secretary, who shall be an officer of the State Government not below the rank of Joint Secretary to that Government.Provided that there shall be at least two women, whether Chairperson, Member or Member-Secretary:

Provided further that there shall be one person belonging to the Scheduled Castes and one person belonging to the Scheduled Tribes, whether Chairperson, Member or Member-Secretary. (3) The Chairperson and other Members shall be appointed from amongst persons- (*a*) who are or have been member of the All India Services or any other civil services of the Union or State or holding a civil post under the Union or State having knowledge and experience in matters relating to food security, policy making and administration in the field of agriculture, civil supplies.nutrition, health or any allied field; (*b*) of eminence in public life with wide knowledge and experience in agriculture, law, human rights, social service, management, nutrition, health, food policy or public administration; or (*c*) who have a proven record of work relating to the improvement of the food and nutrition rights of the poor.

(4) The Chairperson and every other Member shall hold office for a term not exceeding five years from the date on which he enters upon his office and shall be eligible tor reappointment: Provided that no person shall hold office as the Chairperson or other Member after he has attained the age of sixty-five years. (5) The method of appointment and other terms and conditions subject to which the Chairperson, other Members and Member-Secretary of the State Commission may be appointed, and time, place and procedure of meetings of the State Commission (including the quorum at such meetings) and its powers, shall be such as may be prescribed by the State Government.

(6) The State Commission shall undertake the following functions, namely:- (*a*) monitor and evaluate the implementation of the Act, in relation to the State; (*b*) either *suo motu* or on receipt of complaint inquire into violations of entitlements provided above; (*c*) give advice to the State Government on effective implementation of this Act; (d) give advice to the State Government, their agencies, autonomous bodies as well as non- governmental organizations involved in delivery of relevant services, for the effective implementation of food and nutrition related schemes, to enable individuals to fully access their entitlements specified in this Act; (*e*) hear appeals against orders of the District Grievance Redressal Officer; (*f*) prepare annual reports which shall be laid before the State Legislature by the State Government.

(7) The State Government shall make available to the State Commission, such administrative and technical staff, as it may consider necessary for proper functioning of the State Commission. (8) The method of appointment of the staff under sub-section (7), their salaries, allowances and conditions of service shall be such, as may be prescribed by the State Government. (9) The State Government may remove from office the Chairperson or any Member who- (*a*) is, or at any time has been, adjudged as an insolvent; or (*b*) has become physically or mentally incapable of acting as a member; or (*c*) has been convicted of an offence which, in the opinion of the State Government, involves moral turpitude; or (*d*) has acquired such financial or other interest as is likely to affect prejudicially his functions as

a member; or (*e*) has so abused his position as to render his continuation in office detrimental to the public interest.

(*10*) No such Chairperson or Member shall be removed under clause (*d*) or clause (*e*) of sub- section (9) unless he has been given a reasonable opportunity of being heard in the matter. The State Government shall provide for salary and allowances of Chairperson, other Members, Member-Secretary, support staff, and other administrative expenses required for proper functioning of the State Commission. *23A*. The State Government may, if considers it necessary, by notification, designate any statutory commission or a body to exercise the powers and perform the functions of the State Food Commission referred to in section 22.24. Notwithstanding anything contained in sub-section (1) of section 22, two or more States may have a joint State Food Commission for the purposes of this Act with the approval of the Central Government.

The State Commission shall, while inquiring into any matter referred to in clauses (b) and (e) of sub-section (6) of section 22, have all the powers of a Civil court while trying a suit under the Code of Civil Procedure, 1908, and, in particular, in respect of the following matters, namely:- (*a*) summoning and enforcing the attendance of any person and examining him on oath; (*b*) discovery and production of any document; (*c*) receiving evidence on affidavits; (*d*) requisitioning any public record or copy thereof from any court or office; and (*e*) issuing commissions for the examination of witnesses or documents.

The State Commission shall have the power to forward any case to a Magistrate having jurisdiction to try the same and the Magistrate to whom any such case is forwarded shall proceed to hear the complaint against the accused as if the case has been forwarded to him under section 346 of the Code of Criminal Procedure, 1973. No act or proceeding of the State Commission shall be invalid merely by reason of- (a) any vacancy in, or any defect in the constitution of, the State Commission; or (b) any defect in the appointment of a person as the Chairperson or a Member of the State Commission; or (c) any irregularity in the procedure of the State Commission not affecting the merits of the case.

Obligations of Central Government for Food Security

(1) The Central Government shall, for ensuring the regular supply of foodgrains to persons belonging to eligible households, allocate from the central pool the required quantity of food grains to the State Governments under the Targeted Public Distribution System, as per the entitlements under section-3 and at prices specified Schedule 1. (2) The Central Government shall allocate foodgrains in accordance with the number of persons belonging to the eligible households identified in each State under section 15. (3) (4) The Central Government shall provide foodgrains in respect of entitlements under sections 4, 5 and section 6, to the State Governments, at prices specified for the persons belonging to eligible households in Schedule I. (5) Without prejudice to sub- section (1), the Central Government shall;- (*a*) procure foodgrains for the central pool through its own agencies and the State Governments and their agencies; (*b*) allocate foodgrains to the States; (*c*) provide for transportation of food

grains, as per allocation, to the depots designated by the Central Government in each State;

(ca) provide assistance to the State Government in meeting the expenditure incurred by it towards intra-state movement, handling of foodgrains and margins paid to fair price shop dealers, in accordance with such norms and manner as may be prescribed by the Central Government; and (*d*) create and maintain required modern and scientific storage facilities at various levels. In case of short supply of foodgrains from the central pool to a State, the Central Govemment shall provide funds to the extent of short supply to the State Government for meeting obligations under Chapter II in such manner as may be prescribed by the Central Government.

Obligations for State Government for Food Security

(1) The State Government shall be responsible for implementation and monitoring of the schemes of various Ministries and Departments of the Central Government in accordance with guidelines issued by the Central Government for each scheme, and their own schemes, for ensuring food security to the targeted beneficiaries in their State. (2) Under the Targeted Public Distribution System, it shall be the duty of the State Government the Central Government. (3) For foodgrain requirements in respect of entitlements under sections 4, 5 and section 6, it shall be the responsibility of the State Government to take delivery of foodgrains from the designated depots of the Central Government in the State, at the prices specified in Schedule I for persons belonging to eligible households and ensure actual delivery of entitled benefits, as specified in the sections aforesaid. (5) In case of non-supply of the entitled quantities of foodgrains or meals to entitled persons under Chapter II, the State Government shall be responsible for payment of food security allowance specified in section 13. (6) For efficient operations of the Targeted Public Distribution System, every State Government shall,- (*a*) create and maintain scientific storage facilities at the State, District and Block levels, being sufficient to accommodate foodgrains required under the Targeted Public Distribution System and other food based welfare schemes; (*b*) suitably strengthen capacities of their Food and Civil Supplies Corporations and other designated agencies; (*c*) establish institutionalised licensing arrangements for fair price shops in accordance with the relevant provisions of the Public Distribution System (Control) Order, 2001 made under the Essential Commodities Act, 1955, as amended from time to time.

Obligations of Local Authorities

(1)The local authorities shall be responsible for the proper implementation of this Act in their respective areas. (2) Without prejudice to sub-section (1), the State Government may assign, by notification, additional responsibilities for implementation of the Targeted Public Distribution System to the local authority.

In implementing different schemes of the Ministries and Departments of the Central Government and the State Governments, prepared to implement provisions of this Act, the local authority shall be responsible for discharging such duties and responsibilities as may be assigned to them, by notification, by the respective State Governments.

Transparency and Accountability

All Targeted Public Distribution System related records shall be placed in the public domain and kept open for inspection to the public, in such manner as may be prescribed by the State Government.

(1) Every local authority, or any other authority or body, as may be authorised by the State Government, shall conduct or cause to be conducted, periodic social audits on the functioning of fair price shops, Targeted Public Distribution System and other welfare schemes, and cause to publicise its findings and take necessary action, in such manner as may be prescribed by the State Government.

(2) The Central Government may, if it considers necessary, conduct or cause to be conducted social audit through independent agencies having experience in conduct of such audits. (1) For ensuring transparency and proper functioning of the Targeted Public Distribution System and accountability of the functionaries in such system, every State Government shall set up Vigilance Committees as specified in the Public Distribution System (Control) Order, 2001, made under the Essential Commodities Act, 1955, as amended from time to time, at the State, District, Block and fair price shop levels consisting of such persons, as may be prescribed by the State Government giving due representation to the local authorities, the Scheduled Castes, the Scheduled Tribes, women and destitutepersons or persons with disability. (2) The Vigilance Committees shall perform the following functions, namely:- (*a*) regularly supervise the implementation of all schemes under this Act; (*b*) inform the District Grievance Redressal Officer, in writing, of any violation of the provisions of this Act; and (*c*) inform the District Grievance Redressal Officer, in writing, of any malpractice or misappropriation of funds found by it.

Provisions for Advancing Food Security

The Central Government and the State Governments shall, while implementing the provisions of this Act and the schemes for meeting specified entitlements, give special focus to the needs of the vulnerable groups especailly in remote areas and other areas which are difficult to access, hilly and tribal areas for ensuring their food security.

The Central Government, the State Governments and local authorities shall, for the purpose of advancing food and nutritional security, strive to progressively realise the objectives specified in Schedule HI.

Miscellaneous

(1) The provisions of this Act shall not preclude the Central Government or the State Government from continuing or formulating other food based welfare schemes. (2) Notwithstanding anything contained in this Act, the State Govenunent may, continue with or formulate food or nutrition based plans or schemes providing for benefits higher than the benefits provided under this Act, from its own resources. Any public servant or authority found guilty, by the State Commission at the time of deciding any complaint or appeal, of failing to provide the relief recommended by the District Grievance Redressal Officer, without reasonable cause, or willfully

ignoring such recommendation, shall be liable to penalty not exceeding five thousand rupees: Provided that the public servant or the public authority, as the case may be, shall be given a reasonable opportunity of being heard before any penalty is imposed.

(1) For the purpose of adjudging penalty under section 41, the State Commission shall authorise any of its member to be an adjudicating officer for holding an inquiry in the prescribed manner after giving any person concerned a reasonable opportunity of being heard for the purpose of imposing any penalty. (2) While holding an inquiry the adjudicating officer shall have power to summon and enforce the attendance of any person acquainted with the facts and circumstances of the case to give evidence or to produce any document which in the opinion of the adjudicating officer, may be useful for or relevant to the subject matter of the inquiry and if, on such inquiry, he is satisfied that the person has failed to provide the relief recommended by the District Grievance Redressal Officer, without reasonable cause, or wilfully ignored such recommendation, he may impose such penalty as he thinks fit in accordance with the provisions of section 41.

(1) The Central Government may, by notification, direct that the powers exercisable by it (except the power to make rules), in such circumstances and subject to such conditions and limitations, be exercisable also by the State Government or an officer subordinate to the Central Government or the State Government as it may specify in the notification. (2) The State Government may, by notification, direct that the powers exercisable by it (except the power to make rules), in such circumstances and subject to such conditions and limitations, be exercisable also by an officer subordinate to it as it may specify in the notification. The provisions of this Act or the schemes made thereunder shall have effect notwithstanding anything inconsistent therewith contained in any other law for the time being in force or in any instrument having effect by virtue of such law.

(1) If the Central Government is satisfied that it is necessary or expedient so to do, it may, by notification, amend Schedule I or Schedule II or Schedule III and thereupon Schedule I or Schedule II or Schedule III, as the case may be, shall be deemed to have been amended accordingly. (2) A copy of every notification issued under sub-section (1), shall be laid before each House of Parliament as soon as may be after it is issued.

The Central Government may, from time to time, give such directions, as it may consider necessary, to the State Governments for the effective implementation of the provisions of this Act and the State Governments shall comply with such directions. (1) The Central Government may, by notification, and subject to the condition of previous publication, make rules to carry out the provisions of this Act. (2) In particular, and without prejudice to the generality of the foregoing power, such rules may provide for all or any of the following matters, namely:- (*a*) scheme including cost sharing for providing maternity benefit to pregnant women and lactating mothers under clause (*b*) of section 4; (*c*) schemes covering entitlements under sections 4, 5 and section 6 including cost sharing under section 7; (*d*) amount, time and manner of payment of food security allowance to entitled individuals under section 13; (*e*) the norms and manner of providing assistance to the State

Governments in meeting expenditure under clause (ca) of sub-section (5) of section 30; (*P*) manner in which funds shall be provided by the Central Government to the State Governments in case of short supply of foodgrains, under section 31; (*f*) any other matter which is to be, or may be, prescribed or in respect of which provision is to be made by the Central Government by rules. (3) Every rule made by the Central Government under this Act shall be laid, as soon as may be after it is made, before each House of Parliament, while it is in session, for a total period of thirty days which may be comprised in one session or in two or more successive sessions, and if, before the expiry of the session immediately following the session or the successive sessions aforesaid, both Houses agree in making any modification in the rule or both Houses agree that the rule should not be made, the rule shall thereafter have effect only in such modified form or be of no effect, as the case may be; so, however, that any such modification or annulment shall be without prejudice to the validity of anything previously done under that rule.

(1) The State Government may, by notification, and subject to the condition of previous publication, and consistent with this Act and the rules made by the Central Government, make rules to carry out the provisions of this Act.(2) In particular and without prejudice to the generality of the foregoing power, such rules may provide for all or any of the following matters, namely:- (*a*) guidelines for identification of priority households under sub-section (1) of section (*b*) internal grievance redressal mechanism under section 20; (*c*) qualifications for appointment as District Grievance Redressal Officer and its powers under sub-section (2) of section 21; (*b*) method and terms and conditions of appointment of the District Grievance Redressal Officer under sub-section (3) of section 21; (ba) manner and time-limit for hearing complaints by the District Grievance Redressal Officer and the filing of appeals under sub-sections (5) and (7) of section 21; (*c*) method of appointment and the terms and conditions of appointment of Chairperson, other Members and Member-Secretary of the State Commission, procedure for meetings of the Commission and its powers, under sub-section (5) of section 22; (*d*) method of appointment of staff of the State Commission, their salaries, allowances and conditions of service under sub-section (8) of section 22; (*e*) manner in which the Targeted Public Distribution System related records shall be placed in the public domain and kept open for inspection to public under section 35; (j) manner in which the social audit on the functioning of fair price shops, Targeted Public Distribution System and other welfare schemes shall be conducted under section 36; (*g*) composition of Vigilance Committees under sub-section (1) of section 37; (ga) schemes or programmes of the Central Government or the State Governments for utilisation of institutional mechanism under section 51; (*h*) any other matter which is to be, or may be, prescribed or in respect of which provision is to be made by the State Government by rules.

(3) Every rule, notification and guidelines made or issued by the State Government under this Act shall, as soon as may be after it is made or issued, be laid before each House of the State Legislature where there are two Houses, and where there is one House of the State Legislature, before that House. 49. The schemes, guidelines, orders and food standard, existing on the date of commencement of this Act, shall continue to be in force and operate till such schemes, guidelines, orders and food standard are specified under this Act or the rules made thereunder:

Provided that anything done or any action taken under the said schemes, guidelines, orders and food standard shall be deemed to have been done or taken under the corresponding provisions of this Act and shall continue to be in force accordingly unless and until superseded by anything done or by any action taken under this Act.

(1) If any difficulty arises in glvmg effect to the provisrons of this Act, the Central Government may, by order, published in the Official Gazette, make such provisions, not inconsistent with the provisions of this Act, as appear to it to be necessary or expedient for removing the difficulty: Provided that no order shall be made under this section after the expiry of two years from the date of commencement of this Act. (2) Every order made under this section shall be laid, as soon as may be after it is made, before each House of Parliament.

The services of authorities to be appointed or constituted under sections 21 and 22 may be utilised in the implementation of other schemes or programmes of the Central Government or the State Governments, as may be prescribed by the State Government.The Central Government, or as the case may be, the State Government, shall be liable for a claim by any person entitled under this Act, except in the case of war, flood, drought, fire, cyclone or earthquake affecting the regular supply of foodgrains or meals to such person under this Act: Provided that the Central Government may, in consultation with the Planning Commission, declare whether or not any such situation affecting the regular supply of foodgrains or meals to such person has arisen or exist.

SCHEDULE I

[See sections 3(1), 30(1), (4) and 32(2), (3)]

SUBSIDISED PRICES UNDER TARGETED PUBLIC DISTRIBUTION SYSTEM

Eligible households shall be entitled to foodgrains under section 3 at the subsidised price not exceeding rupees 3 per kg for rice, rupees 2 per kg for wheat and rupee 1 per kg for coarse grains for a period of three years from the date of commencement of this Act; and thereafter, at such price, as may be fixed by the Central Government, from time to time, not exceeding,- (i) the minimum support price for wheat and coarse grains; and (ii) the derived minimum support price for rice, as the case may be.

SCHEDULE II

[See sections 4(a), 5(1) and 6]

NUTRITIONAL STANDARDS

Nutritional standards: The nutritional standards for children in the age group of 6 months to 3 years, age group of 3 to 6 years and pregnant and lactating women required to be met by providing 'Take Home Rations,l or nutritious hot cooked meal or ready to eat meal in accordance with the Integrated Child Development Services Scheme and nutritional standards for children in lower and upper primary classes under the Mid Day Meal Scheme are as follows:

Sl.No.	*Category*	*Type of Meal*	*Calories (Kcal)*	*Protein (g)*
1	*2*	*3*	*4*	*5*
1.	Children (6 months to 3 years)	Take Home Ration	500	12-15
2.	Children (3 to 6 years)	Morning Snack and Hot Cooked Meal	500	12-15
3.	Children (6 months to 6 years) who are malnourished	Take Home Ration	800	20-25
4.	Lower primary classes	Hot Cooked Meal	450	12
5.	Upper primary classes	HotCooked Meal	700	20
6.	Pregnant and Lactating mothers	Take Home Ration	600	18-20

Note: 1. Energy Dense Food fortified with micronutrients as per 50 per cent. of Recommended Dietary Allowance.

Note: 2. The provisions of the Food Safety and Standards Act, 2006 and any other law for the time being in force shall apply to Meals referred to in this Schedule.

NB: Nutritional standards are notified to provide balance diet and nutritious foods in terms of the calorie counts, protein value and micronutrients specified.

SCHEDULE III

(See section 39)

PROVISIONS FOR ADVANCING FOOD SECURITY

(1) Revitalisation of Agriculture-

(a) agrarian reforms through measures for securing interests of small and marginal fanners;

(b) increase in investments in agriculture, including research and development, extension services, micro and minor irrigation and power to increase productivity and production;

(c) ensuring livelihood security to farmers by way of remunerative prices, access to inputs, credit, irrigation, power, crop insurance, *etc.*;

(d) prohibiting unwarranted diversion of land and water from food production. (2) Procurement, Storage and Movement related interventions-

(a) incentivising decentralised procurement including procurement of coarse grains;

(b) geographical diversification of procurement operations;

(c) augmentation of adequate decentralised modern and scientific storage;

(d) giving top priority to movement of foodgrains and providing sufficient number of rakes for this purpose, including expanding the line capacity of railways to facilitate foodgrain movement from surplus to consuming regions.

(3) Others: Access to-

(a) safe and adequate drinking water and sanitation;

(b) health care;

(c) nutritional, health and education support to adolescent girls;

(d) adequate pensions for senior citizens, persons with disability and single women.

National Food Security Act, 2013

The **National Food Security Act, 2013** (also **Right to Food Act**) is an Act of the Parliament of Indiawhich aims to provide subsidized food grains to approximately two thirds of India's 1.2 billion people. It was signed into law on September 12, 2013, retroactive to July 5, 2013.

The National Food Security Act, 2013 (NFSA 2013) converts into legal entitlements for existing food security programmes of the Government of India. It includes the Midday Meal Scheme, Integrated Child Development Services scheme and the Public Distribution System. Further, the NFSA 2013 recognizes maternity entitlements. The Midday Meal Scheme and the Integrated Child Development Services Scheme are universal in nature whereas the PDS will reach about two-thirds of the population (75 per cent in rural areas and 50 per cent in urban areas).

Under the provisions of the bill, beneficiaries of the Public Distribution System (or, PDS) are entitled to 5 kilograms (11 lb) per person per month of cereals at the following prices:

- ✰ Rice at 3 (4.7¢ US) per kg
- ✰ Wheat at (3.1¢ US) per kg
- ✰ Coarse grains (millet) at 1 (1.6¢ US) per kg.

Pregnant women, lactating mothers, and certain categories of children are eligible for daily free meals. The bill has been highly controversial. It was introduced into India's parliament on December 22, 2011, promulgated as a presidential ordinance on July 5, 2013, and enacted into law on September 12, 2013.

Salient Features

75 per cent of rural population and 50 per cent of the urban population are entitled for three years from enactment to 5 kilograms (11 lb) food grains per month at 3 (4.7¢ US), 2 (3.1¢ US), 1 (1.6¢ US) per kg for rice, wheat and coarse grains (millet), respectively; The states are responsible for determining eligibility criteria;

Pregnant women and lactating mothers are entitled to a nutritious "take home ration" of 600 Calories and a maternity benefit of at least Rs. 6,000 for six months; Children 6 months to 14 years of age are to receive free hot meals or "take home rations"; The central government will provide funds to states in case of short supplies of food grains; The current food grain allocation of the states will be protected by the central government for at least six months;

The state government will provide a food security allowance to the beneficiaries in case of non-supply of food grains; The Public Distribution System is to be reformed; The eldest woman in the household, 18 years or above, is the head of the household for the issuance of the ration card; There will be state- and district-level redress mechanisms; and

State Food Commissions will be formed for implementation and monitoring of the provisions of the Act.# The cost of the implementation is estimated to be $22 billion(1.25 lac crore), approximately 1.5 per cent of GDP. The poorest who are covered under the Antodaya yojna will remain entitled to the 35 kilograms (77 lb) of grains allotted to them under the mentioned scheme.

Intent

The intent of the National Food Security Bill is spelled out in the Lok Sabha committee report, *The National Food Security Bill, 2011, Twenty Seventh Report,* which states, "Food security means availability of sufficient foodgrains to meet the domestic demand as well as access, at the individual level, to adequate quantities of food at affordable prices." The report adds, "The proposed legislation marks a paradigm shift in addressing the problem of food security – from the current welfare approach to a right based approach. About two thirds(approx 67 per cent) of the population will be entitled to receive subsidized foodgrains under Targeted Public Distribution System. In a country where almost 40 per cent of children are undernourished the importance of the scheme increases significantly."

Scope

The Indian Ministry of Agriculture's Commission on Agricultural Costs and Prices (CACP) has referred to the Bill as the "biggest ever experiment in the world for distributing highly subsidized food by any government through a 'rights based' approach." The Bill extends coverage of the Targeted Public Distribution System, India's principal domestic food aid program, to two thirds of the population, or approximately 820 million people. Initially, the Lok Sabha Standing Committee on Food, Consumer Affairs and Public Distribution estimated a "total requirement of foodgrains, as per the Bill would be 61.55 million [metric] tons in 2012-13." The CACP calculated in May 2013, ".the requirement for average monthly PDS offtake is calculated as 2.3 mt for wheat (27.6 mt annually) and 2.8 mt for rice (33.6 mt annually)." When volumes needed for the Public Distribution System and "Other Welfare Schemes" were aggregated, the CACP estimated rice and wheat requirements to total an "annual requirement of 61.2" million metric tons. However, the final version of the Bill signed into law includes on page 18 an annex, "Schedule IV", which estimates the total food grain allocation as 54.926 million metric tons.

The Standing Committee estimated that the value of additional food subsidies (*i.e.*, on top of the existing Public Distribution System) "during 2012-13 works out to be Rs.2409 crores," that is, 24.09 billion rupees, or about $446 million at the then-current exchange rate, for a total expenditure of 1.122 trillion rupees (or between $20 and $21 billion). However, the Commission on Agricultural Costs and Prices (CACP) calculated, "Currently, the economic cost of FCI for acquiring, storing and distributing foodgrains is about 40 per cent more than the procurement price." The Commission added, The stated expenditure of Rs. 1,20,000 crore annually in NFSB is merely the tip of the iceberg. To support the system and the welfare schemes, additional expenditure is needed for the envisaged administrative set up, scaling up of operations, enhancement of production, investments for storage, movement, processing and market infrastructure *etc.* The existing Food Security Complex of Procurement, Stocking and Distribution- which NFSB perpetuates- would increase the operational expenditure of the Scheme given its creaking infrastructure, leakages and inefficient governance.

The Commission concluded that the total bill for implementation of the Bill " may touch an expenditure of anywhere between Rs. 125,000 to 150,000 crores," *i.e.*, 1.25 to 1.5 trillion rupees. As of the implementation deadline of October 4, 2014, only 11 states had either implemented the Act or declared readiness to do so. On November 28, 2014, the Indian government announced, "Allocation of foodgrains to 11 States/Union Territories (UTs) namely, Bihar, Chandigarh, Chhattisgarh, Delhi, Haryana, Himachal Pradesh, Karnataka, Madhya Pradesh, Maharashtra, Punjab and Rajasthan has started under the Act." and that the "remaining 25 States/UTs have not completed the preparatory measures required for implementation of the Act." The Indian government extended the deadline for implementation of the Act "by another six months, *i.e.* till 04.04.2015."

Commentary

Critics

Criticism of the National Food Security Bill includes accusations of both political motivation and fiscal irresponsibility. One senior opposition politician, Murli Manohar Joshi, went so far as to describe the bill as a measure for "vote security" (for the ruling government coalition) rather than food security. Another political figure, Mulayam Singh Yadav, declared, "It is clearly being brought for elections. Why didn't you bring this bill earlier when poor people were dying because of hunger? Every election, you bring up a measure. There is nothing for the poor."

The report of the 33rd meeting of the Technical Advisory Committee on Monetary Policy stated". Food prices are still elevated and the food security bill will aggravate food price inflation as it will tilt supply towards cereals and away from other farm produce (proteins), which will raise food prices further. Members desired that the Reserve Bank impress on the government the need to address supply side constraints which are causing inflationary pressure, especially on the food front." Dr. Surjit S. Bhalla warned, "The food security bill if implemented honestly, will cost 3 per cent of the GDP in its very first year." The writer Vivek Kaul noted, The government's estimated cost of food security comes at 11.10 per cent of the total receipts. The CACP's estimated cost of food security comes at 21.5 per cent of the total receipts. Bhalla's cost of food security comes at around 28 per cent of the total receipts. Once we express the cost of food security as a percentage of the total estimated receipts of the government, during the current financial year, we see how huge the cost of food security really is.

The Indian Ministry of Agriculture's Commission on Agricultural Costs and Prices warned that enactment of the Bill could be expected to "induce severe imbalance in the production of oilseeds and pulses, "and" will create demand pressures, which will inevitably spillover to market prices of food grains. Furthermore, the higher food subsidy burden on the budget will raise the fiscal deficit, exacerbating macro level inflationary pressures." The Commission argued further that the Bill would restrict private initiative in agriculture, reduce competition in the marketplace due to government domination of the grain market, shift money from investments in agriculture to subsidies, and continue focus on cereals production when shifts in consumer demand patterns indicate a need to focus more on protein, fruits and vegetables.

Advocates

The bill was very widely viewed as a "pet project" of Indian National Congress(INC) President, Sonia Gandhi. Gandhi addressed Parliament the night of the August 2013 Lok Sabha vote on the bill, saying its passage would be a "chance to make history". Former National Advisory Council member and development economist Professor Jean Drèze, reputedly one of the architects of the original, 2011 version of the bill, wrote," the Bill is a form of investment in human capital. It will bring some security in people's lives and make it easier for them to meet their basic needs, protect their health, educate their children, and take risks." Professor Drèze dismissed opposition from business interests, saying, "Corporate hostility does not tell us anything except that the Food Bill does not serve corporate interests. Nobody is claiming that it does, nor is that the purpose of the Bill."

Minister of Consumer Affairs, Food, and Public Distribution K.V. Thomas stated in an interview, This is no mean task, a task being accomplished in the second most populated country in the world. All the while, it has been a satisfying journey. The responsibility is not just of the Central Government but equally of the States/[Union Territories]. I am sure together we can fulfill this dream. The day is not far off, when India will be known the world over for this important step towards eradication of hunger, malnutrition and resultant poverty. By providing food security to 75 per cent of the rural and 50 per cent of the urban population with focus on nutritional needs of children, pregnant and lactating women, the National Food Security Bill will revolutionize food distribution system. In a rebuttal to Dr. Surjit S. Bhalla, three economists responded, " the food subsidy bill should roughly double and come to around 1.35 per cent of GDP, which is still way less than the numbers he put out."

THE NATIONAL FOOD SECURITY MISSION

1. Introduction

The National Food Security Mission (NFSM) which was launched by the Government of India during 2007-08 was implemented in 482 districts of 19 States comprising of 144 districts under Rice in 16 States, 142 districts under Wheat in

9 States and 468 districts under Pulses in 16 States during the XI Plan period. An impact evaluation study of the programme has been carried out by AFC at the instance of the Ministry of Agriculture, Government of India. The sample size for the study comprised of 80 beneficiaries and 20 non-beneficiary farmers each from 30 NFSM districts of 14 Rice growing States, 28 NFSM districts from 9 Wheat growing States and 38 NFSM districts from 14 Pulses growing States. A total of 9600 farmers comprising of 7680 beneficiary farmers and another 1920 non-beneficiary farmers formed the sample for the field survey. The sample includes adequate number of farmers horizontally drawn from across all the interest groups to make it fairly representative in order to assess the true impact of the programme. The data/information were collected from the implementing agencies, farmers - both beneficiary and non-beneficiary and other sources by using various observation schedules/questionnaires containing a set of questions logically related to a problem under study in the survey protocol. The data was analyzed using appropriate analytical tools and necessary inferences have been drawn to arrive at conclusions.

2. Efficacy of Planning

The planning process adopted in the study districts is virtually centralized at the district level. There is hardly any "bottom-up" and "participatory approach" as envisaged under the approach of the Mission. About 31 per cent of the districts were preparing Annual Action Plan at the district level while in the remaining 69 per cent districts, the block level officials, and farmers' organizations are associated in some way in the planning process. However, the institutional mechanism created under ATMA at the Block level is not being utilized to the fullest extent in preparing District/Block level Annual Action Plans. The NFSM Guidelines envisages conduct of base line survey in the implementing districts to determine the status of crop production, its potential and demand so as to use the data/ information for preparation of annual action plan for strategic implementation of various interventions under each component crops. The study revealed that the Baseline Survey had been conducted in the majority of 55 [86 per cent] of the sampled study districts while it was not carried out in only 9 [14 per cent] districts.

The PRIs had been involved in the selection of the beneficiaries in a majority of over 69 per cent of the districts and kept out of picture in a significant number [31 per cent] of the districts in violation of the Mission guidelines. In about 72 per cent of the districts, PRIs were associated in the identification of priority areas for implementation of the programme and 70 per cent of the districts had also engaged PRIs in the execution of local initiatives for achieving better results. The compliance with the requirement of restricting assistance to maximum of 5 ha. to any beneficiary in a single season had been ensured by a large majority of 57 [89 per cent] of the districts and in 7 [11 per cent] of them, no effort was made to check whether the beneficiary farmer had land holdings above 5 ha. in a season. In a majority of 35 [55 per cent] of the districts, there was no system to ensure that the farmer does not get/avail benefit of the same intervention more than once. In the remaining 29 [45 per cent] of the districts, assistance was extended after following the due process of verification of the eligibility of the farmer.

3. Administrative and Technical Set Up

Administrative Set-up at the National, State and District Level

An appropriate organizational structure at National, State and District level has been created under the Mission for its smooth functioning. The General Council constituted at the National level under the chairmanship of the Hon'ble Union Agriculture Minister had met once a year during the period of XI Plan [2007-08 to 2011-12] and played its role very effectively. The National Food Security Mission Executive Committee under the Chairmanship of Secretary, DAC had also done its work well in overseeing the activities of the Mission and approving the State Action Plans.

The State Food Security Mission Executive Committees (SFSMEC) have been constituted under the chairmanship of the respective Chief Secretaries, in all the seventeen participating States. The frequency of the meeting of the Committee was, however, inadequate. The average number of the SFSMEC meetings works out to about 6 during the entire period of 5 years of the XI Plan. SFSMEC may have to meet at least once in every quarter, to oversee the activities of the Mission in the State more closely. All the participating States have also nominated an autonomous agency at the State level which, in the majority of States, is the Agricultural Management and Extension Training Institute [SAMETI]. These agencies have organized several programmes for diverse skill up-gradation of district, block and grassroots workers in different areas such as farming systems approach, participatory management, community mobilization, computer application, *etc.* besides arranging workshops, seminars and training programmes for farmers.

At the District level, a District Food Security Mission Executive Committee [DFSMEC] has been duly constituted and notified in all the sampled districts by the implementing States under the chairmanship of the In-charge of the District Administration. The record of the DFSMEC in convening meetings had, however, been poor. The total meetings held during the entire period of the project during XI Plan in all the study districts works out to about 1 meeting on an average annually. It needs to be made mandatory for holding such meetings at least once every quarter *i.e.* minimum 4 meetings in a year. In most of the sampled districts, the DFSMEC meeting is combined with other review meetings of the Agriculture Department which has resulted in lack of focus and in-depth deliberations, the essence of any Mission mode programme. The records of the meetings were also not being maintained properly. Of the total districts surveyed, the Seed Committees in Mission's 34 [56 per cent] districts had been found constituted while in a significant number of 30 [46 per cent] districts, no separate Seed Committees had been formed. Since seed is a vital input, it is necessary for all the districts to put in place Seed Committees so as to ensure adequate availability of seed varieties of the component crops.

Project Management Team [PMT]

PMTs have since been constituted technically in all the sampled districts but the true concept of the PMT is yet to take off. In the majority of the districts,

Technical Consultants had not been engaged on contract basis, as envisaged under the programme. Under Rice component, the number of districts not engaging Consultants ranged from 60 per cent to 80 per cent, under Wheat from 75 per cent to 80 per cent and under Pulses component from 79 per cent to 92 per cent during different years of the XI Plan. In the absence of consultants/technical experts or utilization of their services properly where ever engaged, the constitution of PMT at the district level has become a mere technical formality. Further, there was hardly any record of meetings/field visits of the Project Management Teams kept in the sampled districts across the districts.

The overall performance of most of the districts in undertaking field visits by the Monitoring Teams during XI Plan with average visits of 12, 49, 58, 57 and 55 per year during 2007-08, 2008-09, 2009-10, 2010-11 and 2011-12 respectively, remained poor. In most of the sampled districts, there was only one Agriculture Officer at the district level. With multifarious duties, the sole officer often finds it difficult to pay much attention to monitoring, evaluation and scrutiny of data received from the NFSM implementing blocks. It is, therefore, felt that at least one dedicated Officer for dealing with NFSM interventions/activities is imperative for a Mission Mode Programme of this nature and scale.

4. Fund Flow and Delivery Mechanism

The districts have not reported any fund constraint in the implementation of the programme. The systems and procedures devised in the transfer of funds are reported to be simple and efficient. The project funds had been released to the respective agencies as per the budgeted cost of the approved activity plan. Funds had been received by a large majority of 88 per cent of the districts on time while delay in receipt of the funds was reported by only 12 per cent of the districts. The average delay was observed to be of 2-3 months. An overwhelming majority [94 per cent] of the districts had been maintaining separate accounts at the district level in respect of funds received from the Mission. A large majority of 92 per cent of the districts had been using 'electronic' banking for fast transfer of funds. Annual accounts of all the sampled districts had been audited by Chartered Accountants since inception *i.e.* 2007-08 to 2012-13.

State-wise Funds Released and Utilized during XI Plan

All the 19 NFSM States including Jammu and Kashmir and Tripura included in the last year of the XI Plan, were released a total amount of Rs.4510 crore during the period of 2007-08 to 2011-12 against which an amount of Rs.4382 crore [97 per cent] was utilized. The overall position of the utilization of funds against releases was satisfactory. About 37 per cent of the States of States had utilized 100 per cent of the funds released, 53 per cent above 90 per cent to 99 per cent, 5 per cent each at 81 per cent and 30 per cent respectively.

District-wise and Component Crop-wise fund utilization

Rice

All the 30 Rice study districts had been released a total amount of Rs.295.58 crore during the entire period of the XI Plan against which the actual utilization was of the order of Rs.260.62 crore. The overall utilization rate at 88.17 per cent of the released funds can be considered as reasonably good. There were, however, district variations. The performance of majority of 18 [60 per cent] of the districts in utilizations of funds was 'good' [above 90 per cent], reasonably good [81 per cent to 90 per cent] of 3 [10 per cent], average [50 per cent to 80 per cent] of 8 [27 per cent] districts and poor [below 50 per cent] of 1 [3 per cent] district.

Wheat

All the 28 Wheat study districts had been released a total amount of Rs.241.07 crore during the entire period of the XI Plan against which the actual utilization was of the order of Rs.211.85 crore. The overall utilization rate at 87.88 per cent of the released funds may be taken as reasonably good. There were, however, district variations. The performance of majority of 16 [60 per cent] of the districts in utilizations of funds was 'good' [above 90 per cent], reasonably good [81 per cent to 90 per cent] of 4 [14 per cent] and average [50 per cent to 80 per cent] of 8 [29 per cent districts. There was no district in poor category [below 50 per cent].

Pulses

All the 38 Pulses study districts had been released a total amount of Rs.236.95 crore during the entire period of the XI Plan against which the actual utilization was of the order of Rs.191.04 crore. The overall utilization rate at 80.62 per cent of the released funds can be categorized as 'average'. The performance of 9 [24 per cent] of the districts in utilizations of funds was good [above 90 per cent], reasonably good [81 per cent to 90 per cent] of next 9 [23 per cent], average [50 per cent to 80 per cent] of the majority of 19 [50 per cent] districts and poor [below 50 per cent] of 1 [3 per cent] district.

Allocation of Mission Funds for MF/SF and Women Farmers

Mission guidelines on the allocation of stipulated 33 per cent funds for marginal farmers [MF], small farmers [SF] and women farmers has been observed in 45 [70 per cent] of the districts. A significant number of 19 [30 per cent] districts remained non-compliant to the prescribed norm.

5. Adherence to Intervention Specific Norms and its Impact

The farmer beneficiaries were queried on several aspects both, to find out the extent to which the stipulated norms for various interventions under each component crop *i.e.* Rice, Wheat and Pulses were adhered to; as well as to elicit their opinion on quality, efficacy and impact of different interventions availed by them; on productivity and income. The primary survey has revealed that the norms laid down in the guidelines were broadly being followed by a large majority of the sampled districts though there were some aberrations and district level variations.

Some of the important aspects are brought out below intervention-wise. Since most of the interventions are common to NFSM-Rice, Wheat and Pulses, the range of per cent age variations are indicated against each parameter in the analysis broadly bringing out the position under each component crop, for a fair understanding of the ground level situation.

Demonstrations

The important guidelines/norms relating to conduct of demonstrations on Improved Package of Practices [IPP], System of Rice Intensification [SRI] and Hybrid Rice Technology [HRT] such as selection of Demo plot in village meetings, conduct of pre-demonstration soil test, pre-training of beneficiaries, plot size, timeliness in sowing, easy accessibility, demarcation of Demo and Check plot, Check plot adjacent to Demo plot *etc.*, had been observed by and large in the majority of the cases.

Over 6 per cent of the beneficiaries of SRI, 5 per cent of HRT and 3 per cent of IPP demonstration reported that there was no clear demarcation between Demo plot and Check plot. In about 2 per cent of the cases, there was reportedly no Check Plot at all. About 9 per cent of the respondents relating to demonstration on IPP and 4 per cent of them in SRI demonstration also reported that the Check plot was not adjacent to Demo plot. Further, the norm of equal sizes of the demo and control plot was observed only in 8 [29 per cent] of the districts in IPP demonstration. The various technical norms of demonstration *viz.* transplanting of young seedlings singly in a square pattern, using organic fertilizers and hand weeding for better growth, use of the recommended age of seedlings, keeping the paddy soil moist during the vegetative growth phase, use of recommended doses of inputs, inter-cropping *etc.* has been largely followed by majority of the sampled farmers. A large majority of 82 per cent of them found significant phenotypic changes occurring in the plant structure as observable characteristics, which is noteworthy.

The demonstration kit was reported to have been received by the beneficiary farmers free of cost except by some farmers from 4 districts under demonstration on IPP. The major portion of seeds contained in 84-85 per cent in case of IPP and SRI demonstration and 71 per cent in respect of Hybrid Rice demonstration, was of certified seed and rest of truthfully labeled seeds. The quality of the inputs [seeds, micronutrients and PP Chemicals] supplied in the kit was rated well by a substantial majority of the respondent farmers. A large majority also found the use of PP Chemicals/Bio-pesticides applied as effective. The other aspects relating to demonstrations such as timeliness in supply of the kit, existence of display board at site, organizing Field Days, visit of Scientists/Experts, *etc.* was largely being followed though there were some aberrations and district level variations. The Resource Conservation. Technologies which has assumed much importance during the recent times, were reported to have been explained during demonstrations only by 67 per cent of the respondents.

A majority of about 63 per cent to 88 per cent of the beneficiary farmers participating in demonstrations under NFSM-Rice, Wheat and Pulses reported facing weed problems. About 28-62 per cent of them had been manually controlling the menace of weeds and 38 per cent to 72 per cent were applying weedicides. The problem of pests/insects was bigger one reported by a large majority of 63 per cent to 92 per cent of the respondent farmers and a majority of them applied pesticides as control measures and 7 per cent to 23 per cent of them let the crop to fend itself. A majority of about 56 per cent to 75 per cent of the sample farmers also reported damage of crops by rodents but a significant number of 22 per cent to 32 per cent had not taken any preventive measure against rodent infestation.

The demonstration programme has been found useful by an overwhelming majority of over 84 to 91 per cent of the respondents while only around 9 to 16 per cent indicated otherwise. A majority of about 76 per cent under IPP, 87 per cent under SRI and 81 per cent under Hybrid Rice Technology demonstrations reported having adopted the new technology learnt fully in their field situations without looking for any further subsidy from the Govt. and the remaining about one fourth [24 per cent] only partially. A predominant majority of the respondent farmers increased the ambit of the programme further by sharing their newly acquired knowledge with fellow farmers and the regular visitors to the site, most of whom, in turn, had introduced the improved package of practices in their farming system.

Impact of Demonstrations on Yield and Income

The primary data analysis reveals that the demonstrations have directly helped in change in the yield of the component crops of the beneficiary farmers. The demonstration on improved package of practices under NFSM-Rice and NFSM-Wheat recorded yield gain of 12 per cent and 11.46 per cent respectively in the Demo plot as against that in the Check plot. In the SRI demonstration plot under NFSM-Rice, an increase in yield of 11 per cent over traditional methods/old varieties was observed. Similarly, yield gain of 10.89 per cent was recorded in the Hybrid Rice Technology Demo plot over the control plot using traditional methods/old varieties. The yield gain of 23.46 per cent in the Demo plot over the Check plot in the Block Technology Demonstrations under NFSM- Pulses was quite impressive. A large majority of the beneficiaries also reported net gain in income between '10 per cent to 20 per cent' due to demonstrations, which is a noteworthy feature.

Supply of Improved Varieties of Seeds [HYV/Hybrid]

The HYV/improved varieties of seed and seed minikits of recently released location specific high yielding varieties had been distributed to the farmers free of cost under the Mission intervention for popularization of varieties and seed multiplication at the farmer's field level. The norms relating to supply of seed at subsidized cost not exceeding 33 per cent of the total seed requirement in a district had been complied with by a majority of the sampled districts and aberrations were observed in 6 districts only. An overwhelming majority of respondents assessed the impact of the improved varieties of seed used by them as 'Good' and were largely satisfied. As regards non-beneficiary farmers, 71 per cent of them were cultivating HYV/improved Varieties and the remaining 29 per cent were growing traditional/ old varieties. Among the not cultivating improved varieties lot, a large majority of 75 per cent attributed to the high cost of the seed as the cause for their not adopting the technology, 15 per cent due to non-availability of quality seeds and the rest 10 per cent to the high cost of cultivation with the HYV/Improved varieties seed.

Yield Gain

The yield obtained by use of HYV and hybrid rice seed recorded a significant increase of 21 per cent and 27 per cent respectively, over the traditional/old varieties. Similar trend was discernible in respect of the beneficiaries of Seed Minikit. The yield obtained with the use of HYV certified seed of wheat registered a significant increase of 16.51 per cent over traditional/old varieties. The yield gain in respect of beneficiaries of Seed Minikit was over 20 per cent, which is notable. The yield obtained with the use of certified seed of pulses recorded a significant increase of 16.51 per cent over traditional/old varieties.

An overwhelming majority of 78-88 per cent of the respondents rated the impact of HYV/improved seed varieties of rice, wheat and pulses, on their productivity and income as 'good' while 12-22 per cent indicated it being 'average'. About 3 per cent of pulses growing farmers, however, rated it as 'poor'. A large majority of the beneficiaries also reported net gain in income between '10 per cent to 20 per cent ' by using improved varieties under the Mission interventions.

Distribution of Micronutrients, Liming Material and Gypsum

Under NFSM, financial assistance is extended for distribution of micronutrients and liming material (in acidic soils) for rice, micronutrients and gypsum (for salt affected soils) for wheat and INM for pulses growers. The beneficiary farmers were by and large satisfied with the quality of the micronutrients/soil ameliorants received by them under the NFSM interventions. An overwhelming majority of over 95 per cent of them found the quality of the input as 'good' and a small proportion of 5 per cent only rated the same as 'poor'. Further, a large majority of 78-92 per cent of the respondent farmers reported micronutrients/soil ameliorants as 'effective', 8-22 per cent indicated it as 'partially effective' and only 4-6 per cent said that it was 'poor'. About 54 per cent to 87 per cent of the respondents rated its impact on income and productivity as 'Good' while only 4 per cent rated as 'poor'. The proportion of beneficiary farmers rating the impact as 'Average' ranged from as low as 9 per cent in the case of lime/gypsum under NFSM-Pulses to a high of 46 per cent in respect of the liming under NFSM-Rice. It is also a noteworthy feature that a large majority of the sampled farmers had reported continuous use of micronutrients/ soil ameliorants as their standard cultivation practice subsequently without looking for any further financial assistance from the Government.

Yield Gain

The micronutrient treated plots under NFSM-Rice and NFSM-Pulses registered a considerable increase in yield of 23 per cent and 26 per cent respectively, over the non-treated plots. Under NFSM- Wheat, the micronutrient treated plots recorded an increase of 12 per cent over the non-treated plots which are again significant. The yield obtained from the liming treated plots under NFSM-Rice registered a substantial increase of 33 per cent over non-treated plots. The yield obtained from the gypsum treated plots under NFSM-Wheat and NFSM-Pulses recorded a significant increase of 16 per cent and 20 per cent respectively, over non-treated plots. A large majority of the beneficiaries also reported net gain in income.

Plant Protection Measures: PP Chemicals/IPM/Weedicides

The beneficiary farmers were reportedly facing weed problems and 49 per cent of them felt the need for an extensive weed management. About 32 per cent of the respondent farmers had already been using weedicides and a large majority of them had preferred hand/manual weeding method. The respondent farmers had been facing weeds problem not only in their own plots but also the threat of the disease from the contiguous fields of other farmers on which they had no control. A large majority of 92 per cent of the respondents reported timely supply of the weedicide material to them.

The beneficiary farmers of rice and pulses were by and large satisfied with the quality of the plant protection chemicals and bio-agents received by them under NFSM. A large majority of 81-94 per cent of beneficiaries of IPP, PP Chemicals (for those not availing full IPM package) and Weedicides under NFSM-Rice and Pulses found the quality of the input as 'good' and a small proportion of 5-6 per cent only rated the same as 'poor'. Further, a large majority of over 81 per cent of the respondent farmers reported the inputs as 'effective' except in the case of Weedicides where only 57 per cent reported in the affirmative and a significant number of them [37 per cent] felt that it was 'partially effective'. About 5-6 per cent of the respondents, however, indicated that the plant protection chemicals were not of much effect.

Impact on Yield

The yield obtained from the PP Chemicals/Bio-Pesticides and weedicides treated plots registered a considerable yield increase in the range of 19 per cent to 28 per cent over the non-treated plots. A substantial proportion of 50 per cent of the respondents reported yield gain over non-treated plot "above 20 per cent ", 31 per cent between '10-20 per cent " and 19 per cent "up to 10 per cent '. Further, a large majority of 87 per cent of the respondents rated the impact of certified seed as 'good', 9 per cent as 'average' and 4 per cent as 'poor', on their productivity and income.

Farmers' Field School (FFS)

The Farmers' Field School [FFS] is an important and effective extension tool for transfer of technology at the grass root level in a participatory mode. Training in FFS is an important intervention eligible for assistance under all the 3 component crops *i.e.* Rice, Wheat and Pulses under NFSM. A large majority of over 88 per cent respondent farmers expressed their satisfaction with the skills of the trainer/resource person conducting FFS while only a small proportion of 12 per cent expressed dissatisfaction on this count. Almost all the topics covering important aspects of the crop husbandry had been included in the FFS programme, in conformity with the Mission guidelines. The norms for conduct of FFS such as close proximity to Demo Plot, easy accessibility, weekly sessions, duration, no. of farmers, distribution of literature/pamphlets, maintenance of records *etc.* had by and large been followed and deviations therein were not much. The season long training was held in about 91 per cent of the cases under Wheat and Pulses while only 63 per cent of the beneficiaries under NFSM-Rice reported having followed such a system, which is a matter of concern. Further, in over 12 per cent cases, it was reported that there was no fixed schedule of training and a significant proportion of about 28-54 per cent of the respondents considered the periodicity and duration of the programme as 'inadequate'. About 11 per cent of the respondents did not support the contention regarding maintenance of the requisite records at FFS.

Only about 41 per cent of the respondents under NFSM-Pulses reported that the FFS training was very useful and majority of 59 per cent said that the programme was useful to them only to some extent. A preponderant majority of 90 per cent of NFSM-Rice and a significant majority of 58 per cent of Wheat beneficiaries, however, indicated that the FFS training was quite useful. Majority of about 58-86 per cent of the farmers had shared their acquired knowledge with fellow farmers some of whom, in turn, had introduced the new learnings' in their field. A large section of the farmers ranging from 48-73 per cent showed their ignorance of any Benchmark Survey having been conducted in their area before the start of FFS programme or an Evaluation Study at the conclusion, as mandated under the NFSM guidelines.

Supply of Farm Implements [FIs]

The Mission provides assistance to the farmers for acquiring various agricultural implements such as Knap Sack Sprayer, Zero Till Seed Drill, Multi Crop Planter, Rotavator, Cono-weeder, *etc.* About 23 per cent to 39 per cent of the respondent farmers only reported that the FIs supplied to them under the programme were of ISI/prescribed standards and 4 per cent to 6 per cent of them specifically stated that the same were not of the required standards. A majority of 56 per cent in case of supply of Zero Till to 71 per cent in respect of supply of Multi Crop Planter reported that they were not aware whether the different types of FIs supplied were of any prescribed standards. On the quality aspect, beneficiary farmers ranging from 82 per cent in case of supply of Multi Crop Planter to 94 per cent in respect of Irrigation Pump Set rated the Machines/FIs as 'good', 2 per cent to 12 per cent as 'average' and 5 per cent -7 per cent as 'ineffective'.

Knap Sack Sprayer, Rotavator, Seed Drill in all the components crops *i.e.* Rice, Wheat and Pulses and Cono-weeder and Power Weeder in Rice have been the most popular implements and are much in demand amongst the farmers in all the sampled districts. Laser Land Leveller has been very popular in the States of Punjab, Haryana and Uttar Pradesh. Multi-crop Planter has been very efficiently used by the rice farmers in the States of Bihar and Jharkhand, wheat farmers mainly in Bihar and UP and pulses growing farmers in the States of Bihar, Maharashtra, Haryana and Madhya Pradesh.

Impact of Farm Implements

The farm mechanization had a positive effect on the farming operation of the respondents in the form of increase in income and productivity and also on the comfort level of the farmers. The use of FIs had brought savings both in time and money to the beneficiary farmers over the traditional method. Apart from savings in time and expenditure, the use of FIs has also ensured farming operations of the respondents well within the limited available time as a spin off. The FIs supplied under the Mission could also have a demonstration affect on the other neighbouring farmers to inspire some of them to adopt the same.

At all study district level, the use of the knap sack sprayer by recipient rice farmers had brought savings in average time required to complete operation in one ha to the extent of 65 per cent as against carrying out the same activity with traditional method thereby affecting a significant saving of 59 per cent in expenditure on an average. Similarly, the use of other farm implements such as Seed Drill/Multi-Crop Planter, Rotavator, Power Weeder and Cono-weeder by beneficiary farmers have resulted in significant savings in time and expenditure to the extent of 43 per cent :24 per cent; 67 per cent :18 per cent; 86 per cent :35 per cent and 60 per cent :52 per cent respectively. On the other hand, the beneficiary wheat farmers reported savings in time and expenditure to the extent of 64 per cent :93 per cent; 60 per cent :45 per cent; 59 per cent :44 per cent; 59 per cent :56 per cent and 57 per cent :61 per cent respectively. In the case of Pulses beneficiary farmers also, significant savings in time and expenditure to the extent of 58 per cent :85 per cent; 51 per cent :56 per cent; 62 per cent :49 per cent and 58 per cent :75 per cent respectively, was reported.

Supply of Irrigation Devices

The Pump Sets supplied to the rice, wheat and pulses growing farmers has created considerable additional irrigation facilities to the extent of 115 per cent, 107 per cent and 124 per cent respectively, indicating a significant impact on the crop yield. Similarly, the Sprinkler Sets supplied to the farmers cultivating wheat and pulses has increased the irrigated area to the extent of 114 per cent and 127 per cent respectively. A large majority of the beneficiaries of irrigation devices also reported net gain in income, which is noteworthy.

The supply of pump sets to a substantial number [29 per cent] of farmers in dark/brown zones of groundwater was against the spirit of the Mission guidelines and the accepted water usage norms. A large proportion of 67 per cent of the farmers further reported that they were not aware of the ground water status of their blocks. Only about 30 per cent of them reported that the sprinkler sets supplied were of ISI Mark/accredited while majority of about 65 per cent showed ignorance on this aspect. About 5-8 per cent of the pump sets and sprinkler sets were reportedly not in working condition at the time of survey. As regards quality aspect of the sprinkler set supplied, a large majority of over 91-94 per cent of the respondent farmers rated it as being of 'good' quality while only 6-9 per cent indicated it as 'poor'. Further, only a miniscule proportion of around 6-7 per cent found the sprinkler sets supplied to them as 'not effective'.

Local Initiatives

The 'local initiatives' intervention had positive impact on all the beneficiaries. The beneficiary farmers had been provided assistance for diverse items such as purchase of tarpaulins, control corn, pesticides, pipe line, seed treatment, termite control and weedicides. About 85 per cent of the beneficiary farmers under local initiatives in all the districts of study were satisfied and 91 per cent felt that the intervention was beneficial to them. The end result was increase in the yield of rice, wheat and pulses crop 'up to 10 per cent ' for majority of around 74-88 per cent and 'above 10 per cent ' for 12-26 per cent of them.

Stakeholders Involved in Delivery of Services

State Agriculture Universities [SAUs]

Only 17 [27 per cent] of the districts had established due links with SAUs for obtaining technical guidance and support. The SAUs have been involved in the FFS training in 3 [18 per cent] districts, field inspections in 4 [24 per cent] districts and supply of seed in one district. Six [35 per cent] districts had been seeking technical guidance and support in holding technology demonstrations. In 4 districts, the SAUs had been helping in imparting training to the farmers for their capacity building.

ATMA Structure at District Level

A majority of 38 [59 per cent] of the districts were also utilizing ATMA forum for implementation of interventions in the form of service delivery of training/ capacity building, farmer awareness and account handling, technical support in demonstration, FFS training, convergence, online field management, popularizing SRI technology, technology dissemination *etc.* A significant number of 25 [39 per cent] districts had not utilized the institutional mechanism created under ATMA.

Involvement of KVKs in Extension Activity of NFSM

A large majority of 47 [73 per cent] of the districts had established links with KVKs to utilize their services in the implementation of various interventions. Their services had mainly been utilized by the concerned districts in the field of seed supply, technical guidance and support, field inspections, FFS training, training/capacity building, preparation and implementation of AAP, popularizing SRI Technology, technology dissemination besides creating awareness about the programme. The performance of KVKs was found to be satisfactory by most of the sampled districts. However, by and large, it was observed that district level agency has failed to utilize the instrument of KVK to its full potential. A good number of 17 [27 per cent] of the districts had not been involving KVKs in the implementation of the Mission programme in any manner.

Involvement of NGOs/SHGs/Farmers' Associations in NFSM Activities

Only 27 [20 per cent] districts had utilized the services of the NGOs for the activities such as supply of inputs, conducting demonstrations, FFS trainings, implementation of local initiatives, *etc.* In 11 [17 per cent] of districts, the support of the FIGs/SHGs/FAs had also been taken in the implementation of above interventions. All the districts had expressed their satisfaction with the performance of the voluntary sector.

Collaboration/Synergy among Line Departments

The study team observed that synergy among the line departments was absolutely lacking in almost all the sampled districts. There is lack of ownership even among officials and also among many farmers. The vertical and horizontal linkage between line department and farmers-researcher-extension linkage has still to go a long way.

Involvement of PRIs in NFSM Activities

The PRIs have been involved in the selection of the beneficiaries in the majority of the 44 [69 per cent] of the districts. In a significant number of 20 [31 per cent] of the districts, the PRIs had been kept out of the selection of the beneficiary farmers for various interventions of the Mission in violation of the guidelines. In a considerable number of 22 [34 per cent] districts, the selection of beneficiaries was not done in a participatory manner by holding meetings in the village. This norm has, however, been observed in the rest of 42 [66 per cent] of the districts.

The active involvement of the PRIs is also considered important under the Mission guidelines to make the planning process realistic and need based. It is observed that while in 46 [72 per cent] of the districts, PRIs were associated in the identification of priority areas for implementation of the programme, they were kept out of the picture in a significant number of 18 [28 per cent] of the districts. Like-wise, as many as 19 [30 per cent] districts did not involve PRIs in the implementation of the local initiatives while the other 45 [70 per cent] of the districts did follow the guidelines and engaged PRIs for achieving better results.

Convergence with Other Government Schemes

The NFSM Programme mandates that the Annual Action Plan to be submitted by the State Governments for funding should explicitly specify the activities to be supported from the resources of other Government schemes as well as from the Mission Programme. Convergence of different schemes/programmes has been reported by 41 [64 per cent] of the districts. The SREP is a mechanism for ensuring convergence of all activities of extension, the resources for which are being provided under different schemes of Centre/State Governments. A majority of the 33 [52 per cent] districts out of the total sampled study districts had reported convergence of Mission interventions with ATMA schemes/programmes.

Monitoring and Reporting Mechanism

The Mission has put in place a strong mechanism of monitoring and evaluation with the involvement of all the implementing agencies and the line departments. The Project Management Teams [PMTs] at the state and district levels have been mandated mainly with the responsibility to undertake field monitoring of various interventions and to provide technical guidance. The feedback received from the sampled districts reveal that PMTs have since been constituted technically in all the sampled districts. The study has, however, revealed that the true concept of the PMT at the District level is yet to take off and the PMTs constituted were mostly redundant.

The State Level Monitoring Teams [SLMT] in respect of all the 17 States had paid a total 330 monitoring visits which works out to an average of 6 visits per year during the entire period of 5 years of the XI Plan. Though there is no minimum stipulation of monitoring visits under the guidelines of the Mission, a minimum of

8 to 10 visits per annum may be desirable as per the practice in the States of Tamil Nadu and Bihar. The overall performance of most of the districts in undertaking field visits by the Monitoring Teams during XI Plan remained poor. Six [9 per cent] districts did not undertake any visit. In case of the remaining 58 [91 per cent] of the districts, 4 [6 per cent] of the districts had undertaken 2-10 visits, 3 [5 per cent] districts 11-20 visits, 17 [27 per cent] districts 21-50 visits, 16 [25 per cent] districts 51-100 visits and 7 [11 per cent] districts undertook 101-200 visits.

A significant number of 23 [36 per cent] of the districts, reported that the staff posted in the NFSM Cell at the state and district levels as also the consultants appointed on contract basis were not adequate to cope up with the work devolved. In the absence of adequate number of staff/consultants or non- engagement of consultants in the Mission Cell, it could not be possible to undertake proper monitoring of various interventions under the component crops, as envisaged under the programme. Furthermore, in most of the sampled districts, the constitution of PMT is a mere technical formality. It was reported by most of the sampled districts that the consultants appointed at the district level were mostly utilized for office work of routine nature and not for the intended specialized technical service and monitoring of NFSM interventions. This apart, there was hardly any record of meetings/field visits of the Project Management Teams kept in the sampled districts across all the States. Equipping the NFSM Cell with adequate staff including consultants will only ensure in the proper implementation of the programme.

Conduct of Baseline Studies

The conduct of baseline survey and feasibility studies in the area of operation was envisaged mainly to determine the status of crop production, its potential and demand. The study has brought out that the Baseline Survey has been conducted in a large majority [89 per cent] of the sampled districts covered under the Mission.

Concurrent Evaluation of NFSM Activities

The NFSM Guidelines also envisages conduct of Concurrent Evaluation by all the implementing States every year to assess the performance of the Mission commensurate with its objectives. The concurrent evaluation has been done only in 18 [28 per cent] of the districts and a large majority of the remaining 46 [72 per cent] of the districts had not undertaken any concurrent evaluation.

Intervention Specific Physical and Financial Achievements against Targets for Various Component Crops

During the XI Plan [2007-08 to 2011-12], there were 19 Mission interventions under NFSM-Rice, 15 under NFSM-Wheat and 20 under NFSM-Pulses for implementation by States and other agencies. The overall performance of all the study districts put together during the entire period of 11th Plan as regards Demonstration on IPP under Rice [100 per cent] and Wheat [97 per cent]; Demonstration on SRI [110 per cent]; Demonstration on Hybrid Rice Technology [99 per cent]; was quite good. The performance of the study districts with regard to conduct of training in

FFS where achievements against target allotted to them was almost fulfilled or was closely achieved with a small deficit, is also remarkable.

Under farm implements category, the overall average performance of all the identified districts in respect of distribution of cono-weeder and other FIs at 188 per cent of the achievement of their targets was outstanding. Similarly, in respect of distribution of knapsack sprayers, the achievement against set targets under all the 3 crops – Rice [175 per cent], Wheat [144 per cent] and Pulses [151 per cent] was really noteworthy. The achievement of States against targets allotted for distributions of irrigation pump sets at 107 per cent, 98 per cent and 109 per cent in respect of Rice, Wheat and Pulses was also good. However, in the matter of distribution of sprinkler sets under Wheat and Pulses, the performance at 88-89 per cent fell far short of the expectations but can certainly be considered as satisfactory. A large majority of the districts in the above categories had achieved targets allotted to them, fully or with a small deficit. The performance of the districts in respect of other farm implements *viz.* seed drill, multi-crop planter, rotavator, power weeder *etc.* fell far short of expectations. The overall financial performance in respect all the above categories of interventions had nearly followed the physical achievement pattern. The performance of all the study districts in respect of other interventions such as micronutrients, liming/gypsum, plant protection chemicals, *etc.* where the achievement level against the targets hovered around 70-80 per cent can be considered as satisfactory in view of the nature of the activity requiring great persuasion to motivate beneficiaries.

Impact of NFSM on Area and Productivity

Rice

All the 30 NFSM study districts had a total area of 44.92 lakh ha. under rice cultivation during the base year of 2006-07 which shrunk to 44.00 lakh ha. in terminal year of 2011-12 of the XI Plan. In the terminal year of 2011-12 of the 5th year of the XI Plan, the overall rice area of the study districts had shown a decline by -1.90 per cent over the base year of 2006-07. Only 12 [40 per cent] districts had enlarged their rice cultivated area. The total rice production in the study districts was 79.92 lakh tonne during the base year of 2006-07 which increased to 100.44 lakh tonne in 2011-12 recording an impressive growth rate of 25.68 per cent. As at the end of the XI Plan, the majority of 22 [73 per cent] of the study districts demonstrated increase in production.

The average yield at all States level, which stood at 1779 kg/ha. during the base year of 2006-07 rose to 2279 kg/ha. in 2011-12 thus recording a yield gain of about 28 per cent over the base year, which was remarkable. There was positive growth rate in 2 years [2010-11 and 2011-12] and negative growth rate in 3 years [2007-08, 2008-09 and 2009-10] during the XI plan, compared to the year 2006-07. Maximum of 10 [33 per cent] of the sampled districts had 'low productivity' [1000-1500 kg/ha] and 3 [10 per cent] districts 'very low productivity' [below 1000 kg/ha], followed by 7 [23 per cent] districts falling under 'medium- low productivity' [1500-2000 kg/ha], 5 [17 per cent] districts each under 'medium productivity' and 'high productivity' [above 2500 kg/ha] respectively.

There were 19 districts [63 per cent] out of 30 sampled rice districts which had yield below their respective State averages during the base year 2006-07 before the launch of NFSM. Of these, 4 districts *viz.* Guntur (AP), West Champaran (Bihar), Gumla (Jharkhand) and Shahdol (MP) which were below their respective State averages during 2006-07 went way ahead of their State averages during the XI Plan period. Out of the remaining 15 districts; except 3 districts namely, Nasik (Maharashtra), Nayagarh and Bolangir (Odisha) which registered a negative growth during XI Plan period as compared to base year 2006-07; all the other 12 districts had registered a significant increase in yield but still continued to be below their respective State averages of XI Plan period yield. Further, 4 districts *viz.* Karbi Anglong (Asom), Rajnandgaon (Chhattisgarh), Thiruvarur and Nagapattinam (Tamil Nadu) which were well above their respective State averages during 2006-07 had fallen below their State averages of XI Plan period yield. Overall, in the majority of 19 [63 per cent] of the sampled districts, the average productivity was below their respective State average yield during the XI Plan period (Average of all the years).

Wheat

All the 28 NFSM districts had a total area of 30.24 lakh ha. under wheat cultivation during the base year of 2006-07 which increased to 33.23 lakh ha. in 2011-12 and registered a growth rate of 9.9 per cent over the base year, which is significant. A majority of 20 [71 per cent] of the districts had enlarged their wheat acreage. The total wheat production in the study districts during the base year of 2006-07 was 87.79 lakh tonne which increased to 105.94 lakh tonne in 2011-12 recording an impressive growth rate of 20.68 per cent. The growth rate in wheat production during all the 5 year period of the XI Plan has shown a steady rising trend all along. The majority of 19 [68 per cent] districts demonstrated a positive growth in wheat production.

The average wheat productivity in relation to the study districts which worked out to 2903.41 kg/ha during the base year of 2006-07 with wide across the States variations, increased to 3187.99 kg ha. in the terminal year of 2011-12 of the 11th Plan. The XI Plan closed with a normal increase of 9.8 per cent in wheat yield. The average yield of all the sampled districts put together showed an increasing trend all along when compared with the base year 2006-07 but, with respect to individual sampled districts during the plan period, there was rising and declining trend compared with the performance of the immediately preceding year. The majority of 15 [54 per cent] NFSM wheat districts had 'high productivity' [above 2500/ha], followed by 6 [21 per cent] districts falling under 'medium productivity' [2000-2500 Kg/ha] range, 3 [11 per cent] districts under 'medium to low productivity' [1500-2000 kg/ha], 2 [7 per cent] districts under 'low productivity' [1000-1500 kg/ha] and the remaining 2 [7 per cent] districts under 'very low productivity' [below 1000 kg/ha].

There were 15 districts [54 per cent] out of 28 sampled wheat districts which had yield below their respective State averages during the base year 2006-07 before the launch of NFSM. Of these, only 1 district namely, Madhubani (Bihar) which had yield below its State average during 2006-07 went way ahead of its State average during the XI Plan period. Out of the remaining 14 districts; except 4 districts namely,

Bhiwani (Haryana), Bikaner (Rajasthan) and Rajgarh and Dindori (Madhya Pradesh) which registered a negative growth during XI Plan period as compared to base year 2006-07; all the other 10 districts had registered a significant increase in yield but still continued to be below their respective State averages of XI Plan period yield. Further, 3 districts *viz.* Banaskantha (Gujarat), Dewas (Madhya Pradesh) and Kushi Nagar (Uttar Pradesh) which were well above their respective State averages during 2006-07 had fallen below their State averages of XI Plan period yield. Overall, in the majority of 17 [61 per cent] of the sampled districts, the average productivity was below their respective State average yield during the XI Plan period (Average of all the years).

Pulses

All the 38 NFSM districts had a total area of 38.24 lakh ha. under pulses cultivation during the base year of 2006-07 which decreased to 34.56 lakh ha. in 2011-12 and registered a negative growth rate of -9.63 per cent over the base year. In the terminal year of 2011-12 of the XI Plan, a majority of 23 [61 per cent] of the districts had recorded a negative growth rate in pulses area. The total pulses production in all the study districts was 23.78 lakh tonne during the base year of 2006-07 which increased to 27.86 lakh tone in 2011-12 recording a promising growth rate of 17.16 per cent.

The average yield at all districts level, which was 621.82 kg/ha. in the base year rose substantially to 806.13 kg/ha. in 2011-12. recording an impressing growth rate of 29.64 per cent even though there was decrease of about 10 per cent in the area. The growth rate in the yield of the pulses during the 5 year period of the XI Plan works out to a negative growth rate of [-] 7.32 per cent, [-] 3.51 per cent, [-] 4.25 per cent during the first three years of the XI Plan and thereafter registering a positive growth rate of 20.39 per cent and 29.64 per cent respectively in the last two years. The majority of 20 [53 per cent] of the sampled districts had 'low productivity' [400-800 kg/ha] and 5 [13 per cent] districts 'very low productivity' [below 400 kg/ha], followed by 11 [29 per cent] districts falling under 'medium productivity' [800-1200 kg/ha]. Only 2 [5 per cent] districts were under the 'high productivity' [above 1200 kg/ha] category.

There were 19 districts [50 per cent] out of 38 sampled pulses districts which had yield below their respective State averages during the base year 2006-07 before the launch of NFSM. Of these, 6 districts namely, Rajnandgaon (Chhattisgarh), Rohtak and Bhiwani (Haryana), Tikamgarh (Madhya Pradesh), Keonjhar (Odisha) and Virudhunagar (Tamil Nadu) which had yield below their respective State averages during 2006-07 went way ahead of their State averages during the XI Plan period. Out of the remaining 13 districts; except 2 districts namely, Nagapattinam (Tamil Nadu) and Purulia (West Bengal) which registered a negative growth during XI Plan period as compared to base year 2006-07; all the other 11 districts had registered a significant increase in yield but still continued to be below their respective State averages of XI Plan period yield. Further, 5 districts namely, Raichur and Mysore (Karnataka); Nasik (Maharashtra), Nayagarh (Odisha) and Thiruvarur (Tamil Nadu) which were well above their respective State averages during 2006-07 had fallen below their State averages of XI Plan period yield. Overall, in the majority of 18 [47 per cent] of the sampled districts, the average productivity was below their respective State average yield during the XI Plan period (Average of all the years).

Pre- and Post-Mission Changes in the Cropping Pattern of Beneficiary Farmes

Rice

The number of sampled farmers growing traditional/old varieties of rice during the base year of 2006-07 showed a decline of 52 per cent post mission [2013-14] while decline in area was to the extent of 30 per cent, as revealed during the primary survey. The loss of the traditional/old varieties was the gain on the HYV/Improved rice varieties. Post Mission, in the year 2013-14, the sampled farmers growing improved/ HYV also registered an impressive growth rate of 57 per cent and 126 per cent in the number of farmers and area respectively, over the pre-Mission condition. Out of the sampled beneficiary farmers, 2 per cent of them had been cultivating other crops in the pre-Mission year of 2006-07 and subsequently switched to rice. The outcome of the Mission had, thus, been very positive in motivating farmers to shift from the low yielding traditional/old varieties to the improved high yielding varieties. Overall 28 per cent decline in the area of traditional/old varieties cultivation in the post-Mission period had taken place and in the majority of 21 [70 per cent] of the districts, the area under traditional/old varieties of rice had decreased. A significant overall increase of 126 per cent in the area under Improved/HYV was noticed in 87 per cent of the districts. Many of the farmers growing traditional/old varieties had adopted improved/HYV under the Mission.

Introduction of hybrid rice cultivation by the beneficiary farmers has been a very positive feature of the Mission. Over 18 per cent of the total sampled farmers adopted hybrid cultivation under the Mission interventions of which a majority of 77 per cent had switched over from traditional/old/HYV varieties to hybrid rice. Increase in the yield has been observed both under old/traditionally grown varieties and the improved/HYV in the post-Mission period compared to pre-Mission year of 2006-07. The overall yield gain in respect of the Improved/HYV in the post-Mission period was of the order of 13.14 per cent. The respondent farmers had thus been benefited in yield gain both in the old/traditionally grown varieties and the Improved/HYV varieties taken up under the Mission interventions. The Improved/HYV rice had 45 per cent yield gain over the old/traditionally grown varieties, which was significant.

On a comparison with control group *i.e.* non-beneficiary farmers, the overall average yield obtained by the beneficiary farmers was more by 12.93 per cent under old/traditionally grown rice varieties, 26.16 per cent under Improved/HYV and 20.37 per cent under Hybrid rice after their participation in the Mission interventions, over the yield obtained by non-beneficiary farmers under the respective varieties. The beneficiary farmers had, thus, an advantage of higher productivity after availing benefits under the Mission interventions over that of the non-beneficiary farmers.

Wheat

The number of sampled farmers growing traditional/old varieties of wheat during the base year of 2006-07 showed a decline of 39 per cent post mission [2013-14] while decline in area was to the extent of 38 per cent. The loss of traditional/old varieties was, however, the gain of the Improved/HYV wheat cultivation. The number of farmers growing improved/HYV registered an impressive growth rate of 40 per cent and 105 per cent in the number of farmers and area, respectively, in the post-Mission over the pre- Mission condition. Out of total sampled farmers growing traditional/old varieties of wheat before their participating in the Mission, 39 per cent opted for Improved/HYV in the post-Mission period [2012-13]. In the majority of 22 [79 per cent] of the districts, farmers had shifted for Improved/HYV cultivation from traditional/old varieties. Overall, there was a decline of 38 per cent in the area of the old/traditionally grown wheat varieties in the districts in the post-Mission era compared to the base year [2006-07].

Increase in the yield has been observed both under old/traditionally grown varieties and the improved/HYV in the post-Mission period compared to pre-Mission year of 2006-07. The wheat varieties grown traditionally, showed a small overall increase of about 4.33 per cent over the pre-Mission period. Similarly, the overall yield gain in respect of the Improved/HYV in the post-Mission period was of the order of 9.63 per cent. The respondent farmers had, thus, been benefited in yield gain both in the old/traditionally grown varieties and the Improved/HYV varieties taken up under the Mission interventions. The Improved/HYV rice had 39 per cent yield gain over the old/traditionally grown varieties, which was significant.

Pulses

An overall increase of 14.39 per cent, as compared to base year 2006-07 position, was witnessed in the area of pulses grown by the beneficiary farmers, which is noteworthy. The average yield taking all the Kharif, Rabi and Summer pulse crops together, recorded 16.40 per cent increase in the post-Mission era over the pre-Mission period [2006-07] which was significant given that the pulses production had been stagnating for decades.

On a comparison with control group *i.e.* non-beneficiary farmers, the average yield obtained by the beneficiary farmers was more than that of the non-beneficiary farmers season-wise and in totality. The increase in yield obtained by the beneficiary farmers over that of the non-beneficiary farmers in Kharif, Rabi and Summer seasons at the end of XI Plan works out to 10.84 per cent, 16.39 per cent and 8.55 per cent, respectively. The overall increase in yield of the beneficiary farmers for all the seasons taken together was 12.03 per cent over that of the non-beneficiary farmers.

Income Generation Patten of Beneficiary Farmes Over the Base Yaer

The contribution of the Mission in raising the productivity and income level of the beneficiary farmers had been significant. The survey findings revealed that there was significant increase of 23.94 per cent in the productivity of rice and consequential income level of farmers, under NFSM programme which is significant. Similarly, in wheat and pulses, the increase was of the order of 15.14 per cent and 15.90 per cent respectively, which is notable. The overall increase of 52 per cent to 74 per cent in the net return under rice, wheat and pulses, attributable mainly to adoption of the improved package of practices by the beneficiary farmers under different interventions of the Mission, was sizeable.

Issues, Suggestions and Recommendations

NFSM, launched in 2007-08, has been successful in achieving targets of food grain production set for the 11th Five Year Plan by addressing major constraints to crop productivity through promotion of relevant technological interventions. The achievements notwithstanding, there still remains a large untapped yield potential reservoir for the exploitation of which Mission has to step up its efforts further. A gap between the potential yield that can be achieved at farmers' field level and what they actually get is very wide. Bridging this yield gap offers an opportunity to produce more even by using the available technologies. The broad issues/constraints in each of the three components of the Mission are discussed below:

A) Rice

The scope for horizontal expansion of area under rice cultivation during Kharif is limited. The only way to sustain production to meet the requirement of increasing population of the country is either to increase the productivity per unit area or increasing gross cropped area [multiple cropping], coverage of area under irrigation and improvement in the productivity levels. There are possibilities of expansion of area under rabi/summer rice provided irrigation facilities are created. Further, about 5 m. ha. area with shallow low land and coastal ecosystem, offers scope for rice production with development of suitable varieties of hybrids of 140-150 days duration and tolerance to salinity. The flood prone areas of Asom, West Bengal, Bihar and Odisha have the potential of cultivation of Boro rice due to favourable soil-water-climatic conditions.

System of Rice Intensification [SRI] and hybrid rice are the principal drivers for accelerating the rice production under the Mission. Despite its having great potential to enhance production and productivity of rice in the country, hybrid rice varieties have not been adopted on large scale as was expected. There is an urgent need to identify new areas in States other than Andhra Pradesh for large scale hybrid seed production. There is a large unmet potential in rainfed rice area which constitutes about 52 per cent of the total rice lands. Developing hybrid seed to meet the region specific needs, raising short duration varieties for rainfed uplands, creation of small-scale water harvesting and supplemental irrigation, combined with better land management will increase the present production level of rice in rainfed areas. The use of SRI techniques coupled with high performing inbred HYVs and hybrids (in potential areas) needs to be widely propagated. Further, these activities should have a targeted yield level nearer or equal to the potential yield of the specific cultivars keeping in view the minimized yield gaps.

B) Wheat

It has been observed that only 32 per cent of the net sown area is cultivated more than once and a large portion of land particularly, in the eastern and central parts of India, covering the states of Madhya Pradesh, Chhattisgarh, Bihar, Jharkhand, West Bengal and Odisha; remains fallow during rabi season. This fallow land can be used for the sowing of wheat during this season through appropriate technologies, inputs and extension services. This, however, will require investment in water-saving techniques both for storage of water and its most economical use like water harvesting and sprinkler irrigation, breeding of short-duration varieties, enabling farmers to grow wheat varieties capable of escaping terminal drought and popularizing zero tillage, besides enhancing capability of farmers for their uptake.

Improving yield level is a more viable and potential option to increase wheat production in the country. The yield gap in wheat production, which is about 30 per cent of the total production, arises on account of various abiotic, biotic and socio-economic problems. Significant potential production in wheat is lost due to attacks by diseases, pests and insects. Development and dissemination of location-specific high-yielding wheat varieties and management of weeds and diseases is an important approach to address the biotic stresses. Specifically for eastern India, the management of sowing time of rainfed wheat should be adequately targeted in cropping system approach adopting conservation agriculture practices. The rainfed districts of eastern and far-eastern India *i.e.* plain areas of north eastern States (NEPZ) may be well targeted for enhanced productivity of wheat.

C) Pulses

The six States of Madhya Pradesh, Rajasthan, Uttar Pradesh, Maharashtra, Andhra Pradesh and Karnataka cover 80 per cent of total pulses production of India. There is an ample opportunity to increase the area under pulses in the other States. It is estimated that additional about 5 million hectares could be realistically brought under Pulses by working out region and farming system specific strategy. In rainfed rice fallows which comprises of about 10 million ha. area, there is a practice of taking only one crop of rice per year. Even if a part of this area could be brought under cultivation for a second crop, it would be a significant increase in pulses production. Promotion of pulses under intercropping, short duration thermo sensitive varieties and development and popularization of late planting in north India would certainly put additional area under pulses.

In order to meet the projected demand of 32 million tonnes of pulses by 2030, as per the Vision 2030 paper prepared by the Indian Institute of Pulses Research, Kanpur, a growth rate of 4.2 per cent has to be ensured. As in the case of cereals, there is scope for a lot of enhancement in pulses productivity. This will, however, require a paradigm shift in research, technology generation and dissemination, popularization of improved crop management practices and commercialization along with capacity building of the stakeholders in frontier areas of research. Genetic enhancement for yield and quality seed would be a critical factor in productivity.

D) Seed Replacement

The wide gaps reported between the potential and realized productivity in most of the crops can be bridged to a large extent by using the seeds of improved varieties. The pace of SRR has, however, not been fast as sufficient quantities of certified seed are not available from all the seed sources put together. Various efforts have been made to ensure availability of good quality seeds of high yielding varieties/hybrids, yet nearly 70-75 per cent of the total seed requirements are met by the farm-saved seed. Further, the released seed varieties are also not reaching farmers in the absence of both information and availability of seed. It is, therefore, imperative to widen the focus on increasing the seed replacement rate of low potential/pest susceptible old varieties by new high yielding varieties with promising yield potential, strengthening of infrastructure facilities for production and distribution of quality seeds and taking up more and more villages under the Seed Villages Programme.

There is a strong need to increase the production of sufficient quantities of quality seeds and its timely distribution among the farmers by coordinating the efforts of all the seed producing and distribution agencies. Production subsidies for certified seeds of inbred HYVs and hybrids also need to be re- introduced. Breeder seeds production of recently released varieties should be adequately channelized with enhanced target of SRR. In order to utilize the rice fallow area of eastern India, promotional incentives for pulse growing may be introduced. Short duration pulses even in the rice field bunds could also be popularized. As pulses are grown mostly in marginal lands by resource poor farmers, assured procurement, contract farming and insitu storage facilities are required for the pulse growers to continue its cultivation in a sustainable manner.

E) Operational Issues

i) Meetings of DFSMEC

The overall record of the DFSMEC in convening meetings had been poor. It needs to be made mandatory for holding such meetings at least once every quarter and without combining with other review meetings of the department for focused and in-depth deliberations.

ii) Monitoring of NFSM Activities

The staff posted in the NFSM Cell at district and state level as also the consultants appointed on contract basis were not found adequate to cope up with the work devolved. In most of the sampled districts, there was only one Agriculture Officer at the district level dealing with several Govt. of India and State Govt. programme. In the absence of adequate number of staff/consultants or non-engagement of consultants in the Mission Cell, it could not be possible to undertake proper monitoring work, as envisaged under the programme. Equipping the NFSM Cell with adequate staff including consultants will only ensure in the proper implementation of the programme. The Project Director at the District level and the State Mission Director may also be appointed exclusively for Mission with fixed tenure of about three years and should be possessing vast nowledge, expertise and ground level experience in agriculture sector.

iii) Management Information System [MIS]

The present MIS needs to be further strengthened. Efficient and effective MIS is essential for proper planning and fulfillment of Mission objectives.

iv) Planning

It was observed that in most of the districts, the planning process adopted is virtually centralized at the district level. There is hardly any "bottom-up" and "participatory approach" as envisaged under the approach of the Mission. It is, therefore, imminent that the present district level action plan is prepared in a participatory manner taking into consideration various agro- climatic conditions and local specific needs in the district based on the block as a basic unit. The views of the farmers, PRIs, other agencies operating in the areas in agricultural field would be helpful in making the plan realistic. The districts should be advised to strictly utilize the institutional mechanism created under ATMA at the Block level to the fullest extent in preparing Annual Action Plans under NFSM for the component crops with due involvement of the Farmers' Advisory Committee [FAC] to have a clear understanding of location-specific needs and problems of the local farmers besides making it truly bottom-up, demand-driven, farmer- centric and participatory.

v) Selection of Beneficiaries

The PRIs should be invariably involved and play an effective role in selection of beneficiaries as well as in implementation of various components/activities of the programme including identification of priority areas, as mandated under the Mission.

vi) Timely availability of Inputs

Ensuring timely availability of fertilizer and other nutrients to the farmers could be one of the critical components for enhancing productivity. The timely availability of inputs, development of organized input market and infrastructure will influence the rate of adoption of knowledge intensive technology. Proper planning for timely availability of inputs and advance tie-up with manufacturers and suppliers is essential for facilitating early procurement and supply. Procurement process for supply of inputs may further be decentralized up to the district level. The Network of outlets for sale of seeds along with other inputs, which is at present inadequate, needs to be increased.

vii) Farm Implements

The high price and limited usage during the sowing/harvesting season discourage farmers from purchasing and using farm equipments and machinery. The setting up of farm machinery centres at different places to lease the farm equipments and machinery to the needy farmers on reasonable rent could be one way to overcome this problem. The Agri-business Centres already opened in many districts could also be encouraged to undertake custom hiring of agriculture machinery so as to improve the access of farmers on adopting. The requirements of the farmers on Farm Implements (FIs) and the brand which they prefer may be assessed with the help of village panchayats every year before commencement of the agricultural season, which needs to be consolidated at the district level for placing orders with manufacturers/suppliers for distribution amongst the farmers. Further, there is also need for a proper tie-up for operation and maintenance of the machineries. The use of farm machineries/implements may be clearly delineated for plains and hills for small and marginal farmers with adequate incentives.

viii) Soil Tests

A proper mechanism should be created for ensuring soil tests in all fields where assistance for micronutrients/soil ameliorants are provided under NFSM. It is essential to provide soil health cards to all farmer beneficiaries. Mobile soil testing laboratory which has already been initiated in some districts, is an innovative way to reach out to farmers, for conducting soil tests in their fields.

ix) Supply of Literature on Package of Practices

Technology package in the form of pamphlets/literatures with targeted yield should be provided during farmer level training and/or in the beginning of each project activity. Climatic zone wise/sub-zone wise customized literature on package of practices could be distributed to farmers for their benefit. Distribution of such literatures through schools to children of farmers studying in higher classes, who can read and transfer the information to their parents in local dialects, may also be an innovative medium to disseminate farming knowledge.

x) Training and Capacity Building

Periodical training of field functionaries in the form of orientation/refresher training programmes is imminent to disseminate technical knowledge and build their capacities for effective implementation of the Mission interventions. Since Human Resource Development is one of the key factors for successful implementation of the programme of this nature and scale, the States/Districts should make all out efforts to depute maximum number of their staff to training to enable the latter to equip themselves with the latest domain knowledge as also the technique of transferring technology to the farmers.

xi) Conduct of FFS and Demonstrations

The targets for conduct of FFS should be corresponding to the availability of technical experts in respective block/district. Farmers' showed reluctance to attend such FFS due to monotony of lectures. Audio-visual aid will create more interest in farmers to join and regularly attend the FFS. Devising of a system of coordination among different agencies *viz.* SAUs/KVKs and drawing up a uniform schedule of visits by the scientists/experts from different institutions at critical stages of plant growth may be more focused and practically useful both in FFS as well as Demonstrations. The visit of crop scientists may be made mandatory at least five times during the crop cycle planting, vegetative stage, flowering, pod development and maturity phases. FFS participants should be given priority for participation in demonstrations, distribution of minikits and other interventions, keeping in view the local situation. Demonstration of Resource Conservation Technologies as well as utility of various farm machineries/implements should be made an integral part of the Demonstration as well as the curriculum in Farmers' Field School [FFS].

xii) Adherence to Intervention Specific Norms

The Districts should put in place a proper control mechanism to ensure compliance with meeting the eligibility criteria of land holding [restricting assistance for a max. 5 ha. per farmer in a season] and availing of assistance only once for the same intervention. The other norms of the scheme such as one FFS for every 1000 ha area, equal size of the check and control plots, existing of check plots, a fixed training schedule for FFS, training and soil tests before demonstration/supply of inputs, organizing field day *etc.* should be scrupulously followed by all the Mission districts.

xiii) Project Management Team (PMT)

In the absence of consultants/technical experts for technology transfer, the constitution of PMT at the district level is a mere technical formality. The consultants, wherever appointed at the district level, were mostly utilized for office work of routine nature and not for the intended specialized technical service. The State Governments and the implementing districts may be advised to ensure that the consultants/technical assistants are appointed immediately, as per the scheme guidelines. The contract staff should further be utilized exclusively for the work of the Mission. There was hardly any record of the monitoring visits of the Project Management Teams in any of the sampled districts across all the States, which may be made mandatory.

xiv) Seed Committees

In most of the sampled districts, the Seed Committees were formed but were virtually dormant and non-functional. The State Governments may be suitably advised to ensure that all the districts make appropriate efforts to make the Seed Committees effective in dealing with the supply of seed to the farmers well in time since shortage of seeds is adversely impacting the production of the crops. The Committee should also decide on the variety to be given in Minikit before the commencement of season.

xv) Publicity and Awareness

It is imperative that the implementing states and districts undertake vigorous awareness campaigns by way of publicizing the features of the Mission through various measures like Advertisements in Print Media, display of video clips on mass media, distribution of relevant literatures, organizing education fair *etc.*, so as to make maximum farmers alert towards the various aspects of the Mission.

xvi) Involvement of NGOs/Farmers' Groups

Extensive efforts are required to be made to involve NGOs, Farmers' Interest Groups and other similar entities in the implementation of the Mission's activities such as supply of inputs, selection of beneficiaries, seed distribution and support in conducting demonstrations, FFS, training programmes *etc.* NGOs being grass root level institutions with excellent rapport with the farmers particularly, Marginal and Small farmers, their involvement will be very useful and lend sustenance to the programme in the long run.

xvii) Collaboration/Synergy among Line Departments

The study team observed that synergy among the line departments was absolutely lacking in almost all the sampled districts. The Block Agriculture Officer is extremely overburdened with several activities under various Government of India and State Govt. programmes. There is a great shortage of field functionaries with top-heavy agricultural administration. Field functionaries, especially at the Block level and below are quite inadequate for technology dissemination. Thus, the vertical and horizontal linkage between line department and farmers-researcher-extension linkage has still to go a long way. Collaboration/Synergy among Line Departments and convergence of schemes concept which is lacking in the programme at present, needs to be considered seriously.

xviii) Involvement of KVKs

Though scientists/experts from SAUs/KVKs were being invited to attend the field days, demonstrations and FFS in a large majority of districts, it was observed that such involvement was largely un-structured and un-planned. The district level agency has virtually failed to utilize the instrument of KVK to its full potential. There is a felt need for associating KVKs actively particularly, in interventions such as Demonstrations, Farmer' Field School *etc.* in addition to their involvement in training programmes and other activities.

Conclusion

The Mission has helped in widening the food basket of the country with sizeable contributions coming from the NFSM districts. The focused and target oriented implementation of mission initiatives has resulted in bumper production of rice, wheat and pulses. The production of wheat has increased from 75.81 million tonnes in pre-NFSM year of 2006-07 to 94.88 million tonnes during 2011-12 *i.e.* an increase of nearly 19 million tonnes against the envisaged target of 8 million tonnes at the end of XI Plan period. Similarly, the total production of rice has increased from 93.36 million tonnes in pre– NFSM year of 2006-07 to 105.31 million tonnes in 2011-12 with an increase of nearly 12 million tonnes against the target of 10 million tonnes. The total production of pulses has also increased from m14.20 million tonnes during 2006-07 to 17.09 million tonnes during 2011-12 with an increase of about 3 million tonnes against the envisaged target of 2 million tonnes. Thus, 34 million tonnes of additional production of total food grains against the target of 20 million tones, were achieved at the end of 11th five year plan against the normal production at the beginning of the 11th five year plan. The Mission has also been successful in achieving its objective of vertical growth by raising productivity of per unit of land.

Significant outputs have been recorded under NFSM since inception of the Mission in 2007-08 up to the terminal year of XI Plan *i.e.* 2011-12. New farm practices have been encouraged through 6.52 lakh demonstrations of improved package of practices on Rice, Wheat and Pulses. As many as 97252 demonstrations on System of Rice Intensification (SRI) as well as 52127 demonstrations on hybrid rice had also been conducted.

There has been an increase in input consumption of seeds, Integrated Nutrient Management (INM), IPM and machinery components under Rice, Wheat and Pulses from 2007-08 to 2011-12 due to awareness generated at the district level towards use of quality seeds, nutrients, plant protection chemicals and farm machinery. The awareness generated through demonstrations and distribution of seed minikits of newly released varieties to the farmers has triggered substantial increase in the use of high yielding varieties of wheat and rice seeds in identified districts. The evaluation study has revealed that about 168 lakh qtls of seeds of high yielding varieties of Rice, Wheat and Pulses and hybrid rice have been distributed. About 112 lakh hectare of area has been treated with soil ameliorants (gypsum/lime/micro nutrients) to restore soil fertility for higher productivity. Further, an area of about 45 lakh ha has been treated under Integrated Pest Management (IPM).

Farmers have enthusiastically responded to NFSM supported farm machinery components, especially to zero till seed drill, rotavator *etc.* which is evident from the increased sale of these machines supported by Mission. Nearly 34 lakh numbers of improved farm implements/machineries of various types including water saving devices such as irrigation pump sets and sprinkler sets have been distributed to help farmers to undertake timely farming operations within the limited available time and to improve irrigation for the crop.

Capacity building of farmers has been encouraged through arranging 47444 Farmers' Field Schools (FFS) at the farm level during the XI Plan period for the benefit of 13.21 lakh farmers since its inception in 2007-08 to 2011-12. FFS has been found to be extremely helpful in orienting the farmers towards adoption of new agricultural technologies.

As revealed by the present study, the programme was very well accepted in the sampled districts in terms of uptake of the recommended technologies by the farmers. The various interventions of the Mission have been instrumental in bringing about significant yield gain to the farmers resulting into increase in their income level. Further, nearly two-thirds of the NFSM districts showed positive production gains. The enhanced investments in agriculture in the NFSM districts have further created various direct and indirect employment opportunities in primary, secondary and tertiary sectors.

Thus, the Mission has made significant achievements in terms of accomplishment of goals set for. Keeping in view the need for sufficient food grains stock to meet the domestic demand and for food security, it is imminent that the present Mission mode programme is continued in the XII Plan by widening its scope and ambit. Building on its strength of targeted interventions backed by direct fund transfers and strong monitoring set up, the Mission has to overcome some of its weaknesses in the form of analytical understanding of some of the laggard Districts.

The success of NFSM activities hinges on ensuring the availability of physical inputs and instruments of production like seeds, machinery, irrigation devices, *etc.* Apart from the above requirements, favourable monsoon and other weather parameters are sine-qua-non for achieving success in fulfilling the targeted production level. A convergence mechanism for getting support of institutions dealing in critical inputs like credit, water, power, and marketing through assured procurement needs to be created for sustaining the gains from the intensive crop development work.

LAND ACQUISITION AND COMPENSATION

Farmers in NCR and beyond are using a section of the 2013 land law to claim their unused land back from the state, and demand fair compensation

Eighty-three year old Jai Narayan is a very angry man. He also has a lot to say. "But no one wants to listen to me," complains the retired schoolteacher from Faridabad. Not the bureaucrats of Haryana, not the judges of the state High Court, not the journalists, at times, not even fellow villagers. We are in Sector 75, Faridabad, standing on what used to be Jai Narayan's farm before the state government acquired it in 2006. Today, it lies barren, surrounded by the menacing multi-storeyed apartment complexes, a sight typical of the National Capital Region's skyline. The sweltering mid-September heat, and the dust raked up by the ongoing constructions in the vicinity, is only making Jai Narayan impatient, and angrier. "Just look around, what does the sarkar need my land for? What can they develop on this patch now? They just want to give it away to big builders, just like all these other plots here".

A five-minute drive from Jai Narayan's patch is the house of Manoj Yadav, 41, who has a similar story. In 2006, about 5,000 farmers from five villages in the region received a notice that HUDA (Haryana Urban Development Authority) would be acquiring 638 acres of their land for "residential and commercial" purposes. The compensation offered for 342 acres in 2009 (some land was released in the interim) at the time was 16 lakhs per acre, an amount that landowners such as Jai Narayan, Manoj and several others in the area felt was much below the market price for their properties. This, they claim, was ascertained by private sale deeds in the area. "There have been sales of land in the area that have fetched owners up to 22 crore per acre. Why is the government then paying us a fraction for the same kind of land?" asks Jai Narayan. He says that while, initially, most owners such as him and Manoj refused compensation — over the differential pricing of land in their area — and resisted giving possession of their land, some did give up along the way. "People thought, 'let us just take whatever is being offered; who knows, one might not even get this later'," says Manoj, adding that only 108 out of the 1,387 farmers in a certain area of the region took compensation. The rest, like him, are still resisting.

Their resistance has not been passive. Since 2008, the group has been going to court several times, to contest the acquisition and demand fair compensation, and has also managed to get stay orders for their land. And last year, when the Right to Fair Compensation and Transparency in Land Acquisition, Rehabilitation and Resettlement Act, 2013, came into force, things started to look better. Like several farmers across the country, especially those in the NCR, Manoj and Jai Narayan figured they could benefit from the contentious Section 24 (2) of the land law. Under this section, in cases where the state had made an award (compensation offer) five years ago, but compensation has not been paid, and/or the physical possession of land has not been taken, the acquisition stands cancelled. Manoj says that for those such as him, where the time lapse between the award and the new Act is 4 years and 9 months — that is, three months short of the requisite time period — there's a provision of enhanced compensation if the amount for a majority of landholdings has not reached the "accounts of beneficiaries".

Under this section, in cases where the state had made an award (compensation offer) five years ago, but compensation has not been paid, and/or the physical possession of land has not been taken, the acquisition stands cancelled. Manoj says that for those such as him, where the time lapse between the award and the new Act is 4 years and 9 months — that is, three months short of the requisite time period — there's a provision of enhanced compensation if the amount for a majority of landholdings has not reached the "accounts of beneficiaries". This is where those such as Manoj believe they have a shot. "Now the authorities have deposited the compensation amount in the court to avoid giving us our due! Also, our matter has been transferred to a committee which includes the same people who were involved in the acquisition. There's need for a neutral body to decide; let them get the CBI to do this," insists Manoj. The mess around land acquisition in Faridabad's Sector 75 is only indicative of the complexity of the system of acquisition in the country, one that is rife with corruption, misinformation and consequently, widespread discontent.

At the policy level, the issue has turned into a perpetual tug-of-war between the present government and the one before it — the new law was followed by three Ordinances and a new Bill, with a set of fresh amendments (including that in Section 24), and consequently, intense opposition. In the latest twist, the government has retracted from its aggressive position — albeit only temporarily — and the 2013 law prevails after the lapse of the Ordinance. Outside policy circles, the discontent over compensation and rehabilitation of displaced populations remains widespread. It is another matter though, that considering the disparities in the value of land and the position of the landowner in the class-caste hierarchy across different states, for some, it is a matter of sheer survival, and for others, it is about maximising gains from one's property.

So, if land acquisition for the Sardar Sarovar Dam has robbed many of their livelihood in Madhya Pradesh, in Haryana, it has given rise to a lot of "heartburn" amongst farmers, who have realised their position in the urban land market. Considering the vast tracts of land lying idle for several years in the country, and the varying degrees of discontent following acquisition, it's not hard to see why the Section 24 clause — that has implications for farmers claiming their unused land back — has become a contentious issue for political parties on both ends, even as it seems to have given a shot in the arm to those who had either filed cases under that section, or are gearing up to do so. "The trend of filing cases under this section is certainly on the rise, at least in the National Capital Region, where land prices are high, owners are more informed of the law and aware of their prospects in negotiating a better deal through litigation," says Deepa Panwar, member of faculty at Symbiosis Law School, Noida, who has done extensive work on the subject.

In areas such as Sonepat, farmers say that compensation offers have indeed increased after owners have agitated, but clearly, it is nowhere close to the market price. Rajvir Singh of Sonepat, who has also filed a case under Section 24 (2) — the acquisition of his land falls in the time period in the clause — concedes that he did take the compensation of 16 lakh, but insists that he did so "under duress". "We are willing to return that amount, in exchange for the market price, which runs into crores". The "heartburn" of those such as Jai Narayan and Rajvir, and their clamour for enhanced compensation or getting their land back, stems from the fact that acquisitions in their region have been rife with corruption and deceit, orchestrated by a builder-state nexus. "This is how it works: the builder buys whatever he can from some of the owners directly. In areas where the sellers refuse the builder price, the state steps in. The state offers a much lesser price, and others whose land is not being acquired in the area get scared. They then sell it to the builder at whatever price he is offering," explains Jai Narayan. The state also transfers land to private entities, who make huge profits from it.

Rajpal Sewli of Sonepat, who received the compensation offer for his land (acquired for the Rajiv Gandhi Educational City in Sonepat) in 2006, narrates another instance of corruption in the process: in many cases, even after the initial notification on acquisition, those who are influential manage to get it released and change the land use [agricultural to non-agricultural]. For those such as Rajpal, who are unable to do so, litigation, — especially under Section 24 — remains one way of exercising their negotiating power with the state. "This is all for private developers, and we have been rendered jobless".

Litigation, however, comes with its own challenges. "A sale deed in the area needs to be produced as evidence for enhanced compensation. Mostly farmers who are closer to urban areas have access to such information, and resources to pursue litigation. While there have been cases where courts have awarded higher compensation, there are hardly any instances where farmers have been able to able to save their land from acquisition and get it back," says Ram Singh, Professor, Delhi School of Economics, author of a comprehensive study on the subject.

Interpretation of a clause such as Section 24 (2) is another challenge, feels Sanjoy Chakravorty, author of The Price. of Land: Acquisition, Conflict, Consequence (OUP, 2013). "What the courts decide is an open question, and will depend on each case. In Singur, for instance, when the Nano project was abandoned, the then Chief Minister asked Tata Motors to return the land leased to the company for 99 years. When Tata Motors refused, the matter went to court. There it was decided that since it was contract, the company had the right to retain the land, even though it was not using it".

While Tata Motors may not be able to stop the farmers from working on the land, in such cases, a lot depends on the terms of contract between the state and the private entity, which the original landowner may not be aware of. But for those doing the rounds of courts, it is a process in which they have invested substantial

Messedup: Acquistion, Corruption and an Alternative

The five years or more limit for releasing acquired land [if farmers have not taken compensation, state has not taken possession] has been set because most acquisitions have happened within those five years. It's a trick.	Closer to urban areas, most people are awate that they can go to court to contest unfair compensation. They have the resources, know the law and are capable of to support their claims.	Land acquisition is changing the way we used to live. There's heartburn amongst farmers over the amount of compensation paid by the state, and private sales. Farming is dying, and this will threaten the country's food security.	The process of land acquisition is fabric in places such as the National Capital Region, where prices are high. Farmers don't have any other skills [aside of farming] and are unable to manage the money	The 2013 Act cannot be used to acquire land, and the high price of land [in and around urban areas] is a big factor. We need to explore alternative formulas such as development rights for the farmers.
MANOJ YADAV, farmer, Sector-75 Faridabad	RAM SINGH, Professor, Delhi School of Economics	SN SOLANKI, Vice President, All India Kisan Sabha, Haryana Unit	DEEPA PANWAR, Member of Faculty Symbiosis Law School, Noida	SANJOY CHAKRAVORTY, Professor, Remple University, USA

money, time, and hope. The fight is not just for money some insist. "We are farmers, and we wouldn't be able to handle the money. We need the land to keep doing what we know best, and we will keep fighting to get it back".

But for those doing the rounds of courts, it is a process in which they have invested substantial money, time, and hope. The fight is not just for money some insist. "We are farmers, and we wouldn't be able to handle the money. We need the land to keep doing what we know best, and we will keep fighting to get it back". With a rise in the negotiating power of the farmer, Chakravorty feels that alternatives such as transfer development rights will need to be explored. Until then, Jai Narayan will keep going to the courts, and stay enraged.

Mechanized Agriculture

A cotton picker at work. The first successful models were introduced in the mid-1940s and each could do the work of 50 hand pickers. Mechanised agriculture is the process of using agricultural machinery to mechanise the work of agriculture, greatly increasing farm worker productivity. In modern times, powered machinery has replaced many jobs formerly carried out bymanual labour or by working animals such as oxen, horses and mules.

The history of agriculture contains many examples of tool use, such as the plough. Mechanization involves the use of an intermediate device between the power source and the work. This intermediate device usually transforms motion, such as rotary to linear, or provides some sort of mechanical advantage, such as speed increase or decrease or leverage. Current mechanised agriculture includes the use of tractors, trucks, combine harvesters, airplanes (crop dusters), helicopters, and other vehicles. Modern farms even sometimes use computers in conjunction with satellite imagery and GPSguidance to increase yields.

List of Agricultural Machinery

Mechanisation was one of the factors responsible for urbanization and industrial economies. Besides improving production efficiency, mechanization encourages large scale production and improves the quality of farm produce. On the other hand, it displaces unskilled farm labor, causes environmental pollution, deforestation and erosion.

Productivity Improving Technologies (Historical)

Threshing machine in 1881. Steam engines were also used to power threshing machines. Today both reaping and threshing are done with a combine harvester

"Better and cheaper than horses" was the theme of many advertisements of the 1910s through 1930s.

Jethro Tull's seed drill (ca. 1701) was a mechanical seed spacing and depth placing device that increased crop yields and saved seed. It was an important factor in the British Agricultural Revolution.

Since the beginning of agriculture threshing was done by hand with a flail, requiring a great deal of labor. Thethreshing machine, which was invented in 1794 but not widely used for several more decades, simplified the operation and allowed the use of animal power. Before the invention of the grain cradle (ca. 1790) an able bodied laborer could reap about one quarter acre of wheat in a day using a sickle. It was estimated that for each of Cyrus McCormick's horse pulled reapers (ca. 1830s) freed up five men for military service in the U.S. Civil War. Later innovations included raking and binding machines. By 1890 two men and two horses could cut, rake and bind 20 acres of wheat per day.

In the 1880s the reaper and threshing machine were combined into the combine harvester. These machines required large teams of horses or mules to pull. Steam power was applied to threshing machines in the late 19th century. There were steam engines that moved around on wheels under their own power for supplying temporary power to stationary threshing machines. These were called road engines, and Henry Ford seeing one as a boy was inspired to build an automobile.

With internal combustion came the first modern tractors in the early 1900s, becoming more popular after the Fordson tractor (ca. 1917). At first reapers and combine harvesters were pulled by tractors, but in the 1930s self powered combines were developed. (Link to a chapter on agricultural mechanisation in the 20th Century at reference).

Advertising for motorized equipment in farm journals during this era did its best to compete against horse-drawn methods with economic arguments, extolling common themes such as that a tractor "eats only when it works", that one tractor could replace many horses, and that mechanisation could allow one man to get more work done per day than he ever had before. The horse population in the U.S. began to decline in the 1920s after the conversion of agriculture and transportation to internal combustion. Peak tractor sales in the U.S. were around 1950. In addition to saving labor, this freed up much land previously used for supporting draft animals. The greatest period of growth in agricultural productivity in the U.S. was from the 1940s to the 1970s, during which time agriculture was benefiting from internal combustion powered tractors and combine harvesters, chemical fertilizers and the green revolution.

Current Status of Future Applications

Asparagus Harvesting

Asparagus are presently harvested by hand with labor costs at 71 per cent of production costs and 44 per cent of selling costs. Asparagus is a difficult crop to harvest since each spear matures at a different speed making it difficult to achieve a uniform harvest. A prototype asparagus harvesting machine - using a light-beam sensor to identify the taller spears - is expected to be available for commercial use.

Chile Pepper Harvesting

As of 2014, prototype chili pepper harvesters are being tested by New Mexico State University. The New Mexico green chile crop is currently hand-picked entirely by field workers.

Orange Harvesting

As of 2010, approximately 10 per cent of the processing orange acreage in Florida is harvested mechanically. Mechanisation has progressed slowly due to the uncertainty of future economic benefits due to competition from Brazil and the transitory damage to orange trees when they are harvested.

Raisin Harvesting

As of 2007, mechanised harvesting of raisins is at 45 per cent; however the rate has slowed due to high raisin demand and prices making the conversion away from hand labor less urgent.

Strawberry Harvesting

Strawberries are a high cost-high value crop with the economics supporting mechanisation. In 2005, picking and hauling costs were estimated at $594 per ton or 51 per cent of the total grower cost. However, the delicate nature of fruit make it an unlikely candidate for mechanisation in the near future. A strawberry harvester developed by Shibuya Seiki and unveiled in Japan in 2013 is able to pick a strawberry every eight seconds. The robot identifies which strawberries are ready to pick by using three separate cameras and then once identified as ready, a mechanized arm snips the fruit free and gently places it in a basket. The robot moves on rails between the rows of strawberries which are generally contained within elevated greenhouses. The machine costs 5 million yen.

Tomato Harvesting

Mechanical harvesting of tomatoes started in 1965 and as of 2010, nearly all processing tomatoes are mechanically harvested. As of 2010, 95 per cent of the U.S. processed tomato crop is produced in California. Although fresh market tomatoes have substantial hand harvesting costs (in 2007, the costs of hand picking and hauling were $86 per ton which is 19 per cent of total grower cost), packing and selling costs were more of a concern (at 44 per cent of total grower cost) making it likely that cost saving efforts would be applied there.

According to a 1977 report by the California Agrarian Action Project, during the summer of 1976 in California, many harvest machines had been equipped with a photo-electric scanner that sorted out green tomatoes among the ripe red ones using infrared lights and color sensors. It worked in lieu of 5,000 hand harvesters causing displacement of innumerable farm laborers as well as wage cuts and shorter work periods. Migrant workers were hit the hardest. To withstand the rigor of the machines, new crop varieties were bred to match the automated pickers. UC Davis Professor G.C. Hanna propagated a thick-skinned tomato called VF-145. But even still, millions were damaged with impact cracks and university breeders produced a more tougher and juiceless "square round" tomato. Small farms were of insufficient size to obtain financing to purchase the equipment and within 10 years, 85 per cent of the state's 4,000 cannery tomato farmers were out of the business. This lead to a concentrated tomato industry in California that "now packed 85 per cent of the nation's tomato products". The monoculture fields fostered rapid pest growth, requiring the use of "more than four million pounds of pesticides each year" which greatly affected the health of the soil, the farm workers, and possibly the consumers.

Crop Insurance in India

Agricultural production and farm incomes in India are frequently affected by natural disasters such as drought, floods, cyclone, storm, landslide, earthquake *etc.* Susceptibility of agriculture to these disasters is compounded by the outbreak of epidemics and man-made disasters such as fire, sale of spurious seeds, fertilizers and pesticides, price crashes, *etc.* All these events severely affect farmers through loss in production and farm income, and are beyond the control of farmers. With growing commercialization of agriculture, the magnitude of loss due to unfavourable eventualities is increasing. In recent times, mechanisms like contract farming and futures trading have been established which are expected to provide some insurance against price fluctuations directly or indirectly. But, agricultural insurance is considered an important mechanism to effectively address the risks to output and income resulting from various natural and manmade events.

Historical Background

i) First Individual Approach Scheme 1972-1978

Different forms of experiments on agricultural insurance on a limited, ad-hoc and scattered scale were started from 1972-73 when the General Insurance Corporation (GIC) of India introduced a Crop Insurance Scheme on H-4 cotton and later included groundnut, wheat and potato. The Scheme was implemented in Andhra Pradesh, Gujarat, Karnataka, Maharashtra, Tamil Nadu and West Bengal. It continued up to 1978-79 and covered only 3,110 farmers for a premium of Rs.4.54 lakh against claims of Rs.37.88 lakh.

ii) Pilot Crop Insurance Scheme (PCIS) 1979-1984

The Pilot Crop Insurance Scheme was launched by the GIC in 1979, which was based on the 'Area Approach' for providing insurance cover against a deficit in crop yield below the threshold level. The Scheme covered cereals, millets, oilseeds, cotton, potato and chickpea and

it was confined to loanee farmers of institutional sources on a voluntary basis. The PCIS 1979 was implemented in 12 states till 1984-85 and covered 6.23 lakh farmers for a premium of Rs.195.01 lakh against claims of Rs.155.68 lakh during the entire period.

iii) Comprehensive Crop Insurance Scheme (CCIS) 1985-99

The Comprehensive Crop Insurance Scheme (CCIS) launched in 1985 was the first nation-wide Scheme. Previous Schemes were either experimental or Pilot Projects, on a small scale and in a scattered manner. This scheme was linked to short-term credit and was based on the 'homogenous area approach'. The Central Government introduced the CCIS during the year 1985-86. Till *Kharif* 1999, the Scheme was adopted by 15 States and 2 Union Territories (UTs). Both, PCIS and CCIS were confined only to farmers who had borrowed seasonal agricultural loans from financial institutions. The main difference between PCIS and CCIS was that PCIS was on voluntary basis while CCIS was compulsory for loanee farmers. The CCIS covered 763 lakh farmers for a premium of Rs.404 crore against claims of Rs.2303 crore. A more comprehensive Scheme, 'National Agricultural Insurance Scheme was launched in 1999 with aim to cover all farmers irrespective of loanee or non-loanee.

National Agricultural Insurance Scheme (NAIS)

The National Agricultural Insurance Scheme (NAIS), with the aim to increase coverage of farmers, crops and risk commitment, was introduced in the country from *Rabi* 1999-2000 replacing the erstwhile Comprehensive Crop Insurance Scheme (CCIS). The main objective of the Scheme was to protect the farmers against the crop losses suffered on account of natural calamities, such as, drought, flood, hailstorm, cyclone, pests and diseases. The Scheme was implemented by the Agriculture Insurance Company of India Ltd. (AIC).

The Scheme was available to all the farmers both loanee and non-loanee irrespective of their size of holding. It envisages coverage of all the food crops (cereals, millets and pulses), oilseeds and annual commercial/horticultural crops, in respect of which past yield data is available for adequate number of years. The Scheme was continued till *Kharif* 2013, however, some States are allowed to implement NAIS during *Rabi* 2013-14 also. The Scheme was optional for States/ Union Territories (UTs) and it had been implemented by the 25 States and 2 Union Territories in one or more seasons. Since the inception of the Scheme 2084.78 lakh farmers for a premium of Rs.8,67,121 lakh against the claim of Rs.25,37,558 lakh were covered until 2012-13. The total area insured was Rs.3137.70 lakh hectares during the same period. The State-wise details of coverage under National Agricultural Insurance Scheme is available at ***Annexure-I***.

Modified National Agricultural Insurance Scheme (MNAIS)

To improve further and make the Scheme easier and more farmer friendly, a proposal on Modified National Agricultural Insurance Scheme (MNAIS) was prepared and was approved by Government of India for implementation on pilot basis in 50 districts from *Rabi* 2010-11 season. During the Five seasons of its implementation in 17 States, the MNAIS covered 45.80 lakh farmers for a premium of Rs.1,08,800 lakh against the claim of Rs.86,400 lakh until *Rabi* 2012-13. The total area insured was 46.79 lakh hectares during the same period. The State-wise details of coverage under Modified National Agricultural Insurance Scheme is available at ***Annexure-II, Table-A.***

Pilot Weather Based Crop Insurance Scheme (WBCIS)

With the objective to bring more farmers under the fold of Crop Insurance, a Pilot Weather Based Crop Insurance Scheme (WBCIS) was launched in 20 States in 2007. Apart from Agriculture Insurance Company of India, some private companies have also been allowed to implement the Scheme. The WBCIS is intended to provide insurance protection to the farmers against adverse weather incidences, such as deficit and excess rainfall, high or low temperature, humidity *etc.* which are deemed to impact adversely the crop production. It has the advantage to settle the claims within shortest possible time. The WBCIS is based on actuarial rates of premium but to make the Scheme attractive, premium actually charged from farmers has been restricted at par with NAIS.

The WBCIS was implemented in 18 States and 469.38 lakh farmers were covered for a premium of Rs.7,51,920 lakh against the claims of Rs. 52,860 lakh under the Scheme from 2007-08 to 2012-13. The total area insured was 632.01 lakh hectares during the same period. The State-wise details of coverage under Pilot Weather Based Crop Insurance

Scheme (WBCIS) is available at ***Annexure-II Table-B***.

Pilot Coconut Palm Insurance Scheme (CPIS)

The Coconut Palm Insurance Scheme (CPIS) was approved for implementation on pilot basis for the years 2009-10 and 2010-11 in the selected areas of Andhra Pradesh, Goa, Karnataka, Kerala, Maharashtra, Odisha and Tamil Nadu. Later on, it was extended to West Bengal also. The pilot was implemented during the years 2011-12 and 2012-13 and continues to be under implementation during the year 2013-14 also. Fifty per cent of the premium is contributed by Government of India, 25 per cent by the concerned State Government and the remaining 25 per cent by the farmer. The CPIS is administered by the Coconut Development Board (CDB).

Under the Scheme, 51,108 farmers were covered for a premium of Rs.167.69 lakh against the claims paid of Rs.214.05 lakh till December 2013. And the total area covered during the same period was 25,938 hectares. The State-wise details of coverage under Pilot Coconut Palm Insurance Scheme (CPIS) is available at ***Annexure-III***.

National Crop Insurance Programme (NCIP)

To make the Crop Insurance Schemes more farmer friendly, a restructured Central Sector Scheme in the name of 'National Crop Insurance Programme' (NCIP) was introduced from Rabi 2013-14. The existing MNAIS, WBCIS and CPIS were merged under this programme with various improvements and changes for implementation throughout the country. However, on the basis of requests received, some States have been allowed to implement NAIS during Rabi 2013-14.

The coverage of NCIP in terms of farmers and area insured has been projected to the level of 50 per cent each from the existing level of about 25 per cent and 20 per cent respectively by the terminal year 2016-17 of the Twelfth Plan. The coverage under CPIS is projected at 25 per cent of coconut growers during 2013-14 with increase of 5 per cent each year during remaining years of Twelfth Five Year Plan.

Issues Related to Crop Insurance

Agriculture in India is varied, diversified and prone to a variety of risks. Most farmers are small and marginal ones. In most areas, agriculture is rain fed, leading to a greater degree of yield variability and risk. Crop Insurance, which aims at addressing yield risk though necessary for a vast majority of farmers - is subject to structural, design and financial problems. Problems of asymmetry of information - moral hazard and adverse selection and co-variability are more pronounced in crop insurance than in other forms of insurance. Consequently, crop insurance schemes face many problems. In response to such problems, Schemes based on the area approach were introduced in the 1980s. More recent Insurance Schemes are based on weather, and adopt an area approach. Due to the vastness of the country, large number of small and marginal farmers, and adoption of area-based approaches, several agencies and organizations are involved in crop insurance programmes. Hence, coordinated efforts are critical for effective implementation of crop insurance scheme.

A four members Committee headed by Dr. P. K. Mishra, former Secretary, Department of Agriculture and Cooperation was constituted in September 2013 to examine the loopholes in the implementation of the Crop Insurance Schemes. The Committee examined the various issues related to Crop Insurance such as:

- Discrepancy in area insured.
- Delay in receiving crop – cutting data and quality and reliability of such data.
- Weather data.
- Crop loan practices.
- Crop insurance premium.
- Settlement of Claims.
- Role of Banks and Agricultural Insurance Companies in the operation of schemes and
- Awareness of farmers regarding various features of the Schemes.

Recommendations for Improvement

After examining the various existing Schemes, the Committee has recommended following measures for improvement in the field of Crop Insurance:

- A web portal should be developed for all States to make land record data be available to financial institutions.
- Premium rates to be revisited for MNAIS.
- Reserve Bank of India (RBI) and National Bank for Agriculture and Rural Development (NABARD) should effectively monitor the compliance of their circulars regarding compulsory crop insurance for loanee farmers.
- Insurance companies and banks should play a pro-active role in insuring effective implementation of crop insurance schemes.
- State Governments should insure the use of GPRS-enabled and camera fitted mobile phones *etc.* while conducting crop cutting experiments.
- An Atlas of critical weather elements for different agro-climatic regions and

- ☆ An Agricultural Insurance Act should be formulated to take care of specific needs of the crop insurance and agricultural insurance in general.

Conclusion

The shift from a social crop insurance program with ad-hoc funding from the Government of India to a market-based crop insurance program with actuarially sound premium rates and product design is a major step forward. The improved product and active involvement of private sector insurance markets are expected to lead to significant benefits for farmers, including faster claims settlement, a more equitable allocation of subsidies and lower basis risk. For the product to be pro-poor, small and marginal farmers must purchase the MNAIS product voluntarily, and insurers and government must experiment with cost-effective ways of increasing outreach.

Use of latest technology such as GPRS-enabled and Camera fitted Mobile Phones may be used to implement Crop Insurance Schemes more effectively. A comprehensive programme of capacity building – in line with the needs of stake holders such as State Government functionaries, insurers and Central Government agencies associated with Crop Insurance Schemes should be organized. Programmes of creating awareness and insurance literacy among farmers should be prepared by Insurance Companies and Banks, in collaboration with the concerned State Governments.

Annexure-I
State-wise details of coverage under National Agricultural Insurance Scheme (NAIS) from Rabi 1999-2000 to Rabi 2012-13 (Rs. in lakh)

Sl.No.	*State/UTs*	*Farmers Insured (No.)*	*Area Insured (H.)*	*Sum Insured*	*Premium Collected*	*Claims Paid*	*Farmers Benefitted (No.)*
1.	Andhra Pradesh	2,85,78,464	4,37,62,477	56,06,783	1,60,398	4,41,990	65,81,020
2.	A&N Island	2,532	4,049	734	17	8	438
3.	Asom	3,38,936	2,52,864	62,238	1,756	1,329	61,056
4.	Bihar	60,36,983	73,77,698	11,90,402	30,383	2,11,347	24,31,008
5.	Chhattisgarh	97,74,546	1,97,81,274	10,22,157	26,108	38,240	16,71,623
6.	Goa	7,893	12,973	279	5	2	702
7.	Gujarat	1,32,98,937	3,05,76,316	35,52,734	1,37,461	4,54,035	48,85,981
8.	Haryana	6,35,688	7,68,865	83,459	2,413	4,024	1,29,217
9.	Himachal Pradesh	3,02,530	2,24,894	47,603 995	995	1,669	1,06,683
10.	J&K	44,520	63,165	9,044	167	123	4,292
11.	Jharkhand	60,88,264	35,92,539	3,25,861	8,078	52,166	21,78,175
12.	Karnataka	1,29,11,992	2,05,74,232	1,58,22,723	46,848	1,77,698	51,33,480
13.	Kerala	4,37,915	3,89,798	76,878	1,646	2,500	75,684
14.	M.P.	2,80,38,512	6,86,84,709	43,68,005	1,24,207	1,72,917	46,77,299
15.	Maharashtra	3,27,51,138	2,84,01,545	20,54,414	81,370	2,25,068	95,33,989
16.	Manipur	19,629	24,372	6,478	160	668	18,946
17.	Meghalaya	29,619	30,925	5,019	245	47	2,668
18.	Mizoram	121	134	23	1	11	119
19.	Odisha	1,44,47,590	1,44,78,255	19,37,570	48,386	1,36,558	26,61,382
20.	Puducherry	37,402	53,434	9,246	176	293	7,037
21.	Rajasthan	15,58,674	3,13,79,980	16,20,309	45,754	2,62,166	52,00,566
22.	Sikkim	1,893	1,324	254	4	1	86
23.	Tamil Nadu	59,61,793	77,82,617	16,11,269	41,164	1,54,814	25,01,663
24.	Tripura	19,573	12,671	2,927	81	58	3,432
25.	UP	2,21,37,563	2,95,55,100	29,76,571	60,756	1,00,973	42,27,898
26.	Uttarakhand	3,86,068	3,62,836	84,106	1,903	4,145	1,18,698
27.	West Bengal	1,11,29,748	56,10,472	11,15,589	46,638	94,707	23,36,616
	Total	**20,84,78,523**	**31,37,69,515**	**2,93,52,675**	**8,67,121**	**25,37,558**	**5,45,49,758**

***Source*:** India. Department of Agriculture and Cooperation, Annual Report, 2013-14, pp. 89-90.

Annexure-II

Table A: State-wise Details of Coverage under Modified National Agricultural Insurance Scheme (MNAIS) from Rabi 2010-11 to Rabi 2012-13

State	Farmers Insured ('000)	Area Insured ('000 Ha.)	Sum Insured (m. Rs)	Gross Premium (m. Rs)	Claims (m. Rs)	Farmers benefitted ('000)	Farmers benefitted (%)
Andhra Pradesh	681	744	29,007	2,008	4,391	300	44
Assam	16	13	489	19	12	2	14
Bihar	504	566	11,612	2,401	564	61	12
Chhattisgarh	0.02	0.03	0.5	0.02	0	0	0
Gujarat	0.40	1	26	3	0	0	0
Haryana	170	285	11,714	398	405	30	18
Jharkhand	45	39	907	88	2	0.4	1
Karnataka	414	693	9,909	1,107	807	114	28
Madhya Pradesh	146	225	2,827	135	10	3	2
Maharashtra	52	50	762	136	0	0	0
Mizoram	1	0.50	10	1	1	1	100
Odisha	76	55	1,623	72	74	13	18
Rajasthan	1,319	1,251	11,462	1,281	601	201	15
Tamil Nadu	212	240	5,141	576	950	104	49
Uttar Pradesh	325	283	8,837	304	78	38	12
Uttarakhand	55	37	930	39	10	8	14
West Bengal	566	196	14,982	2,311	736	91	16
Total	4,580	4,678	1,10,240	10,880	8,640	967	21

***Source*:** India. Department of Agriculture and Cooperation, Report of the Committee to Review the Implementation of Crop Insurance Schemes in India, 2014.

Table - B: State-wise Details of Coverage under Weather Based Crop Insurance Scheme (WBCIS) from Kharif 2007 to Kharif 2013

State	Farmers Insured ('000)	Area Insured ('000 Ha.)	Sum Insured (m. Rs)	Gross Premium (m. Rs)	Claims (m. Rs)	Farmers Benefitted ('000)	Farmers Benefitted (%)
Andhra Pradesh	2,840	4,503	1,12,360	11,297	9,920	2,178	76
Bihar	8,886	9,408	2,15,880	18,704	13,690	6,882	77
Chhattisgarh	214	389	7,460	595	620	155	72
Gujarat	498	413	2,240	224	90	171	34
Haryana	267	427	13,340	1,222	500	144	54
Himachal Pradesh	89	1,000	4,340	500	480	60	68
Jharkhand	358	342	6,560	577	400	294	82
Karnataka	815	1,028	13,640	1,484	1,040	584	72
Kerala	81	57	1730	183	130	44	54
Madhya Pradesh	942	1,662	35,630	3,177	1,720	787	84
Maharashtra	591	679	21,120	2,533	1,800	444	75
Odisha	316	457	11,780	566	320	216	68
Punjab	0.1	0.3	5	0.5	0.1	0.1	75
Rajasthan	30,280	42,046	3,45,770	32,367	21,140	16,692	55
Tamil Nadu	127	185	3,080	295	180	52	41
Uttar Pradesh	447	299	10,140	977	330	203	46
Uttarakhand	84	190	2,640	317	360	44	52
West Bengal	103	115	1,790	173	140	56	55
Total	46,937	63,201	8,09,510	75,192	52,860	29,006	62

Source: India. Department of Agriculture and Cooperation, Report of the Committee to Review the Implementation of Crop Insurance Schemes in India, 2014.

Annexure-III
Statewise Details of Coverage under Pilot Coconut Palm Insurance Scheme (CPIS) till December 2013 (Rs. in lakh)

Sl.No.	*State*	*No. of Farmers*	*Area (Ha)*	*Sum Insured*	*Total Premium*	*Claims Paid*	*Farmers Benefitted*
1.	AndhraPradesh	517	13.33	558.47	2.97	0.00	0
2.	Goa	240	625.26	692.50	3.32	1.63	14
3.	Karnataka	684	635.66	1,035.52	5.84	1.49	36
4.	Kerala	40,330	17,188.20	16,505.84	92.25	193.27	4,338
5.	Odisha	91	21.71	43.53	6.85	0.00	0
6.	Maharashtra	5,963	3,802.78	3,575.18	25.15	8.24	174
7.	Tamil Nadu	2,314	3,308.23	5,680.24	29.28	9.42	129
8.	West Bengal	969	342.66	426.83	2.14	0.00	0
	Total	**51,108**	**25,937.83**	**28,518.12**	**167.79**	**214.05**	**4691**

Source: India. Department of Agriculture and Cooperation, Annual Report, 2013-14.

Drought Today, Trouble Tomorrow

It was nearly impossible to drive through Raichur district of north Karnataka without being lulled into a trance by the unbroken red soil. A warm wind blew hard and free across the arid landscape, and dust rose above the fields. The fields looked just cleared, or recently sown with seeds. The few saplings that were visible looked young, only ankle high. What seemed like the beginning of the kharif season, however, was its withering end; a dry, uneventful harvest. In mid-August, the Karnataka government declared what growers had known for months: the worst drought in 40 years in 27 of the state's 30 districts. The Indian Meteorological Department (IMD) found its grim forecast of deficient rain unfortunately validated. By early September, north Karnataka had received rainfall 43 per cent below normal. It had rained poorly in the region in 2014 too, but this year, 135 of 176 taluks were affected, four times the number last year.

In Lingsugur taluk in Raichur, the dry spell is evident in the bajra stalks, sesame, groundnut and cotton shrubs stunted by thirst. Middle-aged Amresh Gowda and his wife Devamma were uprooting their ruined cotton to plant a variety of fast-growing bajra in one acre, hoping to manage some earnings before the end of September. Their son and daughter, 30 and 21 years old, had left three months ago to Bengaluru to find work. "It's common in our village to migrate to the city for short periods, but it's the first time landed people are going in such large numbers during the Mungaaru (kharif) season," says Devamma. The son returned every few weeks with a part of the wages he saved from construction work. As Devamma cleaned the field, a few cotton pods blew to the side. "You see those cotton buds? We spent nearly Rs.80,000 on seeds, fertiliser and pesticide, and all that grew is the scrap flying around," she says. She slapped her hands to chest. "Thank god we decided against digging a borewell. Then we'd be truly sunk."

A 'Gradual' Problem

Siddalingeshwar Polis Patil, however, dug a borewell. Three, actually. A 72-year-old horticulture farmer, he is well known in Lingsugur for his five convent-educated children who were all going to college, but also for being an early adopter of zerobudget farming techniques, a way of circumventing global markets, debt and price fluctuations by using compost, seeds and cattle urine from one's own farm instead of buying them. He had just returned from a protest in front of the Tahsildar office in Devadurga town, where the Karnataka Farmers Association had demanded subsidies for fodder and electricity. "I can try all kinds of things, but what can I do about water?" asks Patil. Despite a pronounced limp, he jogged speedily through his field to show the three borewells he had dug in the past two years. Each cost him about Rs.2 lakh including pipes, but none struck water. "All I had left was my old well," he says. The well was perfectly dug for ground water to well up, wide enough to harvest the rain and at a low enough point to collect run-off. "But for the first time since I can remember, even the well is fully dry," says Patil. In his three acres of chikoo trees, three of drumstick, and four of mangoes, stand shrivelled young trees. Inadequate rain makes crops more vulnerable to disease as well – Patil's five acres of banana plants bore fruit but a root fungus ate them away.

"Drought is not like rain; it is not sudden," says Patil, walking about in his bone-dry well in Lingsugur. "It will enter our lives slowly, emptying my well first, drying my crop, then my field, starving my cattle, exposing the weak ones to fatal disease, indebting me, then starving me..." He stopped and snorted. "We'll have to think of something before the next stage, won't we?"

As rivers run thin, hydroelectric power generation is down by half. Bengaluru faces 2 hour powercuts today, but rural Karnataka sees electricity for only 5 inconsistent hours a day, rendering pumpsets and lift irrigation useless. In the 13 major multipurpose reservoirs, total water storage is below 40 per cent of capacity. Rural development minister HK Patil says this will be used to meet drinking water needs before irrigation. More than 900 villages are now entirely dependent on tankers for drinking water. After Karnataka declared drought on August 24, chief minister Siddaramiah submitted a memorandum to the Prime Minister asking for central assistance. A nine-member expert team headed by RK Sinha, joint secretary, Union ministry of agriculture, surveyed 16 districts in a week. "The central team is convinced about the severity of the drought," said Siddaramiah on September 11. "We have requested them to allot Rs.3,830 crore from the National Disaster Relief Fund for crop loss relief, meet drinking water needs, and rejuvenate water resources." The team told the press they would submit a report in 10 days.

Not Just Karnataka

Karnataka is the only state to have declared a drought, but Marathwada in Maharashtra has received 51 per cent below normal rain, the worst in the country. Since January, more than 1,000 farmers have committed suicide here, and about 100 in just the past month. Already arid, poorly irrigated and rain dependent, north Karnataka and Marathwada face the second consecutive year of deficient rain and drought-like con- ditions. Last year, Maharashtra announced a package of Rs.4,000 crore for drought-affected farmers, of which more than twothirds has reportedly been distributed. Unless Maharashtra declares a drought this year, however, it cannot access central aid, as Karnataka is seeking to do.

Besides the two states, a Crisil report that calculates the effect of drought on agrarian states found Bihar and Uttar Pradesh also vulnerable. "Monsoon rainfall is crucial for agriculture production and food security, and affects other sectors of the economy," says Dharmakriti Joshi, chief economist at Crisil. The IMD reported an overall 12 per cent shortfall of rain in 40 per cent of the country, the most affected regions being those dependent on rain, or where even the irrigation is rain-fed. According to the agriculture ministry, about 56 per cent of India's net cultivated area is rain-fed. Joshi finds that most impacted crops are jowar, soyabean, tur, maize and cotton – all rain-fed crops grown in drought-prone regions. Even with dismal conditions, most farmers are trying to cope. "Who gives up? Not me," says Hanumaiyya Karigudda in Devadurga, standing in his field where the sesame oilseed saplings were barely ankle-height when they should be nearly 4 feet now. He is borrowing more, sowing seeds of short period crops using whatever water is left, digging more borewells, dipping into his summer fodder stock to feed his cows, selling whose milk he will manage to feed his family. "I try to be creative, try to react as quickly as I can to short spells of rain," he says. When it rained for two days in early September, Hanumaiyya quickly cleared two-thirds of the useless sesame to plant jowar and groundnut, which need less water.

As agriculture becomes undependable, farmers are increasingly depending on cattle. "Few farmers keep cows and bulls these days, but it is a good insurance to tide over bad seasons by selling milk, or selling goats for meat," says Raichurbased oilseed farmer and agriculture researcher Malleshgowda Patil. But cattle are also not immune from rain patterns. As fields get drier, many shepherds spoke about having to travel greater distances to graze their goats and cows; it is a common sight to see 30-40 goats chewing frenzily on the strip of grass on highway dividers. In rain-fed regions like north Karnataka, excessive cash cropping and poor rain has created a shortage of hay (essentially residue stalks of wheat, bajra or corn) and so fodder will have to be bought. And fodder today costs four times it did last year. This explains why the first demand from farmers' groups in both states has been cattle camps or fodder subsidies. Both Maharashtra and Karnataka have made providing drinking water their first priority, followed by creating job opportunities. Water tankers are ferrying water to hundreds of villages. Minister Patil says has allotted Rs.473 crore to gram panchayats to "launch 10,000 works" under the National Rural Employment Guarantee Act, and extended the minimum number of work days from 100 to 150. Maharashtra has authorised the district collectors of Latur, Osmanabad and Beed to set up fodder camps.

Tackling drought, however, cannot be limited to immediate relief. Experts say the more debilitating effects of deficient rain will be slow to take root. Dryness deepens with every shortfall. The Maharashtra government says the number of tankers had increased to 2,130 this year while 48 were deployed at the same time last year. Even if rains pick up now, explained CRISIL's Joshi, the yield will suffer. "Ultimately, delayed rain doesn't help. What's a real worry is that deficient rain has a bearing on the next year's crop too." Despite September's brief showers, the rivers will be thinner, the groundwater over-exploited, and the soil too dry for the rabi (summer) crop next year. "Well-irrigated states like Punjab and Haryana that have been spared this kharif (winter) season might be affected in 2016," he says.

India has suffered weather-related turbulence for years, but now the impact of such events is getting amplified. This year, the drought has meant the highest rural indebtedness India has ever seen, hundreds of suicides, large swathes of unfarmed land, entire villages dependent on water tankers, and thousands of villagers pushing into cities for employment. There is panic in rural areas not only because of lost crop today, but the certainty of worry in the near future. "If you look at the pattern of farmer suicides, it is clear which parts are reeling under the build-up of agrarian stress. It is first about mitigating stress with relief measures, but there are few holistic efforts to offset structural vulnerabilities," says Joshi.

Pune-based water researcher Parineeta Dandekar says chronic water shortage in drought-prone areas like Marathwada has deepened due to single cash cropping (entire villages of water-guzzling sugarcane or rain-fed cotton), and the reluctance to attempt water-conserving techniques like drip irrigation for vegetables or System of Rice Intensification (SRI) for paddy. "Farmers are adopting these now but as the

the water table drops drastically, we can't wait for these ideas to spread organically," says Dandekar. The government focus on increasing output started with a crisis of foodgrain in India – research, state money, politics all were involved in pushing new methods and new seeds. "We are food secure now, and concerted state and research efforts need to focus on the new urgent crisis of water." Political patronage for unsustainable crops like sugarcane and cotton instead of pulses or oilseeds must stop, she adds.

The long-term ability of places like Lingsugur to withstand drought depends on the capacity of reservoir storage in the state, how many farmers have crop insurance, what proportion of farmed land is irrigated, and accessibility of cheap institutional loans. These are chronic issues experts say states must focus on, not only to save farmers and the agrarian sector, but also ease the pressure on cities. As his children all train their eyes on Bengaluru, Patil says, "You may ask why save farmers and farming when our children are trying to be policemen and delivery boys in cities

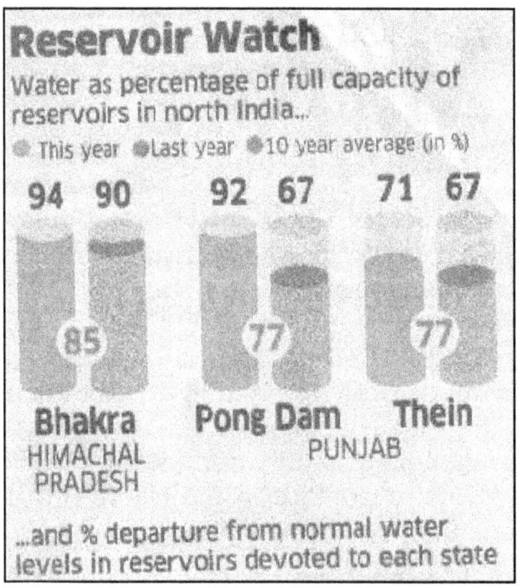

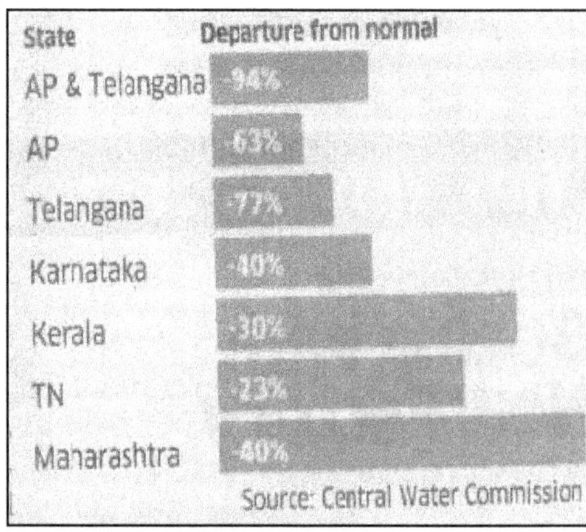

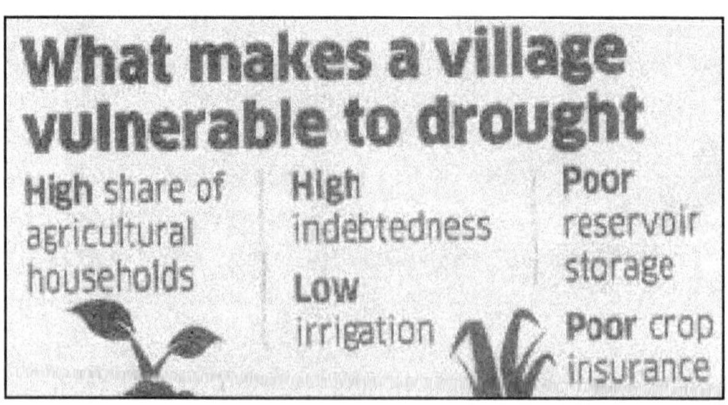

***Source*: Economic Times 15, September 2015.**

anyway. But I was able to send them to school, take education loans and feed them because of my farm." He welcomes the water tankers and relief measures but says, "I will need this attention more than ever tomorrow, when I have run out of energy and ideas to survive the drought."

Indebtedness in Agriculture Sector–Bank Loans are a Double-Edged Weapon

No worthwhile enterprise in any sector can be established and can flourish without institutional production loans. Agriculture is no exception. In the USA and Europe almost all farmers are under debt to banks. Banks there are virtual partners in the business of agriculture. As technology improvements are taking place at a faster rate by the day, farmers require higher and higher doses of finances for updating their resource use patterns and production technologies. This makes them more and more dependent on institutional finance. Perhaps the highest indebted farmer in the world is the American farmer.

In India industrial enterprises and large businesses are the highest-indebted entities. In the secondary and tertiary sectors i.e. industrial enterprises and businesses, the loans obtained are used by and large for productive purposes as per the projects envisaged and evaluated by the banks for financial support. Non-performing assets (NPAs) in these sectors are primarily generated because of projects not succeeding as per plans. In the agriculture sector, if loans obtained are actually used for the purposes for which these are obtained, these investments would yield additional income and repaying capacities. The main problem in agriculture is that of diversion of production loans to meet the consumption needs of the family either directly or to retire the loans obtained earlier from money-lenders to meet the emergency needs or social obligations. Although banks have issued credit cards to farmers, yet the limits do not meet their consumption requirements because these requirements are not embedded into the total entitlements. As a consequence, loans obtained for production purposes in large part get used for consumption purposes. Often production priorities get superseded by consumption necessities. It is therefore of paramount importance that while advancing production credit, reasonable consumption requirements, if any, must be taken care of.

It is bit of a paradox that farm sector financial distress is more common and more sever in the agro-climatically better placed areas compared with states having a not-so-conducive soil-water-climate environment. The Punjab farm sector is much more in the grip of indebtedness and suicides than the agriculturally handicapped states like Bihar, U.P. and Rajasthan. One of the dominant reasons is the higher entitlement of crop loans per unit of land here because of two-crop irrigated lands. Farmers in financial distress borrow money to the limit of entitlement and divert a large part of it for consumption purposes. This leads to a higher level of NPAs in the farm sector. There is therefore a dire need for proper evaluation of the needs for institutional borrowings and continuous monitoring of the expenditure made to meet the objectives specified in the loan applications.

Unfortunately the target-oriented approach has vitiated the credit culture of the country. It is not only in the financial sector for the advancement of loans to various sectors and laying down priorities and earmarking of their share in the total lending, in every sector the concept of targets set from the supply side over the last more than six decades has created a situation similar to the pouring of food on closed mouths. The recipient system in the country has not been paid the needed attention. For instance, in planning targets, the money spent is taken as achievement. No where any impact analysis is done whether the money spent has achieved the intended objectives. Similarly, in priority lending the achievement is the loans advanced only. What happens to the impact of these credit advances on generating additional incomes and repaying capacities is nowhere looked into. As a consequence, the financial institutions have been accumulating NPAs over time and government-owned banks had to be recapitalised again and again at the cost of the exchequer. It is quite a disquieting situation that the premier bank of the country, the State Bank of India, incurred the highest additional NPAs during the first half of the financial year 2013-14. The 40 listed banks of the country had a 38 per cent increase in net NPAs from March 31 to September 30, 2013, from a level of Rs. 93,109 to Rs. 1,28,533 crore. Five major banks –SBI, Bank of Baroda, PNB, CBI and Union Bank – reported an above 30 per cent increase in NPAs in this period. By the end of the financial year the net NPAs are likely to increase to Rs. 1.5 lakh crore. It is the push effect from the supply side that the SBI registered the highest NPAs in agriculture at 9.4 per cent of their lending in the sector compared with 5.7 per cent in both the industrial and services sectors and only 2 per cent in retail. It is all due to the shoddy evaluations under pressure to achieve the targets of priority lending in the sector and virtually no effective monitoring of the end use of the finances advanced to the borrowers. The fact is that the government banks have no solid competence to evaluate the loans and later monitor the use of advances because of lack of knowledge of the sector at the level of bank branches, specially the rural branches, that deal with farm sector loans.

Low interest rates and subventions on timely repayments are very appropriate policy interventions in the farm sector lending, yet it is a double-edged weapon which has also promoted some unscrupulous large farmers get into money-lending to small and marginal farmers as well as non-farming families at higher rates. Although these land owners do not need any loans for themselves, they borrow to the full limits of entitlements and indulge in informal money-lending. It is therefore of paramount importance that where the supply side of the farm credit is being emphasised, the institutional lenders must take equal care of the recipient side through proper evaluation of the loan cases and then effective monitoring of the use of the money advanced for specified purposes. This requires the appointment of agricultural technocrats with an orientation in the business of banking at lending branches in the rural sector. Bankers need to keep in mind that they are the trustees of the depositors' money and recapitalisation of the banks due to escalating NPAs comes from the honest tax payers' money. The probity demands that they realise that they are the superior partners in the business of borrowers and their stakes are much higher as financiers of the business, particularly in the agriculture and rural sectors of the economy.

The Tribune, 10th March, 2014

Beyond Traditional Farming

When Harinder Singh of Bhamri village in Kadian block of Gurdaspur district graduated in 2008, his family wanted him to take up postgraduation and go abroad to improve his prospects. Harinder, however, had other ideas – he insisted that he would stay on in the village and follow the family occupation of farming. His family argued that their five acres of land would not be able to sustain him, especially at a time when argricultural yield was falling and farmers were leaving for other occupations.

Six years later, Harinder has been proved right. In this period he has earned around Rs. 30 lakh and is an example to others in the field. His secret, he says, is "agriculture and subsidiary occupations" that have raised his income far beyond what traditional agriculture would have earned from his small land. "I began with traditional farming on our five acres in 2008, started attending every camp and Kisan Mela being organised by Punjab Agriculture department and Punjab Agriculture University, and collected knowledge about modern farming techniques as well as subsidiary occupation with agriculture," said Harinder.

Not that the fears of Harinder's family were baseless. According to experts, Indian agriculture employs 60 per cent of the population but its contribution to GDP is a mere 14 per cent. In coming years, only an estimated 40 per cent people will remain in agriculture owing to crises in the field and diminishing income, said Kulwant Singh Ahluwalia, a member of the PAU board of management, himself an agriculture expert and progressive farmer, at a Kisan Mela held recently. Around 61 per cent of farmers who participated in a survey said they would readily quit farming should they get alternative jobs in urban areas. Farming, according to them, has become a problem of high production cost and low income – with very little promise of change.

In this scenario, the role of progressive farmers such as Harinder becomes crucial. Their success not only defies the prevalent beliefs but also proves that supplementing traditional agricultural practices with auxiliary or subsidiary occupations is sustainable and yields higher returns. The Punjab government and PAU are stressing this and have made "subsidiary occupations" the theme of Kisan Melas. Subsidiary occupations involve dairy farming, fishery, beekeeping, custom hiring on farm implements and piggery among others, and training programmes for these are conducted through Krishi Vigyan Kendras (KVKs) throughout the state. Harinder, for instance, operated a custom hiring centre for farm implements. "I created a shed on a small portion of my land, purchased farm machinery for Rs. 10 lakh to Rs. 3.30 lakh from state subsidy and the remaining as loan from a bank on soft terms – and started giving out farm implements such as tractor, laser leveller, rotavator and spray pump on rent to farmers who could not afford high-end agricultural machinery and equipment," said Harinder. He earned Rs. 6 lakh in the very first year besides his regular earnings from traditional agriculture. "I rent out a tractor-laser leveller at Rs. 600 per hour, and rotavator and multi-bed planter at Rs. 1,000 per acre," he said. He repaid his loan with a year.

"To raise income from agriculture and horticulture crops, farm machinery is the need of hour of every farmer. Farm mechanisation through upgrade of technology saves around 15 per cent seed, and 30 per cent water and fertiliser each and increases the yield," said Dr Amrik Singh, agriculture development officer, Gurdaspur. Another farmer from Gurdaspur, Maninder Singh, whose family owns four acres, said this is insufficient for a livelihood from the traditional crop cycle of wheat and paddy. Maninder proudly said he earns Rs. 8 to 9 lakh annually by adopting subsidiary occupations. "I farm on three acres and on one acre, I have created a fish pond for fishery, a 3,000 sq ft poultry shed to keep 3,000 birds and a shed for 10 cattle heads including a buffalo and cows," he said. Another farmer, Madan Gopal, owns 2.5 acres and has set up a processing unit for turmeric powder and a cattle shed for 10 animals, besides sowing organic vegetables and pulses. "I earned around Rs. 7 lakh from turmeric cultivation when the rate was good and, at present, earn Rs. 2.5 to 3.0 lakh per acre from it," he said.

VIEWPOINT

- ☆ 61 per cent OF 11,000 : Farmers nationwide preferred some work other than farming. These farmers were interviewed in 274 villages of 137 districts across 18 states.
- ☆ 70 per cent of the farmers said their crops have been destroyed at least once in the past three years.
- ☆ 58 per cent blamed bothe the Centre and state for their problems.
- ☆ 1/10 of poor and small farmers benefit from government schemes and policies, according to most of the farmers surveyed, who felt only rich farmers get the benefits.

Source: Survey by Centre for study of developing societies, commissioned by Bharat Krishak Samaj

Surinder Singh, a foreign-returned farmer, runs a piggery in Paprali village in Ropar district, besides traditional farming. In four years, his small unit of 20 piglets and 20 adult pigs has expanded to 500 ready-to-sell pigs and more than a hundred for breeding. "One can start earning within five to six months as sows deliver piglets twice a year. A sow can deliver 10 to 12 piglets at once. Also, an adult pig of 10 to 11 months is sold between Rs. 35,000 and Rs. 40,000," he said. State director for animal husbandry Dr Harjinderjit Singh Sandha said they were providing subsidy for this venture and there is a a huge demand for pig meat. Discussing the role of subsidiary occupations in enhancing agri-based income, Dr M S Gill, director for extension education, PAU, said it is the need of the hour and the university has been providing farmers with the latest technology through its various KVKs.

BLOSSOMING AGRI STARTUPS*

Entrepreneurs are Looking to Raise Agricultural Productivity with Technology and Automation, Bringing Hope to Drought-prone India.

When Ranjith Mukundan quit his job at Wipro in 2011 after 15 years, farming wasn't on his mind. Bengaluru-born Mukundan and his four co-founders were exploring the Internet of Things space, the idea of connecting ordinary objects to send and receive data. They considered telemedicine, but decided against it since many MNCs were in the business. That's when one of the founders pointed to the problems in dairy management, inclcuding that milk stored in regular containers become non-consumable after a few hours. His uncle was an organic dairy farmer in Tiptur, a town 150km north west of Bengaluru, and had trouble managing his cattle and the supply chain. They liked the idea so Mukundan, now 41, and the team spent the next eight months at the dairy farm and also visited farms in Uttar Pradesh, Madhya Pradesh, Punjab and Odisha. Their company, Stellapps, used their learnings to develop automated dairy solutions, and now works with milk cooperative societies such as Aavin, Milma and Tirumala. Its products are in 250 farms and are used by three lakh dairy farmers. Agriculture in India is a $370 billion sector, but there is little application of technology to improve productivity and lift millions out of poverty. Now, a clutch of entrepreneurs is looking out of their city offices to the countryside to change this.

Bengaluru-based Flybird Agri Innovations places sensors in the soil to detect moisture content and control irrigation. It's installed sensors in 45 villages in Karnataka to help farmers in the drought-hit state optimize irrigation. The company is supported by Chennai-based Villgro, a social impact fund backed by Dell Foundation. Most companies faced resistance when they first set up their equipment. "Our work exposed the malpractices of middlemen," says Mukundan, who is now recruiting sales people in Delhi and Gujarat.The company is looking to expand to allied sectors such as fisheries where perishabil ity is high. Despite the need for technological intervention, there's not much funding coming to startups in this sector.Funding for Indian agriculture startups declined to $56 million in 2015 from $123 million the year before, according to data from Tracxn, a startup activity tracking platform. In 2015, also a drought year, just 20 agriculture-related startups raised money, but $6 billion was invested overall in tech startups that year. Agriculture startups attracted less than 1 per cent of the total.

"The agri funds in the country are mostly early stage investors, and the startups struggle to raise subsequent rounds of funding, thus prolonging their scale up," says emendra Mathur, an investor in agri businesses and former managing director of EAF India Investment Advisory. While overall inflation has been under control, food inflation is growing at 7 per cent, indicating a mismatch of supply and demand.

---

* The Times of India, 24th April 2016.

India imported corn after 16 years, and it has turned from an exporter to an importer of soya. "The inflation translates into an additional `70,000 crore opportunity.The propensity for consumption is also going up. Droughts are an opportunity to develop various seeds and processes that help reduce water consumption without affecting productivity," says Mathur.

Investors tend to view agriculture as a risky proposition. "Most investors don't understand agriculture. The strong gov ernment intervention in the sector along with the vagaries of nature accentuate the business risks associated with startups in the space," says Srikrishna Ramamoorthy, partner at Unitus Seed Fund, an India-focussed fund supported by Gates Foundation that invests in social impact enterprises. It has invested in agri startup Villfarm and is looking for more opportunities. Ramamoorthy says most small farmers are not willing to try technology, and even the large farmers find the equipment too expensive. "Very often, it is not whether the technology is available. The problem is whether there are India-specific solutions," says Mark Kahn, founding partner of Omnivore Partners, an agriculture technology focused fund. It is backed by diversified agri-business firm Godrej Agrovet, SIDBI and SBI among other financial institutions. It has invested in Stellapps and Ecozen, a micro cold storage facility that works on solar power and can be installed even in small farms.

In a country where more than half of the population still survives on agriculture, fruit and vegetables worth $2 billion are wasted because of lack of supply chain management and cold storage facilities."The big cold storage products of the west are not affordable for many Indians," says Kahn. He has investments in 11 agriculture startups, including Skymet, which analyses weather patterns and predicts risk in agriculture for every season. Kahn was earlier head of business strategy and mergers and acquisitions at Godrej Agrovet and found that most agritechnology startups struggle to raise funding. He convinced Godrej to anchor a fund to support more firms and thus Omnivores was founded in 2011 with a corpus of Rs. 260 crore.

The same year also saw the establishment of the SONG Fund by Thomas Hyland, a US citizen who came to India and travelled around the country in trains for a year. He then went on to do a management degree from ISB, Hyderabad, and established Song, backed by US billionaire George Soros, Omidyar Network and Google Capital. "For a country of India's size, there are not enough startups that focus on technologies that addresses the lower middle class and bottom-of-pyramid solutions," says Hyland, who has since started a second fund Aspada. These funds work with the startups for several months before committing to investments. Hyland, for instance, worked with Pune-based SV Agri for eight months before investing. SV works with potato farmers in Maharashtra and Gujarat to improve productivity by providing better seeds, as well as working on supply chain and processing.It supplies potatoes to ITC, Marico and Pepsico for chips. "Not many big companies or startups want to do backward integration when it comes to agriculture because they don't want to get their hands dirty," says Hemant Gaur, founder of SV Agri.

Rohtash Mal, a former CEO of tractor firm Escorts, has established EM3, which provides farm machinery vehicles on demand. "Only 10 per cent of the country's

FUNDING AND DEALS FOR AGRI STARTUPS IN INDIA

Year	Number of deals	Amount ($M)
2016	7	133
2015	20	56
2014	21	123
2013	16	59

farmers can afford farm machinery and ours is a pay-per-use model. We wanted to reach 90 per cent of the farmers," says Mal, who has 15 centres in MP, Bihar and eastern UP. The price is based on the farmhold area measured using satellites. Each centre sends machinery to a 20km radius and Mal says 95 per cent of his customers come back. "Government policies treat agriculture as a poverty alleviation method rather than a means to enhance productivity and raise incomes, and this keeps entrepreneurs out. Mathur says that funds like Omnivore, Villgro and Aspada give him hope, and entrepreneurs are trying to start agri businesses. "The slowdown in e-commerce and the entrepreneurial energy will spur a new wave in rural and agricultural innovation.

Index

A

Adult Male Equivalent Scale (AMES) 132

Aeroponics 200

Agrarian economy 508

Agribusiness 317, 327

Agricultural adjustment Administration (AAA) 296

Agricultural education 117, 317, 341, 342, 502

Agricultural growth 20, 21, 127, 135, 276, 304, 314, 403, 506, 507, 522

Agricultural land 23, 230, 231, 518

Agricultural non-credit societies 182

Agricultural policy 15, 16, 22, 136, 206, 290, 292, 422, 433, 507, 508

Agricultural production 12, 13, 21, 28, 36, 42, 46, 76, 90, 112, 118, 119, 120, 123, 125, 129, 135, 136, 138, 144, 181, 216, 266, 272, 291, 299-304, 319, 332, 352, 353, 355, 359, 365, 392, 423, 424, 497, 505, 506, 507, 517-523, 532, 566

Agricultural production board 299, 365

Agricultural research 24, 28, 29, 47, 108-115, 217, 235, 243, 244, 247, 252, 291, 298, 301, 302, 305, 310, 314, 318-320, 323, 331-333, 340-343, 402, 409, 412, 475, 485, 494, 501, 527

Agricultural research service (ARS) 114, 235

Agricultural scientists' recruitment board (ASRB) 114

Agricultural technology development 24

Agricultural universities 112, 115, 117, 302, 319-325, 341, 343, 475, 526

Agriculture productivity 19, 20

Agri hall of frame 7

Agro-ecosystem 216, 223, 259

Agronomy 1, 113, 401

Agro pastoralism 4, 25, 43

Agro-processing marketing chains 432

All-India Coordinated Research Projects (AICRPs) 112, 117

Animal husbandry 115, 318, 437, 447, 449, 501
Antioxidants 477
APEDA 221, 475
Arable land 124, 273, 275
Awadh peasants protests 101

B

Banjaras 84
BARC 113, 115
Bhabha Atomic Research Centre (BARC) 113
Biodynamic farming 42, 259, 537, 538
Biopharmaceuticals 2
Biotechnology 137
Black revolution 476
Blue revolution 29, 47, 338, 452, 455, 456, 470, 476
British agricultural revolution 30, 37, 632
Bt cotton 239, 245, 248, 251-255, 530, 532, 535, 536, 542, 543, 554
Bt crops 239
Bubbleponics 201

C

Canal irrigation 104, 147, 282, 283
Capital formation 312, 313
Central Food Technological Research Institute (CFTRI) 242
Central Statistics Office (CSO) 490
Certified organic 224
CFTRI 242
Chemical fertilizer 38, 128
China's National Bureau of Statistics 122
CIMMYT 29, 42, 304, 328, 397, 398, 402, 416
Climatic change 61, 67
Coconut Development Board (CDB) 637
Colonial British Era 13, 24, 27, 46
Colonial hydrology 104, 107
Colonialism 8
Commercialisation 417
Commercial seeds 533, 534
Common Agricultural Policy (CAP) 3
Community development programme 347, 351, 352, 507
Comprehensive crop insurance scheme 635
Contract farming 194-197, 206, 422, 429, 485, 486
Contract farming act 485, 486
Conventional hybrid crops 238
Co-operative 179
Co-operative farming societies 182
Corporate farming 205, 421
Cotton Corporation of India (CCI) 487
Cotton export 97
Council for Advancement of People's Action and Rural Technology (CAPART) 358
Council of Scientific and Industrial Research (CSIR) 113, 115
CRIDA 409
Crop-fishery-poultry 209
Crop livestock 208
Cropping pattern 511, 512, 618
Crop production systems 14
Cummings' committee report 320

D

DDT 30, 41
Departmental Commercial Undertaking (DCU) 313
Department of Agricultural Research and Education (DARE) 319, 333
Diversification 260, 261, 426, 571
DNA-based methods 242
Double cropping 413
Drought 355, 643, 644, 651

E

Early common era 8, 24, 26, 45
Early modern era 10, 24, 27, 45
East Punjab holdings 373
Ecological aquaculture 466
Ecological damage 1, 282
Economic impact 88, 297
Economic planning 383
Economic value 17, 19, 465
Environment protection act 239
European economic community (EEC) 450
Evergreen revolution 476

F

Famine commission 28, 141, 142, 290
Famine commission report, 1901 290
FAO 8, 15, 118, 132, 216, 230, 232, 234, 262, 443, 457, 458, 471
Farmers' field school (FFS) 609
Farmer suicide 415, 581
Farming systems trials (FST) 233
Farm wastes 208, 216
Fertilizers 1, 2, 11, 14, 29, 31, 40, 47, 82, 128, 202, 210, 222-225, 259, 262, 289, 301, 312, 323, 334, 359, 360, 363, 370, 371, 386, 397-404, 412-416, 485, 493, 497, 500, 506, 518, 606, 633, 635
FICCI 405
First individual approach scheme 635
Five-year plans 117, 302, 347, 348, 384, 385
Food and Agriculture Organization of the United Nations (FAO) 132
Food consumption 129, 513
ood management 489
ood safety 571
Food security 188, 232, 234, 416, 475, 489, 584-593, 598-603
Foreign direct investment (FDI) 195
Forest research institute 330, 485
Four-field crop rotation 10, 37
Free trade agreements (FTAs) 419

G

Gender revolution 444
Genetically modified (GM) crops 137, 238, 246, 252
Genetically modified organisms (GMOs) 232, 240, 249, 529
Genetic engineering 212, 223, 245, 315
Genetic Engineering Appraisal Authority (GEAC) 242
GIAHS 262, 263
Globally important agricultural heritage system (GIAHS) 262
GM foods and food safety assessment unit (GMFSAU) 240
GM labelling standards 241
Golden revolution 475, 476
Great depression 292, 293, 295-298
Greenhouse 255, 256
Green revolution 2, 11, 15, 24, 29, 30, 36, 42, 47, 108, 110, 115, 182, 216, 228, 236, 253, 259, 264, 304-307, 319, 325-334, 356, 397-416, 423, 445, 456, 459, 476, 499, 506, 507, 519, 543, 569, 574, 584
Green therapy 259
Gregor Mendel 11, 30, 39
Grow more food (GMF) Campaign 301

H

High-yielding varieties (HYVs) 302, 402, 408
High-yielding variety programme 367
Horticulture 19
Human resource development (HRD) 188
Humus farming 211
Hydroponics 196, 199, 200, 257

I

ICRISAT 243, 264, 310, 430, 531, 532
IFOAM 226
Indebtedness 647
Indian Agricultural Research Institute (IARI) 24, 110, 331
Indian Agriculture: Water-Management 12
Indian Central Cotton Committee 111
Indian Council of Medical Research (ICMR) 437
Indian Council of Agricultural Research (ICAR) 24, 28, 47, 108-117, 217, 252, 304, 319, 323, 325, 331-342, 366, 367, 409, 410, 475, 494, 495, 501, 502, 528, 531, 532, 561
Indian Dairy Corporation (IDC) 435
Indian forest service 330
Indian Meteorological Department (IMD) 643
Indian seed system development 525
Industrial agriculture 1, 30, 40, 42
Inland pond culture 470
Institutional Village Linkage Programme (IVLP) 366
Integrated farming system 208
Integrated pest management (IPM) 309, 627
Integrated plant nutrient management (IPNM) 309
Integrated rural development programme 355, 356
Intensive agriculture district programme (IADP) 355, 358, 359, 360, 362, 363, 364, 365
International Center for Maize and Wheat Improvement (CIMMYT) 402
International Rice Research Institute (IRRI) 402, 416
IPNM 309
Irrigation 16, 139

J

Jagirdars 73, 76, 302
John Deere 30, 39

K

Krishi Vigyan Kendras (KVK) 112, 202, 612, 626

L

Land acquisition rehabilitation, and resettlement bill, 2013 579
Land and water management systems 3, 6
Land productivity 271
Land rights 73
Land tax 83, 84, 101
Land tenure system 71, 72, 285
Land use pattern 508, 509
Livestock population trends 451

M

Mahatma Gandhi National Rural Employment Guarantee Scheme (MGNREGS) 269
Mendelian inheritance 39
Milking machine 40
Modified National Agricultural Insurance Scheme (MNAIS) 636, 641
Mollusc farming 468
Mughal period 71, 72, 88
Multi State Co-operative Societies Act (MSCS) 189
Mushroom cultivation 209

N

NAARM 333
NABARD 23, 638
National Agricultural Insurance Scheme (NAIS) 636, 640

National Agricultural Research System (NARS) 108
National Bureau of Plant Genetic Resources (NBPGR) 336
National Co-operative Union of India (NCUI) 188
National Crop Insurance Programme (NCIP) 637
National Development Council (NDC) 383, 385, 391
National food security bill, 2013 584
National Food Security Mission (NFSM) 601
National Horticultural Board (NHB) 475
National Organic Program (NOP) 226
National Programme for Organic Production (NPOP) 221
National Sample Survey (NSS) 438, 491
Natural diet 68
Neolithic revolution 4
NGOs 196, 240, 244, 250, 251, 260, 315, 358, 406, 431, 444, 531, 536, 556, 612, 625, 626
NGT 203
NITI Aayog 381, 388-394, 574

O

OMAFRA 226
Online Procurement Monitoring System (OPMS) 502
Operation Flood 29, 183, 435, 439, 441, 442, 444, 450
Organic agriculture 1, 42, 14, 209-214, 219-223, 226-230, 234-238
Organic aquaculture 467
Organic farming 2, 3, 16, 42, 201, 202, 211-222, 226-233, 259, 260, 264, 423, 424, 431-434, 479, 495, 536-538
Organic Farming Council of Punjab (OFCP) 205, 421
Organic food 211
Organic movement 2, 11, 31, 42, 209, 211, 213, 221
Origins of agriculture 61, 63, 69
Ownership 79

P

Peasants 83
Perennial irrigation 104
Pesticides 1, 2, 11, 14, 31, 41, 82, 128, 211-224, 235-239, 245, 251, 259-262, 309, 312, 324, 334, 369, 370, 401, 404, 412-416, 431, 432, 465, 497, 537, 538, 571, 606, 607, 611, 635
Pilot coconut palm insurance scheme (CPIS) 637
Pilot crop insurance scheme 635
Pilot weather based crop insurance scheme (WBCIS) 636
Pink revolution 476
Planning commission 177, 178, 196, 274, 299, 356, 371, 406, 408, 570, 574, 596
Plant breeding 1, 11, 14, 24, 42, 113, 398, 401, 527, 529, 531, 532
Plant protection 609
Polyculture pond 455
Postharvest management 481
Post Maha Janapadas Period 6, 25
Pre-Mughal period 71
Primary agricultural credit/service societies 181
Project for Intensification of Regional Research on Cotton, Oilseeds and Millets (PIRRCOM) 111, 113
Proto-Dravidian Agriculture 48
Public distribution system 503, 585-593, 595, 598, 599
Public private partnership 531
Punjab Agro Foodgrains Corporation (PAFC) 205, 421
Punjab land revenue act, 1887 375, 377, 379, 380

Q

Quality seed 560-564

R

Rainbow revolution 264, 333
Rainfed farming 310
Raja Ram Jat 87, 88
Recycling 208
Rice-fish-poultry 209
River valley projects 311
Royal commission on agriculture 28, 109, 110, 182, 184, 291, 319, 439
Rural market 19
Russell report 298

S

Seed acquisition 533
Seed bill, 2010 550
Seed industry 527, 567
Seeds Act, 1966 545
Seeds testing laboratory 547
Self help group (SHGs) 188
Sewage farming 203
SIGMEA 43
Silver revolution 476
Soil health card 492, 493, 494
Soilless culture 200
Southern Neolithic 57, 58, 59, 60
Spices and plantation crops 480
State agricultural universities (SAUs) 526
State Food Security Mission Executive Committees (SFSMEC) 603
Supply chain management (SCM) 483
Sustainable agriculture 1, 2, 235, 241, 305, 306, 311, 312, 412, 431, 433, 435, 555
Sustainable aquaculture 463, 475
SWOT analysis 317

T

Tank-and-channel networks 9, 27
Telengana movement 102
The 1919 Reforms 291
Traditional rarming 649
Tube wells rule 570
Twenty point programme 310

U

United States Agency for International Development (USAID) 113, 116, 320, 323, 401
University Grants Commission (UGC) 333
Urban Bias Thesis (UBT) 573

V

Value-addition 314, 317
Vedic period 6, 25, 44
Vermiculture 202
Village community 78
Voelcker report 290

W

Wasteland 568
Water conditions 125
Water traditions 105
Weedicides 609
White revolution 435, 441-446, 450, 476
Women empowerment 589
World bank 21
World food program (WFP) 450
World Trade Organization (WTO) 118, 187, 188, 417, 446, 450, 506, 508, 515, 522, 528, 529
WTO agreement 187, 188

Y

Yellow revolution 476

Z

Zamindars 74-78, 87, 99, 100, 101, 302

www.ingramcontent.com/pod-product-compliance
Ingram Content Group UK Ltd.
Pitfield, Milton Keynes, MK11 3LW, UK
UKHW020223290926
15470UKWH00001B/357

9 789387 057333